One and Two Sample Statistical Inference Summary Tables

Basic Engineering Data Collection and Analysis

Stephen B. Vardeman

Iowa State University

J. Marcus Jobe

Miami University

DUXBURY

THOMSON LEARNING

Australia • Canada • Mexico • Singapore • Spain • United Kingdom • United States

DUXBURY

™

THOMSON LEARNING

Sponsoring Editor: *Carolyn Crockett*
Marketing: *Chris Kelly*
Editorial Assistant: *Ann Day*
Production Editor: *Janet Hill*
Production Service: *Martha Emry*
Permissions: *Sue Ewing*
Cover Design/Illustration: *Denise Davidson*

Interior Design: *John Edeen*
Interior Illustration: *Bob Cordes*
Print Buyer: *Kristina Waller*
Typesetting: *Eigentype Compositors*
Cover Printing: *Phoenix Color Corporation*
Printing/Binding: *Phoenix Color Corporation*

For more information about this or any other Duxbury product, contact:
DUXBURY
511 Forest Lodge Road
Pacific Grove, CA 93950 USA
www.duxbury.com
1-800-423-0563 Thomson Learning Academic Resource Center

www.thomsonrights.com
fax: 1-800-730-2215
phone: 1-800-730-2214

Printed in the United States of America

10 9 8 7 6 5 4 3 2

Library of Congress Cataloging-in-Publication Data

Vardeman, Stephen B.
  Basic engineering data and analysis / Stephen B. Vardeman, J. Marcus Jobe.
        p.   cm.
  ISBN 0-534-36957-X
  1. Engineering–Statistical methods.   2. Engineering–Databases.
  I. Jobe, J. Marcus.   II. Title.

TA340.V35 2001
620′.001′5195–dc21   00-040358

Soli Deo Gloria

Preface

This book is an abridgment and modernization of *Statistics for Engineering Problem Solving* by Stephen Vardeman, which was published in 1994 by PWS Publishing and awarded the (biennial) 1994 Merriam-Wiley Distinguished Author Award by the American Society for Engineering Education recognizing an outstanding new engineering text. The present book preserves the best features of the earlier one, while improving readability and accessibility for engineering students and working engineers, and providing the most essential material in a more compact text.

Basic Engineering Data Collection and Analysis emphasizes real application and implications of statistics in engineering practice. Without compromising mathematical precision, the presentation is carried almost exclusively with references to real cases. Many of these real cases come from student projects from Iowa State University statistics and industrial engineering courses. Others are from our consulting experiences, and some are from engineering journal articles. (Examples bearing only name citations are based on student projects, and we are grateful to those students for the use of their data sets and scenarios.)

We feature the well-proven order and emphasis of presentation from *Statistics for Engineering Problem Solving*. Practical issues of engineering data collection receive early and serious consideration, as do descriptive and graphical methods and the ideas of least squares curve- and surface-fitting and factorial analysis. More emphasis is given to the making of statistical intervals (including prediction and tolerance intervals) than to significance testing. Topics important to engineering practice, such as propagation of error, Shewhart control charts, 2^p factorials and 2^{p-q} fractional factorials are treated thoroughly, instead of being included as supplemental topics intended to make a general statistics text into an "engineering" statistics book. Topics that seem to us less central to common engineering practice (like axiomatic probability and counting) and some slightly more advanced matters (reliability concepts and maximum likelihood model fitting) have been placed in an appendix, where they are available for those instructors who have time to present them but do not interrupt the book's main story line.

· ·

Pedagogical Features

Pedagogical and practical features include:

- Precise exposition

- A logical two-color layout, with examples delineated by a color rule

---

Example 1   Heat Treating Gears

The article "Statistical Analysis: Mack Truck Gear Heat Treating Experiments" by P. Brezler (*Heat Treating*, November, 1986) describes a simple application of engineering statistics. A process engineer was faced with the question, "How should gears be loaded into a continuous carburizing furnace in order to minimize distortion during heat treating?" Various people had various semi-informed opinions about how it should be done—in particular, about whether the gears should be laid flat in stacks or hung on rods passing through the gear bores. But no one really knew the consequences of laying versus hanging.

- Use of computer output

Printout 6   Computations for the Joint Strength Data

```
General Linear Model


Factor     Type Levels  Values
joint      fixed     3 beveled butt    lap
wood       fixed     3      oak pine walnut
```

- Boxing of those formulas students will need to use in exercises

---

Definition 1 identifies $Q(p)$ for all p between $.5/n$ and $(n - .5)/n$. To find $Q(p)$ for such a value of p, one may solve the equation $p = (i - .5)/n$ for i, yielding

Index (i) of the ordered data point that is Q(p)

$$i = np + .5$$

and locate the "$(np + .5)$th ordered data point."

- Margin notes naming formulas and calling attention to some main issues of discussion

---

Purposes of replication

The idea of replication is fundamental in experimentation. **Reproducibility of results** is important in both science and engineering practice. Replication helps establish this, protecting the investigator from unconscious blunders and validating or confirming experimental conclusions.

- Identification of important calculations and final results in Examples

---

To illustrate convention (2) of Definition 1, consider finding the .5 and .93 quantiles of the strength distribution. Since .5 is $\frac{.5-.45}{.55-.45} = .5$ of the way from .45 to .55, linear interpolation gives

$$Q(.5) = (1 - .5)\, Q(.45) + .5\, Q(.55) = .5(9,011) + .5(9,165) = 9,088 \text{ g}$$

The Exercises

There are far more exercises in this text than could ever be assigned over several semesters of teaching from this book. Exercises involving direct application of section material appear at the end of each section, and answers for most of them appear at the end of the book. These give the reader immediate reinforcement that the mechanics and main points of the exposition have been mastered. The rich sets of Chapter Exercises provide more. Beyond additional practice with the computations of the chapter, they add significant insight into how engineering statistics is done and into the engineering implications of the chapter material. These often probe what kinds of analyses might elucidate the main features of a scenario and facilitate substantive engineering progress, and ponder what else might be needed. In most cases, these exercises were written *after* we had analyzed the data and seriously considered what they show in the engineering context. These come from a variety of engineering disciplines, and we expect that instructors will find them to be not only useful for class assignments but also for lecture examples to many different engineering audiences.

Teaching from the Text

A successful ISU classroom-tested, fast-paced introduction to applied engineering statistics can be made by covering most of Chapters 1 through 9 in a single, three-semester hour course (not including those topics designated as "optional" in section or subsection titles). More leisurely single-semester courses can be made, either by skipping the factorial analysis material in Section 4.3 and Chapter 8 altogether, or by covering only Chapters 1 through 6 and Sections 7.5 and 7.6, leaving the rest of the book for self-study as a working engineer finds need of the material.

Instructors who are more comfortable with a traditional "do more probability and do it first, and do factorials last" syllabus will find the additional traditional topics covered with engineering motivation (rather than appeal to cards, coins, and dice!) in Appendix A. For those instructors, an effective order of presentation is the following: Chapters 1 through 3, Appendices A.1 through A.3, Chapter 5, Chapter 6, Section 4.1, Section 9.1, Section 4.2, Section 9.2, Chapter 7, Section 4.3, and Chapter 8.

Ancillaries

Several types of ancillary material are available to support this text.

- The CD packaged with the book provides PowerPoint™ visuals and audio presenting solutions for selected Section Exercises.

- For instructors only, a complete solutions manual is available through the local sales representative or by submitting a request at info@duxbury.com.

- The publisher also maintains a web site supporting instruction using *Basic Engineering Data Collection and Analysis* at www.duxbury.com.

At www.duxbury.com, using the Book Companions and Data Library links, can be found the following:

- Data sets for all exercises

- MINITAB,® JMP,® and Microsoft® Excel help for selected examples from the book

- Formula sheets in PDF and LaTeX formats

- Lists of known errata

Acknowledgments

There are many who deserve thanks for their kind help with this project. People at Duxbury Thomson Learning have been great. We especially thank Carolyn Crockett for her encouragement and vision in putting this project together. Janet Hill has been an excellent Production Editor. We appreciate the help of Seema Atwal with the book's ancillaries, and are truly pleased with the design work overseen by Vernon Boes.

First class help has also come from outside of Duxbury Thomson Learning. Martha Emry of Martha Emry Production Services has simply been dynamite to work with. She is thorough, knowledgeable, possessed of excellent judgment and unbelievably patient. Thanks Martha! And although he didn't work directly on this project, we gratefully acknowledge the meticulous work of Chuck Lerch, who wrote the solutions manual and provided the answer section for *Statistics for Engineering Problem Solving*. We have borrowed liberally from his essentially flawless efforts for answers and solutions carried over to this project. We are also grateful to Jimmy Wright and Victor Chan for their careful work as error checkers. We thank Tom Andrika for his important contributions to the development of the PowerPoint/audio CD supplement. We thank Tiffany Lynn Hagemeyer for her help in preparing the MINITAB, JMP, and Excel data files for download. Andrew Vardeman developed the web site, providing JMP, MINITAB, and Excel help for the text, and we appreciate his contributions to this effort. John Ramberg, University of Arizona; V. A. Samaranayake, University of Missouri at Rolla; Paul Joyce, University of Idaho; James W. Hardin, Texas A & M; and Jagdish K. Patel, University of Missouri at Rolla provided helpful reviews of this book at various stages of completion, and we thank them.

It is our hope that this book proves to be genuinely useful to both engineering students and working engineers, and one that instructors find easy to build their courses around. We'll be glad to receive comments and suggestions at our e-mail addresses.

Steve Vardeman   J. Marcus Jobe
vardeman@iastate.edu   jobejm@muohio.edu

Contents

3   Elementary Descriptive Statistics   66

4   Describing Relationships Between Variables   123

5   Probability: The Mathematics of Randomness

6 Introduction to Formal Statistical Inference   334

7 Inference for Unstructured Multisample Studies   443

1

• • • • • • • • • • • • • • • • • •

Introduction

This chapter lays a foundation for all that follows: It contains a road map for the study of engineering statistics. The subject is defined, its importance is described, some basic terminology is introduced, and the important issue of measurement is discussed. Finally, the role of mathematical models in achieving the objectives of engineering statistics is investigated.

1.1 Engineering Statistics: What and Why

In general terms, what a working engineer does is to design, build, operate, and/or improve physical systems and products. This work is guided by basic mathematical and physical theories learned in an undergraduate engineering curriculum. As the engineer's experience grows, these quantitative and scientific principles work alongside sound engineering judgment. But as technology advances and new systems and products are encountered, the working engineer is inevitably faced with questions for which theory and experience provide little help. When this happens, what is to be done?

On occasion, consultants can be called in, but most often an engineer must independently find out "what makes things tick." It is necessary to **collect and interpret data** that will help in understanding how the new system or product works. Without specific training in data collection and analysis, the engineer's attempts can be haphazard and poorly conceived. Valuable time and resources are then wasted, and sometimes erroneous (or at least unnecessarily ambiguous) conclusions are reached. To avoid this, it is vital for a working engineer to have a toolkit that includes the best possible principles and methods for gathering and interpreting data.

The goal of engineering statistics is to provide the concepts and methods needed by an engineer who faces a problem for which his or her background does not serve as a completely adequate guide. It supplies principles for how to efficiently acquire and process empirical information needed to understand and manipulate engineering systems.

Definition 1

> **Engineering statistics** is the study of how best to
>
> 1. collect engineering data,
> 2. summarize or describe engineering data, and
> 3. draw formal inferences and practical conclusions on the basis of engineering data,
>
> all the while recognizing the reality of variation.

To better understand the definition, it is helpful to consider how the elements of engineering statistics enter into a real problem.

Example 1

Heat Treating Gears

The article "Statistical Analysis: Mack Truck Gear Heat Treating Experiments" by P. Brezler (*Heat Treating*, November, 1986) describes a simple application of engineering statistics. A process engineer was faced with the question, "How should gears be loaded into a continuous carburizing furnace in order to minimize distortion during heat treating?" Various people had various semi-informed opinions about how it should be done—in particular, about whether the gears should be laid flat in stacks or hung on rods passing through the gear bores. But no one really knew the consequences of laying versus hanging.

Data collection

In order to settle the question, the engineer decided to get the facts—to collect some data on "thrust face runout" (a measure of gear distortion) for gears laid and gears hung. Deciding exactly how this data collection should be done required careful thought. There were possible differences in gear raw material lots, machinists and machines that produced the gears, furnace conditions at different times and positions within the furnace, technicians and measurement devices that would produce the final runout measurements, etc. The engineer did not want these differences either to be mistaken for differences between the two loading techniques or to unnecessarily cloud the picture. Avoiding this required care.

Data summarization

In fact, the engineer conducted a well-thought-out and executed study. Table 1.1 shows the runout values obtained for 38 gears laid and 39 gears hung after heat treating. In raw form, the runout values are hardly understandable. They lack organization; it is not possible to simply look at Table 1.1 and tell what is going on. The data needed to be summarized. One thing that was done was to compute some numerical summaries of the data. For example, the process engineer found

$$\text{Mean laid runout} = 12.6$$

$$\text{Mean hung runout} = 17.9$$

Table 1.1
Thrust Face Runouts (.0001 in.)

Gears Laid	Gears Hung
5, 8, 8, 9, 9,	7, 8, 8, 10, 10,
9, 9, 10, 10, 10,	10, 10, 11, 11, 11,
11, 11, 11, 11, 11,	12, 13, 13, 13, 15,
11, 11, 12, 12, 12,	17, 17, 17, 17, 18,
12, 13, 13, 13, 13,	19, 19, 20, 21, 21,
14, 14, 14, 15, 15,	21, 22, 22, 22, 23,
15, 15, 16, 17, 17,	23, 23, 23, 24, 27,
18, 19, 27	27, 28, 31, 36

Further, a simple graphical summarization was made, as shown in Figure 1.1.

Variation   From these summaries of the runouts, several points are obvious. One is that there is variation in the runout values, even within a particular loading method. Variability is an omnipresent fact of life, and all statistical methodology explicitly recognizes this. In the case of the gears, it appears from Figure 1.1 that there is somewhat more variation in the hung values than in the laid values.

But in spite of the variability that complicates comparison between the loading methods, Figure 1.1 and the two group means also carry the message that the laid runouts are on the whole smaller than the hung runouts. By how much? One answer is

$$\text{Mean hung runout} - \text{Mean laid runout} = 5.3$$

Gears laid

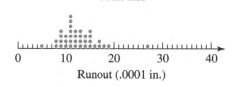

0   10   20   30   40
Runout (.0001 in.)

Gears hung

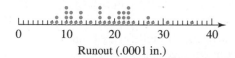

0   10   20   30   40
Runout (.0001 in.)

Figure 1.1   Dot diagrams of runouts

Example 1
(*continued*)

But how "precise" is this figure? Runout values are variable. So is there any assurance that the difference seen in the present means would reappear in further testing? Or is it possibly explainable as simply "stray background noise"? Laying gears is more expensive than hanging them. Can one know whether the extra expense is justified?

Drawing inferences from data

These questions point to the need for methods of formal statistical inference from data and translation of those inferences into practical conclusions. Methods presented in this text can, for example, be used to support the following statements about hanging and laying gears:

1. One can be roughly 90% sure that the difference in long-run mean runouts produced under conditions like those of the engineer's study is in the range

$$3.2 \text{ to } 7.4$$

2. One can be roughly 95% sure that 95% of runouts for gears laid under conditions like those of the engineer's study would fall in the range

$$3.0 \text{ to } 22.2$$

3. One can be roughly 95% sure that 95% of runouts for gears hung under conditions like those of the engineer's study would fall in the range

$$.8 \text{ to } 35.0$$

These are formal quantifications of what was learned from the study of laid and hung gears. To derive practical benefit from statements like these, the process engineer had to combine them with other information, such as the consequences of a given amount of runout and the costs for hanging and laying gears, and had to apply sound engineering judgment. Ultimately, the runout improvement was great enough to justify some extra expense, and the laying method was implemented.

The example shows how the elements of statistics were helpful in solving an engineer's problem. Throughout this text, the intention is to emphasize that the topics discussed are not ends in themselves, but rather tools that engineers can use to help them do their jobs effectively.

Section 1 Exercises

1. Explain why engineering practice is an inherently statistical enterprise.

2. Explain why the concept of variability has a central place in the subject of engineering statistics.

3. Describe the difference between descriptive and (formal) inferential statistics.

1.2   Basic Terminology

Engineering statistics requires learning both new words and new technical meanings for familiar words. This section introduces some common jargon for types of statistical studies, types of data that can arise in those studies, and types of structures those data can have.

1.2.1   Types of Statistical Studies

When an engineer sets about to gather data, he or she must decide how active to be. Will the engineer turn knobs and manipulate process variables or simply let things happen and try to record the salient features?

Definition 2

> An **observational study** is one in which the investigator's role is basically passive. A process or phenomenon is watched and data are recorded, but there is no intervention on the part of the person conducting the study.

Definition 3

> An **experimental study** (or, more simply, an *experiment*) is one in which the investigator's role is active. Process variables are manipulated, and the study environment is regulated.

Most real statistical studies have both observational and experimental features, and these two definitions should be thought of as representing idealized opposite ends of a continuum. On this continuum, the experimental end usually provides the most **efficient and reliable** ways to collect engineering data. It is typically much quicker to manipulate process variables and watch how a system responds to the changes than to passively observe, hoping to notice something interesting or revealing.

Inferring causality

In addition, it is far easier and safer to infer **causality** from an experiment than from an observational study. Real systems are complex. One may observe several instances of good process performance and note that they were all surrounded by circumstances X without being safe in assuming that circumstances X cause good process performance. There may be important variables in the background that are changing and are the true reason for instances of favorable system behavior. These so-called **lurking variables** may govern both process performance and circumstances X. Or it may simply be that many variables change haphazardly without appreciable impact on the system and that by chance, during a limited period of observation, some of these happen to produce X at the same time that good performance occurs. In either case, an engineer's efforts to create X as a means of making things work well will be wasted effort.

On the other hand, in an experiment where the environment is largely regulated except for a few variables the engineer changes in a purposeful way, an inference of causality is much stronger. If circumstances created by the investigator are consistently accompanied by favorable results, one can be reasonably sure that they caused the favorable results.

Example 2

Pelletizing Hexamine Powder

Cyr, Ellson, and Rickard attacked the problem of reducing the fraction of nonconforming fuel pellets produced in the compression of a raw hexamine powder in a pelletizing machine. There were many factors potentially influencing the percentage of nonconforming pellets: among others, Machine Speed, Die Fill Level, Percent Paraffin added to the hexamine, Room Temperature, Humidity at manufacture, Moisture Content, "new" versus "reground" Composition of the mixture being pelletized, and the Roughness of the chute entered by the freshly stamped pellets. Correlating these many factors to process performance through passive observation was hopeless.

The students were, however, able to make significant progress by conducting an experiment. They chose three of the factors that seemed most likely to be important and purposely changed their levels while holding the levels of other factors as close to constant as possible. The important changes they observed in the percentage of acceptable fuel pellets were appropriately attributed to the influence of the system variables they had manipulated.

Besides the distinction between observational and experimental statistical studies, it is helpful to distinguish between studies on the basis of the **intended breadth of application of the results**. Two relevant terms, popularized by the late W. E. Deming, are defined next:

Definition 4

An **enumerative study** is one in which there is a particular, well-defined, finite group of objects under study. Data are collected on some or all of these objects, and conclusions are intended to apply only to these objects.

Definition 5

An **analytical study** is one in which a process or phenomenon is investigated at one point in space and time with the hope that the data collected will be representative of system behavior at other places and times under similar conditions. In this kind of study, there is rarely, if ever, a particular well-defined group of objects to which conclusions are thought to be limited.

Most engineering studies tend to be of the second type, although some important engineering applications do involve enumerative work. One such example is the reliability testing of critical components—e.g., for use in a space shuttle. The interest is in the components actually in hand and how well they can be expected to perform rather than on any broader problem like "the behavior of all components of this type." Acceptance sampling (where incoming lots are checked before taking formal receipt) is another important kind of enumerative study. But as indicated, most engineering studies are analytical in nature.

Example 2
(continued)

The students working on the pelletizing machine were not interested in any particular batch of pellets, but rather in the question of how to make the machine work effectively. They hoped (or tacitly assumed) that what they learned about making fuel pellets would remain valid at later times, at least under shop conditions like those they were facing. Their experimental study was analytical in nature.

Particularly when discussing enumerative studies, the next two definitions are helpful.

Definition 6

A **population** is the entire group of objects about which one wishes to gather information in a statistical study.

Definition 7

A **sample** is the group of objects on which one actually gathers data. In the case of an enumerative investigation, the sample is a subset of the population (and can in some cases include the entire population).

Figure 1.2 shows the relationship between a population and a sample. If a crate of 100 machine parts is delivered to a loading dock and 5 are examined in order to verify the acceptability of the lot, the 100 parts constitute the population of interest, and the 5 parts make up a (single) sample of size 5 from the population. (Notice the word usage here: There is *one* sample, not *five* samples.)

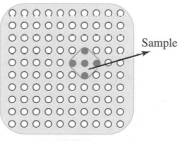

Sample

Population

Figure 1.2   Population and sample

There are several ways in which the meanings of the words *population* and *sample* are often extended. For one, it is common to use them to refer to not only objects under study but also data values associated with those objects. For example, if one thinks of Rockwell hardness values associated with 100 crated machine parts, the 100 hardness values might be called a population (of numbers). Five hardness values corresponding to the parts examined in acceptance sampling could be termed a sample from that population.

Example 2 *(continued)*	Cyr, Ellson, and Rickard identified eight different sets of experimental conditions under which to run the pelletizing machine. Several production runs of fuel pellets were made under each set of conditions, and each of these produced its own percentage of conforming pellets. These eight sets of percentages can be referred to as eight different samples (of numbers).

Also, although strictly speaking there is no concrete population being investigated in an analytical study, it is common to talk in terms of a **conceptual population** in such cases. Phrases like "the population consisting of all widgets that could be produced under these conditions" are sometimes used. We dislike this kind of language, believing that it encourages fuzzy thinking. But it is a common usage, and it is supported by the fact that typically the same mathematics is used when drawing inferences in enumerative and analytical contexts.

1.2.2 Types of Data

Engineers encounter many types of data. One useful distinction concerns the degree to which engineering data are intrinsically numerical.

Definition 8	**Qualitative** or **categorical** data are the values of basically nonnumerical characteristics associated with items in a sample. There can be an order associated with qualitative data, but aggregation and counting are required to produce any meaningful numerical values from such data.

Consider again 5 machine parts constituting a sample from 100 crated parts. If each part can be classified into one of the (ordered) categories (1) conforming, (2) rework, and (3) scrap, and one knows the classifications of the 5 parts, one has 5 qualitative data points. If one aggregates across the 5 and finds 3 conforming, 1 reworkable, and 1 scrap, then numerical summaries have been derived from the original categorical data by counting.

In contrast to categorical data are numerical data.

Definition 9

> **Quantitative** or **numerical** data are the values of numerical characteristics associated with items in a sample. These are typically either **counts** of the number of occurrences of a phenomenon of interest or **measurements** of some physical property of the items.

Returning to the crated machine parts, Rockwell hardness values for 5 selected parts would constitute a set of quantitative measurement data. Counts of visible blemishes on a machined surface for each of the 5 selected parts would make up a set of quantitative count data.

It is sometimes convenient to act as if infinitely precise measurement were possible. From that perspective, measured variables are **continuous** in the sense that their sets of possible values are whole (continuous) intervals of numbers. For example, a convenient idealization might be that the Rockwell hardness of a machine part can lie anywhere in the interval $(0, \infty)$. But of course this is only an idealization. All real measurements are to the nearest unit (whatever that unit may be). This is becoming especially obvious as measurement instruments are increasingly equipped with digital displays. So in reality, when looked at under a strong enough magnifying glass, all numerical data (both measured and count alike) are **discrete** in the sense that they have isolated possible values rather than a continuum of available outcomes. Although $(0, \infty)$ may be mathematically convenient and completely adequate for practical purposes, the real set of possible values for the measured Rockwell hardness of a machine part may be more like $\{.1, .2, .3, \ldots\}$ than like $(0, \infty)$.

Well-known conventional wisdom is that measurement data are preferable to categorical and count data. Statistical methods for measurements are simpler and more informative than methods for qualitative data and counts. Further, there is typically far more to be learned from appropriate measurements than from qualitative data taken on the same physical objects. However, this must sometimes be balanced against the fact that measurement can be more time-consuming (and thus expensive) than the gathering of qualitative data.

Example 3

Pellet Mass Measurements

As a preliminary to their experimental study on the pelletizing process (discussed in Example 2), Cyr, Ellson, and Rickard collected data on a number of aspects of machine behavior. Included was the mass of pellets produced under standard operating conditions. Because a nonconforming pellet is typically one from which some material has broken off during production, pellet mass is indicative of system performance. Informal requirements for (specifications on) pellet mass were from 6.2 to 7.0 grams.

Example 3
(*continued*)

Information on 200 pellets was collected. The students could have simply observed and recorded whether or not a given pellet had mass within the specifications, thereby producing qualitative data. Instead, they took the time necessary to actually measure pellet mass to the nearest .1 gram—thereby collecting measurement data. A graphical summary of their findings is shown in Figure 1.3.

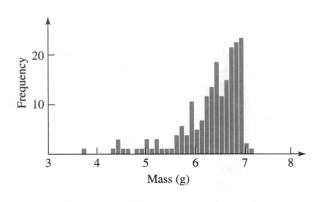

Figure 1.3   Pellet mass measurements

Notice that one can recover from the measurements the conformity/nonconformity information—about 28.5% (57 out of 200) of the pellets had masses that did not meet specifications. But there is much more in Figure 1.3 besides this. The shape of the display can give insights into how the machine is operating and the likely consequences of simple modifications to the pelletizing process. For example, note the **truncated** or chopped-off appearance of the figure. Masses do not trail off on the high side as they do on the low side. The students reasoned that this feature of their data had its origin in the fact that after powder is dispensed into a die, it passes under a paddle that wipes off excess material before a cylinder compresses the powder in the die. The amount initially dispensed to a given die may have a fairly symmetric mound-shaped distribution, but the paddle probably introduces the truncated feature of the display.

Also, from the numerical data displayed in Figure 1.3, one can find a percentage of pellet masses in any interval of interest, not just the interval [6.2, 7.0]. And by mentally sliding the figure to the right, it is even possible to project the likely effects of increasing die size by various amounts.

It is typical in engineering studies to have several response variables of interest. The next definitions present some jargon that is useful in specifying how many variables are involved and how they are related.

Definition 10	**Univariate data** arise when only a single characteristic of each sampled item is observed.

Definition 11	**Multivariate data** arise when observations are made on more than one characteristic of each sampled item. A special case of this involves two characteristics—**bivariate data**.

Definition 12	When multivariate data consist of several determinations of basically the same characteristic (e.g., made with different instruments or at different times), the data are called **repeated measures data**. In the special case of bivariate responses, the term **paired data** is used.

It is important to recognize the multivariate character of data when it is present. Having Rockwell hardness values for 5 of 100 crated machine parts and determinations of the percentage of carbon for 5 other parts is not at all equivalent to having both hardness and carbon content values for a single sample of 5 parts. There are two samples of 5 univariate data points in the first case and a single sample of 5 bivariate data points in the second. The second situation is preferable to the first, because it allows analysis and exploitation of any relationships that might exist between the variables Hardness and Percent Carbon.

Example 4 | Paired Distortion Measurements

In the furnace-loading scenario discussed in Example 1, radial runout measurements were actually made on all $38 + 39 = 77$ gears both before and after heat treating. (Only after-treatment values were given in Table 1.1.) Therefore, the process engineer had two samples (of respective sizes 38 and 39) of paired data. Because of the pairing, the engineer was in the position of being able (if desired) to analyze how post-treatment distortion was correlated with pretreatment distortion.

1.2.3 Types of Data Structures

Statistical engineering studies are sometimes conducted to compare process performance at one set of conditions to a stated standard. Such investigations involve only one sample. But it is far more common for several sets of conditions to be compared with each other, in which case several samples are involved. There are a variety of standard notions of structure or organization for multisample studies. Two of these are briefly discussed in the remainder of this section.

Definition 13

A **(complete) factorial study** is one in which several process variables (and settings of each) are identified as being of interest, and data are collected under each possible combination of settings of the process variables. The process variables are usually called **factors**, and the settings of each variable that are studied are termed **levels** of the factor.

For example, suppose there are four factors of interest—call them A, B, C, and D for convenience. If A has 3 levels, B has 2, C has 2, and D has 4, a study that includes samples collected under each of the $3 \times 2 \times 2 \times 4 = 48$ different possible sets of conditions would be called a $3 \times 2 \times 2 \times 4$ factorial study.

Example 2
(*continued*)

Experimentation with the pelletizing machine produced data with a $2 \times 2 \times 2$ (or 2^3) factorial structure. The factors and respective levels studied were

Die Volume   low volume vs. high volume

Material Flow   current method vs. manual filling

Mixture Type   no binding agent vs. with binder

Combining these then produced eight sets of conditions under which data were collected (see Table 1.2).

Table 1.2
Combinations in a 2^3 Factorial Study

Condition Number	Volume	Flow	Mixture
1	low	current	no binder
2	high	current	no binder
3	low	manual	no binder
4	high	manual	no binder
5	low	current	binder
6	high	current	binder
7	low	manual	binder
8	high	manual	binder

When many factors and/or levels are involved, the number of samples in a full factorial study quickly reaches an impractical size. Engineers often find that they want to collect data for only some of the combinations that would make up a complete factorial study.

Definition 14

A **fractional factorial study** is one in which data are collected for only some of the combinations that would make up a complete factorial study.

One cannot hope to learn as much about how a response is related to a given set of factors from a fractional factorial study as from the corresponding full factorial study. Some information must be lost when only part of all possible sets of conditions are studied. However, some fractional factorial studies will be potentially more informative than others. If only a fixed number of samples can be taken, which samples to take is an issue that needs careful consideration. Sections 8.3 and 8.4 discuss fractional factorials in detail, including how to choose good ones, taking into account what part of the potential information from a full factorial study they can provide.

Example 2
(*continued*)

The experiment actually carried out on the pelletizing process was, as indicated in Table 1.2, a full factorial study. Table 1.3 lists four experimental combinations, forming a well-chosen half of the eight possible combinations. (These are the combinations numbered 2, 3, 5, and 8 in Table 1.2.)

Table 1.3
Half of the 2^3 Factorial

Volume	Flow	Mixture
high	current	no binder
low	manual	no binder
low	current	binder
high	manual	binder

Section 2 Exercises

1. Describe a situation in your field where an observational study might be used to answer a question of real importance. Describe another situation where an experiment might be used.

2. Describe two different contexts in your field where, respectively, qualitative and quantitative data might arise.

3. What kind of information can be derived from a single sample of n bivariate data points (x, y) that can't be derived from two separate samples of, respectively, n data points x and n data points y?

4. Describe a situation in your field where paired data might arise.

5. Consider a study of making paper airplanes, where two different Designs (say, delta versus t wing), two different Papers (say, construction versus typing), and two different Loading Conditions (with a paper clip versus without a paper clip) are of interest in terms of their effects on flight distance. Describe a full factorial and then a fractional factorial data structure that might arise from such a study.

6. Explain why it is safer to infer causality from an experiment than from an observational study.

1.3   Measurement: Its Importance and Difficulty

Success in statistical engineering studies requires the ability to measure. For some physical properties like length, mass, temperature, and so on, methods of measurement are commonplace and obvious. Often, the behavior of an engineering system can be adequately characterized in terms of such properties. But when it cannot, engineers must carefully define what it is about the system that needs observing and then apply ingenuity to create a suitable method of measurement.

Example 5   Measuring Brittleness

A senior design class in metallurgical engineering took on the project of helping a manufacturer improve the performance of a spike-shaped metal part. In its intended application, this part needed to be strong but very brittle. When meeting an obstruction in its path, it had to break off rather than bend, because bending would in turn cause other damage to the machine in which the part functions.

As the class planned a statistical study aimed at finding what variables of manufacture affect part performance, the students came to realize that the company didn't have a good way of assessing part performance. As a necessary step in their study, they developed a measuring device. It looked roughly as in Figure 1.4. A swinging arm with a large mass at its end was brought to a

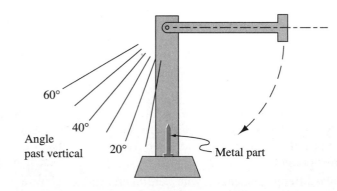

Figure 1.4   A device for measuring brittleness

horizontal position, released, and allowed to swing through a test part firmly fixed in a vertical position at the bottom of its arc of motion. The number of degrees past vertical that the arm traversed after impact with the part provided an effective measure of brittleness.

Example 6   Measuring Wood Joint Strength

Dimond and Dix wanted to conduct a factorial study comparing joint strengths for combinations of three different woods and three glues. They didn't have access to strength-testing equipment and so invented their own. To test a joint, they suspended a large container from one of the pieces of wood involved and poured water into it until the weight was sufficient to break the joint. Knowing the volume of water poured into the container and the density of water, they could determine the force required to break the joint.

Regardless of whether an engineer uses off-the-shelf technology or must fabricate a new device, a number of issues concerning measurement must be considered. These include **validity**, **measurement variation/error**, **accuracy**, and **precision**.

Definition 15   A measurement or measuring method is called **valid** if it usefully or appro-
Validity   priately represents the feature of an object or system that is of engineering importance.

It is impossible to overstate the importance of facing the question of measurement validity before plunging ahead in a statistical engineering study. Collecting engineering data costs money. Expending substantial resources collecting data, only to later decide they don't really help address the problem at hand, is unfortunately all too common.

The point was made in Section 1.1 that when using data, one is quickly faced *Measurement* with the fact that variation is omnipresent. Some of that variation comes about *error* because the objects studied are never exactly alike. But some of it is due to the fact that measurement processes also have their own inherent variability. Given a fine enough scale of measurement, no amount of care will produce exactly the same value over and over in repeated measurement of even a single object. And it is naive to attribute all variation in repeat measurements to bad technique or sloppiness. (Of course, bad technique and sloppiness *can* increase measurement variation beyond that which is unavoidable.)

An exercise suggested by W. J. Youden in his book *Experimentation and Measurement* is helpful in making clear the reality of measurement error. Consider measuring the thickness of the paper in this book. The technique to be used is as follows. The book is to be opened to a page somewhere near the beginning and one somewhere near the end. The stack between the two pages is to be grasped firmly between the thumb and index finger and stack thickness read to the nearest .1 mm using an ordinary ruler. Dividing the stack thickness by the number of sheets in the stack and recording the result to the nearest .0001 mm will then produce a thickness measurement.

Example 7

Book Paper Thickness Measurements

Presented below are ten measurements of the thickness of the paper in Box, Hunter, and Hunter's *Statistics for Experimenters* made one semester by engineering students Wendel and Gulliver.

Wendel: .0807, .0826, .0854, .0817, .0824,
            .0799, .0812, .0807, .0816, .0804

Gulliver: .0972, .0964, .0978, .0971, .0960,
            .0947, .1200, .0991, .0980, .1033

Figure 1.5 shows a graph of these data and clearly reveals that even repeated measurements by one person on one book will vary and also that the patterns of variation for two different individuals can be quite different. (Wendel's values are both smaller and more consistent than Gulliver's.)

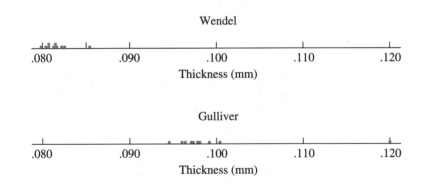

Figure 1.5   Dot diagrams of paper thickness measurements

The variability that is inevitable in measurement can be thought of as having both internal and external components.

Definition 16
Precision

A measurement system is called **precise** if it produces small variation in repeated measurement of the same object.

Precision is the internal consistency of a measurement system; typically, it can be improved only with basic changes in the configuration of the system.

Example 7
(continued)

Ignoring the possibility that some property of Gulliver's book was responsible for his values showing more spread than those of Wendel, it appears that Wendel's measuring technique was more precise than Gulliver's.

The precision of both students' measurements could probably have been improved by giving each a binder clip and a micrometer. The binder clip would provide a relatively constant pressure on the stacks of pages being measured, thereby eliminating the subjectivity and variation involved in grasping the stack firmly between thumb and index finger. For obtaining stack thickness, a micrometer is clearly a more precise instrument than a ruler.

Precision of measurement is important, but for many purposes it alone is not adequate.

Definition 17
Accuracy

A measurement system is called **accurate** (or sometimes, **unbiased**) if on average it produces the true or correct value of a quantity being measured.

Accuracy is the agreement of a measuring system with some external standard. It is a property that can typically be changed without extensive physical change in a measurement method. **Calibration** of a system against a standard (bringing it in line with the standard) can be as simple as comparing system measurements to a standard, developing an appropriate conversion scheme, and thereafter using converted values in place of raw readings from the system.

Example 7
(continued)

It is unknown what the industry-standard measuring methodology would have produced for paper thickness in Wendel's copy of the text. But for the sake of example, suppose that a value of .0850 mm/sheet was appropriate. The fact that Wendel's measurements averaged about .0817 mm/sheet suggests that her future accuracy might be improved by proceeding as before but then multiplying any figure obtained by the ratio of .0850 to .0817—i.e., multiplying by 1.04.

Maintaining the U.S. reference sets for physical measurement is the business of the National Institute of Standards and Technology. It is important business. Poorly calibrated measuring devices may be sufficient for local purposes of comparing local conditions. But to establish the values of quantities in any absolute sense, or to expect local values to have meaning at other places and other times, it is essential to calibrate measurement systems against a constant standard. A millimeter must be the same today in Iowa as it was last week in Alaska.

The possibility of bias or inaccuracy in measuring systems has at least two important implications for planning statistical engineering studies. First, the fact that

Accuracy and statistical studies   measurement systems can lose accuracy over time demands that their performance be monitored over time and that they be recalibrated as needed. The well-known phenomenon of **instrument drift** can ruin an otherwise flawless statistical study. Second, whenever possible, a single system should be used to do all measuring. If several measurement devices or technicians are used, it is hard to know whether the differences observed originate with the variables under study or from differences in devices or technician biases. If the use of several measurement systems is unavoidable, they must be calibrated against a standard (or at least against each other). The following example illustrates the role that human differences can play.

Example 8

Differences Between Technicians in Their Use of a Gauge

Cowan, Renk, Vander Leest, and Yakes worked with a company on the monitoring of a critical dimension of a high-precision metal part produced on a computer-controlled lathe. They encountered large, initially unexplainable variation in this dimension between different shifts at the plant. This variation was eventually traced not to any real shift-to-shift difference in the parts but to an instability in the company's measuring system. A single gauge was in use on all shifts, but different technicians used it quite differently when measuring the critical dimension. The company needed to train the technicians in a single, standardized method of using the gauge.

An analogy that is helpful in understanding the difference between precision and accuracy involves comparing measurement to target shooting. In target shooting, one can be on or off target (accurate or inaccurate) with a small or large cluster of shots (showing precision or imprecision). Figure 1.6 illustrates this analogy.

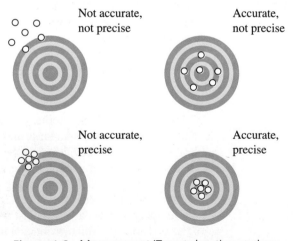

Figure 1.6   Measurement/Target shooting analogy

Good measurement is hard work, but without it data collection is futile. To make progress, engineers must obtain valid measurements, taken by methods whose precision and accuracy are sufficient to let them see important changes in system behavior. Usually, this means that measurement inaccuracy and imprecision must be an order of magnitude smaller than the variation in measured response caused by those changes.

Section 3 Exercises ....

1. Why might it be argued that in terms of producing useful measurements, one must deal first with the issue of validity, then the issue of precision, and only then the issue of accuracy?

2. Often, in order to evaluate a physical quantity (for example, the mean yield of a batch chemical process run according to some standard plant operating procedures), a large number of measurements of the quantity are made and then averaged. Explain which of the three aspects of measurement quality—validity, precision, and accuracy—this averaging of many measurements can be expected to improve and which it cannot.

3. Explain the importance of the stability of the measurement system to the real-world success of a statistical engineering study.

1.4 Mathematical Models, Reality, and Data Analysis

This is not a book on mathematics. Nevertheless, it contains a fair amount of mathematics (that most readers will find to be reasonably elementary—if unfamiliar and initially puzzling). Therefore, it seems wise to try to put the mathematical content of the book in perspective early. In this section, the relationships of mathematics to the physical world and to engineering statistics are discussed.

Mathematical models and reality

Mathematics is a construct of the human mind. While it is of interest to some people in its own right, engineers generally approach mathematics from the point of view that it can be useful in describing and predicting how physical systems behave. Indeed, although they exist only in our minds, mathematical theories are guides in every branch of modern engineering.

Throughout this text, we will frequently use the phrase *mathematical model*.

Definition 18

> A **mathematical model** is a description or summarization of salient features of a real-world system or phenomenon in terms of symbols, equations, numbers, and the like.

Mathematical models are themselves not reality, but they can be extremely effective descriptions of reality. This effectiveness hinges on two somewhat opposing properties of a mathematical model: (1) its degree of **simplicity** and (2) its **predictive**

ability. The most powerful mathematical models are those that simultaneously are simple *and* generate good predictions. A model's simplicity allows one to maneuver within its framework, deriving mathematical consequences of basic assumptions that translate into predictions of process behavior. When these are empirically correct, one has an effective engineering tool.

The elementary "laws" of mechanics are an outstanding example of effective mathematical modeling. For example, the simple mathematical statement that the acceleration due to gravity is constant,

$$a = g$$

yields, after one easy mathematical maneuver (an integration), the prediction that beginning with 0 velocity, after a time t in free fall an object will have velocity

$$v = gt$$

And a second integration gives the prediction that beginning with 0 velocity, a time t in free fall produces displacement

$$d = \frac{1}{2}gt^2$$

The beauty of this is that for most practical purposes, these easy predictions are quite adequate. They agree well with what is observed empirically and can be counted on as an engineer designs, builds, operates, and/or improves physical processes or products.

Mathematics and statistics   But then, how does the notion of mathematical modeling interact with the subject of engineering statistics? There are several ways. For one, data collection and analysis are essential in **fitting or estimating parameters** of mathematical models. To understand this point, consider again the example of a body in free fall. If one postulates that the acceleration due to gravity is constant, there remains the question of what numerical value that constant should have. The parameter g must be evaluated before the model can be used for practical purposes. One does this by gathering data and using them to estimate the parameter.

A standard first college physics lab has traditionally been to empirically evaluate g. The method often used is to release a steel bob down a vertical wire running through a hole in its center and allowing 60-cycle current to arc from the bob through a paper tape to another vertical wire, burning the tape slightly with every arc. A schematic diagram of the apparatus used is shown in Figure 1.7. The vertical positions of the burn marks are bob positions at intervals of $\frac{1}{60}$ of a second. Table 1.4 gives measurements of such positions. (We are grateful to Dr. Frank Peterson of the ISU Physics and Astronomy Department for supplying the tape.) Plotting the bob positions in the table at equally spaced intervals produces the approximately quadratic plot shown in Figure 1.8. Picking a parabola to fit the plotted points involves identifying an appropriate value for g. A method of curve fitting (discussed in Chapter 4) called *least squares* produces a value for g of 9.79m/sec^2, not far from the commonly quoted value of 9.8m/sec^2.

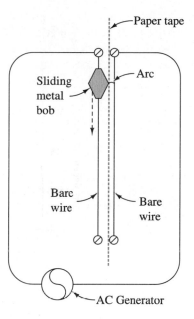

Figure 1.7   A device for measuring g

Table 1.4
Measured Displacements of a Bob in Free Fall

Point Number	Displacement (mm)	Point Number	Displacement (mm)
1	.8	13	223.8
2	4.8	14	260.0
3	10.8	15	299.2
4	20.1	16	340.5
5	31.9	17	385.0
6	45.9	18	432.2
7	63.3	19	481.8
8	83.1	20	534.2
9	105.8	21	589.8
10	131.3	22	647.7
11	159.5	23	708.8
12	190.5		

Notice that (at least before Newton) the data in Table 1.4 might also have been used in another way. The parabolic shape of the plot in Figure 1.8 could have suggested the form of an appropriate model for the motion of a body in free fall. That is, a careful observer viewing the plot of position versus time should conclude that there is an approximately quadratic relationship between position and time (and

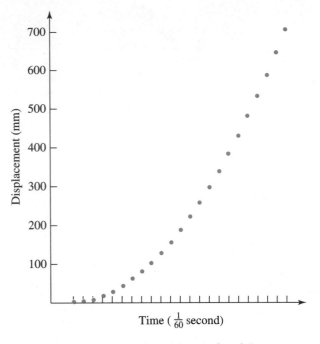

Figure 1.8   Bob positions in free fall

from that proceed via two differentiations to the conclusion that the acceleration due to gravity is roughly constant). This text is full of examples of how helpful it can be to use data both to identify potential forms for empirical models and to then estimate parameters of such models (preparing them for use in prediction).

This discussion has concentrated on the fact that statistics provides raw material for developing realistic mathematical models of real systems. But there is another important way in which statistics and mathematics interact. The mathematical theory of probability provides a framework for quantifying the uncertainty associated with inferences drawn from data.

Definition 19

> **Probability** is the mathematical theory intended to describe situations and phenomena that one would colloquially describe as involving chance.

If, for example, five students arrive at the five different laboratory values of g,

$$9.78, \ 9.82, \ 9.81, \ 9.78, \ 9.79$$

questions naturally arise as to how to use them to state both a best value for g and some measure of precision for the value. The theory of probability provides guidance in addressing these issues. Material in Chapter 6 shows that probability considerations support using the class average of 9.796 to estimate g and attaching to it a precision on the order of plus or minus .02m/sec^2.

We do not assume that the reader has studied the mathematics of probability, so this text will supply a minimal introduction to the subject. But do not lose sight of the fact that probability is not statistics—nor vice versa. Rather, probability is a branch of mathematics and a useful subject in its own right. It is met in a statistics course as a tool because the variation that one sees in real data is closely related conceptually to the notion of chance modeled by the theory of probability.

Section 4 Exercises

1. Explain in your own words the importance of mathematical models to engineering practice.

Chapter 1 Exercises

1. Calibration of measurement equipment is most clearly associated with which of the following concepts: validity, precision, or accuracy? Explain.

2. If factor A has levels 1, 2, and 3, factor B has levels 1 and 2, and factor C has levels 1 and 2, list the combinations of A, B, and C that make up a full factorial arrangement.

3. Explain how paired data might arise in a heat treating study aimed at determining the best way to heat treat parts made from a certain alloy.

4. Losen, Cahoy, and Lewis purchased eight spanner bushings of a particular type from a local machine shop and measured a number of characteristics of these bushings, including their outside diameters. Each of the eight outside diameters was measured once by two student technicians, with the following results. (The units are inches.) Considering both students' measurements, what type of data are given here? Explain.

Bushing	1	2	3	4
Student A	.3690	.3690	.3690	.3700
Student B	.3690	.3695	.3695	.3695
Bushing	5	6	7	8
Student A	.3695	.3700	.3695	.3690
Student B	.3695	.3700	.3700	.3690

5. Describe a situation from your field where a full factorial study might be conducted (name at least three factors, and the levels of each, that would appear in the study).

6. Example 7 concerns the measurement of the thickness of book paper. Variation in measurements is a fact of life. To observe this reality firsthand, measure the thickness of the paper used in this book ten times. Use the method described immediately before Example 7. For each determination, record the measured stack thickness, the number of sheets, and the quotient to four decimal places. If you are using this book in a formal course, be prepared to hand in your results and compare them with the values obtained by others in your class.

7. Exercise 6 illustrates the reality of variation in physical measurement. Another exercise that is similar in spirit, but leads to qualitative data, involves the spinning of U.S. pennies. Spin a penny on a hard surface 20 different times; for each trial, record whether the penny comes to rest with heads or tails showing. Did all the trials have the same outcome? Is the pattern you observed the one you expected to see? If not, do you have any possible explanations?

8. Consider a situation like that of Example 1 (involving the heat treating of gears). Suppose that the original gears can be purchased from a variety of vendors, they can be made out of a variety of materials, they can be heated according to a variety of regimens (involving different times and temperatures), they can be cooled in a number of different ways, and the furnace atmosphere can be adjusted to a variety of different conditions. A number of features of the final gears are of interest, including their flatness, their concentricity, their hardness (both before and after heat treating), and their surface finish.

   (a) What kind of data arise if, for a single set of conditions, the Rockwell hardness of several gears is measured both before and after heat treating? (Use the terminology of Section 1.2.) In the same context, suppose that engineering specifications on flatness require that measured flatness not exceed .40 mm. If flatness is measured for several gears and each gear is simply marked Acceptable or Not Acceptable, what kind of data are generated?

   (b) Describe a three-factor full factorial study that might be carried out in this situation. Name the factors that will be used and describe the levels of each. Write out a list of all the different combinations of levels of the factors that will be studied.

9. Suppose that you wish to determine "the" axial strength of a type of wooden dowel. Why might it be a good idea to test several such dowels in order to arrive at a value for this "physical constant"?

10. Give an example of a 2×3 full factorial data structure that might arise in a student study of the breaking strengths of wooden dowels. (Name the two factors involved, their levels, and write out all six different combinations.) Then make up a data collection form for the study. Plan to record both the breaking strength and whether the break was clean or splintered for each dowel, supposing that three dowels of each type are to be tested.

11. You are a mechanical engineer charged with improving the life-length characteristics of a hydrostatic transmission. You suspect that important variables include such things as the hardnesses, diameters and surface roughnesses of the pistons and the hardnesses, and inside diameters and surface roughnesses of the bores into which the pistons fit. Describe, in general terms, an observational study to try to determine how to improve life. Then describe an experimental study and say why it might be preferable.

12. In the context of Exercise 9, it might make sense to average the strengths you record. Would you expect such an average to be more or less precise than a single measurement as an estimate of the average strength of this kind of dowel? Explain. Argue that such averages can be no more (or less) accurate than the individual measurements that make them up.

13. A toy catapult launches golf balls. There are a number of things that can be altered on the configuration of the catapult: The length of the arm can be changed, the angle the arm makes when it hits the stop can be changed, the pull-back angle can be changed, the weight of the ball launched can be changed, and the place the rubber cord (used to snap the arm forward) is attached to the arm can be changed. An experiment is to be done to determine how these factors affect the distance a ball is launched.

   (a) Describe one three-factor full factorial study that might be carried out. Make out a data collection form that could be used. For each launch, specify the level to be used of each of the three factors and leave a blank for recording the observed value of the response variable. (Suppose two launches will be made for each setup.)

   (b) If each of the five factors mentioned above is included in a full factorial experiment, a minimum of how many different combinations of levels of the five factors will be required? If there is time to make only 16 launches with the device during the available lab period, but you want to vary all five factors, what kind of a data collection plan must you use?

14. As a variation on Exercise 6, you could try using only pages in the first four chapters of the book. If there were to be a noticeable change in the ultimate precision of thickness measurement, what kind of a change would you expect? Try this out by applying the method in Exercise 6 ten times to stacks of pages from only the first four chapters. Is there a noticeable difference in precision of measurement from what is obtained using the whole book?

2

· ·

Data Collection

Data collection is arguably the most important activity of engineering statistics. Often, properly collected data will essentially speak for themselves, making formal inferences rather like frosting on the cake. On the other hand, no amount of cleverness in post-facto data processing will salvage a badly done study. So it makes sense to consider carefully how to go about gathering data.

This chapter begins with a discussion of some general considerations in the collection of engineering data. It turns next to concepts and methods applicable specifically in enumerative contexts, followed by a discussion of both general principles and some specific plans for engineering experimentation. The chapter concludes with advice for the step-by-step planning of a statistical engineering study.

2.1 General Principles in the Collection of Engineering Data

Regardless of the particulars of a statistical engineering study, a number of common general considerations are relevant. Some of these are discussed in this section, organized around the topics of measurement, sampling, and recording.

2.1.1 Measurement

Good measurement is indispensable in any statistical engineering study. An engineer planning a study ought to ensure that data on relevant variables will be collected by well-trained people using measurement equipment of known and adequate quality.

When choosing variables to observe in a statistical study, the concepts of measurement validity and precision, discussed in Section 1.3, must be remembered. One practical point in this regard concerns how directly a measure represents a system property. When a **direct measure** exists, it is preferable to an indirect measure, because it will usually give much better precision.

Example 1

Exhaust Temperature Versus Weight Loss

An engineer working on a drying process for a bulk material was having difficulty determining when a target dryness had been reached. The method being used was monitoring the temperature of hot air being exhausted from the dryer. Exhaust temperature was a valid but very imprecise indicator of moisture content.

Someone suggested measuring the weight loss of the material instead of exhaust temperature. The engineer developed an ingenious method of doing this, at only slightly greater expense. This much more direct measurement greatly improved the quality of the engineer's information.

It is often easier to identify appropriate measures than to carefully and unequivocally define them so that they can be used. For example, suppose a metal cylinder is to be turned on a lathe, and it is agreed that cylinder diameter is of engineering importance. What is meant by the word *diameter*? Should it be measured on one end of the cylinder (and if so, which?) or in the center, or where? In practice, these locations will differ somewhat. Further, when a cylinder is gauged at some chosen location, should it be rolled in the gauge to get a maximum (or minimum) reading, or should it simply be measured as first put into the gauge? The cross sections of real-world cylinders are not exactly circular or uniform, and how the measurement is done will affect how the resulting data look.

It is especially necessary—and difficult—to make careful **operational definitions** where qualitative and count variables are involved. Consider the case of a process engineer responsible for an injection-molding machine producing plastic auto grills. If the number of abrasions appearing on these is of concern and data are to be gathered, how is *abrasion* defined? There are certainly locations on a grill where a flaw is of no consequence. Should those areas be inspected? How big should an abrasion be in order to be included in a count? How (if at all) should an inspector distinguish between abrasions and other imperfections that might appear on a grill? All of these questions must be addressed in an operational definition of "abrasion" before consistent data collection can take place.

Once developed, operational definitions and standard measurement procedures must be communicated to those who will use them. Training of technicians has to be taken seriously. Workers need to understand the importance of adhering to the standard definitions and methods in order to provide consistency. For example, if instructions call for zeroing an instrument before each measurement, it must always be done.

The performance of any measuring equipment used in a study must be known to be adequate—both before beginning and throughout the study. Most large industrial concerns have regular programs for both recalibrating and monitoring the precision of their measuring devices. The second of these activities sometimes goes under the name of **gauge R and R studies**—the two R's being **repeatability** and **reproducibility**. Repeatability is variation observed when a single operator uses the gauge to measure and remeasure one item. Reproducibility is variation in measurement attributable to differences among operators. (A detailed discussion of such studies can be found in Section 2.2.2 of *Statistical Quality Assurance Methods for Engineers* by Vardeman and Jobe.)

Calibration and precision studies should assure the engineer that instrumentation is adequate at the beginning of a statistical study. If the time span involved in the study is appreciable, the *stability* of the instrumentation must be maintained over the study period through checks on calibration and precision.

2.1.2   Sampling

Once it is established how measurement/observation will proceed, the engineer can consider how much to do, who is to do it, where and under what conditions it is to be done, etc. Sections 2.2, 2.3, and 2.4 consider the question of choosing what observations to make, first in enumerative and then in experimental studies. But first, a few general comments about the issues of "How much?", "Who?", and "Where?".

How much data?

The most common question engineers ask about data collection is "How many observations do I need?" Unfortunately, the proper answer to the question is typically "it depends." As you proceed through this book, you should begin to develop some intuition and some rough guides for choosing sample sizes. For the time being, we point out that the only factor on which the answer to the sample size question really depends is the variation in response that one expects (coming both from unit-to-unit variation and from measurement variation).

This makes sense. If objects to be observed were all alike and perfect measurement were possible, then a single observation would suffice for any purpose. But if there is increase either in the measurement noise or in the variation in the system or population under study, the sample size necessary to get a clear picture of reality becomes larger.

However, one feature of the matter of sample size sometimes catches people a bit off guard—the fact that in enumerative studies (provided the population size is large), sample size requirements do not depend on the population size. That is, sample size requirements are not relative to population size, but, rather, are absolute. If a sample size of 5 is adequate to characterize compressive strengths of a lot of 1,000 red clay bricks, then a sample of size 5 would be adequate to characterize compressive strengths for a lot of 100,000 bricks with similar brick-to-brick variability.

Who should collect data?

The "Who?" question of data collection cannot be effectively answered without reference to human nature and behavior. This is true even in a time when automatic data collection devices are proliferating. Humans will continue to supervise these and process the information they generate. Those who collect engineering data must not only be well trained; they must also be convinced that the data they collect will be used and in a way that is in their best interests. Good data must be seen as a help in doing a good job, benefiting an organization, and remaining employed, rather than as pointless or even threatening. If those charged with collecting or releasing data believe that the data will be used against them, it is unrealistic to expect them to produce useful data.

Example 2

Data—An Aid or a Threat?

One of the authors once toured a facility with a company industrial statistician as guide. That person proudly pointed out evidence that data were being collected and effectively used. Upon entering a certain department, the tone of the conversation changed dramatically. Apparently, the workers in that department had been asked to collect data on job errors. The data had pointed unmistakably to poor performance by a particular individual, who was subsequently fired from the company. Thereafter, convincing other workers that data collection is a helpful activity was, needless to say, a challenge.

Perhaps all the alternatives in this situation (like retraining or assignment to a different job) had already been exhausted. But the appropriateness of the firing is not the point here. Rather, the point is that circumstances were allowed to create an atmosphere that was not conducive to the collection and use of data.

Even where those who will gather data are convinced of its importance and are eager to cooperate, care must be exercised. Personal biases (whether conscious or subconscious) must not be allowed to enter the data collection process. Sometimes in a statistical study, hoped-for or predicted best conditions are deliberately or unwittingly given preference over others. If this is a concern, measurements can be made **blind** (i.e., without personnel knowing what set of conditions led to an item being measured). Other techniques for ensuring fair play, having less to do with human behavior, will be discussed in the next two sections.

Where should data be collected?

The "Where?" question of engineering data collection can be answered in general terms: "As close as possible in time and space to the phenomenon being studied." The importance of this principle is most obvious in the routine monitoring of complex manufacturing processes. The performance of one operation in such a process is most effectively monitored at the operation rather than at some later point. If items being produced turn out to be unsatisfactory at the end of the line, it is rarely easy to backtrack and locate the operation responsible. Even if that is accomplished, unnecessary waste has occurred during the time lag between the onset of operation malfunction and its later discovery.

Example 3

IC Chip Manufacturing Process Improvement

The preceding point was illustrated during a visit to a "clean room" where integrated circuit chips are manufactured. These are produced in groups of 50 or so on so-called wafers. Wafers are made by successively putting down a number of appropriately patterned, very thin layers of material on an inert background disk. The person conducting the tour said that at one point, a huge fraction of wafers produced in the room had been nonconforming. After a number of false starts, it was discovered that by appropriate testing (data collection) at the point of application of the second layer, a majority of the eventually nonconforming

Example 3
(continued)

wafers could be identified and eliminated, thus saving the considerable extra expense of further processing. What's more, the need for adjustments to the process was signaled in a timely manner.

2.1.3 Recording

The object of engineering data collection is to get data used. How they are recorded has a major impact on whether this objective is met. A good data recording format can make the difference between success and failure.

Example 4

A Data Collection Disaster

A group of students worked with a maker of molded plastic business signs in an effort to learn what factors affect the shrinkage a sign undergoes as it cools. They considered factors such as Operator, Heating Time, Mold Temperature, Mold Size, Ambient Temperature, and Humidity. Then they planned a partially observational and partially experimental study of the molding process. After spending two days collecting data, they set about to analyze them. The students discovered to their dismay that although they had recorded many features of what went on, they had neglected to record either the size of the plastic sheets before molding or the size of the finished signs. Their considerable effort was entirely wasted. It is likely that this mistake could have been prevented by careful precollection development of a data collection form.

When data are collected in a routine, ongoing, process-monitoring context (as opposed to a one-shot study of limited duration), it is important that they be used to provide effective, timely feedback of information. Increasingly, computer-made graphical displays of data, in real time, are used for this purpose. But it is often possible to achieve this much more cheaply through clever design of a manual data collection form, if the goal of making data recording convenient and immediately useful is kept in sight.

Example 5

Recording Bivariate Data on PVC Bottles

Table 2.1 presents some bivariate data on bottle mass and width of bottom piece resulting from blow molding of PVC plastic bottles (taken from *Modern Methods for Quality Control and Improvement* by Wadsworth, Stephens, and Godfrey). Six consecutive samples of size 3 are represented.

Such bivariate data could be recorded in much the same way as they are listed in Table 2.1. But if it is important to have immediate feedback of information (say, to the operator of a machine), it would be much more effective to use a well-thought-out bivariate **check sheet** like the one in Figure 2.1. On such a sheet, it

Table 2.1
Mass and Bottom Piece Widths of PVC Bottles

Sample	Item	Mass (g)	Width (mm)	Sample	Item	Mass (g)	Width (mm)
1	1	33.01	25.0	4	10	32.80	26.5
1	2	33.08	24.0	4	11	32.86	28.5
1	3	33.24	23.5	4	12	32.89	25.5
2	4	32.93	26.0	5	13	32.73	27.0
2	5	33.17	23.0	5	14	32.57	28.0
2	6	33.07	25.0	5	15	32.65	26.5
3	7	33.01	25.5	6	16	32.43	30.0
3	8	32.82	27.0	6	17	32.54	28.0
3	9	32.91	26.0	6	18	32.61	26.0

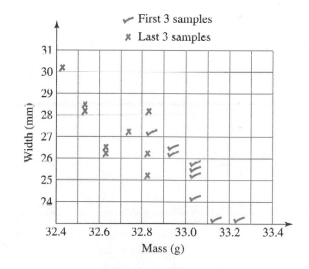

Figure 2.1   Check sheet for the PVC bottle data

is easy to see how the two variables are related. If, as in the figure, the recording symbol is varied over time, it is also easy to track changes in the characteristics over time. In the present case, width seems to be inversely related to mass, which appears to be decreasing over time.

To be useful (regardless of whether data are recorded on a routine basis or in a one-shot mode, automatically or by hand), the recording must carry enough **documentation** that the important circumstances surrounding the study can be reconstructed. In a one-shot experimental study, someone must record responses

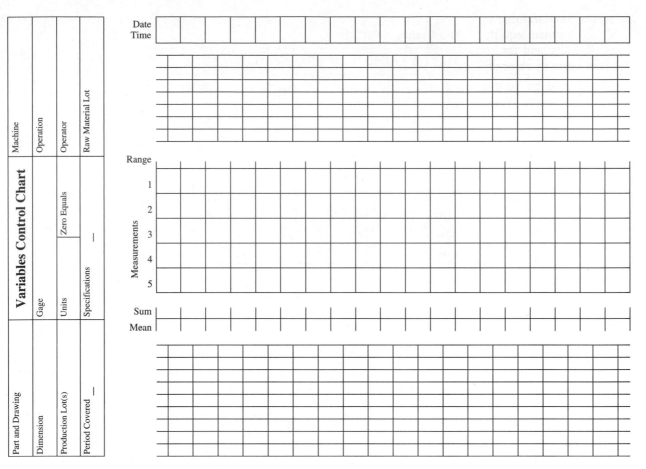

Figure 2.2   Variables control chart form

and values of the experimental variables, and it is also wise to keep track of other variables that might later prove to be of interest. In the context of routine process monitoring, data records will be useful in discovering differences in raw material lots, machines, operators, etc., only if information on these is recorded along with the responses of interest. Figure 2.2 shows a form commonly used for the routine collection of measurements for process monitoring. Notice how thoroughly the user is invited to document the data collection.

Section 1 Exercises

1. Consider the context of a study on making paper airplanes where two different Designs (say delta versus t wing), two different Papers (say construction versus typing), and two different Loading Conditions (with a paper clip versus without a paper clip) are of interest with regard to their impact on flight distance. Give an operational definition of *flight distance* that you might use in such a study.

2. Explain how training operators in the proper use of measurement equipment might affect both the repeatability and the reproducibility of measurements made by an organization.

3. What would be your response to another engineer's comment, "We have great information on

our product—we take 5% samples of every outgoing order, regardless of order size!"?

4. State briefly why it is critical to make careful operational definitions for response variables in statistical engineering studies.

2.2 Sampling in Enumerative Studies

An enumerative study has an identifiable, concrete population of items. This section discusses selecting a sample of the items to include in a statistical investigation.

Using a sample to represent a (typically much larger) population has obvious advantages. Measuring some characteristics of a sample of 30 electrical components from an incoming lot of 10,000 can often be feasible in cases where it would not be feasible to perform a **census** (a study that attempts to include every member of the population). Sometimes testing is destructive, and studying an item renders it unsuitable for subsequent use. Sometimes the timeliness and data quality of a sampling investigation far surpass anything that could be achieved in a census. Data collection technique can become lax or sloppy in a lengthy study. A moderate amount of data, collected under close supervision and put to immediate use, can be very valuable—often more valuable than data from a study that might appear more complete but in fact takes too long.

If a sample is to be used to stand for a population, how that sample is chosen becomes very important. The sample should somehow be representative of the population. The question addressed here is how to achieve this.

Systematic and **judgment-based** methods can in some circumstances yield samples that faithfully portray the important features of a population. If a lot of items is manufactured in a known order, it may be reasonable to select, say, every 20th one for inclusion in a statistical engineering study. Or it may be effective to force the sample to be balanced—in the sense that every operator, machine, and raw material lot (for example) appears in the sample. Or an old hand may be able to look at a physical population and fairly accurately pick out a representative sample.

But there are potential problems with such methods of sample selection. Humans are subject to conscious and subconscious preconceptions and biases. Accordingly, judgment-based samples can produce distorted pictures of populations. Systematic methods can fail badly when unexpected cyclical patterns are present. (For example, suppose one examines every 20th item in a lot according to the order in which the items come off a production line. Suppose further that the items are at one point processed on a machine having five similar heads, each performing the same operation on every fifth item. Examining every 20th item only gives a picture of how one of the heads is behaving. The other four heads could be terribly misadjusted, and there would be no way to find this out.)

Even beyond these problems with judgment-based and systematic methods of sampling, there is the additional difficulty that it is not possible to quantify their properties in any useful way. There is no good way to take information from samples drawn via these methods and make reliable statements of likely margins of error. The method introduced next avoids the deficiencies of systematic and judgment-based sampling.

Definition 1 *Simple random sampling*	A **simple random sample of size** n from a population is a sample selected in such a manner that every collection of n items in the population is a priori equally likely to compose the sample.

Probably the easiest way to think of simple random sampling is that it is conceptually equivalent to drawing n slips of paper out of a hat containing one for each member of the population.

Example 6	Random Sampling Dorm Residents C. Black did a partially enumerative and partially experimental study comparing student reaction times under two different lighting conditions. He decided to recruit subjects from his coed dorm floor, selecting a simple random sample of 20 of these students to recruit. In fact, the selection method he used involved a table of so-called random digits. But he could have just as well written the names of all those living on his floor on standard-sized slips of paper, put them in a bowl, mixed thoroughly, closed his eyes, and selected 20 different slips from the bowl.

Mechanical methods and simple random sampling

Methods for actually carrying out the selection of a simple random sample include **mechanical methods** and **methods using "random digits."** Mechanical methods rely for their effectiveness on symmetry and/or thorough mixing in a physical randomizing device. So to speak, the slips of paper in the hat need to be of the same size and well scrambled before sample selection begins.

The first Vietnam-era U.S. draft lottery was a famous case in which adequate care was not taken to ensure appropriate operation of a mechanical randomizing device. Birthdays were supposed to be assigned priority numbers 1 through 366 in a "random" way. However, it was clear after the fact that balls representing birth dates were placed into a bin by months, and the bin was poorly mixed. When the balls were drawn out, birth dates near the end of the year received a disproportionately large share of the low draft numbers. In the present terminology, the first five dates out of the bin should *not* have been thought of as a simple random sample of size 5. Those who operate games of chance more routinely make it their business to know (via the collection of appropriate data) that their mechanical devices are operating in a more random manner.

Using random digits to do sampling implicitly relies for "randomness" on the appropriateness of the method used to generate those digits. *Physical random processes* like radioactive decay and *pseudorandom number generators* (complicated recursive numerical algorithms) are the most common sources of random digits. Until fairly recently, it was common to record such digits in printed tables. Table B.1 consists of random digits (originally generated by a physical random process). The first five rows of this table are reproduced in Table 2.2 for use in this section.

In making a random digit table, the intention is to use a method guaranteeing that a priori

1. each digit 0 through 9 has the same chance of appearing at any particular location in the table one wants to consider, and

2. knowledge of which digit will occur at a given location provides no help in predicting which one will appear at another.

In a random digit table, condition 1 should typically be reflected in roughly equal representation of the 10 digits, and condition 2 in the lack of obvious internal patterns in the table.

Random digit tables and simple random sampling

For populations that can easily be labeled with consecutive numbers, the following steps can be used to synthetically draw items out of a hat one at a time—to draw a simple random sample using a table like Table 2.2.

Step 1   For a population of N objects, determine the number of digits in N (for example, $N = 1291$ is a four-digit number). Call this number M and assign each item in the population a different M-digit label.

Step 2   Move through the table left to right, top to bottom, M digits at a time, beginning from where you left off in last using the table, and choose objects from the population by means of their associated labels until n have been selected.

Step 3   In moving through the table according to step 2, ignore labels that have not been assigned to items in the population and any that would indicate repeat selection of an item.

Table 2.2
Random Digits

12159	66144	05091	13446	45653	13684	66024	91410	51351	22772
30156	90519	95785	47544	66735	35754	11088	67310	19720	08379
59069	01722	53338	41942	65118	71236	01932	70343	25812	62275
54107	58081	82470	59407	13475	95872	16268	78436	39251	64247
99681	81295	06315	28212	45029	57701	96327	85436	33614	29070

12159 66144 05091 13446 45653 13684 66024 91410 51351 22772
30156 90519 95785 47544 66735 35754 11088 67310 19720 08379

Figure 2.3   Use of a random digit table

As an example of how this works, consider selecting a simple random sample of 25 members of a hypothetical population of 80 objects. One first determines that 80 is an $M = 2$-digit number and therefore labels items in the population as 01, 02, 03, 04, ..., 77, 78, 79, 80 (labels 00 and 81 through 99 are not assigned). Then, if Table 2.2 is being used for the first time, begin in the upper left corner and proceed as indicated in Figure 2.3. Circled numbers represent selected labels, Xs indicate that the corresponding label has not been assigned, and slash marks indicate that the corresponding item has already entered the sample. As the final item enters the sample, the stopping point is marked with a penciled hash mark. Movement through the table is resumed at that point the next time the table is used.

Any predetermined systematic method of moving through the table could be substituted in place of step 2. One could move down columns instead of across rows, for example. It is useful to make the somewhat arbitrary choice of method in step 2 for the sake of classroom consistency.

Statistical or spreadsheet software and simple random sampling

With the wide availability of personal computers, random digit tables have become largely obsolete. That is, random numbers can be generated "on the spot" using statistical or spreadsheet software. In fact, it is even easy to have such software automatically do something equivalent to steps 1 through 3 above, selecting a simple random sample of n of the numbers 1 to N. For example, Printout 1 was produced using the MINITAB™ statistical package. It illustrates the selection of $n = 25$ members of a population of $N = 80$ objects. The numbers 1 through 80 are placed into the first column of a worksheet (using the routine under the "Calc/Make Patterned Data/Simple Set of Numbers" menu). Then 25 of them are selected using MINITAB's pseudorandom number generation capability (located under the "Calc/Random Data/Sample from Columns" menu). Finally, those 25 values (the results beginning with 56 and ending with 72) are printed out (using the routine under the "Manip/Display Data" menu).

Printout 1   Random Selection of 25 Objects from a Population of 80 Objects

```
MTB > Set C1
DATA>   1( 1 : 80 / 1 )1
DATA>   End.
MTB > Sample 25 C1 C2.
MTB > Print C2.

Data Display

C2
      56      74      43      61      80      22      30      67      35       7
      10      69      19      49       8      45       3      37      21      17
       2      12       9      14      72
```

Regardless of how Definition 1 is implemented, several comments about the method are in order. First, it must be admitted that simple random sampling meets the original objective of providing representative samples only in some average or long-run sense. It is possible for the method to produce particular realizations that are horribly unrepresentative of the corresponding population. A simple random sample of 20 out of 80 axles could turn out to consist of those with the smallest diameters. But this doesn't happen often. On the average, a simple random sample will faithfully portray the population. Definition 1 is a statement about a method, not a guarantee of success on a particular application of the method.

Second, it must also be admitted that there is no guarantee that it will be an easy task to make the physical selection of a simple random sample. Imagine the pain of retrieving 5 out of a production run of 1,000 microwave ovens stored in a warehouse. It would probably be a most unpleasant job to locate and gather 5 ovens corresponding to randomly chosen serial numbers to, for example, carry to a testing lab.

But the virtues of simple random sampling usually outweigh its drawbacks. For one thing, it is an **objective method** of sample selection. An engineer using it is protected from conscious and subconscious human bias. In addition, the method **interjects probability** into the selection process in what turns out to be a manageable fashion. As a result, the quality of information from a simple random sample can be quantified. Methods of formal statistical inference, with their resulting conclusions ("I am 95% sure that . . ."), can be applied when simple random sampling is used.

It should be clear from this discussion that there is nothing mysterious or magical about simple random sampling. We sometimes get the feeling while reading student projects (and even some textbooks) that the phrase *random sampling* is used (even in analytical rather than enumerative contexts) to mean "magically OK sampling" or "sampling with magically universally applicable results." Instead, simple random sampling is a concrete methodology for enumerative studies. It is generally about the best one available without a priori having intimate knowledge of the population.

Section 2 Exercises

1. For the sake of exercise, treat the runout values for 38 laid gears (given in Table 1.1) as a population of interest, and using the random digit table (Table B.1), select a simple random sample of 5 of these runouts. Repeat this selection process a total of four different times. (Begin the selection of the first sample at the upper left of the table and proceed left to right and top to bottom.) Are the four samples identical? Are they each what you would call "representative" of the population?

2. Repeat Exercise 1 using statistical or spreadsheet software to do the random sampling.

3. Explain briefly why in an enumerative study, a simple random sample is or is not guaranteed to be representative of the population from which it is drawn.

2.3 Principles for Effective Experimentation

Purposely introducing changes into an engineering system and observing what happens as a result (i.e., experimentation) is a principal way of learning how the system works. Engineers meet such a variety of experimental situations that it is impossible to give advice that will be completely relevant in all cases. But it is possible to raise some general issues, which we do here. The discussion in this section is organized under the headings of

1. taxonomy of variables,

2. handling extraneous variables,

3. comparative study,

4. replication, and

5. allocation of resources.

Then Section 2.4 discusses a few generic experimental frameworks for planning a specific experiment.

2.3.1 Taxonomy of Variables

One of the hard realities of experiment planning is the multidimensional nature of the world. There are typically many characteristics of system performance that the engineer would like to improve and many variables that might influence them. Some terminology is needed to facilitate clear thinking and discussion in light of this complexity.

Definition 2

> A **response variable** in an experiment is one that is monitored as characterizing system performance/behavior.

A response variable is a system output. Some variables that potentially affect a response of interest are **managed** by the experimenter.

Definition 3

> A **supervised (or managed) variable** in an experiment is one over which an investigator exercises power, choosing a setting or settings for use in the study. When a supervised variable is held constant (has only one setting), it is called a **controlled variable**. And when a supervised variable is given several different settings in a study, it is called an **experimental variable**.

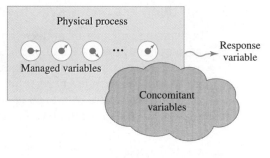

Figure 2.4   Variables in an experiment

Some of the variables that are neither primary responses nor managed in an experiment will nevertheless be observed.

Definition 4	A **concomitant (or accompanying) variable** in an experiment is one that is observed but is neither a primary response variable nor a managed variable. Such a variable can change in reaction to either experimental or unobserved causes and may or may not itself have an impact on a response variable.

Figure 2.4 is an attempt to picture Definitions 2 through 4. In it, the physical process somehow produces values of a response. "Knobs" on the process represent managed variables. Concomitant variables are floating about as part of the experimental environment without being its main focus.

Example 7
(Example 6, Chapter 1, revisited—p. 15)

Variables in a Wood Joint Strength Experiment

Dimond and Dix experimented with three different woods and three different glues, investigating joint strength properties. Their primary interest was in the effects of experimental variables Wood Type and Glue Type on two observed response variables, joint strength in a tension test and joint strength in a shear test.

In addition, they recognized that strengths were probably related to the variables Drying Time and Pressure applied to the joints while drying. Their method of treating the nine wood/glue combinations fairly with respect to the Time and Pressure variables was to manage them as controlled variables, trying to hold them essentially constant for all the joints produced.

Some of the variation the students observed in strengths could also have originated in properties of the particular specimens glued, such as moisture content. In fact, this variable was not observed in the study. But if the students had had some way of measuring it, moisture content might have provided extra insight into how the wood/glue combinations behave. It would have been a potentially informative concomitant variable.

2.3.2   Handling Extraneous Variables

In planning an experiment, there are always variables that could influence the responses but which are not of practical interest to the experimenter. The investigator may recognize some of them as influential but not even think of others. Those that are recognized may fail to be of primary interest because there is no realistic way of exercising control over them or compensating for their effects outside of the experimental environment. So it is of little practical use to know exactly how changes in them affect the system.

But completely ignoring the existence of such **extraneous variables** in experiment planning can needlessly cloud the perception of the effects of factors that *are* of interest. Several methods can be used in an active attempt to avoid this loss of information. These are to manage them (for experimental purposes) as **controlled variables** (recall Definition 3) or as **blocking variables**, or to attempt to balance their effects among process conditions of interest through **randomization**.

Control of extraneous variables

When choosing to control an extraneous variable in an experiment, both the pluses and minuses of that choice should be recognized. On the one hand, the control produces a homogeneous environment in which to study the effects of the primary experimental variables. In some sense, a portion of the background noise has been eliminated, allowing a clearer view of how the system reacts to changes in factors of interest. On the other hand, system behavior at other values of the controlled variable cannot be projected on the firm basis of data. Instead, projections must be based on the basis of *expert opinion* that what is seen experimentally will prove true more generally. Engineering experience is replete with examples where what worked fine in a laboratory (or even a pilot plant) was much less dependable in subsequent experience with a full-scale facility.

Example 7
(continued)

The choice Dimond and Dix made to control Drying Time and the Pressure provided a uniform environment for comparing the nine wood/glue combinations. But strictly speaking, they learned only about joint behavior under their particular experimental Time and Pressure conditions.

To make projections for other conditions, they had to rely on their experience and knowledge of material science to decide how far the patterns they observed were likely to extend. For example, it may have been reasonable to expect what they observed to also hold up for any drying time at least as long as the experimental one, because of expert knowledge that the experimental time was sufficient for the joints to fully set. But such extrapolation is based on other than statistical grounds.

Blocking extraneous variables

An alternative to controlling extraneous variables is to handle them as experimental variables, including them in study planning at several different levels. Notice that this really amounts to applying the notion of control *locally*, by creating not one but several (possibly quite different) homogeneous environments in which to compare levels of the primary experimental variables. The term *blocking* is often used to refer to this technique.

Definition 5

> A **block** of experimental units, experimental times of observation, experimental conditions, etc. is a homogeneous group within which different levels of primary experimental variables can be applied and compared in a relatively uniform environment.

Example 7
(continued)

Consider embellishing a bit on the gluing study of Dimond and Dix. Imagine that the students were uneasy about two issues, the first being the possibility that surface roughness differences in the pieces to be glued might mask the wood/glue combination differences of interest. Suppose also that because of constraints on schedules, the strength testing was going to have to be done in two different sessions a day apart. Measuring techniques or variables like ambient humidity might vary somewhat between such periods. How might such potential problems have been handled?

Blocking is one way. If the specimens of each wood type were separated into relatively rough and relatively smooth groups, the factor Roughness could have then served as an experimental factor. Each of the glues could have been used the same number of times to join both rough and smooth specimens of each species. This would set up comparison of wood/glue combinations separately for rough and for smooth surfaces.

In a similar way, half the testing for each wood/glue/roughness combination might have been done in each testing session. Then, any consistent differences between sessions could be identified and prevented from clouding the comparison of levels of the primary experimental variables. Thus, Testing Period could have also served as a blocking variable in the study.

Experimenters usually hope that by careful planning they can account for the most important extraneous variables via control and blocking. But not all extraneous variables can be supervised. There are an essentially infinite number, most of which cannot even be named. And there is a way to take out insurance against the possibility that major extraneous variables get overlooked and then produce effects that are mistaken for those of the primary experimental variables.

Randomization and extraneous variables

Definition 6

> **Randomization** is the use of a randomizing device or table of random digits at some point where experimental protocol is not already dictated by the specification of values of the supervised variables. Often this means that experimental objects (or units) are divided up between the experimental conditions at random. It can also mean that the order of experimental testing is randomly determined.

The goal of randomization is to average between sets of experimental conditions the effects of all unsupervised extraneous variables. To put it differently, sets of experimental conditions are treated fairly, giving them equal opportunity to shine.

Example 8
(*Example 1, Chapter 1, revisited—p. 2*)

Randomization in a Heat Treating Study

P. Brezler, in his "Heat Treating" article, describes a very simple randomized experiment for comparing the effects on thrust face runout of laying versus hanging gears. The variable Loading Method was the primary experimental variable. Extraneous variables Steel Heat and Machining History were controlled by experimenting on 78 gears from the same heat code, machined as a lot. The 78 gears were broken at random into two groups of 39, one to be laid and the other to be hung. (Note that Table 1.1 gives only 38 data points for the laid group. For reasons not given in the article, one laid gear was dropped from the study.)

Although there is no explicit mention of this in the article, the principle of randomization could have been (and perhaps was) carried a step further by making the runout measurements in a random order. (This means choosing gears 01 through 78 one at a time at random to measure.) The effect of this randomization would have been to protect the investigator from clouding the comparison of heat treating methods with possible unexpected and unintended changes in measurement techniques. Failing to randomize and, for example, making all the laid measurements before the hung measurements, would allow unintended changes in measurement technique to appear in the data as differences between the two loading methods. (Practice with measurement equipment might, for example, increase precision and make later runouts appear to be more uniform than early ones.)

Example 7
(*continued*)

Dimond and Dix took the notion of randomization to heart in their gluing study and, so to speak, randomized everything in sight. In the tension strength testing for a given type of wood, they glued $.5'' \times .5'' \times 3''$ blocks to a $.75'' \times 3.5'' \times 31.5''$ board of the same wood type, as illustrated in Figure 2.5.

Each glue was used for three joints on each type of wood. In order to deal with any unpredicted differences in material properties (e.g., over the extent of the board) or unforeseen differences in loading by the steel strap used to provide pressure on the joints, etc., the students randomized the order in which glue was applied and the blocks placed along the base board. In addition, when it came time to do the strength testing, that was carried out in a randomly determined order.

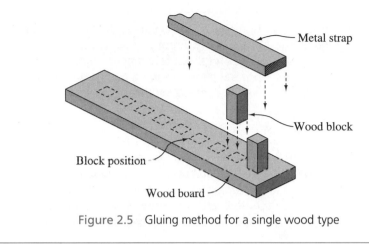

Figure 2.5   Gluing method for a single wood type

Simple random sampling in enumerative studies is only guaranteed to be effective in an average or long-run sense. Similarly, randomization in experiments will not prove effective in averaging the effects of extraneous variables between settings of experimental variables every time it is used. Sometimes an experimenter will be unlucky. But the methodology is objective, effective on the average, and about the best one can do in accounting for those extraneous variables that will not be managed.

2.3.3   Comparative Study

Statistical engineering studies often involve more than a single sample. They usually involve comparison of a number of settings of process variables. This is true not only because there may be many options open to an engineer in a given situation, but for other reasons as well.

Even in experiments where there is only a single new idea or variation on standard practice to be tried out, it is a good idea to make the study **comparative** (and therefore to involve more than one sample). Unless this is done, there is no really firm basis on which to say that any effects observed come from the new conditions under study rather than from unexpected extraneous sources. If standard yield for a chemical process is 63.2% and a few runs of the process with a supposedly improved catalyst produce a mean yield of 64.8%, it is not completely safe to attribute the difference to the catalyst. It could be caused by a number of things, including miscalibration of the measurement system. But suppose a few experimental runs are taken for both the standard and the new catalysts. If these produce two samples with small internal variation and (for example) a difference of 1.6% in mean yields, that difference is more safely attributed to a difference in the catalysts.

Example 8
(*continued*)

In the gear loading study, hanging was the standard method in use at the time of the study. From its records, the company could probably have located some values for thrust face runout to use as a baseline for evaluating the laying method. But the choice to run a comparative study, including both laid and hung gears, put the engineer on firm ground for drawing conclusions about the new method.

A second usage of "control"

In a potentially confusing use of language, the word *control* is sometimes used to mean the practice of including a standard or no-change sample in an experiment for comparison purposes. (Notice that this is not the usage in Definition 3.) When a *control group* is included in a medical study to verify the effectiveness of a new drug, that group is either a standard-treatment or no-treatment group, included to provide a solid basis of comparison for the new treatment.

2.3.4   Replication

In much of what has been said so far, it has been implicit that having more than one observation for a given setting of experimental variables is a good idea.

Definition 7

Replication of a setting of experimental variables means carrying through the whole process of adjusting values for supervised variables, making an experimental "run," and observing the results of that run—more than once. Values of the responses from replications of a setting form the (single) sample corresponding to the setting, which one hopes represents typical process behavior at that setting.

Purposes of replication

The idea of replication is fundamental in experimentation. **Reproducibility of results** is important in both science and engineering practice. Replication helps establish this, protecting the investigator from unconscious blunders and validating or confirming experimental conclusions.

But replication is not only important for establishing that experimental results are reproducible. It is also essential to quantifying the *limits* of that reproducibility—that is, for getting an idea of the size of experimental error. Even under a fixed setting of supervised variables, repeated experimental runs typically will not produce exactly the same observations. The effects of unsupervised variables and measurement errors produce a kind of **baseline variation**, or background noise. Establishing the magnitude of this variation is important. It is only against this background that one can judge whether an apparent effect of an experimental variable is big enough to establish it as clearly real, rather than explainable in terms of background noise.

When planning an experiment, the engineer must think carefully about what kind of repetition will be included. Definition 7 was written specifically to suggest that simply remeasuring an experimental unit does not amount to real replication. Such repetition will capture measurement error, but it ignores the effects of (potentially changing) unsupervised variables. It is a common mistake in logic to seriously underestimate the size of experimental error by failing to adopt a broad enough view of what should be involved in replication, settling instead for what amounts to remeasurement.

Example 9

Replication and Steel Making

A former colleague once related a consulting experience that went approximately as follows. In studying the possible usefulness of a new additive in a type of steel, a metallurgical engineer had one heat (batch) of steel made with the additive and one without. Each of these was poured into ingots. The metallurgist then selected some ingots from both heats, had them cut into pieces, and selected some pieces from the ingots, ultimately measuring a property of interest on these pieces and ending up with a reasonably large amount of data. The data from the heat with additive showed it to be clearly superior to the no-additive heat. As a result, the existing production process was altered (at significant expense) and the new additive incorporated. Unfortunately, it soon became apparent that the alteration to the process had actually degraded the properties of the steel.

The statistician was (only at this point) called in to help figure out what had gone wrong. After all, the experimental results, based on a large amount of data, had been quite convincing, hadn't they?

The key to understanding what had gone wrong was the issue of replication. In a sense, there was none. The metallurgist had essentially just remeasured the same two physical objects (the heats) many times. In the process, he had learned quite a bit about the two particular heats in the study but very little about all heats of the two types. Apparently, extraneous and uncontrolled foundry variables were producing large heat-to-heat variability. The metallurgist had mistaken an effect of this fluctuation for an improvement due to the new additive. The metallurgist had no notion of this possibility because he had not replicated the with-additive and without-additive settings of the experimental variable.

Example 10

Replication and Paper Airplane Testing

Beer, Dusek, and Ehlers completed a project comparing the Kline-Fogelman and Polish Frisbee paper airplane designs on the basis of flight distance under a number of different conditions. In general, it was a carefully done project. However, replication was a point on which their experimental plan was extremely weak. They made a number of trials for each plane under each set of experimental conditions, but only one Kline-Fogelman prototype and one Polish Frisbee prototype were used throughout the study. The students learned quite a bit about the prototypes in hand but possibly much less about the two designs. If their purpose was to pick a winner between the two prototypes, then perhaps the design of their study was appropriate. But if the purpose was to make conclusions about planes

Example 10
(*continued*)

"like" the two used in the study, they needed to make and test several prototypes for each design.

ISU Professor Emeritus L. Wolins calls the problem of identifying what constitutes replication in an experiment the **unit of analysis problem**. There must be replication of the basic experimental unit or object. The agriculturalist who, in order to study pig blood chemistry, takes hundreds of measurements per hour on one pig, has a (highly multivariate) sample of size 1. The pig is the unit of analysis.

Without proper replication, one can only hope to be lucky. If experimental error is small, then accepting conclusions suggested by samples of size 1 will lead to correct conclusions. But the problem is that without replication, one usually has little idea of the size of that experimental error.

2.3.5   Allocation of Resources

Experiments are done by people and organizations that have finite time and money. Allocating those resources and living within the constraints they impose is part of experiment planning. The rest of this section makes several points in this regard.

First, real-world investigations are often most effective when approached **sequentially**, the planning for each stage building upon what has been learned before. The classroom model of planning and/or executing a single experiment is more a result of constraints inherent in our methods of teaching than a realistic representation of how engineering problems are solved. The reality is most often iterative in nature, involving a series of related experiments.

This being the case, one can not use an entire experimental budget on the first pass of a statistical engineering study. Conventional wisdom on this matter is that no more than 20–25% of an experimental budget should be allocated to the first stage of an investigation. This leaves adequate resources for follow-up work built on what is learned initially.

Second, what is easy to do (and therefore usually cheap to do) should not dictate completely what is done in an experiment. In the context of the steel formula development study of Example 9, it seems almost certain that one reason the metallurgist chose to get his "large sample sizes" from pieces of ingots rather than from heats is that it was easy and cheap to get many measurements in that way. But in addition to failing to get absolutely crucial replication and thus botching the study, he probably also grossly overmeasured the two heats.

A final remark is an amplification of the discussion of sample size in Section 2.1. That is, minimum experimental resource requirements are dictated in large part by the magnitude of effects of engineering importance in comparison to the magnitude of experimental error. The larger the effects in comparison to the error (the larger the signal-to-noise ratio), the smaller the sample sizes required, and thus the fewer the resources needed.

Section 3 Exercises ....

1. Consider again the paper airplane study from Exercise 1 of Section 2.1. Describe some variables that you would want to control in such a study. What are the response and experimental variables that would be appropriate in this context? Name a potential concomitant variable here.

2. In general terms, what is the trade-off that must be weighed in deciding whether or not to control a variable in a statistical engineering study?

3. In the paper airplane scenario of Exercise 1 of Section 2.1, if (because of schedule limitations, for example) two different team members will make the flight distance measurements, discuss how the notion of blocking might be used.

4. Again using the paper airplane scenario of Exercise 1 of Section 2.1, suppose that two students are each going to make and fly one airplane of each of the $2^3 = 8$ possible types once. Employ the notion of randomization and Table B.1 and develop schedules for Tom and Juanita to use in their flight testing. Explain how the table was used.

5. Continuing the paper airplane scenario of Exercise 1 of Section 2.1, discuss the pros and cons of Tom and Juanita flying each of their own eight planes twice, as opposed to making and flying two planes of each of the eight types, one time each.

6. Random number tables are sometimes used in the planning of both enumerative and analytical/experimental studies. What are the two different terminologies employed in these different contexts, and what are the different purposes behind the use of the tables?

7. What is blocking supposed to accomplish in an engineering experiment?

8. What are some purposes of replication in a statistical engineering study?

9. Comment briefly on the notion that in order for a statistical engineering study to be statistically proper, one should know before beginning data collection exactly how an entire experimental budget is to be spent. (Is this, in fact, a correct idea?)

....

2.4   Some Common Experimental Plans

In previous sections, experimentation has been discussed in general terms, and the subtlety of considerations that enter the planning of an effective experiment has been illustrated. It should be obvious that any exposition of standard experimental "plans" can amount only to a discussion of standard "skeletons" around which real plans can be built. Nevertheless, it is useful to know something about such skeletons. In this section, so-called completely randomized, randomized complete block, and incomplete block experimental plans are considered.

2.4.1   Completely Randomized Experiments

Definition 8

> A **completely randomized experiment** is one in which all experimental variables are of primary interest (i.e., none are included only for purposes of blocking), and randomization is used at every possible point of choosing the experimental protocol.

Notice that this definition says nothing about how the combinations of settings of experimental variables included in the study are structured. In fact, they may be essentially unstructured or produce data with any of the structures discussed in Section 1.2. That is, there are completely randomized one-factor, factorial, and fractional factorial experiments. The essential point in Definition 8 is that all else is randomized except what is restricted by choice of which combinations of levels of experimental variables are to be used in the study.

Paraphrase of the definition of complete randomization

Although it doesn't really fit every situation (or perhaps even most) in which the term *complete randomization* is appropriate, language like the following is commonly used to capture the intent of Definition 8. "Experimental units (objects) are allocated at random to the treatment combinations (settings of experimental variables). Experimental runs are made in a randomly determined order. And any post-facto measuring of experimental outcomes is also carried out in a random order."

Example 11

Complete Randomization in a Glass Restrengthening Study

Bloyer, Millis, and Schibur studied the restrengthening of damaged glass through etching. They investigated the effects of two experimental factors—the Concentration of hydrofluoric acid in an etching bath and the Time spent in the etching bath—on the resulting strength of damaged glass rods. (The rods had been purposely scratched in a $1''$ region near their centers by sandblasting.) Strengths were measured using a three-point bending method on a 20 kip MTS machine.

The students decided to run a 3×3 factorial experiment. The experimental levels of Concentration were 50%, 75%, and 100% HF, and the levels of Time employed were 30 sec, 60 sec, and 120 sec. There were thus nine treatment combinations, as illustrated in Figure 2.6.

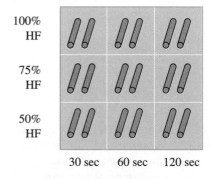

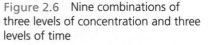

Figure 2.6   Nine combinations of three levels of concentration and three levels of time

The students decided that 18 scratched rods would be allocated—two apiece to each of the nine treatment combinations—for testing. Notice that this could be done at random by labeling the rods 01–18, placing numbered slips of paper in a hat, mixing, drawing out two for 30 sec and 50% concentration, then drawing out two for 30 sec and 75% concentration, etc.

Having determined at random which rods would receive which experimental conditions, the students could again have used the slips of paper to randomly determine an etching order. And a third use of the slips of paper to determine an order of strength testing would have given the students what most people would call a completely randomized 3×3 factorial experiment.

Example 12

Complete Randomization and a Study of the Flight of Golf Balls

G. Gronberg studied drive flight distances for 80, 90, and 100 compression golf balls, using 10 balls of each type in his experiment. Consider what complete randomization would entail in such a study (involving the single factor Compression).

Notice that the paraphrase of Definition 8 is not particularly appropriate to this experimental situation. The levels of the experimental factor are an intrinsic property of the experimental units (balls). There is no way to randomly divide the 30 test balls into three groups and "apply" the treatment levels 80, 90, and 100 compression to them. In fact, about the only obvious point at which randomization could be employed in this scenario is in the choice of an order for hitting the 30 test balls. If one numbered the test balls 01 through 30 and used a table of random digits to pick a hitting order (by choosing balls one at a time without replacement), most people would be willing to call the resulting test a completely randomized one-factor experiment.

Randomization is a good idea. Its virtues have been discussed at some length. So it would be wise to point out that using it can sometimes lead to practically unworkable experimental plans. Dogmatic insistence on complete randomization can in some cases be quite foolish and unrealistic. Changing experimental variables according to a completely randomly determined schedule can sometimes be exceedingly inconvenient (and therefore expensive). If the inconvenience is great and the fear of being misled by the effects of extraneous variables is relatively small, then backing off from complete to partial randomization may be the only reasonable course of action. But when choosing not to randomize, the implications of that choice must be carefully considered.

Example 11
(*continued*)

Consider an embellishment on the glass strengthening scenario, where an experimenter might have access to only a single container to use for a bath and/or have only a limited amount of hydrofluoric acid.

Example 11
(*continued*)

From the discussion of replication in the previous section and present considerations of complete randomization, it would seem that the purest method of conducting the study would be to make a new dilution of HF for each of the rods as its turn comes for testing. But this would be time-consuming and might require more acid than was available.

If the investigator had three containers to use for baths but limited acid, an alternative possibility would be to prepare three different dilutions, one 100%, one 75%, and one 50% dilution. A given dilution could then be used in testing all rods assigned to that concentration. Notice that this alternative allows for a randomized order of testing, but it introduces some question as to whether there is "true" replication.

Taking the resource restriction idea one step further, notice that even if an investigator could afford only enough acid for making one bath, there is a way of proceeding. One could do all 100% concentration testing, then dilute the acid and do all 75% testing, then dilute the acid again and do all 50% testing. The resource restriction would not only affect the "purity" of replication but also prevent complete randomization of the experimental order. Thus, for example, any unintended effects of increased contamination of the acid (as more and more tests were made using it) would show up in the experimental data as indistinguishable from effects of differences in acid concentration.

To choose intelligently between complete randomization (with "true" replication) and the two plans just discussed, the real severity of resource limitations would have to be weighed against the likelihood that extraneous factors would jeopardize the usefulness of experimental results.

2.4.2 Randomized Complete Block Experiments

Definition 9

A **randomized complete block experiment** is one in which at least one experimental variable is a blocking factor (not of primary interest to the investigator); and within each block, every setting of the primary experimental variables appears at least once; and randomization is employed at all possible points where the exact experimental protocol is determined.

A helpful way to think of a randomized complete block experiment is as a collection of completely randomized studies. Each of the blocks yields one of the component studies. Blocking provides the simultaneous advantages of homogeneous environments for studying primary factors and breadth of applicability of the results.

Definition 9 (like Definition 8) says nothing about the structure of the settings of primary experimental variables included in the experiment. Nor does it say anything about the structure of the blocks. It is possible to design experiments where experimental combinations of primary variables have one-factor, factorial, or fractional factorial structure, and at the same time the experimental combinations of blocking variables also have one of these standard structures. The essential points of Definition 9 are the *completeness* of each block (in the sense that it contains each setting of the primary variables) and the *randomization* within each block. The following two examples illustrate that depending upon the specifics of a scenario, Definition 9 can describe a variety of experimental plans.

Example 12
(*continued*)

As actually run, Gronberg's golf ball flight study amounted to a randomized complete block experiment. This is because he hit and recorded flight distances for all 30 balls on six different evenings (over a six-week period). Note that this allowed him to have (six different) homogeneous conditions under which to compare the flight distances of balls having 80, 90, and 100 compression. (The blocks account for possible changes over time in his physical condition and skill level as well as varied environmental conditions.)

Notice the structure of the data set that resulted from the study. The settings of the single primary experimental variable Compression combined with the levels of the single blocking factor Day to produce a 3×6 factorial structure for 18 samples of size 10, as pictured in Figure 2.7.

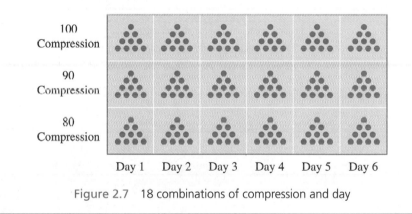

Figure 2.7   18 combinations of compression and day

Example 13
(*Example 2, Chapter 1,
revisited—pp. 6, 13*)

Blocking in a Pelletizing Experiment

Near the end of Section 1.2, the notion of a fractional factorial study was illustrated in the context of a hypothetical experiment on a pelletizing machine. The factors Volume, Flow, and Mixture were of primary interest. Table 1.3 is reproduced here as Table 2.3, listing four (out of eight possible) combinations of two levels each of the primary experimental variables, forming a fractional factorial arrangement.

Consider a situation where two different operators can make four experimental runs each on two consecutive days. Suppose further that Operator and Day are blocking factors, their combinations giving four blocks, within which the four combinations listed in Table 2.3 are run in a random order. This ends

Example 13
(*continued*)

Table 2.3
Half of a 2^3 Factorial

Volume	Flow	Mixture
high	current	no binder
low	manual	no binder
low	current	binder
high	manual	binder

up as a randomized complete block experiment in which the blocks have 2×2 factorial structure and the four combinations of primary experimental factors have a fractional factorial structure.

There are several ways to think of this plan. For one, by temporarily ignoring the structure of the blocks and combinations of primary experimental factors, it can be considered a 4×4 factorial arrangement of samples of size 1, as is illustrated in Figure 2.8. But from another point of view, the combinations under discussion (listed in Table 2.4) have fractional factorial structure of their own, representing a (not particularly clever) choice of 16 out of $2^5 = 32$ different possible combinations of the two-level factors Operator, Day, Volume, Flow, and Mixture. (The lines in Table 2.4 separate the four blocks.) A better use of 16 experimental runs in this situation (at least from the perspective that the combinations in Table 2.4 have their own fractional factorial structure) will be discussed next.

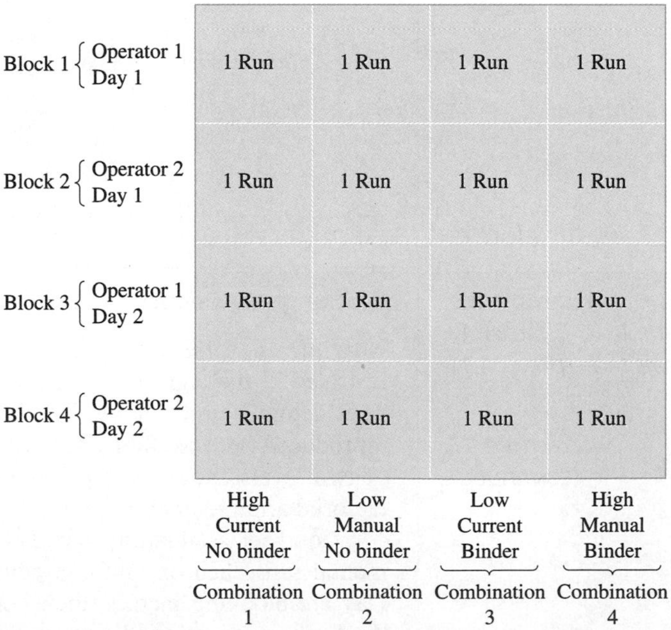

Figure 2.8 16 combinations of blocks and treatments

Table 2.4
Half of a 2^3 Factorial Run Once in Each of Four Blocks

Operator	Day	Volume	Flow	Mixture
1	1	high	current	no binder
1	1	low	manual	no binder
1	1	low	current	binder
1	1	high	manual	binder
2	1	high	current	no binder
2	1	low	manual	no binder
2	1	low	current	binder
2	1	high	manual	binder
1	2	high	current	no binder
1	2	low	manual	no binder
1	2	low	current	binder
1	2	high	manual	binder
2	2	high	current	no binder
2	2	low	manual	no binder
2	2	low	current	binder
2	2	high	manual	binder

2.4.3 Incomplete Block Experiments (*Optional*)

In many experimental situations where blocking seems attractive, physical constraints make it impossible to satisfy Definition 9. This leads to the notion of incomplete blocks.

Definition 10

An **incomplete** (usually randomized) **block experiment** is one in which at least one experimental variable is a blocking factor and the assignment of combinations of levels of primary experimental factors to blocks is such that not every combination appears in every block.

Example 13
(*continued*)

In Section 1.2, the pelletizing machine study examined all eight possible combinations of Volume, Flow, and Mixture. These are listed in Table 2.5. Imagine that only half of these eight combinations can be run on a given day, and there is some fear that daily environmental conditions might strongly affect process performance. How might one proceed?

There are then two blocks (days), each of which will accommodate four runs. Some possibilities for assigning runs to blocks would clearly be poor. For example, running combinations 1 through 4 on the first day and 5 through 8 on

Example 13
(*continued*)

Table 2.5
Combinations in a 2^3 Factorial Study

Combination Number	Volume	Flow	Mixture
1	low	current	no binder
2	high	current	no binder
3	low	manual	no binder
4	high	manual	no binder
5	low	current	binder
6	high	current	binder
7	low	manual	binder
8	high	manual	binder

the second would make it impossible to distinguish the effects of Mixture from any important environmental effects.

What turns out to be a far better possibility is to run, say, the four combinations listed in Table 2.3 (combinations 2, 3, 5, and 8) on one day and the others on the next. This is illustrated in Table 2.6. In a well-defined sense (explained in Chapter 8), this choice of an incomplete block plan minimizes the unavoidable clouding of inferences caused by the fact all eight combinations of levels of Volume, Flow, and Mixture cannot be run on a single day.

As one final variation on the pelletizing scenario, consider an alternative that is superior to the experimental plan outlined in Table 2.4: one that involves incomplete blocks. That is, once again suppose that the two-level primary factors Volume, Flow, and Mixture are to be studied in four blocks of four observations, created by combinations of the two-level blocking factors Operator and Day.

Since a total of 16 experimental runs can be made, all eight combinations of primary experimental factors can be included in the study twice (instead of

Table 2.6
A 2^3 Factorial Run in Two Incomplete Blocks

Day	Volume	Flow	Mixture
2	low	current	no binder
1	high	current	no binder
1	low	manual	no binder
2	high	manual	no binder
1	low	current	binder
2	high	current	binder
2	low	manual	binder
1	high	manual	binder

Table 2.7
A Once-Replicated 2^3 Factorial Run in Four Incomplete Blocks

Operator	Day	Volume	Flow	Mixture
1	1	high	current	no binder
1	1	low	manual	no binder
1	1	low	current	binder
1	1	high	manual	binder
2	1	low	current	no binder
2	1	high	manual	no binder
2	1	high	current	binder
2	1	low	manual	binder
1	2	low	current	no binder
1	2	high	manual	no binder
1	2	high	current	binder
1	2	low	manual	binder
2	2	high	current	no binder
2	2	low	manual	no binder
2	2	low	current	binder
2	2	high	manual	binder

including only four combinations four times apiece). To do this, incomplete blocks are required, but Table 2.7 shows a good incomplete block plan. (Again, blocks are separated by lines.)

Notice the symmetry present in this choice of half of the $2^5 = 32$ different possible combinations of the five experimental factors. For example, a full factorial in Volume, Flow, and Mixture is run on each day, and similarly, each operator runs a full factorial in the primary experimental variables.

It turns out that the study outlined in Table 2.7 gives far more potential for learning about the behavior of the pelletizing process than the one outlined in Table 2.4. But again, a complete discussion of this must wait until Chapter 8.

There may be some reader uneasiness and frustration with the "rabbit out of a hat" nature of the examples of incomplete block experiments, since there has been no discussion of how to go about making up a good incomplete block plan. Both the choosing of an incomplete block plan and corresponding techniques of data analysis are advanced topics that will not be developed until Chapter 8. The purpose here is to simply introduce the possibility of incomplete blocks as a useful option in experimental planning.

Section 4 Exercises

1. What standard name might be applied to the experimental plan you developed for Exercise 4 of Section 2.3?

2. Consider an experimental situation where the three factors A, B, and C each have two levels, and it is desirable to make three experimental runs for each of the possible combinations of levels of the factors.

   (a) Select a completely random order of experimentation. Carefully describe how you use Table B.1 or statistical software to do this. Make an ordered list of combinations of levels of the three factors, prescribing which combination should be run first, second, etc.

   (b) Suppose that because of physical constraints, only eight runs can be made on a given day. Carefully discuss how the concept of blocking could be used in this situation when planning which experimental runs to make on each of three consecutive days. What possible purpose would blocking serve?

   (c) Use Table B.1 or statistical software to randomize the order of experimentation within the blocks you described in part (b). (Make a list of what combinations of levels of the factors are to be run on each day, in what order.)

   How does the method you used here differ from what you did in part (a)?

3. Once more referring to the paper airplane scenario of Exercise 1 of Section 2.1, suppose that only the factors Design and Paper are of interest (all planes will be made without paper clips) but that Tom and Juanita can make and test only two planes apiece. Devise an incomplete block plan for this study that gives each student experience with both designs and both papers. (Which two planes will each make and test?)

4. Again in the paper airplane scenario of Exercise 1 of Section 2.1, suppose that Tom and Juanita each have time to make and test only four airplanes apiece, but that in toto they still wish to test all eight possible types of planes. Develop a sensible plan for doing this. (Which planes should each person test?) You will probably want to be careful to make sure that each person tests two delta wing planes, two construction paper planes, and two paper clip planes. Why is this? Can you arrange your plan so that each person tests each Design/Paper combination, each Design/Loading combination, and each Paper/Loading combination once?

5. What standard name might be applied to the plan you developed in Exercise 4?

2.5 Preparing to Collect Engineering Data

This chapter has raised many of the issues that engineers must consider when planning a statistical study. What is still lacking, however, is a discussion of how to get started. This section first lists and then briefly discusses a series of steps that can be followed in preparing for engineering data collection.

2.5.1 A Series of Steps to Follow

The following is a list of steps that can be used to organize the planning of a statistical engineering study.

PROBLEM DEFINITION

Step 1   Identify the problem to be addressed in general terms.

Step 2   Understand the context of the problem.

Step 3   State in precise terms the objective and scope of the study. (State the questions to be answered.)

STUDY DEFINITION

Step 4   Identify the response variable(s) and appropriate instrumentation.

Step 5   Identify possible factors influencing responses.

Step 6   Decide whether (and if so how) to manage factors that are likely to have effects on the response(s).

Step 7   Develop a detailed data collection protocol and timetable for the first phase of the study.

PHYSICAL PREPARATION

Step 8   Assign responsibility for careful supervision.

Step 9   Identify technicians and provide necessary instruction in the study objectives and methods to be used.

Step 10   Prepare data collection forms and/or equipment.

Step 11   Do a dry run analysis on fictitious data.

Step 12   Write up a "best guess" prediction of the results of the actual study.

These 12 points are listed in a reasonably rational order, but planning any real study may involve departures from the listed order as well as a fair amount of iterating among the steps before they are all accomplished. The need for other steps (like finding funds to pay for a proposed study) will also be apparent in some contexts. Nevertheless, steps 1 through 12 form a framework for getting started.

2.5.2   Problem Definition

Step 1   Identifying the general problem to work on is, for the working engineer, largely a matter of prioritization. An individual engineer's job description and place in an organization usually dictate what problem areas need attention. And far more things could always be done than resources of time and money will permit. So some choice has to be made among the different possibilities.

It is only natural to choose a general topic on the basis of the perceived importance of a problem and the likelihood of solving it (given the available resources). These criteria are somewhat subjective. So, particularly when a project team or other working group must come to consensus before proceeding, even this initial planning step is a nontrivial task. Sometimes it is possible to remove part of the subjectivity and reliance on personal impressions by either examining existing data or commissioning a statistical study of the current state of affairs. For example, suppose members of an engineering project team can name several types of flaws that occur in a mechanical part but disagree about the frequencies or dollar impacts of the flaws. The natural place to begin is to search company records or collect some new data aimed at determining the occurrence rates and/or dollar impacts.

An effective and popular way of summarizing the findings of such a preliminary look at the current situation is through a **Pareto diagram**. This is a bar chart whose vertical axis delineates frequency (or some other measure of impact of system misbehavior) and whose bars, representing problems of various types, have been placed left to right in decreasing order of importance.

Example 14

Maintenance Hours for a Flexible Manufacturing System

Figure 2.9 is an example of a Pareto diagram that represents a breakdown (by craft classification) of the total maintenance hours required in one year on four particular machines in a company's flexible manufacturing system. (This information is excerpted from the ISU M.S. thesis work of M. Patel.) A diagram like Figure 2.9 can be an effective tool for helping to focus attention on the most important problems in an engineering system. Figure 2.9 highlights the fact that (in terms of maintenance hours required) mechanical problems required the most attention, followed by electrical problems.

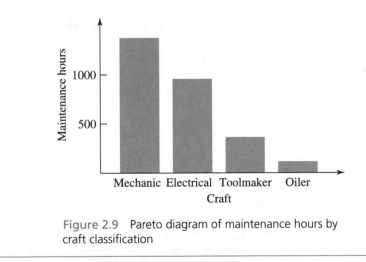

Figure 2.9   Pareto diagram of maintenance hours by craft classification

Step 2   In a statistical engineering study, it is essential to understand the context of the problem. Statistics is no magic substitute for good, hard work learning how a process is configured; what its inputs and environment are; what applicable engineering, scientific, and mathematical theory has to say about its likely behavior; etc. A statistical study is an engineering tool, not a crystal ball. Only when an engineer has studied and asked questions in order to gain expert knowledge about a system is he or she then in a position to decide intelligently what is not known about the system—and thus what data will be of help.

It is often helpful at step 2 to make **flowcharts** describing an ideal process and/or the process as it is currently operating. (Sometimes the comparison of the two is enough in itself to show an engineer how a process should be modified.) During the construction of such a chart, data needs and variables of potential interest can be identified in an organized manner.

Example 15

Work Flow in a Printing Shop

Drake, Lach, and Shadle worked with a printing shop. Before collecting any data, they set about to understand the flow of work through the shop. They made a flowchart similar to Figure 2.10. The flowchart facilitated clear thinking about

Figure 2.10 Flowchart of a printing process

Example 15
(*continued*)
what might go wrong in the printing process and at what points what data could be gathered in order to monitor and improve process performance.

Step 3   After determining the general arena and physical context of a statistical engineering study, it is necessary to agree on a statement of purpose and scope for the study. An engineering project team assigned to work on a wave soldering process for printed circuit boards must understand the steps in that process and then begin to define what part(s) of the process will be included in the study and what the goal(s) of the study will be. Will flux formulation and application, the actual soldering, subsequent cleaning and inspection, and touch-up all be studied? Or will only some part of this list be investigated? Is system throughput the primary concern, or is it instead some aspect of quality or cost? The sharper a statement of purpose and scope can be made at this point, the easier subsequent planning steps will be.

2.5.3   Study Definition

Step 4   Once one has defined in qualitative terms what it is about an engineering system that is of interest, one must decide how to represent that property (or those properties) in precise terms. That is, one must choose a well-defined response variable (or variables) and decide how to measure it (or them). For example, in a manufacturing context, if "throughput" of a system is of interest, should it be measured in pieces/hour, or conforming pieces/hour, or net profit/hour, or net profit/hour/machine, or in some other way?

Sections 1.3 and 2.1 have already discussed issues that arise in measurement and the formation of operational definitions. All that needs to be added here is that these issues must be faced early in the planning of a statistical engineering study. It does little good to carefully plan a study assuming the existence of an adequate piece of measuring equipment, only to later determine that the organization doesn't own a device with adequate precision and that the purchase of one would cost more than the entire project budget.

Step 5   Identification of variables that may affect system response requires expert knowledge of the process under study. Engineers who do not have hands-on experience with a system can sometimes contribute insights gained from experience with similar systems and from basic theory. But it is also wise (in most cases, essential) to include on a project team several people who have first-hand knowledge of the particular process and to talk extensively with those who work with the system on a regular basis.

Typically, the job of identifying factors of potential importance in a statistical engineering study is a group activity, carried out in brainstorming sessions. It is therefore helpful to have tools for lending order to what might otherwise be an inefficient and disorganized process. One tool that has proved effective is variously known as a **cause-and-effect diagram**, or **fishbone diagram**, or **Ishikawa diagram**.

Example 16 | Identifying Potentially Important Variables in a Molding Process

Figure 2.11 shows a cause-and-effect diagram from a study of a molding process for polyurethane automobile steering wheels. It is taken from the paper "Fine Tuning of the Foam System and Optimization of the Process Parameters for the Manufacturing of Polyurethane Steering Wheels Using Reaction Injection Molding by Applying Dr. Taguchi's Method of Design of Experiments" by Vimal Khanna, which appeared in 1985 in the *Third Supplier Symposium on Taguchi Methods*, published by the American Supplier Institute, Inc. Notice how the diagram in Figure 2.11 organizes the huge number of factors possibly affecting

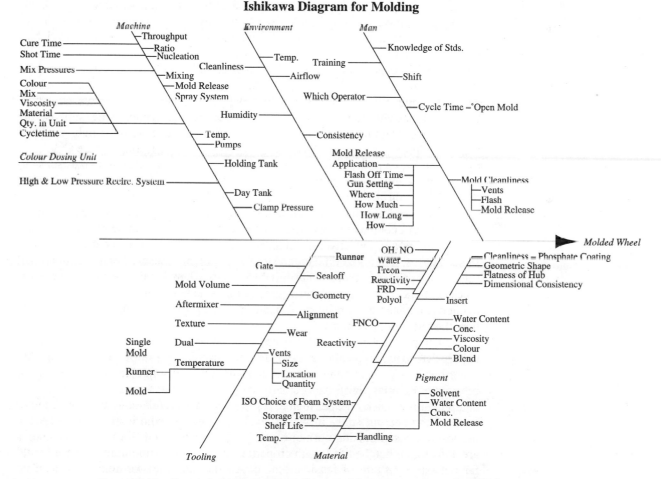

Figure 2.11 Cause and effect diagram for a molding process. From the *Third Symposium on Taguchi Methods*. © Copyright, American Supplier Institute, Dearborn, Michigan (U.S.A.). Reproduced by permission under License No. 930403.

Example 16 *(continued)*	wheel quality. Without some kind of organization, it would be all but impossible to develop anything like a complete list of important factors in a complex situation like this.

Step 6 Armed with (1) a list of variables that might influence the response(s) of interest and some guesses at their relative importance, (2) a solid understanding of the issues raised in Section 2.3, and (3) knowledge of resource and physical constraints and time-frame requirements, one can begin to make decisions about which (if any) variables are to be managed. Experiments have some real advantages over purely observational studies (see Section 1.2). Those must be weighed against possible extra costs and difficulties associated with managing both variables that are of interest and those that are not. The hope is to choose a physically and financially workable set of managed variables in such a way that the aggregate effects of variables not of interest and not managed are not so large as to mask the effects of those variables that *are* of interest.

Step 7 Choosing experimental levels and then combinations for managed variables is part of the task of deciding on a detailed data collection protocol. Levels of controlled and block variables should usually be chosen to be representative of the values that will be met in routine system operation. For example, suppose the amount of contamination in a transmission's hydraulic fluid is thought to affect time to failure when the transmission is subjected to stress testing, where Operating Speed and Pressure are the primary experimental variables. It only makes sense to see that the contamination level(s) during testing are representative of the level(s) that will be typical when the transmission is used in the field.

With regard to primary experimental variables, one should also choose typical levels—with a couple of provisos. Sometimes the goal in an engineering experiment is to compare an innovative, nonstandard way of doing things to current practice. In such cases, it is not good enough simply to look at system behavior with typical settings for primary experimental variables. Also, where primary experimental variables are believed to have relatively small effects on a response, it may be necessary to choose ranges for the primary variables that are wider than normal, to see clearly how they act on the response.

Other physical realities and constraints on data collection may also make it appropriate to use atypical values of managed variables and subsequently extrapolate experimental results to "standard" circumstances. For example, it is costly enough to run studies on **pilot plants** using small quantities of chemical reagents and miniature equipment but much cheaper than experimentation on a full-scale facility. Another kind of engineering study in which levels of primary experimental variables are purposely chosen outside normal ranges is the **accelerated life test**. Such studies are done to predict the life-length properties of products that in normal usage would far outlast any study of feasible length. All that can then be done is to turn up the stress on sample units beyond normal levels, observe performance, and try to extrapolate back to a prediction for behavior under normal usage. (For example, if sensitive electronic equipment performs well under abnormally high temperature and humidity, this could well be expected to imply long useful life under normal temperature and humidity conditions.)

After the experimental levels of individual manipulated variables are chosen, they must be combined to form the experimental patterns (combinations) of managed variables. The range of choices is wide: factorial structures, fractional factorial structures, other standard structures, and patterns tailor-made for a particular problem. (Tailor-made plans will, for example, be needed in situations where particular combinations of factor levels prescribed by standard structures are a priori clearly unsafe or destructive of company property.)

But developing a detailed data collection protocol requires more than even choices of experimental combinations. Experimental order must be decided. Explicit instructions for actually carrying out the testing must be agreed upon and written down in such a way that someone who was not involved in study planning can carry out the data collection. A timetable for initial data collection must be developed. In all of this, it must be remembered that several iterations of data collection and analysis (all within given budget constraints) may be required in order to find a solution to the original engineering problem.

2.5.4 Physical Preparation

Step 8 After a project team has agreed on exactly what is to be done in a statistical study, it can address the details of how to accomplish it and assign responsibility for completion. One team member should be given responsibility for the direct oversight of actual data collection. It is all too common for people who collect the data to say, after the fact, "Oh, I did it the other way . . . I couldn't figure out exactly what you meant here . . . and besides, it was easier the way I did it."

Step 9 Again, technicians who carry out a study planned by an engineering project group often need training in the study objectives and the methods to be used. As discussed in Section 2.1, when people know why they are collecting data and have been carefully shown how to collect them, they will produce better information. Overseeing the data collection process includes making sure that this necessary training takes place.

Steps 10 & 11 The discipline involved in carefully preparing complete data collection forms and doing a dry run data analysis on fictitious values provides opportunities to refine (and even salvage) a study before the expense of data collection is incurred. When carrying out steps 10 and 11, each individual on the team gets a chance to ask, "Will the data be adequate to answer the question at hand? Or are other data needed?" The students referred to in Example 4 (page 30), who failed to measure their primary response variables, learned the importance of these steps the hard way.

Step 12 The final step in this list is writing up a best guess at what the study will show. We first came across this idea in *Statistics for Experimenters* by Box, Hunter, and Hunter. The motivation for it is sound. After a study is complete, it is only human to say, "Of course that's the way things are. We knew that all along." When a careful before-data statement is available to compare to an after-data summarization of findings, it is much easier to see what has been learned and appreciate the value of that learning.

Section 5 Exercises

1. Either take an engineering system and response variable that you are familiar with from your field or consider, for example, the United Airlines passenger flight system and the response variable Customer Satisfaction and make a cause-and-effect diagram showing a variety of variables that may potentially affect the response. How might such a diagram be practically useful?

Chapter 2 Exercises

1. Use Table B.1 and choose a simple random sample of $n = 8$ out of $N = 491$ widgets. Describe carefully how you label the widgets. Begin in the upper left corner of the table. Then use spreadsheet or statistical software to redo the selection.

2. Consider a potential student project concerning the making of popcorn. Possible factors affecting the outcome of popcorn making include at least the following: Brand of corn, Temperature of corn at beginning of cooking, Popping Method (e.g., frying versus hot air popping), Type of Oil used (if frying), Amount of Oil used (if frying), Batch Size, initial Moisture Content of corn, and Person doing the evaluation of a single batch. Using these factors and/or any others that you can think of, answer the following questions about such a project:
   (a) What is a possible response variable in a popcorn project?
   (b) Pick two possible experimental factors in this context and describe a 2×2 factorial data structure in those variables that might arise in such a study.
   (c) Describe how the concept of randomization might be employed.
   (d) Describe how the concept of blocking might be employed.

3. An experiment is to be performed to compare the effects of two different methods for loading gears in a carburizing furnace on the amount of distortion produced in a heat treating process. Thrust face runout will be measured for gears laid and for gears hung while treating.
   (a) 20 gears are to be used in the study. Randomly divide the gears into a group (of 10) to be laid and a group (of 10) to be hung, using either Table B.1 or statistical software. Describe carefully how you do this. If you use the table, begin in the upper left corner.
   (b) What are some purposes of the randomization used in part (a)?

4. A sanitary engineer wishes to compare two methods for determining chlorine content of Cl_2-demand-free water. To do this, eight quite different water samples are split in half, and one determination is made using the MSI method and another using the SIB method. Explain why it could be said that the principle of blocking was used in the engineer's study. Also argue that the resulting data set could be described as consisting of paired measurement data.

5. A research group is testing three different methods of electroplating widgets (say, methods A, B, and C). On a particular day, 18 widgets are available for testing. The effectiveness of electroplating may be strongly affected by the surface texture of the widgets. The engineer running the experiment is able to divide the 18 available widgets into three groups of 6 on the basis of surface texture. (Assume that widgets 1–6 are rough, widgets 7–12 are normal, and widgets 13–18 are smooth.)
   (a) Use Table B.1 or statistical software in an appropriate way and assign each of the treatments to 6 widgets. Carefully explain exactly how you do the assignment of levels of treatments A, B, and C to the widgets.
   (b) If equipment limitations are such that only one widget can be electroplated at once, but it is possible to complete the plating of all 18 widgets on a single day, in exactly what order would you have the widgets plated? Explain where you got this order.
   (c) If, in contrast to the situation in part (b), it is

possible to plate only 9 widgets in a single day, make up an appropriate plan for plating 9 on each of two consecutive days.

(d) If measurements of plating effectiveness are made on each of the 18 widgets, what kind of data structure will result from the scenario in part (b)? From the scenario in part (c)?

6. A company wishes to increase the light intensity of its photoflash cartridge. Two wall thicknesses ($\frac{1}{16}''$ and $\frac{1}{8}''$) and two ignition point placements are under study. Two batches of the basic formulation used in the cartridge are to be made up, each batch large enough to make 12 cartridges. Discuss how you would recommend running this initial phase of experimentation if all cartridges can be made and tested in a short time period by a single technician. Be explicit about any randomization and/or blocking you would employ. Say exactly what kinds of cartridges you would make and test, in what order. Describe the structure of the data that would result from your study.

7. Use Table B.1 or statistical software and
(a) Select a simple random sample of 5 widgets from a production run of 354 such widgets. (If you use the table, begin at the upper left corner and move left to right, top to bottom.)
(b) Select a random order of experimentation for a context where an experimental factor A has two levels; a second factor, B, has three levels; and two experimental runs are going to be made for each of the $2 \times 3 = 6$ different possible combinations of levels of the factors. Carefully describe how you do this.

8. Return to the situation of Exercise 8 of the Chapter 1 Exercises.
(a) Name factors and levels that might be used in a three-factor, full factorial study in this situation. Also name two response variables for the study. Suppose that in accord with good engineering data collection practice, you wish to include some replication in the study. Make up a data collection sheet, listing all the combinations of levels of the factors to be studied, and include blanks where the corresponding

observed values of the two responses could be entered for each experimental run.
(b) Suppose that it is feasible to make the runs listed in your answer to part (a) in a completely randomized order. Use a mechanical method (like slips of paper in a hat) to arrive at a random order of experimentation for your study. Carefully describe the physical steps you follow in developing this order for data collection.

9. Use Table B.1 and
(a) Select a simple random sample of 7 widgets from a production run of 619 widgets (begin at the upper left corner of the table and move left to right, top to bottom). Tell how you labeled the widgets and name which ones make up your sample.
(b) Beginning in the table where you left off in (a), select a second simple random sample of 7 widgets. Is this sample the same as the first? Is there any overlap at all?

10. Redo Exercise 9 using spreadsheet or statistical software.

11. Consider a study comparing the lifetimes (measured in terms of numbers of holes drilled before failure) of two different brands of 8-mm drills in drilling 1045 steel. Suppose that steel bars from three different heats (batches) of steel are available for use in the study, and it is possible that the different heats have differing physical properties. The lifetimes of a total of 15 drills of each brand will be measured, and each of the bars available is large enough to accommodate as much drilling as will be done in the entire study.
(a) Describe how the concept of control could be used to deal with the possibility that different heats might have different physical properties (such as hardnesses).
(b) Name one advantage and one drawback to controlling the heat.
(c) Describe how one might use the concept of blocking to deal with the possibility that different heats might have different physical properties.

3

Elementary Descriptive Statistics

Engineering data are always variable. Given precise enough measurement, even supposedly constant process conditions produce differing responses. Therefore, it is not individual data values that demand an engineer's attention as much as the pattern or **distribution** of those responses. The task of summarizing data is to describe their important distributional characteristics. This chapter discusses simple methods that are helpful in this task.

The chapter begins with some elementary graphical and tabular methods of data summarization. The notion of quantiles of a distribution is then introduced and used to make other useful graphical displays. Next, standard numerical summary measures of location and spread for quantitative data are discussed. Finally comes a brief look at some elementary methods for summarizing qualitative and count data.

3.1 Elementary Graphical and Tabular Treatment of Quantitative Data

Almost always, the place to begin in data analysis is to make appropriate graphical and/or tabular displays. Indeed, where only a few samples are involved, a good picture or table can often tell most of the story about the data. This section discusses the usefulness of dot diagrams, stem-and-leaf plots, frequency tables, histograms, scatterplots, and run charts.

3.1.1 Dot Diagrams and Stem-and-Leaf Plots

When an engineering study produces a small or moderate amount of univariate quantitative data, a **dot diagram**, easily made with pencil and paper, is often quite revealing. A dot diagram shows each observation as a dot placed at a position corresponding to its numerical value along a number line.

Example 1
(*Example 1, Chapter 1, revisited—p. 2*)

Portraying Thrust Face Runouts

Section 1.1 considered a heat treating problem where distortion for gears laid and gears hung was studied. Figure 1.1 has been reproduced here as Figure 3.1. It consists of two dot diagrams, one showing thrust face runout values for gears laid and the other the corresponding values for gears hung, and shows clearly that the laid values are both generally smaller and more consistent than the hung values.

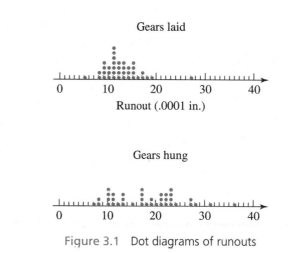

Figure 3.1   Dot diagrams of runouts

Example 2

Portraying Bullet Penetration Depths

Sale and Thom compared penetration depths for several types of .45 caliber bullets fired into oak wood from a distance of 15 feet. Table 3.1 gives the penetration depths (in mm from the target surface to the back of the bullets) for two bullet types. Figure 3.2 presents a corresponding pair of dot diagrams.

Table 3.1
Bullet Penetration Depths (mm)

230 Grain Jacketed Bullets	200 Grain Jacketed Bullets
40.50, 38.35, 56.00, 42.55,	63.80, 64.65, 59.50, 60.70,
38.35, 27.75, 49.85, 43.60,	61.30, 61.50, 59.80, 59.10,
38.75, 51.25, 47.90, 48.15,	62.95, 63.55, 58.65, 71.70,
42.90, 43.85, 37.35, 47.30,	63.30, 62.65, 67.75, 62.30,
41.15, 51.60, 39.75, 41.00	70.40, 64.05, 65.00, 58.00

Example 2
(*continued*)

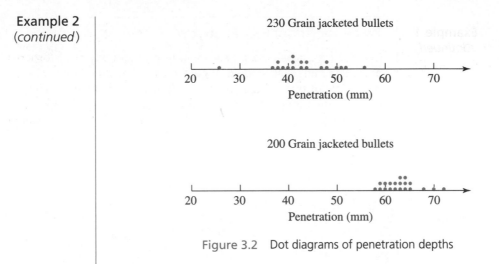

Figure 3.2   Dot diagrams of penetration depths

The dot diagrams show the penetrations of the 200 grain bullets to be both larger and more consistent than those of the 230 grain bullets. (The students had predicted larger penetrations for the lighter bullets on the basis of greater muzzle velocity and smaller surface area on which friction can act. The different consistencies of penetration were neither expected nor explained.)

Dot diagrams give the general feel of a data set but do not always allow the recovery of exactly the values used to make them. A **stem-and-leaf plot** carries much the same visual information as a dot diagram while preserving the original values exactly. A stem-and-leaf plot is made by using the last few digits of each data point to indicate where it falls.

```
0 | 5 8 9 9 9 9
1 | 0 0 1 1 1 1 1 1 1 2 2 2 2 3 3 3 3 4 4 4 5 5 5 5 6 7 7 8 9
2 | 7
3 |
```

```
0 |
0 | 5 8 9 9 9 9
1 | 0 0 1 1 1 1 1 1 1 2 2 2 2 3 3 3 3 4 4 4
1 | 5 5 5 6 7 7 8 9
2 |
2 | 7
3 |
3 |
```

Figure 3.3   Stem-and-leaf plots of laid gear runouts (*Example 1*)

Example 1
(*continued*)

Figure 3.3 gives two possible stem-and-leaf plots for the thrust face runouts of laid gears. In both, the first digit of each observation is represented by the number to the left of the vertical line or "stem" of the diagram. The numbers to the right of the vertical line make up the "leaves" and give the second digits of the observed runouts. The second display shows somewhat more detail than the first by providing "0–4" and "5–9" leaf positions for each possible leading digit, instead of only a single "0–9" leaf for each leading digit.

Example 2
(*continued*)

Figure 3.4 gives two possible stem-and-leaf plots for the penetrations of 200 grain bullets in Table 3.1. On these, it was convenient to use two digits to the left of the decimal point to make the stem and the two following the decimal point to create the leaves. The first display was made by recording the leaf values directly from the table (from left to right and top to bottom). The second display is a better one, obtained by ordering the values that make up each leaf. Notice that both plots give essentially the same visual impression as the second dot diagram in Figure 3.2.

58	.65, .00		58	.00, .65
59	.50, .80, .10		59	.10, .50, .80
60	.70		60	.70
61	.30, .50		61	.30, .50
62	.95, .65, .30		62	.30, .65, .95
63	.80, .55, .30		63	.30, .55, .80
64	.65, .05		64	.05, .65
65	.00		65	.00
66			66	
67	.75		67	.75
68			68	
69			69	
70	.40		70	.40
71	.70		71	.70

Figure 3.4   Stem-and-leaf plots of the 200 grain penetration depths

When comparing two data sets, a useful way to use the stem-and-leaf idea is to make two plots **back-to-back**.

Laid runouts   Hung runouts

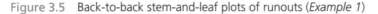

```
                    9 9 9 9 8 8 5 | 0 | 7 8 8
4 4 4 3 3 3 3 2 2 2 2 1 1 1 1 1 1 1 0 0 0 | 1 | 0 0 0 0 1 1 1 2 3 3 3
              9 8 7 7 6 5 5 5 5 | 1 | 5 7 7 7 7 8 9 9
                                 | 2 | 0 1 1 1 2 2 2 3 3 3 3 4
                           7 | 2 | 7 7 8
                                 | 3 | 1
                                 | 3 | 6
```

Figure 3.5   Back-to-back stem-and-leaf plots of runouts (*Example 1*)

Example 1
(*continued*)

Figure 3.5 gives back-to-back stem-and-leaf plots for the data of Table 1.1 (pg. 3). It shows clearly the differences in location and spread of the two data sets.

3.1.2   Frequency Tables and Histograms

Dot diagrams and stem-and-leaf plots are useful devices when mulling over a data set. But they are not commonly used in presentations and reports. In these more formal contexts, frequency tables and histograms are more often used.

A **frequency table** is made by first breaking an interval containing all the data into an appropriate number of smaller intervals of equal length. Then tally marks can be recorded to indicate the number of data points falling into each interval. Finally, frequencies, relative frequencies, and cumulative relative frequencies can be added.

Example 1
(*continued*)

Table 3.2 gives one possible frequency table for the laid gear runouts. The relative frequency values are obtained by dividing the entries in the frequency column

Table 3.2
Frequency Table for Laid Gear Thrust Face Runouts

Runout (.0001 in.)	Tally	Frequency	Relative Frequency	Cumulative Relative Frequency
5–8	|||	3	.079	.079
9–12	⊞ ⊞ ⊞ |||	18	.474	.553
13–16	⊞ ⊞ ||	12	.316	.868
17–20	||||	4	.105	.974
21–24		0	0	.974
25–28			1	.026	1.000
		38	1.000	

by 38, the number of data points. The entries in the cumulative relative frequency column are the ratios of the totals in a given class and all preceding classes to the total number of data points. (Except for round-off, this is the sum of the relative frequencies on the same row and above a given cumulative relative frequency.) The tally column gives the same kind of information about distributional shape that is provided by a dot diagram or a stem-and-leaf plot.

Choosing intervals
for a frequency
table

The choice of intervals to use in making a frequency table is a matter of judgment. Two people will not necessarily choose the same set of intervals. However, there are a number of simple points to keep in mind when choosing them. First, in order to avoid visual distortion when using the tally column of the table to gain an impression of distributional shape, intervals of equal length should be employed. Also, for aesthetic reasons, round numbers are preferable as interval endpoints. Since there is usually aggregation (and therefore some loss of information) involved in the reduction of raw data to tallies, the larger the number of intervals used, the more detailed the information portrayed by the table. On the other hand, if a frequency table is to have value as a summarization of data, it can't be cluttered with too many intervals.

After making a frequency table, it is common to use the organization provided by the table to create a **histogram**. A (frequency or relative frequency) histogram is a kind of bar chart used to portray the shape of a distribution of data points.

Example 2
(continued)

Table 3.3 is a frequency table for the 200 grain bullet penetration depths, and Figure 3.6 is a translation of that table into the form of a histogram.

Table 3.3
Frequency Table for 200 Grain Penetration Depths

Penetration Depth (mm)	Tally	Frequency	Relative Frequency	Cumulative Relative Frequency
58.00–59.99	ЖТ	5	.25	.25
60.00–61.99	|||	3	.15	.40
62.00–63.99	ЖТ		6	.30	.70
64.00–65.99	|||	3	.15	.85
66.00–67.99			1	.05	.90
68.00–69.99		0	0	.90
70.00–71.99	||	2	.10	1.00
		20	1.00	

Example 2
(*continued*)

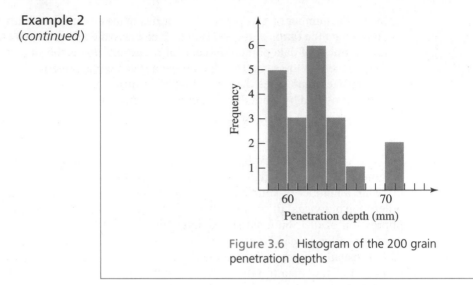

Figure 3.6   Histogram of the 200 grain
penetration depths

The vertical scale in Figure 3.6 is a frequency scale, and the histogram is a **frequency histogram**. By changing to relative frequency on the vertical scale, one can produce a **relative frequency histogram**. In making Figure 3.6, care was taken to

Guidelines for making histograms

1. (continue to) use intervals of equal length,

2. show the entire vertical axis beginning at zero,

3. avoid breaking either axis,

4. keep a uniform scale across a given axis, and

5. center bars of appropriate heights at the midpoints of the (penetration depth) intervals.

Following these guidelines results in a display in which *equal enclosed areas correspond to equal numbers of data points*. Further, data point positioning is clearly indicated by bar positioning on the horizontal axis. If these guidelines are not followed, the resulting bar chart will in one way or another fail to faithfully represent its data set.

Figure 3.7 shows terminology for common **distributional shapes** encountered when making and using dot diagrams, stem-and-leaf plots, and histograms.

The graphical and tabular devices discussed to this point are deceptively simple methods. When routinely and intelligently used, they are powerful engineering tools. The information on location, spread, and shape that is portrayed so clearly on a histogram can give strong hints as to the functioning of the physical process that is generating the data. It can also help suggest physical mechanisms at work in the process.

Examples of engineering interpretations of distribution shape

For example, if data on the diameters of machined metal cylinders purchased from a vendor produce a histogram that is decidedly **bimodal** (or **multimodal**, having several clear humps), this suggests that the machining of the parts was done

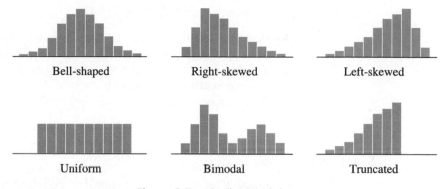

Figure 3.7 Distributional shapes

on more than one machine, or by more than one operator, or at more than one time. The practical consequence of such multichannel machining is a distribution of diameters that has more variation than is typical of a production run of cylinders from a single machine, operator, and setup. As another possibility, if the histogram is **truncated**, this might suggest that the lot of cylinders has been 100% inspected and sorted, removing all cylinders with excessive diameters. Or, upon marking engineering specifications (requirements) for cylinder diameter on the histogram, one may get a picture like that in Figure 3.8. It then becomes obvious that the lathe turning the cylinders needs adjustment in order to increase the typical diameter. But it also becomes clear that the basic process variation is so large that this adjustment will fail to bring essentially all diameters into specifications. Armed with this realization and a knowledge of the economic consequences of parts failing to meet specifications, an engineer can intelligently weigh alternative courses of action: sorting of all incoming parts, demanding that the vendor use more precise equipment, seeking a new vendor, etc.

Investigating the shape of a data set is useful not only because it can lend insight into physical mechanisms but also because shape can be important when determining the appropriateness of methods of formal statistical inference like those discussed later in this book. A methodology appropriate for one distributional shape may not be appropriate for another.

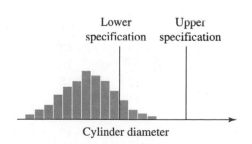

Figure 3.8 Histogram marked with engineering specifications

3.1.3   Scatterplots and Run Charts

Dot diagrams, stem-and-leaf plots, frequency tables, and histograms are univariate tools. But engineering data are often multivariate and *relationships between the variables* are then usually of interest. The familiar device of making a two-dimensional **scatterplot** of data pairs is a simple and effective way of displaying potential relationships between two variables.

Example 3 | Bolt Torques on a Face Plate

Brenny, Christensen, and Schneider measured the torques required to loosen six distinguishable bolts holding the front plate on a type of heavy equipment component. Table 3.4 contains the torques (in ft lb) required for bolts number 3 and 4, respectively, on 34 different components. Figure 3.9 is a scatterplot of the bivariate data from Table 3.4. In this figure, where several points must be plotted at a single location, the number of points occupying the location has been plotted instead of a single dot.

The plot gives at least a weak indication that large torques at position 3 are accompanied by large torques at position 4. In practical terms, this is comforting;

Table 3.4
Torques Required to Loosen Two Bolts on Face Plates (ft lb)

Component	Bolt 3 Torque	Bolt 4 Torque	Component	Bolt 3 Torque	Bolt 4 Torque
1	16	16	18	15	14
2	15	16	19	17	17
3	15	17	20	14	16
4	15	16	21	17	18
5	20	20	22	19	16
6	19	16	23	19	18
7	19	20	24	19	20
8	17	19	25	15	15
9	15	15	26	12	15
10	11	15	27	18	20
11	17	19	28	13	18
12	18	17	29	14	18
13	18	14	30	18	18
14	15	15	31	18	14
15	18	17	32	15	13
16	15	17	33	16	17
17	18	20	34	16	16

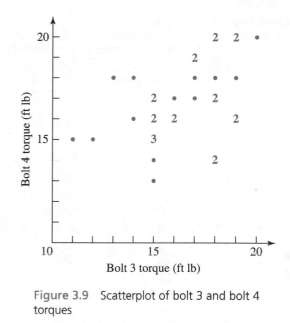

Figure 3.9   Scatterplot of bolt 3 and bolt 4 torques

otherwise, unwanted differential forces might act on the face plate. It is also quite reasonable that bolt 3 and bolt 4 torques be related, since the bolts were tightened by different heads of a single pneumatic wrench operating off a single source of compressed air. It stands to reason that variations in air pressure might affect the tightening of the bolts at the two positions similarly, producing the big-together, small-together pattern seen in Figure 3.9.

The previous example illustrates the point that relationships seen on scatterplots suggest a common physical cause for the behavior of variables and can help reveal that cause.

In the most common version of the scatterplot, the variable on the horizontal axis is a time variable. A scatterplot in which univariate data are plotted against time order of observation is called a **run chart** or **trend chart**. Making run charts is one of the most helpful statistical habits an engineer can develop. Seeing patterns on a run chart leads to thinking about what process variables were changing in concert with the pattern. This can help develop a keener understanding of how process behavior is affected by those variables that change over time.

Example 4

Diameters of Consecutive Parts Turned on a Lathe

Williams and Markowski studied a process for rough turning of the outer diameter on the outer race of a constant velocity joint. Table 3.5 gives the diameters (in inches above nominal) for 30 consecutive joints turned on a particular automatic

Example 4
(continued)

Table 3.5
30 Consecutive Outer Diameters Turned on a Lathe

Joint	Diameter (inches above nominal)	Joint	Diameter (inches above nominal)
1	−.005	16	.015
2	.000	17	.000
3	−.010	18	.000
4	−.030	19	−.015
5	−.010	20	−.015
6	−.025	21	−.005
7	−.030	22	−.015
8	−.035	23	−.015
9	−.025	24	−.010
10	−.025	25	−.015
11	−.025	26	−.035
12	−.035	27	−.025
13	−.040	28	−.020
14	−.035	29	−.025
15	−.035	30	−.015

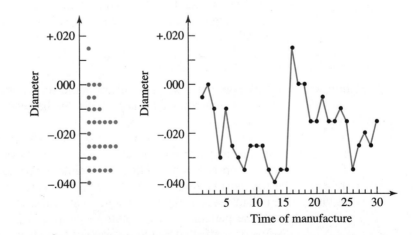

Figure 3.10   Dot diagram and run chart of consecutive outer diameters

lathe. Figure 3.10 gives both a dot diagram and a run chart for the data in the table. In keeping with standard practice, consecutive points on the run chart have been connected with line segments.

Here the dot diagram is not particularly suggestive of the physical mechanisms that generated the data. But the time information added in the run chart is revealing. Moving along in time, the outer diameters tend to get smaller until part 16, where there is a large jump, followed again by a pattern of diameter generally decreasing in time. In fact, upon checking production records, Williams and Markowski found that the lathe had been turned off and allowed to cool down between parts 15 and 16. The pattern seen on the run chart is likely related to the behavior of the lathe's hydraulics. When cold, the hydraulics probably don't do as good a job pushing the cutting tool into the part being turned as when they are warm. Hence, the turned parts become smaller as the lathe warms up. In order to get parts closer to nominal, the aimed-for diameter might be adjusted up by about .020 in. and parts run only after warming up the lathe.

Section 1 Exercises ....

1. The following are percent yields from 40 runs of a chemical process, taken from J. S. Hunter's article "The Technology of Quality" (*RCA Engineer*, May/June 1985):

   65.6, 65.6, 66.2, 66.8, 67.2, 67.5, 67.8, 67.8, 68.0, 68.0, 68.2, 68.3, 68.3, 68.4, 68.9, 69.0, 69.1, 69.2, 69.3, 69.5, 69.5, 69.5, 69.8, 69.9, 70.0, 70.2, 70.4, 70.6, 70.6, 70.7, 70.8, 70.9, 71.3, 71.7, 72.0, 72.6, 72.7, 72.8, 73.5, 74.2

   Make a dot diagram, a stem-and-leaf plot, a frequency table, and a histogram of these data.

2. Make back-to-back stem-and-leaf plots for the two samples in Table 3.1.

3. Osborne, Bishop, and Klein collected manufacturing data on the torques required to loosen bolts holding an assembly on a piece of heavy machinery. The accompanying table shows part of their data concerning two particular bolts. The torques recorded (in ft lb) were taken from 15 different pieces of equipment as they were assembled.
   (a) Make a scatterplot of these paired data. Are there any obvious patterns in the plot?

(b) A trick often employed in the analysis of paired data such as these is to reduce the pairs to differences by subtracting the values of one of the variables from the other. Compute differences (top bolt–bottom bolt) here. Then make and interpret a dot diagram for these values.

Piece	Top Bolt	Bottom Bolt
1	110	125
2	115	115
3	105	125
4	115	115
5	115	120
6	120	120
7	110	115
8	125	125
9	105	110
10	130	110
11	95	120
12	110	115
13	110	120
14	95	115
15	105	105

3.2 Quantiles and Related Graphical Tools

Most readers will be familiar with the concept of a *percentile*. The notion is most famous in the context of reporting scores on educational achievement tests. For example, if a person has scored at the 80th percentile, roughly 80% of those taking the test had worse scores, and roughly 20% had better scores. This concept is also useful in the description of engineering data. However, because it is often more convenient to work in terms of fractions between 0 and 1 rather than in percentages between 0 and 100, slightly different terminology will be used here: "Quantiles," rather than percentiles, will be discussed. After the quantiles of a data set are carefully defined, they are used to create a number of useful tools of descriptive statistics: quantile plots, boxplots, Q-Q plots, and normal plots (a type of theoretical Q-Q plot).

3.2.1 Quantiles and Quantile Plots

Roughly speaking, for a number p between 0 and 1, the p **quantile** of a distribution is a number such that a fraction p of the distribution lies to the left and a fraction $1 - p$ of the distribution lies to the right. However, because of the discreteness of finite data sets, it is necessary to state *exactly* what will be meant by the terminology. Definition 1 gives the precise convention that will be used in this text.

Definition 1

For a data set consisting of n values that when ordered are $x_1 \leq x_2 \leq \cdots \leq x_n$,

1. if $p = \frac{i-.5}{n}$ for a positive integer $i \leq n$, the p **quantile** of the data set is

$$Q(p) = Q\left(\frac{i - .5}{n}\right) = x_i$$

(The ith smallest data point will be called the $\frac{i-.5}{n}$ quantile.)

2. for any number p between $\frac{.5}{n}$ and $\frac{n-.5}{n}$ that is not of the form $\frac{i-.5}{n}$ for an integer i, the p **quantile** of the data set will be obtained by linear interpolation between the two values of $Q(\frac{i-.5}{n})$ with corresponding $\frac{i-.5}{n}$ that bracket p.

In both cases, the notation $Q(p)$ will be used to denote the p quantile.

Definition 1 identifies $Q(p)$ for all p between $.5/n$ and $(n - .5)/n$. To find $Q(p)$ for such a value of p, one may solve the equation $p = (i - .5)/n$ for i, yielding

Index (i) of the ordered data point that is Q(p)

$$i = np + .5$$

and locate the "$(np + .5)$th ordered data point."

Example 5

Quantiles for Dry Breaking Strengths of Paper Towel

Lee, Sebghati, and Straub did a study of the dry breaking strength of several brands of paper towel. Table 3.6 shows ten breaking strengths (in grams) reported by the students for a generic towel. By ordering the strength data and computing values of $\frac{i-.5}{10}$, one can easily find the .05, .15, .25, ..., .85, and .95 quantiles of the breaking strength distribution, as shown in Table 3.7.

Since there are $n = 10$ data points, each one accounts for 10% of the data set. Applying convention (1) in Definition 1 to find (for example) the .35 quantile,

Table 3.6
Ten Paper Towel Breaking
Strengths

Test	Breaking Strength (g)
1	8,577
2	9,471
3	9,011
4	7,583
5	8,572
6	10,688
7	9,614
8	9,614
9	8,527
10	9,165

Table 3.7
Quantiles of the Paper Towel Breaking Strength
Distribution

i	$\frac{i-.5}{10}$	ith Smallest Data Point, $x_i = Q\left(\frac{i-.5}{10}\right)$
1	.05	$7{,}583 = Q(.05)$
2	.15	$8{,}527 = Q(.15)$
3	.25	$8{,}572 = Q(.25)$
4	.35	$8{,}577 = Q(.35)$
5	.45	$9{,}011 = Q(.45)$
6	.55	$9{,}165 = Q(.55)$
7	.65	$9{,}471 = Q(.65)$
8	.75	$9{,}614 = Q(.75)$
9	.85	$9{,}614 = Q(.85)$
10	.95	$10{,}688 = Q(.95)$

Example 5
(*continued*)

the smallest 3 data points and half of the fourth smallest are counted as lying to the left of the desired number, and the largest 6 data points and half of the seventh largest are counted as lying to the right. Thus, the fourth smallest data point must be the .35 quantile, as is shown in Table 3.7.

To illustrate convention (2) of Definition 1, consider finding the .5 and .93 quantiles of the strength distribution. Since .5 is $\frac{.5-.45}{.55-.45} = .5$ of the way from .45 to .55, linear interpolation gives

▶ $$Q(.5) = (1 - .5)\, Q(.45) + .5\, Q(.55) = .5(9,011) + .5(9,165) = 9,088 \text{ g}$$

Then, observing that .93 is $\frac{.93-.85}{.95-.85} = .8$ of the way from .85 to .95, linear interpolation gives

$$Q(.93) = (1 - .8)\, Q(.85) + .8Q(.95) = .2(9,614) + .8(10,688) = 10,473.2 \text{ g}$$

Particular round values of p give quantiles $Q(p)$ that are known by special names.

Definition 2

$Q(.5)$ is called the **median** of a distribution.

Definition 3

$Q(.25)$ and $Q(.75)$ are called the **first (or lower) quartile** and **third (or upper) quartile** of a distribution, respectively.

Example 5
(*continued*)

Referring again to Table 3.7 and the value of $Q(.5)$ previously computed, for the breaking strength distribution

$$\text{Median} = Q(.5) = 9,088 \text{ g}$$
▶ $$\text{1st quartile} = Q(.25) = 8,572 \text{ g}$$
▶ $$\text{3rd quartile} = Q(.75) = 9,614 \text{ g}$$

A way of representing the quantile idea graphically is to make a **quantile plot**.

Definition 4

A **quantile plot** is a plot of $Q(p)$ versus p. For an ordered data set of size n containing values $x_1 \le x_2 \le \cdots \le x_n$, such a display is made by first plotting the points $(\frac{i-.5}{n}, x_i)$ and then connecting consecutive plotted points with straight-line segments.

It is because convention (2) in Definition 1 calls for linear interpolation that straight-line segments enter the picture in making a quantile plot.

Example 5
(*continued*)

Referring again to Table 3.7 for the $\frac{i-.5}{10}$ quantiles of the breaking strength distribution, it is clear that a quantile plot for these data will involve plotting and then connecting consecutive ones of the following ordered pairs.

$$(.05, 7,583) \quad (.15, 8,527) \quad (.25, 8,572)$$
$$(.35, 8,577) \quad (.45, 9,011) \quad (.55, 9,165)$$
$$(.65, 9,471) \quad (.75, 9,614) \quad (.85, 9,614)$$
$$(.95, 10,688)$$

Figure 3.11 gives such a plot.

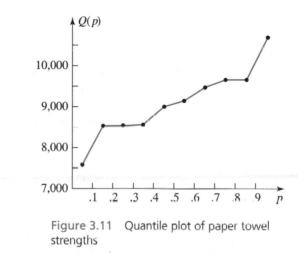

Figure 3.11   Quantile plot of paper towel strengths

A quantile plot allows the user to do some informal visual smoothing of the plot to compensate for any jaggedness. (The tacit assumption is that the underlying data-generating mechanism would itself produce smoother and smoother quantile plots for larger and larger samples.)

3.2.2 Boxplots

Familiarity with the quantile idea is the principal prerequisite for making **boxplots**, an alternative to dot diagrams or histograms. The boxplot carries somewhat less information, but it has the advantage that many can be placed side-by-side on a single page for comparison purposes.

There are several common conventions for making boxplots. The one that will be used here is illustrated in generic fashion in Figure 3.12. A box is made to extend from the first to the third quartiles and is divided by a line at the median. Then the **interquartile range**

Interquartile range

$$IQR = Q(.75) - Q(.25)$$

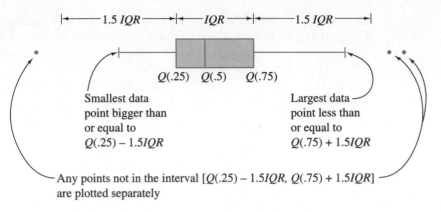

Figure 3.12   Generic boxplot

is calculated and the smallest data point within $1.5IQR$ of $Q(.25)$ and the largest data point within $1.5IQR$ of $Q(.75)$ are determined. Lines called **whiskers** are made to extend out from the box to these values. Typically, most data points will be within the interval $[Q(.25) - 1.5IQR, Q(.75) + 1.5IQR]$. Any that are not then get plotted individually and are thereby identified as *outlying* or unusual.

Example 5
(continued)

Consider making a boxplot for the paper towel breaking strength data. To begin,

$$Q(.25) = 8{,}572 \text{ g}$$
$$Q(.5) = 9{,}088 \text{ g}$$
$$Q(.75) = 9{,}614 \text{ g}$$

So

$$IQR = Q(.75) - Q(.25) = 9{,}614 - 8{,}572 = 1{,}042 \text{ g}$$

and

$$1.5IQR = 1{,}563 \text{ g}$$

Then

$$Q(.75) + 1.5IQR = 9{,}614 + 1{,}563 = 11{,}177 \text{ g}$$

and

$$Q(.25) - 1.5IQR = 8{,}572 - 1{,}563 = 7{,}009 \text{ g}$$

Since all the data points lie in the range 7,009 g to 11,177 g, the boxplot is as shown in Figure 3.13.

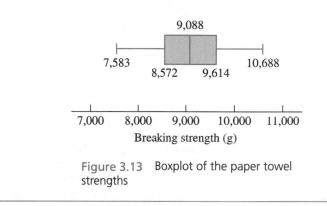

Figure 3.13 Boxplot of the paper towel strengths

A boxplot shows distributional location through the placement of the box and whiskers along a number line. It shows distributional spread through the extent of the box and the whiskers, with the box enclosing the middle 50% of the distribution. Some elements of distributional shape are indicated by the symmetry (or lack thereof) of the box and of the whiskers. And a gap between the end of a whisker and a separately plotted point serves as a reminder that no data values fall in that interval.

Two or more boxplots drawn to the same scale and side by side provide an effective way of comparing samples.

Example 6
(Example 2, page 67, revisited)

More on Bullet Penetration Depths

Table 3.8 contains the raw information needed to find the $\frac{i-.5}{20}$ quantiles for the two distributions of bullet penetration depth introduced in the previous section. For the 230 grain bullet penetration depths, interpolation yields

$$Q(.25) = .5Q(.225) + .5Q(.275) = .5(38.75) + .5(39.75) = 39.25 \text{ mm}$$
$$Q(.5) = .5Q(.475) + .5Q(.525) = .5(42.55) + .5(42.90) = 42.725 \text{ mm}$$
$$Q(.75) = .5Q(.725) + .5Q(.775) = .5(47.90) + .5(48.15) = 48.025 \text{ mm}$$

So

$$IQR = 48.025 - 39.25 = 8.775 \text{ mm}$$
$$1.5IQR = 13.163 \text{ mm}$$
$$Q(.75) + 1.5IQR = 61.188 \text{ mm}$$
$$Q(.25) - 1.5IQR = 26.087 \text{ mm}$$

Example 6
(*continued*)

Similar calculations for the 200 grain bullet penetration depths yield

$$Q(.25) = 60.25 \text{ mm}$$

$$Q(.5) = 62.80 \text{ mm}$$

$$Q(.75) = 64.35 \text{ mm}$$

$$Q(.75) + 1.5IQR = 70.50 \text{ mm}$$

$$Q(.25) - 1.5IQR = 54.10 \text{ mm}$$

Table 3.8
Quantiles of the Bullet Penetration Depth Distributions

i	$\frac{i-.5}{20}$	*i*th Smallest 230 Grain Data Point $= Q(\frac{i-.5}{20})$	*i*th Smallest 200 Grain Data Point $= Q(\frac{i-.5}{20})$
1	.025	27.75	58.00
2	.075	37.35	58.65
3	.125	38.35	59.10
4	.175	38.35	59.50
5	.225	38.75	59.80
6	.275	39.75	60.70
7	.325	40.50	61.30
8	.375	41.00	61.50
9	.425	41.15	62.30
10	.475	42.55	62.65
11	.525	42.90	62.95
12	.575	43.60	63.30
13	.625	43.85	63.55
14	.675	47.30	63.80
15	.725	47.90	64.05
16	.775	48.15	64.65
17	.825	49.85	65.00
18	.875	51.25	67.75
19	.925	51.60	70.40
20	.975	56.00	71.70

Figure 3.14 then shows boxplots placed side by side on the same scale. The plots show the larger and more consistent penetration depths of the 200 grain bullets. They also show the existence of one particularly extreme data point in the 200 grain data set. Further, the relative lengths of the whiskers hint at some skewness (recall the terminology introduced with Figure 3.7) in the data. And all of this is done in a way that is quite uncluttered and compact. Many more of

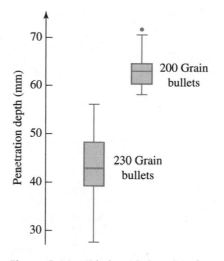

Figure 3.14   Side-by-side boxplots for the bullet penetration depths

these boxes could be added to Figure 3.14 (to compare other bullet types) without visual overload.

3.2.3   *Q-Q* Plots and Comparing Distributional Shapes

It is often important to compare the shapes of two distributions. Comparing histograms is one rough way of doing this. A more sensitive way is to make a single plot based on the quantile functions for the two distributions and exploit the fact that "equal shape" is equivalent to "linearly related quantile functions." Such a plot is called a **quantile-quantile plot** or, more briefly, a *Q-Q* **plot**.

Consider the two small artificial data sets given in Table 3.9. Dot diagrams of these two data sets are given in Figure 3.15. The two data sets have the same shape. But why is this so? One way to look at the equality of the shapes is to note that

$$i\text{th smallest value in data set 2} = 2(i\text{th smallest value in data set 1}) + 1 \qquad (3.1)$$

Then, recognizing ordered data values as quantiles and letting Q_1 and Q_2 stand for the quantile functions of the two respective data sets, it is clear from display (3.1) that

$$Q_2(p) = 2Q_1(p) + 1 \qquad (3.2)$$

Table 3.9
Two Small Artificial Data Sets

Data Set 1	Data Set 2
3, 5, 4, 7, 3	15, 7, 9, 7, 11

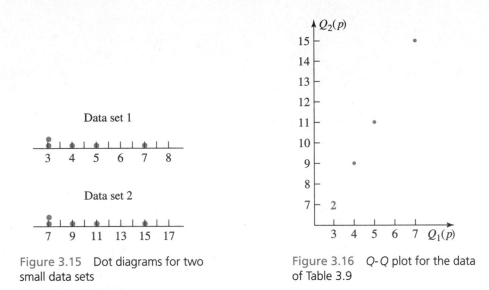

Figure 3.15   Dot diagrams for two small data sets

Figure 3.16   Q-Q plot for the data of Table 3.9

That is, the two data sets have quantile functions that are linearly related. Looking at either display (3.1) or (3.2), it is obvious that a plot of the points

$$\left(Q_1\left(\frac{i-.5}{5}\right), \quad Q_2\left(\frac{i-.5}{5}\right) \right)$$

(for $i = 1, 2, 3, 4, 5$) should be exactly linear. Figure 3.16 illustrates this—in fact Figure 3.16 is a Q-Q plot for the data sets of Table 3.9.

Definition 5

A **Q-Q plot** for two data sets with respective quantile functions Q_1 and Q_2 is a plot of ordered pairs $(Q_1(p), Q_2(p))$ for appropriate values of p. When two data sets of size n are involved, the values of p used to make the plot will be $\frac{i-.5}{n}$ for $i = 1, 2, \ldots, n$. When two data sets of unequal sizes are involved, the values of p used to make the plot will be $\frac{i-.5}{n}$ for $i = 1, 2, \ldots, n$, where n is the size of the smaller set.

To make a Q-Q plot *for two data sets of the same size,*

Steps in making a Q-Q plot

1. order each from the smallest observation to the largest,

2. pair off corresponding values in the two data sets, and

3. plot ordered pairs, with the horizontal coordinates coming from the first data set and the vertical ones from the second.

When data sets of unequal size are involved, the ordered values from the smaller data set must be paired with quantiles of the larger data set obtained by interpolation.

A *Q-Q* plot that is reasonably linear indicates the two distributions involved have similar shapes. When there are significant departures from linearity, the character of those departures reveals the ways in which the shapes differ.

Example 6
(*continued*)

Returning again to the bullet penetration depths, Table 3.8 (page 84) gives the raw material for making a *Q-Q* plot. The depths on each row of that table need only be paired and plotted in order to make the plot given in Figure 3.17.

The scatterplot in Figure 3.17 is not terribly linear when looked at as a whole. However, the points corresponding to the 2nd through 13th smallest values in each data set do look fairly linear, indicating that (except for the *extreme* lower ends) the lower ends of the two distributions have similar shapes.

The horizontal jog the plot takes between the 13th and 14th plotted points indicates that the gap between 43.85 mm and 47.30 mm (for the 230 grain data) is out of proportion to the gap between 63.55 and 63.80 mm (for the 200 grain data). This hints that there was some kind of basic physical difference in the mechanisms that produced the smaller and larger 230 grain penetration depths. Once this kind of indication is discovered, it is a task for ballistics experts or materials people to explain the phenomenon.

Because of the marked departure from linearity produced by the 1st plotted point (27.75, 58.00), there is also a drastic difference in the shapes of the extreme lower ends of the two distributions. In order to move that point back on line with the rest of the plotted points, it would need to be moved to the right or down (i.e., increase the smallest 230 grain observation or decrease the smallest 200 grain observation). That is, *relative to the 200 grain distribution, the 230 grain distribution is long-tailed to the low side.* (Or to put it differently, relative to the 230 grain distribution, the 200 grain distribution is short-tailed to the low side.) Note that the difference in shapes was already evident in the boxplot in Figure 3.14. Again, it would remain for a specialist to explain this difference in distributional shapes

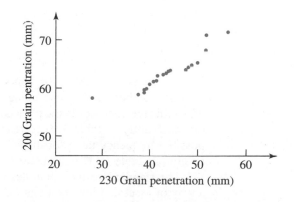

Figure 3.17   *Q-Q* plot for the bullet penetration depths

The Q-Q plotting idea is useful when applied to two data sets, and it is easiest to explain the notion in such an "empirical versus empirical" context. But its greatest usefulness is really when it is applied to one quantile function that represents a data set and a second that represents a *theoretical distribution*.

Definition 6

> A **theoretical Q-Q plot** or **probability plot** for a data set of size n and a theoretical distribution, with respective quantile functions Q_1 and Q_2, is a plot of ordered pairs $(Q_1(p), Q_2(p))$ for appropriate values of p. In this text, the values of p of the form $\frac{i-.5}{n}$ for $i = 1, 2, \ldots, n$ will be used.

Recognizing $Q_1(\frac{i-.5}{n})$ as the ith smallest data point, one sees that a theoretical Q-Q plot is a plot of points with horizontal plotting positions equal to observed data and vertical plotting positions equal to quantiles of the theoretical distribution. That is, with ordered data $x_1 \leq x_2 \leq \cdots \leq x_n$, the points

Ordered pairs making a probability plot

$$\left(x_i, Q_2\left(\frac{i - .5}{n}\right) \right)$$

are plotted. Such a plot allows one to ask, "Does the data set have a shape similar to the theoretical distribution?"

Normal plotting

The most famous version of the theoretical Q-Q plot occurs when quantiles for the **standard normal** or **Gaussian** distribution are employed. This is the familiar bell-shaped distribution. Table 3.10 gives some quantiles of this distribution. In order to find $Q(p)$ for p equal to one of the values .01, .02, ..., .98, .99, locate the entry in the row labeled by the first digit after the decimal place and in the column labeled by the second digit after the decimal place. (For example, $Q(.37) = -.33$.) A simple numerical approximation to the values given in Table 3.10 adequate for most plotting purposes is

Approximate standard normal quantiles

$$Q(p) \approx 4.9(p^{.14} - (1 - p)^{.14}) \tag{3.3}$$

The origin of Table 3.10 is not obvious at this point. It will be explained in Section 5.2, but for the time being consider the following crude argument to the effect that the quantiles in the table correspond to a bell-shaped distribution. Imagine that each entry in Table 3.10 corresponds to a data point in a set of size $n = 99$. A possible frequency table for those 99 data points is given as Table 3.11. The tally column in Table 3.11 shows clearly the bell shape.

The standard normal quantiles can be used to make a theoretical Q-Q plot as a way of assessing how bell-shaped a data set looks. The resulting plot is called a **normal (probability) plot**.

Table 3.10
Standard Normal Quantiles

	.00	.01	.02	.03	.04	.05	.06	.07	.08	.09
.0		−2.33	−2.05	−1.88	−1.75	−1.65	−1.55	−1.48	−1.41	−1.34
.1	−1.28	−1.23	−1.18	−1.13	−1.08	−1.04	−.99	−.95	−.92	−.88
.2	−.84	−.81	−.77	−.74	−.71	−.67	−.64	−.61	−.58	−.55
.3	−.52	−.50	−.47	−.44	−.41	−.39	−.36	−.33	−.31	−.28
.4	−.25	−.23	−.20	−.18	−.15	−.13	−.10	−.08	−.05	−.03
.5	0.00	.03	.05	.08	.10	.13	.15	.18	.20	.23
.6	.25	.28	.31	.33	.36	.39	.41	.44	.47	.50
.7	.52	.55	.58	.61	.64	.67	.71	.74	.77	.81
.8	.84	.88	.92	.95	.99	1.04	1.08	1.13	1.18	1.23
.9	1.28	1.34	1.41	1.48	1.55	1.65	1.75	1.88	2.05	2.33

Table 3.11
A Frequency Table for the Standard Normal Quantiles

Value	Tally	Frequency
−2.80 to −2.30	|	1
−2.29 to −1.79	||	2
−1.78 to −1.28	ⷠⷤ ||	7
−1.27 to −.77	ⷠⷤ ⷠⷤ ||	12
−.76 to −.26	ⷠⷤ ⷠⷤ ⷠⷤ ||	17
−.25 to .25	ⷠⷤ ⷠⷤ ⷠⷤ ⷠⷤ |	21
.26 to .76	ⷠⷤ ⷠⷤ ⷠⷤ ||	17
.77 to 1.27	ⷠⷤ ⷠⷤ ||	12
1.28 to 1.78	ⷠⷤ ||	7
1.79 to 2.29	||	2
2.30 to 2.80	|	1

Example 5
(continued)

Consider again the paper towel strength testing scenario and now the issue of how bell-shaped the data set in Table 3.6 (page 79) is. Table 3.12 was made using Tables 3.7 (page 79) and 3.10; it gives the information needed to produce the theoretical Q-Q plot in Figure 3.18.

Considering the small size of the data set involved, the plot in Figure 3.18 is fairly linear, and so the data set is reasonably bell-shaped. As a practical consequence of this judgment, it is then possible to use the normal probability models discussed in Section 5.2 to describe breaking strength. These could be employed to make breaking strength predictions, and methods of formal statistical inference based on them could be used in the analysis of breaking strength data.

Example 5
(*continued*)

Table 3.12
Breaking Strength and Standard Normal Quantiles

i	$\frac{i-.5}{10}$	$\frac{i-.5}{10}$ Breaking Strength Quantile	$\frac{i-.5}{10}$ Standard Normal Quantile
1	.05	7,583	−1.65
2	.15	8,527	−1.04
3	.25	8,572	−.67
4	.35	8,577	−.39
5	.45	9,011	−.13
6	.55	9,165	.13
7	.65	9,471	.39
8	.75	9,614	.67
9	.85	9,614	1.04
10	.95	10,688	1.65

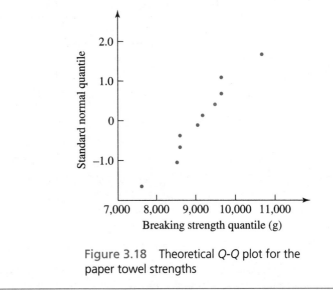

Figure 3.18   Theoretical *Q-Q* plot for the paper towel strengths

Special graph paper, called **normal probability paper** (or just *probability paper*), is available as an alternative way of making normal plots. Instead of plotting points on regular graph paper using vertical plotting positions taken from Table 3.10, points are plotted on probability paper using vertical plotting positions of the form $\frac{i-.5}{n}$. Figure 3.19 is a normal plot of the breaking strength data from Example 5 made on probability paper. Observe that this is virtually identical to the plot in Figure 3.18.

Normal plots are not the only kind of theoretical *Q-Q* plots useful to engineers. Many other types of theoretical distributions are of engineering importance, and each can be used to make theoretical *Q-Q* plots. This point is discussed in more

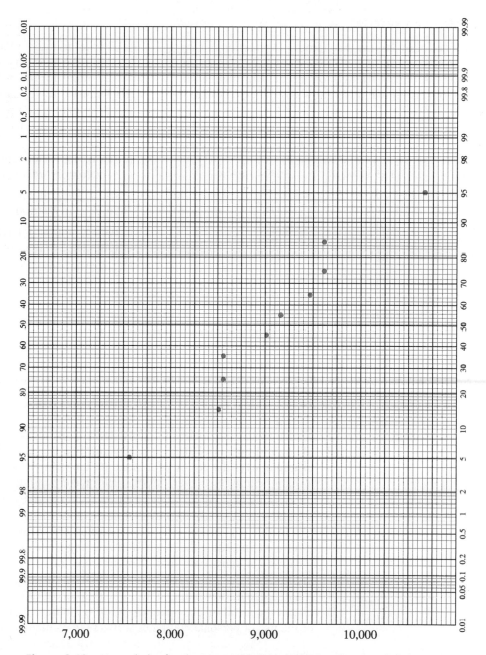

Figure 3.19 Normal plot for the paper towel strengths (made on probability paper, used with permission of the Keuffel and Esser Company)

detail in Section 5.3, but the introduction of theoretical Q-Q plotting here makes it possible to emphasize the relationship between probability plotting and (empirical) Q-Q plotting.

Section 2 Exercises ···

1. The following are data (from *Introduction to Contemporary Statistical Methods* by L. H. Koopmans) on the impact strength of sheets of insulating material cut in two different ways. (The values are in ft lb.)

Lengthwise Cuts	Crosswise Cuts
1.15	.89
.84	.69
.88	.46
.91	.85
.86	.73
.88	.67
.92	.78
.87	.77
.93	.80
.95	.79

(a) Make quantile plots for these two samples. Find the medians, the quartiles, and the .37 quantiles for the two data sets.
(b) Draw (to scale) carefully labeled side-by-side boxplots for comparing the two cutting methods. Discuss what these show about the two methods.
(c) Make and discuss the appearance of a Q-Q plot for comparing the shapes of these two data sets.

2. Make a Q-Q plot for the two small samples in Table 3.13 in Section 3.3.

3. Make and interpret a normal plot for the yield data of Exercise 1 of Section 3.1.

4. Explain the usefulness of theoretical Q-Q plotting.

··

3.3   Standard Numerical Summary Measures

The smooth functioning of most modern technology depends on the reduction of large amounts of data to a few informative numerical summary values. For example, over the period of a month, a lab doing compressive strength testing for a manufacturer's concrete blocks may make hundreds or even thousands of such measurements. But for some purposes, it may be adequate to know that those strengths average 4,618 psi with a range of 2,521 psi (from smallest to largest).

In this section, several standard summary measures for quantitative data are discussed, including the mean, median, range, and standard deviation. Measures of location are considered first, then measures of spread. There follows a discussion of the difference between sample statistics and population parameters and then illustrations of how numerical summaries can be effectively used in simple plots to clarify the results of statistical engineering studies. Finally, there is a brief discussion of the use of personal computer software in elementary data summarization.

3.3.1   Measures of Location

Most people are familiar with the concept of an "average" as being representative of, or in the center of, a data set. Temperatures may vary between different locations in a blast furnace, but an average temperature tells something about a middle or representative temperature. Scores on an exam may vary, but one is relieved to score at least above average.

The word *average*, as used in colloquial speech, has several potential technical meanings. One is the median, $Q(.5)$, which was introduced in the last section. The median divides a data set in half. Roughly half of the area enclosed by the bars of a well-made histogram will lie to either side of the median. As a measure of center, it is completely insensitive to the effects of a few extreme or outlying observations. For example, the small set of data

$$2, 3, 6, 9, 10$$

has median 6, and this remains true even if the value 10 is replaced by 10,000,000 and/or the value 2 is replaced by $-200{,}000$.

The previous section used the median as a center value in the making of boxplots. But the median is not the technical meaning most often attached to the notion of average in statistical analyses. Instead, it is more common to employ the (arithmetic) *mean*.

Definition 7	The **(arithmetic) mean** of a sample of quantitative data (say, $x_1, x_2, \ldots, x_n$) is $$\bar{x} = \frac{1}{n} \sum_{i=1}^{n} x_i$$

The mean is sometimes called the *first moment* or *center of mass* of a distribution, drawing on an analogy to mechanics. Think of placing a unit mass along the number line at the location of each value in a data set—the balance point of the mass distribution is at $\bar{x}$.

Example 7 | **Waste on Bulk Paper Rolls**

Hall, Luethe, Pelszynski, and Ringhofer worked with a company that cuts paper from large rolls purchased in bulk from several suppliers. The company was interested in determining the amount of waste (by weight) on rolls obtained from the various sources. Table 3.13 gives percent waste data, which the students obtained for six and eight rolls, respectively, of paper purchased from two different sources.

The medians and means for the two data sets are easily obtained. For the supplier 1 data,

$$Q(.5) = .5(.65) + .5(.92) = .785\% \text{ waste}$$

and

$$\bar{x} = \frac{1}{6}(.37 + .52 + .65 + .92 + 2.89 + 3.62) = 1.495\% \text{ waste}$$

Example 7
(*continued*)

Table 3.13
Percent Waste by Weight on Bulk Paper Rolls

Supplier 1	Supplier 2
.37, .52, .65,	.89, .99, 1.45, 1.47,
.92, 2.89, 3.62	1.58, 2.27, 2.63, 6.54

For the supplier 2 data,

$$Q(.5) = .5(1.47) + .5(1.58) = 1.525\% \text{ waste}$$

and

$$\bar{x} = \frac{1}{8}(.89 + .99 + 1.45 + 1.47 + 1.58 + 2.27 + 2.63 + 6.54)$$

$$= 2.228\% \text{ waste}$$

Figure 3.20 shows dot diagrams with the medians and means marked. Notice that a comparison of either medians or means for the two suppliers shows the supplier 2 waste to be larger than the supplier 1 waste. But there is a substantial difference between the median and mean values for a given supplier. In both cases, the mean is quite a bit larger than the corresponding median. This reflects the right-skewed nature of both data sets. In both cases, the center of mass of the distribution is pulled strongly to the right by a few extremely large values.

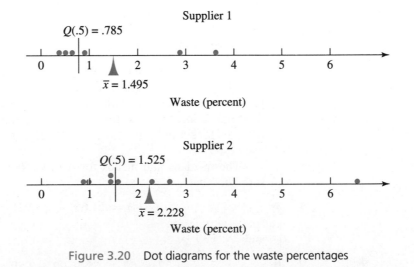

Figure 3.20   Dot diagrams for the waste percentages

Example 7 shows clearly that, in contrast to the median, the mean is a measure of center that can be strongly affected by a few extreme data values. People sometimes say that because of this, one or the other of the two measures is "better." Such statements lack sense. Neither is better; they are simply measures with different properties. And the difference is one that intelligent consumers of statistical information do well to keep in mind. The "average" income of employees at a company paying nine workers each $10,000/year and a president $110,000/year can be described as $10,000/year or $20,000/year, depending upon whether the median or mean is being used.

3.3.2   Measures of Spread

Quantifying the variation in a data set can be as important as measuring its location. In manufacturing, for example, if a characteristic of parts coming off a particular machine is being measured and recorded, the spread of the resulting data gives information about the intrinsic precision or **capability** of the machine. The location of the resulting data is often a function of machine setup or settings of adjustment knobs. Setups can fairly easily be changed, but improvement of intrinsic machine precision usually requires a capital expenditure for a new piece of equipment or overhaul of an existing one.

Although the point wasn't stressed in Section 3.2, the interquartile range, $IQR = Q(.75) - Q(.25)$, is one possible measure of spread for a distribution. It measures the spread of the middle half of a distribution. Therefore, it is insensitive to the possibility of a few extreme values occurring in a data set. A related measure is the **range**, which indicates the spread of the entire distribution.

Definition 8

> The **range** of a data set consisting of ordered values $x_1 \leq x_2 \leq \cdots \leq x_n$ is
>
> $$R = x_n - x_1$$

Notice the word usage here. The word *range* could be used as a verb to say, "The data range from 3 to 21." But to use the word as a noun, one says, "The range is $(21 - 3) = 18$." Since the range depends only on the values of the smallest and largest points in a data set, it is necessarily highly sensitive to extreme (or outlying) values. Because it is easily calculated, it has enjoyed long-standing popularity in industrial settings, particularly as a tool in statistical quality control.

However, most methods of formal statistical inference are based on another measure of distributional spread. A notion of "mean squared deviation" or "root mean squared deviation" is employed to produce measures that are called the **variance** and the **standard deviation**, respectively.

Definition 9

The **sample variance** of a data set consisting of values $x_1, x_2, \ldots, x_n$ is

$$s^2 = \frac{1}{n-1} \sum_{i=1}^{n} (x_i - \bar{x})^2$$

The **sample standard deviation**, s, is the nonnegative square root of the sample variance.

Apart from an exchange of $n - 1$ for n in the divisor, s^2 is an average squared distance of the data points from the central value $\bar{x}$. Thus, s^2 is nonnegative and is 0 only when all data points are exactly alike. The units of s^2 are the squares of the units in which the original data are expressed. Taking the square root of s^2 to obtain s then produces a measure of spread expressed in the original units.

Example 7
(*continued*)

The spreads in the two sets of percentage wastes recorded in Table 3.13 can be expressed in any of the preceding terms. For the supplier 1 data,

$$Q(.25) = .52$$

$$Q(.75) = 2.89$$

and so

$$IQR = 2.89 - .52 = 2.37\% \text{ waste}$$

Also,

$$R = 3.62 - .37 = 3.25\% \text{ waste}$$

Further,

$$s^2 = \frac{1}{6-1}((.37 - 1.495)^2 + (.52 - 1.495)^2 + (.65 - 1.495)^2 + (.92 - 1.495)^2$$

$$+ (2.89 - 1.495)^2 + (3.62 - 1.495)^2)$$

$$= 1.945(\% \text{ waste})^2$$

so that

$$s = \sqrt{1.945} = 1.394\% \text{ waste}$$

Similar calculations for the supplier 2 data yield the values

$$IQR = 1.23\% \text{ waste}$$

and

$$R = 6.54 - .89 = 5.65\% \text{ waste}$$

Further,

$$s^2 = \frac{1}{8-1}((.89 - 2.228)^2 + (.99 - 2.228)^2 + (1.45 - 2.228)^2 + (1.47 - 2.228)^2$$
$$+ (1.58 - 2.228)^2 + (2.27 - 2.228)^2 + (2.63 - 2.228)^2 + (6.54 - 2.228)^2)$$
$$= 3.383(\% \text{ waste})^2$$

so

$$s = 1.839\% \text{ waste}$$

Supplier 2 has the smaller IQR but the larger R and s. This is consistent with Figure 3.20. The central portion of the supplier 2 distribution is tightly packed. But the single extreme data point makes the overall variability larger for the second supplier than for the first.

The calculation of sample variances just illustrated is meant simply to reinforce the fact that s^2 is a kind of mean squared deviation. Of course, the most sensible way to find sample variances in practice is by using either a handheld electronic calculator with a preprogrammed variance function or a statistical package on a personal computer.

The measures of variation, IQR, R, and s, are not directly comparable. Although it is somewhat out of the main flow of this discussion, it is worth interjecting at this point that it is possible to "put R and s on the same scale." This is done by dividing R by an appropriate conversion factor, known to quality control engineers as d_2. Table B.2 contains *control chart constants* and gives values of d_2 for various sample sizes n. For example, to get R and s on the same scale for the supplier 1 data, division of R by 2.534 is in order, since $n = 6$.

Students often have some initial difficulty developing a feel for the meaning of the standard deviation. One possible help in this effort is a famous theorem of a Russian mathematician.

Proposition 1 (*Chebyschev's Theorem*)	For any data set and any number k larger than 1, a fraction of at least $1 - (1/k^2)$ of the data are within ks of $\bar{x}$.

This little theorem says, for example, that at least $\frac{3}{4}$ of a data set is within 2 standard deviations of its mean. And at least $\frac{8}{9}$ of a data set is within 3 standard deviations of its mean. So the theorem promises that if a data set has a small standard deviation, it will be tightly packed about its mean.

Example 7
(*continued*)

Returning to the waste data, consider illustrating the meaning of Chebyschev's theorem with the supplier 1 values. For example, taking $k = 2$, at least $\frac{3}{4} = 1 - \left(\frac{1}{2}\right)^2$ of the 6 data points (i.e., at least 4.5 of them) must be within 2 standard deviations of $\bar{x}$. In fact

$$\bar{x} - 2s = 1.495 - 2(1.394) = -1.294\% \text{ waste}$$

and

$$\bar{x} + 2s = 1.495 + 2(1.394) = 4.284\% \text{ waste}$$

so simple counting shows that all (a fraction of 1.0) of the data are between these two values.

3.3.3   Statistics and Parameters

At this point, it is important to introduce some more basic terminology. Jargon and notation for distributions of samples are somewhat different than for population distributions (and theoretical distributions).

Definition 10

Numerical summarizations of sample data are called (sample) **statistics**. Numerical summarizations of population and theoretical distributions are called (population or model) **parameters**. Typically, Roman letters are used as symbols for statistics, and Greek letters are used to stand for parameters.

As an example, consider the mean. Definition 7 refers specifically to a calculation for a sample. If a data set represents an entire population, then it is common to use the lowercase Greek letter mu (μ) to stand for the **population mean** and to write

Population mean

$$\mu = \frac{1}{N} \sum_{i=1}^{N} x_i \tag{3.4}$$

Comparing this expression to the one in Definition 7, not only is a different symbol used for the mean but also N is used in place of n. It is standard to denote a population size as N and a sample size as n. Chapter 5 gives a definition for the mean of a theoretical distribution. But it is worth saying now that the symbol μ will be used in that context as well as in the context of equation (3.4).

As another example of the usage suggested by Definition 10, consider the variance and standard deviation. Definition 9 refers specifically to the *sample* variance and standard deviation. If a data set represents an entire population, then it is common to use the lowercase Greek sigma squared (σ^2) to stand for the **population variance** and to define

Population variance

$$\sigma^2 = \frac{1}{N} \sum_{i=1}^{N} (x_i - \mu)^2$$

(3.5)

The nonnegative square root of σ^2 is then called the **population standard deviation**, σ. (The division in equation (3.5) is by N, and not the $N - 1$ that might be expected on the basis of Definition 9. There are reasons for this change, but they are not accessible at this point.) Chapter 5 defines a variance and standard deviation for theoretical distributions, and the symbols σ^2 and σ will be used there as well as in the context of equation (3.5).

On one point, this text will deviate from the Roman/Greek symbolism convention laid out in Definition 10: the notation for quantiles. $Q(p)$ will stand for the pth quantile of a distribution, whether it is from a sample, a population, or a theoretical model.

3.3.4 Plots of Summary Statistics

Plots against time

Plotting numerical summary measures in various ways is often helpful in the early analysis of engineering data. For example, plots of summary statistics against time are frequently revealing.

Example 8
(Example 8, Chapter 1, revisited—p. 18)

Monitoring a Critical Dimension of Machined Parts

Cowan, Renk, Vander Leest, and Yakes worked with a company that makes precision metal parts. A critical dimension of one such part was monitored by occasionally selecting and measuring five consecutive pieces and then plotting the sample mean and range. Table 3.14 gives the $\bar{x}$ and R values for 25 consecutive samples of five parts. The values reported are in .0001 in.

Figure 3.21 is a plot of both the means and ranges against order of observation. Looking first at the plot of ranges, no strong trends are obvious, which suggests that the basic short-term variation measured in this critical dimension is stable. The combination of process and measurement precision is neither improving nor degrading with time. The plot of means, however, suggests some kind of physical change. The average dimensions from the second shift on October 27 (samples 9 through 15) are noticeably smaller than the rest of the means. As discussed in Example 8, Chapter 1, it turned out to be the case that the parts produced on that

Table 3.14
Means and Ranges for a Critical Dimension on Samples of $n = 5$ Parts

Sample	Date	Time		$\bar{x}$	R	Sample	Date	Time		$\bar{x}$	R
1	10/27	7:30	AM	3509.4	5	14		10:15		3504.4	4
2		8:30		3509.2	2	15		11:15		3504.6	3
3		9:30		3512.6	3	16	10/28	7:30	AM	3513.0	2
4		10:30		3511.6	4	17		8:30		3512.4	1
5		11:30		3512.0	4	18		9:30		3510.8	5
6		12:30	PM	3513.6	6	19		10:30		3511.8	4
7		1:30		3511.8	3	20		6:15	PM	3512.4	3
8		2:30		3512.2	2	21		7:15		3511.0	4
9		4:15		3500.0	3	22		8:45		3510.6	1
10		5:45		3502.0	2	23		9:45		3510.2	5
11		6:45		3501.4	2	24		10:45		3510.4	2
12		8:15		3504.0	2	25		11:45		3510.8	3
13		9:15		3503.6	3						

Example 8
(*continued*)

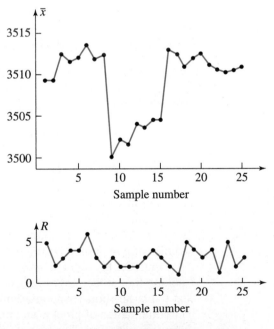

Figure 3.21   Plots of $\bar{x}$ and R over time

shift were not really systematically any different from the others. Instead, the person making the measurements for samples 9 through 15 used the gauge in a fundamentally different way than other employees. The pattern in the $\bar{x}$ values was caused by this change in measurement technique.

Terminology and causes for patterns on plots against Time

Patterns revealed in the plotting of sample statistics against time ought to alert an engineer to look for a physical cause and (typically) a cure. **Systematic variations** or **cycles** in a plot of means can often be related to process variables that come and go on a more or less regular basis. Examples include seasonal or daily variables like ambient temperature or those caused by rotation of gauges or fixtures. **Instability** or variation in excess of that related to basic equipment precision can sometimes be traced to mixed lots of raw material or overadjustment of equipment by operators. **Changes in level** of a process mean can originate in the introduction of new machinery, raw materials, or employee training and (for example) tool wear. **Mixtures** of several patterns of variation on a single plot of some summary statistic against time can sometimes (as in Example 8) be traced to changes in measurement calibration. They are also sometimes produced by consistent differences in machines or streams of raw material.

Plots against process variables

Plots of summary statistics against time are not the only useful ones. Plots against process variables can also be quite informative.

Example 9
(Example 6, Chapter 1, revisited—p. 15)

Plotting the Mean Shear Strength of Wood Joints

In their study of glued wood joint strength, Dimond and Dix obtained the values given in Table 3.15 as mean strengths (over three shear tests) for each combination of three woods and three glues. Figure 3.22 gives a revealing plot of these $3 \times 3 = 9$ different $\bar{x}$'s.

Table 3.15
Mean Joint Strengths for Nine Wood/Glue Combinations

Wood	Glue	$\bar{x}$ Mean Joint Shear Strength (lb)
pine	white	131.7
pine	carpenter's	192.7
pine	cascamite	201.3
fir	white	92.0
fir	carpenter's	146.3
fir	cascamite	156.7
oak	white	257.7
oak	carpenter's	234.3
oak	cascamite	177.7

Example 9
(*continued*)

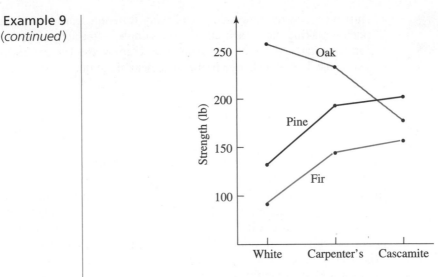

Figure 3.22   Plot of mean joint strength vs. glue type for three woods

From the plot, it is obvious that the gluing properties of pine and fir are quite similar, with pine joints averaging around 40–45 lb stronger. For these two soft woods, cascamite appears slightly better than carpenter's glue, both of which make much better joints than white glue. The gluing properties of oak (a hardwood) are quite different from those of pine and fir. In fact, the glues perform in exactly the opposite ordering for the strength of oak joints. All of this is displayed quite clearly by the simple plot in Figure 3.22.

The two previous examples have illustrated the usefulness of plotting sample statistics against time and against levels of an experimental variable. Other possibilities in specific engineering situations can potentially help the working engineer understand and manipulate the systems on which he or she works.

3.3.5 Summary Statistics and Personal Computer Software

The numerical data summaries introduced in this chapter are relatively simple. For small data sets they can be computed quite easily using only a pocket calculator. However, for large data sets and in cases where subsequent additional calculations or plotting may occur, statistical or spreadsheet software can be convenient.

Printout 1 illustrates the use of the MINITAB statistical package to produce summary statistics for the percent waste data sets in Table 3.13. (The appropriate MINITAB routine is found under the "Stat/Basic Statistics/Display Descriptive Statistics" menu.) The mean, median, and standard deviation values on the printout agree with those produced in Example 7. However, the first and third quartile

Printout 1   Descriptive Statistics for the Percent Waste Data of Table 3.13

Descriptive Statistics

Variable	N	Mean	Median	TrMean	StDev	SE Mean
Supply 1	6	1.495	0.785	1.495	1.394	0.569
Supply 2	8	2.228	1.525	2.228	1.839	0.650

Variable	Minimum	Maximum	Q1	Q3
Supply 1	0.370	3.620	0.483	3.073
Supply 2	0.890	6.540	1.105	2.540

figures on the printout do not match exactly those found earlier. MINITAB simply uses slightly different conventions for those quantities than the ones introduced in Section 3.2.

High-quality statistical packages like MINITAB (and JMP, SAS, SPSS, SYS-TAT, SPLUS, etc.) are widely available. One of them should be on the electronic desktop of every working engineer. Unfortunately, this is not always the case, and engineers often assume that standard spreadsheet software (perhaps augmented with third party plug-ins) provides a workable substitute. Often this is true, but sometimes it is not.

The primary potential problem with using a spreadsheet as a substitute for statistical software concerns numerical accuracy. Spreadsheets can and do on occasion return catastrophically wrong values for even simple statistics. Established vendors of statistical software have many years of experience dealing with subtle numerical issues that arise in the computation of even simple summaries of even small data sets. Most vendors of spreadsheet software seem unaware of or indifferent to these matters. For example, consider the very small data set

$$0, 1, 2$$

The sample variance of these data is easily seen to be 1.0, and essentially any statistical package or spreadsheet will reliably return this value. However, suppose 100,000,000 is added to each of these $n = 3$ values, producing the data set

$$100000000, 100000001, 100000002$$

The actual sample variance is unchanged, and high-quality statistical software will reliably return the value 1.0. However, as of late 1999, the current version of the leading spreadsheet program returned the value 0 for this second sample variance. This is a badly wrong answer to an apparently very simple problem.

So at least until vendors of spreadsheet software choose to integrate an established statistical package into their products, we advise extreme caution in the use of spreadsheets to do statistical computations. A good source of up-to-date information on this issue is the AP Statistics electronic bulletin board found at http://forum.swarthmore.edu/epigone/apstat-l.

Section 3 Exercises

1. Calculate and compare the means, medians, ranges, interquartile ranges, and standard deviations of the two data sets introduced in Exercise 1 of Section 3.2. Discuss the interpretation of these values in the context of comparing the two cutting methods.

2. Are the numerical values you produced in Exercise 1 above most naturally thought of as statistics or as parameters? Explain.

3. Use a statistical package to compute basic summary statistics for the two data sets introduced in Exercise 1 of Section 3.2 and thereby check your answers to Exercise 1 here.

4. Add 1.3 to each of the lengthwise cut impact strengths referred to in Exercise 1 and then recompute the values of the mean, median, range, interquartile range, and standard deviation. How do these compare with the values obtained earlier? Repeat this exercise after multiplying each lengthwise cut impact strength by 2 (instead of adding 1.3).

3.4   Descriptive Statistics for Qualitative and Count Data (*Optional*)

The techniques presented thus far in this chapter are primarily relevant to the analysis of measurement data. As noted in Section 1.2, conventional wisdom is that where they can be obtained, measurement data (or *variables data*) are generally preferable to count and qualitative data (or *attributes data*). Nevertheless, qualitative or count data will sometimes be the primary information available. It is therefore worthwhile to consider their summarization.

This section will cover the reduction of qualitative and count data to per-item or per-inspection-unit figures and the display of those ratios in simple bar charts and plots.

3.4.1   Numerical Summarization of Qualitative and Count Data

Recall from Definitions 8 and 9 in Chapter 1 that aggregation and counting are typically used to produce numerical values from qualitative data. Then, beginning with counts, it is often helpful to calculate rates on a per-item or per-inspection-unit basis.

When each item in a sample of n either does or does not have a characteristic of interest, the notation

Sample fraction of items with a characteristic

$$\hat{p} = \frac{\text{The number of items in the sample with the characteristic}}{n} \tag{3.6}$$

will be used. A given sample can produce many such values of "p hat" if either a single characteristic has many possible categories or many different characteristics are being monitored simultaneously.

Example 10

Defect Classifications of Cable Connectors

Delva, Lynch, and Stephany worked with a manufacturer of cable connectors. Daily samples of 100 connectors of a certain design were taken over 30 production days, and each sampled connector was inspected according to a well-defined (operational) set of rules. Using the information from the inspections, each inspected connector could be classified as belonging to one of the following five mutually exclusive categories:

Category A: having "very serious" defects

Category B: having "serious" defects but no "very serious" defects

Category C: having "moderately serious" defects but no "serious" or "very serious" defects

Category D: having only "minor" defects

Category E: having no defects

Table 3.16 gives counts of sampled connectors falling into the first four categories (the four defect categories) over the 30-day period. Then, using the fact that $30 \times 100 = 3{,}000$ connectors were inspected over this period,

$$\hat{p}_A = 3/3000 = .0010$$

$$\hat{p}_B = 0/3000 = .0000$$

$$\hat{p}_C = 11/3000 = .0037$$

$$\hat{p}_D = 1/3000 = .0003$$

Notice that here $\hat{p}_E = 1 - (\hat{p}_A + \hat{p}_B + \hat{p}_C + \hat{p}_D)$, because categories A through E represent a set of nonoverlapping and exhaustive classifications into which an individual connector must fall, so that the $\hat{p}$'s must total to 1.

Table 3.16
Counts of Connectors Classified into Four Defect Categories

Category	Number of Sampled Connectors
A	3
B	0
C	11
D	1

Example 11

Pneumatic Tool Manufacture

Kraber, Rucker, and Williams worked with a manufacturer of pneumatic tools. Each tool produced is thoroughly inspected before shipping. The students collected some data on several kinds of problems uncovered at final inspection. Table 3.17 gives counts of tools having these problems in a particular production run of 100 tools.

Table 3.17
Counts and Fractions of Tools with Various Problems

Problem	Number of Tools	$\hat{p}$
Type 1 leak	8	.08
Type 2 leak	4	.04
Type 3 leak	3	.03
Missing part 1	2	.02
Missing part 2	1	.01
Missing part 3	2	.02
Bad part 4	1	.01
Bad part 5	2	.02
Bad part 6	1	.01
Wrong part 7	2	.02
Wrong part 8	2	.02

Table 3.17 is a summarization of highly multivariate qualitative data. The categories listed in Table 3.17 are not mutually exclusive; a given tool can fall into more than one of them. Instead of representing different possible values of a single categorical variable (as was the case with the connector categories in Example 10), the categories listed above each amount to 1 (present) of 2 (present and not present) possible values for a different categorical variable. For example, for type 1 leaks, $\hat{p} = .08$, so $1 - \hat{p} = .92$ for the fraction of tools *without* type 1 leaks. The $\hat{p}$ values do not necessarily total to the fraction of tools requiring rework at final inspection. A given faulty tool could be counted in several $\hat{p}$ values.

Another kind of per-item ratio, also based on counts, is sometimes confused with $\hat{p}$. Such a ratio arises when every item in a sample provides an opportunity for a phenomenon of interest to occur, but multiple occurrences are possible and counts are kept of the total number of occurrences. In such cases, the notation

Sample mean occurences per unit or item

$$\hat{u} = \frac{\text{The total number of occurrences}}{\text{The total number of inspection units or sampled items}} \tag{3.7}$$

is used. $\hat{u}$ is really closer in meaning to $\bar{x}$ than to $\hat{p}$, even though it can turn out to be a number between 0 and 1 and is sometimes expressed as a percentage and called a *rate*.

Although the counts totaled in the numerator of expression (3.7) must all be integers, the values totaled to create the denominator need not be. For instance, suppose vinyl floor tiles are being inspected for serious blemishes. If on one occasion inspection of 1 box yields a total of 2 blemishes, on another occasion .5 box yields 0 blemishes, and on still another occasion 2.5 boxes yield a total of 1 blemish, then

$$\hat{u} = \frac{2 + 0 + 1}{1 + .5 + 2.5} = .75 \text{ blemishes/box}$$

Depending on exactly how terms are defined, it may be appropriate to calculate either $\hat{p}$ values or $\hat{u}$ values or both in a single situation.

Example 10
(continued)

It was possible for a single cable connector to have more than one defect of a given severity and, in fact, defects of different severities. For example, Delva, Lynch, and Stephany's records indicate that in the 3,000 connectors inspected, 1 connector had exactly 2 moderately serious defects (along with a single very serious defect), 11 connectors had exactly 1 moderately serious defect (and no others), and 2,988 had no moderately serious defects. So the observed rate of moderately serious defects could be reported as

$$\hat{u} = \frac{2 + 11}{1 + 11 + 2988} = .0043 \text{ moderately serious defects/connector}$$

This is an occurrence rate for moderately serious defects ($\hat{u}$), but not a fraction of connectors having moderately serious defects ($\hat{p}$).

The difference between the statistics $\hat{p}$ and $\hat{u}$ may seem trivial. But it is a point that constantly causes students confusion. Methods of formal statistical inference based on $\hat{p}$ are not the same as those based on $\hat{u}$. The distinction between the two kinds of rates must be kept in mind if those methods are to be applied appropriately.

To carry this warning a step further, note that not every quantity called a *percentage* is even of the form $\hat{p}$ or $\hat{u}$. In a laboratory analysis, a specimen may be declared to be "30% carbon." The 30% cannot be thought of as having the form of $\hat{p}$ in equation (3.6) or $\hat{u}$ in equation (3.7). It is really a single continuous measurement, not a summary statistic. Statistical methods for $\hat{p}$ or $\hat{u}$ have nothing to say about such rates.

3.4.2 Bar Charts and Plots for Qualitative and Count Data

Often, a study will produce several values of $\hat{p}$ or $\hat{u}$ that need to be compared. Bar charts and simple bivariate plots can be a great aid in summarizing these results.

Example 10
(continued)

Figure 3.23 is a bar chart of the fractions of connectors in the categories A through D. It shows clearly that most connectors with defects fall into category C, having moderately serious defects but no serious or very serious defects. This bar chart is a presentation of the behavior of a single categorical variable.

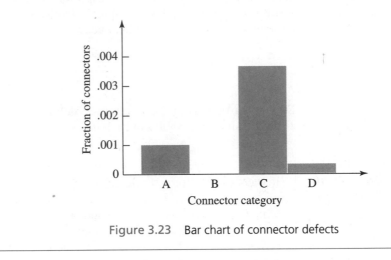

Figure 3.23   Bar chart of connector defects

Example 11
(continued)

Figure 3.24 is a bar chart of the information on tool problems in Table 3.17. It shows leaks to be the most frequently occurring problems on this production run.

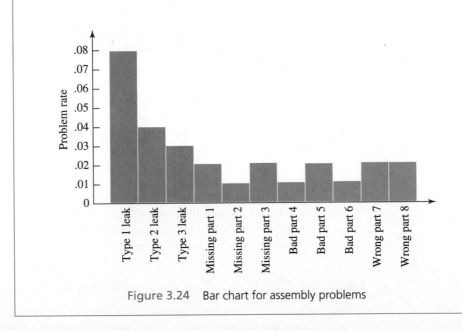

Figure 3.24   Bar chart for assembly problems

Figures 3.23 and 3.24 are both bar charts, but they differ considerably. The first concerns the behavior of a single (ordered) categorical variable—namely, Connector Class. The second concerns the behavior of 11 different present–not present categorical variables, like Type 1 Leak, Missing Part 3, etc. There may be some significance to the shape of Figure 3.23, since categories A through D are arranged in decreasing order of defect severity, and this order was used in the making of the figure. But the shape of Figure 3.24 is essentially arbitrary, since the particular ordering of the tool problem categories used to make the figure is arbitrary. Other equally sensible orderings would give quite different shapes.

The device of **segmenting bars** on a bar chart and letting the segments stand for different categories of a single qualitative variable can be helpful, particularly where several different samples are to be compared.

Example 12

Scrap and Rework in a Turning Operation

The article "Statistical Analysis: Roadmap for Management Action" by H. Rowe (*Quality Progress*, February 1985) describes a statistically based quality-improvement project in the turning of steel shafts. Table 3.18 gives the percentages of reworkable and scrap shafts produced in 18 production runs made during the study.

Figure 3.25 is a corresponding segmented bar graph, with the jobs ordered in time, showing the behavior of both the scrap and rework rates over time. (The total height of any bar represents the sum of the two rates.) The sharp reduction in both scrap and rework between jobs 10 and 11 was produced by overhauling one of the company's lathes. That lathe was identified as needing attention through engineering data analysis early in the plant project.

Table 3.18
Percents Scrap and Rework in a Turning Operation

Job Number	Percent Scrap	Percent Rework	Job Number	Percent Scrap	Percent Rework
1	2	25	10	3	18
2	3	11	11	0	3
3	0	5	12	1	5
4	0	0	13	0	0
5	0	20	14	0	0
6	2	23	15	0	3
7	0	6	16	0	2
8	0	5	17	0	2
9	2	8	18	1	5

Example 12
(*continued*)

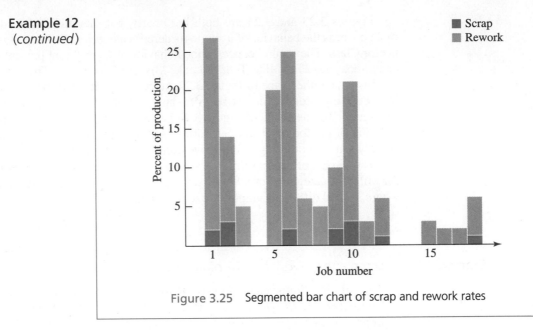

Figure 3.25   Segmented bar chart of scrap and rework rates

In many cases, the simple plotting of $\hat{p}$ or $\hat{u}$ values against time or process variables can make clear the essential message in a set of qualitative or count data.

Example 13

Defects per Truck Found at Final Inspection

In his text *Engineering Statistics and Quality Control*, I. W. Burr illustrates the usefulness of plotting $\hat{u}$ versus time with a set of data on defects found at final inspection at a truck assembly plant. From 95 to 130 trucks were produced daily at the plant; Table 3.19 gives part of Burr's daily defects/truck values. These statistics are plotted in Figure 3.26. The graph shows a marked decrease in quality (increase in $\hat{u}$) over the third and fourth weeks of December, ending with a rate

Table 3.19
Defects Per Truck on 26 Production Days

Date	$\hat{u} = $ Defects/Truck	Date	$\hat{u} = $ Defects/Truck	Date	$\hat{u} = $ Defects/Truck	Date	$\hat{u} = $ Defects/Truck
12/2	1.54	12/11	1.18	12/20	2.32	1/3	1.15
12/3	1.42	12/12	1.39	12/23	1.23	1/6	1.37
12/4	1.57	12/13	1.42	12/24	2.91	1/7	1.79
12/5	1.40	12/16	2.08	12/26	1.77	1/8	1.68
12/6	1.51	12/17	1.85	12/27	1.61	1/9	1.78
12/9	1.08	12/18	1.82	12/30	1.25	1/10	1.84
12/10	1.27	12/19	2.07				

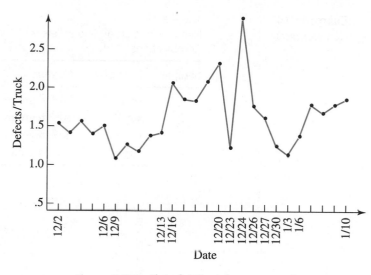

Figure 3.26   Plot of daily defects per truck

of 2.91 defects/truck on Christmas Eve. Apparently, this situation was largely corrected with the passing of the holiday season.

Plots of $\hat{p}$ or $\hat{u}$ against levels of manipulated variables from an experiment are often helpful in understanding the results of that experiment.

Example 14

Plotting Fractions of Conforming Pellets

Greiner, Grim, Larson, and Lukomski experimented with the same pelletizing machine studied by Cyr, Ellson, and Rickard (see Example 2 in Chapter 1). In one part of their study, they ran the machine at an elevated speed and varied the shot size (amount of powder injected into the dies) and the composition of that powder (in terms of the relative amounts of new and reground material). Table 3.20 lists the numbers of conforming pellets produced in a sample of 100 at each of $2 \times 2 = 4$ sets of process conditions. A simple plot of $\hat{p}$ values versus shot size is given in Figure 3.27.

The figure indicates that increasing the shot size is somewhat harmful, but that a substantial improvement in process performance happens when the amount of reground material used in the pellet-making mixture is increased. This makes sense. Reground material had been previously compressed into (nonconforming) pellets. In the process, it had been allowed to absorb some ambient humidity. Both the prior compression and the increased moisture content were potential reasons why this material improved the ability of the process to produce solid, properly shaped pellets.

Example 14
(*continued*)

Table 3.20
Numbers of Conforming Pellets for Four Shot Size/Mixture Combinations

Sample	Shot Size	Mixture	Number Conforming
1	small	20% reground	38
2	small	50% reground	66
3	large	20% reground	29
4	large	50% reground	53

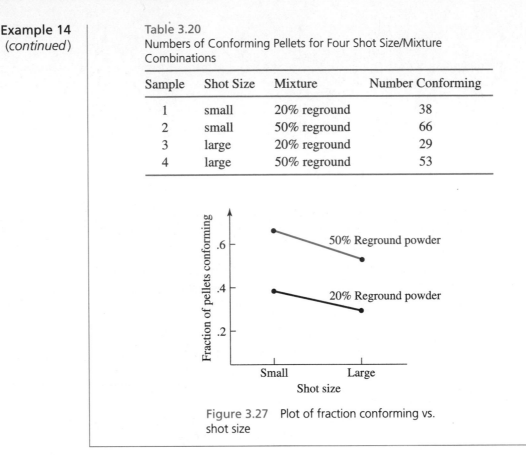

Figure 3.27   Plot of fraction conforming vs. shot size

Section 4 Exercises

1. From your field, give an example of a variable that is a rate (a) of the form $\hat{p}$, (b) of the form $\hat{u}$, and (c) of neither form.

2. Because gauging is easier, it is sometimes tempting to collect qualitative data related to measurements rather than the measurements themselves. For example, in the context of Example 1 in Chapter 1, if gears with runouts exceeding 15 were considered to be nonconforming, it would be possible to derive fractions nonconforming, $\hat{p}$, from simple "go–no go" checking of gears. For the two sets of gears represented in Table 1.1, what would have been the sample fractions nonconforming $\hat{p}$? Give a practical reason why having the values in Table 1.1 might be preferable to knowing only the corresponding $\hat{p}$ values.

3. Consider the measurement of the percentage copper in brass specimens. The resulting data will be a kind of rate data. Are the rates that will be obtained of the type $\hat{p}$, of the type $\hat{u}$, or of neither type? Explain.

Chapter 3 Exercises ....

1. The accompanying values are gains measured on 120 amplifiers designed to produce a 10 dB gain. These data were originally from the *Quality Improvement Tools* workbook set (published by the Juran Institute). They were then used as an example in the article "The Tools of Quality" (*Quality Progress*, September 1990).

   8.1, 10.4, 8.8, 9.7, 7.8, 9.9, 11.7, 8.0, 9.3, 9.0, 8.2, 8.9, 10.1, 9.4, 9.2, 7.9, 9.5, 10.9, 7.8, 8.3, 9.1, 8.4, 9.6, 11.1, 7.9, 8.5, 8.7, 7.8, 10.5, 8.5, 11.5, 8.0, 7.9, 8.3, 8.7, 10.0, 9.4, 9.0, 9.2, 10.7, 9.3, 9.7, 8.7, 8.2, 8.9, 8.6, 9.5, 9.4, 8.8, 8.3, 8.4, 9.1, 10.1, 7.8, 8.1, 8.8, 8.0, 9.2, 8.4, 7.8, 7.9, 8.5, 9.2, 8.7, 10.2, 7.9, 9.8, 8.3, 9.0, 9.6, 9.9, 10.6, 8.6, 9.4, 8.8, 8.2, 10.5, 9.7, 9.1, 8.0, 8.7, 9.8, 8.5, 8.9, 9.1, 8.4, 8.1, 9.5, 8.7, 9.3, 8.1, 10.1, 9.6, 8.3, 8.0, 9.8, 9.0, 8.9, 8.1, 9.7, 8.5, 8.2, 9.0, 10.2, 9.5, 8.3, 8.9, 9.1, 10.3, 8.4, 8.6, 9.2, 8.5, 9.6, 9.0, 10.7, 8.6, 10.0, 8.8, 8.6

   (a) Make a stem-and-leaf plot and a boxplot for these data. How would you describe the shape of this data set? Does the shape of your stem-and-leaf plot (or a corresponding histogram) give you any clue how a high fraction within specifications was achieved?

   (b) Make a normal plot for these data and interpret its shape. (Standard normal quantiles for $p = .0042$ and $p = .9958$ are approximately -2.64 and 2.64, respectively.)

   (c) Although the nominal gain for these amplifiers was to be 10 dB, the design allowed gains from 7.75 dB to 12.2 dB to be considered acceptable. About what fraction, p, of such amplifiers do you expect to meet these engineering specifications?

2. The article "The Lognormal Distribution for Modeling Quality Data When the Mean is Near Zero" by S. Albin (*Journal of Quality Technology*, April 1990) described the operation of a Rutgers University plastics recycling pilot plant. The most important material reclaimed from beverage bottles is PET plastic. A serious impurity is aluminum, which later can clog the filters in extruders when the recycled material is used. The following are the amounts (in ppm by weight of aluminum) found in bihourly samples of PET recovered at the plant over roughly a two-day period.

   291, 222, 125, 79, 145, 119, 244, 118, 182, 63, 30, 140, 101, 102, 87, 183, 60, 191, 119, 511, 120, 172, 70, 30, 90, 115

   (Apparently, the data are recorded in the order in which they were collected, reading left to right, top to bottom.)

   (a) Make a run chart for these data. Are there any obvious time trends? What practical engineering reason is there for looking for such trends?

   (b) Ignoring the time order information, make a stem-and-leaf diagram. Use the hundreds digit to make the stem and the other two digits (separated by commas to indicate the different data points) to make the leaves. After making an initial stem-and-leaf diagram by recording the data in the (time) order given above, make a second one in which the values have been ordered.

   (c) How would you describe the shape of the stem-and-leaf diagram? Is the data set bell-shaped?

   (d) Find the median and the first and third quartiles for the aluminum contents and then find the .58 quantile of the data set.

   (e) Make a boxplot.

   (f) Make a normal plot, using regular graph paper. List the coordinates of the 26 plotted points. Interpret the shape of the plot.

   (g) Try transforming the data by taking natural logarithms and again assess the shape. Is the transformed data set more bell-shaped than the raw data set?

   (h) Find the sample mean, the sample range, and the sample standard deviation for both the original data and the log-transformed values from (g). Is the mean of the transformed values equal to the natural logarithm of the mean of the original data?

3. The accompanying data are three hypothetical samples of size 10 that are supposed to represent measured manganese contents in specimens of 1045 steel (the units are points, or .01%). Suppose that these measurements were made on standard specimens having "true" manganese contents of 80, using three different analytical methods. (Thirty different specimens were involved.)

Method 1
87, 74, 78, 81, 78,
77, 84, 80, 85, 78

Method 2
86, 85, 82, 87, 85,
84, 84, 82, 82, 85

Method 3
84, 83, 78, 79, 85,
82, 82, 81, 82, 79

(a) Make (on the same coordinate system) side-by-side boxplots that you can use to compare the three analytical methods.

(b) Discuss the apparent effectiveness of the three methods in terms of the appearance of your diagram from (a) and in terms of the concepts of accuracy and precision discussed in Section 1.3.

(c) An alternative method of comparing two such analytical methods is to use both methods of analysis once on each of (say) 10 different specimens (10 specimens and 20 measurements). In the terminology of Section 1.2, what kind of data would be generated by such a plan? If one simply wishes to compare the average measurements produced by two analytical methods, which data collection plan (20 specimens and 20 measurements, or 10 specimens and 20 measurements) seems to you most likely to provide the better comparison? Explain.

4. Gaul, Phan, and Shimonek measured the resistances of 15 resistors of $2 \times 5 = 10$ different types. Two different wattage ratings were involved, and five different nominal resistances were used. All measurements were reported to three significant digits. Their data follow.

(a) Make back-to-back stem-and-leaf plots for comparing the $\frac{1}{4}$ watt and $\frac{1}{2}$ watt resistance distributions for each nominal resistance. In a few sentences, summarize what these show.

(b) Make pairs of boxplots for comparing the $\frac{1}{4}$ watt and $\frac{1}{2}$ watt resistance distributions for each nominal resistance.

(c) Make normal plots for the $\frac{1}{2}$ watt nominal 20 ohm and nominal 200 ohm resistors. Interpret these in a sentence or two. From the appearance of the second plot, does it seem that if the nominal 200 ohm resistances were treated as if they had a bell-shaped distribution, the tendency would be to overestimate or to underestimate the fraction of resistances near the nominal value?

$\frac{1}{4}$ Watt Resistors				
20 ohm	75 ohm	100 ohm	150 ohm	200 ohm
19.2	72.9	97.4	148	198
19.2	72.4	95.8	148	196
19.3	72.0	97.7	148	199
19.3	72.5	94.1	148	196
19.1	72.7	95.1	148	196
19.0	72.3	95.4	147	195
19.6	72.9	94.9	148	193
19.2	73.2	98.5	148	196
19.3	71.8	94.8	148	196
19.4	73.4	94.6	147	199
19.4	70.9	98.3	147	194
19.3	72.3	96.0	149	195
19.5	72.5	97.3	148	196
19.2	72.1	96.0	148	195
19.1	72.6	94.8	148	199

$\frac{1}{2}$ Watt Resistors				
20 ohm	*75 ohm*	*100 ohm*	*150 ohm*	*200 ohm*
20.1	73.9	97.2	152	207
19.7	74.2	97.9	151	205
20.2	74.6	96.8	155	214
24.4	72.1	99.2	146	195
20.2	73.8	98.5	148	202
20.1	74.8	95.5	154	211
20.0	75.0	97.2	149	197
20.4	68.6	98.7	150	197
20.3	74.0	96.6	153	199
20.6	71.7	102	149	196
19.9	76.5	103	150	207
19.7	76.2	102	149	210
20.8	72.8	102	145	192
20.4	73.2	100	147	201
20.5	76.7	100	149	257

(d) Compute the sample means and sample standard deviations for all 10 samples. Do these values agree with your qualitative statements made in answer to part (a)?

(e) Make a plot of the 10 sample means computed in part (d), similar to the plot in Figure 3.22. Comment on the appearance of this plot.

5. Blomquist, Kennedy, and Reiter studied the properties of three scales by each weighing a standard 5 g weight, 20 g weight, and 100 g weight twice on each scale. Their data are presented in the accompanying table. Using whatever graphical and numerical data summary methods you find helpful, make sense of these data. Write a several-page discussion of your findings. You will probably want to consider both accuracy and precision and (to the extent possible) make comparisons between scales and between students. Part of your discussion might deal with the concepts of repeatability and reproducibility introduced in Section 2.1. Are the pictures you get of the scale and student performances consistent across the different weights?

5-Gram Weighings			
	Scale 1	*Scale 2*	*Scale 3*
Student 1	5.03, 5.02	5.07, 5.09	4.98, 4.98
Student 2	5.03, 5.01	5.02, 5.07	4.99, 4.98
Student 3	5.06, 5.00	5.10, 5.08	4.98, 4.98

20-Gram Weighings			
	Scale 1	*Scale 2*	*Scale 3*
Student 1	20.04, 20.06	20.04, 20.04	19.94, 19.93
Student 2	20.02, 19.99	20.03, 19.93	19.95, 19.95
Student 3	20.03, 20.02	20.06, 20.03	19.91, 19.96

100-Gram Weighings			
	Scale 1	*Scale 2*	*Scale 3*
Student 1	100.06, 100.35	100.25, 100.08	99.87, 99.88
Student 2	100.05, 100.01	100.10, 100.02	99.87, 99.88
Student 3	100.00, 100.00	100.01, 100.02	99.88, 99.88

6. The accompanying values are the lifetimes (in numbers of 24 mm deep holes drilled in 1045 steel before tool failure) for $n = 12$ D952-II (8 mm) drills. These were read from a graph in "Computer-assisted Prediction of Drill-failure Using In-process Measurements of Thrust Force" by A. Thangaraj and P. K. Wright (*Journal of Engineering for Industry*, May 1988).

47, 145, 172, 86, 122, 110, 172, 52, 194, 116, 149, 48

Write a short report to your engineering manager summarizing what these data indicate about the lifetimes of drills of this type in this kind of application. Use whatever graphical and numerical data summary tools make clear the main features of the data set.

7. Losen, Cahoy, and Lewis purchased eight spanner bushings of a particular type from a local machine shop and measured a number of characteristics of these bushings, including their outside diameters. Each of the eight outside diameters was measured

once by each of two student technicians, with the following results (the units are inches):

Bushing	1	2	3	4
Student A	.3690	.3690	.3690	.3700
Student B	.3690	.3695	.3695	.3695
Bushing	5	6	7	8
Student A	.3695	.3700	.3695	.3690
Student B	.3695	.3700	.3700	.3690

A common device when dealing with paired data like these is to analyze the differences. Subtracting B measurements from A measurements gives the following eight values:

.0000, −.0005, −.0005, .0005, .0000, .0000, −.0005, .0000

(a) Find the first and third quartiles for these differences, and their median.
(b) Find the sample mean and standard deviation for the differences.
(c) Your mean in part (b) should be negative. Interpret this in terms of the original measurement problem.
(d) Suppose you want to make a normal plot of the differences on regular graph paper. Give the coordinates of the lower-left point on such a plot.

8. The accompanying data are the times to failure (in millions of cycles) of high-speed turbine engine bearings made out of two different compounds. These were taken from "Analysis of Single Classification Experiments Based on Censored Samples from the Two-parameter Weibull Distribution" by J. I. McCool (*The Journal of Statistical Planning and Inference*, 1979).

Compound 1
3.03, 5.53, 5.60, 9.30, 9.92,
12.51, 12.95, 15.21, 16.04, 16.84

Compound 2
3.19, 4.26, 4.47, 4.53, 4.67,
4.69, 5.78, 6.79, 9.37, 12.75

(a) Find the .84 quantile of the Compound 1 failure times.
(b) Give the coordinates of the two lower-left points that would appear on a normal plot of the Compound 1 data.
(c) Make back-to-back stem-and-leaf plots for comparing the life length properties of bearings made from Compounds 1 and 2.
(d) Make (to scale) side-by-side boxplots for comparing the life lengths for the two compounds. Mark numbers on the plots indicating the locations of their main features.
(e) Compute the sample means and standard deviations of the two sets of lifetimes.
(f) Describe what your answers to parts (c), (d), and (e) above indicate about the life lengths of these turbine bearings.

9. Heyde, Kuebrick, and Swanson measured the heights of 405 steel punches purchased by a company from a single supplier. The stamping machine in which these are used is designed to use .500 in. punches. Frequencies of the measurements they obtained are shown in the accompanying table.

Punch Height (.001 in.)	Frequency	Punch Height (.001 in.)	Frequency
482	1	496	7
483	0	497	13
484	1	498	24
485	1	499	56
486	0	500	82
487	1	501	97
488	0	502	64
489	1	503	43
490	0	504	3
491	2	505	1
492	0	506	0
493	0	507	0
494	0	508	0
495	6	509	2

(a) Summarize these data, using appropriate graphical and numerical tools. How would you describe the shape of the distribution of punch heights? The specifications for punch heights were in fact .500 in. to .505 in. Does this fact give you any insight as to the origin of the distributional shape observed in the data? Does it appear that the supplier has equipment capable of meeting the engineering specifications on punch height?

(b) In the manufacturing application of these punches, several had to be placed side-by-side on a drum to cut the same piece of material. In this context, why is having small variability in punch height perhaps even more important than having the correct mean punch height?

10. The article "Watch Out for Nonnormal Distributions" by D. C. Jacobs (*Chemical Engineering Progress*, November 1990) contains 100 measured daily purities of oxygen delivered by a single supplier. These are as follows, listed in the time order of their collection (read left to right, top to bottom). The values given are in hundredths of a percent purity above 99.00% (so 63 stands for 99.63%).

63, 61, 67, 58, 55, 50, 55, 56, 52, 64, 73, 57, 63,
81, 64, 54, 57, 59, 60, 68, 58, 57, 67, 56, 66, 60,
49, 79, 60, 62, 60, 49, 62, 56, 69, 75, 52, 56, 61,
58, 66, 67, 56, 55, 66, 55, 69, 60, 69, 70, 65, 56,
73, 65, 68, 59, 62, 58, 62, 66, 57, 60, 66, 54, 64,
62, 64, 64, 50, 50, 72, 85, 68, 58, 68, 80, 60, 60,
53, 49, 55, 80, 64, 59, 53, 73, 55, 54, 60, 60, 58,
50, 53, 48, 78, 72, 51, 60, 49, 67

You will probably want to use a statistical analysis package to help you do the following:

(a) Make a run chart for these data. Are there any obvious time trends? What would be the practical engineering usefulness of early detection of any such time trend?

(b) Now ignore the time order of data collection and represent these data with a stem-and-leaf plot and a histogram. (Use .02% class widths in making your histogram.) Mark on these the supplier's lower specification limit of 99.50%

purity. Describe the shape of the purity distribution.

(c) The author of the article found it useful to reexpress the purities by subtracting 99.30 (remember that the preceding values are in units of .01% above 99.00%) and then taking natural logarithms. Do this with the raw data and make a second stem-and-leaf diagram and a second histogram to portray the shape of the transformed data. Do these figures look more bell-shaped than the ones you made in part (b)?

(d) Make a normal plot for the transformed values from part (c). What does it indicate about the shape of the distribution of the transformed values? (Standard normal quantiles for $p = .005$ and $p = .995$ are approximately -2.58 and 2.58, respectively.)

11. The following are some data taken from the article "Confidence Limits for Weibull Regression with Censored Data" by J. I. McCool (*IEEE Transactions on Reliability*, 1980). They are the ordered failure times (the time units are not given in the paper) for hardened steel specimens subjected to rolling contact fatigue tests at four different values of contact stress.

$.87 \times 10^6$ psi	$.99 \times 10^6$ psi	1.09×10^6 psi	1.18×10^6 psi
1.67	.80	.012	.073
2.20	1.00	.18	.098
2.51	1.37	.20	.117
3.00	2.25	.24	.135
3.90	2.95	.26	.175
4.70	3.70	.32	.262
7.53	6.07	.32	.270
14.7	6.65	.42	.350
27.8	7.05	.44	.386
37.4	7.37	.88	.456

(a) Make side-by-side boxplots for these data. Does it look as if the different stress levels produce life distributions of roughly the same shape? (Engineering experience suggests that different stress levels often change the scale but not the basic shape of life distributions.)

(b) Make Q-Q plots for comparing all six different possible pairs of distributional shapes. Summarize in a few sentences what these indicate about the shapes of the failure time distributions under the different stress levels.

12. Riddle, Peterson, and Harper studied the performance of a rapid-cut industrial shear in a continuous cut mode. They cut nominally 2-in. and 1-in. strips of 14 gauge and 16 gauge steel sheet metal and measured the actual widths of the strips produced by the shear. Their data follow, in units of 10^{-3} in. above nominal.

		Material Thickness	
		14 Gauge	*16 Gauge*
	1 in.	2, 1, 1, 1, 0, 0, −2, −10, −5, 1	−2, −6, −1, −2, −1, −2, −1, −1, −1, −5
Machine Setting			
	2 in.	10, 10, 8, 8, 8, 8, 7, 7, 9, 11	−4, −3, −4, −2, −3, −3, −3, −3, −4, −4

(a) Compute sample means and standard deviations for the four samples. Plot the means in a manner similar to the plot in Figure 3.22. Make a separate plot of this kind for the standard deviations.

(b) Write a short report to an engineering manager to summarize what these data and your summary statistics and plots show about the performance of the industrial shear. How do you recommend that the shear be set up in the future in order to get strips cut from these materials with widths as close as possible to specified dimensions?

13. The accompanying data are some measured resistivity values from *in situ* doped polysilicon specimens taken from the article "LPCVD Process Equipment Evaluation Using Statistical Methods"

by R. Rossi (*Solid State Technology*, 1984). (The units were not given in the article.)

5.55, 5.52, 5.45, 5.53, 5.37, 5.22, 5.62, 5.69, 5.60, 5.58, 5.51, 5.53

(a) Make a dot diagram and a boxplot for these data and compute the statistics $\bar{x}$ and s.

(b) Make a normal plot for these data. How bell-shaped does this data set look? If you were to say that the shape departs from a perfect bell shape, in what specific way does it? (Refer to characteristics of the normal plot to support your answer.)

14. The article "Thermal Endurance of Polyester Enameled Wires Using Twisted Wire Specimens" by H. Goldenberg (*IEEE Transactions on Electrical Insulation*, 1965) contains some data on the lifetimes (in weeks) of wire specimens tested for thermal endurance according to AIEE Standard 57. Several different laboratories were used to make the tests, and the results from two of the laboratories, using a test temperature of 200°C, follow:

Laboratory 1	Laboratory 2
14, 16, 17, 18, 20, 22, 23, 25, 27, 28	27, 28, 29, 29, 29, 30, 31, 31, 33, 34

Consider first only the Laboratory 1 data.

(a) Find the median and the first and third quartiles for the lifetimes and then find the .64 quantile of the data set.

(b) Make and interpret a normal plot for these data. Would you describe this distribution as bell-shaped? If not, in what way(s) does it depart from being bell-shaped? Give the coordinates of the 10 points you plot on regular graph paper.

(c) Find the sample mean, the sample range, and the sample standard deviation for these data.

Now consider comparing the work of the two different laboratories (i.e., consider both data sets).

(d) Make back-to-back stem-and-leaf plots for these two data sets (use two leaves for observations 10–19, two for observations 20–29, etc.)

(e) Make side-by-side boxplots for these two data sets. (Draw these on the same scale.)

(f) Based on your work in parts (d) and (e), which of the two labs would you say produced the more precise results?

(g) Is it possible to tell from your plots in (d) and (e) which lab produced the more accurate results? Why or why not?

15. Agusalim, Ferry, and Hollowaty made some measurements on the thickness of wallboard during its manufacture. The accompanying table shows thicknesses (in inches) of 12 different 4 ft × 8 ft boards (at a single location on the boards) both before and after drying in a kiln. (These boards were nominally .500 in. thick.)

Board	1	2	3	4	5	6
Before Drying	.514	.505	.500	.490	.503	.500
After Drying	.510	.502	.493	.486	.497	.494

Board	7	8	9	10	11	12
Before Drying	.510	.508	.500	.511	.505	.501
After Drying	.502	.505	.488	.486	.491	.498

(a) Make a scatterplot of these data. Does there appear to be a strong relationship between after-drying thickness and before-drying thickness? How might such a relationship be of practical engineering importance in the manufacture of wallboard?

(b) Calculate the 12 before minus after differences in thickness. Find the sample mean and sample standard deviation of these values. How might the mean value be used in running the sheetrock manufacturing process? (Based on the mean value, what is an ideal before-drying thickness for the boards?) If somehow all variability in before-drying thickness could be eliminated, would substantial after-drying variability in thickness remain? Explain in terms of your calculations.

16. The accompanying values are representative of data summarized in a histogram appearing in the article "Influence of Final Recrystallization Heat Treatment on Zircaloy-4 Strip Corrosion" by Foster, Dougherty, Burke, Bates, and Worcester (*Journal of Nuclear Materials*, 1990). Given are $n = 20$ particle diameters observed in a bright-field TEM micrograph of a Zircaloy-4 specimen. The units are $10^{-2}\mu$m.

1.73, 2.47, 2.83, 3.20, 3.20, 3.57, 3.93, 4.30, 4.67, 5.03, 5.03, 5.40, 5.77, 6.13, 6.50, 7.23, 7.60, 8.33, 9.43, 11.27

(a) Compute the mean and standard deviation of these particle diameters.

(b) Make both a dot diagram and a boxplot for these data. Sketch the dot diagram on a ruled scale and make the boxplot below it.

(c) Based on your work in (b), how would you describe the shape of this data set?

(d) Make a normal plot of these data. In what specific way does the distribution depart from being bell-shaped?

(e) It is sometimes useful to find a scale of measurement on which a data set is reasonably bell-shaped. To that end, take the natural logarithms of the raw particle diameters. Normal-plot the log diameters. Does this plot appear to be more linear than your plot in (d)?

17. The data in the accompanying tables are measurements of the latent heat of fusion of ice taken from *Experimental Statistics* (NBS Handbook 91) by M. G. Natrella. The measurements were made (on specimens cooled to $-.072°$C) using two different methods. The first was an electrical method, and the second was a method of mixtures. The units are calories per gram of mass.

(a) Make side-by-side boxplots for comparing the two measurement methods. Does there appear to be any important difference in the precision of the two methods? Is it fair to say that at least one of the methods must be somewhat inaccurate? Explain.

Method A (Electrical)

79.98, 80.04, 80.02, 80.04, 80.03, 80.03, 80.04,
79.97, 80.05, 80.03, 80.02, 80.00, 80.02

Method B (Mixtures)

80.02, 79.94, 79.98, 79.97,
79.97, 80.03, 79.95, 79.97

(b) Compute and compare the sample means and the sample standard deviations for the two methods. How are the comparisons of these numerical quantities already evident on your plot in (a)?

18. T. Babcock did some fatigue life testing on specimens of 1045 steel obtained from three different heats produced by a single steel supplier. The lives till failure of 30 specimens tested on a rotary fatigue strength machine (units are 100 cycles) are

---

Heat 1

313, 100, 235, 250, 457,
11, 315, 584, 249, 204

Heat 2

349, 206, 163, 350, 189,
216, 170, 359, 267, 196

Heat 3

289, 279, 142, 334, 192,
339, 87, 185, 262, 194

(a) Find the median and first and third quartiles for the Heat 1 data. Then find the .62 quantile of the Heat 1 data set.
(b) Make and interpret a normal plot for the Heat 1 data. Would you describe this data set as bell-shaped? If not, in what specific way does the shape depart from the bell shape? (List the

coordinates of the points you plot on regular graph paper.)
(c) Find the sample mean and sample standard deviation of the Heat 1 data.
(d) Make a stem-and-leaf plot for the Heat 1 data using only the leading digits 0, 1, 2, 3, 4 and 5 to the left of the stem (and pairs of final digits to the right).
(e) Now make back-to-back stem-and-leaf plots for the Heat 1 and Heat 2 data. How do the two distributions of fatigue lives compare?
(f) Show the calculations necessary to make box-plots for each of the three data sets above. Then draw these side by side on the same scale to compare the three heats. How would you say that these three heats compare in terms of uniformity of fatigue lives produced? Do you see any clear differences between heats in terms of the average fatigue life produced?

19. Loveland, Rahardja, and Rainey studied a metal turning process used to make some (cylindrical) servo sleeves. Outside diameter measurements made on ten of these sleeves are given here. (Units are 10^{-5} inch above nominal. The "notch" axis of the sleeve was an identifiable axis and the non-notch axis was perpendicular to the notch axis. A dial bore gauge and an air spindler gauge were used.)

Sleeve	1	2	3	4	5
Notch/Dial Bore	130	160	170	310	200
Non-Notch/Dial Bore	150	150	210	160	160
Notch/Air Spindler	40	60	45	0	30
Sleeve	6	7	8	9	10
Notch/Dial Bore	130	200	150	200	140
Non-Notch/Dial Bore	140	220	150	220	160
Notch/Air Spindler	0	25	25	−40	65

(a) What can be learned from the dial bore data that could not be learned from data consisting of the given notch measurements above and ten non-notch measurements on a *different* ten servo sleeves?

(b) The dial bore data might well be termed "paired" data. A common method of analysis for such data is to take differences and study those. Compute the ten "notch minus non-notch" differences for the dial bore values. Make a dot diagram for these and then a boxplot. What physical interpretation does a nonzero mean for such differences have? What physical interpretation does a large variability in these differences have?

(c) Make a scatterplot of the air spindler notch measurements versus the dial bore notch measurements. Does it appear that the air spindler and dial bore measurements are strongly related?

(d) How would you suggest trying to determine which of the two gauges is most precise?

20. Duren, Leng and Patterson studied the drilling of holes in a miniature metal part using two different physical processes (laser drilling and electrical discharge machining). Blueprint specifications on these holes called for them to be drilled at an angle of 45° to the top surface of the part in question. The realized angles measured on 13 parts drilled using each process (26 parts in all) are

---
Laser (Hole A)

42.8, 42.2, 42.7, 43.1, 40.0, 43.5,
42.3, 40.3, 41.3, 48.5, 39.5, 41.1, 42.1

EDM (Hole A)

46.1, 45.3, 45.3, 44.7, 44.2, 44.6,
43.4, 44.6, 44.6, 45.5, 44.4, 44.0, 43.2

(a) Find the median and the first and third quartiles for the Laser data. Then find the .37 quantile of the Laser data set.

(b) Make and interpret a normal plot for the Laser data. Would you describe this distribution as bell-shaped? If not, in what way(s) does it depart from being bell-shaped?

(c) Find the sample mean, the sample range, and the sample standard deviation for the Laser data.

Now consider comparing the two different drilling methods.

(d) Make back-to-back stem-and-leaf plots for the two data sets.

(e) Make side-by-side boxplots for the two data sets. (Draw these on the same scale.)

(f) Based on your work in parts (d) and (e), which of the two processes would you say produced the most consistent results? Which process produced an "average" angle closest to the nominal angle (45°)?

As it turns out, each metal part actually had two holes drilled in it and their angles measured. Below are the measured angles of the second hole drilled in each of the parts made using the Laser process. (The data are listed in the same part order as earlier.)

Laser (Hole B)

43.1, 44.3, 44.5, 46.3, 43.9, 41.9,
43.4, 49.0, 43.5, 47.2, 44.8 ,44.0, 43.9

(g) Taking together the two sets of Laser measurements, how would you describe these values using the terminology of Section 1.2?

(h) Make a scatterplot of the Hole A and Hole B laser data. Does there appear to be a strong relationship between the angles produced in a single part by this drilling method?

(i) Calculate the 13 Hole A minus Hole B differences in measured angles produced using the Laser drilling process. Find the sample mean and sample standard deviation of these values. What do these quantities measure here?

21. Blad, Sobotka, and Zaug did some hardness testing of a metal specimen. They tested it on three different machines, a dial Rockwell tester, a digital Rockwell tester, and a Brinell tester. They made ten measurements with each machine and the values they obtained for Brinell hardness (after

conversion in the case of the Rockwell readings) were

Dial Rockwell

536.6, 539.2, 524.4,
536.6, 526.8, 531.6,
540.5, 534.0, 526.8,
531.6

Digital Rockwell

501.2, 522.0, 531.6,
522.0, 519.4, 523.2,
522.0, 514.2, 506.4,
518.1

Brinell

542.6, 526.0, 520.5,
514.0, 546.6, 512.6,
516.0, 580.4, 600.0,
601.0

Consider first only the Dial Rockwell data.

(a) Find the median and the first and third quartiles for the hardness measurements. Then find the .27 quantile of the data set.

(b) Make and interpret a normal plot for these data. Would you describe this distribution as bell-shaped? If not, in what way(s) does it depart from being bell-shaped?

(c) Find the sample mean, the sample range, and the sample standard deviation for these data.

Now consider comparing the readings from the different testers (i.e., consider all three data sets.)

(d) Make back-to-back stem-and-leaf plots for the two Rockwell data sets. (Use two "leaves" for observations 500–509, two for the observations 510–519, etc.)

(e) Make side-by-side boxplots for all three data sets. (Draw these on the same scale.)

(f) Based on your work in part (e), which of the three machines would you say produced the most precise results?

(g) Is it possible to tell from your plot (e) which machine produced the most accurate results? Why or why not?

22. Ritchey, Bazan, and Buhman did an experiment to compare flight times of several designs of paper helicopters, dropping them from the first to ground floors of the ISU Design Center. The flight times that they reported for two different designs were (the units are seconds)

Design 1	Design 2
2.47, 2.45, 2.43, 2.67, 2.69,	3.42, 3.50, 3.29, 3.51, 3.53,
2.48, 2.44, 2.71, 2.84, 2.84	2.67, 2.69, 3.47, 3.40, 2.87

(a) Find the median and the first and third quartiles for the Design 1 data. Then find the .62 quantile of the Design 1 data set.

(b) Make and interpret a normal plot for the Design 1 data. Would you describe this distribution as bell-shaped? If not, in what way(s) does it depart from being bell-shaped?

(c) Find the sample mean, the sample range, and the sample standard deviation for the Design 1 data. Show some work.

Now consider comparing the two different designs.

(d) Make back-to-back stem-and-leaf plots for the two data sets.

(e) Make side-by-side boxplots for the two data sets. (Draw these on the same scale.)

(f) Based on your work in parts (d) and (e), which of the two designs would you say produced the most consistent results? Which design produced the longest flight times?

(g) It is not really clear from the students' report whether the data came from the dropping of one helicopter of each design ten times, or from the dropping of ten helicopters of each design once. Briefly discuss which of these possibilities is preferable if the object of the study was to identify a superior design. (If necessary, review Section 2.3.4.)

4

● ●

Describing Relationships Between Variables

The methods of Chapter 3 are really quite simple. They require little in the way of calculations and are most obviously relevant to the analysis of a single engineering variable. This chapter provides methods that address the more complicated problem of describing relationships between variables and are computationally more demanding.

The chapter begins with least squares fitting of a line to bivariate quantitative data and the assessment of the goodness of that fit. Then the line-fitting ideas are generalized to the fitting of curves to bivariate data and surfaces to multivariate quantitative data. The next topic is the summarization of data from full factorial studies in terms of so-called factorial effects. Next, the notion of data transformations is discussed. Finally, the chapter closes with a short transitional section that argues that further progress in statistics requires some familiarity with the subject of probability.

● ●

4.1 Fitting a Line by Least Squares

Bivariate data often arise because a quantitative experimental variable x has been varied between several different settings, producing a number of samples of a response variable y. For purposes of summarization, interpolation, limited extrapolation, and/or process optimization/adjustment, it is extremely helpful to have an equation relating y to x. A linear (or straight line) equation

$$y \approx \beta_0 + \beta_1 x \tag{4.1}$$

relating y to x is about the simplest potentially useful equation to consider after making a simple (x, y) scatterplot.

In this section, the principle of least squares is used to fit a line to (x, y) data. The appropriateness of that fit is assessed using the sample correlation and the coefficient of determination. Plotting of residuals is introduced as an important method for further investigation of possible problems with the fitted equation. A discussion of some practical cautions and the use of statistical software in fitting equations to data follows.

4.1.1   Applying the Least Squares Principle

Example 1

Pressing Pressures and Specimen Densities for a Ceramic Compound

Benson, Locher, and Watkins studied the effects of varying pressing pressures on the density of cylindrical specimens made by dry pressing a ceramic compound. A mixture of Al_2O_3, polyvinyl alcohol, and water was prepared, dried overnight, crushed, and sieved to obtain 100 mesh size grains. These were pressed into cylinders at pressures from 2,000 psi to 10,000 psi, and cylinder densities were calculated. Table 4.1 gives the data that were obtained, and a simple scatterplot of these data is given in Figure 4.1.

Table 4.1
Pressing Pressures and Resultant Specimen Densities

x, Pressure (psi)	y, Density (g/cc)
2,000	2.486
2,000	2.479
2,000	2.472
4,000	2.558
4,000	2.570
4,000	2.580
6,000	2.646
6,000	2.657
6,000	2.653
8,000	2.724
8,000	2.774
8,000	2.808
10,000	2.861
10,000	2.879
10,000	2.858

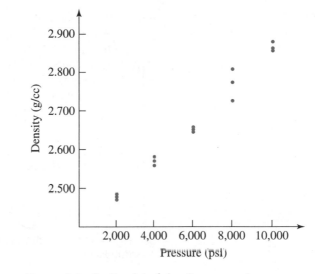

Figure 4.1   Scatterplot of density vs. pressing pressure

It is very easy to imagine sketching a straight line through the plotted points in Figure 4.1. Such a line could then be used to summarize how density depends upon pressing pressure. The principle of **least squares** provides a method of choosing a "best" line to describe the data.

Definition 1

> To apply the **principle of least squares** in the fitting of an equation for y to an n point data set, values of the equation parameters are chosen to minimize
>
> $$\sum_{i=1}^{n} \left(y_i - \hat{y}_i \right)^2 \tag{4.2}$$
>
> where $y_1, y_2, \ldots, y_n$ are the observed responses and $\hat{y}_1, \hat{y}_2, \ldots, \hat{y}_n$ are corresponding responses predicted or fitted by the equation.

In the context of fitting a line to (x, y) data, the prescription offered by Definition 1 amounts to choosing a slope and intercept so as to minimize the sum of squared vertical distances from (x, y) data points to the line in question. This notion is shown in generic fashion in Figure 4.2 for a fictitious five-point data set. (It is the *squares* of the five indicated differences that must be added and minimized.)

Looking at the form of display (4.1), for the fitting of a line,

$$\hat{y} = \beta_0 + \beta_1 x$$

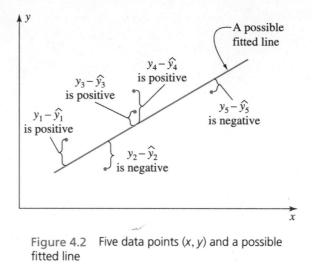

Figure 4.2   Five data points (x, y) and a possible fitted line

Therefore, the expression to be minimized by choice of slope (β_1) and intercept (β_0) is

$$S(\beta_0, \beta_1) = \sum_{i=1}^{n} \left(y_i - (\beta_0 + \beta_1 x_i)\right)^2 \qquad (4.3)$$

The minimization of the function of two variables $S(\beta_0, \beta_1)$ is an exercise in calculus. The partial derivatives of S with respect to β_0 and β_1 may be set equal to zero, and the two resulting equations may be solved simultaneously for β_0 and β_1. The equations produced in this way are

$$n\beta_0 + \left(\sum_{i=1}^{n} x_i\right)\beta_1 = \sum_{i=1}^{n} y_i \qquad (4.4)$$

and

$$\left(\sum_{i=1}^{n} x_i\right)\beta_0 + \left(\sum_{i=1}^{n} x_i^2\right)\beta_1 = \sum_{i=1}^{n} x_i y_i \qquad (4.5)$$

For reasons that are not obvious, equations (4.4) and (4.5) are sometimes called the **normal** (as in perpendicular) **equations** for fitting a line. They are two linear equations in two unknowns and can be fairly easily solved for β_0 and β_1 (provided there are at least two different x_i's in the data set). Simultaneous solution of equations (4.4) and (4.5) produces values of β_1 and β_0 given by

Slope of the least squares line, b_1

$$b_1 = \frac{\sum (x_i - \bar{x})(y_i - \bar{y})}{\sum (x_i - \bar{x})^2}$$

(4.6)

and

Intercept of the least squares line, b_0

$$b_0 = \bar{y} - b_1 \bar{x}$$

(4.7)

Notice the notational convention here. *The particular numerical slope and intercept minimizing $S(\beta_0, \beta_1)$ are denoted (not as β's but) as b_1 and b_0.*

In display (4.6), somewhat standard practice has been followed (and the summation notation abused) by not indicating the variable or range of summation (i, from 1 to n).

Example 1
(continued)

It is possible to verify that the data in Table 4.1 yield the following summary statistics:

$$\sum x_i = 2{,}000 + 2{,}000 + \cdots + 10{,}000 = 90{,}000,$$

$$\text{so } \bar{x} = \frac{90{,}000}{15} = 6{,}000$$

$$\sum (x_i - \bar{x})^2 = (2{,}000 - 6{,}000)^2 + (2{,}000 - 6{,}000)^2 + \cdots +$$

$$(10{,}000 - 6{,}000)^2 = 120{,}000{,}000$$

$$\sum y_i = 2.486 + 2.479 + \cdots + 2.858 = 40.005,$$

$$\text{so } \bar{y} = \frac{40.005}{15} = 2.667$$

$$\sum (y_i - \bar{y})^2 = (2.486 - 2.667)^2 + (2.479 - 2.667)^2 + \cdots +$$

$$(2.858 - 2.667)^2 = .289366$$

$$\sum (x_i - \bar{x})(y_i - \bar{y}) = (2{,}000 - 6{,}000)(2.486 - 2.667) + \cdots +$$

$$(10{,}000 - 6{,}000)(2.858 - 2.667) = 5{,}840$$

Example 1
(*continued*)

Then the least squares slope and intercept, b_1 and b_0, are given via equations (4.6) and (4.7) as

▶
$$b_1 = \frac{5{,}840}{120{,}000{,}000} = .0000486 \text{ (g/cc)/psi}$$

and

▶
$$b_0 = 2.667 - (.0000486)(6{,}000) = 2.375 \text{ g/cc}$$

Figure 4.3 shows the least squares line

▶
$$\hat{y} = 2.375 + .0000487x$$

Interpretation of the slope of the least squares line

sketched on a scatterplot of the (x, y) points from Table 4.1. Note that the slope on this plot, $b_1 \approx .0000487$ (g/cc)/psi, has physical meaning as the (approximate) increase in y (density) that accompanies a unit (1 psi) increase in x (pressure). The intercept on the plot, $b_0 = 2.375$ g/cc, positions the line vertically and is the value at which the line cuts the y axis. But it should probably not be interpreted as the density that would accompany a pressing pressure of $x = 0$ psi. The point is that the reasonably linear-looking relation that the students found for pressures between 2,000 psi and 10,000 psi could well break down at larger or smaller

Extrapolation

pressures. Thinking of b_0 as a 0 pressure density amounts to an extrapolation outside the range of data used to fit the equation, something that ought always to be approached with extreme caution.

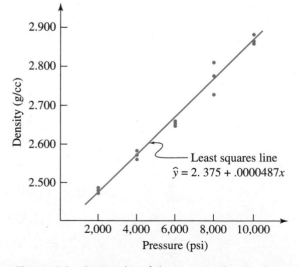

Figure 4.3   Scatterplot of the pressure/density data and the least squares line

As indicated in Definition 1, the value of y on the least squares line corresponding to a given x can be termed a **fitted** or **predicted value**. It can be used to represent likely y behavior at that x.

Example 1
(*continued*)

Consider the problem of determining a typical density corresponding to a pressure of 4,000 psi and one corresponding to 5,000 psi.

First, looking at $x = 4,000$, a simple way of representing a typical y is to note that for the three data points having $x = 4,000$,

$$\bar{y} = \frac{1}{3}(2.558 + 2.570 + 2.580) = 2.5693 \text{ g/cc}$$

and so to use this as a representative value. But assuming that y is indeed approximately linearly related to x, the fitted value

$$\hat{y} = 2.375 + .0000486(4,000) = 2.5697 \text{ g/cc}$$

might be even better for representing average density for 4,000 psi pressure.

Looking then at the situation for $x = 5,000$ psi, there are no data with this x value. The only thing one can do to represent density at that pressure is to ask

Interpolation   whether interpolation is sensible from a physical viewpoint. If so, the fitted value

$$\hat{y} = 2.375 + .0000486(5,000) = 2.6183 \text{ g/cc}$$

can be used to represent density for 5,000 psi pressure.

4.1.2 The Sample Correlation and Coefficient of Determination

Visually, the least squares line in Figure 4.3 seems to do a good job of fitting the plotted points. However, it would be helpful to have methods of quantifying the quality of that fit. One such measure is the **sample correlation**.

Definition 2

The **sample (linear) correlation** between x and y in a sample of n data pairs (x_i, y_i) is

$$r = \frac{\sum (x_i - \bar{x})(y_i - \bar{y})}{\sqrt{\sum (x_i - \bar{x})^2 \cdot \sum (y_i - \bar{y})^2}} \tag{4.8}$$

Interpreting the sample correlation   The sample correlation always lies in the interval from -1 to 1. Further, it is -1 or 1 only when all (x, y) data points fall on a single straight line. Comparison of formulas (4.6) and (4.8) shows that $r = b_1 \left(\sum (x_i - \bar{x})^2 / \sum (y_i - \bar{y})^2 \right)^{1/2}$ so that b_1 *and r have the same sign.* So a sample correlation of -1 means that y decreases linearly in increasing x, while a sample correlation of $+1$ means that y increases linearly in increasing x.

Real data sets do not often exhibit perfect ($+1$ or -1) correlation. Instead r is typically between -1 and 1. But drawing on the facts about how it behaves, people take r as a *measure of the strength of an apparent linear relationship*: r near $+1$ or -1 is interpreted as indicating a relatively strong linear relationship; r near 0 is taken as indicating a lack of linear relationship. The sign of r is thought of as indicating whether y tends to increase or decrease with increased x.

Example 1
(continued)

For the pressure/density data, the summary statistics in the example following display (4.7) (page 127) produces

$$r = \frac{5,840}{\sqrt{(120,000,000)(.289366)}} = .9911$$

This value of r is near $+1$ and indicates clearly the strong positive linear relationship evident in Figures 4.1 and 4.3.

The **coefficient of determination** is another measure of the quality of a fitted equation. It can be applied not only in the present case of the simple fitting of a line to (x, y) data but more widely as well.

Definition 3

The **coefficient of determination** for an equation fitted to an n-point data set via least squares and producing fitted y values $\hat{y}_1, \hat{y}_2, \ldots, \hat{y}_n$ is

$$R^2 = \frac{\sum (y_i - \bar{y})^2 - \sum (y_i - \hat{y}_i)^2}{\sum (y_i - \bar{y})^2} \qquad \textbf{(4.9)}$$

Interpretation of R^2

R^2 may be interpreted as the *fraction of the raw variation in y accounted for using the fitted equation.* That is, provided the fitted equation includes a constant term, $\sum (y_i - \bar{y})^2 \geq \sum (y_i - \hat{y}_i)^2$. Further, $\sum (y_i - \bar{y})^2$ is a measure of raw variability in y, while $\sum (y_i - \hat{y}_i)^2$ is a measure of variation in y remaining after fitting the equation. So the nonnegative difference $\sum (y_i - \bar{y})^2 - \sum (y_i - \hat{y}_i)^2$ is a measure of the variability in y accounted for in the equation-fitting process. R^2 then expresses this difference as a fraction (of the total raw variation).

Example 1
(continued)

Using the fitted line, one can find $\hat{y}$ values for all $n = 15$ data points in the original data set. These are given in Table 4.2.

Table 4.2
Fitted Density Values

x, Pressure	$\hat{y}$, Fitted Density
2,000	2.4723
4,000	2.5697
6,000	2.6670
8,000	2.7643
10,000	2.8617

Then, referring again to Table 4.1,

$$\sum(y_i - \hat{y}_i)^2 = (2.486 - 2.4723)^2 + (2.479 - 2.4723)^2 + (2.472 - 2.4723)^2$$
$$+ (2.558 - 2.5697)^2 + \cdots + (2.879 - 2.8617)^2$$
$$+ (2.858 - 2.8617)^2$$
$$= .005153$$

Further, since $\sum(y_i - \bar{y})^2 = .289366$, from equation (4.9)

$$R^2 = \frac{.289366 - .005153}{.289366} = .9822$$

and the fitted line accounts for over 98% of the raw variability in density, reducing the "unexplained" variation from .289366 to .005153.

R^2 as a squared correlation

The coefficient of determination has a second useful interpretation. For equations that are linear in the parameters (which are the only ones considered in this text), R^2 turns out to be a squared correlation. It is the squared correlation between the observed values y_i and the fitted values $\hat{y}_i$. (Since in the present situation of fitting a line, the $\hat{y}_i$ values are perfectly correlated with the x_i values, R^2 also turns out to be the squared correlation between the y_i and x_i values.)

Example 1
(continued)

For the pressure/density data, the correlation between x and y is

$$r = .9911$$

Example 1
(continued)

Since $\hat{y}$ is perfectly correlated with x, this is also the correlation between $\hat{y}$ and y. But notice as well that

$$r^2 = (.9911)^2 = .9822 = R^2$$

so R^2 is indeed the squared sample correlation between y and $\hat{y}$.

4.1.3 Computing and Using Residuals

When fitting an equation to a set of data, the hope is that the equation extracts the main message of the data, leaving behind (unpredicted by the fitted equation) only the variation in y that is uninterpretable. That is, one hopes that the y_i's will look like the $\hat{y}_i$'s except for small fluctuations explainable only as random variation. A way of assessing whether this view is sensible is through the computation and plotting of **residuals**.

Definition 4

If the fitting of an equation or model to a data set with responses $y_1, y_2, \ldots, y_n$ produces fitted values $\hat{y}_1, \hat{y}_2, \ldots, \hat{y}_n$, then the corresponding **residuals** are the values

$$e_i = y_i - \hat{y}_i$$

If a fitted equation is telling the whole story contained in a data set, then its residuals ought to be patternless. So when they're plotted against time order of observation, values of experimental variables, fitted values, or any other sensible quantities, the plots should look randomly scattered. When they don't, the patterns can themselves suggest what has gone unaccounted for in the fitting and/or how the data summary might be improved.

Example 2

Compressive Strength of Fly Ash Cylinders as a Function of Amount of Ammonium Phosphate Additive

As an exaggerated example of the previous point, consider the naive fitting of a line to some data of B. Roth. Roth studied the compressive strength of concrete-like fly ash cylinders. These were made using varying amounts of ammonium phosphate as an additive. Part of Roth's data are given in Table 4.3. The ammonium phosphate values are expressed as a percentage by weight of the amount of fly ash used.

Table 4.3
Additive Concentrations and Compressive Strengths for Fly Ash Cylinders

x, Ammonium Phosphate (%)	y, Compressive Strength (psi)	x, Ammonium Phosphate (%)	y, Compressive Strength (psi)
0	1221	3	1609
0	1207	3	1627
0	1187	3	1642
1	1555	4	1451
1	1562	4	1472
1	1575	4	1465
2	1827	5	1321
2	1839	5	1289
2	1802	5	1292

Using formulas (4.6) and (4.7), it is possible to show that the least squares line through the (x, y) data in Table 4.3 is

$$\hat{y} = 1498.4 - .6381x \qquad \textbf{(4.10)}$$

Then straightforward substitution into equation (4.10) produces fitted values $\hat{y}_i$ and residuals $e_i = y_i - \hat{y}_i$, as given in Table 4.4. The residuals for this straight-line fit are plotted against x in Figure 4.4.

The distinctly "up-then-back-down-again" curvilinear pattern of the plot in Figure 4.4 is not typical of random scatter. Something has been missed in

Table 4.4
Residuals from a Straight-Line Fit to the Fly Ash Data

x	y	$\hat{y}$	$e = y - \hat{y}$	x	y	$\hat{y}$	$e = y - \hat{y}$
0	1221	1498.4	−277.4	3	1609	1496.5	112.5
0	1207	1498.4	−291.4	3	1627	1496.5	130.5
0	1187	1498.4	−311.4	3	1642	1496.5	145.5
1	1555	1497.8	57.2	4	1451	1495.8	−44.8
1	1562	1497.8	64.2	4	1472	1495.8	−23.8
1	1575	1497.8	77.2	4	1465	1495.8	−30.8
2	1827	1497.2	329.8	5	1321	1495.2	−174.2
2	1839	1497.2	341.8	5	1289	1495.2	−206.2
2	1802	1497.2	304.8	5	1292	1495.2	−203.2

Example 2
(*continued*)

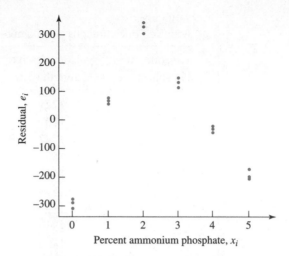

Figure 4.4   Plot of residuals vs. x for a linear fit to the fly ash data

the fitting of a line to Roth's data. Figure 4.5 is a simple scatterplot of Roth's data (which in practice should be made before fitting any curve to such data). It is obvious from the scatterplot that the relationship between the amount of ammonium phosphate and compressive strength is decidedly nonlinear. In fact, a quadratic function would come much closer to fitting the data in Table 4.3.

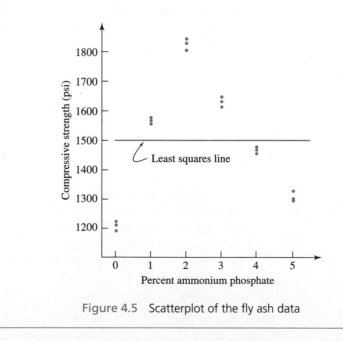

Figure 4.5   Scatterplot of the fly ash data

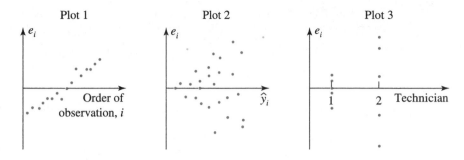

Figure 4.6 Patterns in residual plots

Interpreting patterns on residual plots

Figure 4.6 shows several patterns that can occur in plots of residuals against various variables. Plot 1 of Figure 4.6 shows a trend on a plot of residuals versus time order of observation. The pattern suggests that some variable changing in time is acting on y and has not been accounted for in fitting $\hat{y}$ values. For example, instrument drift (where an instrument reads higher late in a study than it did early on) could produce a pattern like that in Plot 1. Plot 2 shows a fan-shaped pattern on a plot of residuals versus fitted values. Such a pattern indicates that large responses are fitted (and quite possibly produced and/or measured) less consistently than small responses. Plot 3 shows residuals corresponding to observations made by Technician 1 that are on the whole smaller than those made by Technician 2. The suggestion is that Technician 1's work is more precise than that of Technician 2.

Normal-plotting residuals

Another useful way of plotting residuals is to normal-plot them. The idea is that the normal distribution shape is typical of random variation and that normal-plotting of residuals is a way to investigate whether such a distributional shape applies to what is left in the data after fitting an equation or model.

Example 1
(*continued*)

Table 4.5 gives residuals for the fitting of a line to the pressure/density data. The residuals e_i were treated as a sample of 15 numbers and normal-plotted (using the methods of Section 3.2) to produce Figure 4.7.

The central portion of the plot in Figure 4.7 is fairly linear, indicating a generally bell-shaped distribution of residuals. But the plotted point corresponding to the largest residual, and probably the one corresponding to the smallest residual, fail to conform to the linear pattern established by the others. Those residuals seem big in absolute value compared to the others.

From Table 4.5 and the scatterplot in Figure 4.3, one sees that these large residuals both arise from the 8,000 psi condition. And the spread for the three densities at that pressure value does indeed look considerably larger than those at the other pressure values. The normal plot suggests that the pattern of variation at 8,000 psi is genuinely different from those at other pressures. It may be that a different physical compaction mechanism was acting at 8,000 psi than at the other pressures. But it is more likely that there was a problem with laboratory technique, or recording, or the test equipment when the 8,000 psi tests were made.

Example 1
(*continued*)

In any case, the normal plot of residuals helps draw attention to an idiosyncrasy in the data of Table 4.1 that merits further investigation, and perhaps some further data collection.

Table 4.5
Residuals from the Linear Fit to the Pressure/Density Data

x, Pressure	y, Density	$\hat{y}$	$e = y - \hat{y}$
2,000	2.486	2.4723	.0137
2,000	2.479	2.4723	.0067
2,000	2.472	2.4723	−.0003
4,000	2.558	2.5697	−.0117
4,000	2.570	2.5697	.0003
4,000	2.580	2.5697	.0103
6,000	2.646	2.6670	−.0210
6,000	2.657	2.6670	−.0100
6,000	2.653	2.6670	−.0140
8,000	2.724	2.7643	−.0403
8,000	2.774	2.7643	.0097
8,000	2.808	2.7643	.0437
10,000	2.861	2.8617	−.0007
10,000	2.879	2.8617	.0173
10,000	2.858	2.8617	−.0037

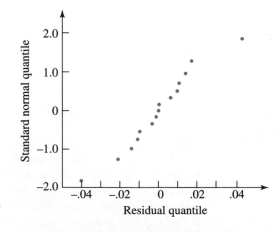

Figure 4.7   Normal plot of residuals from a linear fit to the pressure/density data

4.1.4 Some Cautions

The methods of this section are extremely useful engineering tools when thoughtfully applied. But a few additional comments are in order, warning against some errors in logic that often accompany their use.

r Measures only linear association

The first warning regards the correlation. It must be remembered that r measures only the **linear relationship** between x and y. It is perfectly possible to have a strong *nonlinear* relationship between x and y and yet have a value of r near 0. In fact, Example 2 is an excellent example of this. Compressive strength is strongly related to the ammonium phosphate content. But $r = -.005$, very nearly 0, for the data set in Table 4.3.

Correlation and causation

The second warning is essentially a restatement of one implicit in the early part of Section 1.2: Correlation is not necessarily causation. One may observe a large correlation between x and y in an observational study without it being true that x drives y or vice versa. It may be the case that another variable (say, z) drives the system under study and causes simultaneous changes in both x and y.

The influence of extreme observations

The last warning is that both $R^2(r)$ and least squares fitting can be drastically affected by a few unusual data points. As an example of this, consider the ages and heights of 36 students from an elementary statistics course plotted in Figure 4.8. By the time people reach college age, there is little useful relationship between age and height, but the correlation between ages and heights is .73. This fairly large value is produced by essentially a single data point. If the data point corresponding to the 30-year-old student who happened to be 6 feet 8 inches tall is removed from the data set, the correlation drops to .03.

An engineer's primary insurance against being misled by this kind of phenomenon is the habit of **plotting** data in as many different ways as are necessary to get a feel for how they are structured. Even a simple boxplot of the age data or height

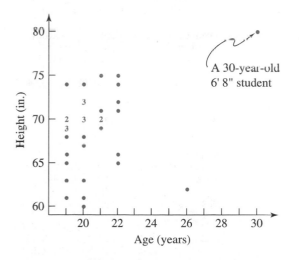

Figure 4.8 Scatterplot of ages and heights of 36 students

data alone would have identified the 30-year-old student in Figure 4.8 as unusual. That would have raised the possibility of that data point strongly influencing both r and any curve that might be fitted via least squares.

4.1.5 Computing

The examples in this section have no doubt left the impression that computations were done "by hand." In practice, such computations are almost always done with a statistical analysis package. The fitting of a line by least squares is done using a **regression program**. Such programs usually also compute R^2 and have an option that allows the computing and plotting of residuals.

It is not the purpose of this text to teach or recommend the use of any particular statistical package, but annotated printouts will occasionally be included to show how MINITAB formats its output. Printout 1 is such a printout for an analysis of the pressure/density data in Table 4.1, paralleling the discussion in this section. (MINITAB's regression routine is found under its "Stat/Regression/Regression" menu.) MINITAB gives its user much more in the way of analysis for least squares curve fitting than has been discussed to this point, so your understanding of Printout 1 will be incomplete. But it should be possible to locate values of the major summary statistics discussed here. The printout shown doesn't include plots, but it's worth noting that the program has options for saving fitted values and residuals for later plotting.

Printout 1 Fitting the Least Squares Line to the Pressure/Density Data

```
Regression Analysis


The regression equation is
density = 2.38 +0.000049 pressure

Predictor        Coef       StDev          T       P
Constant      2.37500     0.01206     197.01   0.000
pressure   0.00004867  0.00000182      26.78   0.000

S = 0.01991     R-Sq = 98.2%     R-Sq(adj) = 98.1%


Analysis of Variance

Source        DF          SS          MS        F       P
Regression     1     0.28421     0.28421   717.06   0.000
Residual Error 13     0.00515     0.00040
Total         14     0.28937

Obs   pressure   density       Fit   StDev Fit    Residual   St Resid
  1       2000   2.48600   2.47233     0.00890     0.01367       0.77
  2       2000   2.47900   2.47233     0.00890     0.00667       0.37
  3       2000   2.47200   2.47233     0.00890    -0.00033      -0.02
  4       4000   2.55800   2.56967     0.00630    -0.01167      -0.62
  5       4000   2.57000   2.56967     0.00630     0.00033       0.02
  6       4000   2.58000   2.56967     0.00630     0.01033       0.55
  7       6000   2.64600   2.66700     0.00514    -0.02100      -1.09
  8       6000   2.65700   2.66700     0.00514    -0.01000      -0.52
```

9	6000	2.65300	2.66700	0.00514	-0.01400	-0.73
10	8000	2.72400	2.76433	0.00630	-0.04033	-2.14R
11	8000	2.77400	2.76433	0.00630	0.00967	0.51
12	8000	2.80800	2.76433	0.00630	0.04367	2.31R
13	10000	2.86100	2.86167	0.00890	-0.00067	-0.04
14	10000	2.87900	2.86167	0.00890	0.01733	0.97
15	10000	2.85800	2.86167	0.00890	-0.00367	-0.21

R denotes an observation with a large standardized residual

At the end of Section 3.3 we warned that using spreadsheet software in place of high-quality statistical software can, without warning, produce spectacularly wrong answers. The example provided at the end of Section 3.3 concerns a badly wrong sample variance of only three numbers. It is important to note that the potential for numerical inaccuracy shown in that example carries over to the rest of the statistical methods discussed in this book, including those of the present section. For example, consider the $n = 6$ hypothetical (x, y) pairs listed in Table 4.6. For fitting a line to these data via least squares, MINITAB correctly produces $R^2 = .997$. But as recently as late 1999, the current version of the leading spreadsheet program returned the ridiculously wrong value, $R^2 = -.81648$. (This data set comes from a posting by Mark Eakin on the "edstat" electronic bulletin board that can be found at http://jse.stat.ncsu.edu/archives/.)

Table 4.6
6 Hypothetical Data Pairs

x	y	x	y
10,000,000.1	1.1	10,000,000.4	3.9
10,000,000.2	1.9	10,000,000.5	4.9
10,000,000.3	3.1	10,000,000.6	6.1

Section 1 Exercises

1. The following is a small set of artificial data. Show the hand calculations necessary to do the indicated tasks.

x	1	2	3	4	5
y	8	8	6	6	4

(a) Obtain the least squares line through these data. Make a scatterplot of the data and sketch this line on that scatterplot.

(b) Obtain the sample correlation between x and y for these data.

(c) Obtain the sample correlation between y and $\hat{y}$ for these data and compare it to your answer to part (b).

(d) Use the formula in Definition 3 and compute R^2 for these data. Compare it to the square of your answers to parts (b) and (c).

(e) Find the five residuals from your fit in part (a). How are they portrayed geometrically on the scatterplot for (a)?

2. Use a computer package and redo the computations and plotting required in Exercise 1. Annotate your output, indicating where on the printout you can

find the equation of the least squares line, the value of r, the value of R^2, and the residuals.

3. The article "Polyglycol Modified Poly (Ethylene Ether Carbonate) Polyols by Molecular Weight Advancement" by R. Harris (*Journal of Applied Polymer Science*, 1990) contains some data on the effect of reaction temperature on the molecular weight of resulting poly polyols. The data for eight experimental runs at temperatures 165°C and above are as follows:

Pot Temperature, x (°C)	Average Molecular Weight, y
165	808
176	940
188	1183
205	1545
220	2012
235	2362
250	2742
260	2935

Use a statistical package to help you complete the following (both the plotting and computations):

(a) What fraction of the observed raw variation in y is accounted for by a linear equation in x?

(b) Fit a linear relationship $y \approx \beta_0 + \beta_1 x$ to these data via least squares. About what change in average molecular weight seems to accompany a 1°C increase in pot temperature (at least over the experimental range of temperatures)?

(c) Compute and plot residuals from the linear relationship fit in (b). Discuss what they suggest about the appropriateness of that fitted equation. (Plot residuals versus x, residuals versus $\hat{y}$, and make a normal plot of them.)

(d) These data came from an experiment where the investigator managed the value of x. There is a fairly glaring weakness in the experimenter's data collection efforts. What is it?

(e) Based on your analysis of these data, what average molecular weight would you predict for an additional reaction run at 188°C? At 200°C? Why would or wouldn't you be willing to make a similar prediction of average molecular weight if the reaction is run at 70°C?

4. Upon changing measurement scales, nonlinear relationships between two variables can sometimes be made linear. The article "The Effect of Experimental Error on the Determination of the Optimum Metal-Cutting Conditions" by Ermer and Wu (*The Journal of Engineering for Industry*, 1967) contains a data set gathered in a study of tool life in a turning operation. The data here are part of that data set.

Cutting Speed, x (sfpm)	Tool Life, y (min)
800	1.00, 0.90, 0.74, 0.66
700	1.00, 1.20, 1.50, 1.60
600	2.35, 2.65, 3.00, 3.60
500	6.40, 7.80, 9.80, 16.50
400	21.50, 24.50, 26.00, 33.00

(a) Plot y versus x and calculate R^2 for fitting a linear function of x to y. Does the relationship $y \approx \beta_0 + \beta_1 x$ look like a reasonable explanation of tool life in terms of cutting speed?

(b) Take natural logs of both x and y and repeat part (a) with these log cutting speeds and log tool lives.

(c) Using the logged variables as in (b), fit a linear relationship between the two variables using least squares. Based on this fitted equation, what tool life would you predict for a cutting speed of 550? What approximate relationship between x and y is implied by a linear approximate relationship between $\ln(x)$ and $\ln(y)$? (Give an equation for this relationship.) By the way, Taylor's equation for tool life is $yx^\alpha = C$.

4.2 Fitting Curves and Surfaces by Least Squares

The basic ideas introduced in Section 4.1 generalize to produce a powerful engineering tool: **multiple linear regression**, which is introduced in this section. (Since the term *regression* may seem obscure, the more descriptive terms **curve fitting** and **surface fitting** will be used here, at least initially.)

This section first covers fitting curves defined by polynomials and other functions that are linear in their parameters to (x, y) data. Next comes the fitting of surfaces to data where a response y depends upon the values of several variables $x_1, x_2, \ldots, x_k$. In both cases, the discussion will stress how useful R^2 and residual plotting are and will consider the question of choosing between possible fitted equations. Lastly, we include some additional practical cautions.

4.2.1 Curve Fitting by Least Squares

In the previous section, a straight line did a reasonable job of describing the pressure/density data. But in the fly ash study, the ammonium phosphate/compressive strength data were very poorly described by a straight line. This section first investigates the possibility of fitting curves more complicated than a straight line to (x, y) data. As an example, an attempt will be made to find a better equation for describing the fly ash data.

A natural generalization of the linear equation

$$y \approx \beta_0 + \beta_1 x \tag{4.11}$$

is the **polynomial equation**

$$y \approx \beta_0 + \beta_1 x + \beta_2 x^2 + \cdots + \beta_k x^k \tag{4.12}$$

The least squares fitting of equation (4.12) to a set of n pairs (x_i, y_i) is conceptually only slightly more difficult than the task of fitting equation (4.11). The function of $k + 1$ variables

$$S(\beta_0, \beta_1, \beta_2, \ldots, \beta_k) = \sum_{i=1}^{n}(y_i - \hat{y}_i)^2 = \sum_{i=1}^{n}(y_i - (\beta_0 + \beta_1 x_i + \beta_2 x_i^2 + \cdots + \beta_k x_i^k))^2$$

must be minimized. Upon setting the partial derivatives of $S(\beta_0, \beta_1, \ldots, \beta_k)$ equal to 0, the set of **normal equations** is obtained for this least squares problem, generalizing the pair of equations (4.4) and (4.5). There are $k + 1$ linear equations in the $k + 1$ unknowns $\beta_0, \beta_1, \ldots, \beta_k$. And typically, they can be solved simultaneously for a single set of values, $b_0, b_1, \ldots, b_k$, minimizing $S(\beta_0, \beta_1, \ldots, \beta_k)$. The mechanics of that solution are carried out using a **multiple linear regression program**.

Example 3
(*Example 2 continued*)

More on the Fly Ash Data of Table 4.3

Return to the fly ash study of B. Roth. A quadratic equation might fit the data better than the linear one. So consider fitting the $k = 2$ version of equation (4.12)

$$y \approx \beta_0 + \beta_1 x + \beta_2 x^2 \qquad (4.13)$$

to the data of Table 4.3. Printout 2 shows the MINITAB run. (After entering x and y values from Table 4.3 into two columns of the worksheet, an additional column was created by squaring the x values.)

Printout 2 Quadratic Fit to the Fly Ash Data

```
Regression Analysis


The regression equation is
y = 1243 + 383 x - 76.7 x**2

Predictor        Coef        StDev          T        P
Constant      1242.89        42.98      28.92    0.000
x              382.67        40.43       9.46    0.000
x**2          -76.661         7.762     -9.88    0.000

S = 82.14       R-Sq = 86.7%      R-Sq(adj) = 84.9%


Analysis of Variance


Source          DF           SS          MS          F        P
Regression       2       658230      329115      48.78    0.000
Residual Error  15       101206        6747
Total           17       759437

Source     DF      Seq SS
x           1          21
x**2        1      658209
```

The fitted quadratic equation is

$$\hat{y} = 1242.9 + 382.7x - 76.7x^2$$

Figure 4.9 shows the fitted curve sketched on a scatterplot of the (x, y) data. Although the quadratic curve is not an altogether satisfactory summary of Roth's data, it does a much better job of following the trend of the data than the line sketched in Figure 4.5.

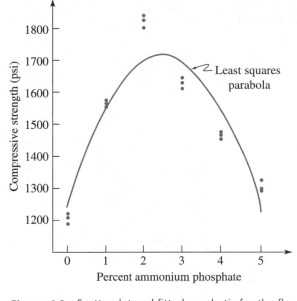

Figure 4.9   Scatterplot and fitted quadratic for the fly ash data

The previous section showed that when fitting a line to (x, y) data, it is helpful to quantify the goodness of that fit using R^2. The coefficient of determination can also be used when fitting a polynomial of form (4.12). Recall once more from Definition 3 that

Coefficient of determination

$$R^2 = \frac{\sum(y_i - \bar{y})^2 - \sum(y_i - \hat{y}_i)^2}{\sum(y_i - \bar{y})^2}$$ **(4.14)**

is the fraction of the raw variability in y accounted for by the fitted equation. Calculation by hand from formula (4.14) is possible, but of course the easiest way to obtain R^2 is to use a computer package.

Example 3
(continued)

Consulting Printout 2, it can be seen that the equation $\hat{y} = 1242.9 + 382.7x - 76.7x^2$ produces $R^2 = .867$. So 86.7% of the raw variability in compressive strength is accounted for using the fitted quadratic. The sample correlation between the observed strengths y_i and fitted strengths $\hat{y}_i$ is $+\sqrt{.867} = .93$.

Comparing what has been done in the present section to what was done in Section 4.1, it is interesting that for the fitting of a line to the fly ash data, R^2 obtained there was only .000 (to three decimal places). The present quadratic is a remarkable improvement over a linear equation for summarizing these data.

A natural question to raise is "What about a cubic version of equation (4.12)?" Printout 3 shows some results of a MINITAB run made to investigate this possibility, and Figure 4.10 shows a scatterplot of the data and a plot of the fitted cubic

Example 3
(continued)

equation. (x values were squared and cubed to provide x, x^2, and x^3 for each y value to use in the fitting.)

Printout 3   Cubic Fit to the Fly Ash Data

```
Regression Analysis


The regression equation is
y = 1188 + 633 x - 214 x**2 + 18.3 x**3

Predictor         Coef       StDev          T        P
Constant       1188.05       28.79      41.27    0.000
x               633.11       55.91      11.32    0.000
x**2           -213.77       27.79      -7.69    0.000
x**3            18.281       3.649       5.01    0.000

S = 50.88       R-Sq = 95.2%      R-Sq(adj) = 94.2%

Analysis of Variance

Source            DF          SS          MS        F        P
Regression         3      723197      241066    93.13    0.000
Residual Error    14       36240        2589
Total             17      759437
```

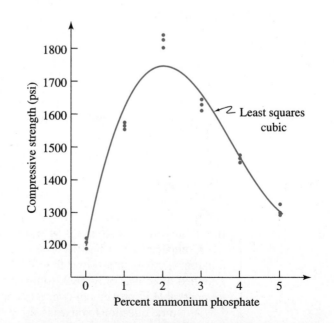

Figure 4.10   Scatterplot and fitted cubic for the fly ash data

R^2 for the cubic equation is .952, somewhat larger than for the quadratic. But it is fairly clear from Figure 4.10 that even a cubic polynomial is not totally satisfactory as a summary of these data. In particular, both the fitted quadratic in Figure 4.9 and the fitted cubic in Figure 4.10 fail to fit the data adequately near an ammonium phosphate level of 2%. Unfortunately, this is where compressive strength is greatest—precisely the area of greatest practical interest.

The example illustrates that R^2 is not the only consideration when it comes to judging the appropriateness of a fitted polynomial. The examination of plots is also important. Not only scatterplots of y versus x with superimposed fitted curves but plots of residuals can be helpful. This can be illustrated on a data set where y is expected to be nearly perfectly quadratic in x.

Example 4 | ### Analysis of the Bob Drop Data of Section 1.4

Consider again the experimental determination of the acceleration due to gravity (through the dropping of the steel bob) data given in Table 1.4 and reproduced here in the first two columns of Table 4.7. Recall that the positions y were recorded at $\frac{1}{60}$ sec intervals beginning at some unknown time t_0 (less than $\frac{1}{60}$ sec) after the bob was released. Since Newtonian mechanics predicts the bob displacement to be

$$\text{displacement} = \frac{gt^2}{2}$$

one expects

$$y \approx \frac{1}{2}g\left(t_0 + \frac{1}{60}(x-1)\right)^2$$

$$= \frac{g}{2}\left(\frac{x}{60}\right)^2 + g\left(t_0 - \frac{1}{60}\right)\left(\frac{x}{60}\right) + \frac{g}{2}\left(t_0 - \frac{1}{60}\right)^2$$

$$= \frac{g}{7200}x^2 + \frac{g}{60}\left(t_0 - \frac{1}{60}\right)x + \frac{g}{2}\left(t_0 - \frac{1}{60}\right)^2 \qquad \textbf{(4.15)}$$

That is, y is expected to be approximately quadratic in x and, indeed, the plot of (x, y) points in Figure 1.8 (p. 22) appears to have that character.

As a slight digression, note that expression (4.15) shows that if a quadratic is fitted to the data in Table 4.7 via least squares,

$$\hat{y} = b_0 + b_1 x + b_2 x^2 \qquad \textbf{(4.16)}$$

is obtained and an experimentally determined value of g (in mm/sec^2) will be

Example 4
(*continued*)

Table 4.7
Data, Fitted Values, and Residuals for a Quadratic Fit to the Bob
Displacement

x, Point Number	y, Displacement	$\hat{y}$, Fitted Displacement	e, Residual
1	.8	.95	−.15
2	4.8	4.56	.24
3	10.8	10.89	−.09
4	20.1	19.93	.17
5	31.9	31.70	.20
6	45.9	46.19	−.29
7	63.3	63.39	−.09
8	83.1	83.31	−.21
9	105.8	105.96	−.16
10	131.3	131.32	−.02
11	159.5	159.40	.10
12	190.5	190.21	.29
13	223.8	223.73	.07
14	260.0	259.97	.03
15	299.2	298.93	.27
16	340.5	340.61	−.11
17	385.0	385.01	−.01
18	432.2	432.13	.07
19	481.8	481.97	−.17
20	534.2	534.53	−.33
21	589.8	589.80	.00
22	647.7	647.80	−.10
23	708.8	708.52	.28

$7200b_2$. This is in fact how the value 9.79 m/sec^2, quoted in Section 1.4, was obtained.

A multiple linear regression program fits equation (4.16) to the bob drop data giving

$$\hat{y} = .0645 - .4716x + 1.3597x^2$$

(from which $g \approx 9790$ mm/sec^2) with R^2 that is 1.0 to 6 decimal places. Residuals for this fit can be calculated using Definition 4 and are also given in Table 4.7. Figure 4.11 is a normal plot of the residuals. It is reasonably linear and thus not remarkable (except for some small suggestion that the largest residual or two may not be as extreme as might be expected, a circumstance that suggests no obvious physical explanation).

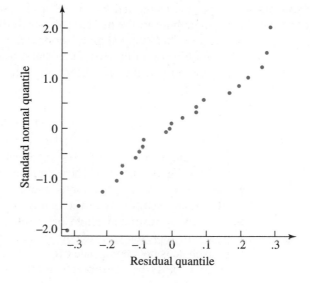

Figure 4.11   Normal plot of the residuals from a quadratic fit to the bob drop data

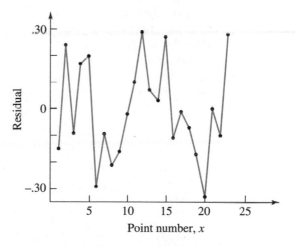

Figure 4.12   Plot of the residuals from the bob drop quadratic fit vs. x

However, a plot of residuals versus x (the time variable) is interesting. Figure 4.12 is such a plot, where successive plotted points have been connected with line segments. There is at least a hint in Figure 4.12 of a **cyclical pattern** in the residuals. Observed displacements are alternately too big, too small, too big, etc. It would be a good idea to look at several more tapes, to see if a cyclical pattern appears consistently, before seriously thinking about its origin. But should the

Example 4
(*continued*)

pattern suggested by Figure 4.12 reappear consistently, it would indicate that something in the mechanism generating the 60 cycle current may cause cycles to be alternately slightly shorter then slightly longer than $\frac{1}{60}$ sec. The practical implication of this would be that if a better determination of g were desired, the regularity of the AC current waveform is one matter to be addressed.

What if a polynomial doesn't fit (x, y) data?

Examples 3 and 4 (respectively) illustrate only partial success and then great success in describing an (x, y) data set by means of a polynomial equation. Situations like Example 3 obviously do sometimes occur, and it is reasonable to wonder what to do when they happen. There are two simple things to keep in mind.

For one, although a polynomial may be unsatisfactory as a global description of a relationship between x and y, it may be quite adequate **locally**—i.e., for a relatively restricted range of x values. For example, in the fly ash study, the quadratic representation of compressive strength as a function of percent ammonium phosphate is not appropriate over the range 0 to 5%. But having identified the region around 2% as being of practical interest, it would make good sense to conduct a follow-up study concentrating on (say) 1.5 to 2.5% ammonium phosphate. It is quite possible that a quadratic fit only to data with $1.5 \leq x \leq 2.5$ would be both adequate and helpful as a summarization of the follow-up data.

The second observation is that the terms $x, x^2, x^3, \ldots, x^k$ in equation (4.12) can be replaced by any (known) functions of x and what we have said here will remain essentially unchanged. The normal equations will still be $k + 1$ linear equations in $\beta_0, \beta_1, \ldots, \beta_k$, and a multiple linear regression program will still produce least squares values $b_0, b_1, \ldots, b_k$. This can be quite useful when there are theoretical reasons to expect a particular (nonlinear but) simple functional relationship between x and y. For example, Taylor's equation for tool life is of the form

$$y \approx \alpha x^\beta$$

for y tool life (e.g., in minutes) and x the cutting speed used (e.g., in sfpm). Taking logarithms,

$$\ln(y) \approx \ln(\alpha) + \beta \ln(x)$$

This is an equation for $\ln(y)$ that is linear in the parameters $\ln(\alpha)$ and β involving the variable $\ln(x)$. So, presented with a set of (x, y) data, empirical values for α and β could be determined by

1. taking logs of both x's and y's,

2. fitting the linear version of (4.12), and

3. identifying $\ln(\alpha)$ with β_0 (and thus α with $\exp(\beta_0)$) and β with β_1.

4.2.2   Surface Fitting by Least Squares

It is a small step from the idea of fitting a line or a polynomial curve to realizing that essentially the same methods can be used to summarize the effects of several different quantitative variables $x_1, x_2, \ldots, x_k$ on some response y. Geometrically the problem is fitting a surface described by an equation

$$y \approx \beta_0 + \beta_1 x_1 + \beta_2 x_2 + \cdots + \beta_k x_k \qquad \textbf{(4.17)}$$

to the data using the least squares principle. This is pictured for a $k = 2$ case in Figure 4.13, where six (x_1, x_2, y) data points are pictured in three dimensions, along with a possible fitted surface of the form (4.17). To fit a surface defined by equation (4.17) to a set of n data points $(x_{1i}, x_{2i}, \ldots, x_{ki}, y_i)$ via least squares, the function of $k + 1$ variables

$$S(\beta_0, \beta_1, \beta_2, \ldots, \beta_k) = \sum_{i=1}^{n} (y_i - \hat{y}_i)^2 = \sum_{i=1}^{n} \big(y_i - (\beta_0 + \beta_1 x_{1i} + \cdots + \beta_k x_{ki})\big)^2$$

must be minimized by choice of the coefficients $\beta_0, \beta_1, \ldots, \beta_k$. Setting partial derivatives with respect to the β's equal to 0 gives normal equations generalizing equations (4.4) and (4.5). The solution of these $k + 1$ linear equations in the $k + 1$ unknowns $\beta_0, \beta_1, \ldots, \beta_k$ is the first task of a multiple linear regression program. The fitted coefficients $b_0, b_1, \ldots, b_k$ that it produces minimize $S(\beta_0, \beta_1, \beta_2, \ldots, \beta_k)$.

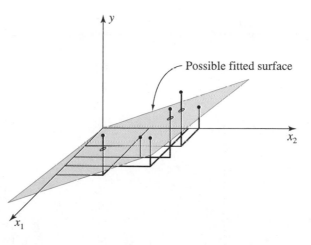

Figure 4.13   Six data points (x_1, x_2, y) and a possible fitted plane

Example 5

Surface Fitting and Brownlee's Stack Loss Data

Table 4.8 contains part of a set of data on the operation of a plant for the oxidation of ammonia to nitric acid that appeared first in Brownlee's *Statistical Theory and Methodology in Science and Engineering*. In plant operation, the nitric oxides produced are absorbed in a countercurrent absorption tower.

The air flow variable, x_1, represents the rate of operation of the plant. The acid concentration variable, x_3, is the percent circulating minus 50 times 10. The response variable, y, is ten times the percentage of ingoing ammonia that escapes from the absorption column unabsorbed (i.e., an inverse measure of overall plant efficiency). For purposes of understanding, predicting, and possibly ultimately optimizing plant performance, it would be useful to have an equation describing how y depends on x_1, x_2, and x_3. Surface fitting via least squares is a method of developing such an empirical equation.

Printout 4 shows results from a MINITAB run made to obtain a fitted equation of the form

$$\hat{y} = b_0 + b_1 x_1 + b_2 x_2 + b_3 x_3$$

Table 4.8
Brownlee's Stack Loss Data

i, Observation Number	x_{1i}, Air Flow	x_{2i}, Cooling Water Inlet Temperature	x_{3i}, Acid Concentration	y_i, Stack Loss
1	80	27	88	37
2	62	22	87	18
3	62	23	87	18
4	62	24	93	19
5	62	24	93	20
6	58	23	87	15
7	58	18	80	14
8	58	18	89	14
9	58	17	88	13
10	58	18	82	11
11	58	19	93	12
12	50	18	89	8
13	50	18	86	7
14	50	19	72	8
15	50	19	79	8
16	50	20	80	9
17	56	20	82	15

The equation produced by the program is

$$\hat{y} = -37.65 + .80x_1 + .58x_2 - .07x_3 \qquad \textbf{(4.18)}$$

Interpreting fitted coefficients from a multiple regression

with $R^2 = .975$. The coefficients in this equation can be thought of as rates of change of stack loss with respect to the individual variables x_1, x_2, and x_3, holding the others fixed. For example, $b_1 = .80$ can be interpreted as the increase in stack loss y that accompanies a one-unit increase in air flow x_1 if inlet temperature x_2 and acid concentration x_3 are held fixed. The signs on the coefficients indicate whether y tends to increase or decrease with increases in the corresponding x. For example, the fact that b_1 is positive indicates that the higher the rate at which the plant is run, the larger y tends to be (i.e., the less efficiently the plant operates). The large value of R^2 is a preliminary indicator that the equation (4.18) is an effective summarization of the data.

Printout 4   Multiple Regression for the Stack Loss Data

```
Regression Analysis


The regression equation is
stack = - 37.7 + 0.798 air + 0.577 water - 0.0671 acid

Predictor        Coef       StDev          T        P
Constant       37.652       4.732      -7.96    0.000
air           0.79769      0.06744      11.83    0.000
water          0.5773       0.1660       3.48    0.004
acid         -0.06706      0.06160      -1.09    0.296

S = 1.253      R-Sq = 97.5%    R-Sq(adj) = 96.9%

Analysis of Variance

Source            DF          SS          MS         F        P
Regression         3      795.83      265.28    169.04    0.000
Residual Error    13       20.40        1.57
Total             16      816.24

Source      DF      Seq SS
air          1      775.48
water        1       18.49
acid         1        1.86

Unusual Observations
Obs       air     stack        Fit   StDev Fit    Residual    St Resid
 10      58.0    11.000     13.506       0.552      -2.506       -2.23R

R denotes an observation with a large standardized residual
```

Although the mechanics of fitting equations of the form (4.17) to multivariate data are relatively straightforward, the **choice and interpretation of appropriate equations** are not so clear-cut. Where many x variables are involved, the number of potential equations of form (4.17) is huge. To make matters worse, there is no completely satisfactory way to plot multivariate $(x_1, x_2, \ldots, x_k, y)$ data to "see" how an equation is fitting. About all that we can do at this point is to (1) offer the broad advice that what is wanted is *the simplest equation that adequately fits the data* and then (2) provide examples of how R^2 and residual plotting can be helpful tools in clearing up the difficulties that arise.

The goal of multiple regression

Example 5
(continued)

In the context of the nitrogen plant, it is sensible to ask whether all three variables, x_1, x_2, and x_3, are required to adequately account for the observed variation in y. For example, the behavior of stack loss might be adequately explained using only one or two of the three x variables. There would be several consequences of practical engineering importance if this were so. For one, in such a case, a simple or **parsimonious** version of equation (4.17) could be used in describing the oxidation process. And if a variable is not needed to predict y, then it is possible that the expense of measuring it might be saved. Or, if a variable doesn't seem to have much impact on y (because it doesn't seem to be essential to include it when writing an equation for y), it may be possible to choose its level on purely economic grounds, without fear of degrading process performance.

As a means of investigating whether indeed some subset of x_1, x_2, and x_3 is adequate to explain stack loss behavior, R^2 values for equations based on all possible subsets of x_1, x_2, and x_3 were obtained and placed in Table 4.9. This shows, for example, that 95% of the raw variability in y can be accounted for using a linear equation in only the air flow variable x_1. Use of both x_1 and the water temperature variable x_2 can account for 97.3% of the raw variability in stack loss. Inclusion of x_3, the acid concentration variable, in an equation already involving x_1 and x_2, increases R^2 only from .973 to .975.

If identifying a simple equation for stack loss that seems to fit the data well is the goal, the message in Table 4.9 would seem to be "Consider an x_1 term first, and then possibly an x_2 term." On the basis of R^2, including an x_3 term in an equation for y seems unnecessary. And in retrospect, this is entirely consistent with the character of the fitted equation (4.18): x_3 varies from 72 to 93 in the original data set, and this means that $\hat{y}$ changes only a total amount

$$.07(93 - 72) \approx 1.5$$

based on changes in x_3. (Remember that $.07 = b_3 =$ the fitted rate of change in y with respect to x_3.) 1.5 is relatively small in comparison to the range in the observed y values.

Once R^2 values have been used to identify potential simplifications of the equation

$$\hat{y} = b_0 + b_1 x_1 + b_2 x_2 + b_3 x_3$$

these can and should go through thorough residual analyses before they are adopted as data summaries. As an example, consider a fitted equation involving

Table 4.9
R^2's for Equations Predicting Stack Loss

Equation Fit	R^2
$y \approx \beta_0 + \beta_1 x_1$	.950
$y \approx \beta_0 + \beta_2 x_2$	.695
$y \approx \beta_0 + \beta_3 x_3$	.165
$y \approx \beta_0 + \beta_1 x_1 + \beta_2 x_2$	.973
$y \approx \beta_0 + \beta_1 x_1 + \beta_3 x_3$	.952
$y \approx \beta_0 + \beta_2 x_2 + \beta_3 x_3$	.706
$y \approx \beta_0 + \beta_1 x_1 + \beta_2 x_2 + \beta_3 x_3$	.975

x_1 and x_2. A multiple linear regression program can be used to produce the fitted equation

$$\hat{y} = -42.00 + .78x_1 + .57x_2 \tag{4.19}$$

Dropping variables from a fitted equation typically changes coefficients

(Notice that b_0, b_1, and b_2 in equation (4.19) differ somewhat from the corresponding values in equation (4.18). That is, equation (4.19) was not obtained from equation (4.18) by simply dropping the last term in the equation. In general, the values of the coefficients b will change depending on which x variables are and are not included in the fitting.)

Residuals for equation (4.19) can be computed and plotted in any number of potentially useful ways. Figure 4.14 shows a normal plot of the residuals and three other plots of the residuals against, respectively, x_1, x_2, and $\hat{y}$. There are no really strong messages carried by the plots in Figure 4.14 except that the data set contains one unusually large x_1 value and one unusually large $\hat{y}$ (which corresponds to the large x_1). But there is enough of a curvilinear "up-then-down-then-back-up-again" pattern in the plot of residuals against x_1 to suggest the possibility of adding an x_1^2 term to the fitted equation (4.19).

You might want to verify that fitting the equation

$$y \approx \beta_0 + \beta_1 x_1 + \beta_2 x_2 + \beta_3 x_1^2$$

to the data of Table 4.8 yields approximately

$$\hat{y} = -15.409 - .069x_1 + .528x_2 + .007x_1^2 \tag{4.20}$$

with corresponding $R^2 = .980$ and residuals that show even less of a pattern than those for the fitted equation (4.19). In particular, the hint of curvature on the plot of residuals versus x_1 for equation (4.19) is not present in the corresponding plot for equation (4.20). Interestingly, looking back over this example, one sees that fitted equation (4.20) has a better R^2 value than even fitted equation (4.18), in

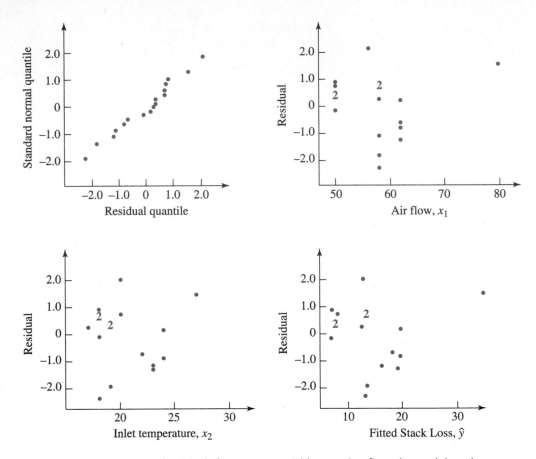

Figure 4.14 Plots of residuals from a two-variable equation fit to the stack loss data ($\hat{y} = -42.00 + .78x_1 + .57x_2$)

Example 5
(continued)

spite of the fact that equation (4.18) involves the process variable x_3 and equation (4.20) does not.

Equation (4.20) is somewhat more complicated than equation (4.19). But because it still really only involves two different input x's and also eliminates the slight pattern seen on the plot of residuals for equation (4.19) versus x_1, it seems an attractive choice for summarizing the stack loss data. A two-dimensional representation of the fitted surface defined by equation (4.20) is given in Figure 4.15. The slight curvature on the plotted curves is a result of the x_1^2 term appearing in equation (4.20). Since most of the data have x_1 from 50 to 62 and x_2 from 17 to 24, the curves carry the message that over these ranges, changes in x_1 seem to produce larger changes in stack loss than do changes in x_2. This conclusion is consistent with the discussion centered around Table 4.9.

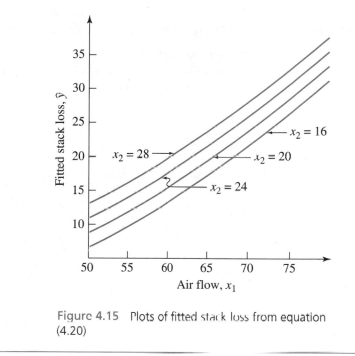

Figure 4.15   Plots of fitted stack loss from equation (4.20)

Common residual plots in multiple regression

The plots of residuals used in Example 5 are typical. They are

1. normal plots of residuals,

2. plots of residuals against all x variables,

3. plots of residuals against $\hat{y}$,

4. plots of residuals against time order of observation, and

5. plots of residuals against variables (like machine number or operator) not used in the fitted equation but potentially of importance.

All of these can be used to help assess the appropriateness of surfaces fit to multivariate data, and they all have the potential to tell an engineer something not previously discovered about a set of data and the process that generated them.

Earlier in this section, there was a discussion of the fact that an "x term" in the equations fitted via least squares can be a known function (e.g., a logarithm) of a basic process variable. In fact, it is frequently helpful to allow an "x term" in equation (4.17) (page 149) to be a known function of *several* basic process variables. The next example illustrates this point.

Example 6

Lift/Drag Ratio for a Three-Surface Configuration

P. Burris studied the effects of the positions relative to the wing of a canard (a forward lifting surface) and tail on the lift/drag ratio for a three-surface configuration. Part of his data are given in Table 4.10, where

x_1 = canard placement in inches above the plane defined by the main wing

x_2 = tail placement in inches above the plane defined by the main wing

(The front-to-rear positions of the three surfaces were constant throughout the study.)

A straightforward least squares fitting of the equation

$$y \approx \beta_0 + \beta_1 x_1 + \beta_2 x_2$$

to these data produces R^2 of only .394. Even the addition of squared terms in both x_1 and x_2, i.e., the fitting of

$$y \approx \beta_0 + \beta_1 x_1 + \beta_2 x_2 + \beta_3 x_1^2 + \beta_4 x_2^2$$

produces an increase in R^2 to only .513. However, Printout 5 shows that fitting the equation

$$y \approx \beta_0 + \beta_1 x_1 + \beta_2 x_2 + \beta_3 x_1 x_2$$

yields $R^2 = .641$ and the fitted relationship

$$\hat{y} = 3.4284 + .5361 x_1 + .3201 x_2 - .5042 x_1 x_2 \qquad \textbf{(4.21)}$$

Table 4.10
Lift/Drag Ratios for 9 Canard/Tail Position Combinations

x_1, Canard Position	x_2, Tail Position	y, Lift/Drag Ratio
−1.2	−1.2	.858
−1.2	0.0	3.156
−1.2	1.2	3.644
0.0	−1.2	4.281
0.0	0.0	3.481
0.0	1.2	3.918
1.2	−1.2	4.136
1.2	0.0	3.364
1.2	1.2	4.018

Printout 5   Multiple Regression for the Lift/Drag Ratio Data

```
Regression Analysis


The regression equation is
y = 3.43 + 0.536 x1 + 0.320 x2 - 0.504 x1*x2

Predictor        Coef        StDev          T        P
Constant       3.4284       0.2613      13.12    0.000
x1             0.5361       0.2667       2.01    0.101
x2             0.3201       0.2667       1.20    0.284
x1*x2         -0.5042       0.2722      -1.85    0.123

S = 0.7839      R-Sq = 64.1%     R-Sq(adj) = 42.5%

Analysis of Variance

Source           DF          SS         MS        F        P
Regression        3      5.4771     1.8257     2.97    0.136
Residual Error    5      3.0724     0.6145
Total             8      8.5495
```

(After reading x_1, x_2, and y values from Table 4.10 into columns of MINITAB's worksheet, $x_1 x_2$ products were created and y fitted to the three predictor variables x_1, x_2, and $x_1 x_2$ in order to create this printout.)

Figure 4.16 shows the nature of the fitted surface (4.21). Raising the canard (increasing x_1) has noticeably different predicted impacts on y, depending on the value of x_2 (the tail position). (It appears that the canard and tail should not be lined up—i.e., x_1 should not be near x_2. For large predicted response, one wants small x_1 for large x_2 and large x_1 for small x_2.) It is the cross-product term $x_1 x_2$ in relationship (4.21) that allows the response curves to have different characters for different x_2 values. Without it, the slices of the fitted $(x_1, x_2, \hat{y})$ surface would be parallel for various x_2, much like the situation in Figure 4.15.

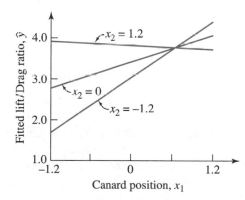

Figure 4.16   Plots of fitted lift/drag from equation (4.21)

Example 6
(*continued*)

Although the main new point of this example has by now been made, it probably should be mentioned that equation (4.21) is not the last word for fitting the data of Table 4.10. Figure 4.17 gives a plot of the residuals for relationship (4.21) versus canard position x_1, and it shows a strong curvilinear pattern. In fact, the fitted equation

$$\hat{y} = 3.9833 + .5361x_1 + .3201x_2 - .4843x_1^2 - .5042x_1x_2 \qquad (4.22)$$

provides $R^2 = .754$ and generally random-looking residuals. It can be verified by plotting $\hat{y}$ versus x_1 curves for several x_2 values that the fitted relationship (4.22) yields nonparallel parabolic slices of the fitted $(x_1, x_2, \hat{y})$ surface, instead of the nonparallel linear slices seen in Figure 4.16.

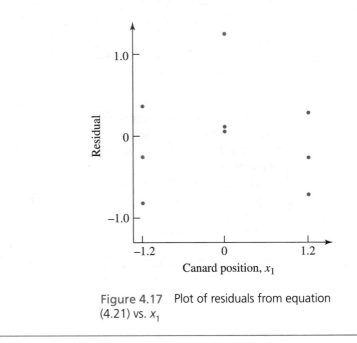

Figure 4.17   Plot of residuals from equation (4.21) vs. x_1

4.2.3   Some Additional Cautions

Least squares fitting of curves and surfaces is of substantial engineering importance—but it must be handled with care and thought. Before leaving the subject until Chapter 9, which explains methods of formal inference associated with it, a few more warnings must be given.

Extrapolation   First, it is necessary to warn of the dangers of extrapolation substantially outside the "range" of the $(x_1, x_2, \ldots, x_k, y)$ data. It is sensible to count on a fitted equation to describe the relation of y to a particular set of inputs $x_1, x_2, \ldots, x_k$ only if they are like the sets used to create the equation. The challenge surface fitting affords is

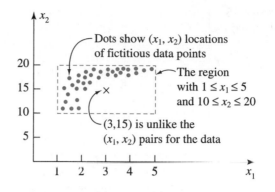

Figure 4.18   Hypothetical plot of (x_1, x_2) pairs

that when several different x variables are involved, it is difficult to tell whether a particular $(x_1, x_2, \ldots, x_k)$ vector is a large extrapolation. About all one can do is check to see that it comes close to matching some *single data point* in the set *on each coordinate* $x_1, x_2, \ldots, x_k$. It is not sufficient that there be some point with x_1 value near the one of interest, another point with x_2 value near the one of interest, etc. For example, having data with $1 \le x_1 \le 5$ and $10 \le x_2 \le 20$ doesn't mean that the (x_1, x_2) pair $(3, 15)$ is necessarily like any of the pairs in the data set. This fact is illustrated in Figure 4.18 for a fictitious set of (x_1, x_2) values.

The influence of outlying data vectors   Another potential pitfall is that the fitting of curves and surfaces via least squares can be strongly affected by a few outlying or extreme data points. One can try to identify such points by examining plots and comparing fits made with and without the suspicious point(s).

Example 5
(*continued*)

Figure 4.14 earlier called attention to the fact that the nitrogen plant data set contains one point with an extreme x_1 value. Figure 4.19 is a scatterplot of (x_1, x_2) pairs for the data in Table 4.8 (page 150). It shows that by most qualitative standards, observation 1 in Table 4.8 is unusual or outlying.

If the fitting of equation (4.20) is redone using only the last 16 data points in Table 4.8, the equation

$$\hat{y} = -56.797 + 1.404x_1 + .601x_2 - .007x_1^2 \tag{4.23}$$

and $R^2 = .942$ are obtained. Using equation (4.23) as a description of stack loss and limiting attention to x_1 in the range 50 to 62 could be considered. But it is possible to verify that though some of the coefficients (the b's) in equations (4.20) and (4.23) differ substantially, the two equations produce comparable $\hat{y}$ values for the 16 data points with x_1 between 50 and 62. In fact, the largest difference in fitted values is about .4. So, since point 1 in Table 4.8 doesn't

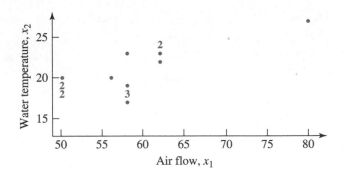

Figure 4.19 Plot of (x_1, x_2) pairs for the stack loss data

radically change predictions made using the fitted equation, it makes sense to leave it in consideration, adopt equation (4.20), and use it to describe stack loss for (x_1, x_2) pairs interior to the pattern of scatter in Figure 4.19.

Replication and surface fitting

A third warning has to do with the notion of replication (first discussed in Section 2.3). It is the fact that the fly ash data of Example 3 has several y's for each x that makes it so clear that even the quadratic and cubic curves sketched in Figures 4.9 and 4.10 are inadequate descriptions of the relationship between phosphate and strength. The fitted curves pass clearly outside the range of what look like believable values of y for some values of x. Without such replication, what is permissible variation about a fitted curve or surface can't be known with confidence. For example, the structure of the lift/drag data set in Example 6 is weak from this viewpoint. There is no replication represented in Table 4.10, so an external value for typical experimental precision would be needed in order to identify a fitted value as obviously incompatible with an observed one.

The nitrogen plant data set of Example 5 was presumably derived from a primarily observational study, where no conscious attempt was made to replicate (x_1, x_2, x_3) settings. However, points number 4 and 5 in Table 4.8 (page 150) do represent the replication of a single (x_1, x_2, x_3) combination and show a difference in observed stack loss of 1. And this makes the residuals for equation (4.20) (which range from -2.0 to 2.3) seem at least not obviously out of line.

Section 9.2 discusses more formal and precise ways of using data from studies with some replication to judge whether or not a fitted curve or surface misses some observed y's too badly. For now, simply note that among replication's many virtues is the fact that it allows more reliable judgments about the appropriateness of a fitted equation than are otherwise possible.

The possibility of overfitting

The fourth caution is that the notion of equation simplicity (*parsimony*) is important for reasons in addition to simplicity of interpretation and reduced expense involved in using the equation. It is also important from the point of view of typically giving smooth interpolation and not **overfitting** a data set. As a hypothetical example,

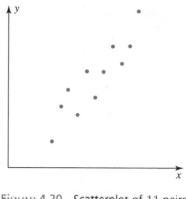

Figure 4.20 Scatterplot of 11 pairs (x, y)

consider the artificial, generally linear (x, y) data plotted in Figure 4.20. It would be possible to run a (wiggly) $k = 10$ version of the polynomial (4.12) through each of these points. But in most physical problems, such a curve would do a much worse job of predicting y at values of x not represented by a data point than would a simple fitted line. A tenth-order polynomial would overfit the data in hand.

Empirical models and engineering   As a final point in this section, consider how the methods discussed here fit into the broad picture of using models for attacking engineering problems. It must be said that physical theories of physics, chemistry, materials, etc. rarely produce equations of the forms (4.12) or (4.17). Sometimes pertinent equations from those theories can be rewritten in such forms, as was possible with Taylor's equation for tool life earlier in this section. But the majority of engineering applications of the methods in this section are to the large number of problems where no commonly known and simple physical theory is available, and a simple **empirical** description of the situation would be helpful. In such cases, the tool of least squares fitting of curves and surfaces can function as a kind of "mathematical French curve," allowing an engineer to develop approximate empirical descriptions of how a response y is related to system inputs $x_1, x_2, \ldots, x_k$.

Section 2 Exercises

1. Return to Exercise 3 of Section 4.1. Fit a quadratic relationship $y \approx \beta_0 + \beta_1 x + \beta_2 x^2$ to the data via least squares. By appropriately plotting residuals and examining R^2 values, determine the advisability of using a quadratic rather than a linear equation to describe the relationship between x and y. If a quadratic fitted equation is used, how does the predicted mean molecular weight at 200°C compare to that obtained in part (e) of the earlier exercise?

2. Here are some data taken from the article "Chemi-thermomechanical Pulp from Mixed High Density Hardwoods" by Miller, Shankar, and Peterson (*Tappi Journal*, 1988). Given are the percent NaOH used as a pretreatment chemical, x_1, the pretreatment time in minutes, x_2, and the resulting value of a specific surface area variable, y (with units of cm³/g), for nine batches of pulp produced from a mixture of hardwoods at a treatment temperature of 75°C in mechanical pulping.

% NaOH, x_1	Time, x_2	Specific Surface Area, y
3.0	30	5.95
3.0	60	5.60
3.0	90	5.44
9.0	30	6.22
9.0	60	5.85
9.0	90	5.61
15.0	30	8.36
15.0	60	7.30
15.0	90	6.43

(a) Fit the approximate relationship $y \approx \beta_0 + \beta_1 x_1 + \beta_2 x_2$ to these data via least squares. Interpret the coefficients b_1 and b_2 in the fitted equation. What fraction of the observed raw variation in y is accounted for using this equation?

(b) Compute and plot residuals for your fitted equation from (a). Discuss what these plots indicate about the adequacy of your fitted equation. (At a minimum, you should plot residuals against all of x_1, x_2, and $\hat{y}$ and normal-plot the residuals.)

(c) Make a plot of y versus x_1 for the nine data points and sketch on that plot the three different linear functions of x_1 produced by setting x_2 first at 30, then 60, and then 90 in your fitted equation from (a). How well do fitted responses appear to match observed responses?

(d) What specific surface area would you predict for an additional batch of pulp of this type produced using a 10% NaOH treatment for a time of 70 minutes? Would you be willing to make a similar prediction for 10% NaOH used for 120 minutes based on your fitted equation? Why or why not?

(e) There are many other possible approximate relationships that might be fitted to these data via least squares, one of which is $y \approx \beta_0 + \beta_1 x_1 + \beta_2 x_2 + \beta_3 x_1 x_2$. Fit this equation to the preceding data and compare the resulting coefficient of determination to the one found in (a). On the basis of these alone, does the use of the more complicated equation seem necessary?

(f) For the equation fit in part (e), repeat the steps of part (c) and compare the plot made here to the one made earlier.

(g) What is an intrinsic weakness of this real published data set?

(h) What terminology (for data structures) introduced in Section 1.2 describes this data set? It turns out that since the data set has this special structure and all nine sample sizes are the same (i.e., are all 1), some special relationships hold between the equation fit in (a) and what you get by separately fitting linear equations in x_1 and then in x_2 to the y data. Fit such one-variable linear equations and compare coefficients and R^2 values to what you obtained in (a). What relationships exist between these?

4.3   Fitted Effects for Factorial Data

The previous two sections have centered on the least squares fitting of equations to data sets where a quantitative response y is presumed to depend on the levels $x_1, x_2, \ldots, x_k$ of *quantitative factors*. In many engineering applications, at least some of the system "knobs" whose effects must be assessed are basically *qualitative* rather than quantitative. When a data set has complete factorial structure (review the meaning of this terminology in Section 1.2), it is still possible to describe it in terms of an equation. This equation involves so-called fitted factorial effects. Sometimes, when a few of these fitted effects dominate the rest, a parsimonious version of this equation can adequately describe the data and have intuitively appealing and understandable interpretations. The use of simple plots and residuals will be discussed, as tools helpful in assessing whether such a simple structure holds.

The discussion begins with the 2-factor case, then considers three (or, by analogy, more) factors. Finally, the special case where each factor has only two levels is discussed.

4.3.1 Fitted Effects for 2-Factor Studies

Example 9 of Chapter 3 (page 101) illustrated how informative a plot of sample means versus levels of one of the factors can be in a 2-factor study. Such plotting is always the place to begin in understanding the story carried by two-way factorial data. In addition, it is helpful to calculate the factor level (marginal) averages of the sample means and the grand average of the sample means. For factor A having I levels and factor B having J levels, the following notation will be used:

Notation for sample means and their averages

$$\bar{y}_{ij} = \text{the sample mean response when factor A is at}$$
$$\text{level } i \text{ and factor B is at level } j$$

$$\bar{y}_{i.} = \frac{1}{J} \sum_{j=1}^{J} \bar{y}_{ij}$$

$$= \text{the average sample mean when factor A is at level } i$$

$$\bar{y}_{.j} = \frac{1}{I} \sum_{i=1}^{I} \bar{y}_{ij}$$

$$= \text{the average sample mean when factor B is at level } j$$

$$\bar{y}_{..} = \frac{1}{IJ} \sum_{i,j} \bar{y}_{ij}$$

$$= \text{the grand average sample mean}$$

The $\bar{y}_{i.}$ and $\bar{y}_{.j}$ are row and column averages when one thinks of the $\bar{y}_{ij}$ laid out in a two-dimensional format, as shown in Figure 4.21.

Example 7

Joint Strengths for Three Different Joint Types in Three Different Woods

Kotlers, MacFarland, and Tomlinson studied the tensile strength of three different types of joints made on three different types of wood. Butt, lap, and beveled joints were made in nominal $1'' \times 4'' \times 12''$ pine, oak, and walnut specimens using a resin glue. The original intention was to test two specimens of each Joint Type/Wood Type combination. But one operator error and one specimen failure not related to its joint removed two of the original data points from consideration and gave the data in Table 4.11. These data have complete 3×3 factorial struc-

Factor B

	Level 1	Level 2		Level J	
Level 1	$\bar{y}_{11}$	$\bar{y}_{12}$	$\cdots$	$\bar{y}_{1J}$	$\bar{y}_{1.}$
Level 2	$\bar{y}_{21}$	$\bar{y}_{22}$	$\cdots$	$\bar{y}_{2J}$	$\bar{y}_{2.}$
Factor A	$\vdots$	$\vdots$		$\vdots$	$\vdots$
Level I	$\bar{y}_{I1}$	$\bar{y}_{I2}$	$\cdots$	$\bar{y}_{IJ}$	$\bar{y}_{I.}$
	$\bar{y}_{.1}$	$\bar{y}_{.2}$	$\cdots$	$\bar{y}_{.J}$	$\bar{y}_{..}$

Figure 4.21 Cell sample means and row, column, and grand average sample means for a two-way factorial

Example 7
(*continued*)

Table 4.11
Measured Strengths of 16 Wood Joints

Specimen	Joint	Wood	y, Stress at Failure (psi)
1	beveled	oak	1518
2	butt	pine	829
3	beveled	walnut	2571
4	butt	oak	1169
5	beveled	oak	1927
6	beveled	pine	1348
7	lap	walnut	1489
8	beveled	walnut	2443
9	butt	walnut	1263
10	lap	oak	1295
11	lap	oak	1561
12	lap	pine	1000
13	butt	pine	596
14	lap	pine	859
15	butt	walnut	1029
16	beveled	pine	1207

Table 4.12
Sample Means for Nine Wood/Joint Combinations

| | | Wood | | | |
		1 (Pine)	*2 (Oak)*	*3 (Walnut)*	
Joint	*1 (Butt)*	$\bar{y}_{11} = 712.5$	$\bar{y}_{12} = 1169.0$	$\bar{y}_{13} = 1146.0$	$\bar{y}_{1.} = 1009.17$
	2 (Beveled)	$\bar{y}_{21} = 1277.5$	$\bar{y}_{22} = 1722.5$	$\bar{y}_{23} = 2507.0$	$\bar{y}_{2.} = 1835.67$
	3 (Lap)	$\bar{y}_{31} = 929.5$	$\bar{y}_{32} = 1428.0$	$\bar{y}_{33} = 1489.0$	$\bar{y}_{3.} = 1282.17$
		$\bar{y}_{.1} = 973.17$	$\bar{y}_{.2} = 1439.83$	$\bar{y}_{.3} = 1714.00$	$\bar{y}_{..} = 1375.67$

*Interaction
Plot*

ture. Collecting y's for the nine different combinations into separate samples and calculating means, the $\bar{y}_{ij}$'s are as presented in tabular form in Table 4.12 and plotted in Figure 4.22. This figure is a so-called **interaction plot** of these means. The qualitative messages given by the plot are as follows:

1. Joint types ordered by strength are "beveled is stronger than lap, which in turn is stronger than butt."

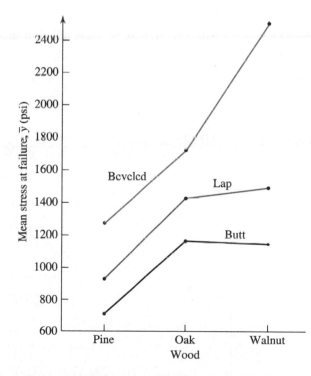

Figure 4.22   Interaction plot of joint strength sample means

Example 7
(*continued*)

2. Woods ordered by overall strength seem to be "walnut is stronger than oak, which in turn is stronger than pine."

3. The strength pattern across woods is not consistent from joint type to joint type (or equivalently, the strength pattern across joints is not consistent from wood type to wood type).

The idea of fitted effects is to invent a way of quantifying such qualitative summaries.

The row and column average means ($\bar{y}_{i.}$'s and $\bar{y}_{.j}$'s, respectively) might be taken as measures of average response behavior at different levels of the factors in question. If so, it then makes sense to use the differences between these and the grand average mean $\bar{y}_{..}$ as measures of the effects of those levels on mean response. This leads to Definition 5.

Definition 5

In a two-way complete factorial study with factors A and B, the **fitted main effect of factor A at its *i*th level** is

$$a_i = \bar{y}_{i.} - \bar{y}_{..}$$

Similarly, the **fitted main effect of factor B at its *j*th level** is

$$b_j = \bar{y}_{.j} - \bar{y}_{..}$$

Example 7
(*continued*)

Simple arithmetic and the $\bar{y}$'s in Table 4.12 yield the fitted main effects for the joint strength study of Kotlers, MacFarland, and Tomlinson. First for factor A (the Joint Type),

a_1 = the Joint Type fitted main effect for butt joints

$\quad$ = 1009.17 − 1375.67

$\quad$ = −366.5 psi

a_2 = the Joint Type fitted main effect for beveled joints

$\quad$ = 1835.67 − 1375.67

$\quad$ = 460.0 psi

a_3 = the Joint Type fitted main effect for lap joints

$\quad$ = 1282.17 − 1375.67

$\quad$ = −93.5 psi

Similarly for factor B (the Wood Type),

$$b_1 = \text{the Wood Type fitted main effect for pine}$$
$$= 973.17 - 1375.67$$
$$= -402.5 \text{ psi}$$
$$b_2 = \text{the Wood Type fitted main effect for oak}$$
$$= 1439.83 - 1375.67$$
$$= 64.17 \text{ psi}$$
$$b_3 = \text{the Wood Type fitted main effect for walnut}$$
$$= 1714.00 - 1375.67$$
$$= 338.33 \text{ psi}$$

These fitted main effects quantify the first two qualitative messages carried by the data and listed as (1) and (2) before Definition 5. For example,

$$a_2 > a_3 > a_1$$

says that beveled joints are strongest and butt joints the weakest. Further, the fact that the a_i's and b_j's are of roughly the same order of magnitude says that the Joint Type and Wood Type factors are of comparable importance in determining tensile strength.

A difference between fitted main effects for a factor amounts to a difference between corresponding row or column averages and quantifies how different response behavior is for those two levels.

Example 7
(continued)

For example, comparing pine and oak wood types,

$$b_1 - b_2 = (\bar{y}_{.1} - \bar{y}_{..}) - (\bar{y}_{.2} - \bar{y}_{..})$$
$$= \bar{y}_{.1} - \bar{y}_{.2}$$
$$= 973.17 - 1439.83$$
$$= -466.67 \text{ psi}$$

which indicates that pine joint average strength is about 467 psi less than oak joint average strength.

In *some* two-factor factorial studies, the fitted main effects as defined in Definition 5 pretty much summarize the story told by the means $\bar{y}_{ij}$, in the sense that

$$\bar{y}_{ij} \approx \bar{y}_{..} + a_i + b_j \qquad \text{for every } i \text{ and } j \qquad \textbf{(4.24)}$$

Display (4.24) implies, for example, that the pattern of mean responses for level 1 of factor A is the same as for level 2 of A. That is, changing levels of factor B (from say j to j') produces the same change in mean response for level 2 as for level 1 (namely, $b_{j'} - b_j$). In fact, if relation (4.24) holds, there are **parallel traces** on an interaction plot of means.

Example 7
(continued)

To illustrate the meaning of expression (4.24), the fitted effects for the Joint Type/Wood Type data have been used to calculate $3 \times 3 = 9$ values of $\bar{y}_{..} + a_i + b_j$ corresponding to the nine experimental combinations. These are given in Table 4.13.

For comparison purposes, the $\bar{y}_{ij}$ from Table 4.12 and the $\bar{y}_{..} + a_i + b_j$ from Table 4.13 are plotted on the same sets of axes in Figure 4.23. Notice the parallel traces for the $\bar{y}_{..} + a_i + b_j$ values for the three different joint types. The traces for

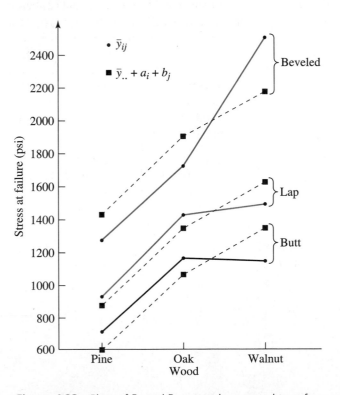

Figure 4.23 Plots of $\bar{y}_{ij}$ and $\bar{y}_{..} + a_i + b_j$ vs. wood type for three joint types

Table 4.13
Values of $\bar{y}_{..} + a_i + b_j$ for the Joint Strength Study

		Wood		
		1 (Pine)	*2 (Oak)*	*3 (Walnut)*
	1 (Butt)	$\bar{y}_{..} + a_1 + b_1 =$ 606.67	$\bar{y}_{..} + a_1 + b_2 =$ 1073.33	$\bar{y}_{..} + a_1 + b_3 =$ 1347.50
Joint	*2 (Beveled)*	$\bar{y}_{..} + a_2 + b_1 =$ 1433.17	$\bar{y}_{..} + a_2 + b_2 =$ 1899.83	$\bar{y}_{..} + a_2 + b_3 =$ 2174.00
	3 (Lap)	$\bar{y}_{..} + a_3 + b_1 =$ 879.67	$\bar{y}_{..} + a_3 + b_2 =$ 1346.33	$\bar{y}_{..} + a_3 + b_3 =$ 1620.50

the $\bar{y}_{ij}$ values for the three different joint types are not parallel (particularly when walnut is considered), so there are apparently substantial differences between the $\bar{y}_{ij}$'s and the $\bar{y}_{..} + a_i + b_j$'s.

When relationship (4.24) fails to hold, the patterns in mean response across levels of one factor depend on the levels of the second factor. In such cases, the differences between the combination means $\bar{y}_{ij}$ and the values $\bar{y}_{..} + a_i + b_j$ can serve as useful *measures of lack of parallelism* on the plots of means, and this leads to another definition.

Definition 6

In a two-way complete factorial study with factors A and B, the **fitted inter-action of factor A at its *i*th level and factor B at its *j*th level** is

$$ ab_{ij} = \bar{y}_{ij} - (\bar{y}_{..} + a_i + b_j) $$

Interpretation of interactions in a two-way factorial study

The fitted interactions in some sense measure how much pattern the combination means $\bar{y}_{ij}$ carry that is not explainable in terms of the factors A and B acting separately. Clearly, when relationship (4.24) holds, the fitted interactions ab_{ij} are all small (nearly 0), and system behavior can be thought of as depending separately on level of A and level of B. In such cases, an important practical consequence is that it is possible to develop recommendations for levels of the two factors independently of each other. For example, one need not recommend one level of A if B is at its level 1 and another if B is at its level 2.

Consider a study of the effects of factors Tool Type and Turning Speed on the metal removal rate for a lathe. If the fitted interactions are small, turning speed recommendations that remain valid for all tool types can be made. However, if the fitted interactions are important, turning speed recommendations might vary according to tool type.

Example 7
(continued)

Again using the Joint Type/Wood Type data, consider calculating the fitted interactions. The raw material for these calculations already exists in Tables 4.12 and 4.13. Simply taking differences between entries in these tables cell-by-cell yields the fitted interactions given in Table 4.14.

It is interesting to compare these fitted interactions to themselves and to the fitted main effects. The largest (in absolute value) fitted interaction (ab_{23}) corresponds to beveled walnut joints. This is consistent with one visual message in Figures 4.22 and 4.23: This Joint Type/Wood Type combination is in some sense most responsible for destroying any nearly parallel structure that might otherwise appear. The fact that (on the whole) the ab_{ij}'s are not as large as the a_i's or b_j's is consistent with a second visual message in Figures 4.22 and 4.23: The lack of parallelism, while important, is not as important as differences in Joint Types or Wood Types.

Table 4.14
Fitted Interactions for the Joint Strength Study

		Wood		
		1 (Pine)	*2 (Oak)*	*3 (Walnut)*
	1 (Butt)	$ab_{11} = 105.83$	$ab_{12} = 95.67$	$ab_{13} = -201.5$
Joint	*2 (Beveled)*	$ab_{21} = -155.66$	$ab_{22} = -177.33$	$ab_{23} = 333.0$
	3 (Lap)	$ab_{31} = 49.83$	$ab_{32} = 81.67$	$ab_{33} = -131.5$

Example 7 has proceeded "by hand." But using a statistical package can make the calculations painless. For example, Printout 6 illustrates that most of the results of Example 7 are readily available in MINITAB's "General Linear Model" routine (found under the "Stat/ANOVA/General Linear Model" menu). Comparing this printout to the example does bring up one point regarding the fitted effects defined in Definitions 5 and 6. Note that the printout provides values of only two (of three) Joint main effects, two (of three) Wood main effects, and four (of nine) Joint × Wood interactions. These are all that are needed, since it is a consequence of Definition 5 that *fitted main effects for a given factor must total to 0*, and it is a consequence of Definition 6 that *fitted interactions must sum to zero across any row or down any column* of the two-way table of factor combinations. The fitted effects not provided by the printout are easily deduced from the ones that are given.

Fitted effects
sum to zero

Printout 6   Computations for the Joint Strength Data

```
General Linear Model

Factor    Type Levels  Values
joint     fixed     3  beveled butt    lap
wood      fixed     3      oak pine walnut
```

Analysis of Variance for strength, using Adjusted SS for Tests

Source	DF	Seq SS	Adj SS	Adj MS	F	P
joint	2	2153879	1881650	940825	32.67	0.000
wood	2	1641095	1481377	740689	25.72	0.001
joint*wood	4	468408	468408	117102	4.07	0.052
Error	7	201614	201614	28802		
Total	15	4464996				

Term	Coef	StDev	T	P
Constant	1375.67	44.22	31.11	0.000
joint				
beveled	460.00	59.63	7.71	0.000
butt	-366.50	63.95	-5.73	0.001
wood				
oak	64.17	63.95	1.00	0.349
pine	-402.50	59.63	-6.75	0.000
joint* wood				
beveled oak	-177.33	85.38	-2.08	0.076
beveled pine	-155.67	82.20	-1.89	0.100
butt oak	95.67	97.07	0.99	0.357
butt pine	105.83	85.38	1.24	0.255

Unusual Observations for strength

Obs	strength	Fit	StDev Fit	Residual	St Resid
4	1169.00	1169.00	169.71	0.00	* X
7	1489.00	1489.00	169.71	0.00	* X

X denotes an observation whose X value gives it large influence.

Least Squares Means for strength

joint	Mean	StDev
beveled	1835.7	69.28
butt	1009.2	80.00
lap	1282.2	80.00
wood		
oak	1439.8	80.00
pine	973.2	69.28
walnut	1714.0	80.00
joint* wood		
beveled oak	1722.5	120.00
beveled pine	1277.5	120.00
beveled walnut	2507.0	120.00
butt oak	1169.0	169.71
butt pine	712.5	120.00
butt walnut	1146.0	120.00
lap oak	1428.0	120.00
lap pine	929.5	120.00
lap walnut	1489.0	169.71

4.3.2 Simpler Descriptions for Some Two-Way Data Sets

Rewriting the equation for ab_{ij} from Definition 6,

$$\bar{y}_{ij} = \bar{y}_{..} + a_i + b_j + ab_{ij} \qquad \textbf{(4.25)}$$

That is, $\bar{y}_{..}$, the fitted main effects, and the fitted interactions provide a decomposition or breakdown of the combination sample means into interpretable pieces. These pieces correspond to an overall effect, the effects of factors acting separately, and the effects of factors acting jointly.

Taking a hint from the equation fitting done in the previous two sections, it makes sense to think of (4.25) as a fitted version of an approximate relationship,

$$y \approx \mu + \alpha_i + \beta_j + \alpha\beta_{ij} \qquad\qquad \textbf{(4.26)}$$

where $\mu, \alpha_1, \alpha_2, \ldots, \alpha_I, \beta_1, \beta_2, \ldots, \beta_J, \alpha\beta_{11}, \ldots, \alpha\beta_{1J}, \alpha\beta_{21}, \ldots, \alpha\beta_{IJ}$ are some constants and the levels of factors A and B associated with a particular response y pick out which of the α_i's, β_j's, and $\alpha\beta_{ij}$'s are appropriate in equation (4.26). By analogy with the previous two sections, the possibility should be considered that a relationship even simpler than equation (4.26) might hold, perhaps not involving $\alpha\beta_{ij}$'s or even α_i's or perhaps β_j's.

It has already been said that when relationship (4.24) is in force, or equivalently

$$ab_{ij} \approx 0 \qquad \text{for every } i \text{ and } j$$

it is possible to understand an observed set of $\bar{y}_{ij}$'s in simplified terms of the factors acting separately. This possibility corresponds to the simplified version of equation (4.26),

$$y \approx \mu + \alpha_i + \beta_j$$

and there are other simplified versions of equation (4.26) that also have appealing interpretations. For example, the simplified version of equation (4.26),

$$y \approx \mu + \alpha_i$$

says that only factor A (not factor B) is important in determining response y. ($\alpha_1, \alpha_2, \ldots, \alpha_I$ still allow for different response behavior for different levels of A.)

Two questions naturally follow on this kind of reasoning: "How is a *reduced* or *simplified* version of equation (4.26) fitted to a data set? And after fitting such an equation, how is the appropriateness of the result determined?" General answers to these questions are subtle. But there is one circumstance in which it is possible to give fairly straightforward answers. That is the case where the data are **balanced**— in the sense that all of the samples (leading to the $\bar{y}_{ij}$'s) have the same size. With balanced data, the fitted effects from Definitions 5 and 6 and simple addition produce fitted responses. And based on such fitted values, the R^2 and residual plotting ideas from the last two sections can be applied here as well. That is, when working with balanced data, least squares fitting of a simplified version of equation (4.26) can be accomplished by

1. calculating fitted effects according to Definitions 5 and 6 and then

2. adding those corresponding to terms in the reduced equation to compute fitted responses, $\hat{y}$.

Residuals are then (as always)

Residuals

$$e = y - \hat{y}$$

(and should look like noise if the simplified equation is an adequate description of the data set). Further, the fraction of raw variation in y accounted for in the fitting process is (as always)

Coefficient of determination

$$R^2 = \frac{\sum(y - \bar{y})^2 - \sum(y - \hat{y})^2}{\sum(y - \bar{y})^2} \qquad (4.27)$$

where the sums are over all observed y's. (Summation notation is being abused even further than usual, by not even subscripting the y's and $\hat{y}$'s.)

Example 8
(Example 12, Chapter 2, revisited—p. 49)

Simplified Description of Two-Way Factorial Golf Ball Flight Data

G. Gronberg tested drive flight distances for golf balls of several different compressions on several different evenings. Table 4.15 gives a small part of the data that he collected, representing 80 and 100 compression flight distances (in yards) from two different evenings. Notice that these data are balanced, all four sample sizes being 10.

Table 4.15
Golf Ball Flight Distances for Four Compression/Evening Combinations

		Evening (B)			
		1		*2*	
		180	192	196	180
		193	190	192	195
	80	197	182	191	197
		189	192	194	192
		187	179	186	193
Compression (A)					
		180	175	190	185
		185	190	195	167
	100	167	185	180	180
		162	180	170	180
		170	185	180	165

Example 8
(continued)

These data have complete two-way factorial structure. The factor Evening is not really of primary interest. Rather, it is a blocking factor, its levels creating homogeneous environments in which to compare 80 and 100 compression flight distances. Figure 4.24 is a graphic using boxplots to represent the four samples and emphasizing the factorial structure.

Calculating sample means corresponding to the four cells in Table 4.15 and then finding fitted effects is straightforward. Table 4.16 displays cell, row, column, and grand average means. And based on those values,

$$a_1 = 189.85 - 184.20 = 5.65 \text{ yards}$$
$$a_2 = 178.55 - 184.20 = -5.65 \text{ yards}$$
$$b_1 = 183.00 - 184.20 = -1.20 \text{ yards}$$
$$b_2 = 185.40 - 184.20 = 1.20 \text{ yards}$$
$$ab_{11} = 188.1 - (184.20 + 5.65 + (-1.20)) = -.55 \text{ yards}$$
$$ab_{12} = 191.6 - (184.20 + 5.65 + 1.20) = .55 \text{ yards}$$
$$ab_{21} = 177.9 - (184.20 + (-5.65) + (-1.20)) = .55 \text{ yards}$$
$$ab_{22} = 179.2 - (184.20 + (-5.65) + 1.20) = -.55 \text{ yards}$$

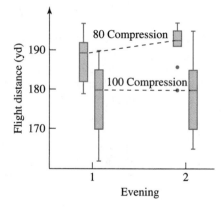

Figure 4.24 Golf ball flight distance boxplots for four combinations of Compression and Evening

Table 4.16
Cell, Row, Column, and Grand Average Means for the Golf Ball Flight Data

		Evening (B)		
		1	*2*	
Compression (A)	*80*	$\bar{y}_{11} = 188.1$	$\bar{y}_{12} = 191.6$	189.85
	100	$\bar{y}_{21} = 177.9$	$\bar{y}_{22} = 179.2$	178.55
		183.00	185.40	184.20

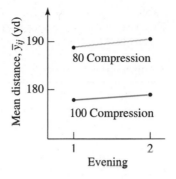

Figure 4.25   Interaction plot for the golf ball flight data

The fitted effects indicate that most of the differences in the cell means in Table 4.16 are understandable in terms of differences between 80 and 100 compression balls. The effect of differences between evenings appears to be on the order of one-fourth the size of the effect of differences between ball compressions. Further, the pattern of flight distances across the two compressions changed relatively little from evening to evening. These facts are portrayed graphically in the interaction plot of Figure 4.25.

The story told by the fitted effects in this example probably agrees with most readers' intuition. There is little reason a priori to expect the relative behaviors of 80 and 100 compression flight distances to change much from evening to evening. But there is slightly more reason to expect the distances to be longer overall on some nights than on others.

It is worth investigating whether the data in Table 4.15 allow the simplest

"Compression effects only"

description, or require the somewhat more complicated

"Compression effects and Evening effects but no interactions"

description, or really demand to be described in terms of

"Compression, Evening, and interaction effects"

To do so, fitted responses are first calculated corresponding to the three different possible corresponding relationships

$$y \approx \mu + \alpha_i \tag{4.28}$$

$$y \approx \mu + \alpha_i + \beta_j \tag{4.29}$$

$$y \approx \mu + \alpha_i + \beta_j + \alpha\beta_{ij} \tag{4.30}$$

Example 8
(*continued*)

Table 4.17
Fitted Responses Corresponding to Equations (4.28), (4.29), and (4.30)

Compression	Evening	For (4.28) $\bar{y}_{..} + a_i = \bar{y}_{i.}$	For (4.29) $\bar{y}_{..} + a_i + b_j$	For (4.30) $\bar{y}_{..} + a_i + b_j + ab_{ij} = \bar{y}_{ij}$
80	1	189.85	188.65	188.10
100	1	178.55	177.35	177.90
80	2	189.85	191.05	191.60
100	2	178.55	179.75	179.20

These are generated using the fitted effects. They are collected in Table 4.17 (not surprisingly, the first and third sets of fitted responses are, respectively, row average and cell means).

Residuals $e = y - \hat{y}$ for fitting the three equations (4.28), (4.29), and (4.30) are obtained by subtracting the appropriate entries in, respectively, the third, fourth, or fifth column of Table 4.17 from each of the data values listed in Table 4.15. For example, 40 residuals for the fitting of the "A main effects only" equation (4.28) would be obtained by subtracting 189.85 from every entry in the upper left cell of Table 4.15, subtracting 178.55 from every entry in the lower left cell, 189.85 from every entry in the upper right cell, and 178.55 from every entry in the lower right cell.

Figure 4.26 provides normal plots of the residuals from the fitting of the three equations (4.28), (4.29), and (4.30). None of the normal plots is especially linear, but at the same time, none of them is grossly nonlinear either. In particular, the first two, corresponding to simplified versions of relationship 4.26, are not significantly worse than the last one, which corresponds to the use of all fitted effects (both main effects and interactions). From the limited viewpoint of producing residuals with an approximately bell-shaped distribution, the fitting of any of the three equations (4.28), (4.29), and (4.30) would appear approximately equally effective.

The calculation of R^2 values for equations (4.28), (4.29), and (4.30) proceeds as follows. First, since the grand average of all 40 flight distances is $\bar{y} = 184.2$ yards (which in this case also turns out to be $\bar{y}_{..}$),

$$\sum(y - \bar{y})^2 = (180 - 184.2)^2 + \cdots + (179 - 184.2)^2$$
$$+ (180 - 184.2)^2 + \cdots + (185 - 184.2)^2$$
$$+ (196 - 184.2)^2 + \cdots + (193 - 184.2)^2$$
$$+ (190 - 184.2)^2 + \cdots + (165 - 184.2)^2$$
$$= 3,492.4$$

(This value can easily be obtained on a pocket calculator by using 39 ($= 40 - 1 = n - 1$) times the sample variance of all 40 flight distances.) Then $\sum(y - \hat{y})^2$

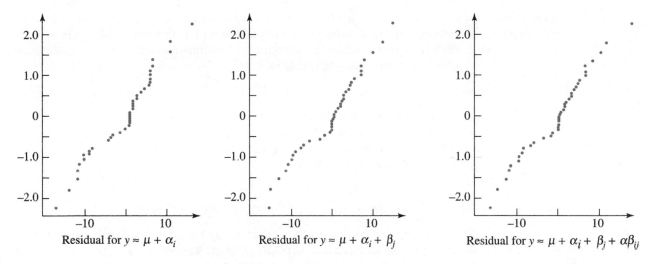

Figure 4.26   Normal plots of residuals from three different equations fitted to the golf data

values for the three equations are obtained as the sums of the squared residuals. For example, using Tables 4.15 and 4.17, for equation (4.29),

$$\sum(y - \hat{y})^2 = (180 - 188.65)^2 + \cdots + (179 - 188.65)^2$$
$$+ (180 - 177.35)^2 + \cdots + (185 - 177.35)^2$$
$$+ (196 - 191.05)^2 + \cdots + (193 - 191.05)^2$$
$$+ (190 - 179.75)^2 + \cdots + (165 - 179.75)^2$$
$$= 2,157.90$$

Finally, equation (4.27) is used. Table 4.18 gives the three values of R^2.

The story told by the R^2 values is consistent with everything else that's been said in this example. None of the values is terribly big, which is consistent with the large within-sample variation in flight distances evident in Figure 4.24. But

Table 4.18
R^2 Values for Fitting Equations
(4.28), (4.29), and (4.30) to
Gronberg's Data

Equation	R^2
$y \approx \mu + \alpha_i$	.366
$y \approx \mu + \alpha_i + \beta_j$	.382
$y \approx \mu + \alpha_i + \beta_j + \alpha\beta_{ij}$	.386

Example 8
(*continued*)

considering A (Compression) main effects does account for some of the observed variation in flight distance, and the addition of B (Evening) main effects adds slightly to the variation accounted for. Introducing interactions into consideration adds little additional accounting power.

The computations in Example 8 are straightforward but tedious. The kind of software used to produce Printout 6 typically allows for the painless fitting of simplified relationships like (4.28), (4.29), and (4.30) and computation (and later plotting) of the associated residuals.

4.3.3 Fitted Effects for Three-Way (and Higher) Factorials

The reasoning that has been applied to two-way factorial data is naturally generalized to complete factorial data structures that are three-way and higher. First, fitted main effects and various kinds of interactions are computed. Then one hopes to discover that a data set can be adequately described in terms of a few of these that are interpretable when taken as a group. This subsection shows how this is carried out for 3-factor situations. Once the pattern has been made clear, the reader can carry it out for situations involving more than three factors, working by analogy.

In order to deal with three-way factorial data, yet more notation is needed. Unfortunately, this involves triple subscripts. For factor A having I levels, factor B having J levels, and factor C having K levels, the following notation will be used:

Notation for sample means and their averages (for three-way factorial data)

$\bar{y}_{ijk}$ = the sample mean response when factor A is at level i, factor B is at level j, and factor C is at level k

$$\bar{y}_{...} = \frac{1}{IJK} \sum_{i,j,k} \bar{y}_{ijk}$$

= the grand average sample mean

$$\bar{y}_{i..} = \frac{1}{JK} \sum_{j,k} \bar{y}_{ijk}$$

= the average sample mean when factor A is at level i

$$\bar{y}_{.j.} = \frac{1}{IK} \sum_{i,k} \bar{y}_{ijk}$$

= the average sample mean when factor B is at level j

$$\bar{y}_{..k} = \frac{1}{IJ} \sum_{i,j} \bar{y}_{ijk}$$

= the average sample mean when factor C is at level k

$$\bar{y}_{ij.} = \frac{1}{K} \sum_k \bar{y}_{ijk}$$

= the average sample mean when factor A is at level i and factor B is at level j

$$\bar{y}_{i.k} = \frac{1}{J} \sum_j \bar{y}_{ijk}$$

= the average sample mean when factor A is at level i and factor C is at level k

$$\bar{y}_{.jk} = \frac{1}{I} \sum_i \bar{y}_{ijk}$$

= the average sample mean when factor B is at level j and factor C is at level k

In these expressions, where a subscript is used as an index of summation, the summation is assumed to extend over all of its I, J, or K possible values.

It is most natural to think of the means from a 3-factor study laid out in three dimensions. Figure 4.27 illustrates this general situation, and the next example employs another common three-dimensional display in a 2^3 context.

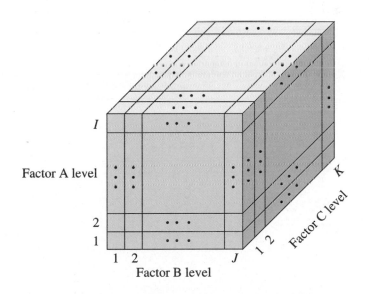

Figure 4.27   *IJK* cells in a three-dimensional table

Example 9

A 2^3 Factorial Experiment on the Strength of a Composite Material

In his article "Application of Two-Cubed Factorial Designs to Process Studies" (*ASQC Technical Supplement Experiments in Industry*, 1985), G. Kinzer discusses a successful 3-factor industrial experiment.

The strength of a proprietary composite material was thought to be related to three process variables, as indicated in Table 4.19. Five specimens were produced under each of the $2^3 = 8$ combinations of factor levels, and their moduli of rupture were measured (in psi) and averaged to produce the means in Table 4.20. (There were also apparently 10 specimens made with an autoclave temperature of 315°F, an autoclave time of 8 hr, and a time span of 8 hr, but this will be ignored for present purposes.)

Cube plot for displaying 2^3 means

A helpful display of these means can be made using the corners of a cube, as in Figure 4.28. Using this three-dimensional picture, one can think of average sample means as averages of $\bar{y}_{ijk}$'s sharing a face or edge of the cube.

Table 4.19
Levels of Three Process Variables in a 2^3 Study of Material Strength

Factor	Process Variable	Level 1	Level 2
A	Autoclave temperature	300°F	330°F
B	Autoclave time	4 hr	12 hr
C	Time span (between product formation and autoclaving)	4 hr	12 hr

Table 4.20
Sample Mean Strengths for 2^3 Treatment Combinations

i, Factor A Level	j, Factor B Level	k, Factor C Level	$\bar{y}_{ijk}$, Sample Mean Strength (psi)
1	1	1	1520
2	1	1	2450
1	2	1	2340
2	2	1	2900
1	1	2	1670
2	1	2	2540
1	2	2	2230
2	2	2	3230

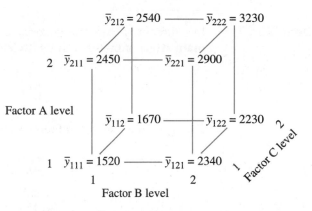

Figure 4.28 2^3 sample mean strengths displayed on a cube plot

For example,

$$\bar{y}_{1..} = \frac{1}{2 \cdot 2}(1520 + 2340 + 1670 + 2230) = 1940 \text{ psi}$$

is the average mean on the bottom face, while

$$\bar{y}_{11.} = \frac{1}{2}(1520 + 1670) = 1595 \text{ psi}$$

is the average mean on the lower left edge. For future reference, all of the average sample means are collected here:

$$
\begin{array}{ll}
\bar{y}_{...} = 2360 \text{ psi} & \\
\bar{y}_{1..} = 1940 \text{ psi} & \bar{y}_{2..} = 2780 \text{ psi} \\
\bar{y}_{.1.} = 2045 \text{ psi} & \bar{y}_{.2.} = 2675 \text{ psi} \\
\bar{y}_{..1} = 2302.5 \text{ psi} & \bar{y}_{..2} = 2417.5 \text{ psi} \\
\bar{y}_{11.} = 1595 \text{ psi} & \bar{y}_{12.} = 2285 \text{ psi} \\
\bar{y}_{21.} = 2495 \text{ psi} & \bar{y}_{22.} = 3065 \text{ psi} \\
\bar{y}_{1.1} = 1930 \text{ psi} & \bar{y}_{1.2} = 1950 \text{ psi} \\
\bar{y}_{2.1} = 2675 \text{ psi} & \bar{y}_{2.2} = 2885 \text{ psi} \\
\bar{y}_{.11} = 1985 \text{ psi} & \bar{y}_{.12} = 2105 \text{ psi} \\
\bar{y}_{.21} = 2620 \text{ psi} & \bar{y}_{.22} = 2730 \text{ psi} \\
\end{array}
$$

Analogy with Definition 5 provides definitions of fitted main effects in a 3-factor study as the differences between factor-level average means and the grand average mean.

Definition 7

In a three-way complete factorial study with factors A, B, and C, the **fitted main effect of factor A at its ith level** is

$$a_i = \bar{y}_{i..} - \bar{y}_{...}$$

The **fitted main effect of factor B at its jth level** is

$$b_j = \bar{y}_{.j.} - \bar{y}_{...}$$

And the **fitted main effect of factor C at its kth level** is

$$c_k = \bar{y}_{..k} - \bar{y}_{...}$$

Using the geometrical representation of factor-level combinations given in Figure 4.28, these fitted effects are averages of $\bar{y}_{ijk}$'s along planes (parallel to one set of faces of the rectangular solid) minus the grand average sample mean.

Next, analogy with Definition 6 produces definitions of fitted two-way interactions in a 3-factor study.

Definition 8

In a three-way complete factorial study with factors A, B, and C, the **fitted 2-factor interaction of factor A at its ith level** and **factor B at its jth level** is

$$ab_{ij} = \bar{y}_{ij.} - (\bar{y}_{...} + a_i + b_j)$$

the **fitted 2-factor interaction of factor A at its ith level** and **factor C at its kth level** is

$$ac_{ik} = \bar{y}_{i.k} - (\bar{y}_{...} + a_i + c_k)$$

and the **fitted 2-factor interaction of factor B at its jth level** and **factor C at its kth level** is

$$bc_{jk} = \bar{y}_{.jk} - (\bar{y}_{...} + b_j + c_k)$$

Interpreting two-way interactions in a three-way study

These fitted 2-factor interactions can be thought of in two equivalent ways:

1. as what one gets as fitted interactions upon averaging across all levels of the factor that is not under consideration to obtain a single two-way table of (average) means and then calculating as per Definition 6 (page 169);

2. as what one gets as averages, across all levels of the factor not under consideration, of the fitted two-factor interactions calculated as per Definition 6, one level of the excluded factor at a time.

Example 9
(continued)

To illustrate the meaning of Definitions 7 and 8, return to the composite material strength study. For example, the fitted A main effects are

$$a_1 = \bar{y}_{1..} - \bar{y}_{...} = 1940 - 2360 = -420 \text{ psi}$$
$$a_2 = \bar{y}_{2..} - \bar{y}_{...} = 2780 - 2360 = 420 \text{ psi}$$

And the fitted AB 2-factor interaction for levels 1 of A and 1 of B is

$$ab_{11} = \bar{y}_{11.} - (\bar{y}_{...} + a_1 + b_1) = 1595 - (2360 + (-420) + (2045 - 2360))$$
$$= -30 \text{ psi}$$

The entire set of fitted effects for the means of Table 4.20 is as follows.

$a_1 = -420$ psi	$b_1 = -315$ psi	$c_1 = -57.5$ psi
$a_2 = 420$ psi	$b_2 = 315$ psi	$c_2 = 57.5$ psi
$ab_{11} = -30$ psi	$ac_{11} = 47.5$ psi	$bc_{11} = -2.5$ psi
$ab_{12} = 30$ psi	$ac_{12} = -47.5$ psi	$bc_{12} = 2.5$ psi
$ab_{21} = 30$ psi	$ac_{21} = -47.5$ psi	$bc_{21} = 2.5$ psi
$ab_{22} = -30$ psi	$ac_{22} = 47.5$ psi	$bc_{22} = -2.5$ psi

Interpretation of three-way interactions

Remember equation (4.25) (page 171). It says that in 2-factor studies, the fitted grand mean, main effects, and two-factor interactions completely describe a factorial set of sample means. Such is not the case in three-factor studies. Instead, a new possibility arises: *3-factor interaction*. Roughly speaking, the fitted three-factor interactions in a 3-factor study measure how much pattern the combination means carry that is not explainable in terms of the factors A, B, and C acting separately and in pairs.

Definition 9

In a three-way complete factorial study with factors A, B, and C, **the fitted 3-factor interaction of A at its *i*th level, B at its *j*th level, and C at its *k*th level** is

$$abc_{ijk} = \bar{y}_{ijk} - (\bar{y}_{...} + a_i + b_j + c_k + ab_{ij} + ac_{ik} + bc_{jk})$$

Example 9
(continued)

To illustrate the meaning of Definition 9, consider again the composite material study. Using the previously calculated fitted main effects and 2-factor interactions,

$$abc_{111} = 1520 - (2360 + (-420) + (-315) + (-57.5) + (-30)$$
$$+ 47.5 + (-2.5)) = -62.5 \text{psi}$$

Similar calculations can be made to verify that the entire set of 3-factor interactions for the means of Table 4.20 is as follows:

$$\begin{array}{ll}
abc_{111} = -62.5 \text{ psi} & abc_{211} = 62.5 \text{ psi} \\
abc_{121} = 62.5 \text{ psi} & abc_{221} = -62.5 \text{ psi} \\
abc_{112} = 62.5 \text{ psi} & abc_{212} = -62.5 \text{ psi} \\
abc_{122} = -62.5 \text{ psi} & abc_{222} = 62.5 \text{ psi}
\end{array}$$

Main effects and 2-factor interactions are more easily interpreted than 3-factor interactions. One insight into their meaning was given immediately before Definition 9. Another is the following. If at the different levels of (say) factor C, the fitted AB interactions are calculated and the fitted AB interactions (the pattern of parallelism or nonparallelism) are essentially the same on all levels of C, then the 3-factor interactions are small (near 0). Otherwise, large 3-factor interactions allow the pattern of AB interaction to change, from one level of C to another.

A second interpretation of three-way interactions

4.3.4   Simpler Descriptions of Some Three-Way Data Sets

Rewriting the equation in Definition 9,

$$\bar{y}_{ijk} = \bar{y}_{...} + a_i + b_j + c_k + ab_{ij} + ac_{ik} + bc_{jk} + abc_{ijk} \qquad \textbf{(4.31)}$$

This is a breakdown of the combination sample means into somewhat interpretable pieces, corresponding to an overall effect, the factors acting separately, the factors acting in pairs, and the factors acting jointly. Display (4.31) may be thought of as a fitted version of an approximate relationship

$$y \approx \mu + \alpha_i + \beta_j + \gamma_k + \alpha\beta_{ij} + \alpha\gamma_{ik} + \beta\gamma_{jk} + \alpha\beta\gamma_{ijk} \qquad \textbf{(4.32)}$$

When beginning the analysis of three-way factorial data, one hopes to discover a simplified version of equation (4.32) that is both interpretable and an adequate description of the data. (Indeed, if it is not possible to do so, little is gained by using the factorial breakdown rather than simply treating the data in question as *IJK unstructured* samples.)

As was the case earlier with two-way factorial data, the process of fitting a simplified version of display (4.32) via least squares is, in general, unfortunately somewhat complicated. But *when all sample sizes are equal* (i.e., the data are balanced), the fitting process can be accomplished by simply adding appropriate fitted effects defined in Definitions 7, 8, and 9. Then the fitted responses lead to residuals that can be used in residual plotting and the calculation of R^2.

Example 9
(*continued*)

Looking over the magnitudes of the fitted effects for Kinzer's composite material strength study, the A and B main effects clearly dwarf the others, suggesting the possibility that the relationship

$$y \approx \mu + \alpha_i + \beta_j \tag{4.33}$$

could be used as a description of the physical system. This relationship doesn't involve factor C at all (either by itself or in combination with A or B) and indicates that responses for a particular AB combination will be comparable for both time spans studied. Further, the fact that display (4.33) doesn't include the $\alpha\beta_{ij}$ term says that factors A and B act on product strength separately, so that their levels can be chosen independently. In geometrical terms corresponding to the cube plot in Figure 4.28, display (4.33) means that observations from the cube's back face will be comparable to corresponding ones on the front face and that parallelism will prevail on both the front and back faces.

Kinzer's article gives only $\bar{y}_{ijk}$ values, not raw data, so a residual analysis and calculation of R^2 are not possible. But because of the balanced nature of the original data set, fitted values are easily obtained. For example, with factor A at level 1 and B at level 1, using the simplified relationship (4.33) and the fitted main effects found earlier produces the fitted value

$$\hat{y} = \bar{y}_{...} + a_1 + b_1 = 2360 + (-420) + (-315) = 1625 \text{ psi}$$

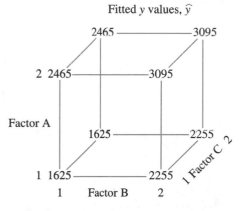

Fitted y values, $\hat{y}$

Figure 4.29   Eight fitted responses for relationship (4.33) and the composite strength study

Example 9
(continued)

All eight fitted values corresponding to equation (4.33) are shown geometrically in Figure 4.29. The fitted values given in the figure might be combined with product requirements and cost information to allow a process engineer to make sound decisions about autoclave temperature, autoclave time, and time span.

In Example 9, the simplified version of display (4.32) was especially interpretable because it involved only main effects. But sometimes even versions of relation (4.32) involving interactions can draw attention to what is going on in a data set.

Example 10

Interactions in a 3-Factor Paper Airplane Experiment

Schmittenberg and Riesterer studied the effects of three factors, each at two levels, on flight distance of paper airplanes. The factors were Plane Design (A) (design 1 versus design 2), Plane Size (B) (large versus small), and Paper Type (C) (heavy versus light). The means of flight distances they obtained for 15 flights of each of the $8 = 2 \times 2 \times 2$ types of planes are given in Figure 4.30.

Calculate the fitted effects corresponding to the $\bar{y}_{ijk}$'s given in Figure 4.30 "by hand." (Printout 7 also gives the fitted effects.) By far the biggest fitted effects (more than three times the size of any others) are the AC interactions. This makes perfect sense. The strongest message in Figure 4.30 is that plane design 1 should be made with light paper and plane design 2 with heavy paper. This is a perfect example of a strong 2-factor interaction in a 3-factor study (where, incidentally, the fitted 3-factor interactions are roughly $\frac{1}{4}$ the size of any other fitted effects). Any simplified version of display (4.32) used to represent this situation would certainly have to include the $\alpha\gamma_{ik}$ term.

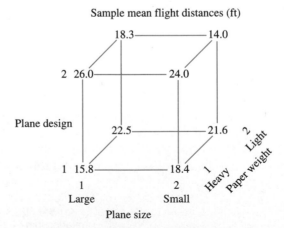

Figure 4.30 2^3 sample mean flight distances displayed on the corners of a cube

Printout 7 Calculation of Fitted Effects for the Airplane Experiment

---

```
General Linear Model


Factor      Type Levels Values
design      fixed    2     1 2
size        fixed    2     1 2
paper       fixed    2     1 2

Analysis of Variance for mean dis, using Adjusted SS for Tests

Source              DF      Seq SS     Adj SS     Adj MS      F       P
design               1       2.000      2.000      2.000     **
size                 1       2.645      2.645      2.645     **
paper                1       7.605      7.605      7.605     **
design*size          1       8.000      8.000      8.000     **
design*paper         1      95.220     95.220     95.220     **
size*paper           1       4.205      4.205      4.205     **
design*size*paper    1       0.180      0.180      0.180     **
Error                0       0.000      0.000      0.000
Total                7     119.855

** Denominator of F-test is zero.

Term                   Coef       StDev        T        P
Constant            20.0750      0.0000        *        *
design
1                   -0.500000    0.000000      *        *
size
1                    0.575000    0.000000      *        *
paper
1                    0.975000    0.000000      *        *
design*size
1      1             -1.00000     0.00000       *        *
design*paper
1      1             -3.45000     0.00000       *        *
size*paper
1      1             -0.725000    0.000000      *        *
design*size*paper
1      1      1      -0.150000    0.000000      *        *
```

4.3.5 Special Devices for 2^p Studies

All of the discussion in this section has been general, in the sense that any value has been permissible for the number of levels for a factor. In particular, all of the definitions of fitted effects in the section work as well for $3 \times 5 \times 7$ studies as they do for $2 \times 2 \times 2$ studies. But from here on in the section, attention will be restricted to 2^p data structures.

Special 2^p factorial notation

Restricting attention to two-level factors affords several conveniences. One is notational. It is possible to reduce the clutter caused by the multiple subscript "ijk" notation, as follows. One level of each factor is designated as a "high" (or "+") level and the other as a "low" (or "−") level. Then the 2^p factorial combinations are labeled with letters corresponding to those factors appearing in the combination at

Table 4.21
Shorthand Names for the 2^3 Factorial Treatment Combinations

Level of Factor A	Level of Factor B	Level of Factor C	Combination Name
1	1	1	(1)
2	1	1	a
1	2	1	b
2	2	1	ab
1	1	2	c
2	1	2	ac
1	2	2	bc
2	2	2	abc

their high levels. For example, if level 2 of each of factors A, B, and C is designated the high level, shorthand names for the $2^3 = 8$ different ABC combinations are as given in Table 4.21. Using these names, for example, $\bar{y}_a$ can stand for a sample mean where factor A is at its high (or second) level and all other factors are at their low (or first) levels.

Special relationship between 2^p effects of a given type

A second convenience special to two-level factorial data structures is the fact that all effects of a given type have the same absolute value. This has already been illustrated in Example 9. For example, looking back, for the data of Table 4.20,

$$a_2 = 420 = -(-420) = -a_1$$

and

$$bc_{22} = -2.5 = bc_{11} = -bc_{12} = -bc_{21}$$

This is always the case for fitted effects in 2^p factorials. In fact, if two fitted effects of the same type are such that an even number of $1 \to 2$ or $2 \to 1$ subscript changes are required to get the second from the first, the fitted effects are equal (e.g., $bc_{22} = bc_{11}$). If an odd number are required, then the second fitted effect is -1 times the first (e.g., $bc_{12} = -bc_{22}$). This fact is so useful because one needs only to do the arithmetic necessary to find one fitted effect of each type and then choose appropriate signs to get all others of that type.

A statistician named Frank Yates is credited with discovering an efficient, mechanical way of generating one fitted effect of each type for a 2^p study. His method is easy to implement "by hand" and produces fitted effects with all "2" subscripts (i.e., corresponding to the "all factors at their high level" combination). The **Yates algorithm** consists of the following steps.

The Yates algorithm for computing fitted 2^p factorial effects

Step 1   Write down the 2^p sample means in a column in what is called **Yates standard order**. Standard order is easily remembered by beginning with (1) and a, then multiplying these two names (algebraically) by b to get b and ab, then multiplying these four names by c to get c, ac, bc, abc, etc.

Step 2   Make up another column of numbers by first adding and then subtracting (first from second) the entries in the previous column in pairs.

Step 3   Follow step 2 a total of p times, and then make up a final column by dividing the entries in the last column by the value 2^p.

The last column (made via step 3) gives fitted effects (all factors at level 2), again in standard order.

Example 9 (*continued*)	Table 4.22 shows the use of the Yates algorithm to calculate fitted effects for the 2^3 composite material study. The entries in the final column of this table are, of course, exactly as listed earlier, and the rest of the fitted effects are easily obtained via appropriate sign changes. This final column is an extremely concise summary of the fitted effects, which quickly reveals which types of fitted effects are larger than others.

Table 4.22
The Yates Algorithm Applied to the Means of Table 4.20

Combination	$\bar{y}$	Cycle 1	Cycle 2	Cycle 3	Cycle 3 $\div$ 8
(1)	1520	3970	9210	18,880	$2360 = \bar{y}_{...}$
a	2450	5240	9670	3,360	$420 = a_2$
b	2340	4210	1490	2,520	$315 = b_2$
ab	2900	5460	1870	−240	$-30 = ab_{22}$
c	1670	930	1270	460	$57.5 = c_2$
ac	2540	560	1250	380	$47.5 = ac_{22}$
bc	2230	870	−370	−20	$-2.5 = bc_{22}$
abc	3230	1000	130	500	$62.5 = abc_{222}$

The Yates algorithm is useful beyond finding fitted effects. For balanced data sets, it is also possible to modify it slightly to find fitted responses, $\hat{y}$, corresponding to a simplified version of a relation like display (4.32). First, the desired (all factors at their high level) fitted effects (using 0's for those types not considered) are written down in reverse standard order. Then, by applying p cycles of the Yates additions and subtractions, the fitted values, $\hat{y}$, are obtained, listed in reverse standard order. (Note that no final division is required in this **reverse Yates algorithm**.)

The reverse Yates algorithm and easy computation of fitted responses

Example 9
(*continued*)

Consider fitting the relationship (4.33) to the balanced data set that led to the means of Table 4.20 via the reverse Yates algorithm. Table 4.23 gives the details. The fitted values in the final column are exactly as shown earlier in Figure 4.29.

Table 4.23
The Reverse Yates Algorithm Applied to Fitting the "A and B Main Effects Only" Equation (4.33) to the Data of Table 4.20

Fitted Effect	Value	Cycle 1	Cycle 2	Cycle 3 ($\hat{y}$)
abc_{222}	0	0	0	$3095 = \hat{y}_{abc}$
bc_{22}	0	0	3095	$2255 = \hat{y}_{bc}$
ac_{22}	0	315	0	$2465 = \hat{y}_{ac}$
c_2	0	2780	2255	$1625 = \hat{y}_{c}$
ab_{22}	0	0	0	$3095 = \hat{y}_{ab}$
b_2	315	0	2465	$2255 = \hat{y}_{b}$
a_2	420	315	0	$2465 = \hat{y}_{a}$
$\bar{y}_{...}$	2360	1940	1625	$1625 = \hat{y}_{(1)}$

The importance of two-level factorials

The restriction to two-level factors that makes these notational and computational devices possible is not as specialized as it may at first seem. When an engineer wishes to study the effects of a large number of factors, even 2^p will be a large number of conditions to investigate. If more than two levels of factors are considered, the sheer size of a complete factorial study quickly becomes unmanageable. Recognizing this, two-level studies are often used for screening to identify a few (from many) process variables for subsequent study at more levels on the basis of their large perceived effects in the screening study. So this 2^p material is in fact quite important to the practice of engineering statistics.

Section 3 Exercises

1. Since the data of Exercise 2 of Section 4.2 have complete factorial structure, it is possible (at least temporarily) to ignore the fact that the two experimental factors are basically quantitative and make a factorial analysis of the data.
   (a) Compute all fitted factorial main effects and interactions for the data of Exercise 2 of Section 4.2. Interpret the relative sizes of these fitted effects, using a interaction plot like Figure 4.22 to facilitate your discussion.
   (b) Compute nine fitted responses for the "main effects only" explanation of y, $y \approx \mu + \alpha_i + \beta_j$.

   Plot these versus level of the NaOH variable, connecting fitted values having the same level of the Time variable with line segments, as in Figure 4.23. Discuss how this plot compares to the two plots of fitted y versus x_1 made in Exercise 2 of Section 4.2.
   (c) Use the fitted values computed in (b) and find a value of R^2 appropriate to the "main effects only" representation of y. How does it compare to the R^2 values from multiple regressions? Also use the fitted values to compute

residuals for this "main effects only" representation. Plot these (versus level of NaOH, level of Time, and $\hat{y}$, and in normal plot form). What do they indicate about the present "no interaction" explanation of specific area?

2. Bachman, Herzberg, and Rich conducted a 2^3 factorial study of fluid flow through thin tubes. They measured the time required for the liquid level in a fluid holding tank to drop from 4 in. to 2 in. for two drain tube diameters and two fluid types. Two different technicians did the measuring. Their data are as follows:

Technician	Diameter (in.)	Fluid	Time (sec)
1	.188	water	21.12, 21.11, 20.80
2	.188	water	21.82, 21.87, 21.78
1	.314	water	6.06, 6.04, 5.92
2	.314	water	6.09, 5.91, 6.01
1	.188	ethylene glycol	51.25, 46.03, 46.09
2	.188	ethylene glycol	45.61, 47.00, 50.71
1	.314	ethylene glycol	7.85, 7.91, 7.97
2	.314	ethylene glycol	7.73, 8.01, 8.32

(a) Compute (using the Yates algorithm or otherwise) the values of all the fitted main effects, two-way interactions, and three-way interactions for these data. Do any simple interpretations of these suggest themselves?

(b) The students actually had some physical theory suggesting that the log of the drain time might be a more convenient response variable than the raw time. Take the logs of the y's and recompute the factorial effects. Does an interpretation of this system in terms of only main effects seem more plausible on the log scale than on the original scale?

(c) Considering the logged drain times as the responses, find fitted values and residuals for a "Diameter and Fluid main effects only" explanation of these data. Compute R^2 appropriate to such a view and compare it to R^2 that results from using all factorial effects to describe log drain time. Make and interpret appropriate residual plots.

(d) Based on the analysis from (c), what change in log drain time seems to accompany a change from .188 in. diameter to .314 in. diameter? What does this translate to in terms of raw drain time? Physical theory suggests that raw time is inversely proportional to the fourth power of drain tube radius. Does your answer here seem compatible with that theory? Why or why not?

3. When analyzing a full factorial data set where the factors involved are quantitative, either the surface-fitting technology of Section 4.2 or the factorial analysis material of Section 4.3 can be applied. What practical engineering advantage does the first offer over the second in such cases?

··

4.4 Transformations and Choice of Measurement Scale (*Optional*)

Sections 4.2 and 4.3 are an introduction to one of the main themes of engineering statistical analysis: the discovery and use of simple structure in complicated situations. Sometimes this can be done by reexpressing variables on some other (nonlinear) scales of measurement besides the ones that first come to mind. That is, sometimes simple structure may not be obvious on initial scales of measurement, but may emerge after some or all variables have been transformed. This section presents several examples where transformations are helpful. In the process, some comments about commonly used types of transformations, and more specific reasons for using them, are offered.

4.4.1 Transformations and Single Samples

In Chapter 5, there are a number of standard theoretical distributions. When one of these standard models can be used to describe a response y, all that is known about the model can be brought to bear in making predictions and inferences regarding y. However, when no standard distributional shape can be found to describe y, it may nevertheless be possible to so describe $g(y)$ for some function $g(\cdot)$.

Example 11	**Discovery Times at an Auto Shop**

Elliot, Kibby, and Meyer studied operations at an auto repair shop. They collected some data on what they called the "discovery time" associated with diagnosing what repairs the mechanics were going to recommend to the car owners. Thirty such discovery times (in minutes) are given in Figure 4.31, in the form of a stem-and-leaf plot.

The stem-and-leaf plot shows these data to be somewhat skewed to the right. Many of the most common methods of statistical inference are based on an assumption that a data-generating mechanism will in the long run produce not skewed, but rather symmetrical and bell-shaped data. Therefore, using these methods to draw inferences and make predictions about discovery times at this shop is highly questionable. However, suppose that some transformation could be applied to produce a bell-shaped distribution of transformed discovery times. The standard methods could be used to draw inferences about transformed discovery times, which could then be translated (by undoing the transformation) to inferences about raw discovery times.

One common transformation that has the effect of shortening the right tail of a distribution is the logarithmic transformation, $g(y) = \ln(y)$. To illustrate its use in the present context, normal plots of both discovery times and log discovery times are given in Figure 4.32. These plots indicate that Elliot, Kibby, and Meyer could not have reasonably applied standard methods of inference to the discovery times, but they could have used the methods with log discovery times. The second normal plot is far more linear than the first.

```
0 | 4  3
0 | 6  5  5  5  6  6  8  9  8  8
1 | 4  0  3  0
1 | 7  5  9  5  6  9  5
2 | 0
2 | 9  5  7  9
3 | 2
3 | 6
```

Figure 4.31   Stem-and-leaf plot of discovery times

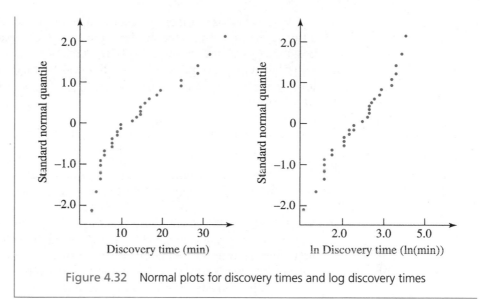

Figure 4.32   Normal plots for discovery times and log discovery times

The logarithmic transformation was useful in the preceding example in reducing the skewness of a response distribution. Some other transformations commonly employed to change the shape of a response distribution in statistical engineering studies are the **power transformations**,

Power transformations

$$g(y) = (y - \gamma)^{\alpha} \tag{4.34}$$

In transformation (4.34), the number γ is often taken as a threshold value, corresponding to a minimum possible response. The number α governs the basic shape of a plot of $g(y)$ versus y. For $\alpha > 1$, transformation (4.34) tends to lengthen the right tail of a distribution for y. For $0 < \alpha < 1$, the transformation tends to shorten the right tail of a distribution for y, the shortening becoming more drastic as α approaches 0 but not as pronounced as that caused by the **logarithmic transformation**

Logarithmic transformation

$$g(y) = \ln(y - \gamma)$$

4.4.2 Transformations and Multiple Samples

Comparing several sets of process conditions is one of the fundamental problems of statistical engineering analysis. It is advantageous to do the comparison on a scale where the samples have comparable variabilities, for at least two reasons. The first is the obvious fact that comparisons then reduce simply to comparisons between response means. Second, standard methods of statistical inference often have well-understood properties only when response variability is comparable for the different sets of conditions.

When response variability is not comparable under different sets of conditions, a transformation can sometimes be applied to all observations to remedy this. This possibility of **transforming to stabilize variance** exists when response variance is roughly a function of response mean. Some theoretical calculations suggest the following guidelines as a place to begin looking for an appropriate variance-stabilizing transformation:

Transformations to stabilize response variance

1. If response standard deviation is approximately proportional to response mean, try a logarithmic transformation.

2. If response standard deviation is approximately proportional to the δ power of the response mean, try transformation (4.34) with $\alpha = 1 - \delta$.

Where several samples (and corresponding $\bar{y}$ and s values) are involved, an empirical way of investigating whether (1) or (2) above might be useful is to plot $\ln(s)$ versus $\ln(\bar{y})$ and see if there is approximate linearity. If so, a slope of roughly 1 makes (1) appropriate, while a slope of $\delta \neq 1$ signals what version of (2) might be helpful.

In addition to this empirical way of identifying a potentially variance-stabilizing transformation, theoretical considerations can sometimes provide guidance. Standard theoretical distributions (like those introduced in Chapter 5) have their own relationships between their (theoretical) means and variances, which can help pick out an appropriate version of (1) or (2) above.

4.4.3   Transformations and Simple Structure in Multifactor Studies

In Section 4.2, Taylor's equation for tool life y in terms of cutting speed x was advantageously reexpressed as a linear equation for $\ln(y)$ in terms of $\ln(x)$. This is just one manifestation of the general fact that many approximate laws of physical science and engineering are **power laws**, expressing one quantity as a product of a constant and powers (some possibly negative) of other quantities. That is, they are of the form

A power law

$$y \approx \alpha x_1^{\beta_1} x_2^{\beta_2} \cdots x_k^{\beta_k} \tag{4.35}$$

Of course, upon taking logarithms in equation (4.35),

$$\ln(y) \approx \ln(\alpha) + \beta_1 \ln(x_1) + \beta_2 \ln(x_2) + \cdots + \beta_k \ln(x_k) \tag{4.36}$$

which immediately suggests the wide usefulness of the logarithmic transformation for both y and x variables in surface-fitting applications involving power laws.

But there is something else in display (4.36) that bears examination: The k functions of the fundamental x variables enter the equation **additively**. In the language of the previous section, there are *no interactions* between the factors whose levels are specified by the variables $x_1, x_2, \ldots, x_k$. This suggests that even in studies involving only seemingly qualitative factors, if a power law for y is at work and the factors act on different fundamental variables x, a logarithmic transformation will tend to create a simple structure. It will do so by eliminating the need for interactions in describing the response.

Example 12

Daniel's Drill Advance Rate Study

In his book *Applications of Statistics to Industrial Experimentation*, Cuthbert Daniel gives an extensive discussion of an unreplicated 2^4 factorial study of the behavior of a new piece of drilling equipment. The response y is a rate of advance of the drill (no units are given), and the experimental factors are Load on the small stone drill (A), Flow Rate through the drill (B), Rotational Speed (C), and Type of Mud used in drilling (D). Daniel's data are given in Table 4.24.

Application of the Yates algorithm to the data in Table 4.24 ($p = 4$ cycles are required, as is division of the results of the last cycle by 2^4) gives the fitted effects:

$$\bar{y}_{\dots} = 6.1550$$

$a_2 = .4563$	$b_2 = 1.6488$	$c_2 = 3.2163$	$d_2 = 1.1425$
$ab_{22} = .0750$	$ac_{22} = .2975$	$ad_{22} = .4213$	
$bc_{22} = .7525$	$bd_{22} = .2213$	$cd_{22} = .7987$	
$abc_{222} = .0838$	$abd_{222} = .2950$	$acd_{222} = .3775$	$bcd_{222} = .0900$
$abcd_{2222} = .2688$			

Looking at the magnitudes of these fitted effects, the candidate relationships

$$y \approx \mu + \beta_j + \gamma_k + \delta_l \tag{4.37}$$

and

$$y \approx \mu + \beta_j + \gamma_k + \delta_l + \beta\gamma_{jk} + \gamma\delta_{kl} \tag{4.38}$$

Table 4.24
Daniel's 2^4 Drill Advance Rate Data

Combination	y	Combination	y
(1)	1.68	d	2.07
a	1.98	ad	2.44
b	3.28	bd	4.09
ab	3.44	abd	4.53
c	4.98	cd	7.77
ac	5.70	acd	9.43
bc	9.97	bcd	11.75
abc	9.07	abcd	16.30

Example 12
(*continued*)

are suggested. (The five largest fitted effects are, in order of decreasing magnitude, the main effects of C, B, and D, and then the two-factor interactions of C with D and B with C.) Fitting equation (4.37) to the balanced data of Table 4.24 produces $R^2 = .875$, and fitting relationship (4.38) produces $R^2 = .948$. But upon closer examination, neither fitted equation turns out to be a very good description of these data.

Figure 4.33 shows a normal plot and a plot against $\hat{y}$ for residuals from a fitted version of equation (4.37). It shows that the fitted version of equation (4.37) produces several disturbingly large residuals and fitted values that are systematically too small for responses that are small and large, but too large for moderate responses. Such a curved plot of residuals versus $\hat{y}$ in general suggests that a nonlinear transformation of y may potentially be effective.

The reader is invited to verify that residual plots for equation (4.38) look even worse than those in Figure 4.33. In particular, it is the bigger responses that are

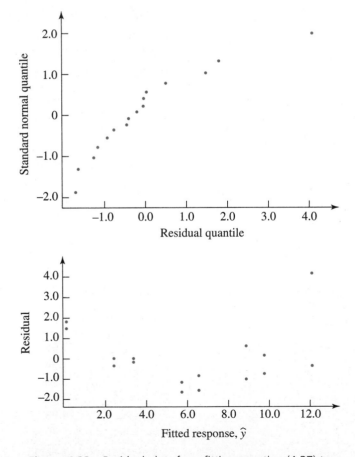

Figure 4.33   Residual plots from fitting equation (4.37) to Daniel's data

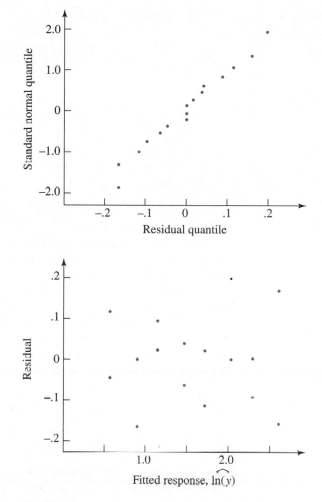

Figure 4.34 Residual plots from fitting equation (4.39)
to Daniel's data

fitted relatively badly by relationship (4.38). This is an unfortunate circumstance, since presumably one study goal is the optimization of response.

But using $y' = \ln(y)$ as a response variable, the situation is much different. The Yates algorithm produces the following fitted effects.

$$\overline{y'}_{....} = 1.5977$$

$$a_2 = .0650 \qquad b_2 = .2900 \qquad c_2 = .5772 \qquad d_2 = .1633$$

$$ab_{22} = -.0172 \qquad ac_{22} = .0052 \qquad ad_{22} = .0334$$

$$bc_{22} = -.0251 \qquad bd_{22} = -.0075 \qquad cd_{22} = .0491$$

$$abc_{222} = .0052 \qquad abd_{222} = .0261 \qquad acd_{222} = .0266 \qquad bcd_{222} = -.0173$$

$$abcd_{2222} = .0193$$

Example 12
(continued)

For the logged drill advance rates, the simple relationship

$$\ln(y) \approx \mu + \beta_j + \gamma_k + \delta_l \qquad (4.39)$$

yields $R^2 = .976$ and absolutely unremarkable residuals. Figure 4.34 shows a normal plot of these and a plot of them against $\widehat{\ln(y)}$.

The point here is that the logarithmic scale appears to be the natural one on which to study drill advance rate. The data can be better described on the log scale without using interaction terms than is possible with interactions on the original scale.

There are sometimes other reasons to consider a logarithmic transformation of a response variable in a multifactor study, besides its potential to produce simple structure. In cases where responses vary over several orders of magnitude, simple curves and surfaces typically don't fit raw y values very well, but they can do a much better job of fitting $\ln(y)$ values (which will usually vary over less than a single order of magnitude). Another potentially helpful property of a log-transformed analysis is that it will never yield physically impossible negative fitted values for a positive variable y. In contrast, an analysis on an original scale of measurement can, rather embarrassingly, do so.

Example 13

A 3^2 Factorial Chemical Process Experiment

The data in Table 4.25 are from an article by Hill and Demler ("More on Planning Experiments to Increase Research Efficiency," *Industrial and Engineering Chemistry*, 1970). The data concern the running of a chemical process where the objective is to achieve high yield y_1 and low filtration time y_2 by choosing settings for Condensation Temperature, x_1, and the Amount of B employed, x_2.

For purposes of this example, consider the second response, filtration time. Fitting the approximate (quadratic) relationship

$$y_2 \approx \beta_0 + \beta_1 x_1 + \beta_2 x_2 + \beta_3 x_1^2 + \beta_4 x_2^2 + \beta_5 x_1 x_2$$

to these data produces the equation

$$\hat{y}_2 = 5179.8 - 56.90x_1 - 146.0x_2 + .1733x_1^2 + 1.222x_2^2 + .6837x_1 x_2 \qquad (4.40)$$

and $R^2 = .866$. Equation (4.40) defines a bowl-shaped surface in three dimensions, which has a minimum at about the set of conditions $x_1 = 103.2°C$ and $x_2 = 30.88$ cc. At first glance, it might seem that the development of equation

Table 4.25
Yields and Filtration Times in a 3^2 Factorial Chemical
Process Study

x_1, Condensation Temperature (°C)	x_2, Amount of B (cc)	y_1, Yield (g)	y_2, Filtration Time (sec)
90	24.4	21.1	150
90	29.3	23.7	10
90	34.2	20.7	8
100	24.4	21.1	35
100	29.3	24.1	8
100	34.2	22.2	7
110	24.4	18.4	18
110	29.3	23.4	8
110	34.2	21.9	10

(4.40) rates as a statistical engineering success story. But there is the embarrassing fact that upon substituting $x_1 = 103.2$ and $x_2 = 30.88$ into equation (4.40), one gets $\hat{y}_2 = -11$ sec, hardly a possible filtration time.

Looking again at the data, it is not hard to see what has gone wrong. The largest response is more than 20 times the smallest. So in order to come close to fitting both the extremely large and more moderate responses, the fitted quadratic surface needs to be very steep—so steep that it is forced to dip below the (x_1, x_2)-plane and produce negative $\hat{y}_2$ values before it can "get turned around" and start to climb again as it moves away from the point of minimum $\hat{y}_2$ toward larger x_1 and x_2.

One cure for the problem of negative predicted filtration times is to use $\ln(y_2)$ as a response variable. Values of $\ln(y_2)$ are given in Table 4.26 to illustrate the moderating effect the logarithm has on the factor of 20 disparity between the largest and smallest filtration times.

Fitting the approximate quadratic relationship

$$\ln(y_2) \approx \beta_0 + \beta_1 x_1 + \beta_2 x_2 + \beta_3 x_1^2 + \beta_4 x_2^2 + \beta_5 x_1 x_2$$

to the $\ln(y_2)$ values produces the equation

$$\widehat{\ln(y_2)} = 99.69 - .8869 x_1 - 3.348 x_2 + .002506 x_1^2 + .03375 x_2^2 + .01196 x_1 x_2$$
$$\textbf{(4.41)}$$

and $R^2 = .975$. Equation (4.41) also represents a bowl-shaped surface in three dimensions and has a minimum approximately at the set of conditions $x_1 = 101.5$°C and $x_2 = 31.6$ cc. The minimum fitted log filtration time is $\widehat{\ln(y_2)} = 1.7582$ ln (sec), which translates to a filtration time of 5.8 sec, a far more sensible value than the negative one given earlier.

Example 13
(*continued*)

Table 4.26
Raw Filtration Times and Corresponding Logged Filtration Times

y_2, Filtration Time (sec)	$\ln(y_2)$, Log Filtration Time (ln(sec))
150	5.0106
10	2.3026
8	2.0794
35	3.5553
8	2.0794
7	1.9459
18	2.8904
8	2.0794
10	2.3026

The taking of logs in this example had two beneficial effects. The first was to cut the ratio of largest response to smallest down to about 2.5 (from over 20), allowing a good fit (as measured by R^2) for a fitted quadratic in two variables, x_1 and x_2. The second was to ensure that minimum predicted filtration time was positive.

Of course, other transformations besides the logarithmic one are also useful in describing the structure of multifactor data sets. Sometimes they are applied to the responses and sometimes to other system variables. As an example of a situation where a power transformation like that specified by equation (4.34) is useful in understanding the structure of a sample of bivariate data, consider the following.

Example 14

Yield Strengths of Copper Deposits and Hall-Petch Theory

In their article "Mechanical Property Testing of Copper Deposits for Printed Circuit Boards" (*Plating and Surface Finishing*, 1988), Lin, Kim, and Weil present some data relating the yield strength of electroless copper deposits to the average grain diameters measured for these deposits. The values in Table 4.27 were deduced from a scatterplot in their paper. These values are plotted in Figure 4.35. They don't seem to promise a simple relationship between grain diameter and yield strength. But in fact, the so called Hall-Petch relationship says that yield strengths of most crystalline materials are proportional to the reciprocal square root of grain diameter. That is, Hall-Petch theory predicts a linear relationship between y and $x^{-.5}$ or between x and y^{-2}. Thus, before trying to further detail the relationship between the two variables, application of transformation (4.34) with $\alpha = -.5$ to x or transformation (4.34) with $\alpha = -2$ to y seems in order. Figure 4.36 shows the partial effectiveness of the reciprocal square root transformation (applied to x) in producing a linear relationship in this context.

Table 4.27
Average Grain Diameters and Yield Strengths for Copper Deposits

x, Average Grain Diameter (μm)	y, Yield Strength (MPa)	x, Average Grain Diameter (μm)	y, Yield Strength (MPa)
.22	330	.48	236
.27	370	.49	224
.33	266	.51	236
.41	270	.90	210

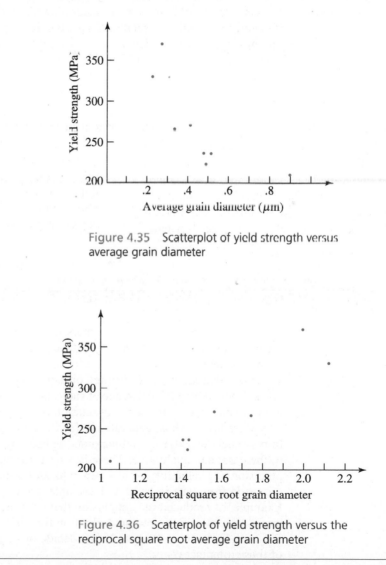

Figure 4.35   Scatterplot of yield strength versus average grain diameter

Figure 4.36   Scatterplot of yield strength versus the reciprocal square root average grain diameter

In the preceding example, a directly applicable and well-known physical theory suggests a natural transformation. Sometimes physical or mathematical considerations that are related to a problem, but do not directly address it, may also suggest some things to try in looking for transformations to produce simple structure. For example, suppose some other property besides yield strength were of interest and thought to be related to grain size. If a relationship with diameter is not obvious, quantifying grain size in terms of cross-sectional area or volume might be considered, and this might lead to squaring or cubing a measured diameter. To take a different example, if some handling characteristic of a car is thought to be related to its speed and a relationship with velocity is not obvious, you might remember that kinetic energy is related to velocity squared, thus being led to square the velocity.

The goal of data transformation   To repeat the main point of this section, the search for appropriate transformations is a quest for measurement scales on which structure is transparent and simple. If the original/untransformed scales are the most natural ones on which to report the findings of a study, the data analysis should be done on the transformed scales but then "untransformed" to state the final results.

Section 4 Exercises

1. What are benefits that can sometimes be derived from transforming data before applying standard statistical techniques?

2. Suppose that a response variable, y, obeys an approximate power law in at least two quantitative variables (say, x_1 and x_2). Will there be important interactions? If the log of y is analyzed instead, will there be important interactions? (In order to make this concrete, you may if you wish consider the relationship $y \approx k x_1^2 x_2^{-3}$. Plot, for at least two different values of x_2, y as a function of x_1. Then plot, for at least two different values of x_2, $\ln(y)$ as a function of x_1. What do these plots show in the way of parallelism?)

4.5   Beyond Descriptive Statistics

We hope that these first four chapters have made you genuinely ready to accept the need for methods of formal statistical inference. Many real data sets have been examined, and many instances of useful structure have been discovered—this in spite of the fact that the structure is often obscured by what might be termed *background noise*. Recognizing the existence of such variation, one realizes that the data in hand are probably not a perfect representation of the population or process from which they were taken. Thus, generalizing from the sample to a broader sphere will have to be somehow hedged. To this point, the hedging has been largely verbal, specific to the case, and qualitative. There is a need for ways to quantitatively express the precision and reliability of any generalizations about a population or process that are made from data in hand. For example, the chemical filtration time problem of Example 13 produced the conclusion that with the temperature set at $101.5°C$ and using 31.6 cc of B, a predicted filtration time is 5.8 sec. But will the next time be 5.8 sec $\pm$ 3 sec or $\pm$ .05 sec? If you decide on $\pm$ *somevalue*, how sure can you be of those tolerances?

In order to quantify precision and reliability for inferences based on samples, the mathematics of probability must be employed. Mathematical descriptions of data generation that are applicable to the original data collection (and any future collection) are necessary. Those mathematical models must explicitly allow for the kind of variation that has been faced in the last two chapters.

The models that are most familiar to engineers do not explicitly account for variation. Rather, they are **deterministic**. For example, Newtonian physics predicts that the displacement of a body in free fall in a time t is exactly $\frac{1}{2}gt^2$. In this statement, there is no explicit allowance for variability. Any observed deviation from the Newtonian predictions is completely unaccounted for. Thus, there is really no logical framework in which to extrapolate from data that don't fit Newtonian predictions exactly.

Stochastic (or probabilistic) models do explicitly incorporate the feature that even measurements generated under the same set of conditions will exhibit variation. Therefore, they can function as descriptions of real-world data collection processes, where many small, unidentifiable causes act to produce the background noise seen in real data sets. Variation is predicted by stochastic or probabilistic models. So they provide a logical framework in which to quantify precision and reliability and to extrapolate from noisy data to contexts larger than the data set in hand.

In the next chapter, some fundamental concepts of probability will be introduced. Then Chapter 6 begins to use probability as a tool in statistical inference.

Section 5 Exercises

1. Read again Section 1.4 and the present one. Then describe in your own words the difference between deterministic and stochastic/probabilistic models. Give an example of a deterministic model that is useful in your field.

Chapter 4 Exercises

1. Nicholson and Bartle studied the effect of the water/cement ratio on 14-day compressive strength for Portland cement concrete. The water/cement ratios (by volume) and compressive strengths of nine concrete specimens are given next.

Water/Cement Ratio, x	14-Day Compressive Strength, y (psi)
.45	2954, 2913, 2923
.50	2743, 2779, 2739
.55	2652, 2607, 2583

(a) Fit a line to the data here via least squares, showing the hand calculations.

(b) Compute the sample correlation between x and y by hand. Interpret this value.

(c) What fraction of the raw variability in y is accounted for in the fitting of a line to the data?

(d) Compute the residuals from your fitted line and make a normal plot of them. Interpret this plot.

(e) What compressive strength would you predict, based on your calculations from (a), for specimens made using a .48 water/cement ratio?

(f) Use a statistical package to find the least squares line, the sample correlation, R^2, and the residuals for this data set.

2. Griffith and Tesdall studied the elapsed time in $\frac{1}{4}$ mile runs of a Camaro Z-28 fitted with different

sizes of carburetor jetting. Their data from six runs of the car follow:

Jetting Size, x	Elapsed Time, y (sec)
66	14.90
68	14.67
70	14.50
72	14.53
74	14.79
76	15.02

(a) What is an obvious weakness in the students' data collection plan?

(b) Fit both a line and a quadratic equation ($y \approx \beta_0 + \beta_1 x + \beta_2 x^2$) to these data via least squares. Plot both of these equations on a scatterplot of the data.

(c) What fractions of the raw variation in elapsed time are accounted for by the two different fitted equations?

(d) Use your fitted quadratic equation to predict an optimal jetting size (allowing fractional sizes).

3. The following are some data taken from "Kinetics of Grain Growth in Powder-formed IN-792: A Nickel-Base Super-alloy" by Huda and Ralph (*Materials Characterization*, September 1990). Three different Temperatures, x_1 (°K), and three different Times, x_2 (min), were used in the heat treating of specimens of a material, and the response

$$y = \text{mean grain diameter } (\mu\text{m})$$

was measured.

Temperature, x_1	Time, x_2	Grain Size, y
1443	20	5
1443	120	6
1443	1320	9
1493	20	14
1493	120	17
1493	1320	25
1543	20	29
1543	120	38
1543	1320	60

(a) What type of data structure did the researchers employ? (Use the terminology of Section 1.2.) What was an obvious weakness in their data collection plan?

(b) Use a regression program to fit the following equations to these data:

$$y \approx \beta_0 + \beta_1 x_1 + \beta_2 x_2$$
$$y \approx \beta_0 + \beta_1 x_1 + \beta_2 \ln(x_2)$$
$$y \approx \beta_0 + \beta_1 x_1 + \beta_2 \ln(x_2) + \beta_3 x_1 \ln(x_2)$$

What are the R^2 values for the three different fitted equations? Compare the three fitted equations in terms of complexity and apparent ability to predict y.

(c) Compute the residuals for the third fitted equation in (b). Plot them against x_1, x_2, and $\hat{y}$. Also normal-plot them. Do any of these plots suggest that the third fitted equation is inadequate as summary of these data? What, if any, possible improvement over the third equation is suggested by these plots?

(d) As a means of understanding the nature of the third fitted equation in (b), make a scatterplot of y vs. x_2 using a logarithmic scale for x_2. On this plot, plot three lines representing $\hat{y}$ as a function of x_2 for the three different values of x_1. Qualitatively, how would a similar plot for the second equation differ from this one?

(e) Using the third equation in (b), what mean grain diameter would you predict for $x_1 = 1500$ and $x_2 = 500$?

(f) It is possible to ignore the fact that the Temperature and Time factors are quantitative and make a factorial analysis of these data. Do so. Begin by making an interaction plot similar to Figure 4.22 for these data. Based on that plot, discuss the apparent relative sizes of the Time and Temperature main effects and the Time × Temperature interactions. Then compute the fitted factorial effects (the fitted main effects and interactions).

4. The article "Cyanoacetamide Accelerators for the Epoxide/Isocyanate Reaction" by Eldin and Renner (*Journal of Applied Polymer Science*, 1990)

reports the results of a 2^3 factorial experiment. Using cyanoacetamides as catalysts for an epoxy/isocyanate reaction, various mechanical properties of a resulting polymer were studied. One of these was

$$y = \text{impact strength (kJ/mm}^2)$$

The three experimental factors employed and their corresponding experimental levels were as follows:

Factor A   Initial Epoxy/Isocyanate Ratio
0.4 (−) vs. 1.2 (+)

Factor B   Flexibilizer Concentration
10 mol % (−) vs. 40 mol % (+)

Factor C   Accelerator Concentration
1/240 mol % (−) vs. 1/30 mol% (+)

(The flexibilizer and accelerator concentrations are relative to the amount of epoxy present initially.) The impact strength data obtained (one observation per combination of levels of the three factors) were as follows:

Combination	y	Combination	y
(1)	6.7	c	6.3
a	11.9	ac	15.1
b	8.5	bc	6.7
ab	16.5	abc	16.4

(a) What is an obvious weakness in the researchers' data collection plan?

(b) Use the Yates algorithm and compute fitted factorial effects corresponding to the "all high" treatment combination (i.e., compute $\bar{y}_{...}$, a_2, b_2, etc.). Interpret these in the context of the original study. (Describe in words which factors and/or combinations of factors appear to have the largest effect(s) on impact strength and interpret the sign or signs.)

(c) Suppose only factor A is judged to be of importance in determining impact strength. What predicted/fitted impact strengths correspond to this judgment? (Find $\hat{y}$ values using the reverse Yates algorithm or otherwise.) Use these eight

values of $\hat{y}$ and compute R^2 for the "A main effects only" description of impact strength. (The formula in Definition 3 works in this context as well as in regression.)

(d) Now recognize that the experimental factors here are quantitative, so methods of curve and surface fitting may be applicable. Fit the equation $y \approx \beta_0 + \beta_1$(epoxy/isocyanate ratio) to the data. What eight values of $\hat{y}$ and value of R^2 accompany this fit?

5. Timp and M-Sidek studied the strength of mechanical pencil lead. They taped pieces of lead to a desk, with various lengths protruding over the edge of the desk. After fitting a small piece of tape on the free end of a lead piece to act as a stop, they loaded it with paper clips until failure. In one part of their study, they tested leads of two different Diameters, used two different Lengths protruding over the edge of the desk, and tested two different lead Hardnesses. That is, they ran a 2^3 factorial study. Their factors and levels were as follows:

Factor A   Diameter   .3 mm (−) vs. .7 mm (+)

Factor B   Length Protruding   3 cm (−) vs.
4.5 cm (+)

Factor C   Hardness   B (−) vs. 2H (+)

and $m = 2$ trials were made at each of the $2^3 = 8$ different sets of conditions. The data the students obtained are given here.

Combination	Number of Clips
(1)	13, 13
a	74, 76
b	9, 10
ab	43, 42
c	16, 15
ac	89, 88
bc	10, 12
abc	54, 55

(a) It appears that analysis of these data in terms of the natural logarithms of the numbers of clips first causing failure is more straightforward than the analysis of the raw numbers of clips. So take natural logs and compute the fitted 2^3 factorial effects. Interpret these. In particular, what (in quantitative terms) does the size of the fitted A main effect say about lead strength? Does lead hardness appear to play a dominant role in determining this kind of breaking strength?

(b) Suppose only the main effects of Diameter are judged to be of importance in determining lead strength. Find a predicted log breaking strength for .7 mm, 2H lead when the length protruding is 4.5 cm. Use this to predict the number of clips required to break such a piece of lead.

(c) What, if any, engineering reasons do you have for expecting the analysis of breaking strength to be more straightforward on the log scale than on the original scale?

6. Ceramic engineering researchers Leigh and Taylor, in their paper "Computer Generated Experimental Designs" (*Ceramic Bulletin*, 1990), studied the packing properties of crushed T-61 tabular alumina powder. The densities of batches of the material were measured under a total of eight different sets of conditions having a 2^3 factorial structure. The following factors and levels were employed in the study:

Factor A   Mesh Size of Powder Particles
            6 mesh $(-)$ vs. 60 mesh $(+)$

Factor B   Volume of Graduated Cylinder
            100 cc $(-)$ vs. 500 cc $(+)$

Factor C   Vibration of Cylinder
            no $(-)$ vs. yes $(+)$

The mean densities (in g/cc) obtained in $m = 5$ determinations for each set of conditions were as follows:

$$\bar{y}_{(1)} = 2.348 \quad \bar{y}_a = 2.080$$
$$\bar{y}_b = 2.298 \quad \bar{y}_{ab} = 1.980$$
$$\bar{y}_c = 2.354 \quad \bar{y}_{ac} = 2.314$$
$$\bar{y}_{bc} = 2.404 \quad \bar{y}_{abc} = 2.374$$

(a) Compute the fitted 2^3 factorial effects (main effects, 2-factor interactions and 3-factor interactions) corresponding to the following set of conditions: 60 mesh, 500 cc, vibrated cylinder.

(b) If your arithmetic for part (a) is correct, you should have found that the largest of the fitted effects (in absolute value) are (respectively) the C main effect, the A main effect, and then the AC 2-factor interaction. (The next largest fitted effect is only about half of the smallest of these, the AC interaction.) Now, suppose you judge these three fitted effects to summarize the main features of the data set. Interpret this data summary (A and C main effects and AC interactions) in the context of this 3-factor study.

(c) Using your fitted effects from (a) and the data summary from (b) (A and C main effects and AC interactions), what fitted response would you have for these conditions: 60 mesh, 500 cc, vibrated cylinder?

(d) Using your fitted effects from (a), what average change in density would you say accompanies the vibration of the graduated cylinder before density determination?

7. The article "An Analysis of Transformations" by Box and Cox (*Journal of the Royal Statistical Society, Series B*, 1964) contains a classical unreplicated 3^3 factorial data set originally taken from an unpublished technical report of Barella and Sust. These researchers studied the behavior of worsted yarns under repeated loading. The response variable was

$$y = \text{the numbers of cycles till failure}$$

for specimens tested with various values of

$$x_1 = \text{length (mm)}$$
$$x_2 = \text{amplitude of the loading cycle (mm)}$$
$$x_3 = \text{load (g)}$$

The researchers' data are given in the accompanying table.

x_1	x_2	x_3	y	x_1	x_2	x_3	y
250	8	40	674	300	9	50	438
250	8	45	370	300	10	40	442
250	8	50	292	300	10	45	332
250	9	40	338	300	10	50	220
250	9	45	266	350	8	40	3,636
250	9	50	210	350	8	45	3,184
250	10	40	170	350	8	50	2,000
250	10	45	118	350	9	40	1,568
250	10	50	90	350	9	45	1,070
300	8	40	1,414	350	9	50	566
300	8	45	1,198	350	10	40	1,140
300	8	50	634	350	10	45	884
300	9	40	1,022	350	10	50	360
300	9	45	620				

(a) To find an equation to represent these data, you might first try to fit multivariable polynomials. Use a regression program and fit a full quadratic equation to these data. That is, fit

$$y \approx \beta_0 + \beta_1 x_1 + \beta_2 x_2 + \beta_3 x_3 + \beta_4 x_1^2 + \beta_5 x_2^2$$
$$+ \beta_6 x_3^2 + \beta_7 x_1 x_2 + \beta_8 x_1 x_3 + \beta_9 x_2 x_3$$

to the data. What fraction of the observed variation in y does it account for? In terms of parsimony (or providing a simple data summary), how does this quadratic equation do as a data summary?

(b) Notice the huge range of values of the response variable. In cases like this, where the response varies over an order of magnitude, taking logarithms of the response often helps produce a simple fitted equation. Here, take (natural) logarithms of all of x_1, x_2, x_3, and y, producing (say) x_1', x_2', x_3', and y', and fit the equation

$$y' \approx \beta_0 + \beta_1 x_1' + \beta_2 x_2' + \beta_3 x_3'$$

to the data. What fraction of the observed variability in $y' = \ln(y)$ does this equation account for? What change in y' seems to accompany a unit (a $1\ln(g)$) increase in x_3'?

(c) To carry the analysis one step further, note that your fitted coefficients for x_1' and x_2' are nearly the negatives of each other. That suggests that y' depends only on the difference between x_1' and x_2'. To see how this works, fit the equation

$$y' \approx \beta_0 + \beta_1 (x_1' - x_2') + \beta_2 x_3'$$

to the data. Compute and plot residuals from this relationship (still on the log scale). How does this relationship appear to do as a data summary? What power law for y (on the original scale) in terms of x_1, x_2, and x_3 (on their original scales) is implied by this last fitted equation? How does this equation compare to the one from (a) in terms of parsimony?

(d) Use your equation from (c) to predict the life of an additional specimen of length 300 mm, at an amplitude of 9 mm, under a load of 45 g. Do the same for an additional specimen of length 325 mm, at an amplitude of 9.5 mm, under a load of 47 g. Why would or wouldn't you be willing to make a similar projection for an additional specimen of length 375 mm, at an amplitude of 10.5 mm, under a load of 51 g?

8. Bauer, Dirks, Palkovic, and Wittmer fired tennis balls out of a "Polish cannon" inclined at an angle of 45°, using three different Propellants and two different Charge Sizes of propellant. They observed the distances traveled in the air by the tennis balls. Their data are given in the accompanying table. (Five trials were made for each Propellant/Charge Size combination and the values given are in feet.)

		Propellant		
		Lighter Fluid	Gasoline	Carburetor Fluid
		58	76	90
		50	79	86
	2.5 ml	53	84	79
		49	73	82
		59	71	86
Charge Size				
		65	96	107
		59	101	102
	5.0 ml	61	94	91
		68	91	95
		67	87	97

Complete a factorial analysis of these data, including a plot of sample means useful for judging the size of Charge Size × Propellant interactions and the computing of fitted main effects and interactions. Write a paragraph summarizing what these data seem to say about how these two variables affect flight distance.

9. The following data are taken from the article "An Analysis of Means for Attribute Data Applied to a 2^4 Factorial Design" by R. Zwickl (*ASQC Electronics Division Technical Supplement*, Fall 1985). They represent numbers of bonds (out of 96) showing evidence of ceramic pull-out on an electronic device called a dual in-line package. (Low numbers are good.) Experimental factors and their levels were:

Factor A   Ceramic Surface
          unglazed (−) vs. glazed (+)

Factor B   Metal Film Thickness
          normal (−) vs. 1.5 times normal (+)

Factor C   Annealing Time
          normal (−) vs. 4 times normal (+)

Factor D   Prebond Clean
          normal clean (−) vs. no clean (+)

The resultant numbers of pull-outs for the 2^4 treatment combinations are given next.

Combination	Pull-Outs	Combination	Pull-Outs
(1)	9	c	13
a	70	ac	55
b	8	bc	7
ab	42	abc	19
d	3	cd	5
ad	6	acd	28
bd	1	bcd	3
abd	7	abcd	6

(a) Use the Yates algorithm and identify dominant effects here.

(b) Based on your analysis from (a), postulate a possible "few effects" explanation for these data. Use the reverse Yates algorithm to find fitted responses for such a simplified description of the system. Use the fitted values to compute residuals. Normal-plot these and plot them against levels of each of the four factors, looking for obvious problems with your representation of system behavior.

(c) Based on your "few effects" description of bond strength, make a recommendation for the future making of these devices. (All else being equal, you should choose what appear to be the least expensive levels of factors.)

10. Exercise 5 of Chapter 3 concerns a replicated 3^3 factorial data set (weighings of three different masses on three different scales by three different students). Use a full-featured statistical package that will compute fitted effects for such data and write a short summary report stating what those fitted effects reveal about the structure of the weighings data.

11. When it is an appropriate description of a two-way factorial data set, what practical engineering advantages does a "main effects only" description offer over a "main effects plus interactions" description?

12. The article referred to in Exercise 4 of Section 4.1 actually considers the effects of both cutting speed and feed rate on tool life. The whole data

set from the article follows. (The data in Section 4.1 are the $x_2 = .01725$ data only.)

Cutting Speed, x_1 (sfpm)	Feed, x_2 (ipr)	Tool Life, y (min)
800	.01725	1.00, 0.90, 0.74, 0.66
700	.01725	1.00, 1.20, 1.50, 1.60
700	.01570	1.75, 1.85, 2.00, 2.20
600	.02200	1.20, 1.50, 1.60, 1.60
600	.01725	2.35, 2.65, 3.00, 3.60
500	.01725	6.40, 7.80, 9.80, 16.50
500	.01570	8.80, 11.00, 11.75, 19.00
450	.02200	4.00, 4.70, 5.30, 6.00
400	.01725	21.50, 24.50, 26.00, 33.00

(a) Taylor's expanded tool life equation is $y x_1^{\alpha_1} x_2^{\alpha_2} = C$. This relationship suggests that $\ln(y)$ may well be approximately linear in both $\ln(x_1)$ and $\ln(x_2)$. Use a multiple linear regression program to fit the relationship

$$\ln(y) \approx \beta_0 + \beta_1 \ln(x_1) + \beta_2 \ln(x_2)$$

to these data. What fraction of the raw variability in $\ln(y)$ is accounted for in the fitting process? What estimates of the parameters α_1, α_2, and C follow from your fitted equation?

(b) Compute and plot residuals (continuing to work on log scales) for the equation you fit in part (a). Make at least plots of residuals versus fitted $\ln(y)$ and both $\ln(x_1)$ and $\ln(x_2)$, and make a normal plot of these residuals. Do these plots reveal any particular problems with the fitted equation?

(c) Use your fitted equation to predict first a log tool life and then a tool life, if in this machining application a cutting speed of 550 and a feed of .01650 is used.

(d) Plot the ordered pairs appearing in the data set in the (x_1, x_2)-plane. Outline a region in the plane where you would feel reasonably safe using the equation you fit in part (a) to predict tool life.

13. K. Casali conducted a gas mileage study on his well-used four-year-old economy car. He drove a 107-mile course a total of eight different times (in comparable weather conditions) at four different speeds, using two different types of gasoline, and ended up with an unreplicated 4×2 factorial study. His data are given in the table below.

Test	Speed (mph)	Gasoline Octane	Gallons Used	Mileage (mpg)
1	65	87	3.2	33.4
2	60	87	3.1	34.5
3	70	87	3.4	31.5
4	55	87	3.0	35.7
5	65	90	3.2	33.4
6	55	90	2.9	36.9
7	70	90	3.3	32.4
8	60	90	3.0	35.7

(a) Make a plot of the mileages that is useful for judging the size of Speed × Octane interactions. Does it look as if the interactions are large in comparison to the main effects?

(b) Compute the fitted main effects and interactions for the mileages, using the formulas of Section 4.3. Make a plot like Figure 4.23 for comparing the observed mileages to fitted mileages computed supposing that there are no Speed × Octane interactions.

(c) Now fit the equation

$$\text{Mileage} \approx \beta_0 + \beta_1(\text{Speed}) + \beta_2(\text{Octane})$$

to the data and plot lines representing the predicted mileages versus Speed for both the 87 octane and the 90 octane gasolines on the same set of axes.

(d) Now fit the equation $\text{Mileage} \approx \beta_0 + \beta_1$ (Speed) separately, first to the 87 octane data and then to the 90 octane data. Plot the two different lines on the same set of axes.

(e) Discuss the different appearances of the plots you made in parts (a) through (d) of this exercise in terms of how well they fit the original data and the different natures of the assumptions involved in producing them.

(f) What was the fundamental weakness in Casali's data collection scheme? A weakness of secondary importance has to do with the fact that tests 1–4 were made ten days earlier than tests 5–8. Why is this a potential problem?

14. The article "Accelerated Testing of Solid Film Lubricants" by Hopkins and Lavik (*Lubrication Engineering*, 1972) contains a nice example of the engineering use of multiple regression. In the study, $m = 3$ sets of journal bearing tests were made on a Mil-L-8937 type film at each combination of three different Loads and three different Speeds. The wear lives of journal bearings, y, in hours, are given next for the tests run by the authors.

Speed, x_1 (rpm)	Load, x_2 (psi)	Wear Life, y (hs)
20	3,000	300.2, 310.8, 333.0
20	6,000	99.6, 136.2, 142.4
20	10,000	20.2, 28.2, 102.7
60	3,000	67.3, 77.9, 93.9
60	6,000	43.0, 44.5, 65.9
60	10,000	10.7, 34.1, 39.1
100	3,000	26.5, 22.3, 34.8
100	6,000	32.8, 25.6, 32.7
100	10,000	2.3, 4.4, 5.8

(a) The authors expected to be able to describe wear life as roughly following the relationship $y x_1 x_2 = C$, but they did not find this relationship to be a completely satisfactory model. So instead, they tried using the more general relationship $y x_1^{\alpha_1} x_2^{\alpha_2} = C$. Use a multiple linear regression program to fit the relationship

$$\ln(y) \approx \beta_0 + \beta_1 \ln(x_1) + \beta_2 \ln(x_2)$$

to these data. What fraction of the raw variability in $\ln(y)$ is accounted for in the fitting process? What estimates of the parameters α_1,

α_2, and C follow from your fitted equation? Using your estimates of α_1, α_2, and C, plot on the same set of (x_1, y) axes the functional relationships between x_1 and y implied by your fitted equation for x_2 equal to 3,000, 6,000, and then 10,000 psi, respectively.

(b) Compute and plot residuals (continuing to work on log scales) for the equation you fit in part (a). Make at least plots of residuals versus fitted $\ln(y)$ and both $\ln(x_1)$ and $\ln(x_2)$, and make a normal plot of these residuals. Do these plots reveal any particular problems with the fitted equation?

(c) Use your fitted equation to predict first a log wear life and then a wear life, if in this application a speed of 20 rpm and a load of 10,000 psi are used.

(d) **(Accelerated life testing)** As a means of trying to make intelligent data-based predictions of wear life at low stress levels (and correspondingly large lifetimes that would be impractical to observe directly), you might (fully recognizing the inherent dangers of the practice) try to extrapolate using the fitted equation. Use your fitted equation to predict first a log wear life and then a wear life if a speed of 15 rpm and load of 1,500 psi are used in this application.

15. The article "Statistical Methods for Controlling the Brown Oxide Process in Multilayer Board Processing" by S. Imadi (*Plating and Surface Finishing*, 1988) discusses an experiment conducted to help a circuit board manufacturer measure the concentration of important components in a chemical bath. Various combinations of levels of

$x_1 = $ % by volume of component A (a proprietary formulation, the major component of which is sodium chlorite)

and

$x_2 = $ % by volume of component B (a proprietary formulation, the major component of which is sodium hydroxide) were set in the chemical bath, and the variables

$y_1 = $ ml of 1N H_2SO_4 used in the first phase of a titration

and

$y_2 = $ ml of 1N H_2SO_4 used in the second phase of a titration

were measured. Part of the original data collected (corresponding to bath conditions free of Na_2CO_3) follow:

x_1	x_2	y_1	y_2
15	25	3.3	.4
20	25	3.4	.4
20	30	4.1	.4
25	30	4.3	.3
25	35	5.0	.5
30	35	5.0	.3
30	40	5.7	.5
35	40	5.8	.4

(a) Fit equations for both y_1 and y_2 linear in both of the variables x_1 and x_2. Does it appear that the variables y_1 and y_2 are adequately described as linear functions of x_1 and x_2?

(b) Solve your two fitted equations from (a) for x_2 (the concentration of primary interest here) in terms of y_1 and y_2. (Eliminate x_1 by solving the first for x_1 in terms of the other three variables and plugging that expression for x_1 into the second equation.) How does this equation seem to do in terms of, so to speak, predicting x_2 from y_1 and y_2 for the original data? Chemical theory in this situation indicated that $x_2 \approx 8(y_1 - y_2)$. Does your equation seem to do better than the one from chemical theory?

(c) A possible alternative to the calculations in (b) is to simply fit an equation for x_2 in terms of y_1 and y_2 directly via least squares. Fit $x_2 \approx \beta_0 + \beta_1 y_1 + \beta_2 y_2$ to the data, using a regression program. Is this equation the same one you found in part (b)?

(d) If you were to compare the equations for x_2 derived in (b) and (c) in terms of the sum of squared differences between the predicted and observed values of x_2, which is guaranteed to be the winner? Why?

16. The article "Nonbloated Burned Clay Aggregate Concrete" by Martin, Ledbetter, Ahmad, and Britton (*Journal of Materials*, 1972) contains data on both composition and resulting physical property test results for a number of different batches of concrete made using burned clay aggregates. The accompanying data are compressive strength measurements, y (made according to ASTM C 39 and recorded in psi), and splitting tensile strength measurements, x (made according to ASTM C 496 and recorded in psi), for ten of the batches used in the study.

Batch	1	2	3	4	5
y	1420	1950	2230	3070	3060
x	207	233	254	328	325

Batch	6	7	8	9	10
y	3110	2650	3130	2960	2760
x	302	258	335	315	302

(a) Make a scatterplot of these data and comment on how linear the relation between x and y appears to be for concretes of this type.

(b) Compute the sample correlation between x and y by hand. Interpret this value.

(c) Fit a line to these data using the least squares principle. Show the necessary hand calculations. Sketch this fitted line on your scatterplot from (a).

(d) About what increase in compressive strength appears to accompany an increase of 1 psi in splitting tensile strength?

(e) What fraction of the raw variability in compressive strength is accounted for in the fitting of a line to the data?

(f) Based on your answer to (c), what measured compressive strength would you predict for a batch of concrete of this type if you were to measure a splitting tensile strength of 245 psi?

(g) Compute the residuals from your fitted line. Plot the residuals against x and against $\hat{y}$. Then make a normal plot of the residuals. What do these plots indicate about the linearity of the relationship between splitting tensile strength and compressive strength?

(h) Use a statistical package to find the least squares line, the sample correlation, R^2, and the residuals for these data.

(i) Fit the quadratic relationship $y \approx \beta_0 + \beta_1 x + \beta_2 x^2$ to the data, using a statistical package. Sketch this fitted parabola on your scatterplot from part (a). Does this fitted quadratic appear to be an important improvement over the line you fit in (c) in terms of describing the relationship of y to x?

(j) How do the R^2 values from parts (h) and (i) compare? Does the increase in R^2 in part (i) speak strongly for the use of the quadratic (as opposed to linear) description of the relationship of y to x for concretes of this type?

(k) If you use the fitted relationship from part (i) to predict y for $x = 245$, how does the prediction compare to your answer for part (f)?

(l) What do the fitted relationships from parts (c) and (i) give for predicted compressive strengths when $x = 400$ psi? Do these compare to each other as well as your answers to parts (f) and (k)? Why would it be unwise to use either of these predictions without further data collection and analysis?

17. In the previous exercise, both x and y were really response variables. As such, they were not subject to direct manipulation by the experimenters. That made it difficult to get several (x, y) pairs with a single x value into the data set. In experimental situations where an engineer gets to choose values of an experimental variable x, why is it useful/important to get several y observations for at least some x's?

18. Chemical engineering graduate student S. Osoka studied the effects of an agitator speed, x_1, and a polymer concentration, x_2, on percent recoveries of pyrite, y_1, and kaolin, y_2, from a step of an ore refining process. (High pyrite recovery and low kaolin recovery rates were desirable.) Data from one set of $n = 9$ experimental runs are given here.

x_1 (rpm)	x_2 (ppm)	y_1 (%)	y_2 (%)
1350	80	77	67
950	80	83	54
600	80	91	70
1350	100	80	52
950	100	87	57
600	100	87	66
1350	120	67	54
950	120	80	52
600	120	81	44

(a) What type of data structure did the researcher use? (Use the terminology of Section 1.2.) What was an obvious weakness in his data collection plan?

(b) Use a regression program to fit the following equations to these data:

$$y_1 \approx \beta_0 + \beta_1 x_1$$
$$y_1 \approx \beta_0 + \beta_2 x_2$$
$$y_1 \approx \beta_0 + \beta_1 x_1 + \beta_2 x_2$$

What are the R^2 values for the three different fitted equations? Compare the three fitted equations in terms of complexity and apparent ability to predict y_1.

(c) Compute the residuals for the third fitted equation in part (b). Plot them against x_1, x_2, and $\hat{y}_1$. Also normal-plot them. Do any of these plots suggest that the third fitted equation is inadequate as a summary of these data?

(d) As a means of understanding the nature of the third fitted equation from part (b), make a scatterplot of y_1 vs. x_2. On this plot, plot three lines representing $\hat{y}_1$ as a function of x_2 for the three different values of x_1 represented in the data set.

(e) Using the third equation from part (b), what pyrite recovery rate would you predict for $x_1 = 1000$ rpm and $x_2 = 110$ ppm?

(f) Consider also a multivariable quadratic description of the dependence of y_1 on x_1 and x_2. That is, fit the equation

$$y_1 \approx \beta_0 + \beta_1 x_1 + \beta_2 x_2 + \beta_3 x_1^2$$
$$+ \beta_4 x_2^2 + \beta_5 x_1 x_2$$

to the data. How does the R^2 value here compare with the ones in part (b)? As a means of understanding this fitted equation, plot on a single set of axes the three different quadratic functions of x_2 obtained by holding x_1 at one of the values in the data set.

(g) It is possible to ignore the fact that the speed and concentration factors are quantitative and to make a factorial analysis of these y_1 data. Do so. Begin by making an interaction plot similar to Figure 4.22 for these data. Based on that plot, discuss the apparent relative sizes of the Speed and Concentration main effects and the Speed × Concentration interactions. Then compute the fitted factorial effects (the fitted main effects and interactions).

(h) If the third equation in part (b) governed y_1, would it lead to Speed × Concentration interactions? What about the equation in part (f)? Explain.

19. The data given in the previous exercise concern both responses y_1 and y_2. The previous analysis dealt with only y_1. Redo all parts of the problem, replacing the response y_1 with y_2 throughout.

20. K. Fellows conducted a 4-factor experiment, with the response variable the flight distance of a paper airplane when propelled from a launcher fabricated specially for the study. This exercise concerns part of the data he collected, constituting a complete 2^4 factorial. The experimental factors involved and levels used were as given here.

Factor A   Plane Design
straight wing (−) vs. t wing (+)

Factor B   Nose Weight
none (−) vs. paper clip (+)

Factor C   Paper Type
notebook (−) vs. construction (+)

Factor D   Wing Tips
straight (−) vs. bent up (+)

The mean flight distances, y (ft), recorded by Fellows for two launches of each plane were as shown in the accompanying table.

(a) Use the Yates algorithm and compute the fitted factorial effects corresponding to the "all high" treatment combination.

(b) Interpret the results of your calculations from (a) in the context of the study. (Describe in words which factors and/or combinations of factors appear to have the largest effect(s) on flight distance. What are the practical implications of these effects?)

Combination	y	Combination	y
(1)	6.25	d	7.00
a	15.50	ad	10.00
b	7.00	bd	10.00
ab	16.50	abd	16.00
c	4.75	cd	4.50
ac	5.50	acd	6.00
bc	4.50	bcd	4.50
abc	6.00	abcd	5.75

(c) Suppose factors B and D are judged to be inert as far as determining flight distance is concerned. (The main effects of B and D and all interactions involving them are negligible.) What fitted/predicted values correspond to this description of flight distance (A and C main effects and AC interactions only)? Use these 16 values of $\hat{y}$ to compute residuals, $y - \hat{y}$. Plot these against $\hat{y}$, levels of A, levels of B, levels of C, and levels of D. Also normal-plot these residuals. Comment on any interpretable patterns in your plots.

(d) Compute R^2 corresponding to the description of flight distance used in part (c). (The formula in Definition 3 works in this context as well as in regression. So does the representation of R^2 as the squared sample correlation between y and $\hat{y}$.) Does it seem that the grand mean, A and C main effects, and AC 2-factor interactions provide an effective summary of flight distance?

21. The data in the accompanying table appear in the text *Quality Control and Industrial Statistics* by Duncan (and were from a paper of L. E. Simon). The data were collected in a study of the effectiveness of armor plate. Armor-piercing bullets were fired at an angle of $40°$ against armor plate of thickness x_1 (in .001 in.) and Brinell hardness number x_2, and the resulting so-called ballistic limit, y (in ft/sec), was measured.

x_1	x_2	y	x_1	x_2	y
253	317	927	253	407	1393
258	321	978	252	426	1401
259	341	1028	246	432	1436
247	350	906	250	469	1327
256	352	1159	242	257	950
246	363	1055	243	302	998
257	365	1335	239	331	1144
262	375	1392	242	355	1080
255	373	1362	244	385	1276
258	391	1374	234	426	1062

(a) Use a regression program to fit the following equations to these data:

$$y \approx \beta_0 + \beta_1 x_1$$
$$y \approx \beta_0 + \beta_2 x_2$$
$$y \approx \beta_0 + \beta_1 x_1 + \beta_2 x_2$$

What are the R^2 values for the three different fitted equations? Compare the three fitted equations in terms of complexity and apparent ability to predict y.

(b) What is the correlation between x_1 and y? The correlation between x_2 and y?

(c) Based on (a) and (b), describe how strongly Thickness and Hardness appear to affect ballistic limit. Review the raw data and speculate as to why the variable with the smaller influence on y seems to be of only minor importance in this data set (although logic says that it must in general have a sizable influence on y).

(d) Compute the residuals for the third fitted equation from (a). Plot them against x_1, x_2, and $\hat{y}$. Also normal-plot them. Do any of these plots suggest that the third fitted equation is seriously deficient as a summary of these data?

(e) Plot the (x_1, x_2) pairs represented in the data set. Why would it be unwise to use any of the fitted equations to predict y for $x_1 = 265$ and $x_2 = 440$?

22. Basgall, Dahl, and Warren experimented with smooth and treaded bicycle tires of different widths. Tires were mounted on the same wheel, placed on a bicycle wind trainer, and accelerated to a velocity of 25 miles per hour. Then pedaling was stopped, and the time required for the wheel to stop rolling was recorded. The sample means, y, of five trials for each of six different tires were as follows:

Tire Width	Tread	Time to Stop, y (sec)
700/19c	smooth	7.30
700/25c	smooth	8.44
700/32c	smooth	9.27
700/19c	treaded	6.63
700/25c	treaded	6.87
700/32c	treaded	7.07

(a) Carefully make an interaction plot of times required to stop, useful for investigating the sizes of Width and Tread main effects and Width × Tread interactions here. Comment briefly on what the plot shows about these effects. Be sure to label the plot very clearly.

(b) Compute the fitted main effects of Width, the fitted main effects of Tread, and the fitted Width × Tread interactions from the y's. Discuss how they quantify features that are evident in your plot from (a).

23. Below are some data read from a graph in the article "Chemical Explosives" by W. B. Sudweeks that appears as Chapter 30 in *Riegel's Handbook of Industrial Chemistry*. The x values are densities (in g/cc) of pentacrythritol tetranitrate (PETN) samples and the y values are corresponding detonation velocities (in km/sec).

x	y	x	y	x	y
.19	2.65	.50	3.95	.91	5.29
.20	2.71	.50	3.87	.91	5.11
.24	2.79	.50	3.57	.95	5.33
.24	3.19	.55	3.84	.95	5.27
.25	2.83	.75	4.70	.97	5.30
.30	3.52	.77	4.19	1.00	5.52
.30	3.41	.80	4.75	1.00	5.46
.32	3.51	.80	4.38	1.00	5.30
.43	3.38	.85	4.83	1.03	5.59
.45	3.13	.85	5.32	1.04	5.71

(a) Make a scatterplot of these data and comment on the apparent linearity (or the lack thereof) of the relationship between y and x.

(b) Compute the sample correlation between y and x. Interpret this value.

(c) Show the "hand" calculations necessary to fit a line to these data by least squares. Then plot your line on the graph from (a).

(d) About what increase in detonation velocity appears to accompany a unit (1 g/cc) increase in PETN density? What increase in detonation velocity would then accompany a .1 g/cc increase in PETN density?

(e) What fraction of the raw variability in detonation velocity is "accounted for" by the fitted line from part (c)?

(f) Based on your analysis, about what detonation velocity would you predict for a PETN density of 0.65 g/cc? If it was your job to produce a PETN explosive charge with a

5.00 km/sec detonation velocity, what PETN density would you employ?

(g) Compute the residuals from your fitted line. Plot them against x and against $\hat{y}$. Then make a normal plot of the residuals. What do these indicate about the linearity of the relationship between y and x?

(h) Use a statistical package and compute the least squares line, the sample correlation, R^2, and the residuals from the least squares line for these data.

24. Some data collected in a study intended to reduce a thread stripping problem in an assembly process follow. Studs screwed into a metal block were stripping out of the block when a nut holding another part on the block was tightened. It was thought that the depth the stud was screwed into the block (the thread engagement) might affect the torque at which the stud stripped out. In the table below, x is the depth (in 10^{-3} inches above .400) and y is the torque at failure (in lbs/in.).

x	y	x	y	x	y	x	y
80	15	40	70	75	70	20	70
76	15	36	65	25	70	40	65
88	25	30	65	30	60	30	75
35	60	0	45	78	25	74	25
75	35	44	50	60	45		

(a) Use a regression program and fit both a linear equation and a quadratic equation to these data. Plot them on a scatterplot of the data. What are the fractions of raw variability in y accounted for by these two equations?

(b) Redo part (a) after dropping the $x = 0$ and $y = 45$ data point from consideration. Do your conclusions about how best to describe the relationship between x and y change appreciably? What does this say about the extent to which a single data point can affect a curve-fitting analysis?

(c) Use your quadratic equation from part (a) and find a thread engagement that provides an optimal predicted failure torque. What would you probably want to do before recommending this depth for use in this assembly process?

25. The textbook *Introduction to Contemporary Statistical Methods* by L. H. Koopmans contains a data set from the testing of automobile tires. A tire under study is mounted on a test trailer and pulled at a standard velocity. Using a braking mechanism, a standard amount of drag (measured in %) is applied to the tire and the force (in pounds) with which it grips the road is measured. The following data are from tests on 19 different tires of the same design made under the same set of road conditions. $x = 0\%$ indicates no braking and $x = 100\%$ indicates the brake is locked.

Drag, x (%)	Grip Force, y (lb)
10	550, 460, 610
20	510, 410, 580
30	470, 360, 480
50	390, 310, 400
70	300, 280, 340
100	250, 200, 200, 200

(a) Make a scatterplot of these data and comment on "how linear" the relation between y and x appears to be.

In fact, physical theory can be called upon to predict that instead of being linear, the relationship between y and x is of the form $y \approx \alpha \exp(\beta x)$ for suitable α and β. Note that if natural logarithms are taken of both sides of this expression, $\ln(y) \approx \ln(\alpha) + \beta x$. Calling $\ln(\alpha)$ by the name β_0 and β by the name β_1, one then has a linear relationship of the form used in Section 4.1.

(b) Make a scatterplot of $y' = \ln(y)$ versus x. Does this plot look more linear than the one in (a)?

(c) Compute the sample correlation between y' and x "by hand." Interpret this value.

(d) Fit a line to the drags and logged grip forces using the least squares principle. Show the necessary hand calculations. Sketch this line on your scatterplot from (b).

(e) About what increase in log grip force appears to accompany an increase in drag of 10% of the total possible? This corresponds to what kind of change in raw grip force?

(f) What fraction of the raw variability in log grip force is accounted for in the fitting of a line to the data in part (d)?

(g) Based on your answer to (d), what log grip force would you predict for a tire of this type under these conditions using 40% of the possible drag? What raw grip force?

(h) Compute the residuals from your fitted line. Plot the residuals against x and against $\hat{y}$. Then make a normal plot of the residuals. What do these plots indicate about the linearity of the relationship between drag and log grip force?

(i) Use a statistical package to find the least squares line, the sample correlation, R^2, and the residuals for these (x, y') data.

26. The article "Laboratory Testing of Asphalt Concrete for Porous Pavements" by Woelfl, Wei, Faulstich, and Litwack (*Journal of Testing and Evaluation*, 1981) studied the effect of asphalt content on the permeability of open-graded asphalt concrete. Four specimens were tested for each of six different asphalt contents, with the following results:

Asphalt Content, x (% by weight)	Permeability, y (in./hr water loss)
3	1189, 840, 1020, 980
4	1440, 1227, 1022, 1293
5	1227, 1180, 980, 1210
6	707, 927, 1067, 822
7	835, 900, 733, 585
8	395, 270, 310, 208

(a) Make a scatterplot of these data and comment on how linear the relation between y and x appears to be. If you focus on asphalt contents between, say, 5% and 7%, does linearity seem to be an adequate description of the relationship between y and x?

Temporarily restrict your attention to the $x = 5, 6$, and 7 data.

(b) Compute the sample correlation between y and x "by hand." Interpret this value.

(c) Fit a line to the asphalt contents and permeabilities using the least squares principle. Show the necessary hand calculations. Sketch this fitted line on your scatterplot from (a).

(d) About what increase in permeability appears to accompany a 1% (by weight) increase in asphalt content?

(e) What fraction of the raw variability in permeability is "accounted for" in the fitting of a line to the $x = 5, 6$, and 7 data in part (c)?

(f) Based on your answer to (c), what measured permeability would you predict for a specimen of this material with an asphalt content of 5.5%?

(g) Compute the residuals from your fitted line. Plot the residuals against x and against $\hat{y}$. Then make a normal plot of the residuals. What do these plots indicate about the linearity of the relationship between asphalt content and permeability?

(h) Use a statistical package and values for $x = 5, 6$, and 7 to find the least squares line, the sample correlation, R^2, and the residuals for these data.

Now consider again the entire data set.

(i) Fit the quadratic relationship $y \approx \beta_0 + \beta_1 x + \beta_2 x^2$ to the data using a statistical package. Sketch this fitted parabola on your second scatterplot from part (a). Does this fitted quadratic appear to be an important improvement over the line you fit in (c) in terms of describing the relationship over the range $3 \le x \le 8$?

(j) Fit the linear relation $y \approx \beta_0 + \beta_1 x$ to the entire data set. How do the R^2 values for this fit and the one in (i) compare? Does the larger R^2 in (i) speak strongly for the use of a quadratic (as opposed to a linear) description of the relationship of y to x in this situation?

(k) If one uses the fitted relationship from (i) to predict y for $x = 5.5$, how does the prediction compare to your answer for (f)?

(l) What do the fitted relationships from (c), (i) and (j) give for predicted permeabilities when $x = 2\%$? Compare these to each other as well as your answers to (f) and (k). Why would it be unwise to use any of these predictions without further data collection?

27. Some data collected by Koh, Morden, and Ogbourne in a study of axial breaking strengths (y) for wooden dowel rods follow. The students tested $m = 4$ different dowels for each of nine combinations of three different diameters (x_1) and three different lengths (x_2).

x_1 (in.)	x_2 (in.)	y (lb)
.125	4	51.5, 37.4, 59.3, 58.5
.125	8	5.2, 6.4, 9.0, 6.3
.125	12	2.5, 3.3, 2.6, 1.9
.1875	4	225.3, 233.9, 211.2, 212.8
.1875	8	47.0, 79.2, 88.7, 70.2
.1875	12	18.4, 22.4, 18.9, 16.6
.250	4	358.8, 309.6, 343.5, 357.8
.250	8	127.1, 158.0, 194.0, 133.0
.250	12	68.9, 40.5, 50.3, 65.6

(a) Make a plot of the 3×3 means, $\bar{y}$, corresponding to the different combinations of diameter and length used in the study, plotting $\bar{y}$ vs. x_2 and connecting the three means for a given diameter with line segments. What does this plot suggest about how successful an equation for y that is linear in x_2 for each fixed x_1 might be in explaining these data?

(b) Replace the strength values with their natural logarithms, $y' = \ln(y)$, and redo the plotting of part (a). Does this second plot suggest that the logarithm of strength might be a linear function of length for fixed diameter?

(c) Fit the following three equations to the data via least squares:

$$y' \approx \beta_0 + \beta_1 x_1,$$
$$y' \approx \beta_0 + \beta_2 x_2,$$
$$y' \approx \beta_0 + \beta_1 x_1 + \beta_2 x_2$$

What are the coefficients of determination for the three fitted equations? Compare the equations in terms of their complexity and their apparent ability to predict y'.

(d) Add three lines to your plot from part (b), showing predicted log strength (from your third fitted equation) as a function of x_2 for the three different values of x_1 included in the study. Use your third fitted equation to predict first a log strength and then a strength for a dowel of diameter .20 in. and length 10 in. Why shouldn't you be willing to use your equation to predict the strength of a rod with diameter .50 in. and length 24 in.?

(e) Compute and plot residuals for the third equation you fit in part (c). Make plots of residuals vs. fitted response and both x_1 and x_2, and normal-plot the residuals. Do these plots suggest any potential inadequacies of the third fitted equation? How might these be remedied?

(f) The students who did this study were strongly suspicious that the ratio $x_3 = x_1^2/x_2$ is the principal determiner of dowel strength. In fact, it is possible to empirically discover the importance of this quantity as follows. Try fitting the equation

$$y' \approx \beta_0 + \beta_1 \ln x_1 + \beta_2 \ln x_2$$

to these data and notice that the fitted coefficients of $\ln x_1$ and $\ln x_2$ are roughly in the ratio of 4 to -2, i.e., 2 to -1. (What does this fitted equation for $\ln(y)$ say about y?) Then plot y vs. x_3 and fit the linear equation

$$y \approx \beta_0 + \beta_3 x_3$$

to these data. Finally, add three curves to your plot from part (a) based on this fitted equation linear in x_3, showing predicted strength as a function of x_2. Make one for each of the three different values of x_1 included in the study.

(g) Since the students' data have a (replicated) 3×3 factorial structure, you can do a factorial analysis as an alternative to the preceding

analysis. Looking again at your plot from (a), does it seem that the interactions of Diameter and Length will be important in describing the raw strengths, y? Compute the fitted factorial effects and comment on the relative sizes of the main effects and interactions.

(h) Redo part (g), referring to the graph from part (b) and working with the logarithms of dowel strength.

28. The paper "Design of a Metal-Cutting Drilling Experiment—A Discrete Two-Variable Problem" by E. Mielnik (*Quality Engineering*, 1993–1994) reports a drilling study run on an aluminum alloy (7075-T6). The thrust (or axial force), y_1, and torque, y_2, required to rotate drills of various diameters x_1 at various feeds (rates of drill penetration into the workpiece) x_2, were measured with the following results:

Diameter, x_1 (in.)	Feed Rate, x_2 (in. rev)	Thrust, y_1 (lb)	Torque, y_2 (ft-lb)
.250	.006	230	1.0
.406	.006	375	2.1
.406	.013	570	3.8
.250	.013	375	2.1
.225	.009	280	1.0
.318	.005	225	1.1
.450	.009	580	3.8
.318	.017	565	3.4
.318	.009	400	2.2
.318	.009	400	2.1
.318	.009	380	2.1
.318	.009	380	1.9

Drilling theory suggests that $y_1 \approx \kappa_1 x_1^a x_2^b$ and $y_2 \approx \kappa_2 x_1^c x_2^d$ for appropriate constants $\kappa_1, \kappa_2, a, b, c$, and d. (Note that upon taking natural logarithms, there are linear relationships between $y_1' = \ln(y_1)$ or $y_2' = \ln(y_2)$ and $x_1' = \ln(x_1)$ and $x_2' = \ln(x_2)$.)

(a) Use a regression program to fit the following equations to these data:

$$y_1' \approx \beta_0 + \beta_1 x_1',$$
$$y_1' \approx \beta_0 + \beta_2 x_2',$$
$$y_1' \approx \beta_0 + \beta_1 x_1' + \beta_2 x_2'$$

What are the R^2 values for the three different fitted equations? Compare the three fitted equations in terms of complexity and apparent ability to predict y_1'.

(b) Compute and plot residuals (continuing to work on log scales) for the third equation you fit in part (a). Make plots of residuals vs. fitted y_1' and both x_1' and x_2', and normal-plot these residuals. Do these plots reveal any particular problems with the fitted equation?

(c) Use your third equation from (a) to predict first a log thrust and then a thrust if a drill of diameter .360 in. and a feed of .011 in./rev are used. Why would it be unwise to make a similar prediction for $x_1 = .450$ and $x_2 = .017$? (*Hint:* Make a plot of the (x_1, x_2) pairs in the data set and locate this second set of conditions on that plot.)

(d) If the third equation fit in part (a) governed y_1, would it lead to Diameter × Feed interactions for y_1 measured on the log scale? To help you answer this question, plot y_1' vs. x_2 (or x_2') for each of $x_1 = .250$, .318, and .406. Does this equation lead to Diameter × Feed interactions for raw y_1?

(e) The first four data points listed in the table constitute a very small complete factorial study (an unreplicated 2 × 2 factorial in the factors Diameter and Feed). Considering only these data points, do a "factorial" analysis of this part of the y_1 data. Begin by making an interaction plot similar to Figure 4.22 for these data. Based on that plot, discuss the apparent relative sizes of the Diameter and Feed main effects on thrust. Then carry out the arithmetic necessary to compute the fitted factorial effects (the main effects and interactions).

(f) Redo part (e), using y_1' as the response variable.

(g) Do your answers to parts (e) and (f) complement those of part (d)? Explain.

29. The article "A Simple Method to Study Dispersion Effects From Non-Necessarily Replicated Data in Industrial Contexts" by Ferrer and Romero (*Quality Engineering*, 1995) describes an unreplicated 2^4 experiment done to improve the adhesive force obtained when gluing on polyurethane sheets as the inner lining of some hollow metal parts. The factors studied were the amount of glue used (A), the predrying temperature (B), the tunnel temperature (C), and the pressure applied (D). The exact levels of the variables employed were not given in the article (presumably for reasons of corporate security). The response variable was the adhesive force, y, in Newtons, and the data reported in the article follow:

Combination	y	Combination	y
(1)	3.80	d	3.29
a	4.34	ad	2.82
b	3.54	bd	4.59
ab	4.59	abd	4.68
c	3.95	cd	2.73
ac	4.83	acd	4.31
bc	4.86	bcd	5.16
abc	5.28	abcd	6.06

(a) Compute the fitted factorial effects corresponding to the "all high" treatment combination.

(b) Interpret the results of your calculations in the context of the study. Which factors and/or combinations of factors appear to have the largest effects on the adhesive force? Suppose that only the A, B, and C main effects and the B × D interactions were judged to be of importance here. Make a corresponding statement to your engineering manager about how the factors impact adhesive force.

(c) Using the reverse Yates algorithm or otherwise, compute fitted/predicted values corresponding to an "A, B, and C main effects and BD interactions" description of adhesive force. Then use these 16 values to compute residuals, $e = y - \hat{y}$. Plot these against $\hat{y}$, and against levels of A, B, C, and D. Also normal-plot them. Comment on any interpretable patterns you see. Particularly in reference to the plot of residuals vs. level of D, what does this graph suggest if one is interested not only in high mean adhesive force but in consistent adhesive force as well?

(d) Find and interpret a value of R^2 corresponding to the description of y used in part (c).

30. The article "Chemical Vapor Deposition of Tungsten Step Coverage and Thickness Uniformity Experiments" by J. Chang (*Thin Solid Films*, 1992) describes an unreplicated 2^4 factorial experiment aimed at understanding the effects of the factors

Factor A   Chamber Pressure (Torr)
            8 (−) vs. 9 (+)

Factor B H_2 Flow (standard cm³/min)
            500 (−) vs. 1000 (+)

Factor C SiH_4 Flow (standard cm³/min)
            15 (−) vs. 25 (+)

Factor D WF_6 Flow (standard cm³/min)
            50 (−) vs. 60 (+)

on a number of response variables in the chemical vapor deposition tungsten films. One response variable reported was the average sheet resistivity, y (mΩ/cm) of the resultant film, and the values reported in the paper follow.

Combination	y	Combination	y
(1)	646	d	666
a	623	ad	597
b	714	bd	718
ab	643	abd	661
c	360	cd	304
ac	359	acd	309
bc	325	bcd	360
abc	318	abcd	318

(a) Use the Yates algorithm and compute the fitted factorial effects corresponding to the "all high" treatment combination. (You will need to employ four cycles in the calculations.)

(b) Interpret the results of your calculations from (a) in the context of the study. (Describe in words which factors and/or combinations of factors appear to have the largest effect(s) on average sheet resistivity. What are the practical implications of these effects?)

(c) Suppose that you judge all factors except C to be "inert" as far as determining sheet resistivity is concerned (the main effects of A, B, and D and all interactions involving them are negligible). What fitted/predicted values correspond to this "C main effects only" description of average sheet resistivity? Use these 16 values to compute residuals, $e = y - \hat{y}$. Plot these against $\hat{y}$, level of A, level of B, level of C, and level of D. Also normal-plot these residuals. Comment on any interpretable patterns in your plots.

(d) Compute an R^2 value corresponding to the description of average sheet resistivity used in part (c). Does it seem that the grand mean and C main effects provide an effective summary of average sheet resistivity? Why?

5

• • • • • • • • • • • • • • • • • • • •

Probability: The Mathematics of Randomness

The theory of probability is the mathematician's description of random variation. This chapter introduces enough probability to serve as a minimum background for making formal statistical inferences.

The chapter begins with a discussion of discrete random variables and their distributions. It next turns to continuous random variables and then probability plotting. Next, the simultaneous modeling of several random variables and the notion of independence are considered. Finally, there is a look at random variables that arise as functions of several others, and how randomness of the input variables is translated to the output variable.

5.1 (Discrete) Random Variables

The concept of a random (or chance) variable is introduced in general terms in this section. Then specialization to discrete cases is considered. The specification of a discrete probability distribution via a probability function or cumulative probability function is discussed. Next, summarization of discrete distributions in terms of (theoretical) means and variances is treated. Then the so-called binomial, geometric, and Poisson distributions are introduced as examples of useful discrete probability models.

5.1.1 Random Variables and Their Distributions

It is usually appropriate to think of a data value as subject to chance influences. In enumerative contexts, chance is commonly introduced into the data collection process through random sampling techniques. Measurement error is nearly always a factor in statistical engineering studies, and the many small, unnameable causes that work to produce it are conveniently thought of as chance phenomena. In analytical contexts, changes in system conditions work to make measured responses vary, and this is most often attributed to chance.

Definition 1

A **random variable** is a quantity that (prior to observation) can be thought of as dependent on chance phenomena. Capital letters near the end of the alphabet are typically used to stand for random variables.

Consider a situation (like that of Example 3 in Chapter 3) where the torques of bolts securing a machine component face plate are to be measured. The next measured value can be considered subject to chance influences and we thus term

$$Z = \text{the next torque recorded}$$

a random variable.

Following Definition 9 in Chapter 1, a distinction was made between discrete and continuous data. That terminology carries over to the present context and inspires two more definitions.

Definition 2

A **discrete random variable** is one that has isolated or separated possible values (rather than a continuum of available outcomes).

Definition 3

A **continuous random variable** is one that can be idealized as having an entire (continuous) interval of numbers as its set of possible values.

Random variables that are basically count variables clearly fall under Definition 2 and are discrete. It could be argued that all measurement variables are discrete—on the basis that all measurements are "to the nearest unit." But it is often mathematically convenient, and adequate for practical purposes, to treat them as continuous.

A random variable is, to some extent, a priori unpredictable. Therefore, in describing or modeling it, the important thing is to specify its set of potential values and the likelihoods associated with those possible values.

Definition 4

To specify a **probability distribution** for a random variable is to give its set of possible values and (in one way or another) consistently assign numbers between 0 and 1—called **probabilities**—as measures of the likelihood that the various numerical values will occur.

The methods used to specify discrete probability distributions are different from those used to specify continuous probability distributions. So the implications of Definition 4 are studied in two steps, beginning in this section with discrete distributions.

5.1.2 Discrete Probability Functions and Cumulative Probability Functions

The tool most often used to describe a discrete probability distribution is the **probability function**.

Definition 5

A **probability function** for a discrete random variable X, having possible values $x_1, x_2, \ldots,$ is a nonnegative function $f(x)$, with $f(x_i)$ giving the probability that X takes the value x_i.

This text will use the notational convention that a capital P followed by an expression or phrase enclosed by brackets will be read "the probability" of that expression. In these terms, a probability function for X is a function f such that

Probability function for the discrete random variable X

$$f(x) = P[X = x]$$

That is, "$f(x)$ is the probability that (the random variable) X takes the value x."

Example 1

A Torque Requirement Random Variable

Consider again Example 3 in Chapter 3, where Brenny, Christensen, and Schneider measured bolt torques on the face plates of a heavy equipment component. With

Z = the next measured torque for bolt 3 (recorded to the nearest integer)

consider treating Z as a discrete random variable and giving a plausible probability function for it.

The relative frequencies for the bolt 3 torque measurements recorded in Table 3.4 on page 74 produce the relative frequency distribution in Table 5.1. This table shows, for example, that over the period the students were collecting data, about 15% of measured torques were 19 ft lb. If it is sensible to believe that the same system of causes that produced the data in Table 3.4 will operate

Example 1
(*continued*)

to produce the next bolt 3 torque, then it also makes sense to base a probability function for Z on the relative frequencies in Table 5.1. That is, the probability distribution specified in Table 5.2 might be used. (In going from the relative frequencies in Table 5.1 to proposed values for $f(z)$ in Table 5.2, there has been some slightly arbitrary rounding. This has been done so that probability values are expressed to two decimal places and now total to exactly 1.00.)

Table 5.1
Relative Frequency Distribution for Measured Bolt 3
Torques

z, Torque (ft lb)	Frequency	Relative Frequency
11	1	$1/34 \approx .02941$
12	1	$1/34 \approx .02941$
13	1	$1/34 \approx .02941$
14	2	$2/34 \approx .05882$
15	9	$9/34 \approx .26471$
16	3	$3/34 \approx .08824$
17	4	$4/34 \approx .11765$
18	7	$7/34 \approx .20588$
19	5	$5/34 \approx .14706$
20	1	$1/34 \approx .02941$
	34	1

Table 5.2
A Probability Function
for Z

Torque z	Probability $f(z)$
11	.03
12	.03
13	.03
14	.06
15	.26
16	.09
17	.12
18	.20
19	.15
20	.03

The appropriateness of the probability function in Table 5.2 for describing Z depends essentially on the physical stability of the bolt-tightening process. But there is a second way in which relative frequencies can become obvious choices for probabilities. For example, think of treating the 34 torques represented in Table 5.1 as a population, from which $n = 1$ item is to be sampled at random, and

$$Y = \text{the torque value selected}$$

The probability distribution of a single value selected at random from a population

Then the probability function in Table 5.2 is also approximately appropriate for Y. This point is not so important in this specific example as it is in general: Where one value is to be selected at random from a population, an appropriate probability distribution is one that is equivalent to the population relative frequency distribution.

This text will usually express probabilities to two decimal places, as in Table 5.2. Computations may be carried to several more decimal places, but final probabilities will typically be reported only to two places. This is because numbers expressed to more than two places tend to look too impressive and be taken too seriously by the uninitiated. Consider for example the statement "There is a .097328 probability of booster engine failure" at a certain missile launch. This may represent the results of some very careful mathematical manipulations and be correct to six decimal places *in the context of the mathematical model used to obtain the value*. But it is doubtful that the model used is a good enough description of physical reality to warrant that much apparent precision. Two-decimal precision is about what is warranted in most engineering applications of simple probability.

Properties of a mathematically valid probability function

The probability function shown in Table 5.2 has two properties that are necessary for the mathematical consistency of a discrete probability distribution. The $f(z)$ values are each in the interval [0, 1] and they total to 1. Negative probabilities or ones larger than 1 would make no practical sense. A probability of 1 is taken as indicating certainty of occurrence and a probability of 0 as indicating certainty of nonoccurrence. Thus, according to the model specified in Table 5.2, since the values of $f(z)$ sum to 1, the occurrence of one of the values 11, 12, 13, 14, 15, 16, 17, 18, 19, and 20 ft lb is certain.

A probability function $f(x)$ gives probabilities of occurrence for individual values. Adding the appropriate values gives probabilities associated with the occurrence of one of a specified type of value for X.

Example 1
(*continued*)

Consider using $f(z)$ defined in Table 5.2 to find

$$P[Z > 17] = P[\text{the next torque exceeds 17}]$$

Adding the $f(z)$ entries corresponding to possible values larger than 17 ft lb,

$$P[Z > 17] = f(18) + f(19) + f(20) = .20 + .15 + .03 = .38$$

The likelihood of the next torque being more than 17 ft lb is about 38%.

Example 1
(*continued*)

If, for example, specifications for torques were 16 ft lb to 21 ft lb, then the likelihood that the next torque measured will be within specifications is

$$P[16 \leq Z \leq 21] = f(16) + f(17) + f(18) + f(19) + f(20) + f(21)$$
$$= .09 + .12 + .20 + .15 + .03 + .00$$
$$= .59$$

In the torque measurement example, the probability function is given in tabular form. In other cases, it is possible to give a formula for $f(x)$.

Example 2

A Random Tool Serial Number

The last step of the pneumatic tool assembly process studied by Kraber, Rucker, and Williams (see Example 11 in Chapter 3) was to apply a serial number plate to the completed tool. Imagine going to the end of the assembly line at exactly 9:00 A.M. next Monday and observing the number plate first applied after 9:00. Suppose that

$$W = \text{the last digit of the serial number observed}$$

Suppose further that tool serial numbers begin with some code special to the tool model and end with consecutively assigned numbers reflecting how many tools of the particular model have been produced. The symmetry of this situation suggests that each possible value of W ($w = 0, 1, \ldots, 9$) is equally likely. That is, a plausible probability function for W is given by the formula

$$f(w) = \begin{cases} .1 & \text{for } w = 0, 1, 2, \ldots, 9 \\ 0 & \text{otherwise} \end{cases}$$

Another way of specifying a discrete probability distribution is sometimes used. That is to specify its **cumulative probability function**.

Definition 6

The **cumulative probability function** for a random variable X is a function $F(x)$ that for each number x gives the probability that X takes that value or a smaller one. In symbols,

$$F(x) = P[X \leq x]$$

Since (for discrete distributions) probabilities are calculated by summing values of $f(x)$, for a discrete distribution,

Cumulative probability function for a discrete variable X

$$F(x) = \sum_{z \leq x} f(z)$$

(The sum is over possible values less than or equal to x.) In this discrete case, the graph of $F(x)$ will be a stair-step graph with jumps located at possible values and equal in size to the probabilities associated with those possible values.

Example 1
(*continued*)

Values of both the probability function and the cumulative probability function for the torque variable Z are given in Table 5.3. Values of $F(z)$ for other z are also easily obtained. For example,

$$F(10.7) = P[Z \leq 10.7] = 0$$

$$F(16.3) = P[Z \leq 16.3] = P[Z \leq 16] = F(16) = .50$$

$$F(32) = P[Z \leq 32] = 1.00$$

A graph of the cumulative probability function for Z is given in Figure 5.1. It shows the stair-step shape characteristic of cumulative probability functions for discrete distributions.

Table 5.3
Values of the Probability Function and Cumulative Probability Function for Z

z, Torque	$f(z) = P[Z = z]$	$F(z) = P[Z \leq z]$
11	.03	.03
12	.03	.06
13	.03	.09
14	.06	.15
15	.26	.41
16	.09	.50
17	.12	.62
18	.20	.82
19	.15	.97
20	.03	1.00

Example 1
(*continued*)

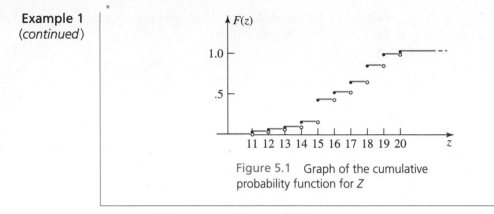

Figure 5.1   Graph of the cumulative probability function for Z

The information about a discrete distribution carried by its cumulative probability function is equivalent to that carried by the corresponding probability function. The cumulative version is sometimes preferred for table making, because round-off problems are more severe when adding several $f(x)$ terms than when taking the difference of two $F(x)$ values to get a probability associated with a consecutive sequence of possible values.

5.1.3   Summarization of Discrete Probability Distributions

Amost all of the devices for describing relative frequency (empirical) distributions in Chapter 3 have versions that can describe (theoretical) probability distributions.

For a discrete random variable with equally spaced possible values, a **probability histogram** gives a picture of the shape of the variable's distribution. It is made by centering a bar of height $f(x)$ over each possible value x. Probability histograms for the random variables Z and W in Examples 1 and 2 are given in Figure 5.2. Interpreting such probability histograms is similar to interpreting relative frequency histograms, except that the areas on them represent (theoretical) probabilities instead of (empirical) fractions of data sets.

It is useful to have a notion of mean value for a discrete random variable (or its probability distribution).

Definition 7

The **mean** or **expected value** of a discrete random variable X (sometimes called the mean of its probability distribution) is

$$EX = \sum_x x\, f(x) \qquad (5.1)$$

EX is read as "the expected value of X," and sometimes the notation μ is used in place of EX.

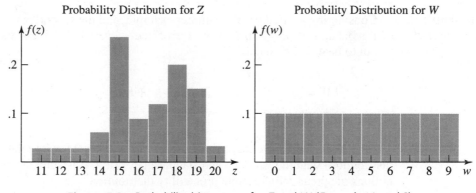

Figure 5.2   Probability histograms for Z and W (*Examples 1 and 2*)

(Remember the warning in Section 3.3 that μ would stand for both the mean of a population and the mean of a probability distribution.)

Example 1
(*continued*)

Returning to the bolt torque example, the expected (or theoretical mean) value of the next torque is

$$EZ = \sum_z z\, f(z)$$

$$= 11(.03) + 12(.03) + 13(.03) + 14(.06) + 15(.26)$$

$$+ 16(.09) + 17(.12) + 18(.20) + 19(.15) + 20(.03)$$

$$= 16.35 \text{ ft lb}$$

This value is essentially the arithmetic mean of the bolt 3 torques listed in Table 3.4. (The slight disagreement in the third decimal place arises only because the relative frequencies in Table 5.1 were rounded slightly to produce Table 5.2.) This kind of agreement provides motivation for using the symbol μ, first seen in Section 3.3, as an alternative to EZ.

The mean of a discrete probability distribution has a balance point interpretation, much like that associated with the arithmetic mean of a data set. Placing (point) masses of sizes $f(x)$ at points x along a number line, EX is the center of mass of that distribution.

Example 2
(*continued*)

Considering again the serial number example, and the second part of Figure 5.2, if a balance point interpretation of expected value is to hold, EW had better turn out to be 4.5. And indeed,

$$EW = 0(.1) + 1(.1) + 2(.1) + \cdots + 8(.1) + 9(.1) = 45(.1) = 4.5$$

It was convenient to measure the spread of a data set (or its relative frequency distribution) with the variance and standard deviation. It is similarly useful to have notions of spread for a discrete probability distribution.

Definition 8

The **variance** of a discrete random variable X (or the variance of its distribution) is

$$\text{Var } X = \sum (x - EX)^2 f(x) \quad \left(= \sum x^2 f(x) - (EX)^2 \right) \qquad (5.2)$$

The **standard deviation** of X is $\sqrt{\text{Var } X}$. Often the notation σ^2 is used in place of Var X, and σ is used in place of $\sqrt{\text{Var } X}$.

The variance of a random variable is its expected (or mean) squared distance from the center of its probability distribution. The use of σ^2 to stand for both the variance of a population and the variance of a probability distribution is motivated on the same grounds as the double use of μ.

Example 1
(*continued*)

The calculations necessary to produce the bolt torque standard deviation are organized in Table 5.4. So

$$\sigma = \sqrt{\text{Var } Z} = \sqrt{4.6275} = 2.15 \text{ ft lb}$$

Except for a small difference due to round-off associated with the creation of Table 5.2, this standard deviation of the random variable Z is numerically the same as the population standard deviation associated with the bolt 3 torques in Table 3.4. (Again, this is consistent with the equivalence between the population relative frequency distribution and the probability distribution for Z.)

Table 5.4
Calculations for Var Z

z	$f(z)$	$(z - 16.35)^2$	$(z - 16.35)^2 f(z)$
11	.03	28.6225	.8587
12	.03	18.9225	.5677
13	.03	11.2225	.3367
14	.06	5.5225	.3314
15	.26	1.8225	.4739
16	.09	.1225	.0110
17	.12	.4225	.0507
18	.20	2.7225	.5445
19	.15	7.0225	1.0534
20	.03	13.3225	.3997
			Var Z = 4.6275

Example 2
(*continued*)

To illustrate the alternative for calculating a variance given in Definition 8, consider finding the variance and standard deviation of the serial number variable W. Table 5.5 shows the calculation of $\sum w^2 f(w)$.

Table 5.5
Calculations for $\sum w^2 f(w)$

w	$f(w)$	$w^2 f(w)$
0	.1	0.0
1	.1	.1
2	.1	.4
3	.1	.9
4	.1	1.6
5	.1	2.5
6	.1	3.6
7	.1	4.9
8	.1	6.4
9	.1	8.1
		28.5

Example 2
(*continued*)

Then

$$\text{Var } W = \sum w^2 f(w) - (EW)^2 = 28.5 - (4.5)^2 = 8.25$$

so that

$$\sqrt{\text{Var } W} = 2.87$$

Comparing the two probability histograms in Figure 5.2, notice that the distribution of W appears to be more spread out than that of Z. Happily, this is reflected in the fact that

$$\sqrt{\text{Var } W} = 2.87 > 2.15 = \sqrt{\text{Var } Z}$$

5.1.4   The Binomial and Geometric Distributions

Discrete probability distributions are sometimes developed from past experience with a particular physical phenomenon (as in Example 1). On the other hand, sometimes an easily manipulated set of mathematical assumptions having the potential to describe a variety of real situations can be put together. When those can be manipulated to derive generic distributions, those distributions can be used to model a number of different random phenomena. One such set of assumptions is that of **independent, identical success-failure trials**.

Independent identical success-failure trials

Many engineering situations involve repetitions of essentially the same "go–no go" (success-failure) scenario, where:

1. There is a *constant chance of a go/success outcome* on each repetition of the scenario (call this probability p).

2. The repetitions are *independent* in the sense that knowing the outcome of any one of them does not change assessments of chance related to any others.

Examples of this kind include the testing of items manufactured consecutively, where each will be classified as either conforming or nonconforming; observing motorists as they pass a traffic checkpoint and noting whether each is traveling at a legal speed or speeding; and measuring the performance of workers in two different workspace configurations and noting whether the performance of each is better in configuration A or configuration B.

In this context, there are two generic kinds of random variables for which deriving appropriate probability distributions is straightforward. The first is the case of a count of the repetitions out of n that yield a go/success result. That is, consider a variable

Binomial random variables

$$X = \text{the number of go/success results in } n \text{ independent identical}$$
$$\text{success-failure trials}$$

X has the **binomial (n, p) distribution**.

Definition 9

The **binomial** (n, p) **distribution** is a discrete probability distribution with probability function

$$f(x) = \begin{cases} \dfrac{n!}{x!\,(n-x)!}\, p^x (1-p)^{n-x} & \text{for } x = 0, 1, \ldots, n \\[2mm] 0 & \text{otherwise} \end{cases}$$ (5.3)

for n a positive integer and $0 < p < 1$.

Equation (5.3) is completely plausible. In it there is one factor of p for each trial producing a go/success outcome and one factor of $(1 - p)$ for each trial producing a no go/failure outcome. And the $n!/x!\,(n - x)!$ term is a count of the number of patterns in which it would be possible to see x go/success outcomes in n trials. The name *binomial* distribution derives from the fact that the values $f(0), f(1), f(2), \ldots, f(n)$ are the terms in the expansion of

$$(p + (1 - p))^n$$

according to the binomial theorem.

Take the time to plot probability histograms for several different binomial distributions. It turns out that for $p < .5$, the resulting histogram is right-skewed. For $p > .5$, the resulting histogram is left-skewed. The skewness increases as p moves away from .5, and it decreases as n is increased. Four binomial probability histograms are pictured in Figure 5.3.

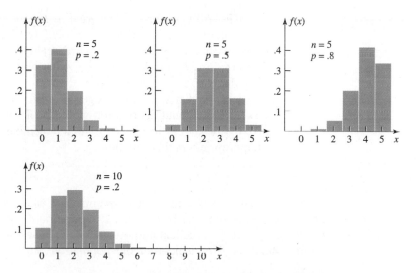

Figure 5.3   Four binomial probability histograms

Example 3

The Binomial Distribution and Counts of Reworkable Shafts

Consider again the situation of Example 12 in Chapter 3 and a study of the performance of a process for turning steel shafts. Early in that study, around 20% of the shafts were typically classified as "reworkable." Suppose that $p = .2$ is indeed a sensible figure for the chance that a given shaft will be reworkable. Suppose further that $n = 10$ shafts will be inspected, and the probability that at least two are classified as reworkable is to be evaluated.

Adopting a model of independent, identical success-failure trials for shaft conditions,

$$U = \text{the number of reworkable shafts in the sample of 10}$$

is a binomial random variable with $n = 10$ and $p = .2$. So

$$
\begin{aligned}
P[\text{at least two reworkable shafts}] &= P[U \geq 2] \\
&= f(2) + f(3) + \cdots + f(10) \\
&= 1 - (f(0) + f(1)) \\
&= 1 - \left(\frac{10!}{0!\,10!}(.2)^0(.8)^{10} + \frac{10!}{1!\,9!}(.2)^1(.8)^9 \right) \\
&= .62
\end{aligned}
$$

(The trick employed here, to avoid plugging into the binomial probability function 9 times by recognizing that the $f(u)$'s have to sum up to 1, is a common and useful one.)

The .62 figure is only as good as the model assumptions that produced it. If an independent, identical success-failure trials description of shaft production fails to accurately portray physical reality, the .62 value is fine mathematics but possibly a poor description of what will actually happen. For instance, say that due to tool wear it is typical to see 40 shafts in specifications, then 10 reworkable shafts, a tool change, 40 shafts in specifications, and so on. In this case, the binomial distribution would be a very poor description of U, and the .62 figure largely irrelevant. (The independence-of-trials assumption would be inappropriate in this situation.)

The binomial distribution and simple random sampling

There is one important circumstance where a model of independent, identical success-failure trials is not exactly appropriate, but a binomial distribution can still be adequate for practical purposes—that is, in describing the results of simple random sampling from a dichotomous population. Suppose a population of size N contains a fraction p of type A objects and a fraction $(1 - p)$ of type B objects. If a simple random sample of n of these items is selected and

$$X = \text{the number of type A items in the sample}$$

strictly speaking, x is not a binomial random variable. But if n is a small fraction of N (say, less than 10%), and p is not too extreme (i.e., is not close to either 0 or 1), X is *approximately* binomial (n, p).

Example 4

Simple Random Sampling from a Lot of Hexamine Pellets

In the pelletizing machine experiment described in Example 14 in Chapter 3, Greiner, Grimm, Larson, and Lukomski found a combination of machine settings that allowed them to produce 66 conforming pellets out of a batch of 100 pellets. Treat that batch of 100 pellets as a population of interest and consider selecting a simple random sample of size $n = 2$ from it.

If one defines the random variable

$$V = \text{the number of conforming pellets in the sample of size 2}$$

the most natural probability distribution for V is obtained as follows. Possible values for V are 0, 1, and 2.

$$f(0) = P[V = 0]$$
$$= P[\text{first pellet selected is nonconforming and} \\ \text{subsequently the second pellet is also nonconforming}]$$
$$f(2) = P[V = 2]$$
$$= P[\text{first pellet selected is conforming and} \\ \text{subsequently the second pellet selected is conforming}]$$
$$f(1) = 1 - (f(0) + f(2))$$

Then think, "In the long run, the first selection will yield a nonconforming pellet about 34 out of 100 times. Considering only cases where this occurs, in the long run the next selection will also yield a nonconforming pellet about 33 out of 99 times." That is, a sensible evaluation of $f(0)$ is

$$f(0) = \frac{34}{100} \cdot \frac{33}{99} = .1133$$

Example 4
(*continued*)

Similarly,

$$f(2) = \frac{66}{100} \cdot \frac{65}{99} = .4333$$

and thus

$$f(1) = 1 - (.1133 + .4333) = 1 - .5467 = .4533$$

Now, V cannot be thought of as arising from exactly independent trials. For example, knowing that the first pellet selected was conforming would reduce most people's assessment of the chance that the second is also conforming from $\frac{66}{100}$ to $\frac{65}{99}$. Nevertheless, for most practical purposes, V can be thought of as *essentially* binomial with $n = 2$ and $p = .66$. To see this, note that

$$\frac{2!}{0!\,2!}(.34)^2(.66)^0 = .1156 \approx f(0)$$

$$\frac{2!}{1!\,1!}(.34)^1(.66)^1 = .4488 \approx f(1)$$

$$\frac{2!}{2!\,0!}(.34)^0(.66)^2 = .4356 \approx f(2)$$

Here, n is a small fraction of N, p is not too extreme, and a binomial distribution is a decent description of a variable arising from simple random sampling.

Calculation of the mean and variance for binomial random variables is greatly simplified by the fact that when the formulas (5.1) and (5.2) are used with the expression for binomial probabilities in equation (5.3), simple formulas result. For X a binomial (n, p) random variable,

Mean of the binomial (n, p) distribution

$$\mu = EX = \sum_{x=0}^{n} x \frac{n!}{x!(n-x)!} p^x (1-p)^{n-x} = np \qquad \textbf{(5.4)}$$

Further, it is the case that

Variance of the binomial (n, p) distribution

$$\sigma^2 = \text{Var } X = \sum_{x=0}^{n} (x - np)^2 \frac{n!}{x!(n-x)!} p^x (1-p)^{n-x} = np(1-p) \qquad \textbf{(5.5)}$$

Example 3 (*continued*)	Returning to the machining of steel shafts, suppose that a binomial distribution with $n = 10$ and $p = .2$ is appropriate as a model for $\qquad U =$ the number of reworkable shafts in the sample of 10 Then, by formulas (5.4) and (5.5), $$EU = (10)(.2) = 2 \text{ shafts}$$ $$\sqrt{\text{Var } U} = \sqrt{10(.2)(.8)} = 1.26 \text{ shafts}$$

A second generic type of random variable associated with a series of independent, identical success-failure trials is

Geometric random variables

$\qquad X =$ the number of trials required to first obtain a go/success result

X has the **geometric** (p) **distribution**.

Definition 10	The **geometric** (p) **distribution** is a discrete probability distribution with probability function $$f(x) = \begin{cases} p(1-p)^{x-1} & \text{for } x = 1, 2, \dots \\ 0 & \text{otherwise} \end{cases} \qquad (5.6)$$ for $0 < p < 1$.

Formula (5.6) makes good intuitive sense. In order for X to take the value x, there must be $x - 1$ consecutive no-go/failure results followed by a go/success. In formula (5.6), there are $x - 1$ terms $(1 - p)$ and one term p. Another way to see that formula (5.6) is plausible is to reason that for X as above and $x = 1, 2, \dots$

$$1 - F(x) = 1 - P[X \leq x]$$
$$= P[X > x]$$
$$= P[x \text{ no-go/failure outcomes in } x \text{ trials}]$$

That is,

Simple relationship for the geometric (p) cumulative probability function

$$1 - F(x) = (1 - p)^x \qquad (5.7)$$

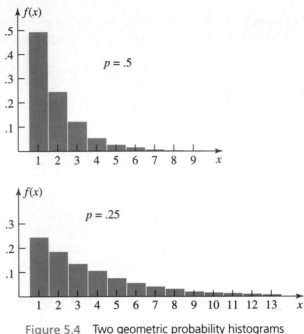

Figure 5.4   Two geometric probability histograms

by using the form of the binomial (x, p) probability function given in equation (5.3). Then for $x = 2, 3, \ldots,$ $f(x) = F(x) - F(x-1) = -(1 - F(x)) + (1 - F(x-1))$. This, combined with equation (5.7), gives equation (5.6).

The name *geometric* derives from the fact that the values $f(1), f(2), f(3), \ldots$ are terms in the geometric infinite series for

$$p \cdot \frac{1}{1 - (1-p)}$$

The geometric distributions are discrete distributions with probability histograms exponentially decaying as x increases. Two different geometric probability histograms are pictured in Figure 5.4.

Example 5 | The Geometric Distribution and Shorts in NiCad Batteries

In "A Case Study of the Use of an Experimental Design in Preventing Shorts in Nickel-Cadmium Cells" (*Journal of Quality Technology*, 1988), Ophir, El-Gad, and Snyder describe a series of experiments conducted in order to reduce the proportion of cells being scrapped by a battery plant because of internal shorts. The experimental program was successful in reducing the percentage of manufactured cells with internal shorts to around 1%.

Suppose that testing begins on a production run in this plant, and let

$$T = \text{the test number at which the first short is discovered}$$

A model for T (appropriate if the independent, identical success-failure trials description is apt) is geometric with $p = .01$. (p is the probability that any particular test yields a shorted cell.) Then, using equation (5.6),

$$P[\text{the first or second cell tested has the first short}] = P[T = 1 \text{ or } T = 2]$$
$$= f(1) + f(2)$$
$$= (.01) + (.01)(1 - .01)$$
$$= .02$$

Or, using equation (5.7),

$$P[\text{at least 50 cells are tested without finding a short}] = P[T > 50]$$
$$= (1 - .01)^{50}$$
$$= .61$$

Like the binomial distributions, the geometric distributions have means and variances that are simple functions of the parameter p. That is, if X is geometric (p),

Mean of the geometric (p) distribution

$$\mu = EX = \sum_{x=1}^{\infty} xp(1 - p)^{x-1} = \frac{1}{p} \qquad (5.8)$$

and

Variance of the geometric (p) distribution

$$\sigma^2 = \text{Var } X = \sum_{x=1}^{\infty} \left(x - \frac{1}{p} \right)^2 p(1 - p)^{x-1} = \frac{1 - p}{p^2} \qquad (5.9)$$

Example 5
(*continued*)

In the context of battery testing, with T as before,

$$ET = \frac{1}{.01} = 100 \text{ batteries}$$

$$\sqrt{\text{Var } T} = \sqrt{\frac{(1 - .01)}{(.01)^2}} = 99.5 \text{ batteries}$$

Example 5
(continued)

Formula (5.8) is an intuitively appealing result. If there is only 1 chance in 100 of encountering a shorted battery at each test, it is sensible to expect to wait through 100 tests on average to encounter the first one.

5.1.5 The Poisson Distributions

As discussed in Section 3.4, it is often important to keep track of the total number of occurrences of some relatively rare phenomenon, where the physical or time unit under observation has the potential to produce many such occurrences. A case of floor tiles has potentially many total blemishes. In a one-second interval, there are potentially a large number of messages that can arrive for routing through a switching center. And a 1 cc sample of glass potentially contains a large number of imperfections.

So probability distributions are needed to describe random *counts of the number of occurrences of a relatively rare phenomenon across a specified interval of time or space*. By far the most commonly used theoretical distributions in this context are the **Poisson distributions**.

Definition 11

The **Poisson (λ) distribution** is a discrete probability distribution with probability function

$$f(x) = \begin{cases} \dfrac{e^{-\lambda}\lambda^{x}}{x!} & \text{for } x = 0, 1, 2, \ldots \\ 0 & \text{otherwise} \end{cases} \quad \textbf{(5.10)}$$

for $\lambda > 0$.

The form of equation (5.10) may initially seem unappealing. But it is one that has sensible mathematical origins, is manageable, and has proved itself empirically useful in many different "rare events" circumstances. One way to arrive at equation (5.10) is to think of a very large number of independent trials (opportunities for occurrence), where the probability of success (occurrence) on any one is very small and the product of the number of trials and the success probability is λ. One is then led to the binomial $(n, \frac{\lambda}{n})$ distribution. In fact, for large n, the binomial $(n, \frac{\lambda}{n})$ probability function approximates the one specified in equation (5.10). So one might think of the Poisson distribution for counts as arising through a mechanism that would present many tiny similar opportunities for independent occurrence or nonoccurrence throughout an interval of time or space.

The Poisson distributions are right-skewed distributions over the values $x = 0, 1, 2, \ldots$, whose probability histograms peak near their respective λ's. Two different Poisson probability histograms are shown in Figure 5.5. λ is both the mean

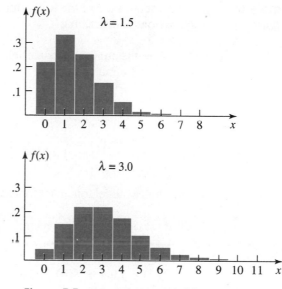

Figure 5.5 Two Poisson probability histograms

and the variance for the Poisson (λ) distribution. That is, if X has the Poisson (λ) distribution, then

Mean of the Poisson (λ) distribution

$$\mu = EX = \sum_{x=0}^{\infty} x \frac{e^{-\lambda}\lambda^x}{x!} = \lambda \tag{5.11}$$

and

Variance of the Poisson (λ) distribution

$$\text{Var } X = \sum_{x=0}^{\infty} (x - \lambda)^2 \frac{e^{-\lambda}\lambda^x}{x!} = \lambda \tag{5.12}$$

Fact (5.11) is helpful in picking out which Poisson distribution might be useful in describing a particular "rare events" situation.

Example 6

The Poisson Distribution and Counts of α-Particles

A classical data set of Rutherford and Geiger, reported in *Philosophical Magazine* in 1910, concerns the numbers of α-particles emitted from a small bar of polonium and colliding with a screen placed near the bar in 2,608 periods of 8 minutes each. The Rutherford and Geiger relative frequency distribution has mean 3.87 and a shape remarkably similar to that of the Poisson probability distribution with mean $\lambda = 3.87$.

Example 6
(*continued*)

In a duplication of the Rutherford/Geiger experiment, a reasonable probability function for describing

S = the number of α-particles striking the screen in an additional 8-minute period

is then

$$f(s) = \begin{cases} \dfrac{e^{-3.87}(3.87)^s}{s!} & \text{for } s = 0, 1, 2, \ldots \\ 0 & \text{otherwise} \end{cases}$$

Using such a model, one has (for example)

P[at least 4 particles are recorded]

$= P[S \geq 4]$

$= f(4) + f(5) + f(6) + \cdots$

$= 1 - (f(0) + f(1) + f(2) + f(3))$

$= 1 - \left(\dfrac{e^{-3.87}(3.87)^0}{0!} + \dfrac{e^{-3.87}(3.87)^1}{1!} + \dfrac{e^{-3.87}(3.87)^2}{2!} + \dfrac{e^{-3.87}(3.87)^3}{3!} \right)$

$= .54$

Example 7

Arrivals at a University Library

Stork, Wohlsdorf, and McArthur collected data on numbers of students entering the ISU library during various periods over a week's time. Their data indicate that between 12:00 and 12:10 P.M. on Monday through Wednesday, an average of around 125 students entered. Consider modeling

M = the number of students entering the ISU library between 12:00 and 12:01 next Tuesday

Using a Poisson distribution to describe M, the reasonable choice of λ would seem to be

$$\lambda = \frac{125 \text{ students}}{10 \text{ minutes}} (1 \text{ minute}) = 12.5 \text{ students}$$

For this choice,

$$EM = \lambda = 12.5 \text{ students}$$

$$\sqrt{\text{Var } M} = \sqrt{\lambda} = \sqrt{12.5} = 3.54 \text{ students}$$

and, for example, the probability that between 10 and 15 students (inclusive) arrive at the library between 12:00 and 12:01 would be evaluated as

$$P[10 \leq M \leq 15] = f(10) + f(11) + f(12) + f(13) + f(14) + f(15)$$

$$= \frac{e^{-12.5}(12.5)^{10}}{10!} + \frac{e^{-12.5}(12.5)^{11}}{11!} + \frac{e^{-12.5}(12.5)^{12}}{12!}$$

$$+ \frac{e^{-12.5}(12.5)^{13}}{13!} + \frac{e^{-12.5}(12.5)^{14}}{14!} + \frac{e^{-12.5}(12.5)^{15}}{15!}$$

$$= .60$$

Section 1 Exercises

1. A discrete random variable X can be described using the probability function

x	2	3	4	5	6
$f(x)$	.1	.2	.3	.3	.1

(a) Make a probability histogram for X. Also plot $F(x)$, the cumulative probability function for X.

(b) Find the mean and standard deviation of X.

2. In an experiment to evaluate a new artificial sweetener, ten subjects are all asked to taste cola from three unmarked glasses, two of which contain regular cola while the third contains cola made with the new sweetener. The subjects are asked to identify the glass whose content is different from the other two. If there is no difference between the taste of sugar and the taste of the new sweetener, the subjects would be just guessing.

(a) Make a table for a probability function for

$$X = \text{the number of subjects correctly identifying the artificially sweetened cola}$$

under this hypothesis of no difference in taste.

(b) If seven of the ten subjects correctly identify the artificial sweetener, is this outcome strong evidence of a taste difference? Explain.

3. Suppose that a small population consists of the $N = 6$ values 2, 3, 4, 4, 5, and 6.

(a) Sketch a relative frequency histogram for this population and compute the population mean, μ, and standard deviation, σ.

(b) Now let $X = $ the value of a single number selected at random from this population. Sketch a probability histogram for this variable X and compute EX and Var X.

(c) Now think of drawing a simple random sample of size $n = 2$ from this small population. Make tables giving the probability distributions of the random variables

$$\overline{X} = \text{the sample mean}$$

$$S^2 = \text{the sample variance}$$

(There are 15 different possible unordered samples of 2 out of 6 items. Each of the 15 possible samples is equally likely to be chosen and has its own corresponding $\bar{x}$ and s^2.) Use the tables and make probability histograms for these random variables. Compute EX and Var $\overline{X}$. How do these compare to μ and σ^2?

4. Sketch probability histograms for the binomial distributions with $n = 5$ and $p = .1, .3, .5, .7,$ and $.9$. On each histogram, mark the location of the mean and indicate the size of the standard deviation.

5. Suppose that an eddy current nondestructive evaluation technique for identifying cracks in critical metal parts has a probability of around .20 of detecting a single crack of length .003 in. in a certain material. Suppose further that $n = 8$ specimens of this material, each containing a single crack of length .003 in., are inspected using this technique. Let W be the number of these cracks that are detected. Use an appropriate probability model and evaluate the following:
   (a) $P[W = 3]$
   (b) $P[W \leq 2]$
   (c) EW
   (d) Var W
   (e) the standard deviation of W

6. In the situation described in Exercise 5, suppose that a series of specimens, each containing a single crack of length .003 in., are inspected. Let Y be the number of specimens inspected in order to obtain the first crack detection. Use an appropriate probability model and evaluate all of the following:
   (a) $P[Y = 5]$
   (b) $P[Y \leq 4]$
   (c) EY
   (d) Var Y
   (e) the standard deviation of Y

7. Sketch probability histograms for the Poisson distributions with means $\lambda = .5, 1.0, 2.0,$ and 4.0. On each histogram, mark the location of the mean and indicate the size of the standard deviation.

8. A process for making plate glass produces an average of four seeds (small bubbles) per 100 square feet. Use Poisson distributions and assess probabilities that
   (a) a particular piece of glass 5 ft × 10 ft will contain more than two seeds.
   (b) a particular piece of glass 5 ft × 5 ft will contain no seeds.

9. Transmission line interruptions in a telecommunications network occur at an average rate of one per day.
   (a) Use a Poisson distribution as a model for

   $$X = \text{the number of interruptions in the next five-day work week}$$

   and assess $P[X = 0]$.
   (b) Now consider the random variable

   $$Y = \text{the number of weeks in the next four in which there are no interruptions}$$

   What is a reasonable probability model for Y? Assess $P[Y = 2]$.

10. Distinguish clearly between the subjects of *probability* and *statistics*. Is one field a subfield of the other?

11. What is the difference between a relative frequency distribution and a probability distribution?

· ·

5.2 Continuous Random Variables

It is often convenient to think of a random variable as not discrete but rather continuous in the sense of having a whole (continuous) interval for its set of possible values. The devices used to describe continuous probability distributions differ from the tools studied in the last section. So the first tasks here are to introduce the notion of a probability density function, to show its relationship to the cumulative probability function for a continuous random variable, and to show how it is used to find the mean and variance for a continuous distribution. After this, several standard continuous distributions useful in engineering applications of probability theory will be discussed. That is, the normal (or Gaussian) exponential and Weibull distributions are presented.

5.2.1 Probability Density Functions and Cumulative Probability Functions

The methods used to specify and describe probability distributions have parallels in mechanics. When considering continuous probability distributions, the analogy to mechanics becomes especially helpful. In mechanics, the properties of a continuous mass distribution are related to the possibly varying density of the mass across its region of location. Amounts of mass in particular regions are obtained from the density by integration.

The concept in probability theory corresponding to mass density in mechanics is *probability density*. To specify a continuous probability distribution, one needs to describe "how thick" the probability is in the various parts of the set of possible values. The formal definition is

Definition 12

> A **probability density function** for a continuous random variable X is a nonnegative function $f(x)$ with
>
> $$\int_{-\infty}^{\infty} f(x)\,dx = 1 \tag{5.13}$$
>
> and such that for all $a \le b$, one is willing to assign $P[a \le X \le b]$ according to
>
> $$P[a \le X \le b] = \int_{a}^{b} f(x)\,dx \tag{5.14}$$

A generic probability density function is pictured in Figure 5.6. In keeping with equations (5.13) and (5.14), the plot of $f(x)$ does not dip below the x axis, the total area under the curve $y = f(x)$ is 1, and areas under the curve above particular intervals give probabilities corresponding to those intervals.

Mechanics analogy for probability density

In direct analogy to what is done in mechanics, if $f(x)$ is indeed the "density of probability" around x, then the probability in an interval of small length dx around x is approximately $f(x)\,dx$. (In mechanics, if $f(x)$ is *mass* density around x, then the *mass* in an interval of small length dx around x is approximately $f(x)\,dx$.) Then to get a probability between a and b, one needs to sum up such $f(x)\,dx$ values. $\int_{a}^{b} f(x)\,dx$ is exactly the limit of $\sum f(x)\,dx$ values as dx gets small. (In mechanics, $\int_{a}^{b} f(x)\,dx$ is the mass between a and b.) So the expression (5.14) is reasonable.

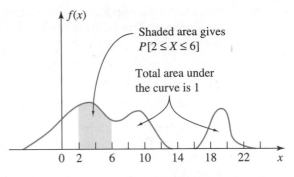

Figure 5.6   A generic probability density function

| Example 8 | The Random Time Until a First Arc in the Bob Drop Experiment |

Consider once again the bob drop experiment first described in Section 1.4 and revisited in Example 4 in Chapter 4. In any use of the apparatus, the bob is almost certainly not released exactly "in sync" with the 60 cycle current that produces the arcs and marks on the paper tape. One could think of a random variable

$$Y = \text{the time elapsed (in seconds) from bob release until the first arc}$$

as continuous with set of possible values $(0, \frac{1}{60})$.

What is a plausible probability density function for Y? The symmetry of this situation suggests that probability density should be constant over the interval $(0, \frac{1}{60})$ and 0 outside the interval. That is, for any two values y_1 and y_2 in $(0, \frac{1}{60})$, the probability that Y takes a value within a small interval around y_1 of length dy (i.e., $f(y_1) \, dy$ approximately) should be the same as the probability that Y takes a value within a small interval around y_2 of the same length dy (i.e., $f(y_2) \, dy$ approximately). This forces $f(y_1) = f(y_2)$, so there must be a constant probability density on $(0, \frac{1}{60})$.

Now if $f(y)$ is to have the form

$$f(y) = \begin{cases} c & \text{for } 0 < y < \frac{1}{60} \\ 0 & \text{otherwise} \end{cases}$$

for some constant c (i.e., is to be as pictured in Figure 5.7), in light of equation (5.13), it must be that

$$1 = \int_{-\infty}^{\infty} f(y) \, dy = \int_{-\infty}^{0} 0 \, dy + \int_{0}^{1/60} c \, dy + \int_{1/60}^{\infty} 0 \, dy = \frac{c}{60}$$

That is, $c = 60$, and thus,

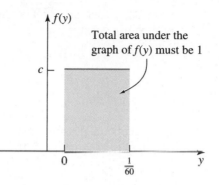

Figure 5.7 Probability density function for Y (time elapsed before arc)

$$f(y) = \begin{cases} 60 & \text{for } 0 < y < \frac{1}{60} \\ 0 & \text{otherwise} \end{cases} \qquad \textbf{(5.15)}$$

If the function specified by equation (5.15) is adopted as a probability density for Y, it is then (for example) possible to calculate that

$$P\left[Y \le \frac{1}{100}\right] = \int_{-\infty}^{1/100} f(y)\,dy = \int_{-\infty}^{0} 0\,dy + \int_{0}^{1/100} 60\,dy = .6$$

One point about continuous probability distributions that may at first seem counterintuitive concerns the probability associated with a continuous random variable assuming a particular prespecified value (say, a). Just as the mass a continuous mass distribution places at a single point is 0, so also is $P[X = a] = 0$ for a continuous random variable X. This follows from equation (5.14), because

For X a continuous random variable, $P[X = a] = 0$

$$P[a \le X \le a] = \int_{a}^{a} f(x)\,dx = 0$$

One consequence of this mathematical curiosity is that when working with continuous random variables, you don't need to worry about whether or not inequality signs you write are strict inequality signs. That is, if X is continuous,

$$P[a \le X \le b] = P[a < X \le b] = P[a \le X < b] = P[a < X < b]$$

Definition 6 gave a perfectly general definition of the cumulative probability function for a random variable (which was specialized in Section 5.1 to the case of a discrete variable). Here equation (5.14) can be used to express the cumulative probability function for a continuous random variable in terms of an integral of its probability density. That is, for X continuous with probability density $f(x)$,

Cumulative probability function for a continuous variable

$$F(x) = P[X \leq x] = \int_{-\infty}^{x} f(t)\, dt \tag{5.16}$$

$F(x)$ is obtained from $f(x)$ by integration, and applying the fundamental theorem of calculus to equation (5.16)

Another relationship between F(x) and f(x)

$$\frac{d}{dx} F(x) = f(x) \tag{5.17}$$

That is, $f(x)$ is obtained from $F(x)$ by differentiation.

Example 8
(continued)

The cumulative probability function for Y, the elapsed time from bob release until first arc, is easily obtained from equation (5.15). For $y \leq 0$,

$$F(y) = P[Y \leq y] = \int_{-\infty}^{y} f(t)\, dt = \int_{-\infty}^{y} 0\, dt = 0$$

and for $0 < y \leq \frac{1}{60}$,

$$F(y) = P[Y \leq y] = \int_{-\infty}^{y} f(t)\, dt = \int_{-\infty}^{0} 0\, dt + \int_{0}^{y} 60\, dt = 0 + 60y = 60y$$

and for $y > \frac{1}{60}$,

$$F(y) = P[Y \leq y] = \int_{-\infty}^{y} f(t)\, dt = \int_{-\infty}^{0} 0\, dt + \int_{0}^{1/60} 60\, dt + \int_{1/60}^{y} 0\, dt = 1$$

That is,

$$F(y) = \begin{cases} 0 & \text{if } y \leq 0 \\ 60y & \text{if } 0 < y \leq 1/60 \\ 1 & \text{if } \frac{1}{60} < y \end{cases}$$

A plot of $F(y)$ is given in Figure 5.8. Comparing Figure 5.8 to Figure 5.7 shows that indeed the graph of $F(y)$ has slope 0 for $y < 0$ and $y > \frac{1}{60}$ and slope 60 for $0 < y < \frac{1}{60}$. That is, $f(y)$ is the derivative of $F(y)$, as promised by equation (5.17).

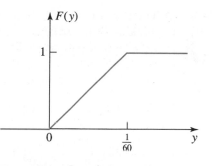

Figure 5.8 Cumulative probability function for Y (time elapsed before arc)

Figure 5.8 is typical of cumulative probability functions for continuous distributions. The graphs of such cumulative probability functions are *continuous* in the sense that they are unbroken curves.

5.2.2 Means and Variances for Continuous Distributions

A plot of the probability density $f(x)$ is a kind of idealized histogram. It has the same kind of visual interpretations that have already been applied to relative frequency histograms and probability histograms. Further, it is possible to define a mean and variance for a continuous probability distribution. These numerical summaries are used in the same way that means and variances are used to describe data sets and discrete probability distributions.

Definition 13

> The **mean** or **expected value** of a continuous random variable X (sometimes called the mean of its probability distribution) is
>
> $$EX = \int_{-\infty}^{\infty} x\, f(x)\, dx \qquad (5.18)$$
>
> As for discrete random variables, the notation μ is sometimes used in place of EX.

Formula (5.18) is perfectly plausible from at least two perspectives. First, the probability in a small interval around x of length dx is approximately $f(x)\,dx$. So multiplying this by x and summing as in Definition 7, one has $\sum x\, f(x)\, dx$, and formula (5.18) is exactly the limit of such sums as dx gets small. And second, in mechanics the center of mass of a continuous mass distribution is of the form given in equation (5.18) except for division by a total mass, which for a probability distribution is 1.

Example 8
(*continued*)

Thinking of the probability density in Figure 5.7 as an idealized histogram and thinking of the balance point interpretation of the mean, it is clear that EY had better turn out to be $\frac{1}{120}$ for the elapsed time variable. Happily, equations (5.18) and (5.15) give

$$\mu = EY = \int_{-\infty}^{\infty} y\, f(y)\, dy = \int_{-\infty}^{0} y \cdot 0 \, dy + \int_{0}^{1/60} y \cdot 60 \, dy + \int_{1/60}^{\infty} y \cdot 0 \, dy$$

$$= 30y^2 \Big|_{0}^{1/60} = \frac{1}{120} \text{ sec}$$

"Continuization" of the formula for the variance of a discrete random variable produces a definition of the variance of a continuous random variable.

Definition 14

The **variance** of a continuous random variable X (sometimes called the variance of its probability distribution) is

$$\text{Var } X = \int_{-\infty}^{\infty} (x - EX)^2 f(x)\, dx \quad \left(= \int_{-\infty}^{\infty} x^2 f(x)\, dx - (EX)^2 \right) \qquad \textbf{(5.19)}$$

The **standard deviation** of X is $\sqrt{\text{Var } X}$. Often the notation σ^2 is used in place of Var X, and σ is used in place of $\sqrt{\text{Var } X}$.

Example 8
(*continued*)

Return for a final time to the bob drop and the random variable Y. Using formula (5.19) and the form of Y's probability density,

$$\sigma^2 = \text{Var } Y = \int_{-\infty}^{0} \left(y - \frac{1}{120} \right)^2 \cdot 0 \, dy + \int_{0}^{1/60} \left(y - \frac{1}{120} \right)^2 \cdot 60 \, dy$$

$$+ \int_{1/60}^{\infty} \left(y - \frac{1}{120} \right)^2 \cdot 0 \, dy = \frac{60 \left(y - \dfrac{1}{120} \right)^3}{3} \Bigg|_{0}^{1/60}$$

$$= \frac{1}{3} \left(\frac{1}{120} \right)^2$$

So the standard deviation of Y is

$$\sigma = \sqrt{\text{Var } Y} = \sqrt{\frac{1}{3} \left(\frac{1}{120} \right)^2} = .0048 \text{ sec}$$

5.2.3   The Normal Probability Distributions

Just as there are a number of standard discrete distributions commonly applied to engineering problems, there are also a number of standard continuous probability distributions. This text has already alluded to the **normal** or **Gaussian distributions** and made use of their properties in producing normal plots. It is now time to introduce them formally.

Definition 15

> The **normal** or **Gaussian** (μ, σ^2) **distribution** is a continuous probability distribution with probability density
>
> $$f(x) = \frac{1}{\sqrt{2\pi\sigma^2}}e^{-(x-\mu)^2/2\sigma^2} \qquad \text{for all } x \qquad\qquad \textbf{(5.20)}$$
>
> for $\sigma > 0$.

It is not necessarily obvious, but formula (5.20) does yield a legitimate probability density, in that the total area under the curve $y = f(x)$ is 1. Further, it is also the case that

Normal distribution mean and variance

$$EX = \int_{-\infty}^{\infty} x \frac{1}{\sqrt{2\pi\sigma^2}}e^{-(x-\mu)^2/2\sigma^2}\,dx = \mu$$

and

$$\text{Var } X = \int_{-\infty}^{\infty} (x-\mu)^2 \frac{1}{\sqrt{2\pi\sigma^2}}e^{-(x-\mu)^2/2\sigma^2}\,dx = \sigma^2$$

That is, the parameters μ and σ^2 used in Definition 15 are indeed, respectively, the mean and variance (as defined in Definitions 13 and 14) of the distribution.

Figure 5.9 is a graph of the probability density specified by formula (5.20). The bell-shaped curve shown there is symmetric about $x = \mu$ and has inflection points at $\mu - \sigma$ and $\mu + \sigma$. The exact form of formula (5.20) has a number of theoretical origins. It is also a form that turns out to be empirically useful in a great variety of applications.

In theory, probabilities for the normal distributions can be found directly by integration using formula (5.20). Indeed, readers with pocket calculators that are preprogrammed to do numerical integration may find it instructive to check some of the calculations in the examples that follow, by straightforward use of formulas (5.14) and (5.20). But the freshman calculus methods of evaluating integrals via antidifferentiation will fail when it comes to the normal densities. They do not have antiderivatives that are expressible in terms of elementary functions. Instead, special normal probability tables are typically used.

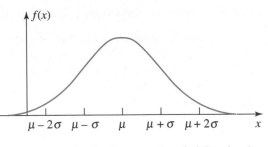

Figure 5.9   Graph of a normal probability density function

The use of tables for evaluating normal probabilities depends on the following relationship. If X is normally distributed with mean μ and variance σ^2,

$$P[a \le X \le b] = \int_a^b \frac{1}{\sqrt{2\pi\sigma^2}} e^{-(x-\mu)^2/2\sigma^2} \, dx = \int_{(a-\mu)/\sigma}^{(b-\mu)/\sigma} \frac{1}{\sqrt{2\pi}} e^{-z^2/2} \, dz \quad \textbf{(5.21)}$$

where the second inequality follows from the change of variable or substitution $z = \frac{x-\mu}{\sigma}$. Equation (5.21) involves an integral of the normal density with $\mu = 0$ and $\sigma = 1$. It says that evaluation of all normal probabilities can be reduced to the evaluation of normal probabilities for that special case.

Definition 16

> The normal distribution with $\mu = 0$ and $\sigma = 1$ is called the **standard normal distribution**.

Relation between normal (μ, σ^2) probabilities and standard normal probabilities

The relationship between normal (μ, σ^2) and standard normal probabilities is illustrated in Figure 5.10. Once one realizes that probabilities for all normal distributions can be had by tabulating probabilities for only the standard normal distribution, it is a relatively simple matter to use techniques of numerical integration to produce a standard normal table. The one that will be used in this text (other forms are possible) is given in Table B.3. It is a table of the **standard normal cumulative probability function**. That is, for values z located on the table's margins, the entries in the table body are

$$\Phi(z) = F(z) = \int_{-\infty}^{z} \frac{1}{\sqrt{2\pi}} e^{-t^2/2} \, dt$$

(Φ is routinely used to stand for the standard normal cumulative probability function, instead of the more generic F.)

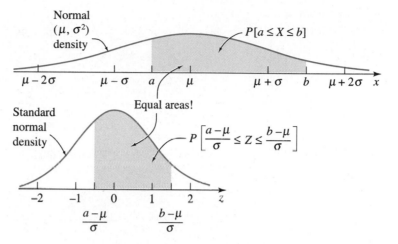

Figure 5.10   Illustration of the relationship between normal (μ, σ^2) and standard normal probabilities

---

Example 9

Standard Normal Probabilities

Suppose that Z is a standard normal random variable. We will find some probabilities for Z using Table B.3.

By a straight table look-up,

$$P[Z < 1.76] = \Phi(1.76) = .96$$

(The tabled value is .9608, but in keeping with the earlier promise to state final probabilities to only two decimal places, the tabled value was rounded to get .96.) After two table look-ups and a subtraction,

$$P[.57 < Z < 1.32] = P[Z < 1.32] - P[Z \leq .57]$$
$$= \Phi(1.32) - \Phi(.57)$$
$$= .9066 - .7157$$
$$= .19$$

And a single table look-up and a subtraction yield a right-tail probability like

$$P[Z > -.89] = 1 - P[Z \leq -.89] = 1 - .1867 = .81$$

As the table was used in these examples, probabilities for values z located on the table's margins were found in the table's body. The process can be run in

Example 9
(continued)

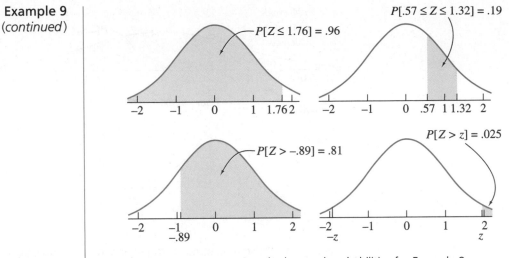

Figure 5.11   Standard normal probabilities for Example 9

reverse. Probabilities located in the table's body can be used to specify values z on the margins. For example, consider locating a value z such that

$$P[-z < Z < z] = .95$$

z will then put probability $\frac{1-.95}{2} = .025$ in the right tail of the standard normal *distribution*—i.e., be such that $\Phi(z) = .975$. Locating .975 in the table body, one sees that $z = 1.96$.

Figure 5.11 illustrates all of the calculations for this example.

The last part of Example 9 amounts to finding the .975 quantile for the standard normal distribution. In fact, the reader is now in a position to understand the origin of Table 3.10 (see page 89). The standard normal quantiles there were found by looking in the body of Table B.3 for the relevant probabilities and then locating corresponding z's on the margins.

In mathematical symbols, for $\Phi(z)$, the standard normal cumulative probability function, and $Q_z(p)$, **the standard normal quantile function,**

$$\left. \begin{array}{l} \Phi(Q_z(p)) = p \\ Q_z(\Phi(z)) = z \end{array} \right\} \tag{5.22}$$

Relationships (5.22) mean that Q_z and Φ are inverse functions. (In fact, the relationship $Q = F^{-1}$ is not just a standard normal phenomenon but is true in general for continuous distributions.)

Relationship (5.21) shows how to use the standard normal cumulative probability function to find general normal probabilities. For X normal (μ, σ^2) and a value

x associated with X, one converts to units of standard deviations above the mean via

z-value for a value x of a normal (μ, σ^2) random variable

$$z = \frac{x - \mu}{\sigma}$$ (5.23)

and then consults the standard normal table using z instead of x.

Example 10

Net Weights of Jars of Baby Food

J. Fisher, in his article "Computer Assisted Net Weight Control" (*Quality Progress*, June 1983), discusses the filling of food containers by weight. In the article, there is a reasonably bell-shaped histogram of individual net weights of jars of strained plums with tapioca. The mean of the values portrayed is about 137.2 g, and the standard deviation is about 1.6 g. The declared (or label) weight on jars of this product is 135.0 g.

Suppose that it is adequate to model

$$W = \text{the next strained plums and tapioca fill weight}$$

with a normal distribution with $\mu = 137.2$ and $\sigma = 1.6$. And further suppose the probability that the next jar filled is below declared weight (i.e., $P[W < 135.0]$) is of interest. Using formula (5.23), $w = 135.0$ is converted to units of standard deviations above μ (converted to a z-value) as

$$z = \frac{135.0 - 137.2}{1.6} = -1.38$$

Then, using Table B.3,

$$P[W < 135.0] = \Phi(-1.38) = .08$$

This model puts the chance of obtaining a below-nominal fill level at about 8%.

As a second example, consider the probability that W is within 1 gram of nominal (i.e., $P[134.0 < W < 136.0]$). Using formula (5.23), both $w_1 = 134.0$ and $w_2 = 136.0$ are converted to z-values or units of standard deviations above the mean as

$$z_1 = \frac{134.0 - 137.2}{1.6} = -2.00$$

$$z_2 = \frac{136.0 - 137.2}{1.6} = -.75$$

Example 10
(*continued*)

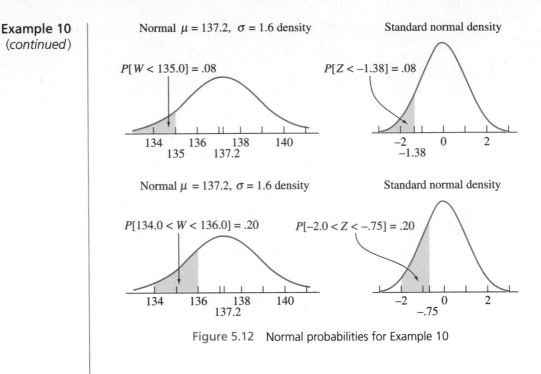

Figure 5.12   Normal probabilities for Example 10

So then

$$P[134.0 < W < 136.0] = \Phi(-.75) - \Phi(-2.00) = .2266 - .0228 = .20$$

The preceding two probabilities and their standard normal counterparts are shown in Figure 5.12.

The calculations for this example have consisted of starting with all of the quantities on the right of formula (5.23) and going from the margin of Table B.3 to its body to find probabilities for W. An important variant on this process is to instead go from the body of the table to its margins to obtain z, and then—given only two of the three quantities on the right of formula (5.23)—to solve for the third.

For example, suppose that it is easy to adjust the aim of the filling process (i.e., the mean μ of W) and one wants to decrease the probability that the next jar is below the declared weight of 135.0 to .01 by increasing μ. What is the minimum μ that will achieve this (assuming that σ remains at 1.6 g)?

Figure 5.13 shows what to do. μ must be chosen in such a way that $w = 135.0$ becomes the .01 quantile of the normal distribution with mean μ and standard deviation $\sigma = 1.6$. Consulting either Table 3.10 or Table B.3, it is easy to determine that the .01 quantile of the standard normal distribution is

$$z = Q_z(.01) = -2.33$$

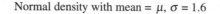

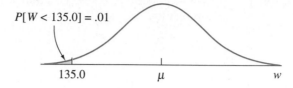

Figure 5.13   Normal distribution and
$P[W < 135.0] = .01$

So in light of equation (5.23) one wants

$$-2.33 = \frac{135.0 - \mu}{1.6}$$

i.e.,

$$\mu = 138.7 \text{ g}$$

An increase of about $138.7 - 137.2 = 1.5$ g in fill level aim is required.

In practical terms, the reduction in $P[W < 135.0]$ is bought at the price of increasing the average *give-away cost* associated with filling jars so that on average they contain much more than the nominal contents. In some applications, this type of cost will be prohibitive. There is another approach open to a process engineer. That is to reduce the variation in fill level through acquiring more precise filling equipment. In terms of equation (5.23), instead of increasing μ one might consider paying the cost associated with reducing σ. The reader is encouraged to verify that a reduction in σ to about .94 g would also produce $P[W < 135.0] = .01$ without any change in μ.

As Example 10 illustrates, equation (5.23) is the fundamental relationship used in problems involving normal distributions. One way or another, three of the four entries in equation (5.23) are specified, and the fourth must be obtained.

5.2.4   The Exponential Distributions (*Optional*)

Section 5.1 discusses the fact that the Poisson distributions are often used as models for the number of occurrences of a relatively rare phenomenon in a specified interval of time. The same mathematical theory that suggests the appropriateness of the Poisson distributions in that context also suggests the usefulness of the **exponential distributions** for describing waiting times until occurrences.

Definition 17

The **exponential (α) distribution** is a continuous probability distribution with probability density

$$f(x) = \begin{cases} \dfrac{1}{\alpha}e^{-x/\alpha} & \text{for } x > 0 \\[2mm] 0 & \text{otherwise} \end{cases} \qquad (5.24)$$

for $\alpha > 0$.

Figure 5.14 shows plots of $f(x)$ for three different values of α. Expression (5.24) is extremely convenient, and it is not at all difficult to show that α is both the mean and the standard deviation of the exponential (α) distribution. That is,

Mean of the exponential (α) distribution

$$\mu = EX = \int_0^\infty x\frac{1}{\alpha}e^{-x/\alpha}\,dx = \alpha$$

and

Variance of the exponential (α) distribution

$$\sigma^2 = \text{Var } X = \int_0^\infty (x - \alpha)^2 \frac{1}{\alpha}e^{-x/\alpha}\,dx = \alpha^2$$

Further, the exponential (α) distribution has a simple cumulative probability function,

Exponential (α) cumulative probability function

$$F(x) = \begin{cases} 0 & \text{if } x \le 0 \\[2mm] 1 - e^{-x/\alpha} & \text{if } x > 0 \end{cases} \qquad (5.25)$$

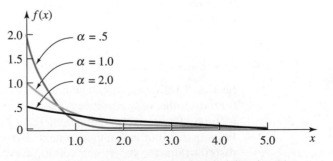

Figure 5.14   Three exponential probability densities

Example 11
(*Example 7 revisited*)

The Exponential Distribution and Arrivals at a University Library

Recall that Stork, Wohlsdorf, and McArthur found the arrival rate of students at the ISU library between 12:00 and 12:10 P.M. early in the week to be about 12.5 students per minute. That translates to a $\frac{1}{12.5} = .08$ min average waiting time between student arrivals.

Consider observing the ISU library entrance beginning at exactly noon next Tuesday and define the random variable

$T =$ the waiting time (in minutes) until the first student passes through the door

A possible model for T is the exponential distribution with $\alpha = .08$. Using it, the probability of waiting more than 10 seconds ($\frac{1}{6}$ min) for the first arrival is

$$P\left[T > \frac{1}{6}\right] = 1 - F\left(\frac{1}{6}\right) = 1 - \left(1 - e^{-1/6(.08)}\right) = .12$$

This result is pictured in Figure 5.15.

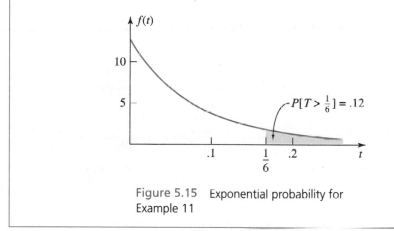

Figure 5.15   Exponential probability for Example 11

Geometric and exponential distributions

The exponential distribution is the continuous analog of the geometric distribution in several respects. For one thing, both the geometric probability function and the exponential probability density decline exponentially in their arguments x. For another, they both possess a kind of **memoryless property**. If the first success in a series of independent identical success-failure trials is known not to have occurred through trial t_0, then the additional number of trials (beyond t_0) needed to produce the first success is a geometric (p) random variable (as was the total number of trials required from the beginning). Similarly, if an exponential (α) waiting time is known not to have been completed by time t_0, then the additional waiting time to completion is exponential (α). This memoryless property is related to the *force-of-mortality function* of the distribution being constant. The force-of-mortality function for a distribution is a concept of reliability theory discussed briefly in Appendix A.4.

5.2.5 The Weibull Distributions (*Optional*)

The Weibull distributions generalize the exponential distributions and provide much more flexibility in terms of distributional shape. They are extremely popular with engineers for describing the strength properties of materials and the life lengths of manufactured devices. The most natural way to specify these distributions is through their cumulative probability functions.

Definition 18

The **Weibull** (α, β) **distribution** is a continuous probability distribution with cumulative probability function

$$F(x) = \begin{cases} 0 & \text{if } x < 0 \\ 1 - e^{-(x/\alpha)^\beta} & \text{if } x \geq 0 \end{cases} \tag{5.26}$$

for parameters $\alpha > 0$ and $\beta > 0$.

Beginning from formula (5.26), it is possible to determine properties of the Weibull distributions. Differentiating formula (5.26) produces the Weibull (α, β) probability density

Weibull (α, β) probability density

$$f(x) = \begin{cases} 0 & \text{if } x < 0 \\ \dfrac{\beta}{\alpha^\beta} x^{\beta-1} e^{-(x/\alpha)^\beta} & \text{if } x > 0 \end{cases} \tag{5.27}$$

This in turn can be shown to yield the mean

Weibull (α, β) mean

$$\mu = EX = \alpha \Gamma \left(1 + \tfrac{1}{\beta} \right) \tag{5.28}$$

and variance

Weibull (α, β) variance

$$\sigma^2 = \text{Var } X = \alpha^2 \left[\Gamma \left(1 + \tfrac{2}{\beta} \right) - \left(\Gamma \left(1 + \tfrac{1}{\beta} \right) \right)^2 \right] \tag{5.29}$$

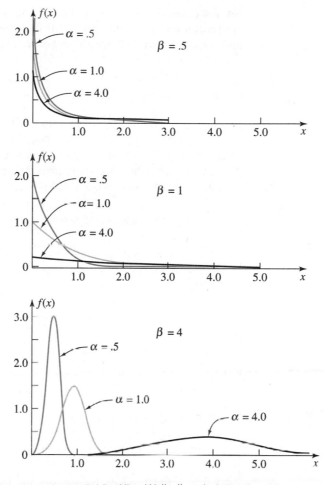

Figure 5.16   Nine Weibull probability densities

where $\Gamma(x) = \int_0^\infty t^{x-1} e^{-t}\, dt$ is the *gamma function* of advanced calculus. (For integer values n, $\Gamma(n) = (n - 1)!$.) These formulas for $f(x)$, μ, and σ^2 are not particularly illuminating. So it is probably most helpful to simply realize that β controls the shape of the Weibull distribution and that α controls the scale. Figure 5.16 shows plots of $f(x)$ for several (α, β) pairs.

Note that $\beta = 1$ gives the special case of the exponential distributions. For small β, the distributions are decidedly right-skewed, but for β larger than about 3.6, they actually become left-skewed. Regarding distribution location, the form of the distribution mean given in equation (5.28) is not terribly revealing. It is perhaps more helpful that the median for the Weibull (α, β) distribution is

Weibull (α, β)
median

$$Q(.5) = \alpha e^{-(.3665/\beta)} \tag{5.30}$$

So, for example, for large shape parameter β the Weibull median is essentially α. And formulas (5.28) through (5.30) show that for fixed β the Weibull mean, median, and standard deviation are all proportional to the scale parameter α.

Example 12

The Weibull Distribution and the Strength of a Ceramic Material

The report "Review of Workshop on Design, Analysis and Reliability Prediction for Ceramics—Part II" by E. Lenoe (*Office of Naval Research Far East Scientific Bulletin*, 1987) suggests that tensile strengths (MPa) of .95 mm rods of HIPped UBE SN-10 with 2.5% yttria material can be described by a Weibull distribution with $\beta = 8.8$ and median 428 MPa. Let

$$S = \text{measured tensile strength of an additional rod (MPa)}$$

Under the assumption that S can be modeled using a Weibull distribution with the suggested characteristics, suppose that $P[S \le 400]$ is needed. Using equation (5.30),

$$428 = \alpha e^{-(.3665/8.8)}$$

Thus, the Weibull scale parameter is

$$\alpha = 446$$

Then, using equation (5.26),

$$P[S \le 400] = 1 - e^{-(400/446)^{8.8}} = .32$$

Figure 5.17 illustrates this probability calculation.

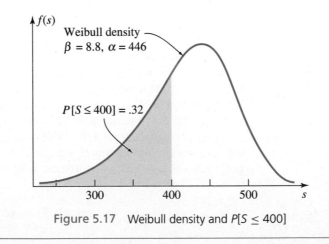

Figure 5.17   Weibull density and $P[S \le 400]$

Section 2 Exercises ....

1. The random number generator supplied on a calculator is not terribly well chosen, in that values it generates are not adequately described by a distribution uniform on the interval (0, 1). Suppose instead that a probability density

$$f(x) = \begin{cases} k(5 - x) & \text{for } 0 < x < 1 \\ 0 & \text{otherwise} \end{cases}$$

   is a more appropriate model for X = the next value produced by this random number generator.
   (a) Find the value of k.
   (b) Sketch the probability density involved here.
   (c) Evaluate $P[.25 < X < .75]$.
   (d) Compute and graph the cumulative probability function for X, $F(x)$.
   (e) Calculate EX and the standard deviation of X.

2. Suppose that Z is a standard normal random variable. Evaluate the following probabilities involving Z:
   (a) $P[Z < .62]$ (b) $P[Z > 1.06]$
   (c) $P[-.37 < Z < .51]$ (d) $P[|Z| \leq .47]$
   (e) $P[|Z| > .93]$ (f) $P[-3.0 < Z < 3.0]$
   Now find numbers # such that the following statements involving Z are true:
   (g) $P[Z \leq \#] = .90$ (h) $P[|Z| < \#] = .90$
   (i) $P[|Z| > \#] = .03$

3. Suppose that X is a normal random variable with mean 43.0 and standard deviation 3.6. Evaluate the following probabilities involving X:
   (a) $P[X < 45.2]$ (b) $P[X \leq 41.7]$
   (c) $P[43.8 < X \leq 47.0]$ (d) $P[|X - 43.0| \leq 2.0]$
   (e) $P[|X - 43.0| > 1.7]$
   Now find numbers # such that the following statements involving X are true:
   (f) $P[X < \#] = .95$ (g) $P[X \geq \#] = .30$
   (h) $P[|X - 43.0| > \#] = .05$

4. The diameters of bearing journals ground on a particular grinder can be described as normally distributed with mean 2.0005 in. and standard deviation .0004 in.
   (a) If engineering specifications on these diameters are 2.0000 in. ± .0005 in., what fraction of these journals are in specifications?

   (b) What adjustment to the grinding process (holding the process standard deviation constant) would increase the fraction of journal diameters that will be in specifications? What appears to be the best possible fraction of journal diameters inside ± .0005 in. specifications, given the $\sigma = .0004$ in. apparent precision of the grinder?
   (c) Suppose consideration was being given to purchasing a more expensive/newer grinder, capable of holding tighter tolerances on the parts it produces. What σ would have to be associated with the new machine in order to guarantee that (when perfectly adjusted so that $\mu = 2.0000$) the grinder would produce diameters with at least 95% meeting 2.0000 in. ± .0005 in. specifications?

5. The mileage to first failure for a model of military personnel carrier can be modeled as exponential with mean 1,000 miles.
   (a) Evaluate the probability that a vehicle of this type gives less than 500 miles of service before first failure. Evaluate the probability that it gives at least 2,000 miles of service before first failure.
   (b) Find the .05 quantile of the distribution of mileage to first failure. Then find the .90 quantile of the distribution.

6. Some data analysis shows that lifetimes, x (in 10^6 revolutions before failure), of certain ball bearings can be modeled as Weibull with $\beta = 2.3$ and $\alpha = 80$.
   (a) Make a plot of the Weibull density (5.27) for this situation. (Plot for x between 0 and 200. Standard statistical software packages like MINITAB will have routines for evaluating this density. In MINITAB look under the "Calc/Probability Distributions/Weibull" menu.)
   (b) What is the median bearing life?
   (c) Find the .05 and .95 quantiles of bearing life.

5.3   Probability Plotting (*Optional*)

Calculated probabilities are only as relevant in a given application as are the distributions used to produce them. It is thus important to have data-based methods to assess the relevance of a given continuous distribution to a given application. The basic logic for making such tools was introduced in Section 3.2. Suppose you have data consisting of n realizations of a random variable X, say $x_1 \leq x_2 \leq \cdots \leq x_n$ and want to know whether a probability density with the same shape as $f(x)$ might adequately describe X. To investigate, it is possible to make and interpret a probability plot consisting of n ordered pairs

Ordered pairs making a probability plot

$$\left(x_i, Q\left(\frac{i - .5}{n}\right)\right)$$

where x_i is the ith smallest data value (the $\left(\frac{i-.5}{n}\right)$ quantile of the data set) and $Q\left(\frac{i-.5}{n}\right)$ is the $\left(\frac{i-.5}{n}\right)$ quantile of the probability distribution specified by $f(x)$.

This section will further discuss the importance of this method. First, some additional points about probability plotting are made in the familiar context where $f(x)$ is the standard normal density (i.e., in the context of normal plotting). Then the general applicability of the idea is illustrated by using it in assessing the appropriateness of exponential and Weibull models. In the course of the discussion, the importance of probability plotting to process capability studies and life data analysis will be indicated.

5.3.1   More on Normal Probability Plots

Definition 15 gives the form of the normal or Gaussian probability density with mean μ and variance σ^2. The discussion that follows the definition shows that all normal distributions have the same essential shape. Thus, a theoretical Q-Q plot using standard normal quantiles can be used to judge whether or not there is *any* normal probability distribution that seems a sensible model.

Example 13

Weights of Circulating U.S. Nickels

Ash, Davison, and Miyagawa studied characteristics of U.S. nickels. They obtained the weights of 100 nickels to the nearest .01 g. They found those to have a mean of 5.002 g and a standard deviation of .055 g. Consider the weight of another nickel taken from a pocket, say, U. It is sensible to think that $EU \approx 5.002$ g and $\sqrt{\mathrm{Var}\, U} \approx .055$ g. Further, it would be extremely convenient if a normal distribution could be used to describe U. Then, for example, normal distribution calculations with $\mu = 5.002$ g and $\sigma = .055$ g could be used to assess

$$P[U > 5.05] = P[\text{the nickel weighs over 5.05 g}]$$

A way of determining whether or not the students' data support the use of a normal model for U is to make a normal probability plot. Table 5.6 presents the data collected by Ash, Davison, and Miyagawa. Table 5.7 shows some of the calculations used to produce the normal probability plot in Figure 5.18.

Table 5.6
Weights of 100 U.S. Nickels

Weight (g)	Frequency	Weight (g)	Frequency
4.81	1	5.00	12
4.86	1	5.01	10
4.88	1	5.02	7
4.89	1	5.03	7
4.91	2	5.04	5
4.92	2	5.05	4
4.93	3	5.06	4
4.94	2	5.07	3
4.95	6	5.08	2
4.96	4	5.09	3
4.97	5	5.10	2
4.98	4	5.11	1
4.99	7	5.13	1

Table 5.7
Example Calculations for a Normal Plot of Nickel Weights

i	$\left(\dfrac{i-.5}{100}\right)$	x_i	$Q_z\left(\dfrac{i-.5}{100}\right)$
1	.005	4.81	-2.576
2	.015	4.86	-2.170
3	.025	4.88	-1.960
4	.035	4.89	-1.812
5	.045	4.91	-1.695
6	.055	4.91	-1.598
7	.065	4.92	-1.514
⋮	⋮	⋮	⋮
98	.975	5.10	1.960
99	.985	5.11	2.170
100	.995	5.13	2.576

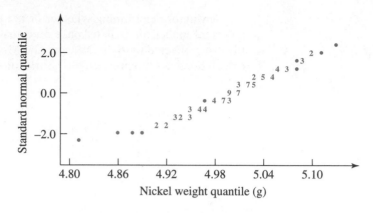

Figure 5.18   Normal plot of nickel weights

At least up to the resolution provided by the graphics in Figure 5.18, the plot is pretty linear for weights above, say, 4.90 g. However, there is some indication that the shape of the lower end of the weight distribution differs from that of a normal distribution. Real nickels seem to be more likely to be light than a normal model would predict. Interestingly enough, the four nickels with weights under 4.90 g were all minted in 1970 or before (these data were collected in 1988). This suggests the possibility that the shape of the lower end of the weight distribution is related to wear patterns and unusual damage (particularly the extreme lower tail represented by the single 1964 coin with weight 4.81 g).

But whatever the origin of the shape in Figure 5.18, its message is clear. For most practical purposes, a normal model for the random variable

$$U = \text{the weight of a nickel taken from a pocket}$$

will suffice. Bear in mind, though, that such a distribution will tend to slightly overstate probabilities associated with larger weights and understate probabilities associated with smaller weights.

Much was made in Section 3.2 of the fact that linearity on a Q-Q plot indicates equality of distribution shape. But to this point, no use has been made of the fact that when there is near-linearity on a Q-Q plot, the *nature of the linear relationship* gives information regarding the relative location and spread of the two distributions involved. This can sometimes provide a way to choose sensible parameters of a theoretical distribution for describing the data set.

For example, a normal probability plot can be used not only to determine whether some normal distribution might describe a random variable but also to graphically pick out *which one* might be used. For a roughly linear normal plot,

Reading a mean and standard deviation from a normal plot

1. the horizontal coordinate corresponding to a vertical coordinate of 0 provides a mean for a normal distribution fit to the data set, and

2. the reciprocal of the slope provides a standard deviation (this is the difference between the horizontal coordinates of points with vertical coordinates differing by 1).

Example 14

Normal Plotting and Thread Lengths of U-bolts

Table 5.8 gives thread lengths produced in the manufacture of some U-bolts for the auto industry. The measurements are in units of .001 in. over nominal. The particular bolts that gave the measurements in Table 5.8 were sampled from a single machine over a 20-minute period.

Figure 5.19 gives a normal plot of the data. It indicates that (allowing for the fact that the relatively crude measurement scale employed is responsible for the discrete/rough appearance of the plot) a normal distribution might well have been a sensible probability model for the random variable

$$L = \text{the actual thread length of an additional U-bolt}$$
$$\text{manufactured in the same time period}$$

The line eye-fit to the plot further suggests appropriate values for the mean and standard deviation: $\mu \approx 10.8$ and $\sigma \approx 2.1$ (Direct calculation with the data in Table 5.8 gives a sample mean and standard deviation of, respectively, $\bar{l} \approx 10.9$ and $s \approx 1.9$.)

Table 5.8
Measured Thread Lengths for 25 U-Bolts

Thread Length (.001 in. over Nominal)	Tally	Frequency
6	|	1
7		0
8	|||	3
9		0
10	||||	4
11	⋕⋕ ⋕⋕	10
12		0
13	⋕⋕ |	6
14	|	1

Example 14
(*continued*)

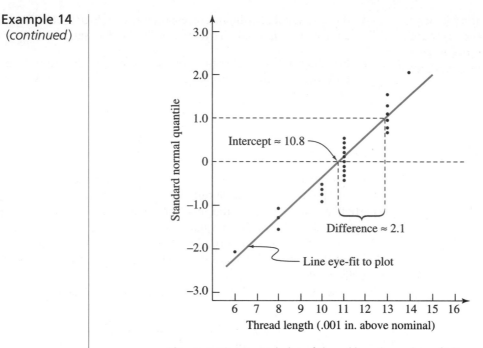

Figure 5.19   Normal plot of thread lengths and eye-fit line

In manufacturing contexts like the previous example, it is common to use the fact that an approximate standard deviation can easily be read from the (reciprocal) slope of a normal plot to obtain a graphical tool for assessing process potential. That is, the primary limitation on the performance of an industrial machine or process is typically the basic precision or short-term variation associated with it. Suppose a dimension of the output of such a process or machine over a short period is approximately normally distributed with standard deviation σ. Then, since for any normal random variable X with mean μ and standard deviation σ,

$$P[\mu - 3\sigma < X < \mu + 3\sigma] > .99$$

*6σ as a process
capability*

it makes some sense to use 6σ $(= (\mu + 3\sigma) - (\mu - 3\sigma))$ as a measure of **process capability**. And it is easy to read such a capability figure off a normal plot. Many companies use specially prepared *process capability analysis forms* (which are in essence pieces of normal probability paper) for this purpose.

Example 14
(*continued*)

Figure 5.20 is a plot of the thread length data from Table 5.8, made on a common capability analysis sheet. Using the plot, it is very easy, even for someone with limited quantitative background (and perhaps even lacking a basic understanding of the concept of a standard deviation), to arrive at the figure

$$\text{Process capability} \approx 16 - 5 = 11(.001 \text{ in.})$$

CAPABILITY ANALYSIS SHEET
R-419-157

Part/Dept./Supplier		Date		(0.003%)
Part Identity		Spec.		+4σ
Operation Indentity		99.73% (±3σ)		
Person Performing Study		99.994% (±4σ)		
Char. Measured *Thread Length*		Unit of Measure *.001" over Nominal*		(0.135%)

1	VALUE					3	4	5	6	7	8	9	10	11	12	13	14	15	16	17		
2	FREQUENCY								1	0	3	0	4	10	0	6	1					
	Follow arrows and perform additions as shown (N ≥ 25)																					
3	EST. ACCUM. FREQ. (EAF)								1	2	5	8	12	26	36	42	49					
4	PLOT POINTS (%) (EAF/2N) × 100								2	4	10	16	24	52	72	84	98					

Figure 5.20   Thread length data plotted on a capability analysis form (used with permission of Reynolds Metals Company)

5.3.2   Probability Plots for Exponential and Weibull Distributions

To illustrate the application of probability plotting to distributions that are not normal (Gaussian), the balance of this section considers its use with first exponential and then general Weibull models.

Example 15	Service Times at a Residence Hall Depot Counter and Exponential Probability Plotting

Jenkins, Milbrath, and Worth studied service times at a residence hall "depot" counter. Figure 5.21 gives the times (in seconds) required to complete 65 different postage stamp sales at the counter.

The shape of the stem-and-leaf diagram is reminiscent of the shape of the exponential probability densities shown in Figure 5.14. So if one defines the random variable

$$T = \text{the next time required to complete a postage stamp sale at the depot counter}$$

an exponential distribution might somehow be used to describe T.

```
0 |
0 | 8  8  8  9  9
1 | 0  0  0  0  0  2  2  2  2  3  3  4  4  4  4  4  4
1 | 5  6  7  7  7  7  8  8  9  9
2 | 0  1  2  2  2  2  3  4
2 | 6  7  8  9  9  9
3 | 0  2  2  2  4  4
3 | 6  6  7  7
4 | 2  3
4 | 5  6  7  8  8
5 |
5 |
6 |
6 |
7 | 0
7 |
8 |
8 | 7
```

Figure 5.21   Stem-and-leaf plot of service times

The exponential distributions introduced in Definition 17 all have the same essential shape. Thus the exponential distribution with $\alpha = 1$ is a convenient representative of that shape. A plot of $\alpha = 1$ exponential quantiles versus corresponding service time quantiles will give a tool for comparing the empirical shape to the theoretical exponential shape.

For an exponential distribution with mean $\alpha = 1$,

$$F(x) = 1 - e^{-x} \qquad \text{for } x > 0$$

So for $0 < p < 1$, setting $F(x) = p$ and solving,

$$x = -\ln(1 - p)$$

That is, $-\ln(1 - p) = Q(p)$, the p quantile of this distribution. Thus, for data $x_1 \leq x_2 \leq \cdots \leq x_n$, an exponential probability plot can be made by plotting the ordered pairs

Points to plot for an exponential probability plot

$$\left(x_i, -\ln\left(1 - \frac{i - .5}{n} \right) \right) \tag{5.31}$$

Figure 5.22 is a plot of the points in display (5.31) for the service time data. It shows remarkable linearity. Except for the fact that the third- and fourth-largest service times (both 48 seconds) appear to be somewhat smaller than might be predicted based on the shape of the exponential distribution, the empirical service time distribution corresponds quite closely to the exponential distribution shape.

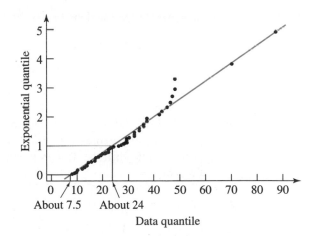

Figure 5.22   Exponential probability plot and eye-fit line for the service times

Example 15
(*continued*)

As was the case in normal-plotting, the character of the linearity in Figure 5.22 also carries some valuable information that can be applied to the modeling of the random variable T. The positioning of the line sketched onto the plot indicates the appropriate location of an exponentially shaped distribution for T, and the slope of the line indicates the appropriate spread for that distribution.

As introduced in Definition 17, the exponential distributions have positive density $f(x)$ for positive x. One might term 0 a *threshold value* for the distributions defined there. In Figure 5.22 the threshold value ($0 = Q(0)$) for the exponential distribution with $\alpha = 1$ corresponds to a service time of roughly 7.5 seconds. This means that to model a variable related to T with a distribution exactly of the form given in Definition 17, it is

$$S = T - 7.5$$

that should be considered.

Further, a change of one unit on the vertical scale in the plot corresponds to a change on the horizontal scale of roughly

$$24 - 7.5 = 16.5 \text{ sec}$$

That is, an exponential model for S ought to have an associated spread that is 16.5 times that of the exponential distribution with $\alpha = 1$.

So ultimately, the data in Figure 5.21 lead via exponential probability plotting to the suggestion that

$S = T - 7.5$

   = the excess of the next time required to complete a postage stamp sale
       over a threshold value of 7.5 seconds

be described with the density

$$f(s) = \begin{cases} \dfrac{1}{16.5} e^{-(s/16.5)} & \text{for } s > 0 \\[2mm] 0 & \text{otherwise} \end{cases} \tag{5.32}$$

Probabilities involving T can be computed by first expressing them in terms of S and then using expression (5.32). If for some reason a density for T itself is desired, simply shift the density in equation (5.32) to the right 7.5 units to obtain the density

$$f(t) = \begin{cases} \dfrac{1}{16.5} e^{-((t-7.5)/16.5)} & \text{for } t > 7.5 \\[2mm] 0 & \text{otherwise} \end{cases}$$

Figure 5.23 shows probability densities for both S and T.

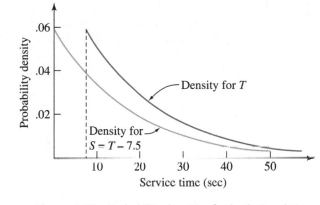

Figure 5.23 Probability densities for both S and T

To summarize the preceding example: Because of the relatively simple form of the exponential $\alpha = 1$ cumulative probability function, it is easy to find quantiles for this distribution. When these are plotted against corresponding quantiles of a data set, an exponential probability plot is obtained. On this plot, linearity indicates exponential shape, the horizontal intercept of a linear plot indicates an appropriate threshold value, and the reciprocal of the slope indicates an appropriate value for the exponential parameter α.

Much the same story can be told for the Weibull distributions for any fixed β. That is, using the form (5.26) of the Weibull cumulative probability function, it is straightforward to argue that for data $x_1 \leq x_2 \leq \cdots \leq x_n$, a plot of the ordered pairs

Points to plot for a fixed β Weibull plot

$$\left(x_i, \left(-\ln\left(1 - \frac{i - .5}{n} \right) \right)^{1/\beta} \right)$$

(5.33)

is a tool for investigating whether a variable might be described using a Weibull-shaped distribution *for the particular β in question*. On such a plot, linearity indicates Weibull shape β, the horizontal intercept indicates an appropriate threshold value, and the reciprocal of the slope indicates an appropriate value for the parameter α.

Although the kind of plot indicated by display (5.33) is easy to make and interpret, it is *not* the most common form of probability plotting associated with the Weibull distributions. In order to plot the points in display (5.33), a value of β is input (and a threshold and scale parameter are read off the graph). In most engineering applications of the Weibull distributions, what is needed (instead of a method that inputs β and can be used to identify a threshold and α) is a method that tacitly inputs the 0 threshold implicit in Definition 18 and can be used to identify α and β. This is particularly true in applications to reliability, where the useful life or time to failure of some device is the variable of interest. It is similarly true in applications to material science, where intrinsically positive material properties like yield strength are under study.

It is possible to develop a probability plotting method that allows identification of values for both α and β in Definition 18. The trick is to work on a log scale. That is, if X is a random variable with the Weibull (α, β) distribution, then for $x > 0$,

$$F(x) = 1 - e^{-(x/\alpha)^{\beta}}$$

so that with $Y = \ln(X)$

$$P[Y \leq y] = P[X \leq e^{y}]$$
$$= 1 - e^{-(e^{y}/\alpha)^{\beta}}$$

So for $0 < p < 1$, setting $p = P[Y \leq y]$ gives

$$p = 1 - e^{-(e^{y}/\alpha)^{\beta}}$$

After some algebra this implies

$$\beta y - \beta \ln(\alpha) = \ln\left(-\ln(1-p)\right) \tag{5.34}$$

Now y is (by design) the p quantile of the distribution of $Y = \ln(X)$. So equation (5.34) says that $\ln(-\ln(1-p))$ is a linear function of $\ln(X)$'s quantile function. The slope of that relationship is β. Further, equation (5.34) shows that when $\ln(-\ln(1-p)) = 0$, the quantile function of $\ln(X)$ has the value $\ln(\alpha)$. So exponentiation of the horizontal intercept gives α. Thus, for data $x_1 \leq x_2 \leq \cdots \leq x_n$, one is led to consider a plot of ordered pairs

Points to plot for a 0-threshold Weibull plot

$$\left(\ln x_i, \ln\left(-\ln\left(1 - \frac{i - .5}{n}\right)\right) \right) \tag{5.35}$$

Reading α and β from a 0-threshold Weibull plot

If data in hand are consistent with a (0-threshold) Weibull (α, β) model, a reasonably linear plot with

1. slope β and

2. horizontal axis intercept equal to $\ln(\alpha)$

may be expected.

Example 16

Electrical Insulation Failure Voltages and Weibull Plotting

The data given in the stem-and-leaf plot of Figure 5.24 are failure voltages (in kv/mm) for a type of electrical cable insulation subjected to increasing voltage

```
3 |
3 | 9.4
4 | 5.3
4 | 9.2, 9.4
5 | 1.3, 2.0, 3.2, 3.2, 4.9
5 | 5.5, 7.1, 7.2, 7.5, 9.2
6 | 1.0, 2.4, 3.8, 4.3
6 | 7.3, 7.7
```

Figure 5.24   Stem-and-leaf plot of insulation failure voltages

stress. They were taken from *Statistical Models and Methods for Lifetime Data* by J. F. Lawless.

Consider the Weibull modeling of

$$R = \text{ the voltage at which one additional specimen}$$
$$\text{of this insulation will fail}$$

Table 5.9 shows some of the calculations needed to use display (5.35) to produce Figure 5.25. The near-linearity of the plot in Figure 5.25 suggests that a (0-threshold) Weibull distribution might indeed be used to describe R. A Weibull shape parameter of roughly

$$\beta \approx \text{ slope of the fitted line} \approx \frac{1 - (-4)}{4.19 - 3.67} \approx 9.6$$

is indicated. Further, a scale parameter α with

$$\ln(\alpha) \approx \text{ horizontal intercept} \approx 4.08$$

and thus

$$\alpha \approx 59$$

appears appropriate.

Example 16
(*continued*)

Table 5.9
Example Calculations for a 0-Threshold Weibull Plot of Failure Voltages

i	$x_i = i$th Smallest Voltage	$\ln(x_i)$	$p = (i - .5)/20$	$\ln(-\ln(1 - p))$
1	39.4	3.67	.025	−3.68
2	45.3	3.81	.075	−2.55
3	49.2	3.90	.125	−2.01
4	49.4	3.90	.175	−1.65
⋮	⋮	⋮	⋮	⋮
19	67.3	4.21	.925	.95
20	67.7	4.22	.975	1.31

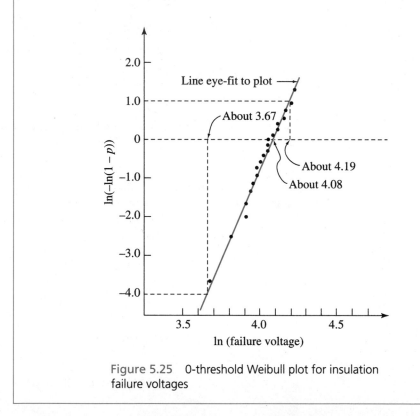

Figure 5.25   0-threshold Weibull plot for insulation failure voltages

Plotting form (5.35) is quite popular in reliability and materials applications. It is common to see such Weibull plots made on special **Weibull paper** (see Figure 5.26). This is graph paper whose scales are constructed so that instead of using plotting positions (5.35) on regular graph paper, one can use plotting positions

$$\left(x_i, \frac{i - .5}{n}\right)$$

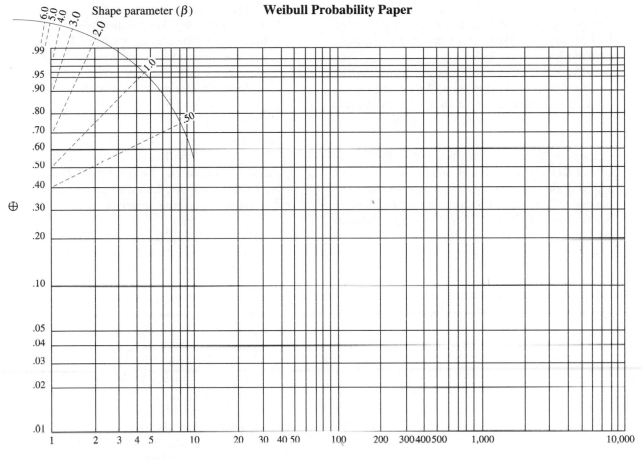

Figure 5.26   Weibull probability paper

for data $x_1 \leq x_2 \leq \cdots \leq x_n$. (The determination of β is even facilitated through the inclusion of the protractor in the upper left corner.) Further, standard statistical packages often have built-in facilities for Weibull plotting of this type.

It should be emphasized that the idea of probability plotting is a quite general one. Its use has been illustrated here only with normal, exponential, and Weibull distributions. But remember that for any probability density $f(x)$, theoretical Q-Q plotting provides a tool for assessing whether the distributional shape portrayed by $f(x)$ might be used in the modeling of a random variable.

Section 3 Exercises

1. What is the practical usefulness of the technique of probability plotting?

2. Explain how an approximate mean μ and standard deviation σ can be read off a plot of standard normal quantiles versus data quantiles.

3. Exercise 3 of Section 3.2 refers to the chemical process yield data of J. S. Hunter given in Exercise 1 of Section 3.1. There you were asked to make a normal plot of those data.

   (a) If you have not already done so, use a computer package to make a version of the normal plot.

   (b) Use your plot to derive an approximate mean and a standard deviation for the chemical process yields.

4. The article "Statistical Investigation of the Fatigue Life of Deep Groove Ball Bearings" by J. Leiblein and M. Zelen (*Journal of Research of the National Bureau of Standards*, 1956) contains the data given below on the lifetimes of 23 ball bearings. The units are 10^6 revolutions before failure.

   > 17.88, 28.92, 33.00, 41.52, 42.12, 45.60,
   > 48.40, 51.84, 51.96, 54.12, 55.56, 67.80,
   > 68.64, 68.64, 68.88, 84.12, 93.12, 98.64,
   > 105.12, 105.84, 127.92, 128.04, 173.40

   (a) Use a normal plot to assess how well a normal distribution fits these data. Then determine if bearing load life can be better represented by a normal distribution if life is expressed on the log scale. (Take the natural logarithms of these data and make a normal plot.) What mean and standard deviation would you use in a normal description of log load life? For these parameters, what are the .05 quantiles of ln(life) and of life?

   (b) Use the method of display (5.35) and investigate whether the Weibull distribution might be used to describe bearing load life. If a Weibull description is sensible, read appropriate parameter values from the plot. Then use the form of the Weibull cumulative probability function given in Section 5.2 to find the .05 quantile of the bearing load life distribution.

5. The data here are from the article "Fiducial Bounds on Reliability for the Two-Parameter Negative Exponential Distribution," by F. Grubbs (*Technometrics*, 1971). They are the mileages at first failure for 19 military personnel carriers.

   > 162, 200, 271, 320, 393, 508, 539, 629,
   > 706, 777, 884, 1008, 1101, 1182, 1462,
   > 1603, 1984, 2355, 2880

   (a) Make a histogram of these data. How would you describe its shape?

   (b) Plot points (5.31) and make an exponential probability plot for these data. Does it appear that the exponential distribution can be used to model the mileage to failure of this kind of vehicle? In Example 15, a threshold service time of 7.5 seconds was suggested by a similar exponential probability plot. Does the present plot give a strong indication of the need for a threshold mileage larger than 0 if an exponential distribution is to be used here?

5.4 Joint Distributions and Independence

Most applications of probability to engineering statistics involve not one but several random variables. In some cases, the application is intrinsically multivariate. It then makes sense to think of more than one process variable as subject to random influences and to evaluate probabilities associated with them in combination. Take, for example, the assembly of a ring bearing with nominal inside diameter 1.00 in. on a rod with nominal diameter .99 in. If

$$X = \text{the ring bearing inside diameter}$$
$$Y = \text{the rod diameter}$$

one might be interested in

$$P[X < Y] = P[\text{there is an interference in assembly}]$$

which involves *both* variables.

But even when a situation is univariate, samples larger than size 1 are essentially always used in engineering applications. The *n* data values in a sample are usually thought of as subject to chance causes and their simultaneous behavior must then be modeled. The methods of Sections 5.1 and 5.2 are capable of dealing with only a single random variable at a time. They must be generalized to create methods for describing several random variables simultaneously.

Entire books are written on various aspects of the simultaneous modeling of many random variables. This section can give only a brief introduction to the topic. Considering first the comparatively simple case of jointly discrete random variables, the topics of joint and marginal probability functions, conditional distributions, and independence are discussed primarily through reference to simple bivariate examples. Then the analogous concepts of joint and marginal probability density functions, conditional distributions, and independence for jointly continuous random variables are introduced. Again, the discussion is carried out primarily through reference to a bivariate example.

5.4.1 Describing Jointly Discrete Random Variables

For several discrete variables the device typically used to specify probabilities is a **joint probability function**. The two-variable version of this is defined next.

Definition 19

A **joint probability function** for discrete random variables X and Y is a nonnegative function $f(x, y)$, giving the probability that (simultaneously) X takes the value x and Y takes the value y. That is,

$$f(x, y) = P[X = x \text{ and } Y = y]$$

Example 17
(Example 1 revisited)

The Joint Probability Distribution of Two Bolt Torques

Return again to the situation of Brenny, Christensen, and Schneider and the measuring of bolt torques on the face plates of a heavy equipment component to the nearest integer. With

$$X = \text{the next torque recorded for bolt 3}$$
$$Y = \text{the next torque recorded for bolt 4}$$

Example 17
(*continued*)

the data displayed in Table 3.4 (see page 74) and Figure 3.9 suggest, for example, that a sensible value for $P[X = 18 \text{ and } Y = 18]$ might be $\frac{1}{34}$, the relative frequency of this pair in the data set. Similarly, the assignments

$$P[X = 18 \text{ and } Y = 17] = \frac{2}{34}$$

$$P[X = 14 \text{ and } Y = 9] = 0$$

also correspond to observed relative frequencies.

If one is willing to accept the whole set of relative frequencies defined by the students' data as defining probabilities for X and Y, these can be collected conveniently in a two-dimensional table specifying a joint probability function for X and Y. This is illustrated in Table 5.10. (To avoid clutter, 0 entries in the table have been left blank.)

Table 5.10
$f(x, y)$ for the Bolt Torque Problem

$y \setminus x$	11	12	13	14	15	16	17	18	19	20
20								2/34	2/34	1/34
19							2/34			
18			1/34	1/34			1/34	1/34	1/34	
17					2/34	1/34	1/34	2/34		
16				1/34	2/34	2/34			2/34	
15	1/34	1/34			3/34					
14					1/34			2/34		
13					1/34					

Properties of a joint probability function for X and Y

The probability function given in tabular form in Table 5.10 has two properties that are necessary for mathematical consistency. These are that the $f(x, y)$ values are each in the interval $[0, 1]$ and that they total to 1. By summing up just some of the $f(x, y)$ values, probabilities associated with X and Y being configured in patterns of interest are obtained.

Example 17
(*continued*)

Consider using the joint distribution given in Table 5.10 to evaluate

$$P[X \geq Y],$$

$$P[|X - Y| \leq 1],$$

$$\text{and } P[X = 17]$$

Take first $P[X \geq Y]$, the probability that the measured bolt 3 torque is at least as big as the measured bolt 4 torque. Figure 5.27 indicates with asterisks which possible combinations of x and y lead to bolt 3 torque at least as large as the bolt 4 torque. Referring to Table 5.10 and adding up those entries corresponding to the cells that contain asterisks,

$$\begin{aligned} P[X \geq Y] &= f(15,13) + f(15,14) + f(15,15) + f(16,16) \\ &\quad + f(17,17) + f(18,14) + f(18,17) + f(18,18) \\ &\quad + f(19,16) + f(19,18) + f(20,20) \\ &= \frac{1}{34} + \frac{1}{34} + \frac{3}{34} + \frac{2}{34} + \cdots + \frac{1}{34} = \frac{17}{34} \end{aligned}$$

Similar reasoning allows evaluation of $P[|X - Y| \leq 1]$—the probability that the bolt 3 and 4 torques are within 1 ft lb of each other. Figure 5.28 shows combinations of x and y with an absolute difference of 0 or 1. Then, adding probabilities corresponding to these combinations,

$$\begin{aligned} P[|X - Y| \leq 1] &= f(15,14) + f(15,15) + f(15,16) + f(16,16) \\ &\quad + f(16,17) + f(17,17) + f(17,18) + f(18,17) \\ &\quad + f(18,18) + f(19,18) + f(19,20) + f(20,20) = \frac{18}{34} \end{aligned}$$

y \ x	11	12	13	14	15	16	17	18	19	20
20										*
19									*	*
18								*	*	*
17							*	*	*	*
16						*	*	*	*	*
15					*	*	*	*	*	*
14				*	*	*	*	*	*	*
13			*	*	*	*	*	*	*	*

Figure 5.27 Combinations of bolt 3 and bolt 4 torques with $x \geq y$

y \ x	11	12	13	14	15	16	17	18	19	20
20									*	*
19								*	*	*
18							*	*	*	
17						*	*	*		
16					*	*	*			
15				*	*	*				
14			*	*	*					
13		*	*	*						

Figure 5.28 Combinations of bolt 3 and bolt 4 torques with $|x - y| \leq 1$

Example 17
(*continued*)

Finally, $P[X = 17]$, the probability that the measured bolt 3 torque is 17 ft lb, is obtained by adding down the $x = 17$ column in Table 5.10. That is,

$$P[X = 17] = f(17, 17) + f(17, 18) + f(17, 19)$$
$$= \frac{1}{34} + \frac{1}{34} + \frac{2}{34}$$
$$= \frac{4}{34}$$

Finding marginal probability functions using a bivariate joint probability function

In bivariate problems like the present one, one can add down columns in a two-way table giving $f(x, y)$ to get values for the probability function of X, $f_X(x)$. And one can add across rows in the same table to get values for the probability function of Y, $f_Y(y)$. One can then write these sums in the *margins* of the two-way table. So it should not be surprising that probability distributions for individual random variables obtained from their joint distribution are called **marginal distributions**. A formal statement of this terminology in the case of two discrete variables is next.

Definition 20

The individual probability functions for discrete random variables X and Y with joint probability function $f(x, y)$ are called **marginal probability functions**. They are obtained by summing $f(x, y)$ values over all possible values of the other variable. In symbols, the marginal probability function for X is

$$f_X(x) = \sum_y f(x, y)$$

and the marginal probability function for Y is

$$f_Y(y) = \sum_x f(x, y)$$

Example 17
(*continued*)

Table 5.11 is a copy of Table 5.10, augmented by the addition of marginal probabilities for X and Y. Separating off the margins from the two-way table produces tables of marginal probabilities in the familiar format of Section 5.1. For example, the marginal probability function of Y is given separately in Table 5.12.

Table 5.11
Joint and Marginal Probabilities for X and Y

$y \backslash x$	11	12	13	14	15	16	17	18	19	20	$f_Y(y)$
20								2/34	2/34	1/34	5/34
19							2/34				2/34
18			1/34	1/34			1/34	1/34	1/34		5/34
17					2/34	1/34	1/34	2/34			6/34
16				1/34	2/34	2/34			2/34		7/34
15	1/34	1/34			3/34						5/34
14					1/34			2/34			3/34
13					1/34						1/34
$f_X(x)$	1/34	1/34	1/34	2/34	9/34	3/34	4/34	7/34	5/34	1/34	

Table 5.12
Marginal
Probability
Function for Y

y	$f_Y(y)$
13	1/34
14	3/34
15	5/34
16	7/34
17	6/34
18	5/34
19	2/34
20	5/34

Getting marginal probability functions from joint probability functions raises the natural question whether the process can be reversed. That is, if $f_X(x)$ and $f_Y(y)$ are known, is there then exactly one choice for $f(x, y)$? The answer to this question is "No." Figure 5.29 shows two quite different bivariate joint distributions that nonetheless possess the same marginal distributions. The marked difference between the distributions in Figure 5.29 has to do with the *joint*, rather than individual, behavior of X and Y.

5.4.2 Conditional Distributions and Independence for Discrete Random Variables

When working with several random variables, it is often useful to think about what is expected of one of the variables, given the values assumed by all others. For

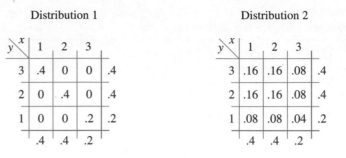

Figure 5.29 Two different joint distributions with the same marginal distributions

example, in the bolt (X) torque situation, a technician who has just loosened bolt 3 and measured the torque as 15 ft lb ought to have expectations for bolt 4 torque (Y) somewhat different from those described by the marginal distribution in Table 5.12. After all, returning to the data in Table 3.4 that led to Table 5.10, the relative frequency distribution of bolt 4 torques for those components with bolt 3 torque of 15 ft lb is as in Table 5.13. Somehow, knowing that $X = 15$ ought to make a probability distribution for Y like the relative frequency distribution in Table 5.13 more relevant than the marginal distribution given in Table 5.12.

Table 5.13
Relative Frequency Distribution for Bolt 4
Torques When Bolt 3 Torque Is 15 ft lb

y, Torque (ft lb)	Relative Frequency
13	1/9
14	1/9
15	3/9
16	2/9
17	2/9

The theory of probability makes allowance for this notion of "distribution of one variable knowing the values of others" through the concept of conditional distributions. The two-variable version of this is defined next.

Definition 21

For discrete random variables X and Y with joint probability function $f(x, y)$, the **conditional probability function of X given $Y = y$** is the function of x

$$f_{X|Y}(x \mid y) = \frac{f(x, y)}{\displaystyle\sum_x f(x, y)}$$

The **conditional probability function of Y given $X = x$** is the function of y

$$f_{Y|X}(y \mid x) = \frac{f(x, y)}{\displaystyle\sum_y f(x, y)}$$

Comparing Definitions 20 and 21

The conditional probability function for X given Y = y

$$f_{X|Y}(x \mid y) = \frac{f(x, y)}{f_Y(y)} \qquad (5.36)$$

and

The conditional probability function for Y given X = x

$$f_{Y|X}(y \mid x) = \frac{f(x, y)}{f_X(x)} \qquad (5.37)$$

Finding conditional distributions from a joint probability function

And formulas (5.36) and (5.37) are perfectly sensible. Equation (5.36) says that starting from $f(x, y)$ given in a two-way table and looking only at the row specified by $Y = y$, the appropriate (conditional) distribution for X is given by the probabilities in that row (the $f(x, y)$ values) divided by their sum ($f_Y(y) = \sum_x f(x, y)$), so that they are renormalized to total to 1. Similarly, equation (5.37) says that looking only at the column specified by $X = x$, the appropriate conditional distribution for Y is given by the probabilities in that column divided by their sum.

Example 17
(continued)

To illustrate the use of equations (5.36) and (5.37), consider several of the conditional distributions associated with the joint distribution for the bolt 3 and bolt 4 torques, beginning with the conditional distribution for Y given that $X = 15$.

From equation (5.37),

$$f_{Y|X}(y \mid 15) = \frac{f(15, y)}{f_X(15)}$$

Referring to Table 5.11, the marginal probability associated with $X = 15$ is $\frac{9}{34}$. So dividing values in the $X = 15$ column of that table by $\frac{9}{34}$, leads to the conditional distribution for Y given in Table 5.14. Comparing this to Table 5.13, indeed formula (5.37) produces a conditional distribution that agrees with intuition.

Example 17
(*continued*)

Table 5.14
The Conditional Probability
Function for Y Given $X = 15$

y	$f_{Y|X}(y \mid 15)$
13	$\left(\dfrac{1}{34}\right) \div \left(\dfrac{9}{34}\right) = \dfrac{1}{9}$
14	$\left(\dfrac{1}{34}\right) \div \left(\dfrac{9}{34}\right) = \dfrac{1}{9}$
15	$\left(\dfrac{3}{34}\right) \div \left(\dfrac{9}{34}\right) = \dfrac{3}{9}$
16	$\left(\dfrac{2}{34}\right) \div \left(\dfrac{9}{34}\right) = \dfrac{2}{9}$
17	$\left(\dfrac{2}{34}\right) \div \left(\dfrac{9}{34}\right) = \dfrac{2}{9}$

Next consider $f_{Y|X}(y \mid 18)$ specified by

$$f_{Y|X}(y \mid 18) = \frac{f(18, y)}{f_X(18)}$$

Consulting Table 5.11 again leads to the conditional distribution for Y given that $X = 18$, shown in Table 5.15. Tables 5.14 and 5.15 confirm that the conditional distributions of Y given $X = 15$ and given $X = 18$ are quite different. For example, knowing that $X = 18$ would on the whole make one expect Y to be larger than when $X = 15$.

Table 5.15
The Conditional
Probability Function for
Y Given $X = 18$

y	$f_{Y|X}(y \mid 18)$
14	2/7
17	2/7
18	1/7
20	2/7

To make sure that the meaning of equation (5.36) is also clear, consider the conditional distribution of the bolt 3 torque (X) given that the bolt 4 torque is 20

$(Y = 20)$. In this situation, equation (5.36) gives

$$f_{X|Y}(x \mid 20) = \frac{f(x, 20)}{f_Y(20)}$$

(Conditional probabilities for X are the values in the $Y = 20$ row of Table 5.11 divided by the marginal $Y = 20$ value.) Thus, $f_{X|Y}(x \mid 20)$ is given in Table 5.16.

Table 5.16
The Conditional Probability
Function for X Given $Y = 20$

x	$f_{X|Y}(x \mid 20)$
18	$\left(\dfrac{2}{34}\right) \div \left(\dfrac{5}{34}\right) = \dfrac{2}{5}$
19	$\left(\dfrac{2}{34}\right) \div \left(\dfrac{5}{34}\right) = \dfrac{2}{5}$
20	$\left(\dfrac{1}{34}\right) \div \left(\dfrac{5}{34}\right) = \dfrac{1}{5}$

The bolt torque example has the feature that the conditional distributions for Y given various possible values for X differ. Further, these are not generally the same as the marginal distribution for Y. X provides some information about Y, in that depending upon its value there are differing probability assessments for Y. Contrast this with the following example.

Example 18

Random Sampling Two Bolt 4 Torques

Suppose that the 34 bolt 4 torques obtained by Brenny, Christensen, and Schneider and given in Table 3.4 are written on slips of paper and placed in a hat. Suppose further that the slips are mixed, one is selected, the corresponding torque is noted, and the slip is replaced. Then the slips are again mixed, another is selected, and the second torque is noted. Define the two random variables

$$U = \text{the value of the first torque selected}$$

and

$$V = \text{the value of the second torque selected}$$

Example 18
(continued)

Intuition dictates that (in contrast to the situation of X and Y in Example 17) the variables U and V don't furnish any information about each other. Regardless of what value U takes, the relative frequency distribution of bolt 4 torques in the hat is appropriate as the (conditional) probability distribution for V, and vice versa. That is, not only do U and V share the common marginal distribution given in Table 5.17 but it is also the case that for all u and v, both

$$f_{U|V}(u \mid v) = f_U(u) \tag{5.38}$$

and

$$f_{V|U}(v \mid u) = f_V(v) \tag{5.39}$$

Equations (5.38) and (5.39) say that the marginal probabilities in Table 5.17 also serve as conditional probabilities. They also specify how joint probabilities for U and V must be structured. That is, rewriting the left-hand side of equation (5.38) using expression (5.36),

$$\frac{f(u, v)}{f_V(v)} = f_U(u)$$

That is,

$$f(u, v) = f_U(u) f_V(v) \tag{5.40}$$

(The same logic applied to equation (5.39) also leads to equation (5.40).) Expression (5.40) says that joint probability values for U and V are obtained by multiplying corresponding marginal probabilities. Table 5.18 gives the joint probability function for U and V.

Table 5.17
The Common Marginal
Probability Function for U
and V

u or v	$f_U(u)$ or $f_V(v)$
13	1/34
14	3/34
15	5/34
16	7/34
17	6/34
18	5/34
19	2/34
20	5/34

Table 5.18
Joint Probabilities for U and V

$v \backslash u$	13	14	15	16	17	18	19	20	$f_V(v)$
20	$\frac{5}{(34)^2}$	$\frac{15}{(34)^2}$	$\frac{25}{(34)^2}$	$\frac{35}{(34)^2}$	$\frac{30}{(34)^2}$	$\frac{25}{(34)^2}$	$\frac{10}{(34)^2}$	$\frac{25}{(34)^2}$	5/34
19	$\frac{2}{(34)^2}$	$\frac{6}{(34)^2}$	$\frac{10}{(34)^2}$	$\frac{14}{(34)^2}$	$\frac{12}{(34)^2}$	$\frac{10}{(34)^2}$	$\frac{4}{(34)^2}$	$\frac{10}{(34)^2}$	2/34
18	$\frac{5}{(34)^2}$	$\frac{15}{(34)^2}$	$\frac{25}{(34)^2}$	$\frac{35}{(34)^2}$	$\frac{30}{(34)^2}$	$\frac{25}{(34)^2}$	$\frac{10}{(34)^2}$	$\frac{25}{(34)^2}$	5/34
17	$\frac{6}{(34)^2}$	$\frac{18}{(34)^2}$	$\frac{30}{(34)^2}$	$\frac{42}{(34)^2}$	$\frac{36}{(34)^2}$	$\frac{30}{(34)^2}$	$\frac{12}{(34)^2}$	$\frac{30}{(34)^2}$	6/34
16	$\frac{7}{(34)^2}$	$\frac{21}{(34)^2}$	$\frac{35}{(34)^2}$	$\frac{49}{(34)^2}$	$\frac{42}{(34)^2}$	$\frac{35}{(34)^2}$	$\frac{14}{(34)^2}$	$\frac{35}{(34)^2}$	7/34
15	$\frac{5}{(34)^2}$	$\frac{15}{(34)^2}$	$\frac{25}{(34)^2}$	$\frac{35}{(34)^2}$	$\frac{30}{(34)^2}$	$\frac{25}{(34)^2}$	$\frac{10}{(34)^2}$	$\frac{25}{(34)^2}$	5/34
14	$\frac{3}{(34)^2}$	$\frac{9}{(34)^2}$	$\frac{15}{(34)^2}$	$\frac{21}{(34)^2}$	$\frac{18}{(34)^2}$	$\frac{15}{(34)^2}$	$\frac{6}{(34)^2}$	$\frac{15}{(34)^2}$	3/34
13	$\frac{1}{(34)^2}$	$\frac{3}{(34)^2}$	$\frac{5}{(34)^2}$	$\frac{7}{(34)^2}$	$\frac{6}{(34)^2}$	$\frac{5}{(34)^2}$	$\frac{2}{(34)^2}$	$\frac{5}{(34)^2}$	1/34
$f_U(u)$	1/34	3/34	5/34	7/34	6/34	5/34	2/34	5/34	

Example 18 suggests that the intuitive notion that several random variables are unrelated might be formalized in terms of all conditional distributions being equal to their corresponding marginal distributions. Equivalently, it might be phrased in terms of joint probabilities being the products of corresponding marginal probabilities. The formal mathematical terminology is that of **independence** of the random variables. The definition for the two-variable case is next.

Definition 22

Discrete random variables X and Y are called **independent** if their joint probability function $f(x, y)$ is the product of their respective marginal probability functions. That is, independence means that

$$f(x, y) = f_X(x)\, f_Y(y) \qquad \text{for all } x, y \qquad \textbf{(5.41)}$$

If formula (5.41) does not hold, the variables X and Y are called **dependent**.

(Formula (5.41) does imply that conditional distributions are all equal to their corresponding marginals, so that the definition does fit its "unrelatedness" motivation.)

U and V in Example 18 are independent, whereas X and Y in Example 17 are dependent. Further, the two joint distributions depicted in Figure 5.29 give an example of a highly dependent joint distribution (the first) and one of independence (the second) that have the same marginals.

Independence of observations in statistical studies

The notion of independence is a fundamental one. When it is sensible to model random variables as independent, *great mathematical simplicity results*. Where engineering data are being collected in an analytical context, and care is taken to make sure that all obvious physical causes of **carryover effects** that might influence successive observations **are minimal**, an assumption of independence between observations is often appropriate. And in enumerative contexts, **relatively small** (compared to the population size) **simple random samples** yield observations that can typically be considered as at least approximately independent.

Example 18
(*continued*)

Again consider putting bolt torques on slips of paper in a hat. The method of torque selection described earlier for producing U and V is not simple random sampling. Simple random sampling as defined in Section 2.2 is *without-replacement* sampling, not the *with-replacement* sampling method used to produce U and V. Indeed, if the first slip is not replaced before the second is selected, the probabilities in Table 5.18 are not appropriate for describing U and V. For example, if no replacement is done, since only one slip is labeled 13 ft lb, one clearly wants

$$f(13, 13) = P[U = 13 \text{ and } V = 13] = 0$$

not the value

$$f(13, 13) = \frac{1}{(34)^2}$$

indicated in Table 5.18. Put differently, if no replacement is done, one clearly wants to use

$$f_{V|U}(13 \mid 13) = 0$$

rather than the value

$$f_{V|U}(13 \mid 13) = f_V(13) = \frac{1}{34}$$

which would be appropriate if sampling is done with replacement. Simple random sampling doesn't lead to exactly independent observations.

But suppose that instead of containing 34 slips labeled with torques, the hat contained 100×34 slips labeled with torques with relative frequencies as in Table 5.17. Then even if sampling is done without replacement, the probabilities developed earlier for U and V (and placed in Table 5.18) remain at least *approximately* valid. For example, with 3,400 slips and using without-replacement sampling,

$$f_{V|U}(13 \mid 13) = \frac{99}{3,399}$$

is appropriate. Then, using the fact that

$$f_{V|U}(v \mid u) = \frac{f(u, v)}{f_U(u)}$$

so that

$$f(u, v) = f_{V|U}(v \mid u)\, f_U(u)$$

without replacement, the assignment

$$f(13, 13) = \frac{99}{3,399} \cdot \frac{1}{34}$$

is appropriate. But the point is that

$$\frac{99}{3,399} \approx \frac{1}{34}$$

and so

$$f(13, 13) \approx \frac{1}{34} \cdot \frac{1}{34}$$

For this hypothetical situation where the population size $N = 3,400$ is much larger than the sample size $n = 2$, independence is a suitable approximate description of observations obtained using simple random sampling.

Where several variables are both independent and have the same marginal distributions, some additional jargon is used.

Definition 23

If random variables $X_1, X_2, \ldots, X_n$ all have the same marginal distribution and are independent, they are termed **iid** or **independent and identically distributed**.

For example, the joint distribution of U and V given in Table 5.18 shows U and V to be iid random variables.

The standard statistical examples of iid random variables are successive measurements taken from a stable process and the results of random sampling with replacement from a single population. The question of whether an iid model is appropriate in a statistical application thus depends on whether or not the data-generating mechanism being studied can be thought of as conceptually equivalent to these.

When can observations be modeled as iid?

5.4.3   Describing Jointly Continuous Random Variables (*Optional*)

All that has been said about joint description of discrete random variables has its analog for continuous variables. Conceptually and computationally, however, the jointly continuous case is more challenging. Probability density functions replace probability functions, and multivariate calculus substitutes for simple arithmetic. So most readers will be best served in the following introduction to multivariate continuous distributions by reading for the main ideas and not getting bogged down in details.

The counterpart of a joint probability function, the device that is commonly used to specify probabilities for several continuous random variables, is a **joint probability density**. The two-variable version of this is defined next.

Definition 24

A **joint probability density** for continuous random variables X and Y is a nonnegative function $f(x, y)$ with

$$\iint f(x, y)\, dx\, dy = 1$$

and such that for any region $\mathcal{R}$, one is willing to assign

$$P[(X, Y) \in \mathcal{R}] = \iint_{\mathcal{R}} f(x, y)\, dx\, dy \qquad \textbf{(5.42)}$$

Instead of summing values of a probability function to find probabilities for a discrete distribution, equation (5.42) says (as in Section 5.2) to integrate a probability density. The new complication here is that the integral is two-dimensional. But it is still possible to draw on intuition developed in mechanics, remembering that this is exactly the sort of thing that is done to specify mass distributions in several dimensions. (Here, mass is probability, and the total mass is 1.)

Example 19
(*Example 15 revisited*)

Residence Hall Depot Counter Service Time and a Continuous Joint Distribution

Consider again the depot service time example. As Section 5.3 showed, the students' data suggest an exponential model with $\alpha = 16.5$ for the random variable

$$S = \text{the excess (over a 7.5 sec threshold) time required}$$
$$\text{to complete the next sale}$$

Imagine that the true value of S will be measured with a (very imprecise) analog stopwatch, producing the random variable

$$R = \text{the measured excess service time}$$

Consider the function of two variables

$$f(s,r) = \begin{cases} \dfrac{1}{16.5}e^{-s/16.5}\dfrac{1}{\sqrt{2\pi(.25)}}\,e^{-(r-s)^2/2(.25)} & \text{for } s > 0 \\ 0 & \text{otherwise} \end{cases} \tag{5.43}$$

as a potential joint probability density for S and R. Figure 5.30 provides a representation of $f(s,r)$, sketched as a surface in three-dimensional space.

As defined in equation (5.43), $f(s,r)$ is nonnegative, and its integral (the volume underneath the surface sketched in Figure 5.30 over the region in the (s,r)-plane where s is positive) is

$$\int\int f(s,r)\,ds\,dr = \int_0^\infty \int_{-\infty}^\infty \frac{1}{16.5\sqrt{2\pi(.25)}}\,e^{-(s/16.5)-((r-s)^2/2(.25))}\,dr\,ds$$

$$= \int_0^\infty \frac{1}{16.5}e^{-s/16.5}\left\{\int_{-\infty}^\infty \frac{1}{\sqrt{2\pi(.25)}}\,e^{-(r-s)^2/2(.25)}\,dr\right\}ds$$

$$= \int_0^\infty \frac{1}{16.5}e^{-s/16.5}\,ds$$

$$= 1$$

(The integral in braces is 1 because it is the integral of a normal density with

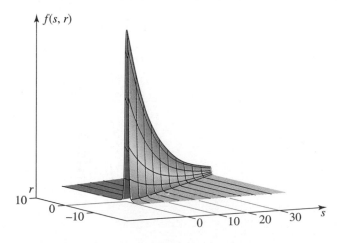

Figure 5.30   A joint probability density for S and R

Example 19
(continued)

mean s and standard deviation .5.) Thus, equation (5.43) specifies a mathematically legitimate joint probability density.

To illustrate the use of a joint probability density in finding probabilities, first consider evaluating $P[R > S]$. Figure 5.31 shows the region in the (s, r)-plane where $f(s, r) > 0$ and $r > s$. It is over this region that one must integrate in order to evaluate $P[R > S]$. Then,

$$P[R > S] = \iint_{\mathcal{R}} f(s, r)\, ds\, dr$$

$$= \int_0^\infty \int_s^\infty f(s, r)\, dr\, ds$$

$$= \int_0^\infty \frac{1}{16.5} e^{-s/16.5} \left\{ \int_s^\infty \frac{1}{\sqrt{2\pi(.25)}} e^{-(r-s)^2/2(.25)}\, dr \right\} ds$$

$$= \int_0^\infty \frac{1}{16.5} e^{-s/16.5} \left\{ \frac{1}{2} \right\} ds$$

$$= \frac{1}{2}$$

(once again using the fact that the integral in braces is a normal (mean s and standard deviation .5) probability).

As a second example, consider the problem of evaluating $P[S > 20]$. Figure 5.32 shows the region over which $f(s, r)$ must be integrated in order to evaluate $P[S > 20]$. Then,

$$P[S > 20] = \iint_{\mathcal{R}} f(s, r)\, ds\, dr$$

$$= \int_{20}^\infty \int_{-\infty}^\infty f(s, r)\, dr\, ds$$

$$= \int_{20}^\infty \frac{1}{16.5} e^{-s/16.5} \left\{ \int_{-\infty}^\infty \frac{1}{\sqrt{2\pi(.25)}} e^{-(r-s)^2/2(.25)}\, dr \right\} ds$$

$$= \int_{20}^\infty \frac{1}{16.5} e^{-s/16.5}\, ds$$

$$= e^{-20/16.5}$$

$$\approx .30$$

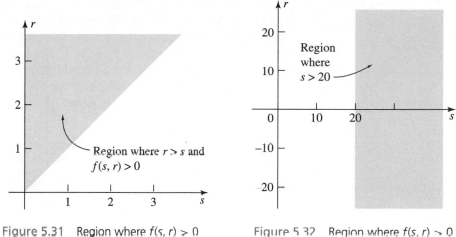

Figure 5.31   Region where $f(s, r) > 0$ and $r > s$

Figure 5.32   Region where $f(s, r) > 0$ and $s > 20$

The last part of the example essentially illustrates the fact that for X and Y with joint density $f(x, y)$,

$$F(x) = P[X \le x] = \int_{-\infty}^{x} \int_{-\infty}^{\infty} f(t, y) \, dy \, dt$$

This is a statement giving the cumulative probability function for X. Differentiation with respect to x shows that a marginal probability density for X is obtained from the joint density by integrating out y. Putting this in the form of a definition gives the following.

Definition 25

The individual probability densities for continuous random variables X and Y with joint probability density $f(x, y)$ are called **marginal probability densities**. They are obtained by integrating $f(x, y)$ over all possible values of the other variable. In symbols, the marginal probability density function for X is

$$f_X(x) = \int_{-\infty}^{\infty} f(x, y) \, dy \qquad (5.44)$$

and the marginal probability density function for Y is

$$f_Y(y) = \int_{-\infty}^{\infty} f(x, y) \, dx \qquad (5.45)$$

Compare Definitions 20 and 25 (page 282). The same kind of thing is done for jointly continuous variables to find marginal distributions as for jointly discrete variables, except that integration is substituted for summation.

Example 19
(*continued*)

Starting with the joint density specified by equation (5.43), it is possible to arrive at reasonably explicit expressions for the marginal densities for S and R. First considering the density of S, Definition 25 declares that for $s > 0$,

$$f_S(s) = \int_{-\infty}^{\infty} \frac{1}{16.5} e^{-s/16.5} \left\{ \frac{1}{\sqrt{2\pi(.25)}} e^{-(r-s)^2/2(.25)} \right\} dr$$

$$= \frac{1}{16.5} e^{-s/16.5}$$

Further, since $f(s, r)$ is 0 for negative s, if $s < 0$,

$$f_S(s) = \int_{-\infty}^{\infty} 0 \, dr = 0$$

That is, the form of $f(s, r)$ was chosen so that (as suggested by Example 15) S has an exponential distribution with mean $\alpha = 16.5$.

The determination of $f_R(r)$ is conceptually no different than the determination of $f_S(s)$, but the details are more complicated. Some work (involving completion of a square in the argument of the exponential function and recognition of an integral as a normal probability) will show the determined reader that for any r,

$$f_R(r) = \int_0^{\infty} \frac{1}{16.5\sqrt{2\pi(.25)}} e^{-(s/16.5)-((r-s)^2/2(.25))} \, ds$$

$$= \frac{1}{16.5} \left(1 - \Phi\left(\frac{1}{33} - 2r \right) \right) \exp\left(\frac{1}{2,178} - \frac{r}{16.5} \right) \qquad \textbf{(5.46)}$$

where, as usual, Φ is the standard normal cumulative probability function. A graph of $f_R(r)$ is given in Figure 5.33.

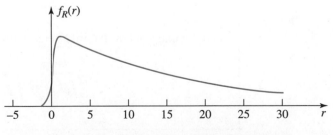

Figure 5.33   Marginal probability density for R

The marginal density for R derived from equation (5.43) does not belong to any standard family of distributions. Indeed, there is generally no guarantee that the process of finding marginal densities from a joint density will produce expressions for the densities even as explicit as that in display (5.46).

5.4.4 Conditional Distributions and Independence for Continuous Random Variables (*Optional*)

In order to motivate the definition for conditional distributions derived from a joint probability density, consider again Definition 21 (page 284). For jointly discrete variables X and Y, the conditional distribution for X given $Y = y$ is specified by holding y fixed and treating $f(x, y)$ as a probability function for X after appropriately renormalizing it—i.e., seeing that its values total to 1. The analogous operation for two jointly continuous variables is described next.

Definition 26

> For continuous random variables X and Y with joint probability density $f(x, y)$, the **conditional probability density function of X given $Y = y$,** is the function of x
>
> $$f_{X|Y}(x \mid y) = \frac{f(x, y)}{\displaystyle\int_{-\infty}^{\infty} f(x, y)\, dx}$$
>
> The **conditional probability density function of Y given $X = x$** is the function of y
>
> $$f_{Y|X}(y \mid x) = \frac{f(x, y)}{\displaystyle\int_{-\infty}^{\infty} f(x, y)\, dy}$$

Definitions 25 and 26 lead to

Conditional probability density for X given Y = y

$$f_{X|Y}(x \mid y) = \frac{f(x, y)}{f_Y(y)} \tag{5.47}$$

and

Conditional probability density for Y given X = x

$$f_{Y|X}(y \mid x) = \frac{f(x, y)}{f_X(x)} \tag{5.48}$$

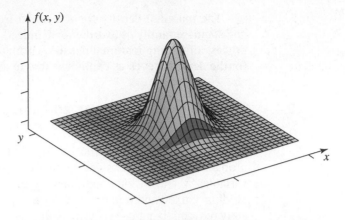

Figure 5.34   A Joint probability density $f(x, y)$ and the shape of a conditional density for X given a value of Y

Geometry of conditional densities

Expressions (5.47) and (5.48) are formally identical to the expressions (5.36) and (5.37) relevant for discrete variables. The geometry indicated by equation (5.47) is that the shape of $f_{X|Y}(x \mid y)$ as a function of x is determined by cutting the $f(x, y)$ surface in a graph like that in Figure 5.34 with the $Y = y$-plane. In Figure 5.34, the divisor in equation (5.47) is the area of the shaded figure above the (x, y)-plane below the $f(x, y)$ surface on the $Y = y$ plane. That division serves to produce a function of x that will integrate to 1. (Of course, there is a corresponding geometric story told for the conditional distribution of Y given $X = x$ in expression (5.48)).

Example 19
(continued)

In the service time example, it is fairly easy to recognize the conditional distribution of R given $S = s$ as having a familiar form. For $s > 0$, applying expression (5.48),

$$f_{R|S}(r \mid s) = \frac{f(s, r)}{f_S(s)} = f(s, r) \div \left(\frac{1}{16.5} e^{-s/16.5} \right)$$

which, using equation (5.43), gives

$$f_{R|S}(r \mid s) = \frac{1}{\sqrt{2\pi(.25)}} e^{-(r-s)^2/2(.25)} \tag{5.49}$$

That is, given that $S = s$, the conditional distribution of R is normal with mean s and standard deviation .5.

This realization is consistent with the bell-shaped cross sections of $f(s, r)$ shown in Figure 5.30. The form of $f_{R|S}(r \mid s)$ given in equation (5.49) says that the measured excess service time is the true excess service time plus a normally distributed measurement error that has mean 0 and standard deviation .5.

It is evident from expression (5.49) (or from the way the positions of the bell-shaped contours on Figure 5.30 vary with s) that the variables S and R ought to be called dependent. After all, knowing that $S = s$ gives the value of R except for a normal error of measurement with mean 0 and standard deviation .5. On the other hand, had it been the case that all conditional distributions of R given $S = s$ were the same (and equal to the marginal distribution of R), S and R should be called independent. The notion of unchanging conditional distributions, all equal to their corresponding marginal, is equivalently and more conveniently expressed in terms of the joint probability density factoring into a product of marginals. The formal version of this for two variables is next.

Definition 27

> Continuous random variables X and Y are called **independent** if their joint probability density function $f(x, y)$ is the product of their respective marginal probability densities. That is, independence means that
>
> $$f(x, y) = f_X(x)\, f_Y(y) \qquad \text{for all } x, y \qquad \textbf{(5.50)}$$
>
> If expression (5.50) does not hold, the variables X and Y are called **dependent**.

Expression (5.50) is formally identical to expression (5.41), which appeared in Definition 22 for discrete variables. The type of factorization given in these expressions provides great mathematical convenience.

It remains in this section to remark that the concept of *iid random variables* introduced in Definition 23 is as relevant to continuous cases as it is to discrete ones. In statistical contexts, it can be appropriate where analytical problems are free of carryover effects and in enumerative problems where (relatively) small simple random samples are being described.

Example 20
(*Example 15 revisited*)

Residence Hall Depot Counter Service Times and iid Variables

Returning once more to the service time example of Jenkins, Milbrath, and Worth, consider the next two excess service times encountered,

$S_1 = $ the first/next excess (over a threshold of 7.5 sec) time required to complete a postage stamp sale at the residence hall service counter

$S_2 = $ the second excess service time

To the extent that the service process is physically stable (i.e., excess service times can be thought of in terms of sampling with replacement from a single population), an iid model seems appropriate for S_1 and S_2. Treating excess service times as

Example 20
(continued)

marginally exponential with mean $\alpha = 16.5$ thus leads to the joint density for S_1 and S_2:

$$f(s_1, s_2) = \begin{cases} \dfrac{1}{(16.5)^2} e^{-(s_1+s_2)/16.5} & \text{if } s_1 > 0 \text{ and } s_2 > 0 \\ 0 & \text{otherwise} \end{cases}$$

Section 4 Exercises

1. Explain in qualitative terms what it means for two random variables X and Y to be independent. What advantage is there when X and Y can be described as independent?

2. Quality audit records are kept on numbers of major and minor failures of circuit packs during burn-in of large electronic switching devices. They indicate that for a device of this type, the random variables

$$X = \text{the number of major failures}$$

and

$$Y = \text{the number of minor failures}$$

can be described at least approximately by the accompanying joint distribution.

y \ x	0	1	2
0	.15	.05	.01
1	.10	.08	.01
2	.10	.14	.02
3	.10	.08	.03
4	.05	.05	.03

(a) Find the marginal probability functions for both X and Y—$f_X(x)$ and $f_Y(y)$.
(b) Are X and Y independent? Explain.
(c) Find the mean and variance of X—EX and Var X.
(d) Find the mean and variance of Y—EY and Var Y.
(e) Find the conditional probability function for Y, given that $X = 0$—i.e., that there are no major circuit pack failures. (That is, find $f_{Y|X}(y \mid 0)$.)

What is the mean of this conditional distribution?

3. A laboratory receives four specimens having identical appearances. However, it is possible that (a single unknown) one of the specimens is contaminated with a toxic material. The lab must test the specimens to find the toxic specimen (if in fact one is contaminated). The testing plan first put forth by the laboratory staff is to test the specimens one at a time, stopping when (and if) a contaminated specimen is found.

Define two random variables

$$X = \text{the number of contaminated specimens}$$

and

$$Y = \text{the number of specimens tested}$$

Let $p = P[X = 0]$ and therefore $P[X = 1] = 1 - p$.

(a) Give the conditional distributions of Y given $X = 0$ and $X = 1$ for the staff's initial testing plan. Then use them to determine the joint probability function of X and Y. (Your joint distribution will involve p, and you may simply fill out tables like the accompanying ones.)

y	$f_{Y|X}(y \mid 0)$
1	
2	
3	
4	

y	$f_{Y|X}(y \mid 1)$.
1	
2	
3	
3	

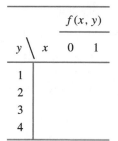

	$f(x, y)$	
$y \backslash x$	0	1
1		
2		
3		
4		

(b) Based on your work in (a), find the marginal distribution of Y. What is EY, the mean number of specimens tested using the staff's original plan?

(c) A second testing method devised by the staff involves testing composite samples of material taken from possibly more than one of the original specimens. By initially testing a composite of all four specimens, then (if the first test reveals the presence of toxic material) following up with a composite of two, and then an appropriate single specimen, it is possible to do the lab's job in one test if $X = 0$, and in three tests if $X = 1$. Suppose that because testing is expensive, it is desirable to hold the number of tests to a minimum. For what values of p is this second method preferable to the first? (*Hint:* What is EY for this second method?)

4. A machine element is made up of a rod and a ring bearing. The rod must fit through the bearing. Model

$$X = \text{the diameter of the rod}$$

and

$$Y = \text{the inside diameter of the ring bearing}$$

as independent random variables, X uniform on (1.97, 2.02) and Y uniform on (2.00, 2.06). ($f_X(x) = 1/.05$ for $1.97 < x < 2.02$, while $f_X(x) = 0$ otherwise. Similarly, $f_Y(y) = 1/.06$ for $2.00 < y < 2.06$, while $f_Y(y) = 0$ otherwise.) With this model, do the following:

(a) Write out the joint probability density for X and Y. (It will be positive only when $1.97 < x < 2.02$ and $2.00 < y < 2.06$.)

(b) Evaluate $P[Y - X < 0]$, the probability of an interference in assembly.

5. Suppose that a pair of random variables have the joint probability density

$$f(x, y) = \begin{cases} 4x(1 - y) & \text{if } 0 \leq x \leq 1 \\ & \text{and } 0 \leq y \leq 1 \\ 0 & \text{otherwise} \end{cases}$$

(a) Find the marginal probability densities for X and Y. What is the mean of X?

(b) Are X and Y independent? Explain.

(c) Evaluate $P[X + 2Y \geq 1]$.

(d) Find the conditional probability density for X given that $Y = .5$. (Find $f_{X|Y}(x \mid .5)$.) What is the mean of this (conditional) distribution?

6. An engineering system consists of two subsystems operating independently of each other. Let

$$X = \text{the time till failure of the first subsystem}$$

and

$$Y = \text{the time till failure of the second subsystem}$$

Suppose that X and Y are independent exponential random variables each with mean $\alpha = 1$ (in appropriate time units).

(a) Write out the joint probability density for X and Y. Be sure to state carefully where the density is positive and where it is 0.

Suppose first that the system is a *series system* (i.e., one that fails when either of the subsystems fail).

(b) The probability that the system is still functioning at time $t > 0$ is then

$$P[X > t \text{ and } Y > t]$$

Find this probability using your answer to (a). (What region in the (x, y)-plane corresponds to the possibility that the system is still functioning at time t?)

(c) If one then defines the random variable

$$T = \text{the time until the system fails}$$

the cumulative probability function for T is

$$F(t) = 1 - P[X > t \text{ and } Y > t]$$

so that your answer to (b) can be used to find the distribution for T. Use your answer to (b) and some differentiation to find the probability density for T. What kind of distribution does T have? What is its mean?

Suppose now that the system is a *parallel system* (i.e., one that fails only when both subsystems fail).

(d) The probability that the system has failed by time t is

$$P[X \le t \text{ and } Y \le t]$$

Find this probability using your answer to part (a).

(e) Now, as before, let T be the time until the system fails. Use your answer to (d) and some differentiation to find the probability density for T. Then calculate the mean of T.

5.5 Functions of Several Random Variables

The last section introduced the mathematics used to simultaneously model several random variables. An important engineering use of that material is in the analysis of system outputs that are functions of random inputs.

This section studies how the variation seen in an output random variable depends upon that of the variables used to produce it. It begins with a few comments on what is possible using exact methods of mathematical analysis. Then the simple and general tool of simulation is introduced. Next, formulas for means and variances of linear combinations of random variables and the related propagation of error formulas are presented. Last is the pervasive central limit effect, which often causes variables to have approximately normal distributions.

5.5.1 The Distribution of a Function of Random Variables

The problem considered in this section is this. Given a joint distribution for the random variables $X, Y, \ldots, Z$ and a function $g(x, y, \ldots, z)$, the object is to predict the behavior of the random variable

$$U = g(X, Y, \ldots, Z) \tag{5.51}$$

In some special simple cases, it is possible to figure out exactly what distribution U inherits from $X, Y, \ldots, Z$.

| Example 21 | The Distribution of the Clearance Between Two Mating Parts with Randomly Determined Dimensions |

Suppose that a steel plate with nominal thickness .15 in. is to rest in a groove of nominal width .155 in., machined on the surface of a steel block. A lot of plates has been made and thicknesses measured, producing the relative fre-

Table 5.19
Relative Frequency Distribution of Plate Thicknesses

Plate Thickness (in.)	Relative Frequency
.148	.4
.149	.3
.150	.3

Table 5.20
Relative Frequency Distribution of Slot Widths

Slot Width (in.)	Relative Frequency
.153	.2
.154	.2
.155	.4
.156	.2

quency distribution in Table 5.19; a relative frequency distribution for the slot widths measured on a lot of machined blocks is given in Table 5.20.

If a plate is randomly selected and a block is separately randomly selected, a natural joint distribution for the random variables

$$X = \text{the plate thickness}$$

$$Y = \text{the slot width}$$

is one of independence, where the marginal distribution of X is given in Table 5.19 and the marginal distribution of Y is given in Table 5.20. That is, Table 5.21 gives a plausible joint probability function for X and Y.

A variable derived from X and Y that is of substantial potential interest is the clearance involved in the plate/block assembly,

$$U = Y - X$$

Notice that taking the extremes represented in Tables 5.19 and 5.20, U is guaranteed to be at least $.153 - .150 = .003$ in. but no more than $.156 - .148 = .008$ in. In fact, much more than this can be said. Looking at Table 5.21, one can see that the diagonals of entries (lower left to upper right) all correspond to the same value of $Y - X$. Adding probabilities on those diagonals produces the distribution of U given in Table 5.22.

Example 21
(*continued*)

Table 5.21
Marginal and Joint Probabilities for X and Y

y \ x	.148	.149	.150	$f_Y(y)$
.156	.08	.06	.06	.2
.155	.16	.12	.12	.4
.154	.08	.06	.06	.2
.153	.08	.06	.06	.2
$f_X(x)$	.4	.3	.3	

Table 5.22
The Probability Function for the
Clearance 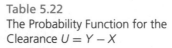 $U = Y - X$

u	$f(u)$
.003	.06
.004	$.12 = .06 + .06$
.005	$.26 = .08 + .06 + .12$
.006	$.26 = .08 + .12 + .06$
.007	$.22 = .16 + .06$
.008	.08

Example 21 involves a very simple discrete joint distribution and a very simple function g—namely, $g(x, y) = y - x$. In general, exact complete solution of the problem of finding the distribution of $U = g(X, Y, \ldots, Z)$ is not practically possible. Happily, for many engineering applications of probability, approximate and/or partial solutions suffice to answer the questions of practical interest. The balance of this section studies methods of producing these approximate and/or partial descriptions of the distribution of U, beginning with a brief look at simulation-based methods.

5.5.2   Simulations to Approximate the Distribution of $U = g(X, Y, \ldots, Z)$

Simulation for independent $X, Y, \ldots, Z$

Many computer programs and packages can be used to produce pseudorandom values, intended to behave as if they were realizations of independent random variables following user-chosen marginal distributions. If the model for $X, Y, \ldots, Z$ is one of independence, it is then a simple matter to generate a simulated value for each of $X, Y, \ldots, Z$ and plug those into g to produce a simulated value for U. If this process is repeated a number of times, a relative frequency distribution for these simulated values of U is developed. One might reasonably use this relative frequency distribution to approximate an exact distribution for U.

Example 22

Uncertainty in the Calculated Efficiency of an Air Solar Collector

The article "Thermal Performance Representation and Testing of Air Solar Collectors" by Bernier and Plett (*Journal of Solar Energy Engineering*, May 1988) considers the testing of air solar collectors. Its analysis of thermal performance based on enthalpy balance leads to the conclusion that under inward leakage conditions, the thermal efficiency of a collector can be expressed as

$$\text{Efficiency} = \frac{M_o C(T_o - T_i) + (M_o - M_i)C(T_i - T_a)}{GA}$$

$$= \frac{C}{GA}\left(M_o T_o - M_i T_i - (M_o - M_i)T_a\right) \tag{5.52}$$

where

C = air specific heat (J/kg°C)

G = global irradiance incident on the plane of the collector (W/m^2)

A = collector gross area (m^2)

M_i = inlet mass flowrate (kg/s)

M_o = outlet mass flowrate (kg/s)

T_a = ambient temperature (°C)

T_i = collector inlet temperature (°C)

T_o = collector outlet temperature (°C)

The authors further give some uncertainty values associated with each of the terms appearing on the right side of equation (5.52) for an example set of measured values of the variables. These are given in Table 5.23.

Table 5.23
Reported Uncertainties in the Measured Inputs
to Collector Efficiency

Variable	Measured Value	Uncertainty
C	1003.8	1.004 (i.e., $\pm$ .1%)
G	1121.4	33.6 (i.e., $\pm$ 3%)
A	1.58	.005
M_i	.0234	.00035 (i.e., $\pm$ 1.5%)
M_o	.0247	.00037 (i.e., $\pm$ 1.5%)
T_a	−13.22	.5
T_i	−6.08	.5
T_o	24.72	.5*

*This value is not given explicitly in the article.

Example 22
(continued)

Plugging the measured values from Table 5.23 into formula (5.52) produces a measured efficiency of about .44. But how good is the .44 value? That is, how do the uncertainties associated with the measured values affect the reliability of the .44 figure? Should you think of the calculated solar collector efficiency as .44 plus or minus .001, or plus or minus .1, or what?

One way of approaching this is to ask the related question, "What would be the standard deviation of *Efficiency* if all of C through T_o were independent random variables with means approximately equal to the measured values and standard deviations related to the uncertainties as, say, half of the uncertainty values?" (This "two sigma" interpretation of uncertainty appears to be at least close to the intention in the original article.)

Printout 1 is from a MINITAB session in which 100 normally distributed realizations of variables C through T_o were generated (using means equal to measured values and standard deviations equal to half of the corresponding uncertainties) and the resulting efficiencies calculated. (The routine under the "Calc/Random Data/Normal" menu was used to generate the realizations of C through T_o. The "Calc/Calculator" menu was used to combine these values according to equation (5.52). Then routines under the "Stat/Basic Statistics/Describe" and "Graph/Character Graphs/Stem-and-Leaf" menus were used to produce the summaries of the simulated efficiencies.) The simulation produced a roughly bell-shaped distribution of calculated efficiencies, possessing a mean value of approximately .437 and standard deviation of about .009. Evidently, if one continues with the understanding that uncertainty means something like "2 standard deviations," an uncertainty of about .02 is appropriate for the nominal efficiency figure of .44.

Printout 1   Simulation of Solar Collector Efficiency

```
Descriptive Statistics


Variable         N       Mean     Median     TrMean      StDev     SE Mean
Efficien       100    0.43729    0.43773    0.43730    0.00949     0.00095


Variable    Minimum    Maximum         Q1         Q3
Efficien    0.41546    0.46088    0.43050    0.44426




Character Stem-and-Leaf Display


Stem-and-leaf of Efficien  N  = 100
Leaf Unit = 0.0010


    5    41 58899
   10    42 22334
   24    42 66666777788999
   39    43 001112233333444
  (21)   43 555556666777889999999
   40    44 00000011122333444
```

```
23    44  555556667788889
 8    45  023344
 2    45  7
 1    46  0
```

The beauty of Example 22 is the ease with which a simulation can be employed to approximate the distribution of U. But the method is so powerful and easy to use that some cautions need to be given about the application of this whole topic before going any further.

Practical cautions   Be careful not to expect more than is sensible from a derived probability distribution ("exact" or approximate) for

$$U = g(X, Y, \ldots, Z)$$

The output distribution can be no more realistic than are the assumptions used to produce it (i.e., the form of the joint distribution and the form of the function $g(x, y, \ldots, z)$). It is all too common for people to apply the methods of this section using a g representing some approximate physical law and U some measurable physical quantity, only to be surprised that the variation in U observed in the real world is *substantially larger* than that predicted by methods of this section. The fault lies not with the methods, but with the naivete of the user. Approximate physical laws are just that, often involving so-called constants that aren't constant, using functional forms that are too simple, and ignoring the influence of variables that aren't obvious or easily measured. Further, although independence of $X, Y, \ldots, Z$ is a very convenient mathematical property, its use is not always justified. When it is inappropriately used as a model assumption, it can produce an inappropriate distribution for U. For these reasons, think of the methods of this section as useful but likely to provide only a *best-case* picture of the variation you should expect to see.

5.5.3 Means and Variances for Linear Combinations of Random Variables

For engineering purposes, it often suffices to know the mean and variance for U given in formula (5.51) (as opposed to knowing the whole distribution of U). When this is the case and g is linear, there are explicit formulas for these.

Proposition 1

If $X, Y, \ldots, Z$ are n independent random variables and $a_0, a_1, a_2, \ldots, a_n$ are $n + 1$ constants, then the random variable $U = a_0 + a_1 X + a_2 Y + \cdots + a_n Z$ has mean

$$EU = a_0 + a_1 EX + a_2 EY + \cdots + a_n EZ \qquad (5.53)$$

and variance

$$\text{Var } U = a_1^2 \text{ Var } X + a_2^2 \text{ Var } Y + \cdots + a_n^2 \text{ Var } Z \qquad \textbf{(5.54)}$$

Formula (5.53) actually holds regardless of whether or not the variables $X, Y, \ldots, Z$ are independent, and although formula (5.54) does depend upon independence, there is a generalization of it that can be used even if the variables are dependent. However, the form of Proposition 1 given here is adequate for present purposes.

One type of application in which Proposition 1 is immediately useful is that of geometrical tolerancing problems, where it is applied with $a_0 = 0$ and the other a_i's equal to plus and minus 1's.

Example 21
(continued)

Consider again the situation of the clearance involved in placing a steel plate in a machined slot on a steel block. With X, Y, and U being (respectively) the plate thickness, slot width, and clearance, means and variances for these variables can be calculated from Tables 5.19, 5.20, and 5.22, respectively. The reader is encouraged to verify that

$$EX = .1489 \quad \text{and} \quad \text{Var } X = 6.9 \times 10^{-7}$$
$$EY = .1546 \quad \text{and} \quad \text{Var } Y = 1.04 \times 10^{-6}$$

Now, since

$$U = Y - X = (-1)X + 1Y$$

Proposition 1 can be applied to conclude that

$$EU = -1EX + 1EY = -.1489 + .1546 = .0057 \text{ in.}$$
$$\text{Var } U = (-1)^2 6.9 \times 10^{-7} + (1)^2 1.04 \times 10^{-6} = 1.73 \times 10^{-6}$$

so that

$$\sqrt{\text{Var } U} = .0013 \text{ in.}$$

It is worth the effort to verify that the mean and standard deviation of the clearance produced using Proposition 1 agree with those obtained using the distribution of U given in Table 5.22 and the formulas for the mean and variance given in Section 5.1. The advantage of using Proposition 1 is that if all that is needed are EU and $\text{Var } U$, there is no need to go through the intermediate step of deriving the distribution of U. The calculations via Proposition 1 use only characteristics of the marginal distributions.

Another particularly important use of Proposition 1 concerns n iid random variables where each a_i is $\frac{1}{n}$. That is, in cases where random variables $X_1, X_2, \ldots, X_n$ are conceptually equivalent to random selections (*with* replacement) from a single numerical population, Proposition 1 tells how the mean and variance of the random variable

$$\overline{X} = \frac{1}{n}X_1 + \frac{1}{n}X_2 + \cdots + \frac{1}{n}X_n$$

are related to the population parameters μ and σ^2. For independent variables $X_1, X_2, \ldots, X_n$ with common mean μ and variance σ^2, Proposition 1 shows that

The mean of an average of n iid random variables

$$E\overline{X} = \frac{1}{n}EX_1 + \frac{1}{n}EX_2 + \cdots + \frac{1}{n}EX_n = n\left(\frac{1}{n}\mu\right) = \mu \qquad \textbf{(5.55)}$$

and

The variance of an average of n iid random variables

$$\operatorname{Var}\overline{X} = \left(\tfrac{1}{n}\right)^2 \operatorname{Var} X_1 + \left(\tfrac{1}{n}\right)^2 \operatorname{Var} X_2 + \cdots + \left(\tfrac{1}{n}\right)^2 \operatorname{Var} X_n$$
$$= n\left(\frac{1}{n}\right)^2 \sigma^2 = \frac{\sigma^2}{n} \qquad \textbf{(5.56)}$$

Since σ^2/n is decreasing in n, equations (5.55) and (5.56) give the reassuring picture of $\overline{X}$ having a probability distribution centered at the population mean μ, with spread that decreases as the sample size increases.

Example 23
(*Example 15 revisited*)

The Expected Value and Standard Deviation for a Sample Mean Service Time

To illustrate the application of formulas (5.55) and (5.56), consider again the stamp sale service time example. Suppose that the exponential model with $\alpha = 16.5$ that was derived in Example 15 for excess service times continues to be appropriate and that several more postage stamp sales are observed and excess service times noted. With

$S_i = $ the excess (over a 7.5 sec threshold) time required to complete the ith additional stamp sale

Example 23
(continued)

consider what means and standard deviations are associated with the probability distributions of the sample average, $\overline{S}$, of first the next 4 and then the next 100 excess service times.

$S_1, S_2, \ldots, S_{100}$ are, to the extent that the service process is physically stable, reasonably modeled as independent, identically distributed, exponential random variables with mean $\alpha = 16.5$. The exponential distribution with mean $\alpha = 16.5$ has variance equal to $\alpha^2 = (16.5)^2$. So, using formulas (5.55) and (5.56), for the first 4 additional service times,

$$E\overline{S} = \alpha = 16.5 \text{ sec}$$

$$\sqrt{\text{Var } \overline{S}} = \sqrt{\frac{\alpha^2}{4}} = 8.25 \text{ sec}$$

Then, for the first 100 additional service times,

$$E\overline{S} = \alpha = 16.5 \text{ sec}$$

$$\sqrt{\text{Var } \overline{S}} = \sqrt{\frac{\alpha^2}{100}} = 1.65 \text{ sec}$$

Notice that going from a sample size of 4 to a sample size of 100 decreases the standard deviation of $\overline{S}$ by a factor of 5 ($= \sqrt{\frac{100}{4}}$).

Relationships (5.55) and (5.56), which perfectly describe the random behavior of $\overline{X}$ under random sampling with replacement, are also approximate descriptions of the behavior of $\overline{X}$ under simple random sampling in enumerative contexts. (Recall Example 18 and the discussion about the approximate independence of observations resulting from simple random sampling of large populations.)

5.5.4 The Propagation of Error Formulas

Proposition 1 gives exact values for the mean and variance of $U = g(X, Y, \ldots, Z)$ only when g is linear. It doesn't seem to say anything about situations involving nonlinear functions like the one specified by the right-hand side of expression (5.52) in the solar collector example. But it is often possible to obtain useful approximations to the mean and variance of U by applying Proposition 1 to a first-order multivariate Taylor expansion of a not-too-nonlinear g. That is, if g is reasonably well-behaved, then for $x, y, \ldots, z$ (respectively) close to $EX, EY, \ldots, EZ$,

$$\left. \begin{aligned} g(x, y, \ldots, z) \approx g(EX, EY, \ldots, EZ) + \frac{\partial g}{\partial x} \cdot (x - EX) + \frac{\partial g}{\partial y} \cdot (y - EY) \\ + \cdots + \frac{\partial g}{\partial z} \cdot (z - EZ) \end{aligned} \right\} \quad \text{(5.57)}$$

where the partial derivatives are evaluated at $(x, y, \ldots, z) = (EX, EY, \ldots, EZ)$. Now the right side of approximation (5.57) is linear in $x, y, \ldots, z$. Thus, if the variances of $X, Y, \ldots, Z$ are small enough so that with high probability, $X, Y, \ldots, Z$ are such that approximation (5.57) is effective, one might think of plugging $X, Y, \ldots, Z$ into expression (5.57) and applying Proposition 1, thus winding up with approximations for the mean and variance of $U = g(X, Y, \ldots, Z)$.

Proposition 2
(The Propagation of Error Formulas)

If $X, Y, \ldots, Z$ are independent random variables and g is well-behaved, for small enough variances $\text{Var } X, \text{Var } Y, \ldots, \text{Var } Z$, the random variable $U = g(X, Y, \ldots, Z)$ has approximate mean

$$EU \approx g(EX, EY, \ldots, EZ) \tag{5.58}$$

and approximate variance

$$\text{Var } U \approx \left(\frac{\partial g}{\partial x}\right)^2 \text{Var } X + \left(\frac{\partial g}{\partial y}\right)^2 \text{Var } Y + \cdots + \left(\frac{\partial g}{\partial z}\right)^2 \text{Var } Z \tag{5.59}$$

Formulas (5.58) and (5.59) are often called the **propagation of error** or **transmission of variance** formulas. They describe how variability or error is propagated or transmitted through an exact mathematical function.

Comparison of Propositions 1 and 2 shows that when g is exactly linear, expressions (5.58) and (5.59) reduce to expressions (5.53) and (5.54), respectively (a_1 through a_n are the partial derivatives of g in the case where $g(x, y, \ldots, z) = a_0 + a_1 x + a_2 y + \cdots + a_n z$.) Proposition 2 is purposely vague about when the approximations (5.58) and (5.59) will be adequate for engineering purposes. Mathematically inclined readers will not have much trouble constructing examples where the approximations are quite poor. But often in engineering applications, expressions (5.58) and (5.59) are at least of the right order of magnitude and certainly better than not having any usable approximations.

Example 24

A Simple Electrical Circuit and the Propagation of Error

Figure 5.35 is a schematic of an assembly of three resistors. If R_1, R_2, and R_3 are the respective resistances of the three resistors making up the assembly, standard theory says that

$$R = \text{the assembly resistance}$$

Example 24
(continued)

is related to R_1, R_2, and R_3 by

$$R = R_1 + \frac{R_2 R_3}{R_2 + R_3} \tag{5.60}$$

A large lot of resistors is manufactured and has a mean resistance of 100 Ω with a standard deviation of resistance of 2 Ω. If three resistors are taken at random from this lot and assembled as in Figure 5.35, consider what formulas (5.58) and (5.59) suggest for an approximate mean and an approximate standard deviation for the resulting assembly resistance.

The g involved here is $g(r_1, r_2, r_3) = r_1 + \dfrac{r_2 r_3}{r_2 + r_3}$, so

$$\frac{\partial g}{\partial r_1} = 1$$

$$\frac{\partial g}{\partial r_2} = \frac{(r_2 + r_3)r_3 - r_2 r_3}{(r_2 + r_3)^2} = \frac{r_3^2}{(r_2 + r_3)^2}$$

$$\frac{\partial g}{\partial r_3} = \frac{(r_2 + r_3)r_2 - r_2 r_3}{(r_2 + r_3)^2} = \frac{r_2^2}{(r_2 + r_3)^2}$$

Also, R_1, R_2, and R_3 are approximately independent with means 100 and standard deviations 2. Then formulas (5.58) and (5.59) suggest that the probability distribution inherited by R has mean

$$ER \approx g(100, 100, 100) = 100 + \frac{(100)(100)}{100 + 100} = 150 \ \Omega$$

and variance

$$\mathrm{Var}\, R \approx (1)^2 (2)^2 + \left(\frac{(100)^2}{(100 + 100)^2} \right)^2 (2)^2 + \left(\frac{(100)^2}{(100 + 100)^2} \right)^2 (2)^2 = 4.5$$

so that the standard deviation inherited by R is

$$\sqrt{\mathrm{Var}\, R} \approx \sqrt{4.5} = 2.12 \ \Omega$$

As something of a check on how good the 150 Ω and 2.12 Ω values are, 1,000 sets of normally distributed R_1, R_2, and R_3 values with the specified population mean and standard deviation were simulated and resulting values of R calculated via formula (5.60). These simulated assembly resistances had $\overline{R} = 149.80$ Ω and a sample standard deviation of 2.14 Ω. A histogram of these values is given in Figure 5.36.

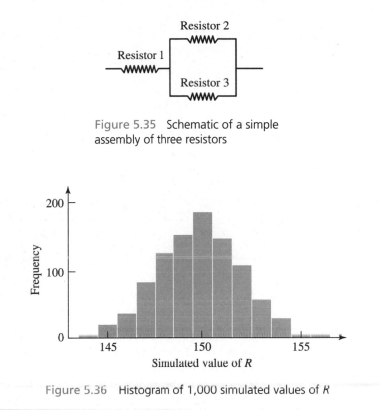

Figure 5.35   Schematic of a simple assembly of three resistors

Figure 5.36   Histogram of 1,000 simulated values of *R*

Example 24 is one to which the cautions following Example 22 (page 307) apply. Suppose you were to actually take a large batch of resistors possessing a mean resistance of 100 Ω and a standard deviation of resistances of 2 Ω, make up a number of assemblies of the type represented in Figure 5.35, and measure the assembly resistances. The standard deviation figures in Example 24 will likely underpredict the variation observed in the assembly resistances.

The propagation of error and simulation methods may do a good job of approximating the (exact) theoretical mean and standard deviation of assembly resistances. But the extent to which the probability model used for assembly resistances can be expected to represent the physical situation is another matter. Equation (5.60) is highly useful, but of necessity only an approximate description of real assemblies. For example, it ignores small but real temperature, inductance, and other second-order physical effects on measured resistance. In addition, although the probability model allows for variation in the resistances of individual components, it does not account for instrument variation or such vagaries of real-world assembly as the quality of contacts achieved when several parts are connected.

In Example 24, the simulation and propagation of error methods produce comparable results. Since the simulation method is so easy to use, why bother to do the calculus and arithmetic necessary to use the propagation of error formulas? One important answer to this question concerns intuition that formula (5.59) provides.

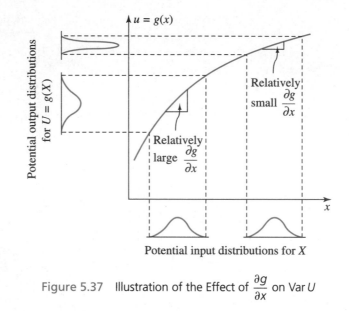

Figure 5.37   Illustration of the Effect of $\dfrac{\partial g}{\partial x}$ on Var U

The effects of the partial derivatives of g on Var U

Consider first the effect that g's partial derivatives have on Var U. Formula (5.59) implies that depending on the size of $\frac{\partial g}{\partial x}$, the variance of X is either inflated or deflated before becoming an ingredient of Var U. And even though formula (5.59) may not be an exact expression, it provides correct intuition. If a given change in x produces a big change in $g(x, y, \ldots, z)$, the impact Var X has on Var U will be greater than if the change in x produces a small change in $g(x, y, \ldots, z)$. Figure 5.37 is a rough illustration of this point. In the case that $U = g(X)$, two different approximately normal distributions for X with different means but a common variance produce radically different spreads in the distribution of U, due to differing rates of change of g (different derivatives).

Partitioning the variance of U

Then, consider the possibility of partitioning the variance of U into interpretable pieces. Formula (5.59) suggests thinking of (for example)

$$\left(\frac{\partial g}{\partial x}\right)^2 \text{Var } X$$

as the contribution the variation in X makes to the variation inherent in U. Comparison of such individual contributions makes it possible to analyze how various potential reductions in input variation can be expected to affect the output variability in U.

Example 22
(continued)

Return to the solar collector example. For means of C through T_o taken to be the measured values in Table 5.23 (page 305), and standard deviations of C through T_o equal to half of the uncertainties listed in the same table, formula

(5.59) might well be applied to the calculated efficiency given in formula (5.52). The squared partial derivatives of *Efficiency* with respect to each of the inputs, times the variances of those inputs, are as given in Table 5.24. Thus, the approximate standard deviation for the efficiency variable provided by formula (5.59) is

$$\sqrt{8.28 \times 10^{-5}} \approx .009$$

which agrees quite well with the value obtained earlier via simulation.

What's given in Table 5.24 that doesn't come out of a simulation is some understanding of the *biggest contributors* to the uncertainty. The largest contribution listed in Table 5.24 corresponds to variable G, followed in order by those corresponding to variables M_o, T_o, and T_i. At least for the values of the means used in this example, it is the uncertainties in those variables that principally produce the uncertainty in *Efficiency*. Knowing this gives direction to efforts to improve measurement methods. Subject to considerations of feasibility and cost, measurement of the variable G deserves first attention, followed by measurement of the variables M_o, T_o, and T_i.

Notice, however, that reduction of the uncertainty in G alone to essentially 0 would still leave a total in Table 5.24 of about 4.01×10^{-5} and thus an approximate standard deviation for *Efficiency* of about $\sqrt{4.01 \times 10^{-5}} \approx .006$. Calculations of this kind emphasize the need for reductions in the uncertainties of M_o, T_o, and T_i as well, if dramatic (order of magnitude) improvements in overall uncertainty are to be realized.

Table 5.24
Contributions to the Output Variation in
Collector Efficiency

Variable	Contributions to Var Efficiency
C	4.73×10^{-8}
G	4.27×10^{-5}
A	4.76×10^{-7}
M_i	5.01×10^{-7}
M_o	1.58×10^{-5}
T_a	3.39×10^{-8}
T_i	1.10×10^{-5}
T_o	1.22×10^{-5}
Total	8.28×10^{-5}

5.5.5   The Central Limit Effect

One of the most frequently used statistics in engineering applications is the sample mean. Formulas (5.55) and (5.56) relate the mean and variance of the probability distribution of the sample mean to those of a single observation when an iid model is appropriate. One of the most useful facts of applied probability is that if the sample size is reasonably large, it is also possible to approximate the *shape* of the probability distribution of $\overline{X}$, independent of the shape of the underlying distribution of individual observations. That is, there is the following fact:

Proposition 3
(*The Central Limit Theorem*)

If $X_1, X_2, \ldots, X_n$ are iid random variables (with mean μ and variance σ^2), then for large n, the variable $\overline{X}$ is approximately normally distributed. (That is, approximate probabilities for $\overline{X}$ can be calculated using the normal distribution with mean μ and variance σ^2/n.)

A proof of Proposition 3 is outside the purposes of this text. But intuition about the effect is fairly easy to develop through an example.

Example 25
(*Example 2 revisited*)

The Central Limit Effect and the Sample Mean of Tool Serial Numbers

Consider again the example from Section 5.1 involving the last digit of essentially randomly selected serial numbers of pneumatic tools. Suppose now that

$W_1 =$ the last digit of the serial number observed next Monday at 9 A.M.

$W_2 =$ the last digit of the serial number observed the following Monday at 9 A.M.

A plausible model for the pair of random variables W_1, W_2 is that they are independent, each with the marginal probability function

$$f(w) = \begin{cases} .1 & \text{if } w = 0, 1, 2, \ldots, 9 \\ 0 & \text{otherwise} \end{cases} \qquad (5.61)$$

that is pictured in Figure 5.38.

Using such a model, it is a straightforward exercise (along the lines of Example 21, page 303) to reason that $\overline{W} = \frac{1}{2}(W_1 + W_2)$ has the probability function given in Table 5.25 and pictured in Figure 5.39.

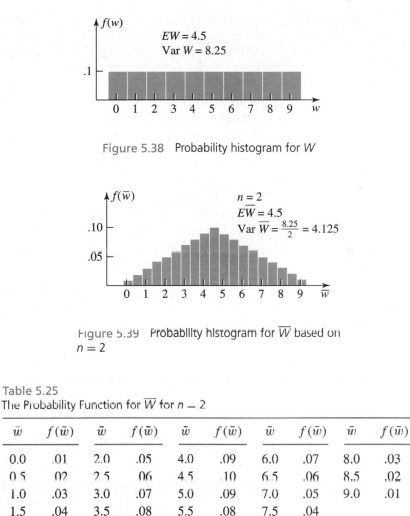

Figure 5.38 Probability histogram for W

Figure 5.39 Probability histogram for $\overline{W}$ based on $n = 2$

Table 5.25
The Probability Function for $\overline{W}$ for $n = 2$

$\bar{w}$	$f(\bar{w})$	$\bar{w}$	$f(\bar{w})$	$\bar{w}$	$f(\bar{w})$	$\bar{w}$	$f(\bar{w})$	$\bar{w}$	$f(\bar{w})$
0.0	.01	2.0	.05	4.0	.09	6.0	.07	8.0	.03
0.5	.02	2.5	.06	4.5	.10	6.5	.06	8.5	.02
1.0	.03	3.0	.07	5.0	.09	7.0	.05	9.0	.01
1.5	.04	3.5	.08	5.5	.08	7.5	.04		

Comparing Figures 5.38 and 5.39, it is clear that even for a completely flat/uniform underlying distribution of W and the small sample size of $n = 2$, the probability distribution of $\overline{W}$ looks far more bell-shaped than the underlying distribution. It is clear why this is so. As you move away from the mean or central value of $\overline{W}$, there are relatively fewer and fewer combinations of w_1 and w_2 that can produce a given value of $\bar{w}$. For example, to observe $\overline{W} = 0$, you must have $W_1 = 0$ and $W_2 = 0$—that is, you must observe not one but two extreme values. On the other hand, there are ten different combinations of w_1 and w_2 that lead to $\overline{W} = 4.5$.

It is possible to use the same kind of logic leading to Table 5.25 to produce exact probability distributions for $\overline{W}$ based on larger sample sizes n. But such

Example 25
(continued)

work is tedious, and for the purpose of indicating roughly how the central limit effect takes over as n gets larger, it is sufficient to approximate the distribution of $\overline{W}$ via simulation for a larger sample size. To this end, 1,000 sets of values for iid variables $W_1, W_2, \ldots, W_8$ (with marginal distribution (5.61)) were simulated and each set averaged to produce 1,000 simulated values of $\overline{W}$ based on $n = 8$. Figure 5.40 is a histogram of these 1,000 values. Notice the bell-shaped character of the plot. (The simulated mean of $\overline{W}$ was $4.508 \approx 4.5 = E\overline{W} = EW$, while the variance of $\overline{W}$ was $1.025 \approx 1.013 = \text{Var } \overline{W} = 8.25/8$, in close agreement with formulas (5.55) and (5.56).)

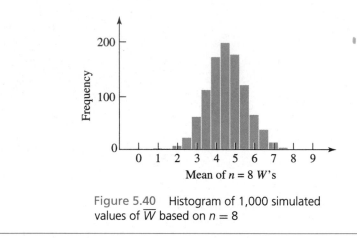

Figure 5.40   Histogram of 1,000 simulated values of $\overline{W}$ based on $n = 8$

Sample size and the central limit effect

What constitutes "large n" in Proposition 3 isn't obvious. The truth of the matter is that what sample size is required before $\overline{X}$ can be treated as essentially normal depends on the shape of the underlying distribution of a single observation. Underlying distributions with decidedly nonnormal shapes require somewhat bigger values of n. But for most engineering purposes, $n \geq 25$ or so is adequate to make $\overline{X}$ essentially normal for the majority of data-generating mechanisms met in practice. (The exceptions are those subject to the occasional production of wildly outlying values.) Indeed, as Example 25 suggests, in many cases $\overline{X}$ is essentially normal for sample sizes much smaller than 25.

The practical usefulness of Proposition 3 is that in many circumstances, only a normal table is needed to evaluate probabilities for sample averages.

Example 23
(continued)

Return one more time to the stamp sale time requirements problem and consider observing and averaging the next $n = 100$ excess service times, to produce

$\overline{S}$ = the sample mean time (over a 7.5 sec threshold) required to complete the next 100 stamp sales

And consider approximating $P[\overline{S} > 17]$.

As discussed before, an iid model with marginal exponential $\alpha = 16.5$ distribution is plausible for the individual excess service times, S. Then

$$E\overline{S} = \alpha = 16.5 \text{ sec}$$

and

$$\sqrt{\text{Var } \overline{S}} = \sqrt{\frac{\alpha^2}{100}} = 1.65 \text{ sec}$$

are appropriate for $\overline{S}$, via formulas (5.55) and (5.56). Further, in view of the fact that $n = 100$ is large, the normal probability table may be used to find approximate probabilities for $\overline{S}$. Figure 5.41 shows an approximate distribution for $\overline{S}$ and the area corresponding to $P[\overline{S} > 17]$.

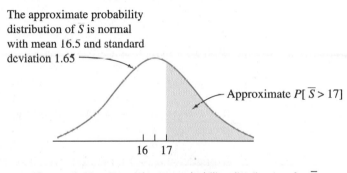

The approximate probability distribution of S is normal with mean 16.5 and standard deviation 1.65

Approximate $P[\overline{S} > 17]$

16 17

Figure 5.41 Approximate probability distribution for $\overline{S}$ and $P[\overline{S} > 17]$

As always, one must convert to z-values before consulting the standard normal table. In this case, the mean and standard deviation to be used are (respectively) 16.5 sec and 1.65 sec. That is, a z-value is calculated as

$$z = \frac{17 - 16.5}{1.65} = .30$$

so

$$P[\overline{S} > 17] \approx P[Z > .30] = 1 - \Phi(.30) = .38$$

The z-value calculated in the example is an application of the general form

z-value for a sample mean

$$z = \frac{\bar{x} - E\overline{X}}{\sqrt{\operatorname{Var}\overline{X}}} = \frac{\bar{x} - \mu}{\dfrac{\sigma}{\sqrt{n}}}$$

(5.62)

appropriate when using the central limit theorem to find approximate probabilities for a sample mean. Formula (5.62) is relevant because by Proposition 3, $\overline{X}$ is approximately normal for large n and formulas (5.55) and (5.56) give its mean and standard deviation.

The final example in this section illustrates how the central limit theorem and some idea of a process or population standard deviation can help guide the choice of sample size in statistical applications.

Example 26
(*Example 10 revisited*)

Sampling a Jar-Filling Process

The process of filling food containers, discussed by J. Fisher in his 1983 "Quality Progress" article, appears (from a histogram in the paper) to have an inherent standard deviation of measured fill weights on the order of 1.6 g. Suppose that in order to calibrate a fill-level adjustment knob on such a process, you set the knob and fill a run of n jars. Their sample mean net contents will then serve as an indication of the process mean fill level corresponding to that knob setting. Suppose further that you would like to choose a sample size, n, large enough that a priori there is an 80% chance the sample mean is within .3 g of the actual process mean.

If the filling process can be thought of as physically stable, it makes sense to model the n observed net weights as iid random variables with (unknown) marginal mean μ and standard deviation $\sigma = 1.6$ g. For large n,

$$\overline{V} = \text{the observed sample average net weight}$$

can be thought of as approximately normal with mean μ and standard deviation $\sigma/\sqrt{n} = 1.6/\sqrt{n}$ (by Proposition 3 and formulas (5.55) and (5.56)).

Now the requirement that $\overline{V}$ be within .3 g of μ can be written as

$$\mu - .3 < \overline{V} < \mu + .3$$

so the problem at hand is to choose n such that

$$P[\mu - .3 < \overline{V} < \mu + .3] = .80$$

Figure 5.42 pictures the situation. The .90 quantile of the standard normal distribution is roughly 1.28—that is, $P[-1.28 < Z < 1.28] = .8$. So evidently Figure 5.42 indicates that $\mu + .3$ should have z-value 1.28. That is, you want

$$1.28 = \frac{(\mu + .3) - \mu}{\frac{1.6}{\sqrt{n}}}$$

or

$$.3 = 1.28 \frac{1.6}{\sqrt{n}}$$

So, solving for n, a sample size of $n \approx 47$ would be required to provide the kind of precision of measurement desired.

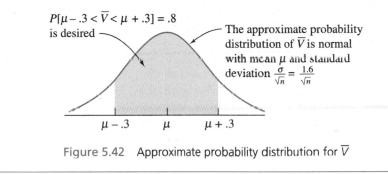

Figure 5.42   Approximate probability distribution for $\overline{V}$

Section 5 Exercises

1. A type of nominal $\frac{3}{4}$ inch plywood is made of five layers. These layers can be thought of as having thicknesses roughly describable as independent random variables with means and standard deviations as follows:

Layer	Mean (in.)	Standard Deviation (in.)
1	.094	.001
2	.156	.002
3	.234	.002
4	.172	.002
5	.094	.001

Find the mean and standard deviation of total thickness associated with the combination of these individual values.

2. The coefficient of linear expansion of brass is to be obtained as a laboratory exercise. For a brass bar that is L_1 meters long at $T_1°C$ and L_2 meters long at $T_2°C$, this coefficient is

$$\alpha = \frac{L_2 - L_1}{L_1(T_2 - T_1)}$$

Suppose that the equipment to be used in the laboratory is thought to have a standard deviation for repeated length measurements of about .00005 m and a standard deviation for repeated temperature measurements of about .1°C.

(a) If using $T_1 \approx 50°C$ and $T_2 \approx 100°C$, $L_1 \approx 1.00000$ m and $L_2 \approx 1.00095$ m are obtained, and it is desired to attach an approximate standard deviation to the derived value of α, find such an approximate standard deviation two different ways. First, use simulation as was done in Printout 1. Then use the propagation of error formula. How well do your two values agree?

(b) In this particular lab exercise, the precision of which measurements (the lengths or the temperatures) is the primary limiting factor in the precision of the derived coefficient of linear expansion? Explain.

(c) Within limits, the larger $T_2 - T_1$, the better the value for α. What (in qualitative terms) is the physical origin of those limits?

3. Consider again the random number generator discussed in Exercise 1 of Section 5.2. Suppose that it is used to generate 25 random numbers and that these may reasonably be thought of as independent random variables with common individual (marginal) distribution as given in Exercise 1 of Section 5.2. Let $\overline{X}$ be the sample mean of these 25 values.

(a) What are the mean and standard deviation of the random variable $\overline{X}$?

(b) What is the approximate probability distribution of $\overline{X}$?

(c) Approximate the probability that $\overline{X}$ exceeds .5.

(d) Approximate the probability that $\overline{X}$ takes a value within .02 of its mean.

(e) Redo parts (a) through (d) using a sample size of 100 instead of 25.

4. Passing a large production run of piston rings through a grinding operation produces edge widths possessing a standard deviation of .0004 in. A simple random sample of rings is to be taken and their edge widths measured, with the intention of using $\overline{X}$ as an estimate of the population mean thickness μ. Approximate the probabilities that $\overline{X}$ is within .0001 in. of μ for samples of size $n = 25$, 100, and 400.

5. A pendulum swinging through small angles approximates simple harmonic motion. The period of the pendulum, τ, is (approximately) given by

$$\tau = 2\pi \sqrt{\frac{L}{g}}$$

where L is the length of the pendulum and g is the acceleration due to gravity. This fact can be used to derive an experimental value for g. Suppose that the length L of about 5 ft can be measured with a standard deviation of about .25 in. (about .0208 foot), and the period τ of about 2.48 sec can be measured with standard deviation of about .1 sec. What is a reasonable standard deviation to attach to a value of g derived using this equipment? Is the precision of the length measurement or the precision of the period measurement the principal limitation on the precision of the derived g?

Chapter 5 Exercises ....

1. Suppose 90% of all students taking a beginning programming course fail to get their first program to run on first submission. Use a binomial distribution and assign probabilities to the possibilities that among a group of six such students,

(a) all fail on their first submissions

(b) at least four fail on their first submissions

(c) less than four fail on their first submissions
Continuing to use this binomial model,

(d) what is the mean number who will fail?

(e) what are the variance and standard deviation of the number who will fail?

2. Suppose that for single launches of a space shuttle, there is a constant probability of O-ring failure (say,

.15). Consider ten future launches, and let X be the number of those involving an O-ring failure. Use an appropriate probability model and evaluate all of the following:

(a) $P[X = 2]$ (b) $P[X \geq 1]$
(c) EX (d) Var X
(e) the standard deviation of X

3. An injection molding process for making auto bumpers leaves an average of 1.3 visual defects per bumper prior to painting. Let Y and Z be the numbers of visual defects on (respectively) the next two bumpers produced. Use an appropriate probability distribution and evaluate the following:

(a) $P[Y = 2]$ (b) $P[Y \geq 1]$
(c) $\sqrt{\text{Var } Y}$
(d) $P[Y + Z \geq 2]$ (*Hint:* What is a sensible distribution for $Y + Z$, the number of blemishes on two bumpers?)

4. Suppose that the random number generator supplied in a pocket calculator actually generates values in such a way that if X is the next value generated, X can be adequately described using a probability density of the form

$$f(x) = \begin{cases} k((x - .5)^2 + 1) & \text{for } 0 < x < 1 \\ 0 & \text{otherwise} \end{cases}$$

(a) Evaluate k and sketch a graph of $f(x)$.
(b) Evaluate $P[X \geq .5]$, $P[X > .5]$, $P[.15 > X \geq .5]$, and $P[|X - .5| \geq .2]$.
(c) Compute EX and Var X.
(d) Compute and graph $F(x)$, the cumulative probability function for X. Read from your graph the .8 quantile of the distribution of X.

5. Suppose that Z is a standard normal random variable. Evaluate the following probabilities involving Z:

(a) $P[Z \leq 1.13]$ (b) $P[Z > -.54]$
(c) $P[-1.02 < Z < .06]$ (d) $P[|Z| \leq .25]$
(e) $P[|Z| > 1.51]$ (f) $P[-3.0 < Z < 3.0]$
Find numbers # such that the following statements about Z are true:
(g) $P[|Z| < \#] = .80$ (h) $P[Z < \#] = .80$
(i) $P[|Z| > \#] = .04$

6. Suppose that X is a normal random variable with mean $\mu = 10.2$ and standard deviation $\sigma = .7$. Evaluate the following probabilities involving X:

(a) $P[X \leq 10.1]$ (b) $P[X > 10.5]$
(c) $P[9.0 < X < 10.3]$ (d) $P[|X - 10.2| \leq .25]$
(e) $P[|X - 10.2| > 1.51]$
Find numbers # such that the following statements about X are true:
(f) $P[|X - 10.2| < \#] = .80$
(g) $P[X < \#] = .80$
(h) $P[|X - 10.2| > \#] = .04$

7. In a grinding operation, there is an upper specification of 3.150 in. on a dimension of a certain part after grinding. Suppose that the standard deviation of this normally distributed dimension for parts of this type ground to any particular mean dimension μ is $\sigma = .002$ in. Suppose further that you desire to have no more than 3% of the parts fail to meet specifications. What is the maximum (minimum machining cost) μ that can be used if this 3% requirement is to be met?

8. A 10 ft cable is made of 50 strands. Suppose that, individually, 10 ft strands have breaking strengths with mean 45 lb and standard deviation 4 lb. Suppose further that the breaking strength of a cable is roughly the sum of the strengths of the strands that make it up.

(a) Find a plausible mean and standard deviation for the breaking strengths of such 10 ft cables.
(b) Evaluate the probability that a 10 ft cable of this type will support a load of 2230 lb. (*Hint:* If $\overline{X}$ is the mean breaking strength of the strands, $\sum(\text{Strengths}) \geq 2230$ is the same as $\overline{X} \geq (\frac{2230}{50})$. Now use the central limit theorem.)

9. The electrical resistivity, ρ, of a piece of wire is a property of the material involved and the temperature at which it is measured. At a given temperature, if a cylindrical piece of wire of length L and cross-sectional area A has resistance R, the material's resistivity is calculated using the formula $\rho = \frac{RA}{L}$. Thus, if a wire's cross section is assumed

to be circular with diameter D, the resistivity at a given temperature is

$$\rho = \frac{R\pi D^2}{4L}$$

In a lab exercise to determine the resistivity of copper at $20°C$, students measure lengths, diameters, and resistances of wire nominally 1.0 m in length (L), 2.0×10^{-3} m in diameter (D), and of resistance (R) $.54 \times 10^{-2}$ Ω. Suppose that it is sensible to describe the measurement precisions in this laboratory with the standard deviations $\sigma_L \approx 10^{-3}$ m, $\sigma_D \approx 10^{-4}$ m, and $\sigma_R \approx 5 \times 10^{-4}$ Ω.

(a) Find an approximate standard deviation that might be used to describe the expected precision for an experimentally derived value of ρ.

(b) Imprecision in which of the measurements is likely to be the largest contributor to imprecision in measured resistivity? Explain.

(c) You should expect that the value derived in (a) underpredict the kind of variation that would be observed in such laboratory exercises over a period of years. Explain why this is so.

10. Suppose that the thickness of sheets of a certain weight of book paper have mean .1 mm and a standard deviation of .003 mm. A particular textbook will be printed on 370 sheets of this paper. Find sensible values for the mean and standard deviation of the thicknesses of copies of this text (excluding, of course, the book's cover).

11. Pairs of resistors are to be connected in parallel and a difference in electrical potential applied across the resistor assembly. Ohm's law predicts that in such a situation, the current flowing in the circuit will be

$$I = V\left(\frac{1}{R_1} + \frac{1}{R_2}\right)$$

where R_1 and R_2 are the two resistances and V the potential applied. Suppose that R_1 and R_2 have

means $\mu_R = 10$ Ω and standard deviations $\sigma_R = .1$ Ω and that V has mean $\mu_V = 9$ volt and $\sigma_V = .2$ volt.

(a) Find an approximate mean and standard deviation for I, treating R_1, R_2, and V as independent random variables.

(b) Based on your work in (a), would you say that the variation in voltage or the combined variations in R_1 and R_2 are the biggest contributors to variation in current? Explain.

12. Students in a materials lab are required to experimentally determine the heat conductivity of aluminum.

(a) If student-derived values are normally distributed about a mean of .5 (cal/(cm)(sec)(°C)) with standard deviation of .03, evaluate the probability that an individual student will obtain a conductivity from .48 to .52.

(b) If student values have the mean and standard deviation given in (a), evaluate the probability that a class of 25 students will produce a sample mean conductivity from .48 to .52.

(c) If student values have the mean and standard deviation given in (a), evaluate the probability that at least 2 of the next 5 values produced by students will be in the range from .48 to .52.

13. Suppose that 10 ft lengths of a certain type of cable have breaking strengths with mean $\mu = 450$ lb and standard deviation $\sigma = 50$ lb.

(a) If five of these cables are used to support a single load L, suppose that the cables are loaded in such a way that support fails if any one of the cables has strength below $\frac{L}{5}$. With $L = 2{,}000$ lb, assess the probability that the support fails, if individual cable strength is normally distributed. Do this in two steps. First find the probability that a particular individual cable fails, then use that to evaluate the desired probability.

(b) Approximate the probability that the sample mean strength of 100 of these cables is below 457 lb.

14. Find EX and Var X for a continuous distribution with probability density

$$f(x) = \begin{cases} .3 & \text{if } 0 < x < 1 \\ .7 & \text{if } 1 < x < 2 \\ 0 & \text{otherwise} \end{cases}$$

15. Suppose that it is adequate to describe the 14-day compressive strengths of test specimens of a certain concrete mixture as normally distributed with mean $\mu = 2{,}930$ psi and standard deviation $\sigma = 20$ psi.
    (a) Assess the probability that the next specimen of this type tested for compressive strength will have strength above 2,945 psi.
    (b) Use your answer to part (a) and assess the probability that in the next four specimens tested, at least one has compressive strength above 2,945 psi.
    (c) Assess the probability that the next 25 specimens tested have a sample mean compressive strength within 5 psi of $\mu = 2{,}930$ psi.
    (d) Suppose that although the particular concrete formula under consideration in this problem is relatively strong, it is difficult to pour in large quantities without serious air pockets developing (which can have important implications for structural integrity). In fact, suppose that using standard methods of pouring, serious air pockets form at an average rate of 1 per 50 cubic yards of poured concrete. Use an appropriate probability distribution and assess the probability that two or more serious air pockets will appear in a 150 cubic yard pour to be made tomorrow.

16. For X with a continuous distribution specified by the probability density

$$f(x) = \begin{cases} .5x & \text{for } 0 < x < 2 \\ 0 & \text{otherwise} \end{cases}$$

find $P[X < 1.0]$ and find the mean, EX.

17. The viscosity of a liquid may be measured by placing it in a cylindrical container and determining the force needed to turn a cylindrical rotor (of nearly the same diameter as the container) at a given velocity in the liquid. The relationship between the viscosity η, force F, area A of the side of the rotor in contact with the liquid, the size L of the gap between the rotor and the inside of the container, and the velocity v at which the rotor surface moves is

$$\eta = \frac{FL}{vA}$$

Suppose that students are to determine an experimental viscosity for SAE no. 10 oil as a laboratory exercise and that appropriate means and standard deviations for the measured variables F, L, v, and A in this laboratory are as follows:

$\mu_F = 151$ N	$\sigma_F = .05$ N
$\mu_A = 1257$ cm^2	$\sigma_A = .2$ cm^2
$\mu_L = .5$ cm	$\sigma_L = .05$ cm
$\mu_v = 30$ cm/sec	$\sigma_v = 1$ cm/sec

(a) Use the propagation of error formulas and find an approximate standard deviation that might serve as a measure of precision for an experimentally derived value of η from this laboratory.
(b) Explain why, if experimental values of η obtained for SAE no. 10 oil in similar laboratory exercises conducted over a number of years at a number of different universities were compared, the approximate standard deviation derived in (a) would be likely to understate the variability actually observed in those values.

18. The heat conductivity, λ, of a cylindrical bar of diameter D and length L, connected between two constant temperature devices of temperatures T_1 and T_2 (respectively), that conducts Q calories in t seconds is

$$\lambda = \frac{4QL}{\pi(T_1 - T_2)tD^2}$$

In a materials laboratory exercise to determine λ for brass, the following means and standard deviations for the variables D, L, T_1, T_2, Q, and t are appropriate, as are the partial derivatives of λ with respect to the various variables (evaluated at the means of the variables):

	D	L	T_1
μ	1.6 cm	100 cm	100°C
σ	.1 cm	.1 cm	1°C
partial	$-.249$	.199	$-.00199$

	T_2	Q	t
μ	0°C	240 cal	600 sec
σ	1°C	10 cal	1 sec
partial	.00199	.000825	.000332

(The units of the partial derivatives are the units of $\lambda(\text{cal}/(\text{cm})(\text{sec})(°\text{C}))$ divided by the units of the variable in question.)

(a) Find an approximate standard deviation to associate with an experimentally derived value of λ.

(b) Which of the variables appears to be the biggest contributor to variation in experimentally determined values of λ? Explain your choice.

19. Suppose that 15% of all daily oxygen purities delivered by an air-products supplier are below 99.5% purity and that it is plausible to think of daily purities as independent random variables. Evaluate the probability that in the next five-day workweek, 1 or less delivered purities will fall below 99.5%.

20. Suppose that the raw daily oxygen purities delivered by an air-products supplier have a standard deviation $\sigma \approx .1$ (percent), and it is plausible to think of daily purities as independent random variables. Approximate the probability that the sample mean $\overline{X}$ of $n = 25$ delivered purities falls within .03 (percent) of the raw daily purity mean, μ.

21. Students are going to measure Young's Modulus for copper by measuring the elongation of a piece of copper wire under a tensile force. For a cylindrical wire of diameter D subjected to a tensile force F, if the initial length (length before applying the force) is L_0 and final length is L_1, Young's Modulus for the material in question is

$$Y = \frac{4FL_0}{\pi D^2 (L_1 - L_0)}$$

The test and measuring equipment used in a particular lab are characterized by the standard deviations

$$\sigma_F \approx 10 \text{ lb} \qquad \sigma_D \approx .001 \text{ in.}$$

$$\sigma_{L_0} = \sigma_{L_1} = .01 \text{ in.}$$

and in the setup employed, $F \approx 300$ lb, $D \approx$ .050 in., $L_0 \approx 10.00$ in., and $L_1 \approx 10.10$ in.

(a) Treating the measured force, diameter, and lengths as independent variables with the preceding means and standard deviations, find an approximate standard deviation to attach to an experimentally derived value of Y. (Partial derivatives of Y at the nominal values of F, D, L_0, and L_1 are approximately $\frac{\partial Y}{\partial F} \approx 5.09 \times 10^4$, $\frac{\partial Y}{\partial D} \approx -6.11 \times 10^8$, $\frac{\partial Y}{\partial L_0} \approx 1.54 \times 10^8$, and $\frac{\partial Y}{\partial L_1} \approx -1.53 \times 10^8$ in the appropriate units.)

(b) Uncertainty in which of the variables is the biggest contributor to uncertainty in Y?

(c) Notice that the equation for Y says that for a particular material (and thus supposedly constant Y), circular wires of constant initial lengths L_0, but of different diameters and subjected to different tensile forces, will undergo elongations $\Delta L = L_1 - L_0$ of approximately

$$\Delta L \approx \kappa \frac{F}{D^2}$$

for κ a constant depending on the material and the initial length. Suppose that you decide to measure ΔL for a factorial arrangement of levels of F and D. Does the equation predict that F and D will or will not have important interactions? Explain.

22. Exercise 6 of Chapter 3 concerns the lifetimes (in numbers of 24 mm deep holes drilled in 1045 steel before failure) of 12 D952-II (8 mm) drills.
    (a) Make a normal plot of the data given in Exercise 6 of Chapter 3. In what specific way does the shape of the data distribution appear to depart from a normal shape?
    (b) The 12 lifetimes have mean $\bar{y} = 117.75$ and standard deviation $s \approx 51.1$. Simply using these in place of μ and σ for the underlying drill life distribution, use the normal table to find an approximate fraction of drill lives below 40 holes.
    (c) Based on your answer to (a), if your answer to (b) is seriously different from the real fraction of drill lives below 40, is it most likely high or low? Explain.

23. Metal fatigue causes cracks to appear on the skin of older aircraft. Assume that it is reasonable to model the number of cracks appearing on a 1 m² surface of planes of a certain model and vintage as Poisson with mean $\lambda = .03$.
    (a) If 1 m² is inspected, assess the probability that at least one crack is present on that surface.
    (b) If 10 m² are inspected, assess the probability that at least one crack (total) is present.
    (c) If ten areas, each of size 1 m², are inspected, assess the probability that exactly one of these has cracks.

24. If a dimension on a mechanical part is normally distributed, how small must the standard deviation be if 95% of such parts are to be within specifications of 2 cm $\pm$ .002 cm when the mean dimension is ideal ($\mu = 2$ cm)?

25. The fact that the "exact" calculation of normal probabilities requires either numerical integration or the use of tables (ultimately generated using numerical integration) has inspired many people to develop approximations to the standard normal cumulative distribution function. Several

of the simpler of these approximations are discussed in the articles "A Simpler Approximation for Areas Under the Standard Normal Curve," by A. Shah (*The American Statistician*, 1985), "Pocket-Calculator Approximation for Areas under the Standard Normal Curve," by R. Norton (*The American Statistician*, 1989), and "Approximations for Hand Calculators Using Small Integer Coefficients," by S. Derenzo (*Mathematics of Computation*, 1977). For $z > 0$, consider the approximations offered in these articles:

$$\Phi(z) \approx g_S(z) = \begin{cases} .5 + \dfrac{z(4.4 - z)}{10} & 0 \leq z \leq 2.2 \\ .99 & 2.2 < z < 2.6 \\ 1.00 & 2.6 \leq z \end{cases}$$

$$\Phi(z) \approx g_N(z) = 1 - \frac{1}{2} \exp\left(-\frac{z^2 + 1.2z^{.8}}{2}\right)$$

$$\Phi(z) \approx g_D(z)$$
$$= 1 - \frac{1}{2} \exp\left(-\frac{(83z + 351)z + 562}{703/z + 165}\right)$$

Evaluate $g_S(z)$, $g_N(z)$, and $g_D(z)$ for $z = .5, 1.0, 1.5, 2.0$, and 2.5. How do these values compare to the corresponding entries in Table B.3?

26. Exercise 25 concerned approximations for normal probabilities. People have also invested a fair amount of effort in finding useful formulas approximating standard normal *quantiles*. One such approximation was given in formula (3.3). A more complicated one, again taken from the article by S. Derenzo mentioned in Exercise 25, is as follows. For $p > .50$, let $y = -\ln(2(1 - p))$ and

$$Q_z(p) \approx \sqrt{\frac{((4y + 100)y + 205)y^2}{((2y + 56)y + 192)y + 131}}$$

For $p < .50$, let $y = -\ln(2p)$ and

$$Q_z(p) \approx -\sqrt{\frac{((4y + 100)y + 205)y^2}{((2y + 56)y + 192)y + 131}}$$

Use these formulas to approximate $Q_z(p)$ for $p = .01, .05, .1, .3, .7, .9, .95,$ and $.99$. How do the values you obtain compare with the corresponding entries in Table 3.10 and the results of using formula (3.3)?

27. The article "Statistical Strength Evaluation of Hot-pressed Si_3N_4" by R. Govila (*Ceramic Bulletin*, 1983) contains summary statistics from an extensive study of the flexural strengths of two high-strength hot-pressed silicon nitrides in $\frac{1}{4}$ point, 4 point bending. The values below are fracture strengths of 30 specimens of one of the materials tested at 20°C. (The units are MPa, and the data were read from a graph in the paper and may therefore individually differ by perhaps as much as 10 MPa from the actual measured values.)

> 514, 533, 543, 547, 584, 619, 653, 684,
> 689, 695, 700, 705, 709, 729, 729, 753,
> 763, 800, 805, 805, 814, 819, 819, 839,
> 839, 849, 879, 900, 919, 979

(a) The materials researcher who collected the original data believed the Weibull distribution to be an adequate model for flexural strength of this material. Make a Weibull probability plot using the method of display (5.35) of Section 5.3 and investigate this possibility. Does a Weibull model fit these data?

(b) Eye-fit a line through your plot from part (a). Use it to help you determine an appropriate shape parameter, β, and an appropriate scale parameter, α, for a Weibull distribution used to describe flexural strength of this material at 20°C. For a Weibull distribution with your fitted values of α and β, what is the median strength? What is a strength exceeded by 80% of such Si_3N_4 specimens? By 90% of such specimens? By 99% of such specimens?

(c) Make normal plots of the raw data and of the logarithms of the raw data. Comparing the three probability plots made in this exercise, is there strong reason to prefer a Weibull model, a normal model, or a lognormal model over

the other two possibilities as a description of the flexural strength?

(d) Eye-fit lines to your plots from part (c). Use them to help you determine appropriate means and standard deviations for normal distributions used to describe flexural strength and the logarithm of flexural strength. Compare the .01, .10, .20, and .50 quantiles of the fitted normal and lognormal distributions for strength to the quantiles you computed in part (b).

28. The article "Using Statistical Thinking to Solve Maintenance Problems" by Brick, Michael, and Morganstein (*Quality Progress*, 1989) contains the following data on lifetimes of sinker rollers. Given are the numbers of 8-hour shifts that 17 sinker rollers (at the bottom of a galvanizing pot and used to direct steel sheet through a coating operation) lasted before failing and requiring replacement.

> 10, 12, 15, 17, 18, 18, 20, 20,
> 21, 21, 23, 25, 27, 29, 29, 30, 35

(a) The authors of the article considered a Weibull distribution to be a likely model for the lifetimes of such rollers. Make a zero-threshold Weibull probability plot for use in assessing the reasonableness of such a description of roller life.

(b) Eye-fit a line to your plot in (a) and use it to estimate parameters for a Weibull distribution for describing roller life.

(c) Use your estimated parameters from (a) and the form of the Weibull cumulative distribution function given in Section 5.2 to estimate the .10 quantile of the roller life distribution.

29. The article "Elementary Probability Plotting for Statistical Data Analysis" by J. King (*Quality Progress*, 1988) contains 24 measurements of deviations from nominal of a distance between two

holes drilled in a steel plate. These are reproduced here. The units are mm.

$$-2, -2, 7, -10, 4, -3, 0, 8, -5, 5, -6, 0,$$
$$2, -2, 1, 3, 3, -4, -6, -13, -7, -2, 2, 2$$

(a) Make a dot diagram for these data and compute $\bar{x}$ and s.

(b) Make a normal plot for these data. Eye-fit a line on the plot and use it to find graphical estimates of a process mean and standard deviation for this deviation from nominal. Compare these graphical estimates with the values you calculated in (a).

(c) Engineering specifications on this deviation from nominal were $\perp 10$ mm. Suppose that $\bar{x}$ and s from (a) are adequate approximations of the process mean and standard deviation for this variable. Use the normal distribution with those parameters and compute a fraction of deviations that fall outside specifications. Does it appear from this exercise that the drilling operation is *capable* (i.e., precise) enough to produce essentially all measured deviations in specifications, at least if properly aimed? Explain.

30. An engineer is responsible for setting up a monitoring system for a critical diameter on a turned metal part produced in his plant. Engineering specifications for the diameter are 1.180 in. $\perp$ .004 in. For ease of communication, the engineer sets up the following nomenclature for measured diameters on these parts:

Green Zone Diameters: 1.178 in. $\leq$ Diameter $\leq$ 1.182 in.

Red Zone Diameters: Diameter $\leq$ 1.176 in. or Diameter $\geq$ 1.184 in.

Yellow Zone Diameters: any other Diameter

Suppose that in fact the diameters of parts coming off the lathe in question can be thought of as independent normal random variables with mean $\mu = 1.181$ in. and standard deviation $\sigma = .002$ in.

(a) Find the probabilities that a given diameter falls into each of the three zones.

(b) Suppose that a technician simply begins measuring diameters on consecutive parts and continues until a Red Zone measurement is found. Assess the probability that more than ten parts must be measured. Also, give the expected number of measurements that must be made.

The engineer decides to use the Green/Yellow/Red gauging system in the following way. Every hour, parts coming off the lathe will be checked. First, a single part will be measured. If it is in the Green Zone, no further action is needed that hour. If the initial part is in the Red Zone, the lathe will be stopped and a supervisor alerted. If the first part is in the Yellow Zone, a second part is measured. If this second measurement is in the Green Zone, no further action is required, but if it is in the Yellow or the Red Zone, the lathe is stopped and a supervisor alerted. It is possible to argue that under this scheme (continuing to suppose that measurements are independent normal variables with mean 1.181 in. and standard deviation .002 in.), the probability that the lathe is stopped in any given hour is .1865.

(c) Use the preceding fact and evaluate the probability that the lathe is stopped exactly twice in 8 consecutive hours. Also, what is the expected number of times the lathe will be stopped in 8 time periods?

31. A random variable X has a cumulative distribution function

$$F(x) = \begin{cases} 0 & \text{for } x \leq 0 \\ \sin(x) & \text{for } 0 < x < \pi/2 \\ 1 & \text{for } \pi/2 < x \end{cases}$$

(a) Find $P[X \leq .32]$.

(b) Give the probability density for X, $f(x)$.

(c) Evaluate EX and Var X.

32. Return to the situation of Exercise 2 of Section 5.4.

Suppose that demerits are assigned to devices of the type considered there according to the formula $D = 5X + Y$.

(a) Find the mean value of D, ED. (Use your answers to (c) and (d) Exercise 2 of Section 5.4 and formula (5.53) of Section 5.5. Formula (5.53) holds whether or not X and Y are independent.)

(b) Find the probability a device of this type scores 7 or less demerits. That is, find $P[D \leq 7]$.

(c) On average, how many of these devices will have to be inspected in order to find one that scores 7 or less demerits? (Use your answer to (b).)

33. Consider jointly continuous random variables X and Y with density

$$f(x, y) = \begin{cases} x + y & \text{for } 0 < x < 1 \text{ and } 0 < y < 1 \\ 0 & \text{otherwise} \end{cases}$$

(a) Find the probability that the product of X and Y is at least $\frac{1}{4}$.

(b) Find the marginal probability density for X. (Notice that Y's is similar.) Use this to find the expected value and standard deviation of X.

(c) Are X and Y independent? Explain.

(d) Compute the mean of $X + Y$. Why can't formula (5.54) of Section 5.5 be used to find the variance of $X + Y$?

34. Return to the situation of Exercise 4 of Section 5.4.

(a) Find EX, Var X, EY, and Var Y using the marginal densities for X and Y.

(b) Use your answer to (a) and Proposition 1 to find the mean and variance of $Y - X$.

35. Visual inspection of integrated circuit chips, even under high magnification, is often less than perfect. Suppose that an inspector has an 80% chance of detecting any given flaw. We will suppose that the inspector never "cries wolf"—that is, sees a flaw where none exists. Then consider the random variables

$X = $ the true number of flaws on a chip

$Y = $ the number of flaws identified by the inspector

(a) What is a sensible conditional distribution for Y given that $X = 5$? Given that $X = 5$, find the (conditional) probability that $Y = 3$.

In general, a sensible conditional probability function for Y given $X = x$ is the binomial probability function with number of trials x and success probability .8. That is, one could use

$$f_{Y|X}(y \mid x) = \begin{cases} \binom{x}{y} .8^y .2^{x-y} & \text{for } y = 0, \\ & \quad 1, 2, \ldots, x \\ 0 & \text{otherwise} \end{cases}$$

Now suppose that X is modeled as Poisson with mean $\lambda = 3$—i.e.,

$$f_X(x) = \begin{cases} \dfrac{e^{-3} 3^x}{x!} & \text{for } x = 0, 1, 2, 3, \ldots \\ 0 & \text{otherwise} \end{cases}$$

Multiplication of the two formulas gives a joint probability function for X and Y.

(b) Find the (marginal) probability that $Y = 0$. (Note that this is obtained by summing $f(x, 0)$ over all possible values of x.)

(c) Find $f_Y(y)$ in general. What (marginal) distribution does Y have?

36. Suppose that cans to be filled with a liquid are circular cylinders. The radii of these cans have mean $\mu_r = 1.00$ in. and standard deviation $\sigma_r = .02$ in. The volumes of liquid dispensed into these cans have mean $\mu_v = 15.10$ in.^{3} and standard deviation $\sigma_v = .05$ in.^{3}.

(a) If the volumes dispensed into the cans are approximately normally distributed, about what fraction will exceed 15.07 in.^{3}?

(b) Approximate the probability that the total volume dispensed into the next 100 cans exceeds 1510.5 in.^{3} (if the total exceeds 1510.5, $\overline{X}$ exceeds 15.105).

(c) Approximate the mean μ_h and standard deviation σ_h of the heights of the liquid in the filled cans. (Recall that the volume of a circular cylinder is $v = \pi r^2 h$, where h is the height of the cylinder.)

(d) Does the variation in bottle radius or the variation in volume of liquid dispensed into the bottles have the biggest impact on the variation in liquid height? Explain.

37. Suppose that a pair of random variables have the joint probability density

$$f(x, y) = \begin{cases} \exp(x - y) & \text{if } 0 \leq x \leq 1 \text{ and } x \leq y \\ 0 & \text{otherwise} \end{cases}$$

(a) Evaluate $P[Y \leq 1.5]$.
(b) Find the marginal probability densities for X and Y.
(c) Are X and Y independent? Explain.
(d) Find the conditional probability density for Y given $X = .25$, $f_{Y|X}(y \mid .25)$. Given that $X = .25$, what is the mean of Y? (*Hint:* Use $f_{Y|X}(y \mid .25)$.)

38. (**Defects per Unit Acceptance Sampling**) Suppose that in the inspection of an incoming product, nonconformities on an inspection unit are counted. If too many are seen, the incoming lot is rejected and returned to the manufacturer. (For concreteness, you might think of blemishes on rolled paper or wire, where an inspection unit consists of a certain length of material from the roll.) Suppose further that the number of nonconformities on a piece of product of any particular size can be modeled as Poisson with an appropriate mean.

(a) Suppose that this rule is followed: "Accept the lot if on a standard size inspection unit, 1 or fewer nonconformities are seen." The *operating characteristic curve* of this acceptance sampling plan is a plot of the probability that the lot is accepted as a function of $\lambda =$ the mean defects per inspection unit. (For $X =$ the number of nonconformities seen, X has Poisson distribution with mean λ and $OC(\lambda) = P[X \leq 1]$.) Make a plot of the operating characteristic curve. List values of the operating characteristic for $\lambda = .25, .5,$ and 1.0.

(b) Suppose that instead of the rule in (a), this rule is followed: "Accept the lot if on 2 standard size inspection units, 2 or fewer total nonconformities are seen." Make a plot of the operating characteristic curve for this second plan and compare it with the plot from part (a). (Note that here, for $X =$ the total number of nonconformities seen, X has a Poisson distribution with mean 2λ and $OC(\lambda) = P[X \leq 2]$.) List values of the operating characteristic for $\lambda = .25, .5,$ and 1.0.

39. A discrete random variable X can be described using the following probability function:

x	1	2	3	4	5
$f(x)$	.61	.24	.10	.04	.01

(a) Make a probability histogram for X. Also plot $F(x)$, the cumulative probability function for X.
(b) Find the mean and standard deviation for the random variable X.
(c) Evaluate $P[X > 3]$ and then find $P[X < 3]$.

40. A classical data set of Rutherford and Geiger (referred to in Example 6) suggests that for a particular experimental setup involving a small bar of polonium, the number of collisions of α particles with a small screen placed near the bar during an 8-minute period can be modeled as a Poisson variable with mean $\lambda = 3.87$. Consider an experimental setup of this type, and let X and Y be (respectively) the numbers of collisions in the next two 8-minute periods. Evaluate the following:

(a) $P[X \geq 2]$ (b) $\sqrt{\text{Var } X}$
(c) $P[X + Y = 6]$ (d) $P[X + Y \geq 3]$
(*Hint for parts (c) and (d):* What is a sensible probability distribution for $X + Y$, the number of collisions in a 16-minute period?)

41. Suppose that X is a continuous random variable with probability density of the form

$$f(x) = \begin{cases} k\left(x^2(1-x)\right) & \text{for } 0 < x < 1 \\ 0 & \text{otherwise} \end{cases}$$

   (a) Evaluate k and sketch a graph of $f(x)$.
   (b) Evaluate $P[X \le .25]$, $P[X \le .75]$, $P[.25 < X \le .75]$, and $P[|X - .5| > .1]$.
   (c) Compute EX and $\sqrt{\text{Var } X}$.
   (d) Compute and graph $F(x)$, the cumulative distribution function for X. Read from your graph the .6 quantile of the distribution of X.

42. Suppose that engineering specifications on the shelf depth of a certain slug to be turned on a CNC lathe are from .0275 in. to .0278 in. and that values of this dimension produced on the lathe can be described using a normal distribution with mean μ and standard deviation σ.
   (a) If $\mu = .0276$ and $\sigma = .0001$, about what fraction of shelf depths are in specifications?
   (b) What machine precision (as measured by σ) would be required in order to produce about 98% of shelf depths within engineering specifications (assuming that μ is at the midpoint of the specifications)?

43. The resistance of an assembly of several resistors connected in series is the sum of the resistances of the individual resistors. Suppose that a large lot of nominal 10 Ω resistors has mean resistance $\mu = 9.91\ \Omega$ and standard deviation of resistances $\sigma = .08\ \Omega$. Suppose that 30 resistors are randomly selected from this lot and connected in series.
   (a) Find a plausible mean and variance for the resistance of the assembly.
   (b) Evaluate the probability that the resistance of the assembly exceeds 298.2 Ω. (*Hint:* If $\overline{X}$ is the mean resistance of the 30 resistors involved, the resistance of the assembly exceeding 298.2 Ω is the same as $\overline{X}$ exceeding 9.94 Ω. Now apply the central limit theorem.)

44. At a small metal fabrication company, steel rods of a particular type cut to length have lengths with standard deviation .005 in.

   (a) If lengths are normally distributed about a mean μ (which can be changed by altering the setup of a jig) and specifications on this length are 33.69 in. $\pm$ .01 in., what appears to be the best possible fraction of the lengths in specifications? What does μ need to be in order to achieve this fraction?
   (b) Suppose now that in an effort to determine the mean length produced using the current setup of the jig, a sample of rods is to be taken and their lengths measured, with the intention of using the value of $\overline{X}$ as an estimate of μ. Approximate the probabilities that $\overline{X}$ is within .0005 in. of μ for samples of size $n = 25, 100$, and 400. Do your calculations for this part of the question depend for their validity on the length distribution being normal? Explain.

45. Suppose that the measurement of the diameters of #10 machine screws produced on a particular machine yields values that are normally distributed with mean μ and standard deviation $\sigma = .03$ mm.
   (a) If $\mu = 4.68$ mm, about what fraction of all measured diameters will fall in the range from 4.65 mm to 4.70 mm?
   (b) Use your value from (a) and an appropriate discrete probability distribution to evaluate the probability (assuming $\mu = 4.68$) that among the next five measurements made, exactly four will fall in the range from 4.65 mm to 4.70 mm.
   (c) Use your value from (a) and an appropriate discrete probability distribution to evaluate the probability (assuming that $\mu = 4.68$) that if one begins sampling and measuring these screws, the first diameter in the range from 4.65 mm to 4.70 mm will be found on the second, third, or fourth screw measured.
   (d) Now suppose that μ is unknown but is to be estimated by $\overline{X}$ obtained from measuring a sample of $n = 25$ screws. Evaluate the probability that the sample mean, $\overline{X}$, takes a value within .01 mm of the long-run (population) mean μ.

(e) What sample size, n, would be required in order to a priori be 90% sure that $\overline{X}$ from n measurements will fall within .005 mm of μ?

46. The random variable $X =$ the number of hours till failure of a disk drive is described using an exponential distribution with mean 15,000 hours.

    (a) Evaluate the probability that a given drive lasts at least 20,000 hours.

    (b) A new computer network has ten of these drives installed on computers in the network. Use your answer to (a) and an assumption of independence of the ten drive lifetimes and evaluate the probability that at least nine of these drives are failure-free through 20,000 hours.

47. Miles, Baumhover, and Miller worked with a company on a packaging problem. Cardboard boxes, nominally 9.5 in. in length were supposed to hold four units of product stacked side by side. They did some measuring and found that in fact the individual product units had widths with mean approximately 2.577 in. and standard deviation approximately .061 in. Further, the boxes had (inside) lengths with mean approximately 9.566 in. and standard deviation approximately .053 in.

    (a) If X_1, X_2, X_3, and X_4 are the actual widths of four of the product units and Y is the actual inside length of a box into which they are to be packed, then the "head space" in the box is $U = Y - (X_1 + X_2 + X_3 + X_4)$. What are a sensible mean and standard deviation for U?

    (b) If X_1, X_2, X_3, X_4, and Y are normally distributed and independent, it turns out that U is also normal. Suppose this is the case. About what fraction of the time should the company expect to experience difficulty packing a box? (What is the probability that the head space as calculated in (a) is negative?)

    (c) If it is your job to recommend a new mean inside length of the boxes and the company wishes to have packing problems in only .5% of the attempts to load four units of product into a box, what is the minimum mean inside length you would recommend? (Assume that standard deviations will remain unchanged.)

Introduction to Formal Statistical Inference

Formal statistical inference uses probability theory to quantify the reliability of data-based conclusions. This chapter introduces the logic involved in several general types of formal statistical inference. Then the most common specific methods for one- and two-sample statistical studies are discussed.

The chapter begins with an introduction to confidence interval estimation, using the important case of large-sample inference for a mean. Then the topic of significance testing is considered, again using the case of large-sample inference for a mean. With the general notions in hand, successive sections treat the standard one- and two-sample confidence interval and significance-testing methods for means, then variances, and then proportions. Finally, the important topics of tolerance and prediction intervals are introduced.

6.1 Large-Sample Confidence Intervals for a Mean

Many important engineering applications of statistics fit the following standard mold. Values for parameters of a data-generating process are unknown. Based on data, the object is

1. identify an interval of values likely to contain an unknown parameter (or a function of one or more parameters) and

2. quantify "how likely" the interval is to cover the correct value.

For example, a piece of equipment that dispenses baby food into jars might produce an unknown mean fill level, μ. Determining a data-based interval likely to contain μ and an evaluation of the reliability of the interval might be important. Or a machine that puts threads on U-bolts might have an inherent variation in thread lengths, describable in terms of a standard deviation, σ. The point of data collection might then be to produce an interval of likely values for σ, together with a statement of how reliable the interval is. Or two different methods of running a pelletizing machine might have different unknown propensities to produce defective pellets, (say, p_1 and p_2). A data-based interval for $p_1 - p_2$, together with an associated statement of reliability, might be needed.

The type of formal statistical inference designed to deal with such problems is called **confidence interval estimation**.

Definition 1	A **confidence interval** for a parameter (or function of one or more parameters) is a data-based interval of numbers thought likely to contain the parameter (or function of one or more parameters) possessing a stated probability-based *confidence* or reliability.

This section discusses how basic probability facts lead to simple large-sample formulas for confidence intervals for a mean, μ. The unusual case where the standard deviation σ is known is treated first. Then parallel reasoning produces a formula for the much more common situation where σ is not known. The section closes with discussions of three practical issues in the application of confidence intervals.

6.1.1   A Large-*n* Confidence Interval for μ Involving σ

The final example in Section 5.5 involved a physically stable filling process known to have a net weight standard deviation of $\sigma = 1.6$ g. Since, for large n, the sample mean of iid random variables is approximately normal, Example 26 of Chapter 5 argued that for $n = 47$ and

$$\bar{x} = \text{the sample mean net fill weight of 47 jars filled by the process (g)}$$

there is an approximately 80% chance that $\bar{x}$ is within .3 gram of μ. This fact is pictured again in Figure 6.1.

Notational conventions   We need to interrupt for a moment to discuss notation. In Chapter 5, capital letters were carefully used as symbols for random variables and corresponding lowercase letters for their possible or observed values. But here a lowercase symbol, $\bar{x}$, has been used for the sample mean *random variable*. This is fairly standard statistical usage, and it is in keeping with the kind of convention used in Chapters 3 and 4. We are thus going to now abandon strict adherence to the capitalization convention introduced in Chapter 5. Random variables will often be symbolized using lowercase letters and the same symbols used for their observed values. The Chapter 5 capitalization convention is especially helpful in learning the basics of probability. But once those basics are mastered, it is common to abuse notation and

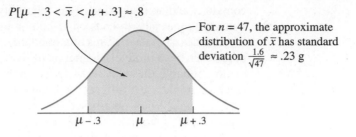

Figure 6.1   Approximate probability distribution for $\bar{x}$ based on $n = 47$

to determine from context whether a random variable or its observed value is being discussed.

The most common way of thinking about a graphic like Figure 6.1 is to think of the possibility that

$$\mu - .3 < \bar{x} < \mu + .3 \qquad (6.1)$$

in terms of whether or not $\bar{x}$ falls in an interval of length $2(.3) = .6$ centered at μ. But the equivalent is to consider whether or not an interval of length .6 centered at $\bar{x}$ falls on top of μ. Algebraically, inequality (6.1) is equivalent to

$$\bar{x} - .3 < \mu < \bar{x} + .3 \qquad (6.2)$$

which shifts attention to this second way of thinking. The fact that expression (6.2) has about an 80% chance of holding true anytime a sample of 47 fill weights is taken suggests that the *random interval*

$$(\bar{x} - .3, \bar{x} + .3) \qquad (6.3)$$

might be used as a confidence interval for μ, with 80% associated reliability or confidence.

Example 1 | A Confidence Interval for a Process Mean Fill Weight

Suppose a sample of $n = 47$ jars produces $\bar{x} = 138.2$ g. Then expression (6.3) suggests that the interval with endpoints

$$138.2 \text{ g} \pm .3 \text{ g}$$

(i.e., the interval from 137.9 g to 138.5 g) be used as an 80% confidence interval for the process mean fill weight.

It is not hard to generalize the logic that led to expression (6.3). Anytime an iid model is appropriate for the elements of a large sample, the central limit theorem implies that the sample mean $\bar{x}$ is approximately normal with mean μ and standard deviation $\sigma/\sqrt{n}$. Then, if for $p > .5$, z is the p quantile of the standard normal distribution, the probability that

$$\mu - z\frac{\sigma}{\sqrt{n}} < \bar{x} < \mu + z\frac{\sigma}{\sqrt{n}} \qquad (6.4)$$

is approximately $1 - 2(1 - p)$. But inequality (6.4) can be rewritten as

$$\bar{x} - z\frac{\sigma}{\sqrt{n}} < \mu < \bar{x} + z\frac{\sigma}{\sqrt{n}} \qquad (6.5)$$

and thought of as the eventuality that the random interval with endpoints

Large-sample known σ confidence limits for μ

$$\boxed{\bar{x} \pm z\frac{\sigma}{\sqrt{n}}} \qquad (6.6)$$

brackets the unknown μ. So an interval with endpoints (6.6) is an approximate confidence interval for μ (with confidence level $1 - 2(1 - p)$).

In an application, z in equation (6.6) is chosen so that the standard normal probability between $-z$ and z corresponds to a desired confidence level. Table 3.10 (of standard normal quantiles) on page 89 or Table B.3 (of standard normal cumulative probabilities) can be used to verify the appropriateness of the entries in Table 6.1. (This table gives values of z for use in expression (6.6) for some common confidence levels.)

Table 6.1
z's for Use in Two-sided
Large-n Intervals for μ

Desired Confidence	z
80%	1.28
90%	1.645
95%	1.96
98%	2.33
99%	2.58

Example 2

Confidence Interval for the Mean Deviation
from Nominal in a Grinding Operation

Dib, Smith, and Thompson studied a grinding process used in the rebuilding of automobile engines. The natural short-term variability associated with the diameters of rod journals on engine crankshafts ground using the process was on the order of $\sigma = .7 \times 10^{-4}$ in. Suppose that the rod journal grinding process can be thought of as physically stable over runs of, say, 50 journals or less. Then if 32 consecutive rod journal diameters have mean deviation from nominal of $\bar{x} = -.16 \times 10^{-4}$ in., it is possible to apply expression (6.6) to make a confidence interval for the current process mean deviation from nominal. Consider a 95% confidence level. Consulting Table 6.1 (or otherwise, realizing that 1.96 is the $p = .975$ quantile of the standard normal distribution), $z = 1.96$ is called for in formula (6.6) (since $.95 = 1 - 2(1 - .975)$). Thus, a 95% confidence interval for the current process mean deviation from nominal journal diameter has endpoints

$$-.16 \times 10^{-4} \pm (1.96)\frac{.7 \times 10^{-4}}{\sqrt{32}}$$

that is, endpoints

$$-.40 \times 10^{-4} \text{ in.} \quad \text{and} \quad .08 \times 10^{-4} \text{ in.}$$

An interval like this one could be of engineering importance in determining the advisability of making an adjustment to the process aim. The interval includes both positive and negative values. So although $\bar{x} < 0$, the information in hand doesn't provide enough precision to tell with any certainty in which direction the grinding process should be adjusted. This, coupled with the fact that potential machine adjustments are probably much coarser than the best-guess misadjustment of $\bar{x} = -.16 \times 10^{-4}$ in., speaks strongly against making a change in the process aim based on the current data.

6.1.2   A Generally Applicable Large-*n* Confidence Interval for μ

Although expression (6.6) provides a mathematically correct confidence interval, the appearance of σ in the formula severely limits its practical usefulness. It is unusual to have to estimate a mean μ when the corresponding σ is known (and can therefore be plugged into a formula). These situations occur primarily in manufacturing situations like those of Examples 1 and 2. Considerable past experience can sometimes give a sensible value for σ, while physical process drifts over time can put the current value of μ in question.

Happily, modification of the line of reasoning that led to expression (6.6) produces a confidence interval formula for μ that depends only on the characteristics of a sample. The argument leading to formula (6.6) depends on the fact that for large n, $\bar{x}$ is approximately normal with mean μ and standard deviation $\sigma/\sqrt{n}$—i.e., that

$$Z = \frac{\bar{x} - \mu}{\frac{\sigma}{\sqrt{n}}} \tag{6.7}$$

is approximately standard normal. The appearance of σ in expression (6.7) is what leads to its appearance in the confidence interval formula (6.6). But a slight generalization of the central limit theorem guarantees that for large n,

$$Z = \frac{\bar{x} - \mu}{\frac{s}{\sqrt{n}}} \tag{6.8}$$

is also approximately standard normal. And the variable (6.8) doesn't involve σ.

Beginning with the fact that (when an iid model for observations is appropriate and n is large) the variable (6.8) is approximately standard normal, the reasoning is much as before. For a positive z,

$$-z < \frac{\bar{x} - \mu}{\frac{s}{\sqrt{n}}} < z$$

is equivalent to

$$\mu - z\frac{s}{\sqrt{n}} < \bar{x} < \mu + z\frac{s}{\sqrt{n}}$$

which in turn is equivalent to

$$\bar{x} - z\frac{s}{\sqrt{n}} < \mu < \bar{x} + z\frac{s}{\sqrt{n}}$$

Thus, the interval with random center $\bar{x}$ and random length $2zs/\sqrt{n}$—i.e., with random endpoints

Large-sample
confidence limits
for μ

$$\boxed{\bar{x} \pm z\frac{s}{\sqrt{n}}} \tag{6.9}$$

can be used as an approximate confidence interval for μ. For a desired confidence, z should be chosen such that the standard normal probability between $-z$ and z corresponds to that confidence level.

Example 3

Breakaway Torques and Hard Disk Failures

F. Willett, in the article "The Case of the Derailed Disk Drives" (*Mechanical Engineering*, 1988), discusses a study done to isolate the cause of "blink code A failure" in a model of Winchester hard disk drive. Included in that article are the data given in Figure 6.2. These are breakaway torques (units are inch ounces) required to loosen the drive's interrupter flag on the stepper motor shaft for 26 disk drives returned to the manufacturer for blink code A failure. For these data, $\bar{x} = 11.5$ in. oz and $s = 5.1$ in. oz.

```
0 | 0  2  3
0 | 7  8  8  9  9
1 | 0  0  0  1  1  2  2  2  3
1 | 5  5  6  6  7  7  7  9
2 | 0
2 |
```

Figure 6.2   Torques required to loosen 26 interrupter flags

If the disk drives that produced the data in Figure 6.2 are thought of as representing the population of drives subject to blink code A failure, it seems reasonable to use an iid model and formula (6.9) to estimate the population mean breakaway torque. Choosing to make a 90% confidence interval for μ, $z = 1.645$ is indicated in Table 6.1. And using formula (6.9), endpoints

$$11.5 \pm 1.645 \frac{5.1}{\sqrt{26}}$$

(i.e., endpoints 9.9 in. oz and 13.1 in. oz) are indicated.

The interval shows that the mean breakaway torque for drives with blink code A failure was substantially below the factory's 33.5 in. oz target value. Recognizing this turned out to be key in finding and eliminating a design flaw in the drives.

6.1.3   Some Additional Comments Concerning Confidence Intervals

Formulas (6.6) and (6.9) have been used to make confidence statements of the type "μ is between a and b." But often a statement like "μ is at least c" or "μ is no more than d" would be of more practical value. For example, an automotive engineer might wish to state, "The mean NO emission for this engine is at most 5 ppm." Or a civil engineer might want to make a statement like "the mean compressive strength for specimens of this type of concrete is at least 4188 psi." That is, practical engineering problems are sometimes best addressed using one-sided confidence intervals.

Making one-sided intervals

There is no real problem in coming up with formulas for one-sided confidence intervals. If you have a workable two-sided formula, all that must be done is to

1. replace the lower limit with $-\infty$ or the upper limit with $+\infty$ and

2. adjust the stated confidence level appropriately upward (this usually means dividing the "unconfidence level" by 2).

This prescription works not only with formulas (6.6) and (6.9) but also with the rest of the two-sided confidence intervals introduced in this chapter.

Example 3
(continued)

For the mean breakaway torque for defective disk drives, consider making a one-sided 90% confidence interval for μ of the form $(-\infty, \#)$, for # an appropriate number. Put slightly differently, consider finding a 90% *upper confidence bound* for μ, (say, #).

Beginning with a two-sided 80% confidence interval for μ, the lower limit can be replaced with $-\infty$ and a one-sided 90% confidence interval determined. That is, using formula (6.9), a 90% upper confidence bound for the mean breakaway torque is

$$\bar{x} + 1.28 \frac{s}{\sqrt{n}} = 11.5 + 1.28 \frac{5.1}{\sqrt{26}} = 12.8 \text{ in. oz}$$

Equivalently, a 90% one-sided confidence interval for μ is $(-\infty, 12.8)$.

The 12.8 in. oz figure here is less than (and closer to the sample mean than) the 13.1 in. oz upper limit from the 90% two-sided interval found earlier. In the one-sided case, $-\infty$ is declared as a lower limit so there is no risk of producing an interval containing only numbers larger than the unknown μ. Thus an upper limit smaller than that for a corresponding two-sided interval can be used.

Interpreting a confidence level

A second issue in the application of confidence intervals is a correct understanding of the technical meaning of the term *confidence*. Unfortunately, there are many possible misunderstandings. So it is important to carefully lay out what confidence does and doesn't mean.

Prior to selecting a sample and plugging into a formula like (6.6) or (6.9), the meaning of a confidence level is obvious. Choosing a (two-sided) 90% confidence level and thus $z = 1.645$ for use in formula (6.9), before the fact of sample selection and calculation, "there is about a 90% chance of winding up with an interval that brackets μ." In symbols, this might be expressed as

$$P\left[\bar{x} - 1.645 \frac{s}{\sqrt{n}} < \mu < \bar{x} + 1.645 \frac{s}{\sqrt{n}}\right] \approx .90$$

But how to think about a confidence level *after* sample selection? This is an entirely different matter. Once numbers have been plugged into a formula like (6.6) or (6.9), the die has already been cast, and the numerical interval is either right or wrong. The practical difficulty is that while which is the case can't be determined, it no longer makes logical sense to attach a probability to the correctness of the interval. For example, it would make no sense to look again at the two-sided interval found in Example 3 and try to say something like "there is a 90% probability that μ is between 9.9 in. oz and 13.1 in. oz." μ is not a random variable. It is a fixed (although unknown) quantity that either is or is not between 9.9 and 13.1. There is no probability left in the situation to be discussed.

So what does it mean that (9.9, 13.1) is a 90% confidence interval for μ? Like it or not, the phrase "90% confidence" refers more to the method used to obtain the interval (9.9, 13.1) than to the interval itself. In coming up with the interval, methodology has been used that would produce numerical intervals bracketing μ in about 90% of repeated applications. But the effectiveness of the particular interval in this application is unknown, and it is not quantifiable in terms of a probability. A person who (in the course of a lifetime) makes many 90% confidence intervals can expect to have a "lifetime success rate" of about 90%. But the effectiveness of any particular application will typically be unknown.

A short statement summarizing this discussion as "the authorized interpretation of confidence" will be useful.

Definition 2
(Interpretation of a Confidence Interval)

To say that a numerical interval (a, b) is (for example) a 90% confidence interval for a parameter is to say that in obtaining it, one has applied methods of data collection and calculation that would produce intervals bracketing the parameter in about 90% of repeated applications. Whether or not the particular interval (a, b) brackets the parameter is unknown and not describable in terms of a probability.

The reader may feel that the statement in Definition 2 is a rather weak meaning for the reliability figure associated with a confidence interval. Nevertheless, the statement in Definition 2 is the correct interpretation and is all that can be rationally expected. And despite the fact that the correct interpretation may initially seem somewhat unappealing, confidence interval methods have proved themselves to be of great practical use.

Sample sizes for estimating μ

As a final consideration in this introduction to confidence intervals, note that formulas like (6.6) and (6.9) can give some crude quantitative answers to the question, "How big must n be?" Using formula (6.9), for example, if you have in mind (1) a desired confidence level, (2) a worst-case expectation for the sample standard deviation, and (3) a desired precision of estimation for μ, it is a simple matter to solve for a corresponding sample size. That is, suppose that the desired confidence level dictates the use of the value z in formula (6.9), s is some likely worst-case value for the sample standard deviation, and you want to have confidence limits (or a limit) of the form $\bar{x} \pm \Delta$. Setting

$$\Delta = z \frac{s}{\sqrt{n}}$$

and solving for n produces the requirement

$$n = \left(\frac{zs}{\Delta}\right)^2$$

Example 3
(*continued*)

Suppose that in the disk drive problem, engineers plan to follow up the analysis of the data in Figure 6.2 with the testing of a number of new drives. This will be done after subjecting them to accelerated (high) temperature conditions, in an effort to understand the mechanism behind the creation of low breakaway torques. Further suppose that the mean breakaway torque for temperature-stressed drives is to be estimated with a two-sided 95% confidence interval and that the torque variability expected in the new temperature-stressed drives is no worse than the $s = 5.1$ in. oz figure obtained from the returned drives. A ± 1 in. oz precision of estimation is desired. Then using the plus-or-minus part of formula (6.9) and remembering Table 6.1, the requirement is

$$1 = 1.96 \frac{5.1}{\sqrt{n}}$$

which, when solved for n, gives

$$n = \left(\frac{(1.96)(5.1)}{1}\right)^2 \approx 100$$

A study involving in the neighborhood of $n = 100$ temperature-stressed new disk drives is indicated. If this figure is impractical, the calculations at least indicate that dropping below this sample size will (unless the variability associated with the stressed new drives is less than that of the returned drives) force a reduction in either the confidence or the precision associated with the final interval.

For two reasons, the kind of calculations in the previous example give somewhat less than an ironclad answer to the question of sample size. The first is that they are only as good as the prediction of the sample standard deviation, s. If s is underpredicted, an n that is not really large enough will result. (By the same token, if one is excessively conservative and overpredicts s, an unnecessarily large sample size will result.) The second issue is that expression (6.9) remains a large-sample formula. If calculations like the preceding ones produce n smaller than, say, 25 or 30, the value should be increased enough to guarantee that formula (6.9) can be applied.

Section 1 Exercises

1. Interpret the statement, "The interval from 6.3 to 7.9 is a 95% confidence interval for the mean μ."

2. In Chapter Exercise 2 of Chapter 3, there is a data set consisting of the aluminum contents of 26 bihourly samples of recycled PET plastic from a recycling facility. Those 26 measurements have $\bar{y} = 142.7$ ppm and $s \approx 98.2$ ppm. Use these facts to respond to the following. (Assume that $n = 26$ is large enough to permit the use of large-sample formulas in this case.)

   (a) Make a 90% two-sided confidence interval for the mean aluminum content of such specimens over the 52-hour study period.

   (b) Make a 95% two-sided confidence interval for the mean aluminum content of such specimens over the 52-hour study period. How does this compare to your answer to part (a)?

   (c) Make a 90% upper confidence bound for the mean aluminum content of such samples over the 52-hour study period. (Find # such that $(-\infty, \#)$ is a 90% confidence interval.) How does this value compare to the upper endpoint of your interval from part (a)?

   (d) Make a 95% upper confidence bound for the mean aluminum content of such samples over the 52-hour study period. How does this value compare to your answer to part (c)?

   (e) Interpret your interval from (a) for someone with little statistical background. (Speak in the context of the recycling study and use Definition 2 as your guide.)

3. Return to the context of Exercise 2. Suppose that in order to monitor for possible process changes, future samples of PET will be taken. If it is desirable to estimate the mean aluminum content with ± 20 ppm precision and 90% confidence, what future sample size do you recommend?

4. DuToit, Hansen, and Osborne measured the diameters of some no. 10 machine screws with two different calipers (digital and vernier scale). Part of their data are recorded here. Given in the small frequency table are the measurements obtained on 50 screws by one of the students using the digital calipers.

Diameter (mm)	Frequency
4.52	1
4.66	4
4.67	7
4.68	7
4.69	14
4.70	9
4.71	4
4.72	4

   (a) Compute the sample mean and standard deviation for these data.

   (b) Use your sample values from (a) and make a 98% two-sided confidence interval for the mean diameter of such screws as measured by this student with these calipers.

   (c) Repeat part (b) using 99% confidence. How does this interval compare with the one from (b)?

   (d) Use your values from (a) and find a 98% lower confidence bound for the mean diameter. (Find a number # such that $(\#, \infty)$ is a 98% confidence interval.) How does this value compare to the lower endpoint of your interval from (b)?

   (e) Repeat (d) using 99% confidence. How does the value computed here compare to your answer to (d)?

   (f) Interpret your interval from (b) for someone with little statistical background. (Speak in the context of the diameter measurement study and use Definition 2 as your guide.)

6.2   Large-Sample Significance Tests for a Mean

The last section illustrated how probability can enable confidence interval estimation. This section makes a parallel introduction of significance testing.

The goal of significance testing

Significance testing amounts to using data to quantitatively assess the plausibility of a trial value of a parameter (or function of one or more parameters). This trial value typically embodies a status quo/"pre-data" view. For example, a process engineer might employ significance testing to assess the plausibility of an ideal value of 138 g as the current process mean fill level of baby food jars. Or two different methods of running a pelletizing machine might have unknown propensities to produce defective pellets, (say, p_1 and p_2), and significance testing could be used to assess the plausibility of $p_1 - p_2 = 0$—i.e., that the two methods are equally effective.

This section describes how basic probability facts lead to simple large sample significance tests for a mean, μ. It introduces significance testing terminology in the case where the standard deviation σ is known. Next, a five-step format for summarizing significance testing is presented. Then the more common situation of significance testing for μ where σ is not known is considered. The section closes with two discussions about practical issues in the application of significance-testing logic.

6.2.1   Large-*n* Significance Tests for μ Involving σ

Recall once more Example 26 in Chapter 5, where a physically stable filling process is known to have $\sigma = 1.6$ g for net weight. Suppose further that with a declared (label) weight of 135 g, process engineers have set a target mean net fill weight at $135 + 3\sigma = 139.8$ g. Finally, suppose that in a routine check of filling-process performance, intended to detect any change of the process mean from its target value, a sample of $n = 25$ jars produces $\bar{x} = 139.0$ g. What does this value have to say about the plausibility of the current process mean actually being at the target of 139.8 g?

The central limit theorem can be called on here. If indeed the current process mean is at 139.8 g, $\bar{x}$ has an approximately normal distribution with mean 139.8 g and standard deviation $\sigma/\sqrt{n} = 1.6/\sqrt{25} = .32$ g, as pictured in Figure 6.3 along with the observed value of $\bar{x} = 139.0$ g.

Figure 6.4 shows the standard normal picture that corresponds to Figure 6.3. It is based on the fact that if the current process mean is on target at 139.8 g, then the fact that $\bar{x}$ is approximately normal with mean μ and standard deviation $\sigma/\sqrt{n} = .32$ g implies that

$$Z = \frac{\bar{x} - 139.8}{\dfrac{\sigma}{\sqrt{n}}} = \frac{\bar{x} - 139.8}{.32} \tag{6.10}$$

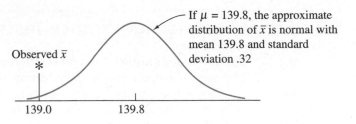

Figure 6.3 Approximate probability distribution for $\bar{x}$ if $\mu = 139.8$, and the observed value of $\bar{x} = 139.0$

is approximately standard normal. The observed $\bar{x} = 139.0$ g in Figure 6.3 has corresponding observed $z = -2.5$ in Figure 6.4.

It is obvious from either Figure 6.3 or Figure 6.4 that if the process mean is on target at 139.8 g (and thus the figures are correct), a fairly extreme/rare $\bar{x}$, or equivalently z, has been observed. Of course, extreme/rare things occasionally happen. But the nature of the observed $\bar{x}$ (or z) might instead be considered as making the possibility that the process is on target implausible.

The figures even suggest a way of quantifying their own implausibility—through calculating a probability associated with values of $\bar{x}$ (or Z) at least as extreme as the one actually observed. Now "at least as extreme" must be defined in relation to the original purpose of data collection—to detect either a decrease of μ below target or an increase above target. Not only are values $\bar{x} \leq 139.0$ g ($z \leq -2.5$) as extreme as that observed but so also are values $\bar{x} \geq 140.6$ g ($z \geq 2.5$). (The first kind of $\bar{x}$ suggests a decrease in μ, and the second suggests an increase.) That is, the implausibility of being on target might be quantified by noting that if this were so, only a fraction

$$\Phi(-2.5) + \big(1 - \Phi(2.5)\big) = .01$$

of all samples would produce a value of $\bar{x}$ (or Z) as extreme as the one actually observed. Put in those terms, the data seem to speak rather convincingly against the process being on target.

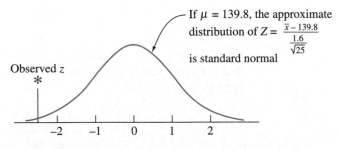

Figure 6.4 The standard normal picture corresponding to Figure 6.3

The argument that has just been made is an application of typical significance-testing logic. In order to make the pattern of thought obvious, it is useful to isolate some elements of it in definition form. This is done next, beginning with a formal restatement of the overall purpose.

Definition 3

> Statistical **significance testing** is the use of data in the quantitative assessment of the plausibility of some trial value for a parameter (or function of one or more parameters).

Logically, significance testing begins with the specification of the trial or hypothesized value. Special jargon and notation exist for the statement of this value.

Definition 4

> A **null hypothesis** is a statement of the form
>
> $$\text{Parameter} = \#$$
>
> or
>
> $$\text{Function of parameters} = \#$$
>
> (for some number, #) that forms the basis of investigation in a significance test. A null hypothesis is usually formed to embody a status quo/"pre-data" view of the parameter (or function of the parameter(s)). It is typically denoted as H_0.

The notion of a null hypothesis is so central to significance testing that it is common to use the term **hypothesis testing** in place of *significance testing*. The "null" part of the phrase "null hypothesis" refers to the fact that null hypotheses are statements of *no difference*, or equality. For example, in the context of the filling operation, standard usage would be to write

$$H_0 \colon \mu = 139.8 \tag{6.11}$$

meaning that there is no difference between μ and the target value of 139.8 g.

After formulating a null hypothesis, what kinds of departures from it are of interest must be specified.

Definition 5

> An **alternative hypothesis** is a statement that stands in opposition to the null hypothesis. It specifies what forms of departure from the null hypothesis are of concern. An alternative hypothesis is typically denoted as H_a. It is of the

same form as the corresponding null hypothesis, except that the equality sign is replaced by $\neq$, $>$, or $<$.

Often, the alternative hypothesis is based on an investigator's suspicions and/or hopes about the true state of affairs, amounting to a kind of *research hypothesis* that the investigator hopes to establish. For example, if an engineer tests what is intended to be a device for improving automotive gas mileage, a null hypothesis expressing "no mileage change" and an alternative hypothesis expressing "mileage improvement" would be appropriate.

Definitions 4 and 5 together imply that for the case of testing about a single mean, the three possible pairs of null and alternative hypotheses are

$$\text{H}_0: \mu = \# \qquad \text{H}_0: \mu = \# \qquad \text{H}_0: \mu = \#$$
$$\text{H}_a: \mu > \# \qquad \text{H}_a: \mu < \# \qquad \text{H}_a: \mu \neq \#$$

In the example of the filling operation, there is a need to detect both the possibility of consistently underfilled ($\mu < 139.8$ g) and the possibility of consistently overfilled ($\mu > 139.8$ g) jars. Thus, an appropriate alternative hypothesis is

$$\text{H}_a: \mu \neq 139.8 \qquad\qquad \textbf{(6.12)}$$

Once null and alternative hypotheses have been established, it is necessary to lay out carefully how the data will be used to evaluate the plausibility of the null hypothesis. This involves specifying a statistic to be calculated, a probability distribution appropriate for it if the null hypothesis is true, and what kinds of observed values will make the null hypothesis seem implausible.

Definition 6

A **test statistic** is the particular form of numerical data summarization used in a significance test. The formula for the test statistic typically involves the number appearing in the null hypothesis.

Definition 7

A **reference (or null) distribution** for a test statistic is the probability distribution describing the test statistic, provided the null hypothesis is in fact true.

The values of the test statistic considered to cast doubt on the validity of the null hypothesis are specified after looking at the form of the alternative hypothesis. Roughly speaking, values are identified that are more likely to occur if the alternative hypothesis is true than if the null hypothesis holds.

The discussion of the filling process scenario has vacillated between using $\bar{x}$ and its standardized version Z given in equation (6.10) for a test statistic. Equation (6.10) is a specialized form of the general (large-n, known σ) test statistic for μ,

Large-sample known σ test statistic for μ

$$Z = \frac{\bar{x} - \#}{\frac{\sigma}{\sqrt{n}}}$$ **(6.13)**

for the present scenario, where the hypothesized value of μ is 139.8, $n = 25$, and $\sigma = 1.6$. It is most convenient to think of the test statistic for this kind of problem in the standardized form shown in equation (6.13) rather than as $\bar{x}$ itself. Using form (6.13), the reference distribution will always be the same—namely, standard normal.

Continuing with the filling example, note that if instead of the null hypothesis (6.11), the alternative hypothesis (6.12) is operating, observed $\bar{x}$'s much larger or much smaller than 139.8 will tend to result. Such $\bar{x}$'s will then, via equation (6.13), translate respectively to large or small (that is, large negative numbers in this case) observed values of Z—i.e., large values $|z|$. Such observed values render the null hypothesis implausible.

Having specified how data will be used to judge the plausibility of the null hypothesis, it remains to collect them, plug them into the formula for the test statistic, and (using the calculated value and the reference distribution) arrive at a quantitative assessment of the plausibility of H_0. There is jargon for the form this will take.

Definition 8

The **observed level of significance** or **p-value** in a significance test is the probability that the reference distribution assigns to the set of possible values of the test statistic that are at least as extreme as the one actually observed (in terms of casting doubt on the null hypothesis).

Small p-values are evidence against H_0

The smaller the observed level of significance, the stronger the evidence against the validity of the null hypothesis. In the context of the filling operation, with an observed value of the test statistic of

$$z = -2.5$$

the p-value or observed level of significance is

$$\Phi(-2.5) + \left(1 - \Phi(2.5)\right) = .01$$

which gives fairly strong evidence against the possibility that the process mean is on target.

6.2.2   A Five-Step Format for Summarizing Significance Tests

Five-step significance testing format

It is helpful to lay down a step-by-step format for organizing write-ups of significance tests. The one that will be used in this text includes the following five steps:

Step 1   State the null hypothesis.

Step 2   State the alternative hypothesis.

Step 3   State the test criteria. That is, give the formula for the test statistic (plugging in only a hypothesized value from the null hypothesis, but not any sample information) and the reference distribution. Then state in general terms what observed values of the test statistic will constitute evidence against the null hypothesis.

Step 4   Show the sample-based calculations.

Step 5   Report an observed level of significance and (to the extent possible) state its implications in the context of the real engineering problem.

Example 4

A Significance Test Regarding a Process Mean Fill Level

The five-step significance-testing format can be used to write up the preceding discussion of the filling process.

1. $H_0: \mu = 139.8$.

2. $H_a: \mu \neq 139.8$.

3. The test statistic is

$$Z = \frac{\bar{x} - 139.8}{\dfrac{\sigma}{\sqrt{n}}}$$

The reference distribution is standard normal, and large observed values $|z|$ will constitute evidence against H_0.

4. The sample gives

$$z = \frac{139.0 - 139.8}{\dfrac{1.6}{\sqrt{100}}} = -2.5$$

5. The observed level of significance is

$$P[\text{a standard normal variable} \leq -2.5]$$
$$+ P[\text{a standard normal variable} \geq 2.5]$$
$$= P\,[|\text{a standard normal variable}| \geq 2.5]$$
$$= .01$$

> This is reasonably strong evidence that the process mean fill level is not on target.

6.2.3 Generally Applicable Large-n Significance Tests for μ

The significance-testing method used to carry the discussion thus far is easy to discuss and understand but of limited practical use. The problem with it is that statistic (6.13) involves the parameter σ. As remarked in Section 6.1, there are few engineering contexts where one needs to make inferences regarding μ but knows the corresponding σ. Happily, because of the same probability fact that made it possible to produce a large-sample confidence interval formula for μ free of σ, it is also possible to do large-n significance testing for μ without having to supply σ.

For observations that are describable as essentially equivalent to random selections with replacement from a single population with mean μ and variance σ^2, if n is large,

$$Z = \frac{\bar{x} - \mu}{\frac{s}{\sqrt{n}}}$$

is approximately standard normal. This means that for large n, to test

$$H_0 : \mu = \#$$

a widely applicable method will simply be to use the logic already introduced but with the statistic

Large-sample test statistic for μ

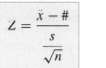

$$Z = \frac{\bar{x} - \#}{\frac{s}{\sqrt{n}}} \tag{6.14}$$

in place of statistic (6.13).

Example 5
(*Example 3 revisited*)

Significance Testing and Hard Disk Failures

Consider again the problem of disk drive blink code A failure. Breakaway torques set at the factory on the interrupter flag connection to the stepper motor shaft averaged 33.5 in. oz, and there was suspicion that blink code A failure was associated with reduced breakaway torque. Recall that a sample of $n = 26$ failed drives had breakaway torques (given in Figure 6.2) with $\bar{x} = 11.5$ in. oz and $s = 5.1$ in. oz.

Consider the situation of an engineer wishing to judge the extent to which the data in hand debunk the possibility that drives experiencing blink code A failure

Example 5
(continued)

have mean breakaway torque equal to the factory-set mean value of 33.5 in. oz. The five-step significance-testing format can be used.

1. $H_0: \mu = 33.5$.

2. $H_a: \mu < 33.5$.
   (Here the alternative hypothesis is directional, amounting to a research hypothesis based on the engineer's suspicions about the relationship between drive failure and breakaway torque.)

3. The test statistic is

$$Z = \frac{\bar{x} - 33.5}{\dfrac{s}{\sqrt{n}}}$$

The reference distribution is standard normal, and small observed values z will constitute evidence against the validity of H_0. (Means less than 33.5 will tend to produce $\bar{x}$'s of the same nature and therefore small—i.e., large negative—z's.)

4. The sample gives

$$z = \frac{11.5 - 33.5}{\dfrac{5.1}{\sqrt{26}}} = -22.0$$

5. The observed level of significance is

$$P[\text{a standard normal variable} < -22.0] \approx 0$$

The sample provides overwhelming evidence that failed drives have a mean breakaway torque below the factory-set level.

It is important not to make too much of a logical jump here to an incorrect conclusion that this work constitutes the complete solution to the real engineering problem. Drives returned for blink code A failure have substandard breakaway torques. But in the absence of evidence to the contrary, it is possible that they are no different in that respect from nonfailing drives currently in the field. And even if reduced breakaway torque is at fault, a real-world fix of the drive failure problem requires the identification and prevention of the physical mechanism producing it. This is not to say the significance test lacks importance, but rather to remind the reader that it is but one of many tools an engineer uses to do a job.

6.2.4 Significance Testing and Formal Statistical Decision Making (*Optional*)

The basic logic introduced in this section is sometimes applied in a decision-making context, where data are being counted on to provide guidance in choosing between two rival courses of action. In such cases, a decision-making framework is often built into the formal statistical analysis in an explicit way, and some additional terminology and patterns of thought are standard.

In some decision-making contexts, it is possible to conceive of two different possible decisions or courses of action as being related to a null and an alternative hypothesis. For example, in the filling-process scenario, $H_0: \mu = 139.8$ might correspond to the course of action "leave the process alone," and $H_a: \mu \neq 139.8$ could correspond to the course of action "adjust the process." When such a correspondence holds, two different errors are possible in the decision-making process.

Definition 9	When significance testing is used in a decision-making context, deciding in favor of H_a when in fact H_0 is true is called a **type I error**.

Definition 10	When significance testing is used in a decision-making context, deciding in favor of H_0 when in fact H_a is true is called a **type II error**.

The content of these two definitions is represented in the 2×2 table pictured in Figure 6.5. In the filling-process problem, a type I error would be adjusting an on-target process. A type II error would be failing to adjust an off-target process.

Significance testing is harnessed and used to come to a decision by choosing a critical value and, if the observed level of significance is smaller than the critical value (thus making the null hypothesis correspondingly implausible), deciding in favor of H_a. Otherwise, the course of action corresponding to H_0 is followed. The critical value for the observed level of significance ends up being the a priori

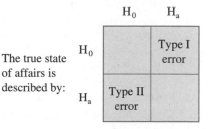

Figure 6.5 Four potential outcomes in a decision problem

probability the decision maker runs of deciding in favor of H_a, calculated supposing H_0 to be true. There is special terminology for this concept.

Definition 11	When significance testing is used in a decision-making context, a critical value separating those large observed levels of significance for which H_0 will be accepted from those small observed levels of significance for which H_0 will be rejected in favor of H_a is called the **type I error probability** or the **significance level**. The symbol α is usually used to stand for the type I error probability.

It is standard practice to use small numbers, like .1, .05, or even .01, for α. This puts some inertia in favor of H_0 into the decision-making process. (Such a practice guarantees that *type I* errors won't be made very often. But at the same time, it creates an asymmetry in the treatment of H_0 and H_a that is not always justified.)

Definition 10 and Figure 6.5 make it clear that type I errors are not the only undesirable possibility. The possibility of type II errors must also be considered.

Definition 12	When significance testing is used in a decision-making context, the probability—calculated supposing a particular parameter value described by H_a holds—that the observed level of significance is bigger than α (i.e., H_0 is not rejected) is called a **type II error probability**. The symbol β is usually used to stand for a type II error probability.

For most of the testing methods studied in this book, calculation of β's is more than the limited introduction to probability given in Chapter 5 will support. But the job can be handled for the simple known-σ situation that was used to introduce the topic of significance testing. And making a few such calculations will provide some intuition consistent with what, qualitatively at least, holds in general.

Example 4 *(continued)*	Again consider the filling process and testing $H_0: \mu = 139.8$ vs. $H_a: \mu \neq 139.8$. This time suppose that significance testing based on $n = 25$ will be used tomorrow to decide whether or not to adjust the process. Type II error probabilities, calculated supposing $\mu = 139.5$ and $\mu = 139.2$ for tests using $\alpha = .05$ and $\alpha = .2$, will be compared. First consider $\alpha = .05$. The decision will be made in favor of H_0 if the *p*-value exceeds .05. That is, the decision will be in favor of the null hypothesis if the observed value of Z given in equation (6.10) (generalized in formula (6.13)) is such that

$$|z| < 1.96$$

i.e., if

$$139.8 - 1.96(.32) < \bar{x} < 139.8 + 1.96(.32)$$

i.e., if

$$139.2 < \bar{x} < 140.4 \qquad\qquad \textbf{(6.15)}$$

Now if μ described by H_a given in display (6.12) is the true process mean, $\bar{x}$ is not approximately normal with mean 139.8 and standard deviation .32, but rather approximately normal with mean μ and standard deviation .32. So for such a μ, expression (6.15) and Definition 12 show that the corresponding β will be the probability the corresponding normal distribution assigns to the possibility that $139.2 < \bar{x} < 140.4$. This is pictured in Figure 6.6 for the two means $\mu = 139.5$ and $\mu = 139.2$.

It is an easy matter to calculate z-values corresponding to $\bar{x} = 139.2$ and $\bar{x} = 140.4$ using means of 139.5 and 139.2 and a standard deviation of .32 and to consult a standard normal table in order to verify the correctness of the two β's marked in Figure 6.6.

Parallel reasoning for the situation with $\alpha = .2$ is as follows. The decision will be in favor of H_0 if the p-value exceeds .2. That is, the decision will be in favor of H_0 if $|z| < 1.28$—i.e., if

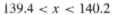

$$139.4 < x < 140.2$$

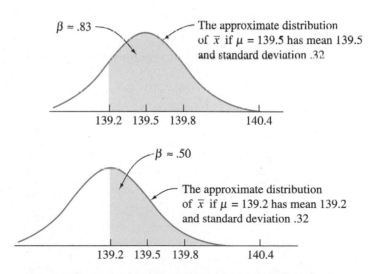

Figure 6.6 Approximate probability distributions for $\bar{x}$ for two different values of μ described by H_a and the corresponding β's, when $\alpha = .05$

Example 4
(continued)

If μ described by H_a is the true process mean, $\bar{x}$ is approximately normal with mean μ and standard deviation .32. So the corresponding β will be the probability this normal distribution assigns to the possibility that $139.4 < \bar{x} < 140.2$. This is pictured in Figure 6.7 for the two means $\mu = 139.5$ and $\mu = 139.2$, having corresponding type II error probabilities $\beta = .61$ and $\beta = .27$.

The calculations represented by the two figures are collected in Table 6.2. Notice two features of the table. First, the β values for $\alpha = .05$ are larger than those for $\alpha = .2$. If one wants to run only a 5% chance of (incorrectly) deciding to adjust an on-target process, the price to be paid is a larger probability of failure to recognize an off-target condition. Secondly, the β values for $\mu = 139.2$ are smaller than the β values for $\mu = 139.5$. The further the filling process is from being on target, the less likely it is that the off-target condition will fail to be detected.

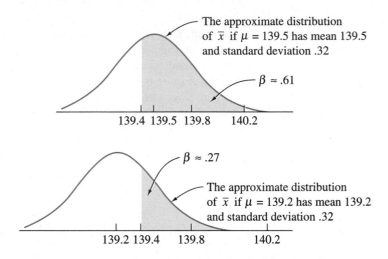

Figure 6.7 Approximate probability distributions for $\bar{x}$ for two different values of μ described by H_a and the corresponding β's, when $\alpha = .2$

Table 6.2
$n = 25$ type II error probabilities (β)

		μ	
		139.2	139.5
α	.05	.50	.83
	.2	.27	.61

The story told by Table 6.2 applies in qualitative terms to all uses of significance testing in decision-making contexts. The further H_0 is from being true, the smaller the corresponding β. And small α's imply large β's and vice versa.

The effect of sample size on β's

There is one other element of this general picture that plays an important role in the determination of error probabilities. That is the matter of sample size. If a sample size can be increased, for a given α, the corresponding β's can be reduced. Redo the calculations of the previous example, this time supposing that $n = 100$ rather than 25. Table 6.3 shows the type II error probabilities that should result, and comparison with Table 6.2 serves to indicate the sample-size effect in the filling-process example.

Analogy between testing and a criminal trial

An analogy helpful in understanding the standard logic applied when significance testing is employed in decision-making involves thinking of the process of coming to a decision as a sort of legal proceeding, like a criminal trial. In a criminal trial, there are two opposing hypotheses, namely

$$H_0 : \text{The defendant is innocent}$$

$$H_a : \text{The defendant is guilty}$$

Evidence, playing a role similar to the data used in testing, is gathered and used to decide between the two hypotheses. Two types of potential error exist in a criminal trial: the possibility of convicting an innocent person (parallel to the type I error) and the possibility of acquitting a guilty person (similar to the type II error). A criminal trial is a situation where the two types of error are definitely thought of as having differing consequences, and the two hypotheses are treated asymmetrically. The a priori presumption in a criminal trial is in favor of H_0, the defendant's innocence. In order to keep the chance of a false conviction small (i.e., keep α small), overwhelming evidence is required for conviction, in much the same way that if small α is used in testing, extreme values of the test statistic are needed in order to indicate rejection of H_0. One consequence of this method of operation in criminal trials is that there is a substantial chance that a guilty individual will be acquitted, in the same way that small α's produce big β's in testing contexts.

This significance testing/criminal trial parallel is useful, but do not make more of it than is justified. Not all significance-testing applications are properly thought of in this light. And few engineering scenarios are simple enough to reduce to a "decide between H_0 and H_a" choice. Sensible applications of significance testing are

Table 6.3
$n = 100$ Type II Error
Probabilities (β)

		μ	
		139.2	139.5
α	.05	.04	.53
	.2	.01	.28

often only steps of "evidence evaluation" in a many-faceted, data-based detective job necessary to solve an engineering problem. And even when a real problem can be reduced to a simple "decide between H$_0$ and H$_a$" framework, it need not be the case that the "choose a small α" logic is appropriate. In some engineering contexts, the practical consequences of a type II error are such that rational decision-making strikes a balance between the opposing goals of small α and small β's.

6.2.5 Some Comments Concerning Significance Testing and Estimation

Confidence interval estimation and significance testing are the two most commonly used forms of formal statistical inference. These having been introduced, it is appropriate to offer some comparative comments about their practical usefulness and, in the process, admit to an *estimation orientation* that will be reflected in much of the rest of this book's treatment of formal inference.

More often than not, engineers need to know "What is the value of the parameter?" rather than "Is the parameter equal to some hypothesized value?" And it is confidence interval estimation, not significance testing, that is designed to answer the first question. A confidence interval for a mean breakaway torque of from 9.9 in. oz to 13.1 in. oz says what values of μ seem plausible. A tiny observed level of significance in testing H$_0$: $\mu = 33.5$ says only that the data speak clearly against the possibility that $\mu = 33.5$, but it doesn't give any clue to the likely value of μ.

The fact that significance testing doesn't produce any useful indication of what parameter values are plausible is sometimes obscured by careless interpretation of semistandard jargon. For example, it is common in some fields to term p-values less than .05 "statistically significant" and ones less than .01 "highly significant." The danger in this kind of usage is that "significant" can be incorrectly heard to mean "of great practical consequence" and the p-value incorrectly interpreted as a measure of how much a parameter differs from a value stated in a null hypothesis. One reason this interpretation doesn't follow is that the observed level of significance in a test depends not only on how far H$_0$ appears to be from being correct but on the sample size as well. Given a large enough sample size, any departure from H$_0$, whether of practical importance or not, can be shown to be "highly significant."

"Statistical significance" and practical importance

Example 6

Statistical Significance and Practical Importance in a Regulatory Agency Test

A good example of the previous points involves the newspaper article in Figure 6.8. Apparently the Pass Master manufacturer did enough physical mileage testing (used a large enough n) to produce a p-value less than .05 for testing a null hypothesis of no mileage improvement. That is, a "statistically significant" result was obtained.

But the size of the actual mileage improvement reported is only "small but real," amounting to about .8 mpg. Whether or not this improvement is of *practical importance* is a matter largely separate from the significance-testing

WASHINGTON (AP)—A gadget that cuts off a car's air conditioner when the vehicle accelerates has become the first product aimed at cutting gasoline consumption to win government endorsement.

The device, marketed under the name "Pass Master," can provide a "small but real fuel economy benefit," the Environmental Protection Agency said Wednesday.

Motorists could realize up to 4 percent fuel reduction while using their air conditioners on cars equipped with the device, the agency said. That would translate into .8-miles-per-gallon improvement for a car that normally gets 20 miles to the gallon with the air conditioner on.

The agency cautioned that the 4 percent figure was a maximum amount and could be less depending on a motorist's driving habits, the type of car and the type of air conditioner.

But still the Pass Master, which sells for less than $15, is the first of 40 products to pass the EPA's tests as making any "statistically significant" improvement in a car's mileage.

Figure 6.8 Article from *The Lafayette Journal and Courier*, Page D-3, August 28, 1980. Reprinted by permission of the Associated Press. © 1980 the Associated Press.

result. And an engineer equipped with a confidence interval for the mean mileage improvement is in a better position to judge this than is one who knows only that the *p*-value was less than .05.

Example 5
(continued)

To illustrate the effect that sample size has on observed level of significance, return to the breakaway torque problem and consider two hypothetical samples, one based on $n = 25$ and the other on $n = 100$ but both giving $\bar{x} = 32.5$ in. oz and $s = 5.1$ in. oz.

For testing H_0: $\mu = 33.5$ with H_a: $\mu < 33.5$, the first hypothetical sample gives

$$z = \frac{32.5 - 33.5}{\frac{5.1}{\sqrt{25}}} = -.98$$

with associated observed level of significance

$$\Phi(-.98) = .16$$

The second hypothetical sample gives

$$z = \frac{32.5 - 33.5}{\frac{5.1}{\sqrt{100}}} = -1.96$$

Example 5
(*continued*)

with corresponding p-value

$$\Phi(-1.96) = .02$$

Because the second sample size is larger, the second sample gives stronger evidence that the mean breakaway torque is below 33.5 in. oz. But the best data-based guess at the difference between μ and 33.5 is $\bar{x} - 33.5 = -1.0$ in. oz *in both cases*. And it is the size of the difference between μ and 33.5 that is of primary engineering importance.

It is further useful to realize that in addition to doing its primary job of providing an interval of plausible values for a parameter, a confidence interval itself also provides some significance-testing information. For example, a 95% confidence interval for a parameter contains all those values of the parameter for which significance tests using the data in hand would produce p-values bigger than 5%. (Those values not covered by the interval would have associated p-values smaller than 5%.)

Example 5
(*continued*)

Recall from Section 6.1 that a 90% one-sided confidence interval for the mean breakaway torque for failed drives is $(-\infty, 12.8)$. This means that for any value, #, larger than 12.8 in. oz, a significance test of $H_0: \mu = \#$ with $H_a: \mu < \#$ would produce a p-value less than .1. So clearly, the observed level of significance corresponding to the null hypothesis $H_0: \mu = 33.5$ is less than .1 . (In fact, as was seen earlier in this section, the p-value is 0 to two decimal places.) Put more loosely, the interval $(-\infty, 12.8)$ is a long way from containing 33.5 in. oz and therefore makes such a value of μ quite implausible.

The discussion here could well raise the question "What practical role remains for significance testing?" Some legitimate answers to this question are

1. In an almost negative way, p-values can help an engineer gauge the extent to which data in hand are inconclusive. When observed levels of significance are large, more information is needed in order to arrive at any definitive judgment.

2. Sometimes legal requirements force the use of significance testing in a compliance or effectiveness demonstration. (This was the case in Figure 6.8, where before the Pass Master could be marketed, some mileage improvement had to be legally demonstrated.)

3. There are cases where the use of significance testing in a decision-making framework is necessary and appropriate. (An example is acceptance sampling: Based on information from a sample of items from a large lot, one must determine whether or not to receive shipment of the lot.)

So, properly understood and handled, significance testing does have its place in engineering practice. Thus, although the rest of this book features estimation over significance testing, methods of significance testing will not be completely ignored.

Section 2 Exercises

1. In the aluminum contamination study discussed in Exercise 2 of Section 6.1 and in Chapter Exercise 2 of Chapter 3, it was desirable to have mean aluminum content for samples of recycled plastic below 200 ppm. Use the five-step significance-testing format and determine the strength of the evidence in the data that in fact this contamination goal has been violated. (You will want to begin with $H_0: \mu = 200$ ppm and use $H_a: \mu > 200$ ppm.)

2. Heyde, Kuebrick, and Swanson measured the heights of 405 steel punches of a particular type. These were all from a single manufacturer and were supposed to have heights of .500 in. (The stamping machine in which these are used is designed to use .500 in. punches.) The students' measurements had $\bar{x} = .5002$ in. and $s = .0026$ in. (The raw data are given in Chapter Exercise 9 of Chapter 3.)

   (a) Use the five-step format and test the hypothesis that the mean height of such punches is "on spec" (i.e., is .500 in.).

   (b) Make a 98% two-sided confidence interval for the mean height of such punches produced by this manufacturer under conditions similar to those existing when the students' punches were manufactured. Is your interval consistent with the outcome of the test in part (a)? Explain.

   (c) In the students' application, the mean height of the punches did not tell the whole story about how they worked in the stamping machine. Several of these punches had to be placed side by side and used to stamp the same piece of material. In this context, what other feature of the height distribution is almost certainly of practical importance?

3. Discuss, in the context of Exercise 2, part (a), the potential difference between statistical significance and practical importance.

4. In the context of the machine screw diameter study of Exercise 4 of Section 6.1, suppose that the nominal diameter of such screws is 4.70 mm. Use the five-step significance-testing format and assess the strength of the evidence provided by the data that the long-run mean measured diameter differs from nominal. (You will want to begin with $H_0: \mu = 4.70$ mm and use $H_a: \mu \neq 4.70$ mm.)

5. Discuss, in the context of Exercise 4, the potential difference between statistical significance and practical importance.

6.3 One- and Two-Sample Inference for Means

Sections 6.1 and 6.2 introduced the basic concepts of confidence interval estimation and significance testing. There are thousands of specific methods of these two types. This book can only discuss a small fraction that are particularly well known and useful to engineers. The next three sections consider the most elementary of these—some of those that are applicable to one- and two-sample studies—beginning in this section with methods of formal inference for means.

Inferences for a single mean, based not on the large samples of Sections 6.1 and 6.2 but instead on small samples, are considered first. In the process, it is necessary to introduce the so-called (Student) t probability distributions. Presented next are methods of formal inference for paired data. The section concludes with discussions of both large- and small-n methods for data-based comparison of two means based on independent samples.

6.3.1 Small-Sample Inference for a Single Mean

The most important practical limitation on the use of the methods of the previous two sections is the requirement that n must be large. That restriction comes from the fact that without it, there is no way to conclude that

$$\frac{\bar{x} - \mu}{\dfrac{s}{\sqrt{n}}} \tag{6.16}$$

is approximately standard normal. So if, for example, one mechanically uses the large-n confidence interval formula

$$\bar{x} \pm z \frac{s}{\sqrt{n}} \tag{6.17}$$

with a small sample, there is no way of assessing what actual level of confidence should be declared. That is, for small n, using $z = 1.96$ in formula (6.17) generally doesn't produce 95% confidence intervals. And without a further condition, there is neither any way to tell what confidence might be associated with $z = 1.96$ nor any way to tell how to choose z in order to produce a 95% confidence level.

There is one important special circumstance in which it is possible to reason in a way parallel to the work in Sections 6.1 and 6.2 and arrive at inference methods for means based on small sample sizes. That is the situation where it is sensible to model the observations as iid normal random variables. The normal observations case is convenient because although the variable (6.16) is not standard normal, it does have a recognized, tabled distribution. This is the **Student t distribution**.

Definition 13

The **(Student) t distribution with degrees of freedom parameter** ν is a continuous probability distribution with probability density

$$f(t) = \frac{\Gamma\left(\dfrac{\nu+1}{2}\right)}{\Gamma\left(\dfrac{\nu}{2}\right)\sqrt{\pi\nu}} \left(1 + \frac{t^2}{\nu}\right)^{-(\nu+1)/2} \qquad \text{for all } t \tag{6.18}$$

If a random variable has the probability density given by formula (6.18), it is said to have a t_ν distribution.

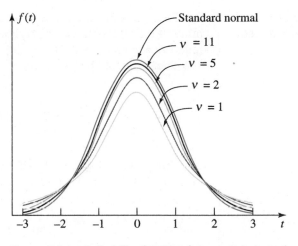

Figure 6.9 t Probability densities for $\nu = 1, 2, 5,$ and 11 and the standard normal density

The word *Student* in Definition 13 was the pen name of the statistician who first came upon formula (6.18). Expression (6.18) is rather formidable looking. No direct computations with it will actually be required in this book. But, it is useful to have expression (6.18) available in order to sketch several t probability densities, to get a feel for their shape. Figure 6.9 pictures the t densities for degrees of freedom $\nu = 1$, 2, 5, and 11, along with the standard normal density.

The message carried by Figure 6.9 is that the t probability densities are bell shaped and symmetric about 0. They are flatter than the standard normal density but are increasingly like it as ν gets larger. In fact, for most practical purposes, for ν larger than about 30, the t distribution with ν degrees of freedom and the standard normal distribution are indistinguishable.

t distributions and the standard normal distribution

Probabilities for the t distributions are not typically found using the density in expression (6.18), as no simple antiderivative for $f(t)$ exists. Instead, it is common to use tables (or statistical software) to evaluate common t distribution quantiles and to get at least crude bounds on the types of probabilities needed in significance testing. Table B.4 is a typical table of t quantiles. Across the top of the table are several cumulative probabilities. Down the left side are values of the degrees of freedom parameter, ν. In the body of the table are corresponding quantiles. Notice also that the last line of the table is a "$\nu = \infty$" (i.e., standard normal) line.

Example 7

Use of a Table of t Distribution Quantiles

Suppose that T is a random variable having a t distribution with $\nu = 5$ degrees of freedom. Consider first finding the .95 quantile of T's distribution, then seeing what Table B.4 reveals about $P[T < -1.9]$ and then about $P[|T| > 2.3]$.

Example 7
(continued)

First, looking at the $\nu = 5$ row of Table B.4 under the cumulative probability .95, 2.015 is found in the body of the table. That is, $Q(.95) = 2.015$ or
(equivalently) $P[T \leq 2.015] = .95$.

Then note that by symmetry,

$$P[T < -1.9] = P[T > 1.9] = 1 - P[T \leq 1.9]$$

Looking at the $\nu = 5$ row of Table B.4, 1.9 is between the .90 and .95 quantiles
of the t_5 distribution. That is,

$$.90 < P[T \leq 1.9] \leq .95$$

so finally

$$.05 < P[T < -1.9] < .10$$

Lastly, again by symmetry,

$$P[|T| > 2.3] = P[T < -2.3] + P[T > 2.3] = 2P[T > 2.3]$$
$$= 2(1 - P[T \leq 2.3])$$

Then, from the $\nu = 5$ row of Table B.4, 2.3 is seen to be between the .95 and
.975 quantiles of the t_5 distribution. That is,

$$.95 < P[T \leq 2.3] < .975$$

so

$$.05 < P[|T| > 2.3] < .10$$

The three calculations of this example are pictured in Figure 6.10.

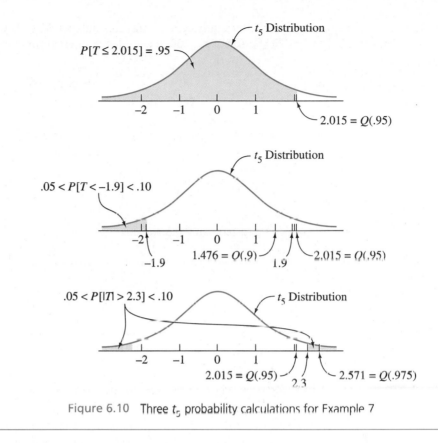

Figure 6.10 Three t_5 probability calculations for Example 7

The connection between expressions (6.18) and (6.16) that allows the development of small-n inference methods for normal observations is that if an iid normal model is appropriate,

$$T = \frac{\bar{x} - \mu}{\frac{s}{\sqrt{n}}} \qquad (6.19)$$

has the t distribution with $\nu = n - 1$ degrees of freedom. (This is consistent with the basic fact used in the previous two sections. That is, for large n, ν is large, so the t_ν distribution is approximately standard normal; and for large n, the variable (6.19) has already been treated as approximately standard normal.)

Since the variable (6.19) can under appropriate circumstances be treated as a t_{n-1} random variable, we are in a position to work in exact analogy to what was done in Sections 6.1 and 6.2 to find methods for confidence interval estimation and significance testing. That is, if a data-generating mechanism can be thought of as essentially equivalent to drawing independent observations from a single normal distribution, a two-sided confidence interval for μ has endpoints

Normal distribution confidence limits for μ

$$\bar{x} \pm t \frac{s}{\sqrt{n}} \qquad (6.20)$$

where t is chosen such that the t_{n-1} distribution assigns probability corresponding to the desired confidence level to the interval between $-t$ and t. Further, the null hypothesis

$$H_0: \mu = \#$$

can be tested using the statistic

Normal distribution test statistic for μ

$$T = \frac{\bar{x} - \#}{\frac{s}{\sqrt{n}}} \qquad (6.21)$$

and a t_{n-1} reference distribution.

Operationally, the only difference between the inference methods indicated here and the large-sample methods of the previous two sections is the exchange of standard normal quantiles and probabilities for ones corresponding to the t_{n-1} distribution. *Conceptually*, however, the nominal confidence and significance properties here are practically relevant only under the extra condition of a reasonably normal underlying distribution. Before applying either expression (6.20) or (6.21) in practice, it is advisable to investigate the appropriateness of a normal model assumption.

Example 8 | Small-Sample Confidence Limits for a Mean Spring Lifetime

Part of a data set of W. Armstrong (appearing in *Analysis of Survival Data* by Cox and Oakes) gives numbers of cycles to failure of ten springs of a particular type under a stress of 950 N/mm^2. These spring-life observations are given in Table 6.4, in units of 1,000 cycles.

Table 6.4
Cycles to Failure of Ten
Springs under 950 N/mm^2
Stress (10^3 cycles)

Spring Lifetimes
225, 171, 198, 189, 189
135, 162, 135, 117, 162

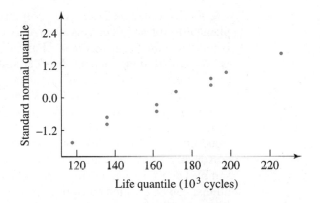

Figure 6.11   Normal plot of spring lifetimes

An important question here might be "What is the average spring lifetime under conditions of 950 N/mm^2 stress?" Since only $n = 10$ observations are available, the large-sample method of Section 6.1 is not applicable. Instead, only the method indicated by expression (6.20) is a possible option. For it to be appropriate, lifetimes must be normally distributed.

Without a relevant base of experience in materials, it is difficult to speculate a priori about the appropriateness of a normal lifetime model in this context. But at least it is possible to examine the data in Table 6.4 themselves for evidence of strong departure from normality. Figure 6.11 is a normal plot for the data. It shows that in fact no such evidence exists.

For the ten lifetimes, $\bar{x} = 168.3$ ($\times 10^3$ cycles) and $s = 33.1$ ($\times 10^3$ cycles). So to estimate the mean spring lifetime, these values may be used in expression (6.20), along with an appropriately chosen value of t. Using, for example, a 90% confidence level and a two-sided interval, t should be chosen as the .95 quantile of the t distribution with $\nu = n - 1 = 9$ degrees of freedom. That is, one uses the t_9 distribution and chooses $t > 0$ such that

$$P[-t < \text{a } t_9 \text{ random variable} < t] = .90$$

Consulting Table B.4, the choice $t = 1.833$ is in order. So a two-sided 90% confidence interval for μ has endpoints

$$168.3 \pm 1.833 \frac{33.1}{\sqrt{10}}$$

i.e.,

$$168.3 \pm 19.2$$

i.e.,

$$149.1 \times 10^3 \text{ cycles} \quad \text{and} \quad 187.5 \times 10^3 \text{ cycles}$$

As illustrated in Example 8, normal-plotting the data as a rough check on the plausibility of an underlying normal distribution is a sound practice, and one that is used repeatedly in this text. However, it is important not to expect more than is justified from the method. It is certainly preferable to use it rather than making an unexamined leap to a possibly inappropriate normal assumption. But it is also true that when used with small samples, the method doesn't often provide definitive indications as to whether a normal model can be used. Small samples from normal distributions will often have only marginally linear-looking normal plots. At the same time, small samples from even quite nonnormal distributions can often have reasonably linear normal plots. In short, because of sampling variability, small samples don't carry much information about underlying distributional shape. About all that can be counted on from a small-sample preliminary normal plot, like that in Example 8, is a warning in case of gross departure from normality associated with an underlying distributional shape that is much heavier in the tails than a normal distribution (i.e., producing more extreme values than a normal shape would).

What is a "nonlinear" normal plot?

It is a good idea to make the effort to (so to speak) calibrate normal-plot perceptions if they are going to be used as a tool for checking a model. One way to do this is to use simulation and generate a number of samples of the size in question from a standard normal distribution and normal-plot these. Then the shape of the normal plot of the data in hand can be compared to the simulations to get some feeling as to whether any nonlinearity it exhibits is really unusual. To illustrate, Figure 6.12 shows normal plots for several simulated samples of size $n = 10$ from the standard normal distribution. Comparing Figures 6.11 and 6.12, it is clear that indeed the spring-life data carry no strong indication of nonnormality.

Small sample tests for μ

Example 8 shows the use of the confidence interval formula (6.20) but not the significance testing method (6.21). Since the small-sample method is exactly analogous to the large-sample method of Section 6.2 (except for the substitution of the t distribution for the standard normal distribution), and the source from which the data were taken doesn't indicate any particular value of μ belonging naturally in a null hypothesis, the use of the method indicated in expression (6.21) by itself will not be illustrated at this point. (There is, however, an application of the testing method to paired differences in Example 9.)

6.3.2 Inference for the Mean of Paired Differences

An important type of application of the foregoing methods of confidence interval estimation and significance testing is to *paired data*. In many engineering problems, it is natural to make two measurements of essentially the same kind, but differing in timing or physical location, on a single sample of physical objects. The goal in such situations is often to investigate the possibility of consistent differences between the two measurements. (Review the discussion of paired data terminology in Section 1.2.)

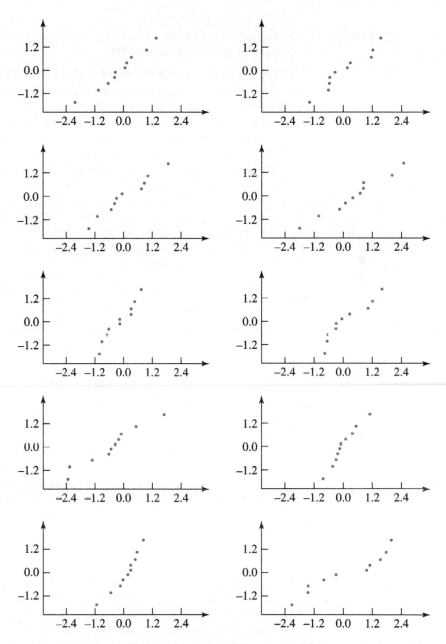

Figure 6.12   Normal plots of samples of size $n = 10$ from a standard normal distribution (data quantiles on the horizontal axes)

| Example 9 | Comparing Leading-Edge and Trailing-Edge Measurements on a Shaped Wood Product |

Drake, Hones, and Mulholland worked with a company on the monitoring of the operation of an end-cut router in the manufacture of a wood product. They measured a critical dimension of a number of pieces of a particular type as they came off the router. Both a leading-edge and a trailing-edge measurement were made on each piece. The design for the piece in question specified that both leading-edge and trailing-edge values were to have a target value of .172 in. Table 6.5 gives leading- and trailing-edge measurements taken by the students on five consecutive pieces.

Table 6.5
Leading-Edge and Trailing-Edge Dimensions for Five Workpieces

Piece	Leading-Edge Measurement (in.)	Trailing-Edge Measurement (in.)
1	.168	.169
2	.170	.168
3	.165	.168
4	.165	.168
5	.170	.169

In this situation, the correspondence between leading- and trailing-edge dimensions was at least as critical to proper fit in a later assembly operation as was the conformance of the individual dimensions to the nominal value of .172 in. This was thus a paired-data situation, where one issue of concern was the possibility of a consistent difference between leading- and trailing-edge dimensions that might be traced to a machine misadjustment or unwise method of router operation.

In situations like Example 9, one simple method of investigating the possibility of a consistent difference between paired data is to first reduce the two measurements on each physical object to a single difference between them. Then the methods of confidence interval estimation and significance testing studied thus far may be applied to the differences. That is, after reducing paired data to differences $d_1, d_2, \ldots, d_n$, if n (the number of data pairs) is large, endpoints of a confidence interval for the underlying mean difference, μ_d, are

Large-sample confidence limits for μ_d

$$\bar{d} \pm z \frac{s_d}{\sqrt{n}}$$

(6.22)

where s_d is the sample standard deviation of $d_1, d_2, \ldots, d_n$. Similarly, the null hypothesis

$$H_0\colon \mu_d = \#$$

(6.23)

can be tested using the test statistic

Large-sample test statistic for μ_d

$$Z = \frac{\bar{d} - \#}{\dfrac{s_d}{\sqrt{n}}}$$

(6.24)

and a standard normal reference distribution.

If n is small, in order to come up with methods of formal inference, an underlying normal distribution *of differences* must be plausible. If that is the case, a confidence interval for μ_d has endpoints

Normal distribution confidence limits for μ_d

$$\bar{d} \pm t\frac{s_d}{\sqrt{n}}$$

(6.25)

and the null hypothesis (6.23) can be tested using the test statistic

Normal distribution test statistic for μ_d

$$T - \frac{\bar{d} - \#}{\dfrac{s_d}{\sqrt{n}}}$$

(6.26)

and a t_{n-1} reference distribution.

Example 9
(continued)

To illustrate this method of paired differences, consider testing the null hypothesis $H_0\colon \mu_d = 0$ and making a 95% confidence interval for any consistent difference between leading- and trailing-edge dimensions, μ_d, based on the data in Table 6.5.

Begin by reducing the $n = 5$ paired observations in Table 6.5 to differences

$$d = \text{leading-edge dimension} - \text{trailing-edge dimension}$$

appearing in Table 6.6. Figure 6.13 is a normal plot of the $n = 5$ differences in Table 6.6. A little experimenting with normal plots of simulated samples of size $n = 5$ from a normal distribution will convince you that the lack of linearity in Figure 6.13 would in no way be atypical of normal data. This, together with the fact that normal distributions are very often appropriate for describ-

Example 9
(continued)

Table 6.6
Five Differences in Leading- and Trailing-Edge Measurements

Piece	$d =$ Difference in Dimensions (in.)	
1	$-.001$	$(= .168 - .169)$
2	$.002$	$(= .170 - .168)$
3	$-.003$	$(= .165 - .168)$
4	$-.003$	$(= .165 - .168)$
5	$.001$	$(= .170 - .169)$

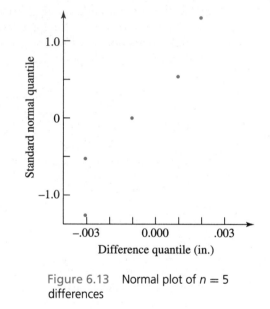

Figure 6.13   Normal plot of $n = 5$ differences

ing machined dimensions of mass-produced parts, suggests the conclusion that the methods represented by expressions (6.25) and (6.26) are in order in this example.

The differences in Table 6.6 have $\bar{d} = -.0008$ in. and $s_d = .0023$ in. So, first investigating the plausibility of a "no consistent difference" hypothesis in a five-step significance testing format, gives the following:

1. $H_0\colon \mu_d = 0.$

2. $H_a\colon \mu_d \neq 0.$
(There is a priori no reason to adopt a one-sided alternative hypothesis.)

3. The test statistic will be

$$T = \frac{\bar{d} - 0}{\frac{s_d}{\sqrt{n}}}$$

The reference distribution will be the t distribution with $\nu = n - 1 = 4$ degrees of freedom. Large observed $|t|$ will count as evidence against H_0 and in favor of H_a.

4. The sample gives

$$t = \frac{-.0008}{\frac{.0023}{\sqrt{5}}} = -.78$$

5. The observed level of significance is $P[|\text{a } t_4 \text{ random variable}| \geq .78]$, which can be seen from Table B.4 to be larger than $2(.10) = .2$. The data in hand are not convincing in favor of a systematic difference between leading- and trailing-edge measurements.

Consulting Table B.4 for the .975 quantile of the t_4 distribution, $t = 2.776$ is the appropriate multiplier for use in expression (6.25) for 95% confidence. That is, a two-sided 95% confidence interval for the mean difference between the leading- and trailing-edge dimensions has endpoints

$$-.0008 \pm 2.776 \frac{.0023}{\sqrt{5}}$$

i.e.,

$$-.0008 \text{ in.} \pm .0029 \text{ in.} \tag{6.27}$$

i.e.,

$$-.0037 \text{ in.} \quad \text{and} \quad .0021 \text{ in.}$$

This confidence interval for μ_d implicitly says (since 0 is in the calculated interval) that the observed level of significance for testing H_0: $\mu_d = 0$ is more than .05 ($= 1 - .95$). Put slightly differently, it is clear from display (6.27) that the imprecision represented by the plus-or-minus part of the expression is large enough to make it believable that the perceived difference, $\bar{d} = -.0008$, is just a result of sampling variability.

Large-sample inference for μ_d

Example 9 treats a small-sample problem. No example for large n is included here, because after the taking of differences just illustrated, such an example would reduce to a rehash of things in Sections 6.1 and 6.2. In fact, since for large n the t distribution with $\nu = n - 1$ degrees of freedom becomes essentially standard normal, one could even imitate Example 9 for large n and get into no logical problems. So at this point, it makes sense to move on from consideration of the paired-difference method.

6.3.3 Large-Sample Comparisons of Two Means (Based on Independent Samples)

One of the principles of effective engineering data collection discussed in Section 2.3 was *comparative study*. The idea of paired differences provides inference methods of a very special kind for comparison, where one sample of items in some sense provides its own basis for comparison. Methods that can be used to compare two means where two different "unrelated" samples form the basis of inference are studied next, beginning with large-sample methods.

Example 10	Comparing the Packing Properties of Molded and Crushed Pieces of a Solid

A company research effort involved finding a workable geometry for molded pieces of a solid. One comparison made was between the weight of molded pieces of a particular geometry, that could be poured into a standard container, and the weight of irregularly shaped pieces (obtained through crushing), that could be poured into the same container. A series of 24 attempts to pack both molded and crushed pieces of the solid produced the data (in grams) that are given in Figure 6.14 in the form of back-to-back stem-and-leaf diagrams.

Notice that although the same number of molded and crushed weights are represented in the figure, there are two distinctly different samples represented. This is in no way comparable to the paired-difference situation treated in Example 9, and a different method of statistical inference is appropriate.

In situations like Example 10, it is useful to adopt subscript notation for both the parameters and the statistics—for example, letting μ_1 and μ_2 stand for underlying distributional means corresponding to the first and second conditions and $\bar{x}_1$ and $\bar{x}_2$ stand for corresponding sample means. Now if the two data-generating mechanisms are conceptually essentially equivalent to sampling with replacement from two distributions, Section 5.5 says that $\bar{x}_1$ has mean μ_1 and variance σ_1^2/n_1, and $\bar{x}_2$ has mean μ_2 and variance σ_2^2/n_2.

The difference in sample means $\bar{x}_1 - \bar{x}_2$ is a natural statistic to use in comparing μ_1 and μ_2. Proposition 1 in Chapter 5 (see page 307) implies that if it is reasonable

Molded		Crushed
7.9	11	
4.5, 3.6, 1.2	12	
9.8, 8.9, 7.9, 7.1, 6.1, 5.7, 5.1	12	
2.3, 1.3, 0.0	13	
8.0, 7.0, 6.5, 6.3, 6.2	13	
2.2, 0.1	14	
	14	
2.1, 1.2, 0.2	15	
	15	
	16	1.8
	16	5.8, 9.6
	17	1.3, 2.0, 2.4, 3.3, 3.4, 3.7
	17	6.6, 9.8
	18	0.2, 0.9, 3.3, 3.8, 4.9
	18	5.5, 6.5, 7.1, 7.3, 9.1, 9.8
	19	0.0, 1.0
	19	

Figure 6.14 Back-to-back stem-and-leaf plots of packing weights for molded and crushed pieces

to think of the two samples as separately chosen/independent, the random variable has

$$E(\bar{x}_1 - \bar{x}_2) = \mu_1 - \mu_2$$

and

$$\text{Var}(\bar{x}_1 - \bar{x}_2) = \frac{\sigma_1^2}{n_1} + \frac{\sigma_2^2}{n_2}$$

If, in addition, n_1 and n_2 are large (so that $\bar{x}_1$ and $\bar{x}_2$ are each approximately normal), $\bar{x}_1 - \bar{x}_2$ is approximately normal—i.e.,

$$Z = \frac{\bar{x}_1 - \bar{x}_2 - (\mu_1 - \mu_2)}{\sqrt{\dfrac{\sigma_1^2}{n_1} + \dfrac{\sigma_2^2}{n_2}}} \tag{6.28}$$

has an approximately standard normal probability distribution.

It is possible to begin with the fact that the variable (6.28) is approximately standard normal and end up with confidence interval and significance-testing methods for $\mu_1 - \mu_2$ by using logic exactly parallel to that in the "known-σ" parts of Sections 6.1 and 6.2. But practically, it is far more useful to begin instead with an expression that is free of the parameters σ_1 and σ_2. Happily, for large n_1 and n_2, not only is the variable (6.28) approximately standard normal but so is

$$Z = \frac{\bar{x}_1 - \bar{x}_2 - (\mu_1 - \mu_2)}{\sqrt{\dfrac{s_1^2}{n_1} + \dfrac{s_2^2}{n_2}}} \tag{6.29}$$

Then the standard logic of Section 6.1 shows that a two-sided large-sample confidence interval for the difference $\mu_1 - \mu_2$ based on two independent samples has endpoints

Large-sample confidence limits for $\mu_1 - \mu_2$

$$\bar{x}_1 - \bar{x}_2 \pm z \sqrt{\frac{s_1^2}{n_1} + \frac{s_2^2}{n_2}} \tag{6.30}$$

where z is chosen such that the probability that the standard normal distribution assigns to the interval between $-z$ and z corresponds to the desired confidence. And the logic of Section 6.2 shows that under the same conditions,

$$H_0\!: \mu_1 - \mu_2 = \#$$

can be tested using the statistic

Large-sample test statistic for $\mu_1 - \mu_2$

$$Z = \frac{\bar{x}_1 - \bar{x}_2 - \#}{\sqrt{\dfrac{s_1^2}{n_1} + \dfrac{s_2^2}{n_2}}} \tag{6.31}$$

and a standard normal reference distribution.

Example 10
(continued)

In the molding problem, the crushed pieces were a priori expected to pack better than the molded pieces (that for other purposes are more convenient). Consider testing the statistical significance of the difference in mean weights and also making a 95% one-sided confidence interval for the difference (declaring that the crushed mean weight minus the molded mean weight is at least some number).

The sample sizes here ($n_1 = n_2 = 24$) are borderline for being called large. It would be preferable to have a few more observations of each type. Lacking them, we will go ahead and use the methods of expressions (6.30) and (6.31) but remain properly cautious of the results should they in any way produce a "close call" in engineering or business terms.

Arbitrarily labeling "crushed" condition 1 and "molded" condition 2 and calculating from the data in Figure 6.14 that $\bar{x}_1 = 179.55$ g, $s_1 = 8.34$ g, $\bar{x}_2 = 132.97$ g, and $s_2 = 9.31$ g, the five-step testing format produces the following summary:

1. $H_0: \mu_1 - \mu_2 = 0$.

2. $H_a: \mu_1 - \mu_2 > 0$.
   (The research hypothesis here is that the crushed mean exceeds the molded mean so that the difference, taken in this order, is positive.)

3. The test statistic is

$$Z - \frac{\bar{x}_1 - \bar{x}_2 - 0}{\sqrt{\dfrac{s_1^2}{n_1} + \dfrac{s_2^2}{n_2}}}$$

The reference distribution is standard normal, and large observed values z will constitute evidence against H_0 and in favor of H_a.

4. The samples give

$$z = \frac{179.55 - 132.97 - 0}{\sqrt{\dfrac{(8.34)^2}{24} + \dfrac{(9.31)^2}{24}}} - 18.3$$

5. The observed level of significance is $P[$a standard normal variable $\geq 18.3] \approx 0$. The data present overwhelming evidence that $\mu_1 - \mu_2 > 0$— i.e., that the mean packed weight of crushed pieces exceeds that of the molded pieces.

Then turning to a one-sided confidence interval for $\mu_1 - \mu_2$, note that only the lower endpoint given in display (6.30) will be used. So $z = 1.645$ will be appropriate. That is, with 95% confidence, we conclude that the difference in means (crushed minus molded) exceeds

$$(179.55 - 132.97) - 1.645\sqrt{\frac{(8.34)^2}{24} + \frac{(9.31)^2}{24}}$$

i.e., exceeds

$$46.58 - 4.20 = 42.38 \text{ g}$$

Example 10
(*continued*)

Or differently put, a 95% one-sided confidence interval for $\mu_1 - \mu_2$ is

$$(42.38, \infty)$$

Students are sometimes uneasy about the arbitrary choice involved in labeling the two conditions in a two-sample study. The fact is that either one can be used. As long as a given choice is followed through consistently, the real-world conclusions reached will be completely unaffected by the choice. In Example 10, if the molded condition is labeled number 1 and the crushed condition number 2, an appropriate one-sided confidence for the molded mean minus the crushed mean is

$$(-\infty, -42.38)$$

This has the same meaning in practical terms as the interval in the example.

The present methods apply where single measurements are made on each element of two different samples. This stands in contrast to problems of paired data (where there are bivariate observations on a single sample). In the woodworking case of Example 9, the data were paired because both leading-edge and trailing-edge measurements were made on each piece. If leading-edge measurements were taken from one group of items and trailing-edge measurements from another, a two-sample (not a paired difference) analysis would be in order.

6.3.4 Small-Sample Comparisons of Two Means (Based on Independent Samples from Normal Distributions)

The last inference methods presented in this section are those for the difference in two means in cases where at least one of n_1 and n_2 is small. All of the discussion for this problem is limited to cases where observations are normal. And in fact, the most straightforward methods are for cases where, in addition, the two underlying standard deviations are comparable. The discussion begins with these.

Graphical check on the plausibility of the model

A way of making at least a rough check on the plausibility of "normal distributions with a common variance" model assumptions in an application is to normal-plot two samples on the same set of axes, checking not only for approximate linearity but also for approximate equality of slope.

Example 8
(*continued*)

The data of W. Armstrong on spring lifetimes (appearing in the book by Cox and Oakes) not only concern spring longevity at a 950 N/mm^2 stress level but also longevity at a 900 N/mm^2 stress level. Table 6.7 repeats the 950 N/mm^2 data from before and gives the lifetimes of ten springs at the 900 N/mm^2 stress level as well.

Table 6.7
Spring Lifetimes under Two Different Levels of Stress
(10^3 cycles)

950 N/mm² Stress	900 N/mm² Stress
225, 171, 198, 189, 189	216, 162, 153, 216, 225
135, 162, 135, 117, 162	216, 306, 225, 243, 189

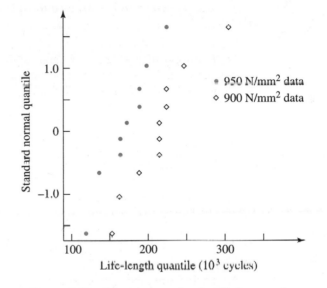

Figure 6.15   Normal plots of spring lifetimes under two different levels of stress

Figure 6.15 consists of normal plots for the two samples made on a single set of axes. In light of the kind of variation in linearity and slope exhibited in Figure 6.12 by the normal plots for samples of this size ($n = 10$) from a single normal distribution, there is certainly no strong evidence in Figure 6.15 against the appropriateness of an "equal variances, normal distributions" model for spring lifetimes.

If the assumption that $\sigma_1 = \sigma_2$ is used, then the common value is called σ, and it makes sense that both s_1 and s_2 will approximate σ. That suggests that they should somehow be combined into a single estimate of the basic, baseline variation. As it turns out, mathematical convenience dictates a particular method of combining or **pooling** the individual s's to arrive at a single estimate of σ.

Definition 14

If two numerical samples of respective sizes n_1 and n_2 produce respective sample variances s_1^2 and s_2^2, the **pooled sample variance**, s_P^2, is the weighted average of s_1^2 and s_2^2 where the weights are the sample sizes minus 1. That is,

$$s_P^2 = \frac{(n_1 - 1)s_1^2 + (n_2 - 1)s_2^2}{(n_1 - 1) + (n_2 - 1)} = \frac{(n_1 - 1)s_1^2 + (n_2 - 1)s_2^2}{n_1 + n_2 - 2} \qquad (6.32)$$

The **pooled sample standard deviation**, s_P, is the square root of s_P^2.

s_P is a kind of average of s_1 and s_2 that is guaranteed to fall between the two values s_1 and s_2. Its exact form is dictated more by considerations of mathematical convenience than by obvious intuition.

Example 8
(continued)

In the spring-life case, making the arbitrary choice to call the 900 N/mm^2 stress level condition 1 and the 950 N/mm^2 stress level condition 2, $s_1 = 42.9$ (10^3 cycles) and $s_2 = 33.1$ (10^3 cycles). So pooling the two sample variances via formula (6.32) produces

$$s_P^2 = \frac{(10 - 1)(42.9)^2 + (10 - 1)(33.1)^2}{(10 - 1) + (10 - 1)} = 1,468(10^3 \text{ cycles})^2$$

Then, taking the square root,

$$s_P = \sqrt{1,468} = 38.3(10^3 \text{ cycles})$$

In the argument leading to large-sample inference methods for $\mu_1 - \mu_2$, the quantity given in expression (6.28),

$$Z = \frac{\bar{x}_1 - \bar{x}_2 - (\mu_1 - \mu_2)}{\sqrt{\dfrac{\sigma_1^2}{n_1} + \dfrac{\sigma_2^2}{n_2}}}$$

was briefly considered. In the $\sigma_1 = \sigma_2 = \sigma$ context, this can be rewritten as

$$Z = \frac{\bar{x}_1 - \bar{x}_2 - (\mu_1 - \mu_2)}{\sigma\sqrt{\dfrac{1}{n_1} + \dfrac{1}{n_2}}} \qquad (6.33)$$

One could use the fact that expression (6.33) is standard normal to produce methods for confidence interval estimation and significance testing. But for use, these would require the input of the parameter σ. So instead of beginning with expression (6.28) or (6.33), it is standard to replace σ in expression (6.33) with s_P and begin with the quantity

$$T = \frac{(\bar{x}_1 - \bar{x}_2) - (\mu_1 - \mu_2)}{s_P\sqrt{\dfrac{1}{n_1} + \dfrac{1}{n_2}}} \tag{6.34}$$

Expression (6.34) is crafted exactly so that under the present model assumptions, the variable (6.34) has a well-known, tabled probability distribution: the t distribution with $\nu = (n_1 - 1) + (n_2 - 1) = n_1 + n_2 - 2$ degrees of freedom. (Notice that the $n_1 - 1$ degrees of freedom associated with the first sample add together with the $n_2 - 1$ degrees of freedom associated with the second to produce $n_1 + n_2 - 2$ overall.) This probability fact, again via the kind of reasoning developed in Sections 6.1 and 6.2, produces inference methods for $\mu_1 - \mu_2$. That is, a two-sided confidence interval for the difference $\mu_1 - \mu_2$, based on independent samples from normal distributions with a common variance, has endpoints

Normal distributions
($\sigma_1 = \sigma_2$) confidence
limits for $\mu_1 - \mu_2$

$$\bar{x}_1 - \bar{x}_2 \pm t s_P \sqrt{\frac{1}{n_1} + \frac{1}{n_2}} \tag{6.35}$$

where t is chosen such that the probability that the $t_{n_1+n_2-2}$ distribution assigns to the interval between $-t$ and t corresponds to the desired confidence. And under the same conditions,

$$H_0: \mu_1 - \mu_2 = \#$$

can be tested using the statistic

Normal distributions
($\sigma_1 = \sigma_2$) test
statistic for $\mu_1 - \mu_2$

$$T = \frac{\bar{x}_1 - \bar{x}_2 - \#}{s_P\sqrt{\dfrac{1}{n_1} + \dfrac{1}{n_2}}} \tag{6.36}$$

and a $t_{n_1+n_2-2}$ reference distribution.

Example 8 (*continued*)	We return to the spring-life case to illustrate small-sample inference for two means. First consider testing the hypothesis of equal mean lifetimes with an alternative of increased lifetime accompanying a reduction in stress level. Then

Example 8
(*continued*)

consider making a two-sided 95% confidence interval for the difference in mean lifetimes.

Continuing to call the 900 N/mm^2 stress level condition 1 and the 950 N/mm^2 stress level condition 2, from Table 6.7 $\bar{x}_1 = 215.1$ and $\bar{x}_2 = 168.3$, while (from before) $s_P = 38.3$. The five-step significance-testing format then gives the following:

1. $H_0: \mu_1 - \mu_2 = 0$.

2. $H_a: \mu_1 - \mu_2 > 0$.
   (The engineering expectation is that condition 1 produces the larger lifetimes.)

3. The test statistic is $T = \dfrac{\bar{x}_1 - \bar{x}_2 - 0}{s_P\sqrt{\dfrac{1}{n_1} + \dfrac{1}{n_2}}}$

   The reference distribution is t with $10 + 10 - 2 = 18$ degrees of freedom, and large observed t will count as evidence against H_0.

4. The samples give

$$t = \frac{215.1 - 168.3 - 0}{38.3\sqrt{\dfrac{1}{10} + \dfrac{1}{10}}} = 2.7$$

5. The observed level of significance is $P[\text{a } t_{18} \text{ random variable} \geq 2.7]$, which (according to Table B.4) is between .01 and .005. This is strong evidence that the lower stress level is associated with larger mean spring lifetimes.

Then, if the expression (6.35) is used to produce a two-sided 95% confidence interval, the choice of t as the .975 quantile of the t_{18} distribution is in order. Endpoints of the confidence interval for $\mu_1 - \mu_2$ are

$$(215.1 - 168.3) \pm 2.101(38.3)\sqrt{\frac{1}{10} + \frac{1}{10}}$$

i.e.,

$$46.8 \pm 36.0$$

i.e.,

$$10.8 \times 10^3 \text{ cycles} \quad \text{and} \quad 82.8 \times 10^3 \text{ cycles}$$

The data in Table 6.7 provide enough information to establish convincingly that increased stress is associated with reduced mean spring life. But although the apparent size of that reduction when moving from the 900 N/mm^2 level (condition 1) to the 950 N/mm^2 level (condition 2) is 46.8×10^3 cycles, the variability present in the data is large enough (and the sample sizes small enough) that only a precision of $\pm 36.0 \times 10^3$ cycles can be attached to the figure 46.8×10^3 cycles.

Small-sample inference for $\mu_1 - \mu_2$ without the $\sigma_1 = \sigma_2$ assumption

There is no completely satisfactory answer to the question of how to do inference for $\mu_1 - \mu_2$ when it is not sensible to assume that $\sigma_1 = \sigma_2$. The most widely accepted (but approximate) method for the problem is one due to Satterthwaite that is related to the large-sample formula (6.30). That is, while endpoints (6.30) are not appropriate when n_1 or n_2 is small (they don't produce actual confidence levels near the nominal one), a modification of them is appropriate. Let

Satterthwaite's "estimated degrees of freedom"

$$\hat{v} = \frac{\left(\dfrac{s_1^2}{n_1} + \dfrac{s_2^2}{n_2}\right)^2}{\dfrac{s_1^4}{(n_1 - 1)n_1^2} + \dfrac{s_2^4}{(n_2 - 1)n_2^2}} \tag{6.37}$$

and for a desired confidence level, suppose that $\hat{t}$ is such that the t distribution with $\hat{v}$ degrees of freedom assigns that probability to the interval between $-\hat{t}$ and $\hat{t}$. Then the two endpoints

Satterthwaite (approximate) normal distribution confidence limits for $\mu_1 - \mu_2$

$$\bar{x}_1 - \bar{x}_2 \pm \hat{t}\sqrt{\frac{s_1^2}{n_1} + \frac{s_2^2}{n_2}} \tag{6.38}$$

can serve as confidence limits for $\mu_1 - \mu_2$ with a confidence level approximating the desired one. (One of the two limits (6.38) may be used as a single confidence bound with the two-sided unconfidence level halved.)

Example 8
(*continued*)

Armstrong collected spring lifetime data at stress levels besides the 900 and 950 N/mm^2 levels used thus far in this example. Ten springs tested at 850 N/mm^2 had lifetimes with $\bar{x} = 348.1$ and $s = 57.9$ (both in 10^3 cycles) and a reasonably linear normal plot. But taking the 850, 900, and 950 N/mm^2 data together, there is a clear trend to smaller and more consistent lifetimes as stress is increased. In light of this fact, should mean lifetimes at the 850 and 950 N/mm^2 stress levels be compared, use of a constant variance assumption seems questionable.

Example 8
(*continued*)

Consider then what the Satterthwaite method (6.38) gives for two-sided approximate 95% confidence limits for the difference in 850 and 950 N/mm^2 mean lifetimes. Equation (6.37) gives

$$\hat{v} = \frac{\left(\dfrac{(57.9)^2}{10} + \dfrac{(33.1)^2}{10} \right)^2}{\dfrac{(57.9)^4}{9(100)} + \dfrac{(33.1)^4}{9(100)}} = 14.3$$

and (rounding "degrees of freedom" down) the .975 quantile of the t_{14} distribution is 2.145. So the 95% limits (6.38) for the (850 N/mm^2 minus 950 N/mm^2) difference in mean lifetimes ($\mu_{850} - \mu_{950}$) are

$$348.1 - 168.3 \pm 2.145 \sqrt{\frac{(57.9)^2}{10} + \frac{(33.1)^2}{10}}$$

i.e.,

$$179.8 \pm 45.2$$

i.e.,

$$134.6 \times 10^3 \text{ cycles} \quad \text{and} \quad 225.0 \times 10^3 \text{ cycles}$$

The inference methods represented by displays (6.35), (6.36), and (6.38) are the last of the standard one- and two-sample methods for means. In the next two sections, parallel methods for variances and proportions are considered. But before leaving this section to consider those methods, a final comment is appropriate about the small-sample methods.

This discussion has emphasized that, strictly speaking, the nominal properties (in terms of coverage probabilities for confidence intervals and relevant p-value declarations for significance tests) of the small-sample methods depend on the appropriateness of exactly normal underlying distributions and (in the cases of the methods (6.35) and (6.36)) exactly equal variances. On the other hand, when actually applying the methods, rather crude probability-plotting checks have been used for verifying (only) that the models are roughly plausible. According to conventional statistical wisdom, the small-sample methods presented here are remarkably robust to all but gross departures from the model assumptions. That is, as long as the model assumptions are at least roughly a description of reality, the nominal confidence levels and p-values will not be ridiculously incorrect. (For example, a nominally 90% confidence interval method might in reality be only an 80% method, but it will not be only a 20% confidence interval method.) So the kind of plotting that has been illustrated here is often taken as adequate precaution against unjustified application of the small-sample inference methods for means.

Section 3 Exercises

1. What is the practical consequence of using a "normal distribution" confidence interval formula when in fact the underlying data-generating mechanism cannot be adequately described using a normal distribution? Say something more specific/informative than "an error might be made," or "the interval might not be valid." (What, for example, can be said about the real confidence level that ought to be associated with a nominally 90% confidence interval in such a situation?)

2. Consider again the situation of Exercise 3 of Section 3.1. (It concerns the torques required to loosen two particular bolts holding an assembly on a piece of machinery.)

   (a) What model assumptions are needed in order to do inference for the mean top-bolt torque here? Make a plot to investigate the necessary distributional assumption.

   (b) Assess the strength of the evidence in the data that the mean top-bolt torque differs from a target value of 100 ft lb.

   (c) Make a two-sided 98% confidence interval for the mean top-bolt torque.

   (d) What model assumptions are needed in order to compare top-bolt and bottom-bolt torques here? Make a plot for investigating the necessary distributional assumption.

   (e) Assess the strength of the evidence that there is a mean increase in required torque as one moves from the top to the bottom bolts.

   (f) Give a 98% two-sided confidence interval for the mean difference in torques between the top and bottom bolts.

3. The machine screw measurement study of DuToit, Hansen, and Osborne referred to in Exercise 4 of Section 6.1 involved measurement of diameters of each of 50 screws with both digital and vernier-scale calipers. For the student referred to in that exercise, the differences in measured diameters (digital minus vernier, with units of mm) had the following frequency distribution:

Difference	−.03	−.02	−.01	.00	.01	.02
Frequency	1	3	11	19	10	6

   (a) Make a 90% two-sided confidence interval for the mean difference in digital and vernier readings for this student.

   (b) Assess the strength of the evidence provided by these differences to the effect that there is a systematic difference in the readings produced by the two calipers (at least when employed by this student).

   (c) Briefly discuss why your answers to parts (a) and (b) of this exercise are compatible. (Discuss how the outcome of part (b) could easily have been anticipated from the outcome of part (a).)

4. B. Choi tested the stopping properties of various bike tires on various surfaces. For one thing, he tested both treaded and smooth tires on dry concrete. The lengths of skid marks produced in his study under these two conditions were as follows (in cm).

Treaded	Smooth
365, 374, 376	341, 348, 349
391, 401, 402	355, 375, 391

   (a) In order to make formal inferences about $\mu_{\text{Treaded}} - \mu_{\text{Smooth}}$ based on these data, what must you be willing to use for model assumptions? Make a plot to investigate the reasonableness of those assumptions.

   (b) Proceed under the necessary model assumptions to assess the strength of Choi's evidence of a difference in mean skid lengths.

   (c) Make a 95% two-sided confidence interval for $\mu_{\text{Treaded}} - \mu_{\text{Smooth}}$ assuming that treaded and smooth skid marks have the same variability.

   (d) Use the Satterthwaite method and make an approximate 95% two-sided confidence interval for $\mu_{\text{Treaded}} - \mu_{\text{Smooth}}$ assuming only that skid mark lengths for both types of tires are normally distributed.

6.4 One- and Two-Sample Inference for Variances

This text has repeatedly indicated that engineers must often pay close attention to the measurement, the prediction, and sometimes the physical reduction of variability associated with a system response. Accordingly, it makes sense to consider inference for a single variance and inference for comparing two variances. In doing so, two more standard families of probability distributions—the χ^2 distributions and the F distributions—will be introduced.

6.4.1 Inference for the Variance of a Normal Distribution

The key step in developing most of the formal inference methods discussed in this chapter has been to find a random quantity involving both the parameter (or function of parameters) of interest and sample-based quantities that under appropriate assumptions can be shown to have some well-known distribution. Inference methods for a single variance rely on a type of continuous probability distribution that has not yet been discussed in this book: the χ^2 distributions.

Definition 15

The χ^2 **(Chi-squared) distribution with degrees of freedom parameter**, ν, is a continuous probability distribution with probability density

$$f(x) = \begin{cases} \dfrac{1}{2^{\nu/2}\Gamma\left(\dfrac{\nu}{2}\right)} x^{(\nu/2)-1} e^{-x/2} & \text{for } x > 0 \\ 0 & \text{otherwise} \end{cases} \tag{6.39}$$

If a random variable has the probability density given by formula (6.39), it is said to have the χ^2_ν distribution.

Form (6.39) is not terribly inviting, but neither is it unmanageable. For instance, it is easy enough to use it to make the kind of plots in Figure 6.16 for comparing the shapes of the χ^2_ν distributions for various choices of ν.

The χ^2_ν distribution has mean ν and variance 2ν. For $\nu = 2$, it is exactly the exponential distribution with mean 2. For large ν, the χ^2_ν distributions look increasingly bell-shaped (and can in fact be approximated by normal distributions with matching means and variances). Rather than using form (6.39) to find χ^2 probabilities, it is more common to use tables of χ^2 quantiles. Table B.5 is one such table. Across the top of the table are several cumulative probabilities. Down the left side of the table are values of the degrees of freedom parameter, ν. In the body of the table are corresponding quantiles.

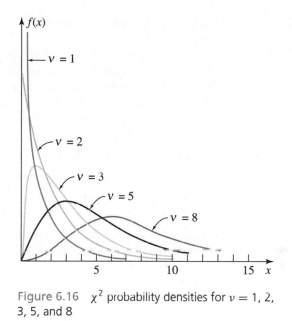

Figure 6.16 χ^2 probability densities for $\nu = 1, 2,$ 3, 5, and 8

Example 11

Using the χ^2 table, Table B.5

Use of a Table of χ^2 Distribution Quantiles

Suppose that V is a random variable with a χ_3^2 distribution. Consider first finding the .95 quantile of V's distribution and then seeing what Table B.5 says about $P[V < .4]$ and $P[V > 10.0]$.

First, looking at the $\nu = 3$ row of Table B.5 under the cumulative probability .95, one finds 7.815 in the body of the table. That is, $Q(.95) = 7.815$, or (equivalently) $P[V \leq 7.815] = .95$. Then note that again using the $\nu = 3$ line of Table B.5, .4 lies between the .05 and .10 quantiles of the χ_3^2 distribution. Thus,

$$.05 < P[V < .4] < .10$$

Finally, since 10.0 lies between the ($\nu = 3$ line) entries of the table corresponding to cumulative probabilities .975 and .99 (i.e., the .975 and .99 quantiles of the χ_3^2 distribution), one may reason that

$$.01 < P[V > 10.0] < .025$$

The χ^2 distributions are of interest here because of a probability fact concerning the behavior of the random variable s^2 if the observations from which it is calculated are iid normal random variables. Under such assumptions,

$$X^2 = \frac{(n-1)s^2}{\sigma^2} \tag{6.40}$$

has a χ^2_{n-1} distribution. This fact is what is needed to identify inference methods for σ.

That is, given a desired confidence level concerning σ, one can choose χ^2 quantiles (say, L and U) such that the probability that a χ^2_{n-1} random variable will take a value between L and U corresponds to that confidence level. (Typically, L and U are chosen to "split the 'unconfidence' between the upper and lower χ^2_{n-1} tails"—for example, using the .05 and .95 χ^2_{n-1} quantiles for L and U, respectively, if 90% confidence is of interest.) Then, because the variable (6.40) has a χ^2_{n-1} distribution, the probability that

$$L < \frac{(n-1)s^2}{\sigma^2} < U \qquad \textbf{(6.41)}$$

corresponds to the desired confidence level. But expression (6.41) is algebraically equivalent to the eventuality that

$$\frac{(n-1)s^2}{U} < \sigma^2 < \frac{(n-1)s^2}{L}$$

This then means that when an engineering data-generating mechanism can be thought of as essentially equivalent to random sampling from a normal distribution, a two-sided confidence interval for σ^2 has endpoints

Normal distribution confidence limits for σ^2

$$\boxed{\frac{(n-1)s^2}{U} \quad \text{and} \quad \frac{(n-1)s^2}{L}} \qquad \textbf{(6.42)}$$

where L and U are such that the χ^2_{n-1} probability assigned to the interval (L, U) corresponds to the desired confidence.

Further, there is an obvious significance-testing method for σ^2. That is, subject to the same modeling limitations needed to support the confidence interval method,

$$H_0 : \sigma^2 = \#$$

can be tested using the statistic

Normal distribution test statistic for σ^2

$$\boxed{X^2 = \frac{(n-1)s^2}{\#}} \qquad \textbf{(6.43)}$$

and a χ^2_{n-1} reference distribution.

One feature of the testing methodology that needs comment concerns the com-

p-values for testing $H_0 : \sigma^2 = \#$

puting of p-values in the case that the alternative hypothesis is of the form H_a: $\sigma^2 \neq \#$. (p-values for the one-sided alternative hypotheses H_a: $\sigma^2 < \#$ and H_a: $\sigma^2 > \#$ are, respectively, the left and right χ^2_{n-1} tail areas beyond the observed value of X^2.) The fact that the χ^2 distributions have no point of symmetry leaves some doubt for two-sided significance testing as to how an observed value of X^2 should be translated into a (two-sided) p-value. The convention that will be used here is as follows. If the observed value is larger than the χ^2_{n-1} median, the (two-sided) p-value will be twice the χ^2_{n-1} probability to the right of the observed value. If the observed value of X^2 is smaller than the χ^2_{n-1} median, the (two-sided) p-value will be twice the χ^2_{n-1} probability to the left of the observed value.

Confidence limits for functions of σ^2   Knowing that display (6.42) gives endpoints for a confidence interval for σ^2 also leads to confidence intervals for functions of σ^2. The square roots of the values in display (6.42) give endpoints for a confidence interval for the standard deviation, σ. And six times the square roots of the values in display (6.42) could be used as endpoints of a confidence interval for the "6σ" **capability of a process**.

Example 12

Inference for the Capability of a CNC Lathe

Cowan, Renk, Vander Leest, and Yakes worked with a manufacturer of high-precision metal parts on a project involving a computer numerically controlled (CNC) lathe. A critical dimension of one particular part produced on the lathe had engineering specifications of the form

$$\text{Nominal dimension} + .0020 \text{ in.}$$

An important practical issue in such situations is whether or not the machine is capable of meeting specifications of this type. One way of addressing this is to collect data and do inference for the intrinsic machine short-term variability, represented as a standard deviation. Table 6.8 gives values of the critical dimension measured on 20 parts machined on the lathe in question over a three-hour period. The units are .0001 in. over nominal.

Table 6.8
Measurements of a Dimension on 20 Parts
Machined on a CNC Lathe

Measured Dimension (.0001 in. over nominal)	Frequency
8	1
9	1
10	10
11	4
12	3
13	1

Example 12
(*continued*)

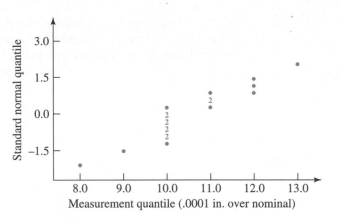

Figure 6.17   Normal plot of measurements on 20 parts machined on a CNC lathe

Suppose one takes the $\pm.0020$ in. engineering specifications as a statement of worst acceptable "$\pm3\sigma$" machine capability, accordingly uses the data in Table 6.8, and (since $\frac{.0020}{3} \approx .0007$) tests $H_0: \sigma = .0007$. The relevance of the methods represented by displays (6.42) and (6.43) depends on the appropriateness of a normal distribution as a description of the critical dimension (as machined in the three-hour period in question). In this regard, note that (after allowing for the fact of the obvious discreteness of measurement introduced by gauging read to .0001 in.) the normal plot of the data from Table 6.8 shown in Figure 6.17 is not distressing in its departure from linearity. Further, at least over periods where manufacturing processes like the one in question are physically stable, normal distributions often prove to be quite adequate models for measured dimensions of mass-produced parts. Other evidence available on the machining process indicated that for practical purposes, the machining process was stable over the three-hour period in question. So one may proceed to use the normal-based methods, with no strong reason to doubt their relevance.

Direct calculation with the data of Table 6.8 shows that $s = 1.1 \times 10^{-4}$ in. So, using the five-step significance-testing format produces the following:

1. $H_0: \sigma = .0007$.

2. $H_a: \sigma > .0007$.
   (The most practical concern is the possibility that the machine is not capable of holding to the stated tolerances, and this is described in terms of σ larger than standard.)

3. The test statistic is

$$X^2 = \frac{(n-1)s^2}{(.0007)^2}$$

The reference distribution is χ^2 with $\nu = (20 - 1) = 19$ degrees of freedom, and large observed values x^2 (resulting from large values of s^2) will constitute evidence against H_0.

4. The sample gives

$$x^2 = \frac{(20 - 1)(.00011)^2}{(.0007)^2} = .5$$

5. The observed level of significance is $P[a \; \chi^2_{19} \text{ random variable} \geq .5]$. Now .5 is smaller than the .005 quantile of the χ^2_{19} distribution, so the p-value exceeds .995. There is nothing in the data in hand to indicate that the machine is incapable of holding to the given tolerances.

Consider, too, making a one-sided 99% confidence interval of the form $(0, \#)$ for 3σ. According to Table B.5, the .01 quantile of the χ^2_{19} distribution is $L = 7.633$. So using display (6.42), a 99% upper confidence bound for 3σ is

$$3\sqrt{\frac{(20 - 1)(1.1 \times 10^{-4} \text{ in.})^2}{7.633}} = 5.0 \times 10^{-4} \text{ in.}$$

When this is compared to the $\pm 20 \times 10^{-4}$ in. engineering requirement, it shows that the lathe in question is clearly capable of producing the kind of precision specified for the given dimension.

6.4.2 Inference for the Ratio of Two Variances (Based on Independent Samples from Normal Distributions)

To move from inference for a single variance to inference for comparing two variances requires the introduction of yet another new family of probability distributions: (Snedecor's) F distributions.

Definition 16

The **(Snedecor) F distribution with numerator and denominator degrees of freedom parameters ν_1 and ν_2** is a continuous probability distribution with probability density

$$f(x) = \begin{cases} \dfrac{\Gamma\left(\dfrac{\nu_1 + \nu_2}{2}\right)\left(\dfrac{\nu_1}{\nu_2}\right)^{\nu_1/2} x^{(\nu_1/2)-1}}{\Gamma\left(\dfrac{\nu_1}{2}\right)\Gamma\left(\dfrac{\nu_2}{2}\right)\left(1 + \dfrac{\nu_1 x}{\nu_2}\right)^{(\nu_1+\nu_2)/2}} & \text{for } x > 0 \\ 0 & \text{otherwise} \end{cases} \qquad \textbf{(6.44)}$$

If a random variable has the probability density given by formula (6.44), it is said to have the F_{ν_1, ν_2} distribution.

As Figure 6.18 reveals, the F distributions are strongly right-skewed distributions, whose densities achieve their maximum values at arguments somewhat less than 1. Roughly speaking, the smaller the values ν_1 and ν_2, the more asymmetric and spread out is the corresponding F distribution.

Direct use of formula (6.44) to find probabilities for the F distributions requires numerical integration methods. For purposes of applying the F distributions in statistical inference, the typical path is to instead make use of either statistical software or some fairly abbreviated tables of F distribution quantiles. Tables B.6 are tables of F quantiles. The body of a particular one of these tables, for a single p, gives the F distribution p quantiles for various combinations of ν_1 (the numerator degrees of freedom) and ν_2 (the denominator degrees of freedom). The values of ν_1 are given across the top margin of the table and the values of ν_2 down the left margin.

Using the F distribution tables, Tables B.6

Tables B.6 give only p quantiles for p larger than .5. Often F distribution quantiles for p smaller than .5 are needed as well. Rather than making up tables of such values, it is standard practice to instead make use of a computational trick. By using a relationship between F_{ν_1, ν_2} and F_{ν_2, ν_1} quantiles, quantiles for small p can be determined. If one lets Q_{ν_1, ν_2} stand for the F_{ν_1, ν_2} quantile function and Q_{ν_2, ν_1} stand for the quantile function for the F_{ν_2, ν_1} distribution,

Relationship between F_{ν_1, ν_2} and F_{ν_2, ν_1} quantiles

$$Q_{\nu_1, \nu_2}(p) = \frac{1}{Q_{\nu_2, \nu_1}(1 - p)}$$ **(6.45)**

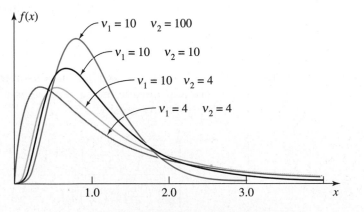

Figure 6.18 Four different F probability densities

Fact (6.45) means that a small lower percentage point of an F distribution may be obtained by taking the reciprocal of a corresponding small upper percentage point of the F distribution with degrees of freedom reversed.

Example 13

Use of Tables of F Distribution Quantiles

Suppose that V is an $F_{3,5}$ random variable. Consider finding the .95 and .01 quantiles of V's distribution and then seeing what Tables B.6 reveal about $P[V > 4.0]$ and $P[V < .3]$.

First, a direct look-up in the $p = .95$ table of quantiles, in the $v_1 = 3$ column and $v_2 = 5$ row, produces the number 5.41. That is, $Q(.95) = 5.41$, or (equivalently) $P[V < 5.41] = .95$.

To find the $p = .01$ quantile of the $F_{3,5}$ distribution, expression (6.45) must be used. That is,

$$Q_{3,5}(.01) = \frac{1}{Q_{5,3}(.99)}$$

so that using the $v_1 = 5$ column and $v_2 = 3$ row of the table of F .99 quantiles, one has

$$Q_{3,5}(.01) = \frac{1}{28.24} = .04$$

Next, considering $P[V > 4.0]$, one finds (using the $v_1 = 3$ columns and $v_2 = 5$ rows of Tables B.6) that 4.0 lies between the .90 and .95 quantiles of the $F_{3,5}$ distribution. That is,

$$.90 < P[V \leq 4.0] < .95$$

so that

$$.05 < P[V > 4.0] < .10$$

Finally, considering $P[V < .3]$, note that none of the entries in Tables B.6 is less than 1.00. So to place the value .3 in the $F_{3,5}$ distribution, one must locate its reciprocal, $3.33(= 1/.3)$, in the $F_{5,3}$ distribution and then make use of expression (6.45). Using the $v_1 = 5$ columns and $v_2 = 3$ rows of Tables B.6, one finds that 3.33 is between the .75 and .90 quantiles of the $F_{5,3}$ distribution. So by expression (6.45), .3 is between the .1 and .25 quantiles of the $F_{3,5}$ distribution, and

$$.10 < P[V < .3] < .25$$

The extra effort required to find small F distribution quantiles is an artifact of standard table-making practice, rather than being any intrinsic extra difficulty associated with the F distributions. One way to eliminate the difficulty entirely is to use standard statistical software or a statistical calculator to find F quantiles.

The F distributions are of use here because a probability fact ties the behavior of ratios of independent sample variances based on samples from normal distributions to the variances σ_1^2 and σ_2^2 of those underlying distributions. That is, when s_1^2 and s_2^2 come from independent samples from normal distributions, the variable

$$F = \frac{s_1^2}{\sigma_1^2} \cdot \frac{\sigma_2^2}{s_2^2} \qquad (6.46)$$

has an F_{n_1-1,n_2-1} distribution. (s_1^2 has $n_1 - 1$ associated degrees of freedom and is in the numerator of this expression, while s_2^2 has $n_2 - 1$ associated degrees of freedom and is in the denominator, providing motivation for the language introduced in Definition 16.)

This fact is exactly what is needed to produce formal inference methods for the ratio σ_1^2/σ_2^2. For example, it is possible to pick appropriate F quantiles L and U such that the probability that the variable (6.46) falls between L and U corresponds to a desired confidence level. (Typically, L and U are chosen to "split the 'unconfidence' " between the upper and lower F_{n_1-1,n_2-1} tails.) But

$$L < \frac{s_1^2}{\sigma_1^2} \cdot \frac{\sigma_2^2}{s_2^2} < U$$

is algebraically equivalent to

$$\frac{1}{U} \cdot \frac{s_1^2}{s_2^2} < \frac{\sigma_1^2}{\sigma_2^2} < \frac{1}{L} \cdot \frac{s_1^2}{s_2^2}$$

That is, when a data-generating mechanism can be thought of as essentially equivalent to independent random sampling from two normal distributions, a two-sided confidence interval for σ_1^2/σ_2^2 has endpoints

Normal distributions confidence limits for σ_1^2/σ_2^2

$$\frac{s_1^2}{U \cdot s_2^2} \quad \text{and} \quad \frac{s_1^2}{L \cdot s_2^2} \qquad (6.47)$$

where L and U are (F_{n_1-1,n_2-1} quantiles) such that the F_{n_1-1,n_2-1} probability assigned to the interval (L, U) corresponds to the desired confidence.

In addition, there is an obvious significance-testing method for σ_1^2/σ_2^2. That is, subject to the same modeling limitations as needed to support the confidence interval method,

$$H_0: \frac{\sigma_1^2}{\sigma_2^2} = \#\tag{6.48}$$

can be tested using the statistic

Normal distributions test statistic for σ_1^2/σ_2^2

$$F = \frac{s_1^2/s_2^2}{\#}\tag{6.49}$$

and an F_{n_1-1,n_2-1} reference distribution. (The choice of $\# = 1$ in displays (6.48) and (6.49), so that the null hypothesis is one of equality of variances, is the only one commonly used in practice.) p-values for the one-sided alternative hypotheses $H_a: \sigma_1^2/\sigma_2^2 < \#$ and $H_a: \sigma_1^2/\sigma_2^2 > \#$ are (respectively) the left and right F_{n_1-1,n_2-1} tail areas beyond the observed values of the test statistic. For the two-sided alternative hypothesis $H_a: \sigma_1^2/\sigma_2^2 \neq \#$, the standard convention is to report twice the F_{n_1-1,n_2-1} probability to the right of the observed f if $f > 1$ and to report twice the F_{n_1-1,n_2-1} probability to the left of the observed f if $f < 1$.

p-values for testing

$H_0: \frac{\sigma_1^2}{\sigma_2^2} = \#$

Example 14

Comparing Uniformity of Hardness Test Results for Two Types of Steel

Condon, Smith, and Woodford did some hardness testing on specimens of 4% carbon steel. Part of their data are given in Table 6.9, where Rockwell hardness measurements for ten specimens from a lot of heat-treated steel specimens and five specimens from a lot of cold-rolled steel specimens are represented.

Consider comparing measured hardness *uniformity* for these two steel types (rather than mean hardness, as might have been done in Section 6.3). Figure 6.19 shows side-by-side dot diagrams for the two samples and suggests that there is a larger variability associated with the heat-treated specimens than with the cold-rolled specimens. The two normal plots in Figure 6.20 indicate no obvious problems with a model assumption of normal underlying distributions.

Table 6.9
Rockwell Hardness Measurements for Steel Specimens
of Two Types

Heat-Treated	Cold-Rolled
32.8, 44.9, 34.4, 37.0, 23.6, 29.1, 39.5, 30.1, 29.2, 19.2	21.0, 24.5, 19.9, 14.8, 18.8

Example 14
(continued)

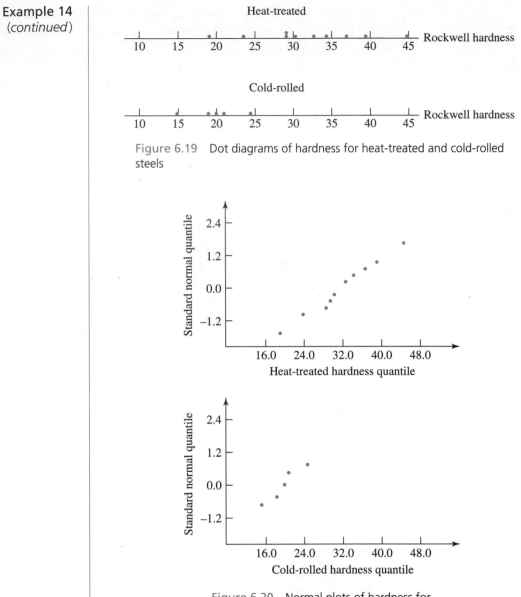

Figure 6.19 Dot diagrams of hardness for heat-treated and cold-rolled steels

Figure 6.20 Normal plots of hardness for heat-treated and cold-rolled steels

Then, arbitrarily choosing to call the heat-treated condition number 1 and the cold-rolled condition 2, $s_1 = 7.52$ and $s_2 = 3.52$, and a five-step significance test of equality of variances based on the variable (6.49) proceeds as follows:

1. $H_0: \dfrac{\sigma_1^2}{\sigma_2^2} = 1.$

2. $H_a: \dfrac{\sigma_1^2}{\sigma_2^2} \neq 1.$

   (If there is any materials-related reason to pick a one-sided alternative hypothesis here, the authors don't know it.)

3. The test statistic is

$$F = \frac{s_1^2}{s_2^2}$$

   The reference distribution is the $F_{9,4}$ distribution, and both large observed f and small observed f will constitute evidence against H_0.

4. The samples give

$$f = \frac{(7.52)^2}{(3.52)^2} = 4.6$$

5. Since the observed f is larger than 1, for the two-sided alternative, the p-value is

$$2P[\text{an } F_{9,4} \text{ random variable} \geq 4.6]$$

   From Tables B.6, 4.6 is between the $F_{9,4}$ distribution .9 and .95 quantiles, so the observed level of significance is between .1 and .2. This makes it moderately (but not completely) implausible that the heat-treated and cold-rolled variabilities are the same.

In an effort to pin down the relative sizes of the heat-treated and cold-rolled hardness variabilities, the square roots of the expressions in display (6.47) may be used to give a 90% two-sided confidence interval for σ_1/σ_2. Now the .95 quantile of the $F_{9,4}$ distribution is 6.0, while the .95 quantile of the $F_{4,9}$ distribution is 3.63, implying that the .05 quantile of the $F_{9,4}$ distribution is $\frac{1}{3.63}$. Thus, a 90% confidence interval for the ratio of standard deviations σ_1/σ_2 has endpoints

$$\sqrt{\frac{(7.52)^2}{6.0(3.52)^2}} \quad \text{and} \quad \sqrt{\frac{(7.52)^2}{(1/3.63)(3.52)^2}}$$

That is,

$$.87 \quad \text{and} \quad 4.07$$

Example 14
(continued)

The fact that the interval (.87, 4.07) covers values both smaller and larger than 1 indicates that the data in hand do not provide definitive evidence even as to which of the two variabilities in material hardness is larger.

One of the most important engineering applications of the inference methods represented by displays (6.47) through (6.49) is in the comparison of inherent precisions for different pieces of equipment and for different methods of operating a single piece of equipment.

Example 15

Comparing Uniformities of Operation of Two Ream Cutters

Abassi, Afinson, Shezad, and Yeo worked with a company that cuts rolls of paper into sheets. The uniformity of the sheet lengths is important, because the better the uniformity, the closer the average sheet length can be set to the nominal value without producing undersized sheets, thereby reducing the company's giveaway costs. The students compared the uniformity of sheets cut on a ream cutter having a manual brake to the uniformity of sheets cut on a ream cutter that had an automatic brake. The basis of that comparison was estimated standard deviations of sheet lengths cut by the two machines—just the kind of information used to frame formal inferences in this section. The students estimated $\sigma_{\text{manual}}/\sigma_{\text{automatic}}$ to be on the order of 1.5 and predicted a period of two years or less for the recovery of the capital improvement cost of equipping all the company's ream cutters with automatic brakes.

The methods of this section are, strictly speaking, normal distribution methods. It is worthwhile to ask, "How essential is this normal distribution restriction to the predictable behavior of these inference methods for one and two variances?" There is a remark at the end of Section 6.3 to the effect that the methods presented there for *means* are fairly robust to moderate violation of the section's model assumptions. Unfortunately, such is *not* the case for the methods for *variances* presented here.

Caveats about inferences for variances

These are methods whose nominal confidence levels and *p*-values can be fairly badly misleading unless the normal models are good ones. This makes the kind of careful data scrutiny that has been implemented in the examples (in the form of normal-plotting) essential to the responsible use of the methods of this section. And it suggests that since normal-plotting itself isn't typically terribly revealing unless the sample size involved is moderate to large, formal inferences for variances will be most safely made on the basis of moderate to large normal-looking samples.

The importance of the "normal distribution(s)" restriction to the predictable operation of the methods of this section is not the only reason to prefer large sample sizes for inferences on variances. A little experience with the formulas in this section will convince the reader that (even granting the appropriateness of normal models) small samples often do not prove adequate to answer practical questions about variances. χ^2 and F confidence intervals for variances and variance ratios based on small samples can be so big as to be of little practical value, and the engineer will typically be driven to large sample sizes in order to solve variance-related real-world problems. This is not in any way a failing of the present methods. It is simply a warning and quantification of the fact that learning about variances requires more data than (for example) learning about means.

Section 4 Exercises

1. Return to data on Choi's bicycle stopping distance given in Exercise 4 of Section 6.3.
   (a) Operating under the assumption that treaded tires produce normally distributed stopping distances, give a two-sided 95% confidence interval for the standard deviation of treaded tire stopping distances.
   (b) Operating under the assumption that smooth tires produce normally distributed stopping distances, give a 99% upper confidence bound for the standard deviation of smooth tire stopping distances.
   (c) Operating under the assumption that both treaded and smooth tires produce normally distributed stopping distances, assess the strength of Choi's evidence that treaded and smooth stopping distances differ in their variability. (Use $H_0: \sigma_{\text{Treaded}} = \sigma_{\text{Smooth}}$ and $H_a: \sigma_{\text{Treaded}} \neq \sigma_{\text{Smooth}}$ and show the whole five-step format.)
   (d) Operating under the assumption that both treaded and smooth tires produce normally distributed stopping distances, give a 90% two-sided confidence interval for the ratio $\sigma_{\text{Treaded}}/\sigma_{\text{Smooth}}$.

2. Consider again the situation of Exercise 3 of Section 3.1 and Exercise 2 of Section 6.3. (It concerns the torques required to loosen two particular bolts holding an assembly on a piece of machinery.)
   (a) Operating under the assumption that top-bolt torques are normally distributed, give a 95% lower confidence bound for the standard deviation of the top-bolt torques.
   (b) Translate your answer to part (a) into a 95% lower confidence bound on the "6σ process capability" of the top-bolt tightening process.
   (c) It is not appropriate to use the methods (6.47) through (6.49) and the data given in Exercise 3 of Section 3.1 to compare the consistency of top-bolt and bottom-bolt torques. Why?

6.5 One- and Two-Sample Inference for Proportions

The methods of formal statistical inference in the previous four sections are useful in the analysis of quantitative data. Occasionally, however, engineering studies produce only qualitative data, and one is faced with the problem of making properly hedged inferences from such data. This section considers how the sample fraction $\hat{p}$ (defined in Section 3.4) can be used as the basis for formal statistical inferences. It begins with the use of $\hat{p}$ from a single sample to make formal inferences about a single system or population. The section then treats the use of sample proportions from two samples to make inferences comparing two systems or populations.

6.5.1 Inference for a Single Proportion

Recall from display (3.6) (page 104) that the notation $\hat{p}$ is used for the fraction of a sample that possesses a characteristic of engineering interest. A sample of pellets produced by a pelletizing machine might prove individually conforming or nonconforming, and $\hat{p}$ could be the sample fraction conforming. Or in another case, a sample of turned steel shafts might individually prove acceptable, reworkable, or scrap; $\hat{p}$ could be the sample fraction reworkable.

If formal statistical inferences are to be based on $\hat{p}$, one must think of the physical situation in such a way that $\hat{p}$ is related to some parameter characterizing it. Accordingly, this section considers scenarios where $\hat{p}$ is derived from an *independent identical success/failure trials* data-generating mechanism. (See again Section 5.1.4 to review this terminology.) Applications will include inferences about physically stable processes, where p is a system's propensity to produce an item with the characteristic of interest. And they will include inferences drawn about population proportions p in enumerative contexts involving large populations. For example, the methods of this section can be used both to make inferences about the routine operation of a physically stable pelletizing machine and also to make inferences about the fraction of nonconforming machine parts contained in a specific lot of 10,000 such parts.

Review of the material on independent success/failure trials (and particularly the binomial distributions) in Section 5.1.4 should convince the reader that

$$X = n\hat{p} = \text{the number of items in the sample with the characteristic of interest}$$

has the binomial (n, p) distribution. The sample fraction $\hat{p}$ is just a scale change away from $X = n\hat{p}$, so facts about the distribution of X have immediate counterparts regarding the distribution of $\hat{p}$. For example, Section 5.1.4 stated that the mean and variance for the binomial (n, p) distribution are (respectively) np and $np(1 - p)$. This (together with Proposition 1 in Chapter 5) implies that $\hat{p}$ has

Mean of the sample proportion

$$E\hat{p} = E\left(\frac{X}{n}\right) = \frac{1}{n}EX = \frac{1}{n} \cdot np = p \qquad (6.50)$$

and

Variance of the sample proportion

$$\text{Var}\,\hat{p} = \text{Var}\left(\frac{X}{n}\right) = \left(\frac{1}{n}\right)^2 \text{Var}\,X = \frac{np(1 - p)}{n^2} = \frac{p(1 - p)}{n} \qquad (6.51)$$

Equations (6.50) and (6.51) provide a reassuring picture of the behavior of the statistic $\hat{p}$. They show that the probability distribution of $\hat{p}$ is centered at the underlying parameter p, with a variability that decreases as n increases.

Example 16
(Example 3, Chapter 5, revisited—page 234)

Means and Standard Deviations of Sample Fractions of Reworkable Shafts

Return again to the case of the performance of a process for turning steel shafts. Assume for the time being that the process is physically stable and that the likelihood that a given shaft is reworkable is $p = .20$. Consider $\hat{p}$, the sample fraction of reworkable shafts in samples of first $n = 4$ and then $n = 100$ shafts.

Expressions (6.50) and (6.51) show that for the $n = 4$ sample size,

$$E\hat{p} = p = .2$$

$$\sqrt{\text{Var }\hat{p}} = \sqrt{\frac{p(1-p)}{n}} = \sqrt{\frac{(.2)(.8)}{4}} = .2$$

Similarly, for the $n = 100$ sample size,

$$E\hat{p} = p = .2$$

$$\sqrt{\text{Var }\hat{p}} = \sqrt{\frac{(.2)(.8)}{100}} = .04$$

Comparing the two standard deviations, it is clear that the effect of a change in sample size from $n = 4$ to $n = 100$ is to produce a factor of 5 ($= \sqrt{100/4}$) decrease in the standard deviation of $\hat{p}$, while the distribution of $\hat{p}$ is centered at p for both sample sizes.

Approximate normality of the sample proportion

The basic new insight needed to provide large-sample inference methods based on $\hat{p}$ is the fact that for large n, the binomial (n, p) distribution (and therefore also the distribution of $\hat{p}$) is approximately normal. That is, for large n, approximate probabilities for $X = n\hat{p}$ (or $\hat{p}$) can be found using the normal distribution with mean $\mu = np$ (or $\mu = p$) and variance $\sigma^2 = np(1 - p)$ (or $\sigma^2 = \frac{p(1-p)}{n}$).

Example 16
(continued)

In the shaft-turning example, consider the probability that for a sample of $n = 100$ shafts, $\hat{p} \geq .25$. Notice that $\hat{p} \geq .25$ is equivalent here to the eventuality that $n\hat{p} \geq 25$. So in theory the form of the binomial probability function given in Definition 9 of Chapter 5 could be used and the desired probability could be evaluated exactly as

$$P[\hat{p} \geq .25] = P[X \geq 25] = f(25) + f(26) + \cdots + f(99) + f(100)$$

But instead of making such laborious calculations, it is common (and typically adequate for practical purposes) to settle instead for a normal approximation to probabilities such as this one.

Example 16
(*continued*)

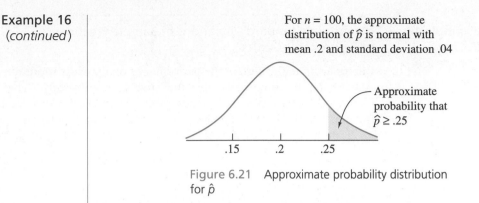

For $n = 100$, the approximate distribution of $\hat{p}$ is normal with mean .2 and standard deviation .04

Approximate probability that $\hat{p} \geq .25$

.15   .2   .25

Figure 6.21 Approximate probability distribution for $\hat{p}$

Figure 6.21 shows the normal distribution with mean $\mu = p = .2$ and standard deviation $\sigma = \sqrt{p(1-p)/n} = .04$ and the corresponding probability assigned to the interval $[.25, \infty)$. Conversion of .25 to a z-value and then an approximate probability proceeds as follows:

$$z = \frac{.25 - E\,\hat{p}}{\sqrt{\text{Var}\,\hat{p}}} = \frac{.25 - .2}{.04} = 1.25$$

so

$$P[\hat{p} \geq .25] \approx 1 - \Phi(1.25) = .1056 \approx .11$$

The exact value of $P[\hat{p} \geq .25]$ (calculated to four decimal places using the binomial probability function) is .1314. (This can, for example, be obtained using the MINITAB routine under the "Calc/Probability Distributions/Binomial" menu.)

The statement that for large n, the random variable $\hat{p}$ is approximately normal is actually a version of the central limit theorem. For a given n, the approximation is best for moderate p (i.e., p near .5), and a common rule of thumb is to require that both the expected number of successes and the expected number of failures be at least 5 before making use of a normal approximation to the binomial (n, p) distribution. This is a requirement that

$$np \geq 5 \quad \text{and} \quad n(1-p) \geq 5$$

which amounts to a requirement that

Conditions for the normal approximation to the binomial

$$\boxed{5 \leq np \leq n - 5} \tag{6.52}$$

(Notice that in Example 16, $np = 100(.2) = 20$ and $5 \leq 20 \leq 95$.)

An alternative, and typically somewhat stricter rule of thumb (which comes from a requirement that the mean of the binomial distribution be at least 3 standard deviations from both 0 and n) is to require that

Another set of conditions for the normal approximation to the binomial

$$9 \le (n+9)p \le n \tag{6.53}$$

before using the normal approximation. (Again in Example 16, $(n+9)p = (100 + 9)(.2) = 21.8$ and $9 \le 21.8 \le 100$.)

The approximate normality of $\hat{p}$ for large n implies that for large n,

$$Z = \frac{\hat{p} - p}{\sqrt{\dfrac{p(1-p)}{n}}} \tag{6.54}$$

is approximately standard normal. This and the reasoning of Section 6.2 then imply that the null hypothesis

$$\mathrm{H_0}\colon p = \#$$

can be tested using the statistic

Large-sample test statistic for p

$$Z = \frac{\hat{p} - \#}{\sqrt{\dfrac{\#(1 - \#)}{n}}} \tag{6.55}$$

and a standard normal reference distribution. Further, reasoning parallel to that in Section 6.1 (beginning with the fact that the variable (6.54) is approximately standard normal), leads to the conclusion that an interval with endpoints

$$\hat{p} \pm z\sqrt{\frac{p(1-p)}{n}} \tag{6.56}$$

(where z is chosen such that the standard normal probability between $-z$ and z corresponds to a desired confidence) is a mathematically valid two-sided confidence interval for p.

However, the endpoints indicated by expression (6.54) are of no practical use as they stand, since they involve the unknown parameter p. There are two standard ways of remedying this situation. One draws its motivation from the simple plot of $p(1-p)$ shown in Figure 6.22. That is, from Figure 6.22 it is easy to see that $p(1-p) \le (.5)^2 = .25$, so the plus-or-minus part of formula (6.56) has (for $z > 0$)

$$z\sqrt{\frac{p(1-p)}{n}} \le z\frac{1}{2\sqrt{n}}$$

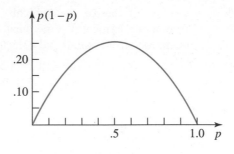

Figure 6.22 Plot of $p(1-p)$ versus p

Thus, modifying the endpoints in formula (6.56) by replacing the plus-or-minus part with $\pm z/2\sqrt{n}$ produces an interval that is guaranteed to be as wide as necessary to give the desired approximate confidence level. That is, the interval with endpoints

Large-sample conservative confidence limits for p

$$\hat{p} \pm z\frac{1}{2\sqrt{n}} \qquad \textbf{(6.57)}$$

where z is chosen such that the standard normal probability between $-z$ and z corresponds to a desired confidence, is a practically usable large-n, two-sided, *conservative* confidence interval for p. (Appropriate use of only one of the endpoints in display (6.57) gives a one-sided confidence interval.)

The other common method of dealing with the fact that the endpoints in formula (6.56) are of no practical use is to begin the search for a formula from a point other than the approximate standard normal distribution of the variable (6.54). For large n, not only is the variable (6.54) approximately standard normal, but so is

$$Z = \frac{\hat{p} - p}{\sqrt{\dfrac{\hat{p}(1 - \hat{p})}{n}}} \qquad \textbf{(6.58)}$$

And the denominator of the quantity (6.58) (which amounts to an estimated standard deviation for $\hat{p}$) is free of the parameter p. So when manipulations parallel to those in Section 6.1 are applied to expression (6.58), the conclusion is that the interval with endpoints

Large-sample confidence limits for p

$$\hat{p} \pm z\sqrt{\frac{\hat{p}(1 - \hat{p})}{n}} \qquad \textbf{(6.59)}$$

can be used as a two-sided, large-n confidence interval for p with confidence level corresponding to the standard normal probability assigned to the interval between $-z$ and z. (One-sided confidence limits are obtained in the usual way, using only one of the endpoints in display (6.59) and appropriately adjusting the confidence level.)

Example 17   Inference for the Fraction of Dry Cells with Internal Shorts

The article "A Case Study of the Use of an Experimental Design in Preventing Shorts in Nickel-Cadmium Cells" by Ophir, El-Gad, and Snyder (*Journal of Quality Technology*, 1988) describes a series of experiments conducted to find how to reduce the proportion of cells scrapped by a battery plant because of internal shorts. At the beginning of the study, about 6% of the cells produced were being scrapped because of internal shorts.

Among a sample of 235 cells made under a particular trial set of plant operating conditions, 9 cells had shorts. Consider what formal inferences can be drawn about the set of operating conditions based on such data. $\hat{p} = \frac{9}{235} = .038$, so two-sided 95% confidence limits for p, are by expression (6.59)

$$.038 \pm 1.96\sqrt{\frac{(.038)(1 - .038)}{235}}$$

i.e.,

$$.038 \pm .025$$

i.e.,

$$.013 \quad \text{and} \quad .063 \tag{6.60}$$

Notice that according to display (6.60), although $\hat{p} = .038 < .06$ (and thus indicates that the trial conditions were an improvement over the standard ones), the case for this is not airtight. The data in hand allow some possibility that p for the trial conditions even exceeds .06. And the ambiguity is further emphasized if the conservative formula (6.57) is used in place of expression (6.59). Instead of 95% confidence endpoints of $.038 \pm .025$, formula (6.57) gives endpoints $.038 \pm .064$.

To illustrate the significance-testing method represented by expression (6.55), consider testing with an alternative hypothesis that the trial plant conditions are an improvement over the standard ones. One then has the following summary:

1. $H_0: p = .06$.

2. $H_a: p < .06$.

3. The test statistic is

$$Z = \frac{\hat{p} - .06}{\sqrt{\frac{(.06)(1 - .06)}{n}}}$$

The reference distribution is standard normal, and small observed values z will count as evidence against H_0.

Example 17
(continued)

4. The sample gives

$$z = \frac{.038 - .06}{\sqrt{\dfrac{(.06)(1 - .06)}{235}}} = -1.42$$

5. The observed level of significance is then

$$\Phi(-1.42) = .08$$

This is strong but not overwhelming evidence that the trial plant conditions are an improvement on the standard ones.

It needs to be emphasized again that these inferences depend for their practical relevance on the appropriateness of the "stable process/independent, identical trials" model for the battery-making process and extend only as far as that description continues to make sense. It is important that the experience reported in the article was gained under (presumably physically stable) regular production, so there is reason to hope that a single "independent, identical trials" model can describe both experimental and future process behavior.

Sample size determination for estimating p

Section 6.1 illustrated the fact that the form of the large-n confidence interval for a mean can be used to guide sample-size choices for estimating μ. The same is true regarding the estimation of p. If one (1) has in mind a desired confidence level, (2) plans to use expression (6.57) or has in mind a worst-case (largest) expectation for $\hat{p}(1 - \hat{p})$ in expression (6.59), and (3) has a desired precision of estimation of p, it is a simple matter to solve for a corresponding sample size. That is, suppose that the desired confidence level dictates the use of the value z in formula (6.57) and one wants to have confidence limits (or a limit) of the form $\hat{p} \pm \Delta$. Setting

$$\Delta = z \frac{1}{2\sqrt{n}}$$

and solving for n produces the requirement

$$n = \left(\frac{z}{2\Delta}\right)^2$$

Example 17
(continued)

Return to the nicad battery case and suppose that for some reason a better fix on the implications of the new operating conditions was desired. In fact, suppose that p is to be estimated with a two-sided conservative 95% confidence interval, and $\pm.01$ (fraction defective) precision of estimation is desired. Then, using the plus-or-minus part of expression (6.57) (or equivalently, the plus-or-minus part of expression (6.59) under the worst-case scenario that $\hat{p} = .5$), one is led to set

$$.01 = 1.96\frac{1}{2\sqrt{n}}$$

From this, a sample size of

$$n \approx 9,604$$

is required.

In most engineering contexts this sample size is impractically large. Rethinking the calculation by planning the use of expression (6.59) and adopting the point of view that, say, 10% is a worst-case expectation for $\hat{p}$ (and thus $.1(1 - .1) = .09$ is a worst-case expectation for $\hat{p}(1 - \hat{p})$), one might be led instead to set

$$.01 = 1.96\sqrt{\frac{(.1)(1 - .1)}{n}}$$

However, solving for n, one has

$$n \approx 3,458$$

which is still beyond what is typically practical.

The moral of these calculations is that something has to give. The kind of large confidence and somewhat precise estimation requirements set at the beginning here cannot typically be simultaneously satisfied using a realistic sample size. One or the other of the requirements must be relaxed.

Cautions concerning inference based on sample proportions   The sample-size conclusions just illustrated are typical, and they justify two important points about the use of qualitative data. First, qualitative data carry less information than corresponding numbers of quantitative data (and therefore usually require very large samples to produce definitive inferences). This makes measurements generally preferable to qualitative observations in engineering applications. Second, if inferences about p based on even large values of n are often disappointing in their precision or reliability, there is little practical motivation to consider small-sample inference for p in a beginning text like this.

6.5.2 Inference for the Difference Between Two Proportions (Based on Independent Samples)

Two separately derived sample proportions $\hat{p}_1$ and $\hat{p}_2$, representing different processes or populations, can enable formal comparison of those processes or populations. The logic behind those methods of inference concerns the difference $\hat{p}_1 - \hat{p}_2$. If

1. the "independent, identical success-failure trials" description applies separately to the mechanisms that generate two samples,

2. the two samples are reasonably described as independent, and

3. both n_1 and n_2 are large,

a very simple approximate description of the distribution of $\hat{p}_1 - \hat{p}_2$ results.

Assuming $\hat{p}_1$ and $\hat{p}_2$ are independent, Proposition 1 in Chapter 5 and the discussion in this section concerning the mean and variance of a single sample proportion imply that $\hat{p}_1 - \hat{p}_2$ has

Mean of a difference in sample proportions

$$E(\hat{p}_1 - \hat{p}_2) = E\hat{p}_1 + (-1)E\hat{p}_2 = p_1 - p_2 \qquad \text{(6.61)}$$

and

Variance of a difference in sample proportions

$$\text{Var}(\hat{p}_1 - \hat{p}_2) = (1)^2 \, \text{Var} \, \hat{p}_1 + (-1)^2 \, \text{Var} \, \hat{p}_2 = \frac{p_1(1 - p_1)}{n_1} + \frac{p_2(1 - p_2)}{n_2} \qquad \text{(6.62)}$$

Approximate normality of $\hat{p}_1 - \hat{p}_2$

Then the approximate normality of $\hat{p}_1$ and $\hat{p}_2$ for large sample sizes turns out to imply the approximate normality of the difference $\hat{p}_1 - \hat{p}_2$.

Example 16
(continued)

Consider again the turning of steel shafts, and imagine that two different, physically stable lathes produce reworkable shafts at respective rates of 20 and 25%. Then suppose that samples of (respectively) $n_1 = 50$ and $n_2 = 50$ shafts produced by the machines are taken, and the reworkable sample fractions $\hat{p}_1$ and $\hat{p}_2$ are found. Consider approximating the probability that $\hat{p}_1 \geq \hat{p}_2$ (i.e., that $\hat{p}_1 - \hat{p}_2 \geq 0$).

Using expressions (6.61) and (6.62), the variable $\hat{p}_1 - \hat{p}_2$ has

$$E(\hat{p}_1 - \hat{p}_2) = .20 - .25 = -.05$$

and

$$\sqrt{\text{Var}(\hat{p}_1 - \hat{p}_2)} = \sqrt{\frac{(.20)(1 - .20)}{50} + \frac{(.25)(1 - .25)}{50}} = \sqrt{.00695} = .083$$

Figure 6.23 shows the approximately normal distribution of $\hat{p}_1 - \hat{p}_2$ and the area corresponding to $P[\hat{p}_1 - \hat{p}_2 \geq 0]$. The z-value corresponding to $\hat{p}_1 - \hat{p}_2 = 0$ is

$$z = \frac{0 - E(\hat{p}_1 - \hat{p}_2)}{\sqrt{\text{Var}(\hat{p}_1 - \hat{p}_2)}} = \frac{0 - (-.05)}{.083} = .60$$

so that

$$P[\hat{p}_1 - \hat{p}_2 \geq 0] = 1 - \Phi(.60) = .27$$

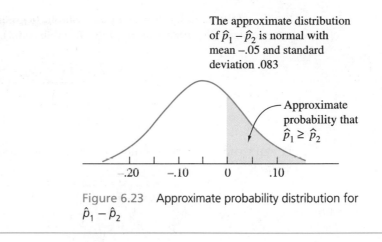

Figure 6.23   Approximate probability distribution for $\hat{p}_1 - \hat{p}_2$

The large-sample approximate normality of $\hat{p}_1 - \hat{p}_2$ translates to the realization that

$$Z = \frac{\hat{p}_1 - \hat{p}_2 - (p_1 - p_2)}{\sqrt{\dfrac{p_1(1 - p_1)}{n_1} + \dfrac{p_2(1 - p_2)}{n_2}}} \tag{6.63}$$

is approximately standard normal, and this observation forms the basis for inference concerning $p_1 - p_2$. First consider confidence interval estimation for $p_1 - p_2$. The familiar argument of Section 6.1 (beginning with the quantity (6.63)) shows

$$\hat{p}_1 - \hat{p}_2 + z \sqrt{\frac{p_1(1 - p_1)}{n_1} + \frac{p_2(1 - p_2)}{n_2}} \tag{6.64}$$

to be a mathematically correct but practically unusable formula for endpoints of a confidence interval for $p_1 - p_2$. Conservative modification of expression (6.64), via replacement of both $p_1(1 - p_1)$ and $p_2(1 - p_2)$ with .25, shows that the two-sided interval with endpoints

Large-sample conservative confidence limits for $p_1 - p_2$

$$\hat{p}_1 - \hat{p}_2 \pm z \cdot \frac{1}{2}\sqrt{\frac{1}{n_1} + \frac{1}{n_2}} \tag{6.65}$$

is a large-sample, two-sided, conservative confidence interval for $p_1 - p_2$ with confidence at least that corresponding to the standard normal probability between $-z$ and z. (One-sided intervals are obtained from expression (6.65) in the usual way.)

In addition, in by now familiar fashion, beginning with the fact that for large sample sizes, the modification of the variable (6.63),

$$Z = \frac{\hat{p}_1 - \hat{p}_2 - (p_1 - p_2)}{\sqrt{\dfrac{\hat{p}_1(1 - \hat{p}_1)}{n_1} + \dfrac{\hat{p}_2(1 - \hat{p}_2)}{n_2}}} \tag{6.66}$$

is approximately standard normal leads to the conclusion that the interval with endpoints

Large-sample confidence limits for $p_1 - p_2$

$$\hat{p}_1 - \hat{p}_2 \pm z\sqrt{\frac{\hat{p}_1(1 - \hat{p}_1)}{n_1} + \frac{\hat{p}_2(1 - \hat{p}_2)}{n_2}} \tag{6.67}$$

is a large-sample, two-sided confidence interval for $p_1 - p_2$ with confidence corresponding to the standard normal probability assigned to the interval between $-z$ and z. (Again, use of only one of the endpoints in display (6.67) gives a one-sided confidence interval.)

Example 18
(Example 14, Chapter 3, revisited—page 111)

Comparing Fractions Conforming for Two Methods of Operating a Pelletizing Process

Greiner, Grim, Larson, and Lukomski studied a number of different methods of running a pelletizing process. Two of these involved a mix with 20% reground powder with respectively small (condition 1) and large (condition 2) shot sizes. Of $n_1 = n_2 = 100$ pellets produced under these two sets of conditions, sample fractions $\hat{p}_1 = .38$ and $\hat{p}_2 = .29$ of the pellets conformed to specifications. Consider making a 90% confidence interval for comparing the two methods of process operation (i.e., an interval for $p_1 - p_2$).

Use of expression (6.67) shows that the interval with endpoints

$$.38 - .29 \pm 1.645\sqrt{\frac{(.38)(1 - .38)}{100} + \frac{(.29)(1 - .29)}{100}}$$

i.e.,

$$.09 \pm .109$$

i.e.,

$$-.019 \quad \text{and} \quad .199 \tag{6.68}$$

is a 90% confidence interval for $p_1 - p_2$, the difference in long-run fractions of conforming pellets that would be produced under the two sets of conditions. Notice that although appearances are that condition 1 has the higher associated likelihood of producing a conforming pellet, the case for this made by the data in hand is not airtight. The interval (6.68) allows some possibility that $p_1 - p_2 < 0$—i.e., that p_2 actually exceeds p_1. (The conservative interval indicated by expression (6.65) has endpoints of the form $.09 \pm .116$ and thus tells a similar story.)

The usual significance-testing method for $p_1 - p_2$ concerns the null hypothesis

$$H_0: p_1 - p_2 = 0 \qquad (6.69)$$

i.e., the hypothesis that the parameters p_1 and p_2 are equal. Notice that if $p_1 - p_2$ and the common value is denoted as p, expression (6.63) can be rewritten as

$$Z = \frac{\hat{p}_1 - \hat{p}_2}{\sqrt{p(1-p)}\sqrt{\dfrac{1}{n_1} + \dfrac{1}{n_2}}} \qquad (6.70)$$

The variable (6.70) cannot serve as a test statistic for the null hypothesis (6.69), since it involves the unknown hypothesized common value of p_1 and p_2. What is done to modify the variable (6.70) to arrive at a usable test statistic, is to replace p with a sample-based estimate, obtained by pooling together the two samples. That is, let

Pooled estimator of a common p

$$\hat{p} = \frac{n_1 \hat{p}_1 + n_2 \hat{p}_2}{n_1 + n_2} \qquad (6.71)$$

($\hat{p}$ is the total number of items in the two samples with the characteristic of interest divided by the total number of items in the two samples). Then a significance test of hypothesis (6.69) can be carried out using the test statistic

Large-sample test statistic for $H_0: p_1 - p_2 = 0$

$$Z = \frac{\hat{p}_1 - \hat{p}_2}{\sqrt{\hat{p}(1-\hat{p})}\sqrt{\dfrac{1}{n_1} + \dfrac{1}{n_2}}} \qquad (6.72)$$

If $H_0: p_1 - p_2 = 0$ is true, Z in equation (6.72) is approximately standard normal, so a standard normal reference distribution is in order.

Example 18
(continued)

As further confirmation of the fact that in the pelletizing problem sample fractions of $\hat{p}_1 = .38$ and $\hat{p}_2 = .29$ based on samples of size $n_1 = n_2 = 100$ are not completely convincing evidence of a real difference in process performance for small and large shot sizes, consider testing $H_0\colon p_1 - p_2 = 0$ with $H_a\colon p_1 - p_2 \neq 0$. As a preliminary step, from expression (6.71),

$$\hat{p} = \frac{100(.38) + 100(.29)}{100 + 100} = \frac{67}{200} = .335$$

Then the five-step summary gives the following:

1. $H_0\colon p_1 - p_2 = 0$.
2. $H_a\colon p_1 - p_2 \neq 0$.
3. The test statistic is

$$Z = \frac{\hat{p}_1 - \hat{p}_2}{\sqrt{\hat{p}(1 - \hat{p})}\sqrt{\dfrac{1}{n_1} + \dfrac{1}{n_2}}}$$

The reference distribution is standard normal, and large observed values $|z|$ will constitute evidence against H_0.

4. The samples give

$$z = \frac{.38 - .29}{\sqrt{(.335)(1 - .335)}\sqrt{\dfrac{1}{100} + \dfrac{1}{100}}} = 1.35$$

5. The p-value is $P[|a \text{ standard normal variable}| \geq 1.35]$. That is, the p-value is

$$\Phi(-1.35) + \big(1 - \Phi(1.35)\big) = .18$$

The data furnish only fairly weak evidence of a real difference in long-run fractions of conforming pellets for the two shot sizes.

The kind of results seen in Example 18 may take some getting used to. Even with sample sizes as large as 100, sample fractions differing by nearly .1 are still not necessarily conclusive evidence of a difference in p_1 and p_2. But this is just another manifestation of the point that *individual qualitative observations carry disappointingly little information.*

A final reminder of the large-sample nature of the methods presented here is in order. The methods here all rely (for the agreement of nominal and actual confidence levels or the validity of their p-values) on the adequacy of normal approximations to binomial distributions. The approximations are workable provided expression (6.52) or (6.53) holds. When testing $H_0: p = \#$, it is easy to plug both n and $\#$ into expression (6.52) or (6.53) before putting great stock in normal-based p-values. But when estimating p or $p_1 - p_2$ or testing $H_0: p_1 - p_2 = 0$, no parallel check is obvious. So it is not completely clear how to screen potential applications for ones where the nominal confidence levels or p-values are possibly misleading. What is often done is to plug both n and $\hat{p}$ (or both n_1 and $\hat{p}_1$ and n_2 and $\hat{p}_2$) into expression (6.52) or (6.53) and verify that the inequalities hold before trusting nominal (normal-based) confidence levels and p-values. Since these random quantities are only approximations to the corresponding nonrandom quantities, one will occasionally be misled regarding the appropriateness of the normal approximations by such empirical checks. But they are better than automatic application, protected by no check at all.

Section 5 Exercises

1. Consider the situation of Example 14 of Chapter 3, and in particular the results for the 50% reground mixture.
   (a) Make and interpret 95% one-sided and two-sided confidence intervals for the fraction of conforming pellets that would be produced using the 50% mixture and the small shot size. (For the one-sided interval, give a lower confidence bound.) Use both methods of dealing with the fact that σ_p is not known and compare the resulting pairs of intervals.
   (b) If records show that past pelletizing performance was such that 55% of the pellets produced were conforming, does the value in Table 3.20 constitute strong evidence that the conditions of 50% reground mixture and small shot-size provide an improvement in yield? Show the five-step format.
   (c) Compare the small and large shot-size conditions using a 95% two-sided confidence interval for the difference in fractions conforming. Interpret the interval in the context of the example.
   (d) Assess the strength of the evidence given in Table 3.20 that the shot size affects the fraction of pellets conforming (when the 50% reground mixture is used).

2. In estimating a proportion p, a two-sided interval $\hat{p} \pm \Delta$ is used. Suppose that 95% confidence and $\Delta \leq .01$ are desired. About what sample size will be needed to guarantee this?

3. Specifications on the punch heights referred to in Chapter Exercise 9 of Chapter 3 were .500 in. to .505 in. In the sample of 405 punches measured by Hyde, Kuebrick, and Swanson, there were only 290 punches meeting these specifications. Suppose that the 405 punches can be thought of as a random sample of all such punches manufactured by the supplier under standard manufacturing conditions. Give an approximate 99% two-sided confidence interval for the standard fraction of nonconforming punches of this type produced by the punch supplier.

4. Consider two hypothetical machines producing a particular widget. If samples of $n_1 = 25$ and $n_2 = 25$ widgets produced by the respective machines have fractions nonconforming $\hat{p}_1 = .2$ and $\hat{p}_2 = .32$, is this strong evidence of a difference in machine nonconforming rates? What does this suggest about the kind of sample sizes typically needed in order to reach definitive conclusions based on attributes or qualitative data?

6.6 Prediction and Tolerance Intervals

Methods of confidence interval estimation and significance testing concern the problem of reasoning from sample information to statements about underlying *parameters* of the data generation, such as μ, σ, and p. These are extremely important engineering tools, but they often fail to directly address the question of real interest. Sometimes what is really needed as the ultimate product of a statistical analysis is not a statement about a parameter but rather an indication of reasonable bounds on other *individual values* generated by the process under study. For example, suppose you are about to purchase a new car. For some purposes, knowing that "the mean EPA mileage for this model is likely in the range 25 mpg $\pm$ .5 mpg" is not nearly as useful as knowing that "the EPA mileage figure for the particular car you are ordering is likely in the range 25 mpg $\pm$ 3 mpg." Both of these statements may be quite accurate, but they serve different purposes. The first statement is one about a *mean* mileage and the second is about an *individual* mileage. And it is only statements of the first type that have been directly treated thus far.

This section indicates what is possible in the way of formal statistical inferences, not for parameters but rather for individual values generated by a stable data-generating mechanism. There are two types of formal inference methods aimed in this general direction—statistical prediction interval methods and statistical tolerance interval methods—and both types will be discussed. The section begins with prediction intervals for a normal distribution. Then tolerance intervals for a normal distribution are considered. Finally, there is a discussion of how it is possible to use minimum and/or maximum values in a sample to create prediction and tolerance intervals for even nonnormal underlying distributions.

6.6.1 Prediction Intervals for a Normal Distribution

One fruitful way to phrase the question of inference for additional individual values produced by a process is the following: How might data in hand, $x_1, x_2, \ldots, x_n$, be used to create a numerical interval likely to bracket **one additional (as yet unobserved) value**, x_{n+1}, from the same data-generating mechanism? How, for example, might mileage tests on ten cars of a particular model be used to predict the results of the same test applied to an eleventh?

If the underlying distribution is adequately described as normal with mean μ and variance σ^2, there is a simple line of reasoning based on the random variable

$$\bar{x} - x_{n+1} \tag{6.73}$$

that leads to an answer to this question. That is, the random variable in expression (6.73) has, by the methods of Section 5.5 (Proposition 1 in particular),

$$E(\bar{x} - x_{n+1}) = E\bar{x} + (-1)Ex_{n+1} = \mu - \mu = 0 \tag{6.74}$$

and

$$\mathrm{Var}(\bar{x} - x_{n+1}) = (1)^2 \,\mathrm{Var}\,\bar{x} + (-1)^2 \,\mathrm{Var}\,x_{n+1} = \frac{\sigma^2}{n} + \sigma^2 = \left(1 + \frac{1}{n}\right)\sigma^2 \quad \textbf{(6.75)}$$

Further, it turns out that the difference (6.73) is normally distributed, so the variable

$$Z = \frac{(\bar{x} - x_{n+1}) - 0}{\sigma\sqrt{1 + \dfrac{1}{n}}} \qquad \textbf{(6.76)}$$

is standard normal. And taking one more step, if s^2 is the usual sample variance of $x_1, x_2, \ldots, x_n$, substituting s for σ in expression (6.76) produces a variable

$$T = \frac{(\bar{x} - x_{n+1}) - 0}{s\sqrt{1 + \dfrac{1}{n}}} \qquad \textbf{(6.77)}$$

which has a t distribution with $\nu = n - 1$ degrees of freedom.

Now (upon identifying x_{n+1} with μ and $\sqrt{1 + (1/n)}$ with $\sqrt{1/n}$), the variable (6.77) is formally similar to the t-distributed variable used to derive a small-sample confidence interval for μ. In fact, algebraic steps parallel to those used in the first part of Section 6.3 show that if $t > 0$ is such that the t_{n-1} distribution assigns, say, .95 probability to the interval between $-t$ and t, there is then .95 probability that

$$\bar{x} - ts\sqrt{1 + \frac{1}{n}} < x_{n+1} < \bar{x} + ts\sqrt{1 + \frac{1}{n}}$$

This reasoning suggests in general that the interval with endpoints

Normal distribution prediction limits for a single additional observation

$$\boxed{\bar{x} \pm ts\sqrt{1 + \frac{1}{n}}} \qquad \textbf{(6.78)}$$

can be used as a two-sided interval to predict x_{n+1} and that the probability-based reliability figure attached to the interval should be the t_{n-1} probability assigned to the interval from $-t$ to t. The interval (6.78) is a called a **prediction interval** with associated confidence the t_{n-1} probability assigned to the interval from $-t$ to t. In general, the language indicated in Definition 17 will be used.

Definition 17	A **prediction interval** for a single additional observation is a data-based interval of numbers thought likely to contain the observation, possessing a stated probability-based confidence or reliability.

It is the fact that a finite sample gives only a somewhat clouded picture of a distribution that prevents the making of a normal distribution prediction interval from being a trivial matter of probability calculations like those in Section 5.2. That is, suppose there were enough data to "know" the mean, μ, and variance, σ^2, of a normal distribution. Then, since 1.96 is the .975 standard normal quantile, the interval with endpoints

$$\mu - 1.96\sigma \quad \text{and} \quad \mu + 1.96\sigma \tag{6.79}$$

has a 95% chance of bracketing the next value generated by the distribution. The fact that (when based only on small samples), the knowledge of μ and σ is noisy forces expression (6.79) to be abandoned for an interval like (6.78). It is thus comforting that for large n and 95% confidence, formula (6.78) produces an interval with endpoints approximating those in display (6.79). That is, for large n and 95% confidence, $t \approx 1.96$, $\sqrt{1 + (1/n)} \approx 1$, and one expects that typically $\bar{x} \approx \mu$ and $s \approx \sigma$, so that expressions (6.78) and (6.79) will essentially agree. The beauty of expression (6.78) is that it allows in a rational fashion for the uncertainties involved in the $\mu \approx \bar{x}$ and $\sigma \approx s$ approximations.

Example 19 (*Example 8 revisited*)	Predicting a Spring Lifetime Recall from Section 6.3 that $n = 10$ spring lifetimes under 950 N/mm^2 stress conditions given in Table 6.4 (page 366) produced a fairly linear normal plot, $\bar{x} = 168.3$ ($\times 10^3$ cycles) and $s = 33.1$ ($\times 10^3$ cycles). Consider now predicting the lifetime of an additional spring of this type (under the same test conditions) with 90% confidence. Using $\nu = 10 - 1 = 9$ degrees of freedom, the .95 quantile of the t distribution is (from Table B.4) 1.833. So, employing expression (6.78), there are two-sided 90% prediction limits for an additional spring lifetime $$168.3 \pm 1.833(33.1)\sqrt{1 + \frac{1}{10}}$$ i.e., $$104.7 \times 10^3 \text{ cycles} \quad \text{and} \quad 231.9 \times 10^3 \text{ cycles} \tag{6.80}$$

The interval indicated by display (6.80) is not at all the same as the confidence interval for μ found in Example 8. The limits of

$$149.1 \times 10^3 \text{ cycles} \quad \text{and} \quad 187.5 \times 10^3 \text{ cycles}$$

found on page 367 apply to the mean spring lifetime, μ, not to an additional observation x_{11} as the ones in display (6.80) do.

Example 20

Predicting the Weight of a Newly Minted Penny

The delightful book *Experimentation and Measurement* by W. J. Youden (published as NBS Special Publication 672 by the U.S. Department of Commerce) contains a data set giving the weights of $n = 100$ newly minted U.S. pennies measured to 10^{-4} g but reported only to the nearest .02 g. These data are reproduced in Table 6.10. Figure 6.24 is a normal plot of these data and shows that a normal distribution is a plausible model for weights of newly minted pennies.

Further, calculation with the values in Table 6.10 shows that for the penny weights, $\bar{x} = 3.108$ g and $s = .043$ g. Then interpolation in Table B.4 shows the .9 quantile of the t_{99} distribution to be about 1.290, so that using only the "plus" part of expression (6.78), a one-sided 90% prediction interval of the form $(-\infty, \#)$ for the weight of a single additional penny has upper endpoint

$$3.108 + 1.290(.043)\sqrt{1 + \frac{1}{100}}$$

i.e.,

$$3.164 \text{ g} \tag{6.81}$$

Table 6.10
Weights of 100 Newly Minted U.S. Pennies

Penny Weight (g)	Frequency	Penny Weight (g)	Frequency
2.99	1	3.11	24
3.01	4	3.13	17
3.03	4	3.15	13
3.05	4	3.17	6
3.07	7	3.19	2
3.09	17	3.21	1

Example 20
(continued)

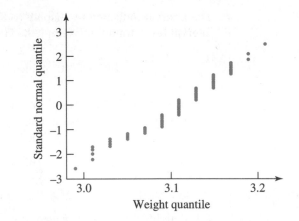

Figure 6.24   Normal plot of the penny weights

This example illustrates at least two important points. First, the two-sided prediction limits in display (6.78) can be modified to get a one-sided limit exactly as two-sided confidence limits can be modified to get a one-sided limit. Second, the calculation represented by the result (6.81) is, because $n = 100$ is a fairly large sample size, only marginally different from what one would get assuming $\mu = 3.108$ g exactly and $\sigma = .043$ g exactly. That is, since the .9 normal quantile is 1.282, "knowing" μ and σ leads to an upper prediction limit of

$$\mu + 1.282\sigma = 3.108 + (1.282)(.043) = 3.163 \text{ g} \qquad \textbf{(6.82)}$$

The fact that the result (6.81) is slightly larger than the final result in display (6.82) reflects the small uncertainty involved in the use of $\bar{x}$ in place of μ and s in place of σ.

Cautions about "prediction"

The name "prediction interval" probably has some suggested meanings that should be dismissed before going any further. *Prediction* suggests the future and thus potentially different conditions. But no such meaning should be associated with statistical prediction intervals. The assumption behind formula (6.78) is that $x_1, x_2, \ldots, x_n$ and x_{n+1} are *all* generated according to the *same* underlying distribution. If (for example, because of potential physical changes in a system during a time lapse between the generation of $x_1, x_2, \ldots, x_n$ and the generation of x_{n+1}) no single stable process model for the generation of all $n + 1$ observations is appropriate, then neither is formula (6.78). Statistical inference is not a crystal ball for foretelling an erratic and patternless future. It is rather a methodology for quantifying the extent of knowledge about a pattern of variation existing in a consistent present. It has implications in other times and at other places only if that same pattern of variation can be expected to repeat itself in those conditions.

It is also appropriate to comment on the meaning of the confidence or reliability figure attached to a prediction interval. Since a prediction interval is doing a different job than the confidence intervals of previous sections, the meaning of *confidence* given in Definition 2 doesn't quite apply here.

Prior to the generation of any of $x_1, x_2, \ldots, x_n, x_{n+1}$, planned use of expression (6.78) gives a guaranteed probability of success in bracketing x_{n+1}. And after all of $x_1, x_2, \ldots, x_n, x_{n+1}$ have been generated, one has either been completely successful or completely unsuccessful in bracketing x_{n+1}. But it is not altogether obvious how to think about "confidence" of prediction when $x_1, x_2, \ldots, x_n$ are in hand, but prior to the generation of x_{n+1}. For example, in the context of Example 19, having used sample data to arrive at the prediction limits in display (6.80)—i.e.,

$$104.7 \times 10^3 \text{ cycles} \quad \text{to} \quad 231.9 \times 10^3 \text{ cycles}$$

since x_{11} is a random variable, it would make sense to contemplate

$$P[104.7 \times 10^3 \leq x_{11} \leq 231.9 \times 10^3]$$

However, there is no guarantee on this probability nor any way to determine it. In particular, it is *not* necessarily .9 (the confidence level associated with the prediction interval). That is, there is no practical way to employ probability to describe the likely effectiveness of a numerical prediction interval. One is thus left with the interpretation of confidence of prediction given in Definition 18.

Definition 18 *(Interpretation of a Prediction Interval)*	To say that a numerical interval (a, b) is (for example) a 90% prediction interval for an additional observation x_{n+1} is to say that in obtaining it, methods of data collection and calculation have been applied that would produce intervals bracketing an $(n + 1)$th observation in about 90% of repeated applications of the entire process of (1) selecting the sample $x_1, \ldots, x_n$, (2) calculating an interval, and (3) generating a single additional observation x_{n+1}. Whether or not x_{n+1} will fall into the numerical interval (a, b) is not known, and although there is some probability associated with that eventuality, it is not possible to evaluate it. And in particular, it need not be 90%.

When using a 90% prediction interval method, although some samples $x_1, \ldots, x_n$ produce numerical intervals with probability less than .9 of bracketing x_{n+1} and others produce numerical intervals with probability more than .9, the average for all samples $x_1, \ldots, x_n$ does turn out to be .9. The practical problem is simply that with data $x_1, \ldots, x_n$ in hand, you don't know whether you are above, below, or at the .9 figure.

6.6.2 Tolerance Intervals for a Normal Distribution

The emphasis, when making a prediction interval of the type just discussed, is on a *single* additional observation beyond those n already in hand. But in some practical engineering problems, many additional items are of interest. In such cases, one may wish to declare a data-based interval likely to encompass **most measurements from the rest of these items**.

Prediction intervals are not designed for the purpose of encompassing most of the measurements from the additional items of interest. The paragraph following Definition 18 argues that only on average is the fraction of a normal distribution bracketed by a 90% prediction interval equal to 90%. So a crude analysis (identifying the mean fraction bracketed with the median fraction bracketed) then suggests that the probability that the actual fraction bracketed is at least 90% is only about .5. That is, a 90% prediction interval is not constructed to be big enough for the present purpose. What is needed instead is a **statistical tolerance interval**.

Definition 19

> A **statistical tolerance interval for a fraction** p **of an underlying distribution** is a data-based interval thought likely to contain at least a fraction p and possessing a stated (usually large) probability-based confidence or reliability.

The derivation of normal distribution tolerance interval formulas requires probability background well beyond what has been developed in this text. But results of that work look about as would be expected. It is possible, for a desired confidence level and fraction p of an underlying normal distribution, to find a corresponding constant τ_2 such that the two-sided interval with endpoints

Two-sided normal distribution tolerance limits

$$\bar{x} \pm \tau_2 s \tag{6.83}$$

is a tolerance interval for a fraction p of the underlying distribution. The τ_2 appearing in expression (6.83) is, for common (large) confidence levels, larger than the multiplier $t\sqrt{1 + (1/n)}$ appearing in expression (6.78) for two-sided confidence of prediction p. On the other hand, as n gets large, both τ_2 from expression (6.83) and $t\sqrt{1 + (1/n)}$ from expression (6.78) tend to the $(\frac{1+p}{2})$ standard normal quantile. Table B.7A gives some values of τ_2 for 95% and 99% confidence and $p = .9, .95,$ and .99. (The use of this table will be demonstrated shortly.)

The factors τ_2 are not used to make one-sided tolerance intervals. Instead, another set of constants that will here be called τ_1 values have been developed. They are such that for a given confidence and fraction p of an underlying normal distribution, both of the one-sided intervals

A one-sided normal tolerance interval

$$(-\infty, \bar{x} + \tau_1 s) \tag{6.84}$$

and

$$(\bar{x} - \tau_1 s, \infty) \tag{6.85}$$

are tolerance intervals for a fraction p of the distribution. τ_1 appearing in intervals (6.84) and (6.85) is, for common confidence levels, larger than the multiplier $t\sqrt{1 + (1/n)}$ appearing in expression (6.78) for one-sided confidence of prediction p. And as n gets large, both τ_1 from expression (6.84) or (6.85) and $t\sqrt{1 + (1/n)}$ from expression (6.78) tend to the standard normal p quantile. Table B.7B gives some values of τ_1.

Example 19
(continued)

Consider making a two-sided 95% tolerance interval for 90% of additional spring lifetimes based on the data of Table 6.4. As earlier, for these data, $\bar{x} = 168.3$ ($\times 10^3$ cycles) and $s = 33.1$ ($\times 10^3$ cycles). Then consulting Table B.7A, since $n = 10$, $\tau_2 = 2.856$ is appropriate for use in expression (6.83). That is, two-sided 95% tolerance limits for 90% of additional spring lifetimes are

$$168.3 \pm 2.856 \, (33.1)$$

i.e.,

$$73.8 \times 10^3 \text{ cycles} \quad \text{and} \quad 262.8 \times 10^3 \text{ cycles} \tag{6.86}$$

It is obvious from comparing displays (6.80) and (6.86) that the effect of moving from the prediction of a single additional spring lifetime to attempting to bracket most of a large number of additional lifetimes is to increase the size of the declared interval.

Example 20
(continued)

Consider again the new penny weights given in Table 6.10 and now the problem of making a one-sided 95% tolerance interval of the form $(-\infty, \#)$ for the weights of 90% of additional pennies. Remembering that for the penny weights, $\bar{x} = 3.108$ g and $s = .043$ g, and using Table B.7B for $n = 100$, the desired upper tolerance bound for 90% of the penny weights is

$$3.108 + 1.527(.043) = 3.174 \text{ g}$$

As expected, this is larger (more conservative) than the value of 3.164 g given in display (6.81) as a one-sided 90% prediction limit for a single additional penny weight.

The correct interpretation of the confidence level for a tolerance interval should be fairly easy to grasp. Prior to the generation of $x_1, x_2, \ldots, x_n$, planned use of expression (6.83), (6.84), or (6.85) gives a guaranteed probability of success in bracketing a fraction of at least p of the underlying distribution. But after observing $x_1, \ldots, x_n$ and making a numerical interval, it is impossible to know whether the attempt has or has not been successful. Thus the following interpretation:

Definition 20
(*Interpretation of a Tolerance Interval*)

To say that a numerical interval (a, b) is (for example) a 90% tolerance interval for a fraction p of an underlying distribution is to say that in obtaining it, methods of data collection and calculation have been applied that would produce intervals bracketing a fraction of at least p of the underlying distribution in about 90% of repeated applications (of generation of $x_1, \ldots, x_n$ and subsequent calculation). Whether or not the numerical interval (a, b) actually contains at least a fraction p is unknown and not describable in terms of a probability.

6.6.3 Prediction and Tolerance Intervals Based on Minimum and/or Maximum Values in a Sample

Formulas (6.78), (6.83), (6.84), and (6.85) for prediction and tolerance limits are definitely normal distribution formulas. So what if an engineering data-generation process is stable but does not produce normally distributed observations? How, if at all, can prediction or tolerance limits be made? Two kinds of answers to these questions will be illustrated in this text. The first employs the transformation idea presented in Section 4.4, and the second involves the use of minimum and/or maximum sample values to establish prediction and/or tolerance bounds.

First (as observed in Section 4.4) if a response variable y fails to be normally distributed, it may still be possible to find some transformation g (essentially specifying a revised scale of measurement) such that $g(y)$ is normal. Then normal-based methods might be applied to $g(y)$ and answers of interest translated back into statements about y.

Example 21
(*Example 11, Chapter 4, revisited—page 192*)

Prediction and Tolerance Intervals for Discovery Times Obtained Using a Transformation

Section 5.3 argued that the auto service discovery time data of Elliot, Kibby, and Meyer given in Figure 4.31 (see page 192) are not themselves normal-looking, but that their natural logarithms are. This, together with the facts that the $n = 30$ natural logarithms have $\bar{x} = 2.46$ and $s = .68$, can be used to make prediction or tolerance intervals for log discovery times.

For example, using expression (6.78) to make a two-sided 99% prediction interval for an additional log discovery time produces endpoints

$$2.46 \pm 2.756(.68)\sqrt{1 + \frac{1}{30}}$$

i.e.,

$$.55 \ln \min \quad \text{and} \quad 4.37 \ln \min \qquad \textbf{(6.87)}$$

And using expression (6.83) to make, for example, a 95% tolerance interval for 99% of additional log discovery times produces endpoints

$$2.46 \pm 3.355(.68)$$

i.e.,

$$.18 \ln \min \quad \text{and} \quad 4.74 \ln \min \qquad \textbf{(6.88)}$$

Then the intervals specified in displays (6.87) and (6.88) for log discovery times have, via exponentiation, their counterparts for raw discovery times. That is, exponentiation of the values in display (6.87) gives a 99% prediction interval for another discovery time of from

$$1.7 \min \quad \text{to} \quad 79.0 \min$$

And exponentiation of the values in display (6.88) gives a 95% tolerance interval for 99% of additional discovery times of from

$$1.2 \min \quad \text{to} \quad 114.4 \min$$

When it is not possible to find a transformation that will allow normal-based methods to be used, prediction and tolerance interval formulas derived for other standard families of distributions (e.g., the Weibull family) can sometimes be appropriate. (The book *Statistical Intervals: A Guide for Practitioners*, by Hahn and Meeker, is a good place to look for these methods.) What can be done here is to point out that intervals from the smallest observation and/or to the largest value in a sample can be used as prediction and/or tolerance intervals for *any* underlying continuous distribution.

That is, if $x_1, x_2, \ldots, x_n$ are values in a sample and $\min(x_1, \ldots, x_n)$ and $\max(x_1, \ldots, x_n)$ are (respectively) the smallest and largest values among x_1, $x_2, \ldots, x_n$, consider the use of the intervals

Interval based on the sample maximum

$$(-\infty, \max(x_1, \ldots, x_n)) \tag{6.89}$$

and

Interval based on the sample minimum

$$(\min(x_1, \ldots, x_n), \infty) \tag{6.90}$$

and

Interval based on the sample minimum and maximum

$$(\min(x_1, \ldots, x_n), \max(x_1, \ldots, x_n)) \tag{6.91}$$

as prediction or tolerance intervals. Independent of exactly what underlying continuous distribution is operating, if the generation of $x_1, x_2, \ldots, x_n$ (and if relevant, x_{n+1}) can be described as a stable process, it is possible to evaluate the confidence levels associated with intervals (6.89), (6.90), and (6.91).

Consider first intervals (6.89) or (6.90) used as one-sided prediction intervals for a single additional observation x_{n+1}. The associated confidence level is

Prediction confidence for a one-sided interval

$$\text{One-sided prediction confidence level} = \frac{n}{n+1} \tag{6.92}$$

Then, considering interval (6.91) as a two-sided prediction interval for a single additional observation x_{n+1}, the associated confidence level is

Prediction confidence for a two-sided interval

$$\text{Two-sided prediction confidence level} = \frac{n-1}{n+1} \tag{6.93}$$

The confidence levels for intervals (6.89), (6.90), and (6.91) as tolerance intervals must of necessity involve p, the fraction of the underlying distribution one hopes to bracket. The fact is that using interval (6.89) or (6.90) as a one-sided tolerance interval for a fraction p of an underlying distribution, the associated confidence level is

Confidence level for a one-sided tolerance interval

$$\text{One-sided confidence level} = 1 - p^n \tag{6.94}$$

And when interval (6.91) is used as a tolerance interval for a fraction p of an underlying distribution, the appropriate associated confidence is

Confidence level for a two-sided tolerance interval

$$\text{Two-sided confidence level} = 1 - p^n - n(1 - p)p^{n-1} \qquad \textbf{(6.95)}$$

Example 19
(continued)

Return one more time to the spring-life scenario, and consider the use of interval (6.91) as first a prediction interval and then a tolerance interval for 90% of additional spring lifetimes. Notice in Table 6.4 (page 366) that the smallest and largest of the observed spring lifetimes are, respectively,

$$\min(x_1, \ldots, x_{10}) = 117 \times 10^3 \text{ cycles}$$

and

$$\max(x_1, \ldots, x_{10}) = 225 \times 10^3 \text{ cycles}$$

so the numerical interval under consideration is the one with endpoints 117 ($\times 10^3$ cycles) and 225 ($\times 10^3$ cycles).

Then expression (6.93) means that this interval can be used as a prediction interval with

$$\text{Prediction confidence} = \frac{10}{10 + 1} \quad \frac{1}{11} = \frac{9}{11} = 82\%$$

And expression (6.95) says that as a tolerance interval for a fraction $p = .9$ of many additional spring lifetimes, the interval can be used with associated confidence

$$\text{Confidence} = 1 - (.9)^{10} - 10(1 - .9)(.9)^9 = 26\%$$

Example 20
(continued)

Looking for a final time at the penny weight data in Table 6.10, consider the use of interval (6.89) as first a prediction interval and then a tolerance interval for 99% of additional penny weights. Notice that in Table 6.10, the largest of the $n = 100$ weights is 3.21 g, so

$$\max(x_1, \ldots, x_{100}) = 3.21 \text{ g}$$

Example 20
(*continued*)

Then expression (6.92) says that when used as an upper prediction limit for a single additional penny weight, the prediction confidence associated with 3.21 g is

$$\text{Prediction confidence} = \frac{100}{100 + 1} = 99\%$$

And expression (6.94) shows that as a tolerance interval for 99% of many additional penny weights, the interval $(-\infty, 3.21)$ has associated confidence

$$\text{Confidence} = 1 - (.99)^{100} = 63\%$$

A little experience with formulas (6.92), (6.93), (6.94), and (6.95) will convince the reader that the intervals (6.89), (6.90), and (6.91) often carry disappointingly small confidence coefficients. Usually (but not always), you can do better in terms of high confidence and short intervals if (possibly after transformation) the normal distribution methods discussed earlier can be applied. But the beauty of intervals (6.89), (6.90), and (6.91) is that they are both widely applicable (in even nonnormal contexts) and extremely simple.

Prediction and tolerance interval methods are very useful engineering tools. Historically, they probably haven't been used as much as they should be for lack of accessible textbook material on the methods. We hope the reader is now aware of the existence of the methods as the appropriate form of formal inference when the focus is on individual values generated by a process rather than on process parameters. When the few particular methods discussed here don't prove adequate for practical purposes, the reader should look into the topic further, beginning with the book by Hahn and Meeker mentioned earlier.

Section 6 Exercises

1. Confidence, prediction, and tolerance intervals are all intended to do different jobs. What are these jobs? Consider the differing situations of an official of the EPA, a consumer about to purchase a single car, and a design engineer trying to equip a certain model with a gas tank large enough that most cars produced will have highway cruising ranges of at least 350 miles. Argue that depending on the point of view adopted, a lower confidence bound for a mean mileage, a lower prediction bound for an individual mileage, or a lower tolerance bound for most mileages would be of interest.

2. The 900 N/mm^2 stress spring lifetime data in Table 6.7 used in Example 8 have a fairly linear normal plot.

   (a) Make a two-sided 90% prediction interval for an additional spring lifetime under this stress.
   (b) Make a two-sided 95% tolerance interval for 90% of all spring lifetimes under this stress.
   (c) How do the intervals from (a) and (b) compare? (Consider both size and interpretation.)
   (d) There is a two-sided 90% confidence interval for the mean spring lifetime under this stress given in Example 8. How do your intervals from (a) and (b) compare to the interval in Example 8? (Consider both size and interpretation.)
   (e) Make a 90% lower prediction bound for an additional spring lifetime under this stress.

(f) Make a 95% lower tolerance bound for 90% of all spring lifetimes under this stress.

3. The natural logarithms of the aluminum contents discussed in Exercise 2 of Chapter 3 have a reasonably bell-shaped relative frequency distribution. Further, these 26 log aluminum contents have sample mean 4.9 and sample standard deviation .59. Use this information to respond to the following:

(a) Give a two-sided 99% tolerance interval for 90% of additional log aluminum contents at the Rutgers recycling facility. Then translate this interval into a 99% tolerance interval for 90% of additional raw aluminum contents.

(b) Make a 90% prediction interval for one additional log aluminum content and translate it into a prediction interval for a single additional aluminum content.

(c) How do the intervals from (a) and (b) compare?

4. Again in the context of Chapter Exercise 2 of Chapter 3, if the interval from 30 ppm to 511 ppm is used as a prediction interval for a single additional aluminum content measurement from the study period, what associated prediction confidence level can be stated? What confidence can be associated with this interval as a tolerance interval for 90% of all such aluminum content measurements?

Chapter 6 Exercises

1. Consider the breaking strength data of Table 3.6. Notice that the normal plot of these data given as Figure 3.18 is reasonably linear. It may thus be sensible to suppose that breaking strengths for generic towel of this type (as measured by the students) are adequately modeled as normal. Under this assumption,

(a) Make and interpret 95% two-sided and one-sided confidence intervals for the mean breaking strength of generic towels (make a one-sided interval of the form $(\#, \infty)$).

(b) Make and interpret 95% two-sided and one-sided prediction intervals for a single additional generic towel breaking strength (for the one-sided interval, give the lower prediction bound).

(c) Make and interpret 95% two-sided and one-sided tolerance intervals for 99% of generic towel breaking strengths (for the one-sided interval, give the lower tolerance bound).

(d) Make and interpret 95% two-sided and one-sided confidence intervals for σ, the standard deviation of generic towel breaking strengths.

(e) Put yourself in the position of a quality control inspector, concerned that the mean breaking strength not fall under 9,500 g. Assess the strength of the evidence in the data that the mean generic towel strength is in fact below the 9,500 g target. (Show the whole five-step significance-testing format.)

(f) Now put yourself in the place of a quality control inspector concerned that the breaking strength be reasonably consistent—i.e., that σ be small. Suppose in fact it is desirable that σ be no more than 400 g. Use the significance-testing format and assess the strength of the evidence given in the data that in fact σ exceeds the target standard deviation.

2. Consider the situation of Example 1 in Chapter 1.

(a) Use the five-step significance-testing format to assess the strength of the evidence collected in this study to the effect that the laying method is superior to the hanging method in terms of mean runouts produced.

(b) Make and interpret 90% two-sided and one-sided confidence intervals for the improvement in mean runout produced by the laying method over the hanging method (for the one-sided interval, give a lower bound for $\mu_{\text{hung}} - \mu_{\text{laid}}$).

(c) Make and interpret a 90% two-sided confidence interval for the mean runout for laid gears.

(d) What is it about Figure 1.1 that makes it questionable whether "normal distribution" prediction and tolerance interval formulas ought to be used to describe runouts for laid gears? Suppose instead that you used the methods of Section 6.6.3 to make prediction and tolerance intervals for laid gear runouts. What confidence could be associated with the largest observed laid runout as an upper prediction bound for a single additional laid runout? What confidence could be associated with the largest observed laid runout as an upper tolerance bound for 95% of additional laid gear runouts?

3. Consider the situation of Example 1 in Chapter 4. In particular, limit attention to those densities obtained under the 2,000 and 4,000 psi pressures. (One can view the six corresponding densities as two samples of size $n_1 = n_2 = 3$.)

(a) Assess the strength of the evidence that increasing pressure increases the mean density of the resulting cylinders. Use the five-step significance-testing format.

(b) Give a 99% lower confidence bound for the increase in mean density associated with the change from 2,000 to 4,000 psi conditions.

(c) Assess the strength of the evidence (in the six density values) that the variability in density differs for the 2,000 and 4,000 psi conditions (i.e., that $\sigma_{2,000} \neq \sigma_{4,000}$).

(d) Give a 90% two-sided confidence interval for the ratio of density standard deviations for the two pressures.

(e) What model assumptions stand behind the formal inferences you made in parts (a) through (d) above?

4. Simple counting with the data of Chapter Exercise 2 in Chapter 3 shows that 18 out of the 26 PET samples had aluminum contents above 100 ppm. Give a two-sided approximate 95% confidence interval for the fraction of all such samples with aluminum contents above 100 ppm.

5. Losen, Cahoy, and Lewis measured the lengths of some spanner bushings of a particular type purchased from a local machine supply shop. The lengths obtained by one of the students were as follows (the units are inches):

1.1375, 1.1390, 1.1420, 1.1430, 1.1410, 1.1360, 1.1395, 1.1380, 1.1350, 1.1370, 1.1345, 1.1340, 1.1405, 1.1340, 1.1380, 1.1355

(a) If you were to, for example, make a confidence interval for the population mean measured length of these bushings via the formulas in Section 6.3, what model assumption must you employ? Make a probability plot to assess the reasonableness of the assumption.

(b) Make a 90% two-sided confidence interval for the mean measured length for bushings of this type measured by this student.

(c) Give an upper bound for the mean length with 90% associated confidence.

(d) Make a 90% two-sided prediction interval for a single additional measured bushing length.

(e) Make a 95% two-sided tolerance interval for 99% of additional measured bushing lengths.

(f) Consider the statistical interval derived from the minimum and maximum sample values— namely, (1.1340, 1.1430). What confidence level should be associated with this interval as a prediction interval for a single additional bushing length? What confidence level should be associated with this interval as a tolerance interval for 99% of additional bushing lengths?

6. The study mentioned in Exercise 5 also included measurement of the outside diameters of the 16 bushings. Two of the students measured each of the bushings, with the results given here.

Bushing	1	2	3	4
Student A	.3690	.3690	.3690	.3700
Student B	.3690	.3695	.3695	.3695
Bushing	**5**	**6**	**7**	**8**
Student A	.3695	.3700	.3695	.3690
Student B	.3695	.3700	.3700	.3690

Bushing	9	10	11	12
Student A	.3690	.3695	.3690	.3690
Student B	.3700	.3690	.3695	.3695

Bushing	13	14	15	16
Student A	.3695	.3700	.3690	.3690
Student B	.3690	.3695	.3690	.3690

(a) If you want to compare the two students' average measurements, the methods of formulas (6.35), (6.36), and (6.38) are not appropriate. Why?

(b) Make a 95% two-sided confidence interval for the mean difference in outside diameter measurements for the two students.

7. Find the following quantiles using the tables of Appendix B:

   (a) the .90 quantile of the t_5 distribution
   (b) the .10 quantile of the t_5 distribution
   (c) the .95 quantile of the χ_7^2 distribution
   (d) the .05 quantile of the χ_7^2 distribution
   (e) the .95 quantile of the F distribution with numerator degrees of freedom 8 and denominator degrees of freedom 4
   (f) the .05 quantile of the F distribution with numerator degrees of freedom 8 and denominator degrees of freedom 4

8. Find the following quantiles using the tables of Appendix B:

   (a) the .99 quantile of the t_{13} distribution
   (b) the .01 quantile of the t_{13} distribution
   (c) the .975 quantile of the χ_3^2 distribution
   (d) the .025 quantile of the χ_3^2 distribution
   (e) the .75 quantile of the F distribution with numerator degrees of freedom 6 and denominator degrees of freedom 12
   (f) the .25 quantile of the F distribution with numerator degrees of freedom 6 and denominator degrees of freedom 12

9. Ho, Lewer, Peterson, and Riegel worked with the lack of flatness in a particular kind of manufactured steel disk. Fifty different parts of this type were measured for what the students called "wobble," with the results that the 50 (positive) values

obtained had mean $\bar{x} = .0287$ in. and standard deviation $s = .0119$ in.

(a) Give a 95% two-sided confidence interval for the mean wobble of all such disks.

(b) Give a lower bound for the mean wobble possessing a 95% confidence level.

(c) Suppose that these disks are ordered with the requirement that the mean wobble not exceed .025 in. Assess the strength of the evidence in the students' data to the effect that the requirement is being violated. Show the whole five-step format.

(d) Is the requirement of part (c) the same as an upper specification of .025 in. on individual wobbles? Explain. Is it possible for a lot with many individual wobbles exceeding .025 in. to meet the requirement of part (c)?

(e) Of the measured wobbles, 19 were .030 in. or more. Use this fact and make an approximate 90% two-sided confidence interval for the fraction of all such disks with wobbles of at least .030 in.

10. T. Johnson tested properties of several brands of 10 lb test monofilament fishing line. Part of his study involved measuring the stretch of a fixed length of line under a 3.5 kg load. Test results for three pieces of two of the brands follow. The units are cm.

Brand B	Brand D
.86, .88, .88	1.06, 1.02, 1.04

(a) Considering first only Brand B, use "normal distribution" model assumptions and give a 90% upper prediction bound for the stretch of an additional piece of Brand B line.

(b) Again considering only Brand B, use "normal distribution" model assumptions and give a 95% upper tolerance bound for stretch measurements of 90% of such pieces of Brand B line.

(c) Again considering only Brand B, use "normal distribution" model assumptions and give 90% two-sided confidence intervals for the mean and for the standard deviation of the Brand B stretch distribution.

(d) Compare the Brand B and Brand D standard deviations of stretch using an appropriate 90% two-sided confidence interval.

(e) Compare the Brand B and Brand D mean stretch values using an appropriate 90% two-sided confidence interval. Does this interval give clear indication of a difference in mean stretch values for the two brands?

(f) Carry out a formal significance test of the hypothesis that the two brands have the same mean stretch values (use a two-sided alternative hypothesis). Does the conclusion you reach here agree with your answer to part (e)?

11. The accompanying data are $n = 10$ daily measurements of the purity (in percent) of oxygen being delivered by a certain industrial air products supplier. (These data are similar to some given in a November 1990 article in *Chemical Engineering Progress* and used in Chapter Exercise 10 of Chapter 3.)

$$
\begin{array}{ccccc}
99.77 & 99.66 & 99.61 & 99.59 & 99.55 \\
99.64 & 99.53 & 99.68 & 99.49 & 99.58
\end{array}
$$

(a) Make a normal plot of these data. What does the normal plot reveal about the shape of the purity distribution? ("It is not bell-shaped" is not an adequate answer. Say how its shape departs from the normal shape.)

(b) What statistical "problems" are caused by lack of a normal distribution shape for data such as these?

As a way to deal with problems like those from part (b), you might try transforming the original data. Next are values of $y' = \ln(y - 99.3)$ corresponding to each of the original data values y, and some summary statistics for the transformed values.

$$
\begin{array}{ccccc}
-.76 & -1.02 & -1.17 & -1.24 & -1.39 \\
-1.08 & -1.47 & -.97 & -1.66 & -1.27
\end{array}
$$

$$
\bar{y}' = -1.203 \quad \text{and} \quad s_{y'} = .263
$$

(c) Make a normal plot of the transformed values and verify that it is very linear.

(d) Make a 95% two-sided prediction interval for the next transformed purity delivered by this supplier. What does this "untransform" to in terms of raw purity?

(e) Make a 99% two-sided tolerance interval for 95% of additional transformed purities from this supplier. What does this "untransform" to in terms of raw purity?

(f) Suppose that the air products supplier advertises a median purity of at least 99.5%. This corresponds to a median (and therefore mean) transformed value of at least -1.61. Test the supplier's claim ($H_0: \mu_{y'} = -1.61$) against the possibility that the purity is substandard. Show and carefully label all five steps.

12. Chapter Exercise 6 of Chapter 3 contains a data set on the lifetimes (in numbers of 24 mm deep holes drilled in 1045 steel before tool failure) of 12 D952-II (8 mm) drills. The data there have mean $\bar{y} = 117.75$ and $s = 51.1$ holes drilled. Suppose that a normal distribution can be used to roughly describe drill lifetimes.

(a) Give a 90% lower confidence bound for the mean lifetime of drills of this type in this kind of industrial application.

(b) Based on your answer to (a), do you think a hypothesis test of $H_0: \mu = 100$ versus $H_a: \mu > 100$ would have a large p-value or a small p-value? Explain.

(c) Give a 90% lower prediction bound for the next life length of a drill of this type in this kind of industrial application.

(d) Give two-sided tolerance limits with 95% confidence for 90% of all life lengths for drills of this type in this kind of industrial application.

(e) Give two-sided 90% confidence limits for the standard deviation of life lengths for drills of this type in this kind of industrial application.

13. M. Murphy recorded the mileages he obtained while commuting to school in his nine-year-old economy car. He kept track of the mileage for ten

different tankfuls of fuel, involving gasoline of two different octanes. His data follow.

87 Octane	90 Octane
26.43, 27.61, 28.71, 28.94, 29.30	30.57, 30.91, 31.21, 31.77, 32.86

(a) Make normal plots for these two samples of size 5 on the same set of axes. Does the "equal variances, normal distributions" model appear reasonable for describing this situation?

(b) Find s_P for these data. What is this quantity measuring in the present context?

(c) Give a 95% two-sided confidence interval for the difference in mean mileages obtainable under these circumstances using the fuels of the two different octanes. From the nature of this confidence interval, would you expect to find a large p-value or a small p-value when testing $H_0: \mu_{87} = \mu_{90}$ versus $H_a: \mu_{87} \neq \mu_{90}$?

(d) Conduct a significance test of $H_0: \mu_{87} = \mu_{90}$ against the alternative that the higher-octane gasoline provides a higher mean mileage.

(e) Give 95% lower prediction bounds for the next mileages experienced, using first 87 octane fuel and then 90 octane fuel.

(f) Give 95% lower tolerance bounds for 95% of additional mileages experienced, using first 87 octane fuel and then 90 octane fuel.

14. Eastman, Frye, and Schnepf worked with a company that mass-produces plastic bags. They focused on start-up problems of a particular machine that could be operated at either a high speed or a low speed. One part of the data they collected consisted of counts of faulty bags produced in the first 250 manufactured after changing a roll of plastic feedstock. The counts they obtained for both low- and high-speed operation of the machine were 147 faulty ($\hat{p}_H = \frac{147}{250}$) under high-speed operation and 12 faulty under low-speed operation ($\hat{p}_L = \frac{12}{250}$). Suppose that it is sensible to think of the machine as operating in a physically stable fashion during the production of the first 250 bags after changing

a roll of plastic, with a constant probability (p_H or p_L) of any particular bag produced being faulty.

(a) Give a 95% upper confidence bound for p_H.

(b) Give a 95% upper confidence bound for p_L.

(c) Compare p_H and p_L using an appropriate two-sided 95% confidence interval. Does this interval provide a clear indication of a difference in the effectiveness of the machine at start-up when run at the two speeds? What kind of a p-value (big or small) would you expect to find in a test of $H_0: p_H = p_L$ versus $H_a: p_H \neq p_L$?

(d) Use the five-step format and test $H_0: p_H = p_L$ versus $H_a: p_H \neq p_L$.

15. Hamilton, Seavey, and Stucker measured resistances, diameters, and lengths for seven copper wires at two different temperatures and used these to compute experimental resistivities for copper at these two temperatures. Their data follow. The units are 10^{-8} Ωm.

Wire	0.0°C	21.8°C
1	1.52	1.72
2	1.44	1.56
3	1.52	1.68
4	1.52	1.64
5	1.56	1.69
6	1.49	1.71
7	1.56	1.72

(a) Suppose that primary interest here centers on the difference between resistivities at the two different temperatures. Make a normal plot of the seven observed differences. Does it appear that a normal distribution description of the observed difference in resistivities at these two temperatures is plausible?

(b) Give a 90% two-sided confidence interval for the mean difference in resistivity measurements for copper wire of this type at 21.8°C and 0.0°C.

(c) Give a 90% two-sided prediction interval for an additional difference in resistivity measurements for copper wire of this type at 21.8°C and 0.0°C.

16. The students referred to in Exercise 15 also measured the resistivities for seven aluminum wires at the same temperatures. The 21.8°C measurements that they obtained follow:

$$2.65, 2.83, 2.69, 2.73, 2.53, 2.65, 2.69$$

(a) Give a 99% two-sided confidence interval for the mean resistivity value derived from such experimental determinations.

(b) Give a 95% two-sided prediction interval for the next resistivity value that would be derived from such an experimental determination.

(c) Give a 95% two-sided tolerance interval for 99% of resistivity values derived from such experimental determinations.

(d) Give a 95% two-sided confidence interval for the standard deviation of resistivity values derived from such experimental determinations.

(e) How strong is the evidence that there is a real difference in the precisions with which the aluminum resistivities and the copper resistivities can be measured at 21.8°C? (Carry out a significance test of $H_0: \sigma_{\text{copper}} = \sigma_{\text{aluminum}}$ versus $H_a: \sigma_{\text{copper}} \neq \sigma_{\text{aluminum}}$ using the data of this problem and the 21.8°C data of Exercise 15.)

(f) Again using the data of this exercise and Exercise 15, give a 90% two-sided confidence interval for the ratio $\sigma_{\text{copper}}/\sigma_{\text{aluminum}}$.

17. **(The Stein Two-Stage Estimation Procedure)** One of the most common of all questions faced by engineers planning a data-based study is how much data to collect. The last part of Example 3 illustrates a rather crude method of producing an answer to the sample-size question when estimation of a single mean is involved. In fact, in such circumstances, a more careful two-stage procedure due to Charles Stein can sometimes be used to find appropriate sample sizes.

Suppose that one wishes to use an interval of the form $\bar{x} \pm \Delta$ with a particular confidence coefficient to estimate the mean μ of a normal distribution. If it is desirable to have $\Delta \leq$ # for some number # and one can collect data in two stages, it is possible to choose an overall sample size to satisfy these criteria as follows. After taking a small or moderate initial sample of size n_1 (n_1 must be at least 2 and is typically at least 4 or 5), one computes the sample standard deviation of the initial data—say, s_1. Then if t is the appropriate $t_{n_1 - 1}$ distribution quantile for producing the desired (one- or two-sided) confidence, it is necessary to find the smallest integer n such that

$$n \geq \left(\frac{ts_1}{\#} \right)^2$$

If this integer is larger than n_1, then $n_2 = n - n_1$ additional observations are taken. (Otherwise, $n_2 = 0$.) Finally, with $\bar{x}$ the sample mean of all the observations (from both the initial and any subsequent sample), the formula $\bar{x} \pm ts_1/\sqrt{n_1 + n_2}$ (with t still based on $n_1 - 1$ degrees of freedom) is used to estimate μ.

Suppose that in estimating the mean resistance of a production run of resistors, it is desirable to have the two-sided confidence level be 95% and the "± part" of the interval no longer than .5 Ω.

(a) If an initial sample of $n_1 = 5$ resistors produces a sample standard deviation of 1.27 Ω, how many (if any) additional resistors should be sampled in order to meet the stated goals?

(b) If all of the $n_1 + n_2$ resistors taken together produce the sample mean $\bar{x} = 102.8$ Ω, what confidence interval for μ should be declared?

18. Example 15 of Chapter 5 concerns some data on service times at a residence hall depot counter. The data portrayed in Figure 5.21 are decidedly nonnormal-looking, so prediction and tolerance interval formulas based on normal distributions are not appropriate for use with these data. However, the largest of the $n = 65$ observed service times in that figure is 87 sec.

(a) What prediction confidence level can be associated with 87 sec as an upper prediction bound for a single additional service time?

(b) What confidence level can be associated with 87 sec as an upper tolerance bound for 95% of service times?

19. Caliste, Duffie, and Rodriguez studied the process of keymaking using a manual machine at a local lumber yard. The records of two different employees who made keys during the study period were as follows. Employee 1 made a total of 54 different keys, 5 of which were returned as not fitting their locks. Employee 2 made a total of 73 different keys, 22 of which were returned as not fitting their locks.

(a) Give approximate 95% two-sided confidence intervals for the long-run fractions of faulty keys produced by these two different employees.

(b) Give an approximate 95% two-sided confidence interval for the difference in long-run fractions of faulty keys produced by these two different employees.

(c) Assess the strength of the evidence provided in these two samples of a real difference in the keymaking proficiencies of these two employees. (Test $H_0: p_1 = p_2$ using a two-sided alternative hypothesis.)

20. The article "Optimizing Heat Treatment with Factorial Design" by T. Lim (*JOM*, 1989) discusses the improvement of a heat-treating process for gears through the use of factorial experimentation. To compare the performance of the heat-treating process under the original settings of process variables to that using the "improved" settings (identified through factorial experimentation), $n_1 = n_2 = 10$ gears were treated under both sets of conditions. Then measures of flatness, y_1 (in mm of distortion), and concentricity, y_2 (again in mm of distortion), were made on each of the gears. The data shown were read from graphs in the article (and may in some cases differ by perhaps $\pm.002$ mm from the original measurements).

Improved settings		
Gear	y_1 (mm)	y_2 (mm)
1A	.036	.050
2A	.040	.054
3A	.026	.043
4A	.051	.071
5A	.034	.043
6A	.050	.058
7A	.059	.061
8A	.055	.048
9A	.051	.060
10A	.050	.033

Original settings		
Gear	y_1 (mm)	y_2 (mm)
1B	.056	.070
2B	.064	.062
3B	.070	.075
4B	.037	.060
5B	.054	.071
6B	.060	.070
7B	.065	.060
8B	.060	.060
9B	.051	.070
10B	.062	.070

(a) What assumptions are necessary in order to make inferences regarding the parameters of the y_1 (or y_2) distribution for the improved settings of the process variables?

(b) Make a normal plot for the improved settings' y_1 values. Does it appear that it is reasonable to treat the improved settings' flatness distribution as normal? Explain.

(c) Suppose that the improved settings' flatness distribution is normal, and do the following:
 (i) Give a 90% two-sided confidence interval for the mean flatness distortion value for gears of this type.
 (ii) Give a 90% two-sided prediction interval for an additional flatness distortion value.

(iii) Give a 95% two-sided tolerance interval for 90% of additional flatness distortion values.

(iv) Give a 90% two-sided confidence interval for the standard deviation of flatness distortion values for gears of this type.

(d) Repeat parts (b) and (c) using the improved settings' concentricity values, y_2, instead of flatness.

(e) Explain why it is not possible to base formal inferences (tests and confidence intervals), for comparing the standard deviations of the y_1 and y_2 distributions for the improved process settings, on the sample standard deviations of the y_1 and y_2 measurements from gears 1A through 10A.

(f) What assumptions are necessary in order to make comparisons between parameters of the y_1 (or y_2) distributions for the original and improved settings of the process variables?

(g) Make normal plots of the y_1 data for the original settings and for the improved settings on the same set of axes. Does an "equal variances, normal distributions" model appear tenable here? Explain.

(h) Supposing that the flatness distortion distributions for the original and improved process settings are adequately described as normal with a common standard deviation, do the following.

(i) Use an appropriate significance test to assess the strength of the evidence in the data to the effect that the improved settings produce a reduction in mean flatness distortion.

(ii) Give a 90% lower confidence bound on the reduction in mean flatness distortion provided by the improved process settings.

(i) Repeat parts (g) and (h) using the y_2 values and concentricity instead of flatness.

21. R. Behne measured air pressure in car tires in a student parking lot. Shown here is one summary of the data he reported. Any tire with pressure reading more than 3 psi below its recommended value was considered underinflated, while any tire with pressure reading more than 3 psi above its recommended value was considered overinflated. The counts in the accompanying table are the numbers of cars (out of 25 checked) falling into the four possible categories.

| | | Underinflated tires | |
		None	At Least One Tire
Overinflated tires	None	6	5
	At Least One Tire	10	4

(a) Behne's sample was in all likelihood a convenience sample (as opposed to a genuinely simple random sample) of the cars in the large lot. Does it make sense to argue in this case that the data can be treated as if the sample were a simple random sample? On what basis? Explain.

(b) Give a two-sided 90% confidence interval for the fraction of all cars in the lot with at least one underinflated tire.

(c) Give a two-sided 90% confidence interval for the fraction of the cars in the lot with at least one overinflated tire.

(d) Give a 90% lower confidence bound on the fraction of cars in the lot with at least one misinflated tire.

(e) Why can't the data here be used with formula (6.67) of Section 6.5 to make a confidence interval for the difference in the fraction of cars with at least one underinflated tire and the fraction with at least one overinflated tire?

22. The article "A Recursive Partitioning Method for the Selection of Quality Assurance Tests" by Raz and Bousum (*Quality Engineering*, 1990) contains some data on the fractions of torque converters manufactured in a particular facility failing a final inspection (and thus requiring some rework). For a particular family of four-element converters, about 39% of 442 converters tested were out of specifications on a high-speed operation inlet flow test.

(a) If plant conditions tomorrow are like those under which the 442 converters were manufactured, give a two-sided 98% confidence interval for the probability that a given converter manufactured will fail the high-speed inlet flow test.

(b) Suppose that a process change is instituted in an effort to reduce the fraction of converters failing the high-speed inlet flow test. If only 32 out of the first 100 converters manufactured fail the high-speed inlet flow test, is this convincing evidence that a real process improvement has been accomplished? (Give and interpret a 90% two-sided confidence interval for the change in test failure probability)

23. Return to the situation of Chapter Exercise 1 in Chapter 3 and the measured gains of 120 amplifiers. The nominal/design value of the gain was 10.0 dB; 16 of the 120 amplifiers measured had gains above nominal. Give a 95% two-sided confidence interval for the fraction of all such amplifiers with above-nominal gains.

24. The article "Multi-functional Pneumatic Gripper Operating Under Constant Input Actuation Air Pressure" by J. Przybyl (*Journal of Engineering Technology*, 1988) discusses the performance of a 6-digit pneumatic robotic gripper. One part of the article concerns the gripping pressure (measured by manometers) delivered to objects of different shapes for fixed input air pressures. The data given here are the measurements (in psi) reported for an actuation pressure of 40 psi for (respectively) a 1.7 in. × 1.5 in. × 3.5 in. rectangular bar and a circular bar of radius 1.0 in. and length 3.5 in.

Rectangular Bar	Circular Bar
76	84
82	87
85	94
88	80
82	92

(a) Compare the variabilities of the gripping pressures delivered to the two different objects using an appropriate 98% two-sided confidence interval. Does there appear to be much evidence in the data of a difference between these? Explain.

(b) Supposing that the variabilities of gripping pressure delivered by the gripper to the two different objects are comparable, give a 95% two-sided confidence interval for the difference in mean gripping pressures delivered.

(c) The data here came from the operation of a single prototype gripper. Why would you expect to see more variation in measured gripping pressures than that represented here if each measurement in a sample were made on a different gripper? Strictly speaking, to what do the inferences in (a) and (b) apply? To the single prototype gripper or to all grippers of this design? Discuss this issue.

25. A sample of 95 U-bolts produced by a small company has thread lengths with a mean of $\bar{x} = 10.1$ (.001 in. above nominal) and $s = 3.2$ (.001 in.).

(a) Give a 95% two-sided confidence interval for the mean thread length (measured in .001 in. above nominal). Judging from this interval, would you expect a small or a large p-value when testing $H_0: \mu = 0$ versus $H_a: \mu \neq 0$? Explain.

(b) Use the five-step format of Section 6.2 and assess the strength of the evidence provided by the data to the effect that the population mean thread length exceeds nominal.

26. D. Kim did some crude tensile strength testing on pieces of some nominally .012 in. diameter wire of various lengths. Below are Kim's measured strengths (kg) for pieces of wire of lengths 25 cm and 30 cm.

25 cm Lengths	30 cm Lengths
4.00, 4.65, 4.70, 4.50	4.10, 4.50, 3.80, 4.60
4.40, 4.50, 4.50, 4.20	4.20, 4.60, 4.60, 3.90

(a) If one is to make a confidence interval for the mean measured strength of 25 cm pieces of this wire using the methods of Section 6.3, what model assumption must be employed? Make a probability plot useful in assessing the reasonableness of the assumption.

(b) Make a 95% two-sided confidence interval for the mean measured strength of 25 cm pieces of this wire.

(c) Give a 95% lower confidence bound for the mean measured strength of 25 cm pieces.

(d) Make a 95% two-sided prediction interval for a single additional measured strength for a 25 cm piece of wire.

(e) Make a 99% two-sided tolerance interval for 95% of additional measured strengths of 25 cm pieces of this wire.

(f) Consider the statistical interval derived from the minimum and maximum sample values for the 25 cm lengths—namely, (4.00, 4.70). What confidence should be associated with this interval as a prediction interval for a single additional measured strength? What confidence should be associated with this interval as a tolerance interval for 95% of additional measured strengths for 25 cm pieces of this wire?

(g) In order to make formal inferences about $\mu_{25} - \mu_{30}$ based on these data, what must you be willing to use for model assumptions? Make a plot useful for investigating the reasonableness of those assumptions.

(h) Proceed under the assumptions discussed in part (g) and assess the strength of the evidence provided by Kim's data to the effect that an increase in specimen length produces a decrease in measured strength.

(i) Proceed under the necessary model assumptions to give a 98% two-sided confidence interval for $\mu_{25} - \mu_{30}$.

27. The article "Influence of Final Recrystallization Heat Treatment on Zircaloy-4 Strip Corrosion" by Foster, Dougherty, Burke, Bates, and Worcester (*Journal of Nuclear Materials*, 1990) reported some summary statistics from the measurement of

the diameters of 821 particles observed in a bright field TEM micrograph of a Zircaloy-4 specimen. The sample mean diameter was $\bar{x} = .055$ μm, and the sample standard deviation of the diameters was $s = .028$ μm.

(a) The engineering researchers wished to establish from their observation of this single specimen the impact of a certain combination of specimen lot and heat-treating regimen on particle size. Briefly discuss why data such as the ones summarized have serious limitations for this purpose. (*Hints:* The apparent "sample size" here is huge. But of what is there a sample? How widely do the researchers want their results to apply? Given this desire, is the "real" sample size really so large?)

(b) Use the sample information and give a 98% two-sided confidence interval for the mean diameter of particles in this particular Zircaloy-4 specimen.

(c) Suppose that a standard method of heat treating for such specimens is believed to produce a mean particle diameter of .057 μm. Assess the strength of the evidence contained in the sample of diameter measurements to the effect that the specimen's mean particle diameter is different from the standard. Show the whole five-step format.

(d) Discuss, in the context of part (c), the potential difference between the mean diameter being statistically different from .057 μm and there being a difference between μ and .057 that is of practical importance.

28. Return to Kim's tensile strength data given in Exercise 26.

(a) Operating under the assumption that measured tensile strengths of 25 cm lengths of the wire studied are normally distributed, give a two-sided 98% confidence interval for the standard deviation of measured strengths.

(b) Operating under the assumption that measured tensile strengths of 30 cm lengths of the wire studied are normally distributed, give a 95% upper confidence bound for the standard deviation of measured strengths.

(c) Operating under the assumption that both 25 and 30 cm lengths of the wire have normally distributed measured tensile strengths, assess the strength of Kim's evidence that 25 and 30 cm lengths differ in variability of their measured tensile strengths. (Use $H_0: \sigma_{25} = \sigma_{30}$ and $H_a: \sigma_{25} \neq \sigma_{30}$ and show the whole five-step format.)

(d) Operating under the assumption that both 25 and 30 cm lengths produce normally distributed tensile strengths, give a 98% two-sided confidence interval for the ratio σ_{25}/σ_{30}.

29. Find the following quantiles:
(a) the .99 quantile of the χ_4^2 distribution
(b) the .025 quantile of the χ_4^2 distribution
(c) the .99 quantile of the F distribution with numerator degrees of freedom 3 and denominator degrees of freedom 15
(d) the .25 quantile of the F distribution with numerator degrees of freedom 3 and denominator degrees of freedom 15

30. The digital and vernier caliper measurements of no. 10 machine screw diameters summarized in Exercise 3 of Section 6.3 are such that for 19 out of 50 of the screws, there was no difference in the measurements. Based on these results, give a 95% confidence interval for the long-run fraction of such measurements by the student technician that would produce agreement between the digital and vernier caliper measurements.

31. Duren, Leng, and Patterson studied the drilling of holes in a miniature metal part using electrical discharge machining. Blueprint specifications on a certain hole called for diameters of $.0210 \pm .0003$ in. The diameters of this hole were measured on 50 parts with plug gauges and produced $\bar{x} = .02046$ and $s = .00178$. Assume that the holes the students measured were representative of the output of a physically stable drilling process.
(a) Give a 95% two-sided confidence interval for the mean diameter of holes drilled by this process.
(b) Give a 95% lower confidence bound for the mean diameter of the holes drilled by this process. (Find a number, #, so that $(\#, \infty)$

is a 95% confidence interval.) How does this number compare to the lower end point of your interval from (a)?
(c) Repeat (a) using 90% confidence. How does this interval compare with the one from (a)?
(d) Repeat (b) using 90% confidence. How does this bound compare to the one found in (b)?
(e) Interpret your interval from (a) for someone with little statistical background. (Speak in the context of the drilling study and use the "authorized interpretation" of confidence as your guide.)
(f) Based on your confidence intervals, would you expect the p-value in a test of $H_0: \mu = .0210$ versus $H_a: \mu \neq .0210$ to be small? Explain.
(g) Based on your confidence intervals, would you expect the p-value in a test of $H_0: \mu = .0210$ versus $H_a: \mu > .0210$ to be small? Explain.
(h) Consider again your answer to part (a). A colleague sees your calculations and says, "Oh, so 95% of the measured diameters would be in that range?" What do you say to this person?
(i) Use the five step significance-testing format of Section 6.2 and assess the strength of the evidence provided by the data to the effect that the process mean diameter differs from the mid-specification of .0210. (Begin with $H_0: \mu = .0210$ and use $H_a: \mu \neq .0210$.
(j) Thus far in this exercise, inference for the mean hole diameter has been of interest. Explain why in practice the variability of diameters is also important. The methods of Sections 6.1 are not designed for analyzing distributional spread. Where in Chapter 6 can you find inference methods for this feature?

32. Return to Babcock's fatigue life testing data in Chapter Exercise 18 of Chapter 3 and for now focus on the fatigue life data for heat 1.
(a) In order to do inference based on this small sample, what model assumptions must you employ? What does a normal plot say about the appropriateness of these assumptions?

(b) Give a 90% two-sided confidence interval for the mean fatigue life of such specimens from this heat.

(c) Give a 90% lower confidence bound for the mean fatigue life of such specimens from this heat.

(d) If you are interested in quantifying the variability in fatigue lives produced by this heat of steel, inference for σ becomes relevant. Give a 95% two-sided confidence interval for σ based on display (6.42) of the text.

(e) Make a 90% two-sided prediction interval for a single additional fatigue life for a specimen from this heat.

(f) Make a 95% two-sided tolerance interval for 90% of additional fatigue lives for specimens from this heat. How does this interval compare to your interval from (e)?

(g) Now consider the statistical interval derived from the minimum and maximum sample values from heat 1, namely (11, 548). What confidence should be associated with this interval as a prediction interval for a single additional fatigue life from this heat? What confidence should be associated with the interval as a tolerance interval for 90% of additional fatigue lives?

Now consider both the data for heat 1 and the data for heat 3.

(h) In order to make formal inferences about $\mu_1 - \mu_3$ based on these data, what must be assumed about fatigue lives for specimens from these two heats? Make a plot useful for investigating the reasonableness of these assumptions.

(i) Under the appropriate assumptions (state them), give a 95% two-sided confidence interval for $\mu_1 - \mu_3$.

33. Consider the Notch/Dial Bore and Notch/Air Spindler measurements on ten servo sleeves recorded in Chapter Exercise 19 in Chapter 3.

(a) If one wishes to compare the dial bore gauge and the air spindler gauge measurements, the methods of formulas (6.35), (6.36), and (6.38) are not appropriate. Why?

(b) What assumption must you make in order to do formal inference on the mean difference in dial bore and air spindler gauge measurements? Make a plot useful for assessing the reasonableness of this assumption. Comment on what it indicates in this problem.

(c) Make the necessary assumptions about the dial bore and air spindler measurements and assess the strength of the evidence in the data of a systematic difference between the two gauges.

(d) Make a 95% two-sided confidence interval for the mean difference in dial bore and air spindler measurements.

(e) Briefly discuss how your answers for parts (c) and (d) of this problem are consistent.

34. Chapter Exercise 20 in Chapter 3 concerned the drilling of holes in miniature metal parts using laser drilling and electrical discharge machining. Return to that problem and consider first only the EDM values.

(a) In order to use the methods of inference of Section 6.3 with these data, what model assumptions must be made? Make a plot useful for investigating the appropriateness of those assumptions. Comment on the shape of that plot and what it says about the appropriateness of the model assumptions.

(b) Give a 99% two-sided confidence interval for the mean angle produced by the EDM drilling of this hole.

(c) Give a 99% upper confidence bound for the mean angle produced by the EDM drilling of this hole.

(d) Give a 95% two-sided confidence interval for the standard deviation of angles produced by the EDM drilling of this hole.

(e) Make a 99% two-sided prediction interval for the next measured angle produced by the EDM drilling of this hole.

(f) Make a 95% two-sided tolerance interval for 99% of angles produced by the EDM drilling of this hole.

(g) Consider the statistical interval derived from the minimum and maximum sample EDM values, namely (43.2, 46.1). What confidence should be associated with this interval as a prediction interval for a single additional measured angle? What confidence should be associated with this interval as a tolerance interval for 99% of additional measured angles?

Now consider both the EDM and initial set of Laser values in Chapter Exercise 20 of Chapter 3 (two sets of 13 parts).

(h) In order to make formal inferences about $\mu_{Laser} - \mu_{EDM}$ based on these data, what must you be willing to use for model assumptions? Make a plot useful for investigating the reasonableness of those assumptions.

(i) Proceed under appropriate assumptions to assess the strength of the evidence provided by the data that there is a difference in the mean angles produced by the two drilling methods.

(j) Give a 95% two-sided confidence interval for $\mu_{Laser} - \mu_{EDM}$.

(k) Give a 90% two-sided confidence interval for comparing the standard deviations of angles produced by Laser and EDM drilling of this hole.

Now consider both sets of Laser measurements given in Chapter Exercise 20 of Chapter 3. (Holes A and B are on the same 13 parts.)

(l) If you wished to compare the mean angle measurements for the two holes, the formulas used in (i) and (j) are not appropriate. Why?

(m) Make a 90% two-sided confidence interval for the mean difference in angles for the two holes made with the laser equipment.

(n) Assess the strength of the evidence provided by these data that there is a systematic difference in the angles of the holes made with the laser equipment.

(o) Briefly discuss why your answers to parts (m) and (n) of this exercise are compatible. (Discuss how the outcome of part (n) could have been anticipated from the outcome of part (m).)

35. A so-called "tilttable" test was run in order to determine the angles at which certain vehicles experience lift-off of one set of wheels and begin to roll over on their sides. "Tilttable ratios" (which are the tangents of the angles at which lift-off occurred) were measured for two minivans of different makes four times each with the following results.

Van 1	Van 2
1.096, 1.093,	.962, .970,
1.090, 1.093	.967, .966

(a) If you were to make a confidence interval for the long-run mean measured tilttable ratio for Van 1 (under conditions like those experienced during the testing) using the methods of Section 6.3, what model assumption must be made?

(b) Make a 95% two-sided confidence interval for the mean measured tilttable ratio for Van 1 under conditions like those experienced during the testing.

(c) Give a 95% lower confidence bound for the mean measured tilttable ratio for Van 1.

(d) Give a 95% lower confidence bound for the standard deviation of tilttable ratios for Van 1.

(e) Make a 95% two-sided prediction interval for a single additional measured tilttable ratio for Van 1 under conditions such as those experienced during testing.

(f) Make a 99% two-sided tolerance interval for 95% of additional measured tilttable ratios for Van 1.

(g) Consider the statistical interval derived from the minimum and maximum sample values for Van 1, namely (1.090, 1.096). What confidence should be associated with this interval as a prediction interval for a single additional measured tilttable ratio? What confidence should be associated with this interval as a tolerance interval for 95% of additional tilttable test results for Van 1?

Now consider the data for both vans.

(h) In order to make formal inferences about $\mu_1 - \mu_2$ based on these data, what must you be willing to use for model assumptions?

(i) Proceed under the necessary assumptions to assess the strength of the evidence provided by the data that there is a difference in mean measured tilttable ratios for the two vans.

(j) Proceed under the necessary model assumptions to give a 90% two-sided confidence interval for $\mu_1 - \mu_2$.

(k) Proceed under the necessary model assumptions to give a 90% two-sided confidence interval for σ_1/σ_2.

....

Chapter 6 Summary Tables

The methods presented in Chapter 6 can seem overwhelming in their variety. It is sometimes helpful to have a summary of them. The tables here give such a summary and can be used to help you locate methods appropriate in a particular problem or application.

Table 1
Inference Methods for Individual Values

Inference For	Assumptions	Interval	Section
x_{n+1} (a single additional value)		$(\min(x_1, \ldots, x_n), \max(x_1, \ldots, x_n))$ or $(\min(x_1, \ldots, x_n), \infty)$ or $(-\infty, \max(x_1, \ldots, x_n))$	6.6
	observations normal	$\bar{x} \pm ts\sqrt{1 + \dfrac{1}{n}}$	6.6
most of the distribution		$(\min(x_1, \ldots, x_n), \max(x_1, \ldots, x_n))$ or $(\min(x_1, \ldots, x_n), \infty)$ or $(-\infty, \max(x_1, \ldots, x_n))$	6.6
	observations normal	$\bar{x} \pm \tau_2 s$ or $(\bar{x} - \tau_1 s, \infty)$ or $(-\infty, \bar{x} + \tau_1 s)$	6.6

Table 2
Inference Methods for One and Two Means

Inference For	Sample Size	Assumptions	H_0, Test Stat, Reference	Interval	Section
μ (one mean)	large n		$H_0: \mu = \#$ $Z = \dfrac{\bar{x} - \#}{s/\sqrt{n}}$ standard normal	$\bar{x} \pm z\dfrac{s}{\sqrt{n}}$	6.1, 6.2
	small n	observations normal	$H_0: \mu = \#$ $T = \dfrac{\bar{x} - \#}{s/\sqrt{n}}$ t with $\nu = n - 1$	$\bar{x} \pm t\dfrac{s}{\sqrt{n}}$	6.3
$\mu_1 - \mu_2$ (difference in means)	large n_1, n_2	independent samples	$H_0: \mu_1 - \mu_2 = \#$ $Z = \dfrac{\bar{x}_1 - \bar{x}_2 - \#}{\sqrt{\dfrac{s_1^2}{n_1} + \dfrac{s_2^2}{n_2}}}$ standard normal	$\bar{x}_1 - \bar{x}_2 \pm z\sqrt{\dfrac{s_1^2}{n_1} + \dfrac{s_2^2}{n_2}}$	6.3
	small n_1 or n_2	independent normal samples $\sigma_1 = \sigma_2$	$H_0: \mu_1 - \mu_2 = \#$ $T = \dfrac{\bar{x}_1 - \bar{x}_2 - \#}{s_P\sqrt{\dfrac{1}{n_1} + \dfrac{1}{n_2}}}$ t with $\nu = n_1 + n_2 - 2$	$\bar{x}_1 - \bar{x}_2 \pm ts_P\sqrt{\dfrac{1}{n_1} + \dfrac{1}{n_2}}$	6.3
		possibly $\sigma_1 \neq \sigma_2$		$\bar{x}_1 - \bar{x}_2 \pm \hat{t}\sqrt{\dfrac{s_1^2}{n_1} + \dfrac{s_2^2}{n_2}}$ use random $\hat{\nu}$ given in (6.37)	6.3
μ_d (mean difference)	large n	(paired data)	$H_0: \mu_d = \#$ $Z = \dfrac{\bar{d} - \#}{s_d/\sqrt{n}}$ standard normal	$\bar{d} \pm z\dfrac{s_d}{\sqrt{n}}$	6.3
	small n	(paired data) normal differences	$H_0: \mu_d = \#$ $T = \dfrac{\bar{d} - \#}{s_d/\sqrt{n}}$ t with $\nu = n - 1$	$\bar{d} \pm t\dfrac{s_d}{\sqrt{n}}$	6.3

Table 3
Inference Methods for Variances

Inference For	Assumptions	H_0, Test Stat, Reference	Interval	Section
σ^2 (one variance)	observations normal	$H_0: \sigma^2 = \#$ $X^2 = \dfrac{(n-1)s^2}{\#}$ χ^2 with $\nu = n - 1$	$\dfrac{(n-1)s^2}{U}$ and/or $\dfrac{(n-1)s^2}{L}$	6.4
σ_1^2/σ_2^2 (variance ratio)	observations normal independent samples	$H_0: \dfrac{\sigma_1^2}{\sigma_2^2} = \#$ $F = \dfrac{s_1^2/s_2^2}{\#}$ F with $\nu_1 = n_1 - 1$ and $\nu_2 = n_2 - 1$	$\dfrac{s_1^2}{U \cdot s_2^2}$ and/or $\dfrac{s_1^2}{L \cdot s_2^2}$	6.4

Table 4
Inference Methods for Proportions

Inference For	Sample Size	Assumptions	H_0, Test Stat, Reference	Interval	Section
p (one proportion)	large n		$H_0: p = \#$ $Z = \dfrac{\hat{p} - \#}{\sqrt{\dfrac{\#(1 - \#)}{n}}}$ standard normal	$\hat{p} \pm z\sqrt{\dfrac{\hat{p}(1 - \hat{p})}{n}}$ or $\hat{p} \pm z\dfrac{1}{2\sqrt{n}}$	6.5
$p_1 - p_2$ difference in proportions	large n_1, n_2	independent samples	$H_0: p_1 - p_2 = 0$ $Z = \dfrac{\hat{p}_1 - \hat{p}_2}{\sqrt{\hat{p}(1-\hat{p})}\sqrt{\frac{1}{n_1} + \frac{1}{n_2}}}$ use $\hat{p}$ given in (6.71) standard normal	$\hat{p}_1 - \hat{p}_2 \pm z\sqrt{\dfrac{\hat{p}_1(1 - \hat{p}_1)}{n_1} + \dfrac{\hat{p}_2(1 - \hat{p}_2)}{n_2}}$ or $\hat{p}_1 - \hat{p}_2 \pm z \cdot \dfrac{1}{2}\sqrt{\dfrac{1}{n_1} + \dfrac{1}{n_2}}$	6.5

7

• •

Inference for Unstructured Multisample Studies

Chapter 6 introduced the basics of formal statistical inference in one and two-sample studies. This chapter begins to consider formal inference for multisample studies, with a look at methods that make no explicit use of structure relating the samples (beyond time order of data collection). That is, the study of inference methods specifically crafted for use in factorial and fractional factorial studies and in curve- and surface-fitting analyses will be delayed until subsequent chapters.

The chapter opens with a discussion of the standard one-way model typically used in the analysis of measurement data from multisample studies and of the role of residuals in judging its appropriateness. The making of confidence intervals in multisample contexts is then considered, including both individual and simultaneous confidence interval methods. The one-way analysis of variance (ANOVA) test for the hypothesis of equality of several means and a related method of estimating variance components are introduced next. The chapter then covers the basics of Shewhart control (or process monitoring) charts. The $\bar{x}$, R, and s control charts for measurement data are studied. The chapter then closes with a section on p charts and u charts for attributes data.

• •

7.1 The One-Way Normal Model

Statistical engineering studies often produce samples taken under not one or two, but rather many different sets of conditions. So although the inference methods of Chapter 6 are a start, they are not a complete statistical toolkit for engineering problem solving. Methods of formal inference appropriate to multisample studies are also needed.

This section begins to provide such methods. First the reader is reminded of the usefulness of some of the simple graphical tools of Chapter 3 for making informal comparisons in multisample studies. Next the "equal variances, normal distributions" model is introduced. The role of residuals in evaluating the reasonableness of that model in an application is explained and emphasized. The section then proceeds to introduce the notion of combining several sample variances to produce a single pooled estimate of baseline variation. Finally, there is a discussion of how standardized residuals can be helpful when sample sizes vary considerably.

7.1.1 Graphical Comparison of Several Samples of Measurement Data

Any thoughtful analysis of several samples of engineering measurement data should begin with the making of graphical representations of those data. Where samples are small, side-by-side dot diagrams are the most natural graphical tool. Where sample sizes are moderate to large (say, at least six or so data points per sample), side-by-side boxplots are effective.

Example 1 | Comparing Compressive Strengths for Eight Different Concrete Formulas

Armstrong, Babb, and Campen did compressive strength testing on 16 different concrete formulas. Part of their data are given in Table 7.1, where eight different

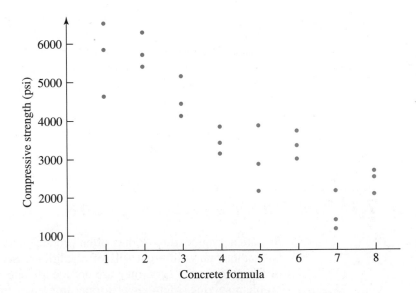

Figure 7.1 Side-by-side dot diagrams for eight samples of compressive strengths

Table 7.1
Compressive Strengths for 24 Concrete Specimens

Specimen	Concrete Formula	28-Day Compressive Strength (psi)
1	1	5,800
2	1	4,598
3	1	6,508
4	2	5,659
5	2	6,225
6	2	5,376
7	3	5,093
8	3	4,386
9	3	4,103
10	4	3,395
11	4	3,820
12	4	3,112
13	5	3,820
14	5	2,829
15	5	2,122
16	6	2,971
17	6	3,678
18	6	3,325
19	7	2,122
20	7	1,372
21	7	1,160
22	8	2,051
23	8	2,631
24	8	2,490

formulas are represented. (The only differences between formulas 1 through 8 are their water/cement ratios. Formula 1 had the lowest water/cement ratio, and the ratio increased with formula number in the progression .40, .44, .49, .53, .58, .62, .66, .71. Of course, knowing these water/cement ratios suggests that a curve-fitting analysis might be useful with these data, but for the time being this possibility will be ignored.)

Making side-by-side dot diagrams for these eight samples of sizes $n_1 = n_2 = n_3 = n_4 = n_5 = n_6 = n_7 = n_8 = 3$ amounts to making a scatterplot of compressive strength versus formula number. Such a plot is shown in Figure 7.1. The general message conveyed by Figure 7.1 is that there are clear differences in mean compressive strengths between the formulas but that the variabilities in compressive strengths are roughly comparable for the eight different formulas.

Example 2

Comparing Empirical Spring Constants for Three Different Types of Springs

Hunwardsen, Springer, and Wattonville did some testing of three different types of steel springs. They made experimental determinations of spring constants for $n_1 = 7$ springs of type 1 (a 4 in. design with a theoretical spring constant of 1.86), $n_2 = 6$ springs of type 2 (a 6 in. design with a theoretical spring constant of 2.63), and $n_3 = 6$ springs of type 3 (a 4 in. design with a theoretical spring constant of 2.12), using an 8.8 lb load. The students' experimental values are given in Table 7.2.

These samples are just barely large enough to produce meaningful boxplots. Figure 7.2 gives a side-by-side boxplot representation of these data. The primary qualitative message carried by Figure 7.2 is that there is a substantial difference in empirical spring constants between the 6 in. spring type and the two 4 in. spring types but that no such difference between the two 4 in. spring types is obvious. Of course, the information in Table 7.2 could also be presented in side-by-side dot diagram form, as in Figure 7.3.

Table 7.2
Empirical Spring Constants

Type 1 Springs	Type 2 Springs	Type 3 Springs
1.99, 2.06, 1.99	2.85, 2.74, 2.74	2.10, 2.01, 1.93
1.94, 2.05, 1.88	2.63, 2.74, 2.80	2.02, 2.10, 2.05
2.30		

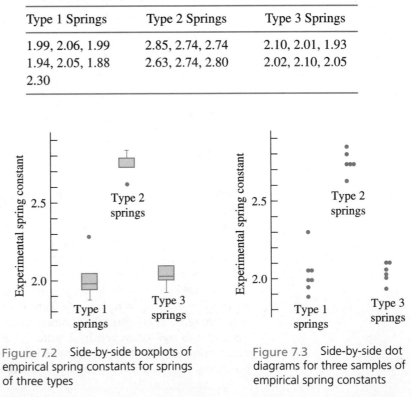

Figure 7.2   Side-by-side boxplots of empirical spring constants for springs of three types

Figure 7.3   Side-by-side dot diagrams for three samples of empirical spring constants

Methods of formal statistical inference are meant to sharpen and quantify the impressions that one gets when making a descriptive analysis of data. But an intelligent graphical look at data and a correct application of formal inference methods rarely tell completely different stories. Indeed, the methods of formal inference offered here for simple, unstructured multisample studies are *confirmatory*—in cases like Examples 1 and 2, they should confirm what is clear from a descriptive or *exploratory* look at the data.

7.1.2 The One-Way (Normal) Multisample Model, Fitted Values, and Residuals

Chapter 6 emphasized repeatedly that to make one- and two-sample inferences, one must adopt a model for data generation that is both manageable and plausible. The present situation is no different, and standard inference methods for unstructured multisample studies are based on a natural extension of the model used in Section 6.3 to support small-sample comparison of two means. The present discussion will be carried out under the assumption that r samples of respective sizes

One-way normal model assumptions

$n_1, n_2, \ldots, n_r$ are independent samples from normal underlying distributions with a common variance—say, σ^2. Just as in Section 6.3 the $r = 2$ version of this **one-way** (as opposed, for example, to several-way factorial) **model** led to useful inference methods for $\mu_1 - \mu_2$, this general version will support a variety of useful inference methods for r-sample studies. Figure 7.4 shows a number of different normal distributions with a common standard deviation. It represents essentially what must be generating measured responses if the methods of this chapter are to be applied.

In addition to a description of the one-way model in words and the pictorial representation given in Figure 7.4, it is helpful to have a description of the model in symbols. This and the next three sections will employ the notation

$$y_{ij} = \text{the } j\text{th observation in sample } i$$

The model equation used to specify the one-way model is then

One-way model statement in symbols

$$y_{ij} = \mu_i + \epsilon_{ij} \tag{7.1}$$

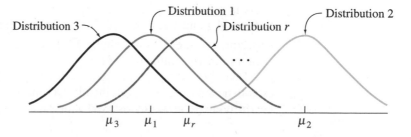

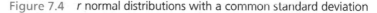

Figure 7.4 *r* normal distributions with a common standard deviation

where μ_i is the ith underlying mean and the quantities $\epsilon_{11}, \epsilon_{12}, \ldots, \epsilon_{1n_1}, \epsilon_{21}, \epsilon_{22}, \ldots,$ $\epsilon_{2n_2}, \ldots, \epsilon_{r1}, \epsilon_{r2}, \ldots, \epsilon_{rn_r}$ are independent normal random variables with mean 0 and variance σ^2. (In this statement, the means $\mu_1, \mu_2, \ldots, \mu_r$ and the variance σ^2 are typically unknown parameters.)

Equation (7.1) says exactly what is conveyed by Figure 7.4 and the statement of the one-way assumptions in words. But it says it in a way that is suggestive of another useful pattern of thinking, reminiscent of the "residual" notion that was used extensively in Sections 4.1, 4.2, and 4.3. That is, equation (7.1) says that an observation in sample i is made up of the corresponding underlying mean plus some random noise, namely

$$\epsilon_{ij} = y_{ij} - \mu_i$$

This is a theoretical counterpart of an empirical notion met in Chapter 4. There, it was useful to decompose data into fitted values and the corresponding residuals.

In the present situation, since any structure relating the r different samples is specifically being ignored, it may not be obvious how to apply the notions of fitted values and residuals. But a plausible meaning for

$$\hat{y}_{ij} = \text{the fitted value corresponding to } y_{ij}$$

in the present context is the ith sample mean

ith sample mean

$$\bar{y}_i = \frac{1}{n_i} \sum_{j=1}^{n_i} y_{ij}$$

That is,

Fitted values for the one-way model

$$\hat{y}_{ij} = \bar{y}_i \tag{7.2}$$

(This is not only intuitively plausible but also consistent with what was done in Sections 4.1 and 4.2. If one fits the approximate relationship $y_{ij} \approx \mu_i$ to the data via least squares—i.e., by minimizing $\sum_{ij}(y_{ij} - \mu_i)^2$ over choices of $\mu_1, \mu_2, \ldots, \mu_r$— each minimizing value of μ_i is $\bar{y}_i$.)

Taking equation (7.2) to specify fitted values for an r-sample study, the pattern established in Chapter 4 (specifically, Definition 4, page 132) then says that residuals are differences between observed values and sample means. That is, with

$$e_{ij} = \text{the residual corresponding to } y_{ij}$$

one has

$$e_{ij} = y_{ij} - \hat{y}_{ij} = y_{ij} - \bar{y}_i \qquad (7.3)$$

Rearranging display (7.3) gives the relationship

$$y_{ij} = \hat{y}_{ij} + e_{ij} = \bar{y}_i + e_{ij} \qquad (7.4)$$

which is an empirical counterpart of the theoretical statement (7.1). In fact, combining equations (7.1) and (7.4) into a single statement gives

$$y_{ij} = \mu_i + \epsilon_{ij} = \bar{y}_i + e_{ij} \qquad (7.5)$$

This is a specific instance of a pattern of thinking that runs through all of the common normal-distribution-based methods of analysis for multisample studies. In words, equation (7.5) says

$$\text{Observation} = \text{deterministic response} + \text{noise} = \text{fitted value} + \text{residual} \qquad (7.6)$$

and display (7.6) is a paradigm that provides a unified way of approaching the majority of the analysis methods presented in the rest of this book.

The decompositions (7.5) and (7.6) suggest that

1. the fitted values ($\hat{y}_{ij} = \bar{y}_i$) are meant to approximate the deterministic part of a system response (μ_i), and

2. the residuals (e_{ij}) are therefore meant to approximate the corresponding noise in the response (ϵ_{ij}).

The fact that the ϵ_{ij} in equation (7.1) are assumed to be iid normal $(0, \sigma^2)$ random variables then suggests that the e_{ij} ought to look at least approximately like a random sample from a normal distribution.

So the normal-plotting of an entire set of residuals (as in Chapter 4) is a way of checking on the reasonableness of the one-way model. The plotting of residuals against (1) fitted values, (2) time order of observation, or (3) any other potentially relevant variable—hoping (as in Chapter 4) to see only random scatter—are other ways of investigating the appropriateness of the model assumptions.

These kinds of plotting, which combine residuals from all r samples, are often especially useful in practice. When r is large at all, budget constraints on total data collection costs often force the individual sample sizes $n_1, n_2, \ldots, n_r$ to be fairly small. This makes it fruitless to investigate "single variance, normal distributions" model assumptions using (for example) sample-by-sample normal plots. (Of course, where all of $n_1, n_2, \ldots, n_r$ are of a decent size, a sample-by-sample approach can be effective.)

Example 1
(*continued*)

Returning again to the concrete strength study, consider investigating the reasonableness of model (7.1) for this case. Figure 7.1 is a first step in this investigation. As remarked earlier, it conveys the visual impression that at least the "equal variances" part of the one-way model assumptions is plausible. Next, it makes sense to compute some summary statistics and examine them, particularly the sample standard deviations. Table 7.3 gives sample sizes, sample means, and sample standard deviations for the data in Table 7.1.

At first glance, it might seem worrisome that in this table s_1 is more than three times the size of s_8. But the sample sizes here are so small that a largest ratio of

Table 7.3
Summary Statistics for the Concrete Strength Study

i, Concrete Formula	n_i, Sample Size	$\bar{y}_i$, Sample Mean (psi)	s_i, Sample Standard Deviation (psi)
1	$n_1 = 3$	$\bar{y}_1 = 5{,}635.3$	$s_1 = 965.6$
2	$n_2 = 3$	$\bar{y}_2 = 5{,}753.3$	$s_2 = 432.3$
3	$n_3 = 3$	$\bar{y}_3 = 4{,}527.3$	$s_3 = 509.9$
4	$n_4 = 3$	$\bar{y}_4 = 3{,}442.3$	$s_4 = 356.4$
5	$n_5 = 3$	$\bar{y}_5 = 2{,}923.7$	$s_5 = 852.9$
6	$n_6 = 3$	$\bar{y}_6 = 3{,}324.7$	$s_6 = 353.5$
7	$n_7 = 3$	$\bar{y}_7 = 1{,}551.3$	$s_7 = 505.5$
8	$n_8 = 3$	$\bar{y}_8 = 2{,}390.7$	$s_8 = 302.5$

Table 7.4
Example Computations of Residuals for the Concrete Strength Study

Specimen	i, Concrete Formula	y_{ij}, Compressive Strength (psi)	$\hat{y}_{ij} = \bar{y}_i$, Fitted Value	e_{ij}, Residual
1	1	5,800	5,635.3	164.7
2	1	4,598	5,635.3	−1,037.3
3	1	6,508	5,635.3	872.7
4	2	5,659	5,753.3	−94.3
5	2	6,225	5,753.3	471.7
⋮	⋮	⋮	⋮	⋮
22	8	2,051	2,390.7	−339.7
23	8	2,631	2,390.7	240.3
24	8	2,490	2,390.7	99.3

sample standard deviations on the order of 3.2 is hardly unusual (for $r = 8$ samples of size 3 from a normal distribution). Note from the F tables (Tables B.6) that for samples of size 3, even if only 2 (rather than 8) sample standard deviations were involved, a ratio of sample variances of $(965.6/302.5)^2 \approx 10.2$ would yield a p-value between .10 and .20 for testing the null hypothesis of equal variances with a two-sided alternative. The sample standard deviations in Table 7.3 really carry no strong indication that the one-way model is inappropriate.

Since the individual sample sizes are so small, trying to see anything useful in eight separate normal plots of the samples is hopeless. But some insight can be gained by calculating and plotting all $8 \times 3 = 24$ residuals. Some of the calculations necessary to compute residuals for the data in Table 7.1 (using the fitted values appearing as sample means in Table 7.3) are shown in Table 7.4. Figures 7.5 and 7.6 are, respectively, a plot of residuals versus fitted y (e_{ij} versus $\bar{y}_{ij}$) and a normal plot of all 24 residuals.

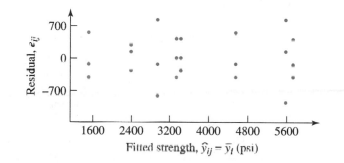

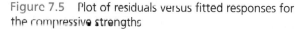

Figure 7.5 Plot of residuals versus fitted responses for the compressive strengths

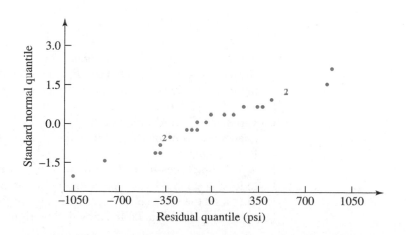

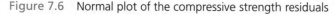

Figure 7.6 Normal plot of the compressive strength residuals

Example 1
(continued)

Figure 7.5 gives no indication of any kind of strong dependence of σ on μ (which would violate the "constant variance" restriction). And the plot in Figure 7.6 is reasonably linear, thus identifying no obvious difficulty with the assumption of normal distributions. In all, it seems from examination of both the raw data and the residuals that analysis of the data in Table 7.1 on the basis of model (7.1) is perfectly sensible.

Example 2
(continued)

The spring testing data can also be examined with the potential use of the one-way normal model (7.1) in mind. Figures 7.2 and 7.3 indicate reasonably comparable variabilities of experimental spring constants for the $r = 3$ different spring types. The single very large value (for spring type 1) causes some doubt both in terms of this judgment and also (by virtue of its position on its boxplot as an outlying value) regarding a "normal distribution" description of type 1 experimental constants. Summary statistics for these samples are given in Table 7.5.

Table 7.5
Summary Statistics for the Empirical
Spring Constants

i, Spring Type	n_i	$\bar{y}_i$	s_i
1	7	2.030	.134
2	6	2.750	.074
3	6	2.035	.064

Without the single extreme value of 2.30, the first sample standard deviation would be .068, completely in line with those of the second and third samples. But even the observed ratio of largest to smallest sample variance (namely $(.134/.064)^2 = 4.38$) is not a compelling reason to abandon a one-way model description of the spring constants. (A look at the F tables with $\nu_1 = 6$ and $\nu_2 = 5$ shows that 4.38 is between the $F_{6,5}$ distribution .9 and .95 quantiles. So even if there were only two rather than three samples involved, a variance ratio of 4.38 would yield a p-value between .1 and .2 for (two-sided) testing of equality of variances.) Before letting the single type 1 empirical spring constant of 2.30 force abandonment of the highly tractable model (7.1) some additional investigation is warranted.

Sample sizes $n_1 = 7$ and $n_2 = n_3 = 6$ are large enough that it makes sense to look at sample-by-sample normal plots of the spring constant data. Such plots, drawn on the same set of axes, are shown in Figure 7.7. Further, use of the fitted values ($\bar{y}_i$) listed in Table 7.5 with the original data given in Table 7.2 produces

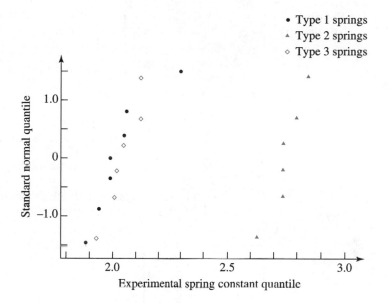

Figure 7.7   Normal plots of empirical spring constants for springs of three types

Table 7.6
Example Computations of Residuals for the Spring Constant Study

i, Spring Type	j, Observation Number	y_{ij}, Spring Constant	$\hat{y}_{ij} = \bar{y}_i$, Sample Mean	e_{ij}, Residual
1	1	1.99	2.030	−.040
⋮	⋮	⋮	⋮	⋮
1	7	2.30	2.030	.270
2	1	2.85	2.750	.100
⋮	⋮	⋮	⋮	⋮
2	6	2.80	2.750	.050
3	1	2.10	2.035	.065
⋮	⋮	⋮	⋮	⋮
3	6	2.05	2.035	.015

19 residuals, as partially illustrated in Table 7.6. Then Figures 7.8 and 7.9, respectively, show a plot of residuals versus fitted responses and a normal plot of all 19 residuals.

Example 2
(*continued*)

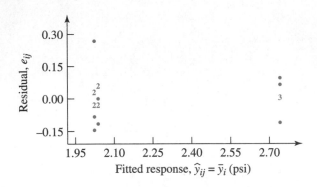

Figure 7.8 Plot of residuals versus fitted responses for the empirical spring constants

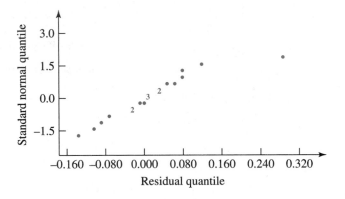

Figure 7.9 Normal plot of the spring constant residuals

But Figures 7.8 and 7.9 again draw attention to the largest type 1 empirical spring constant. Compared to the other measured values, 2.30 is simply too large (and thus produces a residual that is too large compared to all the rest) to permit serious use of model (7.1) with the spring constant data. Barring the possibility that checking of original data sheets would show the 2.30 value to be an arithmetic blunder or gross error of measurement (which could be corrected or legitimately force elimination of the 2.30 value from consideration), it appears that the use of model (7.1) with the $r = 3$ spring types could produce inferences with true (and unknown) properties quite different from their nominal properties.

One might, of course, limit attention to spring types 2 and 3. There is nothing in the second or third samples to render the "equal variances, normal distributions" model untenable for those two spring types. But the pattern of variation for springs of type 1 appears to be detectably different from that for springs of types 2 and 3, and the one-way model is not appropriate when all three types are considered.

7.1.3   A Pooled Estimate of Variance for Multisample Studies

The "equal variances, normal distributions" model (7.1) has as a fundamental parameter, σ, the standard deviation associated with responses from any of conditions $1, 2, 3, \ldots, r$. Similar to what was done in the $r = 2$ situation of Section 6.3, it is typical in multisample studies to **pool** the r sample variances to arrive at a single estimate of σ derived from all r samples.

Definition 1

If r numerical samples of respective sizes $n_1, n_2, \ldots, n_r$ produce sample variances $s_1^2, s_2^2, \ldots, s_r^2$, the **pooled sample variance**, s_{P}^2, is the weighted average of the sample variances, where the weights are the sample sizes minus 1. That is,

$$s_{\mathrm{P}}^2 = \frac{(n_1 - 1)s_1^2 + (n_2 - 1)s_2^2 + \cdots + (n_r - 1)s_r^2}{(n_1 - 1) + (n_2 - 1) + \cdots + (n_r - 1)} \tag{7.7}$$

The **pooled sample standard deviation**, s_{P}, is the square root of s_{P}^2.

Definition 1 is just Definition 14 in Chapter 6 restated for the case of more than two samples. As was the case for s_{P} based on two samples, s_{P} is guaranteed to lie between the largest and smallest of the s_i and is a mathematically convenient form of compromise value.

Equation (7.7) can be rewritten in a number of equivalent forms. For one thing, letting

The total number of observations in an r-sample study

$$n = \textstyle\sum_{i=1}^{r} n_i = \text{the total number of observations in all } r \text{ samples}$$

it is common to rewrite the denominator on the right of equation (7.7) as

$$\sum_{i=1}^{r}(n_i - 1) = \sum_{i=1}^{r} n_i - \sum_{i=1}^{r} 1 = n - r$$

And noting that the ith sample variance is

$$s_i^2 = \frac{1}{n_i - 1} \sum_{j=1}^{n_i} (y_{ij} - \bar{y}_i)^2$$

the numerator on the right of equation (7.7) is

$$\sum_{i=1}^{r}(n_i - 1)\left(\frac{1}{(n_i - 1)}\sum_{j=1}^{n_i}(y_{ij} - \bar{y}_i)^2\right) = \sum_{i=1}^{r}\sum_{j=1}^{n_i}(y_{ij} - \bar{y}_i)^2 \qquad (7.8)$$

$$= \sum_{i=1}^{r}\sum_{j=1}^{n_i}e_{ij}^2 \qquad (7.9)$$

Alternative formulas for s_P^2

So one can define s_P^2 in terms of the right-hand side of equation (7.8) or (7.9) divided by $n - r$.

Example 1
(*continued*)

For the compressive strength data, each of $n_1, n_2, \ldots, n_8$ are 3, and s_1 through s_8 are given in Table 7.3. So using equation (7.7),

$$s_P^2 = \frac{(3 - 1)(965.6)^2 + (3 - 1)(432.3)^2 + \cdots + (3 - 1)(302.5)^2}{(3 - 1) + (3 - 1) + \cdots + (3 - 1)}$$

$$= \frac{2[(965.6)^2 + (432.3)^2 + \cdots + (302.5)^2]}{16}$$

$$= \frac{2,705,705}{8}$$

$$= 338,213 \ (\text{psi})^2$$

and thus

$$s_P = \sqrt{338,213} = 581.6 \text{ psi}$$

One estimates that if a large number of specimens of any one of formulas 1 through 8 were tested, a standard deviation of compressive strengths on the order of 582 psi would be obtained.

The meaning of s_P

s_P is an estimate of the intrinsic or baseline variation present in a response variable at a fixed set of conditions, calculated supposing that the baseline variation is constant across the conditions under which the samples were collected. When that supposition is reasonable, the pooling idea allows a number of individually unreliable small-sample estimates to be combined into a single, relatively more reliable combined estimate. It is a fundamental measure that figures prominently in a variety of useful methods of formal inference.

On occasion, it is helpful to have not only a single number as a data-based best guess at σ^2 but a confidence interval as well. Under model restrictions (7.1), the variable

$$\frac{(n-r)s_P^2}{\sigma^2}$$

has a χ_{n-r}^2 distribution. Thus, in a manner exactly parallel to the derivation in Section 6.4, a two-sided confidence interval for σ^2 has endpoints

Confidence limits for the one-way model variance

$$\frac{(n-r)s_P^2}{U} \quad \text{and} \quad \frac{(n-r)s_P^2}{L} \tag{7.10}$$

where L and U are such that the χ_{n-r}^2 probability assigned to the interval (L, U) is the desired confidence level. And, of course, a one-sided interval is available by using only one of the endpoints (7.10) and choosing U or L such that the χ_{n-r}^2 probability assigned to the interval $(0, U)$ or (L, ∞) is the desired confidence.

Example 1
(continued)

In the concrete compressive strength case, consider the use of display (7.10) in making a two-sided 90% confidence interval for σ. Since $n - r = 16$ degrees of freedom are associated with s_P^2, one consults Table B.5 for the .05 and .95 quantiles of the χ_{16}^2 distribution. These are 7.962 and 26.296, respectively. Thus, from display (7.10), a confidence interval for σ^2 has endpoints

$$\frac{16(581.6)^2}{26.296} \quad \text{and} \quad \frac{16(581.6)^2}{7.962}$$

So a two-sided 90% confidence interval for σ has endpoints

$$\sqrt{\frac{16(581.6)^2}{26.296}} \quad \text{and} \quad \sqrt{\frac{16(581.6)^2}{7.962}}$$

that is,

$$453.7\text{psi} \quad \text{and} \quad 824.5\text{psi}$$

7.1.4 Standardized Residuals

In discussing the use of residuals, the reasoning has been that the e_{ij} are meant to approximate the corresponding random errors ϵ_{ij}. Since the model assumptions are that the ϵ_{ij} are iid normal variables, the e_{ij} ought to look approximately like iid normal variables. This is sensible rough-and-ready reasoning, adequate for many circumstances. But strictly speaking, the e_{ij} are neither independent nor identically distributed, and it can be important to recognize this.

As an extreme example of the dependence of the residuals for a given sample i, consider a case where $n_i = 2$. Since

$$e_{ij} = y_{ij} - \bar{y}_i$$

one immediately knows that $e_{i1} = -e_{i2}$. So e_{i1} and e_{i2} are clearly dependent.

One can further apply Proposition 1 of Chapter 5 to show that if the sample sizes n_i are varied, the residuals don't have the same variance (and therefore can't be identically distributed). That is, since

$$e_{ij} = y_{ij} - \bar{y}_i = \left(\frac{n_i - 1}{n_i}\right) y_{ij} - \frac{1}{n_i} \sum_{j' \neq j} y_{ij'}$$

it is the case that

$$\operatorname{Var} e_{ij} = \left(\frac{n_i - 1}{n_i}\right)^2 \sigma^2 + \left(-\frac{1}{n_i}\right)^2 (n_i - 1)\sigma^2 = \frac{n_i - 1}{n_i} \sigma^2 \qquad \textbf{(7.11)}$$

So, for example, residuals from a sample of size $n_i = 2$ have variance $\sigma^2/2$, while those from a sample of size $n_i = 100$ have variance $99\sigma^2/100$, and *one ought to expect residuals from larger samples to be somewhat bigger in magnitude than those from small samples.*

A way of addressing at least the issue that residuals need not have a common variance is through the use of **standardized residuals**.

Definition 2 If a residual e has variance $a \cdot \sigma^2$ for some positive constant a, and s is some estimate of σ, the **standardized residual** corresponding to e is

$$e^* = \frac{e}{s\sqrt{a}} \qquad \textbf{(7.12)}$$

The division by $s\sqrt{a}$ in equation (7.12) is a division by an estimated standard deviation of e. It serves, so to speak, to put all of the residuals on the same scale.

Plotting with standardized residuals

Standardized residuals for the one-way model

$$e_{ij}^* = \frac{e_{ij}}{s_{\mathrm{P}}\sqrt{\dfrac{n_i - 1}{n_i}}} \tag{7.13}$$

is a somewhat more refined way of judging the adequacy of the one-way model than the plotting of raw residuals e_{ij} illustrated in Examples 1 and 2. When all n_i are the same, as in Example 1, the plotting of the standardized residuals in equation (7.13) is completely equivalent to plotting with the raw residuals. And as a practical matter, unless some n_i are very small and others are very large, the standardization used in equation (7.13) typically doesn't have much effect on the appearance of residual plots.

Example 2
(continued)

In the spring constant study, allowing for the fact that sample 1 is larger than the other two (and thus according to the model (7.1) should produce larger residuals) doesn't materially change the outcome of the residual analysis. To see this, note that using the summary statistics in Table 7.5,

$$s_{\mathrm{P}}^2 = \frac{(7 - 1)(.134)^2 + (6 - 1)(.074)^2 + (6 - 1)(.064)^2}{(7 - 1) + (6 - 1) + (6 - 1)} = .0097$$

so that

$$s_{\mathrm{P}} = \sqrt{.0097} = .099$$

Then using equation (7.13), each residual from sample 1 should be divided by

$$.099\sqrt{\frac{7 - 1}{7}} = .0913$$

to get standardized residuals, while each residual from the second and third samples should be divided by

$$.099\sqrt{\frac{6 - 1}{6}} = .0900$$

Clearly, .0913 and .0900 are not much different, and the division before plotting has little effect on the appearance of residual plots. By way of example, a normal plot of all 19 standardized residuals is given in Figure 7.10. Verify its similarity to the normal plot of all 19 raw residuals given in Figure 7.9 on page 454.

Example 2
(*continued*)

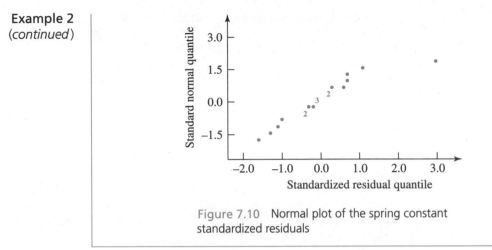

Figure 7.10 Normal plot of the spring constant standardized residuals

The notion of standardized residuals is often introduced only in the context of curve- and surface-fitting analyses, where the variances of residuals $e = (y - \hat{y})$ depend not only on the sizes of the samples involved but also on the associated values of the independent or predictor variables $(x_1, x_2, \ldots,$ etc.). The concept has been introduced here, not only because it can be of importance in the present situation if the sample sizes vary widely but also because it is particularly easy to motivate the idea in the present context.

Section 1 Exercises

1. Return again to the data of Example 1 in Chapter 4. These may be viewed as simply $r = 5$ samples of $m = 3$ densities. (For the time being, ignore the fact that the pressure variable is quantitative and that curve fitting seems a most natural method of analysis to apply to these data.)
   (a) Compute and make a normal plot of the residuals for the one-way model. What does the plot indicate about the appropriateness of the one-way model assumptions here?
   (b) Using the five samples, find s_P, the pooled estimate of σ. What does this value measure in this context? Give a two-sided 90% confidence interval for σ based on s_P.

2. In an ISU engineering research project, so called "tilttable tests" were done in order to determine the angles at which vehicles experience lift-off of the "high-side" wheels and begin to roll over. So called "tilttable ratios" (which are the tangents of angles at which lift-off occurs) were measured for four different vans with the following results:

Van #1	Van #2	Van #3	Van #4
1.096, 1.093	.962, .970	1.010, 1.024	1.002, 1.001
1.090, 1.093	.967, .966	1.021, 1.020	1.002, 1.004
		1.022	

(Notice that Van #3 was tested five times while the others were tested four times each.) Vans #1 and #2 were minivans, and Vans #3 and #4 were full-size vans.
   (a) Compute and normal-plot residuals as a crude means of investigating the appropriateness of the one-way model assumptions for tilttable ratios. Comment on the appearance of your plot.
   (b) Redo part (a) using standardized residuals.
   (c) Compute a pooled estimate of the standard deviation based on these four samples. What is s_P supposed to be measuring in this example? Give a two-sided 95% confidence interval for σ based on s_P.

··

7.2 Simple Confidence Intervals in Multisample Studies

Section 6.3 illustrates how useful confidence intervals for means and differences in means can be in one- and two-sample studies. Estimating an individual mean and comparing a pair of means are every bit as important when there are r samples as they are when there are only one or two. The methods of Chapter 6 can be applied in r-sample studies by simply limiting attention to one or two of the samples at a time. But since individual sample sizes in multisample studies are often small, such a strategy of inference often turns out to be relatively uninformative. Under the one-way model assumptions discussed in the previous section, it is possible to base inference methods on the pooled standard deviation, s_P. Those tend to be relatively more informative than the direct application of the formulas from Section 6.3 in the present context.

This section first considers the confidence interval estimation of a single mean and of the difference between two means under the "equal variances, normal distributions" model. There follows a discussion of confidence intervals for any linear combination of underlying means. Finally, the section closes with some comments concerning the notions of individual and simultaneous confidence levels.

7.2.1 Intervals for Means and for Comparing Means

The primary drawback to applying the formulas from Section 6.3 in a multisample context is that typical small sample sizes lead to small degrees of freedom, large t multipliers in the plus-or-minus parts of the interval formulas, and thus long intervals. But based on the one-way model assumptions, confidence interval formulas can be developed that tend to produce shorter intervals.

That is, in a development parallel to that in Section 6.3, under the one-way normal model,

$$T = \frac{\bar{y}_i - \mu_i}{\dfrac{s_P}{\sqrt{n_i}}}$$

has a t_{n-r} distribution. Hence, a two-sided confidence interval for the ith mean, μ_i, has endpoints

Confidence limits for μ_i based on the one-way model

$$\boxed{\bar{y}_i \pm t \frac{s_P}{\sqrt{n_i}}} \qquad (7.14)$$

where the associated confidence is the probability assigned to the interval from $-t$ to t by the t_{n-r} distribution. This is exactly formula (6.20) from Section 6.3, except that s_P has replaced s_i and the degrees of freedom have been adjusted from $n_i - 1$ to $n - r$.

In the same way, for conditions i and i', the variable

$$T = \frac{\bar{y}_i - \bar{y}_{i'} - (\mu_i - \mu_{i'})}{s_{\text{P}}\sqrt{\dfrac{1}{n_i} + \dfrac{1}{n_{i'}}}}$$

has a t_{n-r} distribution. Hence, a two-sided confidence interval for $\mu_i - \mu_{i'}$ has endpoints

Confidence limits for $\mu_i - \mu_{i'}$ based on the one-way model

$$\bar{y}_i - \bar{y}_{i'} \pm t s_{\text{P}}\sqrt{\frac{1}{n_i} + \frac{1}{n_{i'}}} \qquad\qquad (7.15)$$

where the associated confidence is the probability assigned to the interval from $-t$ to t by the t_{n-r} distribution. Display (7.15) is essentially formula (6.35) of Section 6.3, except that s_{P} is calculated based on r samples instead of two and the degrees of freedom are $n - r$ instead of $n_i + n_{i'} - 2$.

Of course, use of only one endpoint from formula (7.14) or (7.15) produces a one-sided confidence interval with associated confidence corresponding to the t_{n-r} probability assigned to the interval $(-\infty, t)$ (for $t > 0$). The virtues of formulas (7.14) and (7.15) (in comparison to the corresponding formulas from Section 6.3) are that (when appropriate) for a given confidence, they will tend to produce shorter intervals than their Chapter 6 counterparts.

Example 3
(*Example 1 revisited*)

Confidence Intervals for Individual, and Differences of, Mean Concrete Compressive Strengths

Return to the concrete strength study of Armstrong, Babb, and Campen. Consider making first a 90% two-sided confidence interval for the mean compressive strength of an individual concrete formula and then a 90% two-sided confidence interval for the difference in mean compressive strengths for two different formulas. Since $n = 24$ and $r = 8$, there are $n - r = 16$ degrees of freedom associated with $s_{\text{P}} = 581.6$. So the .95 quantile of the t_{16} distribution, namely 1.746, is appropriate for use in both formulas (7.14) and (7.15).

Turning first to the estimation of a single mean compressive strength, since each n_i is 3, the plus-or-minus part of formula (7.14) gives

$$t\frac{s_{\text{P}}}{\sqrt{n_i}} = 1.746\frac{581.6}{\sqrt{3}} = 586.3 \text{ psi}$$

So ±586.3 psi precision could be attached to any one of the sample means in Table 7.7 as an estimate of the corresponding formula's mean strength. For example, since $\bar{y}_3 = 4{,}527.3$ psi, a 90% two-sided confidence interval for μ_3 has endpoints

$$4{,}527.3 \pm 586.3$$

that is,

$$3{,}941.0 \text{ psi} \quad \text{and} \quad 5{,}113.6 \text{ psi}$$

In parallel fashion, consider estimation of the difference in two mean compressive strengths with 90% confidence. Again, since each n_i is 3, the plus-or-minus part of formula (7.15) gives

$$t s_{\text{P}} \sqrt{\frac{1}{n_i} + \frac{1}{n_{i'}}} = 1.746(581.6)\sqrt{\frac{1}{3} + \frac{1}{3}} = 829.1 \text{ psi}$$

Thus, ± 829.1 psi precision could be attached to any difference between sample means in Table 7.7 as an estimate of the corresponding difference in formula mean strengths. For instance, since $\bar{y}_3 = 4{,}527.3$ psi and $\bar{y}_7 = 1{,}551.3$ psi, a 90% two-sided confidence interval for $\mu_3 - \mu_7$ has endpoints

$$(4{,}527.3 - 1{,}551.3) \pm 829.1$$

That is,

$$2{,}146.9 \text{ psi} \quad \text{and} \quad 3{,}805.1 \text{ psi}$$

Table 7.7
Concrete Formula Sample Mean Strengths

Concrete Formula	Sample Mean Strength (psi)
1	5,635.3
2	5,753.3
3	4,527.3
4	3,442.3
5	2,923.7
6	3,324.7
7	1,551.3
8	2,390.7

The use of $n - r = 16$ degrees of freedom in Example 3 instead of $n_i - 1 = 2$ and $n_i + n_{i'} - 2 = 4$ reflects the reduction in uncertainty associated with s_{P} as an estimate of σ as compared to that of s_i and of s_P based on only two samples. That reduction is, of course, bought at the price of restriction to problems where the "equal variances" model is tenable.

7.2.2   Intervals for General Linear Combinations of Means

There is an important and simple generalization of the formulas (7.14) and (7.15) that is easy to state and motivate at this point. Its most common applications are in the context of factorial studies. But it is pedagogically most sound to introduce the method in the unstructured r-sample context, so that the logic behind it is clear and is seen not to be limited to factorial analyses. The basic notion is that μ_i and $\mu_i - \mu_{i'}$ are particular linear combinations of the r means $\mu_1, \mu_2, \ldots, \mu_r$, and the same logic that produces confidence intervals for μ_i and $\mu_i - \mu_{i'}$ will produce a confidence interval for any linear combination of the r means.

That is, suppose that for constants $c_1, c_2, \ldots, c_r$, the quantity

A linear combination of population means

$$L = c_1\mu_1 + c_2\mu_2 + \cdots + c_r\mu_r \tag{7.16}$$

is of engineering interest. (Note that, for example, if all c_i's except c_3 are 0 and $c_3 = 1$, $L = \mu_3$, the mean response from condition 3. Similarly, if $c_3 = 1$, $c_5 = -1$, and all other c_i's are 0, $L = \mu_3 - \mu_5$, the difference in mean responses from conditions 3 and 5.) A natural data-based way to approximate L is to replace the theoretical or underlying means, μ_i, with empirical or sample means, $\bar{y}_i$. That is, define an estimator of L by

A linear combination of sample means

$$\hat{L} = c_1\bar{y}_1 + c_2\bar{y}_2 + \cdots + c_r\bar{y}_r \tag{7.17}$$

(Clearly, if $L = \mu_3$, then $\hat{L} = \bar{y}_3$, while if $L = \mu_3 - \mu_5$, then $\hat{L} = \bar{y}_3 - \bar{y}_5$.)

The one-way model assumptions make it very easy to describe the distribution of $\hat{L}$ given in equation (7.17). Since $E\bar{y}_i = \mu_i$ and $\mathrm{Var}\,\bar{y}_i = \sigma^2/n_i$, one can appeal again to Proposition 1 of Chapter 5 (page 307) and conclude that

$$
\begin{aligned}
E\hat{L} &= c_1 E\bar{y}_1 + c_2 E\bar{y}_2 + \cdots + c_r E\bar{y}_r \\
&= c_1\mu_1 + c_2\mu_2 + \cdots + c_r\mu_r \\
&= L
\end{aligned}
$$

and

$$
\begin{aligned}
\mathrm{Var}\,\hat{L} &= c_1^2\,\mathrm{Var}\,\bar{y}_1 + c_2^2\,\mathrm{Var}\,\bar{y}_2 + \cdots + c_r^2\,\mathrm{Var}\,\bar{y}_r \\
&= c_1^2\frac{\sigma^2}{n_1} + c_2^2\frac{\sigma^2}{n_2} + \cdots + c_r^2\frac{\sigma^2}{n_r}
\end{aligned}
$$

$$= \sigma^2 \left(\frac{c_1^2}{n_1} + \frac{c_2^2}{n_2} + \cdots + \frac{c_r^2}{n_r} \right)$$

The one-way model restrictions imply that the $\bar{y}_i$ are independent and normal and, in turn, that $\hat{L}$ is normal. So the standardized version of $\hat{L}$,

$$Z = \frac{\hat{L} - E\hat{L}}{\sqrt{\text{Var } \hat{L}}} = \frac{\hat{L} - L}{\sigma \sqrt{\frac{c_1^2}{n_1} + \frac{c_2^2}{n_2} + \cdots + \frac{c_r^2}{n_r}}} \tag{7.18}$$

is standard normal. The usual manipulations beginning with this fact would produce an unusable confidence interval for L involving the unknown parameter σ. A way to reason to something of practical importance is to begin not with the variable (7.18), but with

$$T = \frac{\hat{L} - L}{s_P \sqrt{\frac{c_1^2}{n_1} + \frac{c_2^2}{n_2} + \cdots + \frac{c_r^2}{n_r}}} \tag{7.19}$$

instead. The fact is that under the current assumptions, the variable (7.19) has a t_{n-r} distribution. And this leads in the standard way to the fact that the interval with endpoints

Confidence limits for a linear combination of means

$$\hat{L} \pm t s_P \sqrt{\frac{c_1^2}{n_1} + \frac{c_2^2}{n_2} + \cdots + \frac{c_r^2}{n_r}} \tag{7.20}$$

can be used as a two-sided confidence interval for L with associated confidence the t_{n-r} probability assigned to the interval between $-t$ and t. Further, a one-sided confidence interval for L can be obtained by using only one of the endpoints in display (7.20) and appropriately adjusting the confidence level upward by reducing the unconfidence in half.

It is worthwhile to verify that the general formula (7.20) reduces to the formula (7.14) if a single c_i is 1 and all others are 0. And if one c_i is 1, one other is -1, and all others are 0, the general formula (7.20) reduces to formula (7.15).

Example 4 | Comparing Absorbency Properties for Three Brands of Paper Towels

D. Speltz did some absorbency testing for several brands of paper towels. His study included (among others) a generic brand and two national brands. $n_1 = n_2 = n_3 = 5$ tests were made on towels of each of these $r = 3$ brands, and the numbers of milliliters of water (out of a possible 100) not absorbed out of a

Example 4
(continued)

graduated cylinder were recorded. Some summary statistics for the tests on these brands are given in Table 7.8. Plots (not shown here) of the raw absorbency values and residuals indicate no problems with the use of the one-way model in the analysis of the absorbency data.

One question of practical interest is "On average, do the national brands absorb more than the generic brand?" A way of quantifying this is to ask for a two-sided 95% confidence interval for

$$L = \mu_1 - \frac{1}{2}(\mu_2 + \mu_3) \tag{7.21}$$

the difference between the average liquid left by the generic brand and the arithmetic mean of the national brand averages.

With L as in equation (7.21), formula (7.17) shows that

$$\hat{L} = 93.2 - \frac{1}{2}(81.0) - \frac{1}{2}(83.8) = 10.8 \text{ ml}$$

is an estimate of the increased absorbency offered by the national brands. Using the standard deviations given in Table 7.8,

$$s_P^2 = \frac{(5-1)(.8)^2 + (5-1)(.7)^2 + (5-1)(.8)^2}{(5-1) + (5-1) + (5-1)} = .59$$

and thus

$$s_P = \sqrt{.59} = .77 \text{ ml}$$

Now $n - r = 15 - 3 = 12$ degrees of freedom are associated with s_P, and the .975 quantile of the t_{12} distribution for use in (7.20) is 2.179. In addition, since $c_1 = 1$, $c_2 = -\frac{1}{2}$, and $c_3 = -\frac{1}{2}$ and all three sample sizes are 5,

$$\sqrt{\frac{c_1^2}{n_1} + \frac{c_2^2}{n_2} + \frac{c_3^2}{n_3}} = \sqrt{\frac{1}{5} + \frac{\left(-\frac{1}{2}\right)^2}{5} + \frac{\left(-\frac{1}{2}\right)^2}{5}} = .55$$

Table 7.8
Summary Statistics for Absorbencies of Three
Brands of Paper Towels

Brand	i	n_i	$\bar{y}_i$	s_i
Generic	1	5	93.2 ml	.8 ml
National B	2	5	81.0 ml	.7 ml
National V	3	5	83.8 ml	.8 ml

So finally, endpoints for a two-sided 95% confidence interval for L given in equation (7.21) are

$$10.8 \pm 2.179(.77)(.55)$$

that is,

$$10.8 \pm .9$$

i.e.,

$$9.9 \text{ ml} \quad \text{and} \quad 11.7 \text{ ml} \tag{7.22}$$

The interval indicated in display (7.22) shows definitively the substantial advantage in absorbency held by the national brands over the generic, particularly in view of the fact that the amount actually absorbed by the generic brand appears to average only about 6.8 ml ($= 100 \text{ ml} - 93.2 \text{ ml}$).

Example 5

A Confidence Interval for a Main Effect in a 2^2 Factorial Brick Fracture Strength Study

Graves, Lundeen, and Micheli studied the fracture strength properties of brick bars. They included several experimental variables in their study, including both bar composition and heat-treating regimen. Part of their data are given in Table 7.9. Modulus of rupture values under a bending load are given in psi for $n_1 = n_2 = n_3 = n_4 = 3$ bars of $r = 4$ types.

Table 7.9
Modulus of Rupture Measurements for Brick Bars in a 2^2 Factorial Study

i, Bar Type	% Water in Mix	Heat-Treating Regimen	MOR (psi)
1	17	slow cool	4911, 5998, 5676
2	19	slow cool	4387, 5388, 5007
3	17	fast cool	3824, 3140, 3502
4	19	fast cool	4768, 3672, 3242

Notice that the data represented in Table 7.9 have a 2×2 complete factorial structure. Indeed, returning to Section 4.3 (in particular, to Definition 5, page 166),

Example 5
(*continued*)

it becomes clear that the fitted main effect of the factor Heat-Treating Regimen at its slow cool level is

$$\frac{1}{2}(\bar{y}_1 + \bar{y}_2) - \frac{1}{4}(\bar{y}_1 + \bar{y}_2 + \bar{y}_3 + \bar{y}_4) \qquad (7.23)$$

But the variable (7.23) is the $\hat{L}$ for the linear combination of mean strengths μ_1, μ_2, μ_3, and μ_4 given by

$$L = \frac{1}{4}\mu_1 + \frac{1}{4}\mu_2 - \frac{1}{4}\mu_3 - \frac{1}{4}\mu_4 \qquad (7.24)$$

So subject to the relevance of the "equal variances, normal distributions" description of modulus of rupture for fired brick clay bodies of these four types, formula (7.20) can be applied to develop a precision figure to attach to the fitted effect (7.23).

Table 7.10 gives summary statistics for the data of Table 7.9. Using the values in Table 7.10 leads to

$$\hat{L} = \frac{1}{2}(\bar{y}_1 + \bar{y}_2) - \frac{1}{4}(\bar{y}_1 + \bar{y}_2 + \bar{y}_3 + \bar{y}_4)$$

$$= \frac{1}{4}\bar{y}_1 + \frac{1}{4}\bar{y}_2 - \frac{1}{4}\bar{y}_3 - \frac{1}{4}\bar{y}_4$$

$$= \frac{1}{4}(5,528.3 + 4,927.3 - 3,488.7 - 3,894.0)$$

$$= 768.2 \text{ psi}$$

and

$$s_{\mathrm{P}} = \sqrt{\frac{(3-1)(558.3)^2 + (3-1)(505.2)^2 + (3-1)(342.2)^2 + (3-1)(786.8)^2}{(3-1) + (3-1) + (3-1) + (3-1)}}$$

$$= 570.8 \text{ psi}$$

Table 7.10
Summary Statistics for the
Modulus of Rupture Measurements

i, Bar Type	$\bar{y}_i$	s_i
1	5,528.3	558.3
2	4,927.3	505.2
3	3,488.7	342.2
4	3,894.0	786.8

Further, $n - r = 12 - 4 = 8$ degrees of freedom are associated with s_p. There-fore, if one wanted (for example) a two-sided 98% confidence interval for L given in equation (7.24), the necessary .99 quantile of the t_8 distribution is 2.896. Then, since

$$\sqrt{\frac{\left(\frac{1}{4}\right)^2}{3} + \frac{\left(\frac{1}{4}\right)^2}{3} + \frac{\left(-\frac{1}{4}\right)^2}{3} + \frac{\left(-\frac{1}{4}\right)^2}{3}} = .2887$$

a two-sided 98% confidence interval for L has endpoints

$$768.2 \pm 2.896(570.8)(.2887)$$

that is,

$$291.1 \text{ psi} \quad \text{and} \quad 1{,}245.4 \text{ psi} \tag{7.25}$$

Display (7.25) establishes convincingly the effectiveness of a slow cool regimen in increasing MOR. It says that the differences in sample mean MOR values for slow- and fast-cooled bricks are not simply reflecting background variation. In fact, multiplying the endpoints in display (7.25) each by 2 in order to get a confidence interval for

$$2L = \frac{1}{2}(\mu_1 + \mu_2) - \frac{1}{2}(\mu_3 + \mu_4)$$

shows that (when averaged over 17% and 19% water mixtures) the slow, cool regimen seems to offer an increase in MOR in the range from

$$582.2 \text{ psi} \quad \text{to} \quad 2{,}490.8 \text{ psi}$$

7.2.3 Individual and Simultaneous Confidence Levels

This section has introduced a variety of confidence intervals for multisample studies. In a particular application, several of these might be used, perhaps several times each. For example, even in the relatively simple context of Example 4 (the paper towel absorbency study), it would be reasonable to desire confidence intervals for each of

$$\mu_1, \mu_2, \mu_3, \mu_1 - \mu_2, \mu_1 - \mu_3, \mu_2 - \mu_3, \text{ and } \mu_1 - \frac{1}{2}(\mu_2 + \mu_3)$$

Since many confidence statements are often made in multisample studies, it is important to reflect on the meaning of a confidence level and realize that it is attached to one interval at a time. If many 90% confidence intervals are made, the 90% figure applies to the intervals **individually**. One is "90% sure" of the first interval, *separately* "90% sure" of the second, *separately* "90% sure" of the third, and so on. It is not at all clear how to arrive at a reliability figure for the intervals jointly or simultaneously (i.e., an a priori probability that all the intervals are effective). But it is fairly obvious that it must be less than 90%. That is, the **simultaneous or joint confidence** (the overall reliability figure) to be associated with a group of intervals is generally not easy to determine, but it is typically less (and sometimes much less) than the *individual* confidence level(s) associated with the intervals one at a time.

There are at least three different approaches to be taken once the difference between simultaneous and individual confidence levels is recognized. The most obvious option is to make individual confidence intervals and be careful to interpret them as such (being careful to recognize that as the number of intervals one makes increases, so does the likelihood that among them are one or more intervals that fail to cover the quantities they are meant to locate).

A second way of handling the issue of simultaneous versus individual confidence is to use very large individual confidence levels for the separate intervals and then employ a somewhat crude inequality to find at least a minimum value for the simultaneous confidence associated with an entire group of intervals. That is, if k confidence intervals have associated confidences $\gamma_1, \gamma_2, \ldots, \gamma_k$, the **Bonferroni inequality** says that the simultaneous or joint confidence that all k intervals are effective (say, γ) satisfies

The Bonferroni inequality

$$\gamma \geq 1 - \left((1 - \gamma_1) + (1 - \gamma_2) + \cdots + (1 - \gamma_k)\right) \qquad (7.26)$$

(Basically, this statement says that the joint "unconfidence" associated with k intervals $(1 - \gamma)$ is no larger than the sum of the k individual unconfidences. For example, five intervals with individual 99% confidence levels have a joint or simultaneous confidence level of at least 95%.)

The third way of approaching the issue of simultaneous confidence is to develop and employ methods that for some specific, useful set of unknown quantities provide intervals with a known level of simultaneous confidence. There are whole books full of such simultaneous inference methods. In the next section, two of the better known and simplest of these are discussed.

Section 2 Exercises

1. Return to the situation of Exercise 1 of Section 7.1 (and the pressure/density data of Example 1 in Chapter 4).
   (a) Individual two-sided confidence intervals for the five different means here would be of the form $\bar{y}_i \pm \Delta$ for a number Δ. If 95% individual confidence is desired, what is Δ? If all five of these intervals are made, what does the Bonferroni inequality guarantee for a minimum joint or simultaneous confidence?
   (b) Individual two-sided confidence intervals for the differences in the five different means

would be of the form $\bar{y}_i - \bar{y}_{i'} \pm \Delta$ for a number Δ. If 95% individual confidence is desired, what is Δ?

(c) Note that if mean density is a linear function of pressure over the range of pressures from 2,000 to 6,000 psi, then $\mu_{4000} - \mu_{2000} = \mu_{6000} - \mu_{4000}$, that is $L = \mu_{6000} - 2\mu_{4000} + \mu_{2000}$ has the value 0. Give 95% two-sided confidence limits for this L. What does your interval indicate about the linearity of the pressure/density relationship?

2. Return to the tilttable testing problem of Exercise 2 of Section 7.1.

(a) Make (individual) 99% two-sided confidence intervals for the four different mean tilttable ratios for the four vans, μ_1, μ_2, μ_3 and μ_4. What does the Bonferroni inequality guarantee for a minimum joint or simultaneous confidence for these four intervals?

(b) Individual confidence intervals for the differences between particular pairs of mean tilttable ratios are of the form $\bar{y}_i - \bar{y}_{i'} \pm \Delta$ for appropriate values of Δ. Find values of Δ if individual 99% two-sided intervals are desired, first for pairs of means with samples of size 4 and then for pairs of means where one sample size is 4 and the other is 5.

(c) It might be of interest to compare the average of the tilttable ratios for the minivans to that of the full-size vans. Give a 99% two-sided confidence interval for the quantity $\frac{1}{2}(\mu_1 + \mu_2) - \frac{1}{2}(\mu_3 + \mu_4)$.

3. Explain the difference between several intervals having associated 95% individual confidences and having associated 95% simultaneous confidence.

7.3 Two Simultaneous Confidence Interval Methods

As Section 7.2 illustrated, there are several kinds of confidence intervals for means and linear combinations of means that could be made in a multisample study. The issue of individual versus simultaneous confidence was also raised, but only the use of the Bonferroni inequality was given as a means of controlling a simultaneous confidence level.

This section presents two methods for making a number of confidence intervals and in the process maintaining a desired simultaneous confidence. The first of these is due to Pillai and Ramachandran; it provides a guaranteed simultaneous confidence in the estimation of all r individual underlying means. The second is Tukey's method for the simultaneous confidence interval estimation of all differences in pairs of underlying means.

7.3.1 The Pillai-Ramachandran Method

One of the things typically of interest in an r-sample statistical engineering study is the **estimation of all r individual mean responses** $\mu_1, \mu_2, \ldots, \mu_r$. If the individual confidence interval formula of Section 7.2,

$$\bar{y}_i \pm t \frac{s_P}{\sqrt{n_i}} \tag{7.27}$$

is applied r times to estimate these means, the only handle one has on the corresponding simultaneous confidence is given by the Bonferroni inequality (7.26). This fairly crude tool says that if $r = 8$ and one wants 95% simultaneous confidence, individual "unconfidences" of $\frac{.05}{8} = .00625$ (i.e., individual confidences of 99.375%) for the eight different intervals will suffice to produce the desired simultaneous confidence.

Another approach to the setting of simultaneous confidence limits on all of $\mu_1, \mu_2, \ldots, \mu_r$ is to replace t in formula (7.27) with a multiplier derived specifically for the purpose of providing an exact, stated, *simultaneous* confidence in the estimation of all the means. Such multipliers were derived by Pillai and Ramachandran, where either all of the intervals for the r means are two-sided or all are one-sided. That is, Table B.8A gives values of constants k_2^* such that the r two-sided intervals with respective endpoints

P-R two-sided simultaneous 95% confidence limits for r means

$$\bar{y}_i \pm k_2^* \frac{s_P}{\sqrt{n_i}} \tag{7.28}$$

have simultaneous 95% confidence as intervals for the means $\mu_1, \mu_2, \ldots, \mu_r$. (These values k_2^* are in fact .95 quantiles of the *Studentized maximum modulus distributions*.)

Table B.8B gives values of some other constants k_1^* such that if for each i from 1 to r, an interval of the form

P-R one-sided simultaneous 95% confidence intervals for r means

$$\left(-\infty, \bar{y}_i + k_1^* \frac{s_P}{\sqrt{n_i}} \right) \tag{7.29}$$

or of the form

P-R one-sided simultaneous 95% confidence intervals for r means

$$\left(\bar{y}_i - k_1^* \frac{s_P}{\sqrt{n_i}}, \infty \right) \tag{7.30}$$

is made as a confidence interval for μ_i, the simultaneous confidence associated with the collection of r one-sided intervals is 95%. (These k_1^* values are in fact .95 quantiles of the *Studentized extreme deviate distributions*.)

In this book, the use of r intervals of one of the forms (7.28) through (7.30) will be called the P-R method of simultaneous confidence intervals. In order to apply the P-R method, one must find (using interpolation as needed) the entry in Tables B.8 in the column corresponding to the number of samples, r, and the row corresponding to the degrees of freedom associated with s_P, namely $\nu = n - r$.

Example 6
(*Example 3 revisited*)

Simultaneous Confidence Intervals
for Eight Mean Concrete Compressive Strengths

Consider again the concrete strength study of Armstrong, Babb, and Campen. Recall that tests on $n_i = 3$ specimens of each of $r = 8$ different concrete formulas gave $s_P = 581.6$ psi. Using formula (7.27) and remembering that there are $n - r = 16$ degrees of freedom associated with s_P, one has endpoints for 95% two-sided intervals for the formula mean compressive strengths

$$\bar{y}_i \pm 2.120 \frac{581.6}{\sqrt{3}}$$

that is,

$$y_i \pm 711.9 \text{ psi} \tag{7.31}$$

In contrast to intervals (7.31), consider the use of formula (7.28) to produce $r = 8$ two-sided intervals for the formula mean strengths with *simultaneous* 95% confidence. Table B.8A shows that $k_7^* = 3.099$ is appropriate in this application. So each concrete formula mean compressive strength, μ_i, should be estimated using

$$\bar{y}_i \pm 3.099 \frac{581.6}{\sqrt{3}}$$

that is,

$$\bar{y}_i \pm 1{,}040.6 \text{ psi} \tag{7.32}$$

Expressions (7.31) and (7.32) provide two-sided intervals for the eight mean compressive strengths. If one-sided intervals of the form $(\#, \infty)$ were desired instead, consulting the t table for the .95 quantile of the t_{16} distribution and use of formula (7.27) shows that the values

$$\bar{y}_i - 1.746 \frac{581.6}{\sqrt{3}}$$

that is,

$$\bar{y}_i - 586.3 \text{ psi} \tag{7.33}$$

are individual 95% lower confidence bounds for the formula mean compressive strengths, μ_i. At the same time, consulting Table B.8B shows that for

Example 6
(continued)

simultaneous 95% confidence, use of $k_1^* = 2.779$ in formula (7.30) is appropriate, and the values

$$\bar{y}_i - 2.779 \frac{581.6}{\sqrt{3}}$$

that is,

$$\bar{y}_i - 933.2 \text{ psi} \qquad\qquad \textbf{(7.34)}$$

are simultaneous 95% lower confidence bounds for the formula mean compressive strengths, μ_i.

Comparing intervals (7.31) with intervals (7.32) and bounds (7.33) with bounds (7.34) shows clearly the impact of requiring simultaneous rather than individual confidence. *For a given nominal confidence level, the simultaneous intervals must be bigger (more conservative) than the corresponding individual intervals.*

It is common practice to summarize the information about mean responses gained in a multisample study in a plot of sample means versus sample numbers, enhanced with "error bars" around the sample means to indicate the uncertainty associated with locating the means. There are various conventions for the making of these bars. When looking at such a plot, one typically forms an overall visual impression. Therefore, it is our opinion that error bars derived from the P-R simultaneous confidence limits of display (7.28) are the most sensible representation of what is known about a group of r means. For example, Figure 7.11 is a graphical representation of the eight formula sample mean strengths given in Table 7.7 with $\pm 1{,}040.6$ psi error bars, as indicated by expression (7.32).

When looking at a display like Figure 7.11, it is important to remember that what is represented is the precision of knowledge about the *mean* strengths, rather than any kind of predictions for individual compressive strengths. In this regard, the similarity of the spread of the samples on the side-by-side dot diagram given as Figure 7.1 and the size of the error bars here is coincidental. As sample sizes increase, spreads on displays of individual measurements like Figure 7.1 will tend to stabilize (representing the spreads of the underlying distributions), while the lengths of error bars associated with means will shrink to 0 as increased information gives sharper and sharper evidence about the underlying means.

In any case, Figure 7.11 shows clearly that the information in the data is quite adequate to establish the existence of differences in formula mean compressive strengths.

7.3.2   Tukey's Method

A second set of quantities often of interest in an r-sample study consists of the **differences in all $\frac{r(r-1)}{2}$ pairs of mean responses** μ_i **and** $\mu_{i'}$. Section 7.2 argued

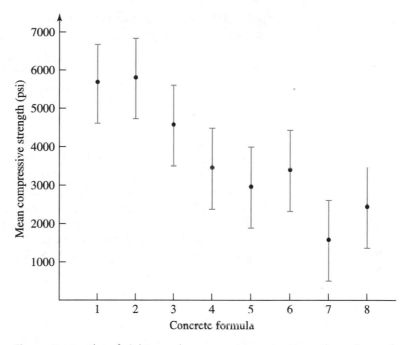

Figure 7.11 Plot of eight sample mean compressive strengths, enhanced with error bars derived from P-R simultaneous confidence limits

that a single difference in mean responses, $\mu_i - \mu_{i'}$, can be estimated using an interval with endpoints

$$\bar{y}_i - \bar{y}_{i'} \pm t s_P \sqrt{\frac{1}{n_i} + \frac{1}{n_{i'}}} \tag{7.35}$$

where the associated confidence level is an individual one. But if, for example, $r = 8$, there are 28 different two-at-a-time comparisons of underlying means to be considered (μ_1 versus μ_2, μ_1 versus μ_3, ..., μ_1 versus μ_8, μ_2 versus μ_3, ..., and μ_7 versus μ_8). If one wishes to guarantee a reasonable simultaneous confidence level for all these comparisons via the crude Bonferroni idea, a huge individual confidence level is required for the intervals (7.35). For example, the Bonferroni inequality requires 99.82% individual confidence for 28 intervals in order to guarantee simultaneous 95% confidence.

A better approach to the setting of simultaneous confidence limits on all of the differences $\mu_i - \mu_{i'}$ is to replace t in formula (7.35) with a multiplier derived specifically for the purpose of providing exact, stated, simultaneous confidence in the estimation of all such differences. J. Tukey first pointed out that it is possible to provide such multipliers using quantiles of the *Studentized range distributions*.

Tables B.9A and B.9B give values of constants q^* such that the set of two-sided intervals with endpoints

Tukey's two-sided simultaneous confidence limits for all differences in r means

$$\bar{y}_i - \bar{y}_{i'} \pm \frac{q^*}{\sqrt{2}} s_\text{P} \sqrt{\frac{1}{n_i} + \frac{1}{n_{i'}}} \qquad (7.36)$$

has simultaneous confidence at least 95% or 99% (depending on whether $Q(.95)$ is read from Table B.9A or $Q(.99)$ is read from Table B.9B) in the estimation of all differences $\mu_i - \mu_{i'}$. If all the sample sizes $n_1, n_2, \ldots, n_r$ are equal, the 95% or 99% nominal simultaneous confidence figure is exact, while if the sample sizes are not all equal, the true value is at least as big as the nominal value.

In order to apply Tukey's method, one must find (using interpolation as needed) the column in Tables B.9 corresponding to the number of samples/means to be compared and the row corresponding to the degrees of freedom associated with s_P, (namely, $\nu = n - r$).

Example 6
(continued)

Consider the making of confidence intervals for differences in formula mean compressive strengths. If a 95% two-sided individual confidence interval is desired for a *specific* difference $\mu_i - \mu_{i'}$, formula (7.35) shows that appropriate endpoints are

$$\bar{y}_i - \bar{y}_{i'} \pm 2.120(581.6)\sqrt{\frac{1}{3} + \frac{1}{3}}$$

that is,

$$\bar{y}_i - \bar{y}_{i'} \pm 1{,}006.7 \text{ psi} \qquad (7.37)$$

On the other hand, if one plans to estimate *all* differences in mean compressive strengths with *simultaneous* 95% confidence, by formula (7.36) Tukey two-sided intervals with endpoints

$$\bar{y}_i - \bar{y}_{i'} \pm \frac{4.90}{\sqrt{2}}(581.6)\sqrt{\frac{1}{3} + \frac{1}{3}}$$

that is,

$$\bar{y}_i - \bar{y}_{i'} \pm 1{,}645.4 \text{ psi} \qquad (7.38)$$

are in order (4.90 is the value in the $r = 8$ column and $\nu = 16$ row of Table B.9A.)

In keeping with the fact that the confidence level associated with the intervals (7.38) is a simultaneous one, the Tukey intervals are wider than those indicated in formula (7.37).

The plus-or-minus part of display (7.38) is not as big as twice the plus-or-minus part of expression (7.32). Thus, when looking at Figure 7.11, it is not necessary that the error bars around two means fail to overlap before it is safe to judge the corresponding underlying means to be detectably different. Rather, it is only necessary that the two sample means differ by the plus-or-minus part of formula (7.36)—1,645.4 psi in the present situation.

This section has mentioned only two of many existing methods of simultaneous confidence interval estimation for multisample studies. These should serve to indicate the general character of such methods and illustrate the implications of a simultaneous (as opposed to individual) confidence guarantee.

One final word of caution has to do with the theoretical justification of all of the methods found in this section. It is the "equal variances, normal distributions" model that supports these engineering tools. If any real faith is to be put in the nominal confidence levels attached to the P-R and Tukey methods presented here, that faith should be based on evidence (typically gathered, at least to some extent, as illustrated in Section 7.1) that the standard one-way normal model is a sensible description of a physical situation.

Section 3 Exercises

1. Return to the situation of Exercises 1 of Sections 7.1 and 7.2 (and the pressure/density data of Example 1 in Chapter 4).
   (a) Using the P-R method, what Δ can be employed to make two-sided intervals of the form $\bar{y}_i \pm \Delta$ for all five mean densities, possessing simultaneous 95% confidence? How does this Δ compare to the one computed in part (a) of Exercise 1 of Section 7.2?
   (b) Using the Tukey method, what Δ can be employed to make two-sided intervals of the form $\bar{y}_i - \bar{y}_{i'} \pm \Delta$ for all differences in the five mean densities, possessing simultaneous 95% confidence? How does this Δ compare to the one computed in part (b) of Exercise 1 of Section 7.2?

2. Return to the tilttable study of Exercises 2 of Sections 7.1 and 7.2.

   (a) Use the P-R method of simultaneous confidence intervals and make simultaneous 95% two-sided confidence intervals for the four mean tilttable ratios.
   (b) Simultaneous confidence intervals for the differences in all pairs of mean tilttable ratios are of the form $\bar{y}_i - \bar{y}_{i'} \pm \Delta$. Find appropriate values of Δ if simultaneous 99% two-sided intervals are desired, first for pairs of means with samples of size 4 and then for pairs of means where one sample size is 4 and the other is 5. How do these compare to the intervals you found in part (b) of Exercise 2 of Section 7.2? Why is it reasonable that the Δ's should be related in this way?

7.4 One-Way Analysis of Variance (ANOVA)

This book's approach to inference in multisample studies has to this point been completely "interval-oriented." But there are also significance-testing methods that are appropriate to the multiple-sample context. This section considers some of these and the issues raised by their introduction. It begins with some general comments regarding significance testing in r-sample studies. Then the one-way analysis of variance (ANOVA) test for the equality of r means is discussed. Next, the one-way ANOVA table and the organization and intuition that it provides are presented. Finally, there is a brief look at the one-way random effects model and ANOVA-based inference for its parameters.

7.4.1 Significance Testing and Multisample Studies

Just as there are many quantities one might want to estimate in a multisample study, there are potentially many issues of statistical significance to be judged. For instance, one might desire p-values for hypotheses like

$$H_0: \mu_3 = 7 \tag{7.39}$$

$$H_0: \mu_3 - \mu_7 = 0 \tag{7.40}$$

$$H_0: \mu_1 - \frac{1}{2}(\mu_2 + \mu_3) = 0 \tag{7.41}$$

The confidence interval methods discussed in Section 7.2 have their significance-testing analogs for treating hypotheses that, like all three of these, involve linear combinations of the means $\mu_1, \mu_2, \ldots, \mu_r$.

In general (under the standard one-way model), if

$$L = c_1\mu_1 + c_2\mu_2 + \cdots + c_r\mu_r$$

the hypothesis

$$H_0: L = \# \tag{7.42}$$

can be tested using the test statistic

$$T = \frac{\hat{L} - \#}{s_P\sqrt{\dfrac{c_1^2}{n_1} + \dfrac{c_2^2}{n_2} + \cdots + \dfrac{c_r^2}{n_r}}} \tag{7.43}$$

and a t_{n-r} reference distribution. This fact specializes to cover hypotheses of types (7.39) to (7.41) by appropriate choice of the c_i and #.

But the significance-testing method most often associated with the one-way normal model is not for hypotheses of the type (7.42). Instead, the most common method concerns the hypothesis that all r underlying means have the same value. In symbols, this is

$$H_0: \mu_1 = \mu_2 = \cdots = \mu_r \qquad (7.44)$$

Given that one is working under the assumptions of the one-way model to begin with, hypothesis (7.44) amounts to a statement that all r underlying distributions are essentially the same—or "There are no differences between treatments."

Hypothesis (7.44) can be thought of in terms of the simultaneous equality of $\frac{r(r-1)}{2}$ pairs of means—that is, as equivalent to the statement that simultaneously

$$\mu_1 - \mu_2 = 0, \quad \mu_1 - \mu_3 = 0, \quad \ldots, \quad \mu_1 - \mu_r = 0,$$
$$\mu_2 - \mu_3 = 0, \quad \ldots, \quad \text{and} \quad \mu_{r-1} - \mu_r = 0$$

And this fact should remind the reader of the ideas about simultaneous confidence intervals from the previous section (specifically, Tukey's method). In fact, one way of judging the statistical significance of an r-sample data set in reference to hypothesis (7.44) is to apply Tukey's method of simultaneous interval estimation and note whether or not all the intervals for differences in means include 0. If they all do, the associated p-value is larger than 1 minus the simultaneous confidence level. If not all of the intervals include 0, the associated p-value is smaller than 1 minus the simultaneous confidence level. (If simultaneous 95% intervals all include 0, no differences between means are definitively established, and the corresponding p-value exceeds .05.)

We admit a bias toward estimation over testing per se. A consequence of this bias is a fondness for deriving a rough idea of a p-value for hypothesis (7.44) as a byproduct of Tukey's method. But a most famous significance-testing method for hypothesis (7.44) also deserves discussion: the one-way analysis of variance test. (At this point it may seem strange that a test about means has a name apparently emphasizing variance. The motivation for this jargon is that the test is associated with a very helpful way of thinking about partitioning the overall variability that is encountered in a response variable.)

7.4.2   The One-Way ANOVA *F* Test

The standard method of testing the hypothesis (7.44)

$$H_0: \mu_1 = \mu_2 = \cdots = \mu_r$$

of no differences among r means against

$$H_a: \text{not } H_0 \qquad (7.45)$$

is based essentially on a comparison of a measure of variability among the sample means to the pooled sample variance, s_P^2. In order to fully describe this method some additional notational conventions are needed.

Repeatedly in the balance of this book, it will be convenient to have symbols for the summary measures of Section 3.3 (sample means and variances) applied to the data from multisample studies, *ignoring the fact that there are r different samples involved*. Already the unsubscripted letter n has been used to stand for $n_1 + n_2 + \cdots + n_r$, the number of observations in hand ignoring the fact that r samples are involved. This kind of convention will now be formally extended to include statistics calculated from the n responses. For emphasis, this will be stated in definition form.

Definition 3 (*A Notational Convention* *for Multisample Studies*)	In multisample studies, symbols for sample sizes and sample statistics appearing without subscript indices or dots will be understood to be calculated from all responses in hand, obtained by combining all samples.

So n will stand for the total number of data points (even in an r-sample study), $\bar{y}$ for the grand sample average of response y, and s^2 for a grand sample variance calculated completely ignoring sample boundaries.

For present purposes (of writing down a test statistic for testing hypothesis (7.44)), one needs to make use of $\bar{y}$, the grand sample average. It is important to recognize that $\bar{y}$ and

The (unweighted)
average of r sample
means

$$\bar{y}_. = \frac{1}{r}\sum_{i=1}^{r}\bar{y}_i \tag{7.46}$$

are not necessarily the same unless all sample sizes are equal. That is, when sample sizes vary, $\bar{y}$ is the (unweighted) arithmetic average of the raw data values y_{ij} but is a weighted average of the sample means $\bar{y}_i$. On the other hand, $\bar{y}_.$ is the (unweighted) arithmetic mean of the sample means $\bar{y}_i$ but is a weighted average of the raw data values y_{ij}. For example, in the simple case that $r = 2$, $n_1 = 2$, and $n_2 = 3$,

$$\bar{y} = \frac{1}{5}(y_{11} + y_{12} + y_{21} + y_{22} + y_{23}) = \frac{2}{5}\bar{y}_1 + \frac{3}{5}\bar{y}_2$$

while

$$\bar{y}_. = \frac{1}{2}(\bar{y}_1 + \bar{y}_2) = \frac{1}{4}y_{11} + \frac{1}{4}y_{12} + \frac{1}{6}y_{21} + \frac{1}{6}y_{22} + \frac{1}{6}y_{23}$$

and, in general, $\bar{y}$ and $\bar{y}_.$ will not be the same.

Now, under the hypothesis (7.44), that $\mu_1 = \mu_2 = \cdots = \mu_r$, $\bar{y}$ is a natural estimate of the common mean. (All underlying distributions are the same, so the data in hand are reasonably thought of not as r different samples, but rather as a single sample of size n.) Then the differences $\bar{y}_i - \bar{y}$ are indicators of possible differences among the μ_i. It is convenient to summarize the size of these differences $\bar{y}_i - \bar{y}$ in terms of a kind of total of their squares—namely,

$$\sum_{i=1}^{r} n_i (\bar{y}_i - \bar{y})^2 \tag{7.47}$$

One can think of statistic (7.47) either as a weighted sum of the quantities $(\bar{y}_i - \bar{y})^2$ or as an unweighted sum, where there is a term in the sum for each raw data point and therefore n_i of the type $(\bar{y}_i - \bar{y})^2$. The quantity (7.47) is a measure of the **between-sample variation** in the data. For a given set of sample sizes, the larger it is, the more variation there is between the sample means $\bar{y}_i$.

In order to produce a test statistic for hypothesis (7.44), one simply divides the measure (7.47) by $(r-1)s_P^2$, giving

One-way ANOVA test statistic for equality of r means

$$F = \frac{\dfrac{1}{r-1} \sum_{i=1}^{r} n_i (\bar{y}_i - \bar{y})^2}{s_P^2} \tag{7.48}$$

The fact is that if $H_0: \mu_1 = \mu_2 = \cdots = \mu_r$ is true, the one-way model assumptions imply that this statistic has an $F_{r-1, n-r}$ distribution. So the hypothesis of equality of r means can be tested using the statistic in equation (7.48) with an $F_{r-1, n-r}$ reference distribution, where large observed values of F are taken as evidence against H_0 in favor of H_a: not H_0.

Example 7
(*Example 1 revisited*)

Returning again to the concrete compressive strength study of Armstrong, Babb, and Campen, $\bar{y} = 3{,}693.6$ and the 8 sample means $\bar{y}_i$ have differences from this value given in Table 7.11.

Then since each $n_i = 3$, in this situation,

$$\sum_{i=1}^{r} n_i (\bar{y}_i - \bar{y})^2 = 3(1{,}941.7)^2 + 3(2{,}059.7)^2 + \cdots$$

$$+ 3(-2{,}142.3)^2 + 3(-1{,}302.9)^2$$

$$= 47{,}360{,}780 \text{ (psi)}^2$$

Example 7
(*continued*)

Table 7.11
Sample Means and Their
Deviations from $\bar{y}$ in the Concrete
Strength Study

i, Formula	$\bar{y}_i$	$\bar{y}_i - \bar{y}$
1	5,635.3	1,941.7
2	5,753.3	2,059.7
3	4,527.3	833.7
4	3,442.3	−251.3
5	2,923.7	−769.9
6	3,324.7	−368.9
7	1,551.3	−2,142.3
8	2,390.7	−1,302.9

In order to use this figure to judge statistical significance, one standardizes via equation (7.48) to arrive at the observed value of the test statistic

$$f = \frac{\frac{1}{8-1}(47{,}360{,}780)}{(581.6)^2} = 20.0$$

It is easy to verify from Tables B.6 that 20.0 is larger than the .999 quantile of the $F_{7,16}$ distribution. So

$$p\text{-value} = P[\text{an } F_{7,16} \text{ random variable} \geq 20.0] < .001$$

That is, the data provide overwhelming evidence that $\mu_1, \mu_2, \ldots, \mu_8$ are not all equal.

For pedagogical reasons, the one-way ANOVA test has been presented after discussing interval-oriented methods of inference for r-sample studies. But if it is to be used in applications, the testing method typically belongs chronologically before estimation. That is, the ANOVA test can serve as a *screening device* to determine whether the data in hand are adequate to differentiate conclusively between the means, or whether more data are needed.

7.4.3 The One-Way ANOVA Identity and Table

Associated with the ANOVA test statistic is some strong intuition related to the partitioning of observed variability. This is related to an algebraic identity that is stated here in the form of a proposition.

Proposition 1	For any n numbers y_{ij}

*One-way
ANOVA
identity*

$$(n-1)s^2 = \sum_{i=1}^{r} n_i(\bar{y}_i - \bar{y})^2 + (n-r)s_P^2 \qquad \textbf{(7.49)}$$

or in other symbols,

*A second statement
of the one-way
ANOVA identity*

$$\sum_{i,j}(y_{ij} - \bar{y})^2 = \sum_{i=1}^{r} n_i(\bar{y}_i - \bar{y})^2 + \sum_{i=1}^{r}\sum_{j=1}^{n_i}(y_{ij} - \bar{y}_i)^2 \qquad \textbf{(7.50)}$$

Proposition 1 should begin to shed some light on the phrase "analysis of variance." It says that an overall measure of variability in the response y, namely,

$$(n-1)s^2 = \sum_{i,j}(y_{ij} - \bar{y})^2$$

can be partitioned or decomposed algebraically into two parts. One,

$$\sum_{i=1}^{r} n_i(\bar{y}_i - \bar{y})^2$$

can be thought of as measuring variation between the samples or "treatments," and the other,

$$(n-r)s_P^2 = \sum_{i=1}^{r}\sum_{j=1}^{n_i}(y_{ij} - \bar{y}_i)^2$$

measures variation within the samples (and in fact consists of the sum of the squared residuals). The F statistic (7.48), developed for testing $H_0: \mu_1 = \mu_2 = \cdots = \mu_r$, has a numerator related to the first of these and a denominator related to the second. So using the ANOVA F statistic amounts to a kind of analyzing of the raw variability in y.

In recognition of their prominence in the calculation of the one-way ANOVA F statistic and their usefulness as descriptive statistics in their own right, the three sums (of squares) appearing in formulas (7.49) and (7.50) are usually given special names and shorthand. These are stated here in definition form.

Definition 4	In a multisample study, $(n-1)s^2$, the sum of squared differences between the raw data values and the grand sample mean, will be called the **total sum of squares** and denoted as *SSTot*.

Definition 5	In an unstructured multisample study, $\sum n_i(\bar{y}_i - \bar{y})^2$ will be called the **treatment sum of squares** and denoted as *SSTr*.

Definition 6	In a multisample study, the sum of squared residuals, $\sum(y - \hat{y})^2$ (which is $(n-r)s_P^2$ in the unstructured situation) will be called the **error sum of squares** and denoted as *SSE*.

In the new notation introduced in these definitions, Proposition 1 states that in an unstructured multisample context,

A third statement of the one-way ANOVA identity

$$SSTot = SSTr + SSE \qquad (7.51)$$

Partially as a means of organizing calculation of the F statistic given in formula (7.48) and partially because it reinforces and extends the variance partitioning insight provided by formulas (7.49), (7.50), and (7.51), it is useful to make an **ANOVA table**. There are many forms of ANOVA tables corresponding to various multisample analyses. The form most relevant to the present situation is given in symbolic form as Table 7.12.

The column headings in Table 7.12 are Source (of variation), <u>S</u>um of <u>S</u>quares (corresponding to the source), <u>d</u>egrees of <u>f</u>reedom (corresponding to the source), <u>M</u>ean <u>S</u>quare (corresponding to the source), and <u>F</u> (for testing the significance of the source in contributing to the overall observed variability). The entries in the Source column of the table are shown here as being Treatments, Error, and Total. But the name Treatments is sometimes replaced by Between (Samples), and the

Table 7.12
General Form of the One-Way ANOVA Table

ANOVA Table (for testing $H_0: \mu_1 = \mu_2 = \cdots = \mu_r$)				
Source	*SS*	*df*	*MS*	*F*
Treatments	*SSTr*	$r-1$	$SSTr/(r-1)$	*MSTr/MSE*
Error	*SSE*	$n-r$	$SSE/(n-r)$	
Total	*SSTot*	$n-1$		

name Error is sometimes replaced by Within (Samples) or Residual. The first two entries in the *SS* column must sum to the third, as indicated in equation (7.51). Similarly, the Treatments and Error degrees of freedom add to the Total degrees of freedom, $(n - 1)$. Notice that the entries in the df column are those attached to the numerator and denominator, respectively, of the test statistic in equation (7.48). The ratios of sums of squares to degrees of freedom are called mean squares, here the mean square for treatments (*MSTr*) and the mean square for error (*MSE*). Verify that in the present context, $MSE = s_P^2$ and *MSTr* is the numerator of the F statistic given in equation (7.48). So the single ratio appearing in the F column is the observed value of F for testing $H_0 : \mu_1 = \mu_2 = \cdots = \mu_r$.

Example 7
(continued)

Consider once more the concrete strength study. It is possible to return to the raw data given in Table 7.1 and find that $\bar{y} = 3{,}693.6$, so

$$SSTot = (n - 1)s^2$$
$$= (5{,}800 - 3{,}693.6)^2 + (4{,}598 - 3{,}693.6)^2 + (6{,}508 - 3{,}693.6)^2$$
$$+ \cdots + (2{,}631 - 3{,}693.6)^2 + (2{,}490 - 3{,}693.6)^2$$
$$= 52{,}772{,}190 \text{ (psi)}^2$$

Further, as in Section 7.1, $s_P^2 = 338{,}213.1$ (psi)^2 and $n - r = 16$, so

$$SSE = (n - r)s_P^2 = 5{,}411{,}410 \text{ (psi)}^2$$

And from earlier in this section,

$$SSTr = \sum_{i=1}^{r} n_i(\bar{y}_i - \bar{y})^2 = 47{,}360{,}780$$

Then, plugging these and appropriate degrees of freedom values into the general form of the one-way ANOVA table produces the table for the concrete compressive strength study, presented here as Table 7.13.

Table 7.13
One-Way ANOVA Table for the Concrete Strength Study

| **ANOVA Table (for testing $H_0 : \mu_1 = \mu_2 = \cdots = \mu_8$)** | | | | |
Source	*SS*	*df*	*MS*	*F*
Treatments	47,360,780	7	6,765,826	20.0
Error	5,411,410	16	338,213	
Total	52,772,190	23		

Example 7
(*continued*)

Notice that, as promised by the one-way ANOVA identity, the sum of the treatment and error sums of squares is the total sum of squares. Also, Table 7.13 serves as a helpful summary of the testing process, showing at a glance the observed value of F, the appropriate degrees of freedom, and $s_P^2 = MSE$.

The computations here are by no means impossible to do "by hand." But the most sensible way to handle them is to employ a statistical package. Printout 1 shows the results of using MINTAB to create an ANOVA table. (The routine under MINITAB's "Stat/ANOVA/One-way" menu was used.)

Printout 1   ANOVA Table for a One-Way Analysis
             of the Concrete Strength Data

```
One-way Analysis of Variance

Analysis of Variance for strength
Source      DF        SS        MS        F        P
formula      7  47360781   6765826    20.00    0.000
Error       16   5411409    338213
Total       23  52772190

                                   Individual 95% CIs For Mean
                                   Based on Pooled StDev
Level     N      Mean     StDev    -----+---------+---------+---------+-
1         3    5635.3     965.6                            (---*----)
2         3    5753.3     432.3                            (---*---)
3         3    4527.3     509.9                     (---*----)
4         3    3442.3     356.4               (----*---)
5         3    2923.7     852.9            (---*----)
6         3    3324.7     353.5              (----*---)
7         3    1551.3     505.5   (----*---)
8         3    2390.7     302.5        (----*---)
                                   -----+---------+---------+---------+-
Pooled StDev =    581.6            1600      3200      4800      6400
```

You may recall having used a breakdown of a "raw variation in the data" earlier in this text (namely, in Chapter 4). In fact, there is a direct connection between the present discussion and the discussion and use of R^2 in Sections 4.1, 4.2, and 4.3. (See Definition 3 in Chapter 4 and its use throughout those three sections.) In the present notation, the coefficient of determination defined as a descriptive measure in Section 4.1 is

The coefficient of determination in general sums of squares notation

$$R^2 = \frac{SSTot - SSE}{SSTot} \tag{7.52}$$

(Fitted values for the present situation are the sample means and SSE is the sum of squared residuals here, just as it was earlier.) Expression (7.52) is a perfectly general recasting of the definition of R^2 into "SS" notation. In the present one-way context, the one-way identity (7.51) makes it possible to rewrite the numerator of the right-hand side of formula (7.52) as *SSTr*. So in an unstructured *r*-sample study (where the fitted values are the sample means)

The coefficient of determination in a one-way analysis

$$R^2 = \frac{SSTr}{SSTot} \tag{7.53}$$

That is, the first entry in the *SS* column of the ANOVA table divided by the total entry of that column can be taken as "the fraction of the raw variability in *y* accounted for in the process of fitting the equation $y_{ij} \approx \mu_i$ to the data."

Example 7
(continued)

In the concrete compressive strength study, a look at Table 7.13 and equation (7.53) shows that

$$R^2 = \frac{SSTr}{SSTot} = \frac{47,360,780}{52,772,190} = .897$$

That is, another way to describe these data is to say that differences between concrete formulas account for nearly 90% of the raw variability observed in compressive strength.

So the ANOVA breakdown of variability not only facilitates the testing of $H_0 \colon \mu_1 = \mu_2 = \cdots = \mu_r$ but it also makes direct connection with the earlier descriptive analyses of what part of the raw variability is accounted for in fitting a model equation.

7.4.4 Random Effects Models and Analyses *(Optional)*

On occasion, the *r* particular conditions leading to the *r* samples in a multisample study are not so much of interest in and of themselves, as they are of interest as representing a wider set of conditions. For example, in the nondestructive testing of critical metal parts, if $n_i = 3$ mechanical wave travel-time measurements are made on each of $r = 6$ parts selected from a large lot of such parts, the six particular parts are of interest primarily as they provide information on the whole lot.

In such situations, rather than focusing formal inference on the particular *r* means actually represented in the data (i.e., $\mu_1, \mu_2, \ldots, \mu_r$), it is more natural to make inferences about *the mechanism that generates the means* μ_i. And it is possible, under appropriate model assumptions, to use the ANOVA ideas introduced in this section in this way. The balance of this section is concerned with how this is done.

The most commonly used probability model for the analysis of *r*-sample data, where the *r* conditions actually studied represent a much wider set of conditions of interest, is a variation on the one-way model of this chapter called the **one-way random effects model**. It is built on the usual one-way assumptions that

$$y_{ij} = \mu_i + \epsilon_{ij} \tag{7.54}$$

Random effects model assumptions

where the ϵ_{ij} are iid normal $(0, \sigma^2)$ random variables. But it doesn't treat the means μ_i as parameters/unknown constants. Instead, the means $\mu_1, \mu_2, \ldots, \mu_r$ are treated as (unobservable) *random variables* independent of the ϵ_{ij}'s and themselves iid according to some normal distribution with an unknown mean μ and unknown variance σ_τ^2. The random variables μ_i are now called **random (treatment) effects**, and the variances σ^2 and σ_τ^2 are called **variance components**. The objects of formal inference become μ (the mean of the random effects) and the two variance components σ^2 and σ_τ^2.

Example 8

Magnesium Contents at Different Locations on an Alloy Rod and the Random Effects Model

Youden's *Experimentation and Measurement* contains an interesting data set concerned with the magnesium contents of different parts of a long rod of magnesium alloy. A single ingot had been drawn into a rod of about 100 m in length, with a square cross section about 4.5 cm on a side. $r = 5$ flat test pieces 1.2 cm thick were cut from the rod (after it had been cut into 100 bars and 5 of these randomly selected to represent the rod), and multiple magnesium determinations were made on the 5 specimens. $n_i = 10$ of the resulting measurements for each specimen are given in Table 7.14. (There were actually other observations made not listed in Table 7.14. And some additional structure in Youden's original data

Table 7.14
Measured Magnesium Contents for Five Alloy Specimens

Specimen 1	Specimen 2	Specimen 3	Specimen 4	Specimen 5
76	69	73	73	70
71	71	69	75	66
70	68	68	69	68
67	71	69	72	68
71	66	70	69	64
65	68	70	69	70
67	71	65	72	69
71	69	67	63	67
66	70	67	69	69
68	68	64	69	67
$\bar{y}_1 = 69.2$	$\bar{y}_2 = 69.1$	$\bar{y}_3 = 68.2$	$\bar{y}_4 = 70.0$	$\bar{y}_5 = 67.8$
$s_1 = 3.3$	$s_2 = 1.7$	$s_3 = 2.6$	$s_4 = 3.3$	$s_5 = 1.9$

will also be ignored for present purposes.) The units of measurement in Table 7.14 are .001% magnesium.

In this example, on the order of 8,300 test specimens could be cut from the 100 m rod. The purpose of creating the rod was to provide secondary standards for field calibration of chemical analysis instruments. That is, laboratories purchasing pieces of this rod could use them as being of "known" magnesium content to calibrate their instruments. As such, the practical issues at stake here are not primarily how the $r = 5$ particular test specimens analyzed compare. Rather, the issues are what the overall magnesium content is and whether or not the rod is consistent enough in content along its length to be of any use as a calibration tool. A random effects model and inference for the mean effect μ and the variance components are quite natural in this situation. Here, σ_τ^2 represents the variation in magnesium content among the potentially 8,300 different test specimens, and σ^2 represents measurement error plus variation in magnesium content within the 1.2 cm thick specimens, test location to test location.

When all of the r sample sizes n_i are the same (say, equal to m), it turns out to be quite easy to do some diagnostic checking of the aptness of the normal random effects model (7.54) and make subsequent inferences about μ, σ^2, and σ_τ^2. So this discussion will be limited to cases of equal sample sizes.

As far as investigation of the reasonableness of the model restrictions on the distribution of the μ_i and inference for μ are concerned, a key observation is that

$$\bar{y}_i = \frac{1}{m} \sum_{j=1}^{m} (\mu_i + \epsilon_{ij}) = \mu_i + \bar{\epsilon}_i$$

(where, of course, $\bar{\epsilon}_i$ is the sample mean of $\epsilon_{i1}, \ldots, \epsilon_{im}$). Under the random effects model (7.54), these $\bar{y}_i = \mu_i + \bar{\epsilon}_i$ are iid normal variables with mean μ and variance $\sigma_\tau^2 + \sigma^2/m$. So normal-plotting the $\bar{y}_i$ is a sensible method of at least indirectly investigating the appropriateness of the normal distribution assumption for the μ_i. In addition, the fact that the model says the $\bar{y}_i$ are independent normal variables with mean μ and a common variance suggests that the small-sample inference methods from Section 6.3 should simply be applied to the sample means $\bar{y}_i$ in order to do inference for μ. In doing so, the "sample size" involved is the number of $\bar{y}_i$'s—namely, r.

Example 8
(continued)

For the magnesium alloy rod, the $r = 5$ sample means are in Table 7.14. Figure 7.12 gives a normal plot of those five values, showing no obvious problems with a normal random effects model for specimen magnesium contents.

To find a 95% two-sided confidence interval for μ, we calculate as follows (treating the five values $\bar{y}_i$ as "observations"). The sample mean (of $\bar{y}_i$'s) is

$$\bar{y}_. = \frac{1}{5} \sum_{i=1}^{5} \bar{y}_i = 68.86$$

Example 8
(*continued*)

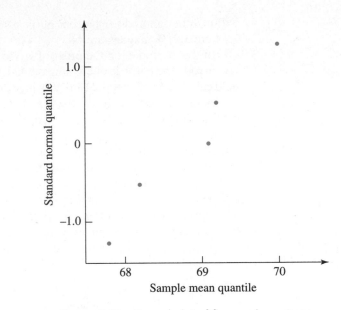

Figure 7.12 Normal plot of five specimen mean magnesium contents

and the sample variance (of $\bar{y}_i$'s) is

$$\frac{1}{5-1}\sum_{i=1}^{5}(\bar{y}_i - \bar{y}_.)^2 = .76$$

so that the sample standard deviation (of $\bar{y}_i$'s) is

$$\sqrt{\frac{1}{5-1}\sum_{i=1}^{5}(\bar{y}_i - \bar{y}_.)^2} = .87$$

Applying the small-sample confidence interval formula for a single mean from Section 6.3 (since $r - 1 = 4$ degrees of freedom are appropriate), a two-sided 95% confidence for μ has endpoints

$$68.86 \pm 2.776\frac{.87}{\sqrt{5}}$$

that is,

$$67.78 \times 10^{-3}\% \quad \text{and} \quad 69.94 \times 10^{-3}\%$$

These limits provide a notion of precision appropriate for the number $68.86 \times 10^{-3}\%$ as an estimate of the rod's mean magnesium content.

It is useful to write out in symbols what was just done to get a confidence interval for μ. That is, a sample variance of $\bar{y}_i$'s was used. This is

$$\frac{1}{r-1}\sum_{i=1}^{r}(\bar{y}_i - \bar{y}_{.})^2 = \frac{1}{m(r-1)}\sum_{i=1}^{r}m(\bar{y}_i - \bar{y})^2 = \frac{1}{m(r-1)}SSTr = \frac{1}{m}MSTr$$

because all n_i are m and $\bar{y}_{.} = \bar{y}$ in this case. But this means that under the assumptions of the one-way normal random effects model, a two-sided confidence interval for μ has endpoints

Balanced data confidence limits for the overall mean in the one-way random effects model

$$\boxed{\bar{y}_{.} \pm t\sqrt{\frac{MSTr}{mr}}} \tag{7.55}$$

where t is such that the probability the t_{r-1} distribution assigns to the interval between $-t$ and t is the desired confidence. One-sided intervals are obtained in the usual way, by employing only one of the endpoints in display (7.55).

7.4.5 ANOVA Based Inference for Variance Components (*Optional*)

Turning attention to the variance components in the random effects model (7.54), first note that as far as diagnostic checking of the assumption that the ϵ_{ij} are iid normal variables and inference for $\sigma^2 -$ Var ϵ_{ij} are concerned, all of the methods of Section 7.1 remain in force. If one thinks of holding the μ_i fixed in formula (7.54), it is clear that (conditional on the μ_i) the random effects model treats the r samples as random samples from normal distributions with a common variance. So before doing inference for σ^2 (or σ_τ^2 for that matter) via usual normal theory formulas, it is advisable to do the kind of sample-by-sample normal-plotting and plotting of residuals illustrated in Section 7.1. And if it is then plausible that the ϵ_{ij} are iid normal $(0, \sigma^2)$ variables, formula (7.10) of Section 7.1 can be used to produce a confidence interval for σ^2, and significance testing for σ^2 can be done based on the fact that $r(m-1)s_P^2/\sigma^2$ has a $\chi_{r(m-1)}^2$ distribution.

Inference for σ_τ^2 borrows from things already discussed but also provides a new wrinkle or two of its own. First, significance testing for

$$H_0 : \sigma_\tau^2 = 0 \tag{7.56}$$

is made possible by the observation that if H_0 is true, then (just as when $H_0 : \mu_1 = \mu_2 = \cdots = \mu_r$ in the case where the μ_i are not random effects but fixed parameters) the $n = mr$ observations are all coming from a single normal distribution. So

ANOVA test statistic for $H_0 : \sigma_\tau^2 = 0$ in the one-way random effects model

$$F = \frac{MSTr}{MSE} \tag{7.57}$$

has an $F_{r-1, n-r}$ distribution under the assumptions of the random effects model (7.54) when the null hypothesis (7.56) holds. Thus, the same one-way ANOVA F test used to test $H_0: \mu_1 = \mu_2 = \cdots = \mu_r$ when the means μ_i are considered fixed parameters can also be used to test $H_0: \sigma_\tau^2 = 0$ under the assumptions of the random effects model.

As far as estimation goes, it doesn't turn out to be possible to give a simple confidence interval formula for σ_τ^2 directly. But what can be done in a straightforward fashion is to give both a natural ANOVA-based single-number estimate of σ_τ^2 and a confidence interval for the ratio σ_τ^2/σ^2. To accomplish the first of these, consider the mean values of random variables $MSTr$ and $MSE \ (= s_P^2)$ under the assumptions of the random effects model. Not too surprisingly,

$$E(MSE) = Es_P^2 = \sigma^2 \tag{7.58}$$

(After all, s_P^2 has been used to approximate σ^2. That the "center" of the probability distribution of s_P^2 is σ^2 should therefore seem only reassuring.) And further,

$$E(MSTr) = \sigma^2 + m\sigma_\tau^2 \tag{7.59}$$

Then, from equations (7.58) and (7.59),

$$\frac{1}{m}\big(E(MSTr) - E(MSE)\big) = \sigma_\tau^2$$

or

$$E\frac{1}{m}(MSTr - MSE) = \sigma_\tau^2 \tag{7.60}$$

So equation (7.60) suggests that the random variable

$$\frac{1}{m}(MSTr - MSE) \tag{7.61}$$

is one whose distribution is centered about the variance component σ_τ^2 and thus is a natural ANOVA-based estimator of σ_τ^2. The variable in display (7.61) is potentially negative. When that occurs, common practice is to estimate σ_τ^2 by 0. So the variable actually used to estimate σ_τ^2 is

An ANOVA-based estimator of the treatment variance

$$\hat{\sigma}_\tau^2 = \max\left(0, \frac{1}{m}(MSTr - MSE)\right) \tag{7.62}$$

Facts (7.58) and (7.60), which motivate this method of estimating σ_τ^2, are important enough that they are often included as entries in an Expected Mean Square column added to the one-way ANOVA table when testing $H_0: \sigma_\tau^2 = 0$.

Although no elementary confidence interval for σ_τ^2 is known, it is possible to give one for the ratio σ_τ^2/σ^2. A basic probability fact is that under the assumptions of the random effects model (7.54),

$$F = \frac{\dfrac{MSTr}{\sigma^2 + m\sigma_\tau^2}}{\dfrac{MSE}{\sigma^2}}$$

has an $F_{r-1,\,n-r}$ distribution. Some algebraic manipulations beginning from this fact show that the interval with endpoints

Confidence limits for σ_τ^2/σ^2 in the one-way random effects model

$$\frac{1}{m}\left(\frac{MSTr}{U \cdot MSE} - 1\right) \quad \text{and} \quad \frac{1}{m}\left(\frac{MSTr}{L \cdot MSE} - 1\right) \qquad (7.63)$$

can be used as a two-sided confidence interval for σ_τ^2/σ^2, where the associated confidence is the probability the $F_{r-1,\,n-r}$ distribution assigns to the interval (L, U). One-sided intervals for σ_τ^2/σ^2 can be had by using only one of the endpoints and choosing L or U such that the probability assigned by the $F_{r-1,\,n-r}$ distribution to (L, ∞) or $(0, U)$ is the desired confidence.

Example 8
(continued)

Consider again the measured magnesium contents for specimens cut from the 100 m alloy rod. Some normal plotting shows the "single variance normal c_{ij}" part of the model assumptions (7.54) to be at least not obviously flawed. Sample-by-sample normal plots show fair linearity (at least after allowing for the discreteness introduced in the data by the measurement scale used), except perhaps for sample 4, with its five identical values. The five sample standard deviations are roughly of the same order of magnitude, and the normal plot of residuals in Figure 7.13 is pleasantly linear. So it is sensible to consider formal inference for σ^2 and σ_τ^2 based on the normal theory model.

Table 7.15 is an ANOVA table for the data of Table 7.14. From Table 7.15, the p-value for testing $H_0: \sigma_\tau^2 = 0$ is the $F_{4,45}$ probability to the right of 1.10. According to Tables B.6, this is larger than .25, giving very weak evidence of detectable variation between specimen mean magnesium contents.

The *EMS* column in Table 7.15 is based on relationships (7.58) and (7.59) and is a reminder first that $MSE = s_P^2 = 6.88$ serves as an estimate of σ^2. So multiple magnesium determinations on a given specimen would be estimated to have a standard deviation on the order of $\sqrt{6.88} = 2.6 \times 10^{-3}\%$. Then the expected mean squares further suggest that σ_τ^2 be estimated by

$$\hat{\sigma}_\tau^2 = \frac{1}{10}(MSTr - MSE) = \frac{1}{10}(7.58 - 6.88) = .07$$

Example 8
(*continued*)

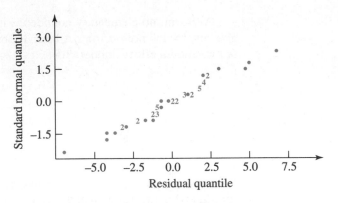

Figure 7.13 Normal plot of residuals for the magnesium content study

Table 7.15
ANOVA Table for the Magnesium Content Study

ANOVA Table (for testing $H_0: \sigma_\tau^2 = 0$)					
Source	*SS*	*df*	*MS*	*EMS*	*F*
Treatments	30.32	4	7.58	$\sigma^2 + 10\sigma_\tau^2$	1.10
Error	309.70	45	6.88	σ^2	
Total	340.02	49			

as in equation (7.62). So an estimate of σ_τ is

$$\sqrt{.07} = .26 \times 10^{-3}\%$$

That is, the standard deviation of specimen mean magnesium contents is estimated to be on the order of $\frac{1}{10}$ of the standard deviation associated with multiple measurements on a single specimen.

A confidence interval for σ^2 could be made using formula (7.10) of Section 7.1. That will not be done here, but formula (7.63) will be used to make a one-sided 90% confidence interval of the form (0, #) for σ_τ/σ. The .90 quantile of the $F_{45,4}$ distribution is about 3.80, so the .10 quantile of the $F_{4,45}$ distribution is about $\frac{1}{3.80}$. Then taking the root of the second endpoint given in display (7.63), a 90% upper confidence bound for σ_τ/σ is

$$\sqrt{\frac{1}{10}\left(\frac{7.58}{\left(\frac{1}{3.80}\right)6.88} - 1\right)} = .56$$

The bottom line here is that σ_τ is small compared to σ and is not even clearly other than 0. Most of the variation in the data of Table 7.14 is associated with the making of multiple measurements on a single specimen. Of course, this is good news if the rod is to be cut up and distributed as pieces having known magnesium contents and thus useful for measurement instrument calibration.

Section 4 Exercises

1. Return to the situation in Exercises 1 of Sections 7.1 through 7.3 (and the pressure/density data of Example 1 in Chapter 4).
   (a) In part (b) of Exercise 1 of Section 7.3, you were asked to make simultaneous confidence intervals for all differences in the $r = 5$ mean densities. From your intervals, what kind of a p-value (small or large) do you expect to find when testing the equality of these means? Explain.
   (b) Make an ANOVA table (in the form of Table 7.12) for the data of Example 1 in Chapter 4. You should do the calculations by hand first and then check your arithmetic using a statistical computer package. Then use the calculations to find both R^2 for the one-way model and also the observed level of significance for an F test of the null hypothesis that all five pressures produce the same mean density.

2. Return to the tilttable study of Exercises 2 of Sections 7.1 through 7.3.
   (a) In part (b) of Exercise 2 of Section 7.3, you were asked to make simultaneous confidence intervals for all differences in the $r = 4$ mean tilttable ratios. From your intervals, what kind of a p-value (small or large) do you expect to find when testing the equality of these means? Explain.
   (b) Make an ANOVA table (in the form of Table 7.12) for the data of Exercise 2 of Section 7.1. Then find both R^2 for the one-way model and also the observed level of significance for an F test of the null hypothesis that all four vans have the same mean tilttable ratio.

3. The following data are taken from the paper "Zero-Force Travel-Time Parameters for Ultrasonic Head-

Waves in Railroad Rail" by Bray and Leon-Salamanca (*Materials Evaluation*, 1985). Given are measurements in nanoseconds of the travel time (in excess of 36.1 μs) of a certain type of mechanical wave induced by mechanical stress in railroad rails. Three measurements were made on each of six different rails.

Rail	Travel Time (nanoseconds above 36.1 μs)
1	55, 53, 54
2	26, 37, 32
3	78, 91, 85
4	92, 100, 96
5	49, 51, 50
6	80, 85, 83

   (a) Make plots to check the appropriateness of a one-way random effects analysis of these data. What do these suggest?
   (b) Ignoring any possible problems with the standard assumptions of the random effects model revealed in (a), make an ANOVA table for these data (like Table 7.15) and find estimates of σ and σ_τ. What, in the context of this problem, do these two estimates measure?
   (c) Find and interpret a two-sided 90% confidence interval for the ratio σ_τ / σ.

4. The following are some general questions about the random effects analyses:
   (a) Explain in general terms when a random effects analysis is appropriate for use with multisample data.
   (b) Consider a scenario where $r = 5$ different technicians employed by a company each make

$m = 2$ measurements of the diameter of a particular widget using a particular gauge in a study of how technician differences show up in diameter data the company collects. Under what circumstances would a random effects analysis of the resulting data be appropriate?

(c) Suppose that the following ANOVA table was made in a random effects analysis of data like those described in part (b). Give estimates of the standard deviation associated with repeat diameter measurements for a given technician (σ) and then for the standard deviation of long-run mean measurements for various technicians (σ_τ). The sums of squares are in units of square inches.

ANOVA Table

Source	SS	df	MS	F
Technician	.0000136	4	.0000034	1.42
Error	.0000120	5	.0000024	
Total	.0000256	9		

7.5 Shewhart Control Charts for Measurement Data

This text has repeatedly made use of the phrase "stable process" and emphasized that unless data generation has associated with it a single, repeatable pattern of variation, there is no way to move from data in hand to predictions and inferences. The notion that "baseline" or "inherent" variation evident in the output of a process is a principal limitation on system performance has also been stressed. But no tools have yet been presented that are specifically crafted for evaluating the extent to which a data-generation mechanism can be thought of as stable, or for determining the size of the baseline variation of a process.

W. Shewhart, working in the late 1920s and early 1930s at Bell Laboratories, developed an extremely simple yet effective device for doing these jobs. This tool has become known as the *Shewhart control chart*. (Actually, the nonstandard name *Shewhart monitoring chart* is far more descriptive. It also avoids the connotations of automatic/feedback process adjustment that the word *control* may carry for readers familiar with the field of engineering control.)

This section and the next introduce the topic of Shewhart control charts, beginning here with charts for measurement data. This section begins with some generalities, discussing Shewhart's conceptualization of process variability. Then the specific instances of Shewhart control charts for means, ranges, and standard deviations are considered in turn. Finally, the section closes with comments about the place of control charts in the improvement of modern industrial processes.

7.5.1 Generalities about Shewhart Control Charts

Stability of an engineering data-generating process refers to a consistency or repeatability over time. When one thinks of empirically assessing the stability of a process, it is therefore clear that samples of data taken from it at different points in time will be needed.

Example 9

Monitoring the Lengths of Sheets Cut on a Ream Cutter

Shervheim and Snider worked with a company on the cutting of a rolled material into sheets using a ream cutter. Every two minutes they sampled five consecutive sheets and measured their lengths. Part of the students' length data are given in Table 7.16, in units of $\frac{1}{64}$ inch over a certain reference length.

One of the goals of the study was to investigate the stability of the cutting process over time. The kind of multisample data the students collected, where the samples were separated and ordered in time, are ideal for that purpose.

Table 7.16
Lengths of 22 Samples of Five Sheets
Cut on a Ream Cutter

Sample	Time	Excess Length
1	12:40	9, 10, 7, 8, 10
2	12:42	6, 10, 8, 8, 10
3	12:44	11, 10, 9, 5, 11
4	12:46	10, 9, 9, 8, 7
5	12:48	7, 5, 11, 9, 5
6	12:50	9, 9, 10, 7, 9
7	12:52	10, 8, 6, 11, 8
8	12:54	7, 10, 8, 8, 9
9	12:56	10, 9, 9, 5, 12
10	12:58	8, 10, 6, 8, 10
11	1:00	8, 10, 4, 7, 8
12	1.02	8, 10, 10, 6, 9
13	1:04	10, 8, 6, 7, 10
14	1:06	8, 6, 10, 8, 8
15	1:08	13, 5, 8, 8, 13
16	1:10	10, 4, 9, 10, 8
17	1:12	7, 7, 9, 7, 8
18	1:14	9, 7, 7, 9, 6
19	1:16	5, 10, 5, 8, 10
20	1:18	9, 6, 8, 9, 11
21	1:20	6, 10, 11, 5, 6
22	1:22	15, 3, 7, 9, 11

Data (like those in Table 7.16) collected for purposes of assessing process stability will often be r samples of some fixed sample size m, lacking any structure except for the fact that they were taken in a particular time order. So Shewhart control charting is at home in this chapter that treats inference methods for unstructured multisample studies.

Shewhart's fundamental qualitative insight regarding variation seen in process data over time is that

Shewhart's partition of process variation

$$\text{Overall process variation} = \text{baseline variation} + \text{variation that can be eliminated} \qquad \textbf{(7.64)}$$

Shewhart conceived of **baseline variation** as that which will remain even under the most careful process monitoring and appropriate physical interventions—an inherent property of a particular system configuration, which cannot be reduced without basic changes in the physical process or how it is run. This is variation due to **common (universal) causes** or **system causes**. Other terms used for it are **random variation** and **short-term variation**. In the context of the cutting operation of Example 9, this kind of variation might be seen in consecutive sheet lengths cut on a single ream cutter, from a single roll of material, without any intervening operator adjustments, following a particular plant standard method of machine operation, etc. It is variation that comes from hundreds of small unnameable, unidentifiable physical causes. When only this kind of variation is acting, it is reasonable to call a process "stable."

The second component of overall process variation is **variation that can potentially be eliminated** by appropriate physical intervention. This kind of variation has been called variation due to **special** or **assignable causes**, **nonrandom variation**, and **long-term variation**. In the sheet-cutting example, this might be variation in sheet length brought about by undesirable changes in tension on the material being cut, roller slippage on the cutter, unwarranted operator adjustments to the machine, eccentricities associated with how a particular incoming roll of material was wound, etc. Shewhart reasoned that being able to separate the two kinds of variation is a prerequisite to ensuring good process performance. It provides a basis for knowing when to intervene and find and eliminate the cause of any assignable variation, thereby producing process stability.

Shewhart control charts

Shewhart's method for separating the two components of overall variation in equation (7.64) is graphical and based on the following logic. First, periodically taken samples are reduced to appropriate summary statistics, and the summary statistics are plotted against time order of observation. To this simple time-plotting of summary statistics, Shewhart added the notion that lines be drawn on the chart to separate values that are consistent with a "baseline variation only" view of process performance from those that are not. Shewhart called these lines of demarcation **control limits**. When all plotted points fall within the control limits, the process is judged to be stable, subject only to chance causes. But when a point falls outside the limits, physical investigation and intervention is called for, to eliminate any assignable cause of variation. Figure 7.14 is a plot of a generic control chart for a summary statistic, w. It shows upper and lower control limits (*UCL* and *LCL*), some plotted values, and one "out of control" point.

There are any number of charts that fit the general pattern of Figure 7.14. For example, common possibilities relevant in the sheet-cutting case of Example 9 include control charts for the sample mean, sample range, and sample standard

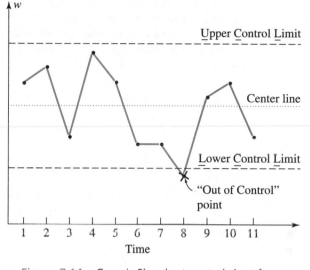

Figure 7.14   Generic Shewhart control chart for a statistic w

deviation of sheet lengths. These will presently be discussed in detail. But first, some additional generalities still need to be considered.

Setting control limits   For one thing, there remains the matter of how to set the position of the control limits. Shewhart argued that probability theory can be applied and appropriate stable-process/iid-observations distributions developed for the plotted statistics. Then small upper and lower percentage points for these can be used to establish control limits. As an example, the central limit material in Section 5.5 should have conditioned the reader to think of sample means as approximately normal with mean μ and standard deviation $\sigma/\sqrt{m}$, where μ and σ describe individual observations and m is the sample size. So for plotting sample means, the upper and lower control limits might be set at small upper and lower percentage points of the normal distribution with mean μ and standard deviation $\sigma/\sqrt{m}$, where μ and σ are a process mean and short-term standard deviation, respectively.

"Standards given" contexts   Two different circumstances are possible regarding the origin of values for process parameters used to produce control limits. In some applications, values of process parameters (and therefore, parameters for the "stable process" distribution of the plotted statistic) and thus control limits are provided from outside the data producing the charted values. Such circumstances will be called "**standards given**" situations. For emphasis, the meaning of this term is stated here in definition form.

Definition 7

> When control limits are derived from data, requirements, or knowledge of the behavior of a process that are outside the information contained in the samples whose summary statistics are to be plotted, the charting is said to be done with **standards given**.

For example, suppose that in the sheet-cutting context of Example 9, past experience with the ream cutter indicates that a process short-term standard deviation of $\sigma = 1.9$ ($\frac{1}{64}$ in.) is appropriate when the cutter is operating as it should. Further, suppose that legal and other considerations have led to the establishment of a target process mean of $\mu = 10.0$ ($\frac{1}{64}$ in. above the reference length). Then control limits based on these values and applied to data collected tomorrow would be "standards given" control limits.

"Standards given" charting and hypothesis testing

One way to think about a "standards given" control chart is as a graphical means of repeatedly testing the hypothesis

$$H_0: \text{Process parameters are at their standard values} \qquad \textbf{(7.65)}$$

When a plotted point lies inside control limits, one is directed to a decision in favor of hypothesis (7.65) for the time period in question. A point plotting outside limits makes hypothesis (7.65) untenable at the time represented by the sample.

Retrospective contexts

In contrast to "standards given" applications, there are situations in which no external values for process parameters are used. Instead, a single set of samples taken from the process is used to both develop a plausible set of parameters for the process and also to judge the stability of the process over the period represented by the data. The terms **retrospective** or **"as past data"** will be used in this text for such control charting applications.

Definition 8

When control limits are derived from the same samples whose summary statistics are plotted, the charting is said to be done **retrospectively** or "**as past data**."

In the context of Example 9, control limits derived from the data in Table 7.16 and applied to summary statistics for those same data would be "as past data" control limits for assessing the cutting process stability over the period from 12:40 through 1:22 on the day the data were taken.

Retrospective charting and hypothesis testing

A way of thinking about a retrospective control chart is a graphical means of testing the hypothesis

$$H_0: \text{A single set of process parameters was acting throughout the time period studied} \qquad \textbf{(7.66)}$$

When a point or points plot outside of control limits derived from the whole data set, the hypothesis (7.66) of process stability over the period represented by the data becomes untenable.

7.5.2 "Standards Given" $\bar{x}$ Control Charts

The single most famous and frequently used Shewhart control chart is the one where sample mean measurements are plotted. Control charts are typically named by the symbols used for the plotted statistics. So the following discussion concerns **Shewhart $\bar{x}$ charts**. In using this terminology (and other notation from the statistical quality control field), this text must choose a path through notational conflicts that exist between the most common usages in control charting and those for other multisample analyses. The options that will be exercised here must be explained.

Notational conventions for x charting

In the first place, to this point in Chapter 7 (also in Chapter 4, for that matter) the symbol y has been used for the basic response variable in a multisample statistical engineering study, $\bar{y}_i$ for a sample mean, and $\bar{\bar{y}}$ and $\bar{y}$ for unweighted and weighted averages of the $\bar{y}_i$, respectively. In contrast, in Chapters 3 and 6, where the discussion centered primarily on one- and two-sample studies, x was used as the basic response variable and $\bar{x}$ (or $\bar{x}_i$ in the case of two-sample studies) to stand for a sample mean. Standard usage in Shewhart control charting is to use the x and $\bar{x}$ ($\bar{x}_i$) convention, and the precedent is so strong that this section will adopt it as well. In addition, historical momentum in control charting dictates that rather than using $\bar{\bar{x}}$ notation,

Average sample mean (quality control notation)

$$\bar{\bar{x}} = \frac{1}{r} \sum_{i=1}^{r} \bar{x}_i \qquad (7.67)$$

is used for the average of sample means. But this "bar bar" or "double bar" notation is used in this book *only* in this section.

Something must also be said about notation for sample sizes. It is universal to use the notation n_i for an individual sample size. But there is some conflict when all sample sizes n_i have a common value. The convention in this chapter has been to use m for such a common value and n for $\sum n_i$. Standard quality control notation is to instead use n for a common sample size. In this matter, we will continue to use the conventions established thus far in Chapter 7, believing that to do otherwise invites too much confusion. But the reader is hereby alerted to the fact that the m used here is usually going to appear as n in other treatments of control charting.

Having dealt with the notational problems, we turn to the making of a "standards given" Shewhart $\bar{x}$ chart based on samples of size m. An iid model for observations from a process with mean μ and standard deviation σ produces

$$E\bar{x} = \mu \qquad (7.68)$$

and

$$\sqrt{\text{Var}\,\bar{x}} = \frac{\sigma}{\sqrt{m}} \qquad (7.69)$$

and often an approximately normal distribution for $\bar{x}$. The fact that essentially all of the probability of a normal distribution is within 3 standard deviations of its mean led Shewhart to suggest that given process standards μ and σ, $\bar{x}$ chart control limits could be set at

"Standards given"
control limits for $\bar{x}$

$$LCL_{\bar{x}} = \mu - 3\frac{\sigma}{\sqrt{m}} \quad \text{and} \quad UCL_{\bar{x}} = \mu + 3\frac{\sigma}{\sqrt{m}} \qquad \textbf{(7.70)}$$

Additionally, he suggested drawing a *center line* on an $\bar{x}$ chart at the standard mean μ.

Limits (7.70) have proved themselves of great utility even in cases where m is fairly small and there is no reason to expect a normal distribution for observations in a sampling period. Formulas (7.68) and (7.69) hold regardless of whether a process distribution is normal, and the 3-sigma (of the plotted statistic $\bar{x}$) control limits in display (7.70) tend to bracket most of the distribution of $\bar{x}$ under nearly any circumstances. (Indeed, a crude but universal analysis, based on a probability version of the Chebyschev theorem stated in Section 3.3 for relative frequency distributions, guarantees that limits (7.70) will bracket at least $\frac{8}{9}$ of the distribution of $\bar{x}$ in any stable process context.)

Example 9
(continued)

Consider the use of process standards $\mu = 10$ and $\sigma = 1.9$ in $\bar{x}$ charting based on the data given in Table 7.16 (recall the values there are in units of $\frac{1}{64}$ in. over a reference length). With these standard values for μ and σ, since the $r = 22$

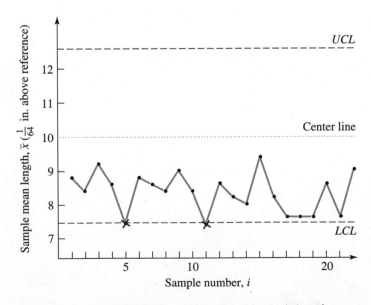

Figure 7.15 "Standards given" Shewhart $\bar{x}$ control chart for cut sheet lengths

samples are all of size $m = 5$, formulas (7.70) indicate control limits

$$UCL_{\bar{x}} = 10 + 3\frac{1.9}{\sqrt{5}} = 12.55 \quad \text{and} \quad LCL_{\bar{x}} = 10 - 3\frac{1.9}{\sqrt{5}} = 7.45$$

along with a center line drawn at $\mu = 10$. Table 7.17 gives some sample-by-sample summary statistics for the data of Table 7.16, including the sample means $\bar{x}_i$. Figure 7.15 is a "standards given" Shewhart $\bar{x}$ chart for the same data.

Figure 7.15 shows two points plotting below the lower control limit: the means for samples 5 and 11. But it is perfectly obvious from the plot what was going on in the data of Table 7.16 to produce the "out of control" points and corresponding debunking of hypothesis (7.65). Not one of the $r = 22$ plotted

Table 7.17
Sample-by-Sample Summary Statistics
for 22 Samples of Sheet Lengths

i, Sample	$\bar{x}_i$	s_i	R_i
1	8.8	1.30	3
2	8.4	1.67	4
3	9.2	2.49	6
4	8.6	1.14	3
5	7.4	2.61	6
6	8.8	1.10	3
7	8.6	1.95	5
8	8.4	1.14	3
9	9.0	2.55	7
10	8.4	1.67	4
11	7.4	2.19	6
12	8.6	1.67	4
13	8.2	1.79	4
14	8.0	1.41	4
15	9.4	3.51	8
16	8.2	2.49	6
17	7.6	.89	2
18	7.6	1.34	3
19	7.6	2.51	5
20	8.6	1.82	5
21	7.6	2.70	6
22	9.0	4.47	12
	$\sum \bar{x} = 183.4$	$\sum s = 44.41$	$\sum R = 109$

Example 9
(continued)

sample means lies at or above 10. If an average sheet length of $\mu = 10$ was truly desired, a simple adjustment was needed, to increase sheet lengths roughly

$$10 - \bar{\bar{x}} = 10 - 8.3 = 1.7 \left(\tfrac{1}{64} \text{ in.} \right)$$

The true process mean operating to produce the data was clearly below the standard mean.

7.5.3 Retrospective $\bar{x}$ Control Charts

Retrospective (or "as past data") control limits for $\bar{x}$ come about by replacing μ and σ in formulas (7.70) with estimates made from data in hand, under the provisional assumption that the process was stable over the period represented by the data. That is, in calculating such estimates, a single set of parameters is presumed to be adequate to describe process behavior during the study period. Notice that supposing process stability the present situation is exactly the one met in the ANOVA material of Section 7.4 under the hypothesis of equality of r means. So one way to think about a retrospective $\bar{x}$ chart is as a graphical test of the constancy of the process mean over time. Further, the analogy with the material of Section 7.4 suggests natural estimates of μ and σ for use in formulas (7.70).

In Section 7.4, $\bar{y}$ was used to approximate a hypothesized common value of $\mu_1, \mu_2, \ldots, \mu_r$. In the present notation, this suggests replacing μ in formulas (7.70) with $\bar{\bar{x}}$. Regarding an estimate of σ for use in formulas (7.70), analogy with all that has gone before in this chapter suggests s_P. And indeed, s_P is a perfectly rational choice. But it is not one that is commonly used. Historical precedent/accident in the quality control field has made other estimates much more widely used. These must therefore be discussed, not so much because they are better than s_P, but because they represent standard practice.

The most common way of approximating a supposedly constant σ in control charting contexts is based on probability facts about the range, R, of a sample of m observations from a normal distribution. It is possible to derive the probability density for R defined in Definition 8 in Chapter 3 (see page 95), supposing m iid normal variables with mean μ and standard deviation σ are involved. That density will not be given in this book. But it is useful to know that the mean of that distribution is (for a given sample size m) proportional to σ. The constant of proportionality is typically called d_2, and in symbols,

$$ER = d_2 \sigma \tag{7.71}$$

or equivalently,

$$\sigma = \frac{ER}{d_2} \tag{7.72}$$

Values of d_2 for various m are given in Table B.2. (Return to the comments preceding Proposition 1 in Section 3.3 and recognize that what was cryptic there should now make sense.)

Statements (7.71) and (7.72) are theoretical. The way they find practical relevance is to think that under the hypothesis that the process standard deviation is constant, the sample mean of sample ranges

Average sample range

$$\overline{R} = \frac{1}{r} \sum_{i=1}^{r} R_i \quad (7.73)$$

can be expected to approximate the theoretical mean range, ER. That is, from statement (7.72), it seems that

A range-based estimator of σ

$$\hat{\sigma} = \frac{\overline{R}}{d_2} \quad (7.74)$$

is a plausible way to estimate σ. On theoretical grounds, $\overline{R}/d_2$ is inferior to s_p, but it has the weight of historical precedent behind it, and it is simple to calculate (an important virtue before the advent of widespread computing power).

A second estimator of σ with quality control origins comes about by making the same kind of argument that led to statistic (7.74), beginning not with R but instead with s. That is, the fact that it is possible to derive a χ^2_{m-1} probability density for $(m-1)s^2/\sigma^2$ if s^2 is based on m iid normal (μ, σ^2) random variables has been used extensively (beginning in Section 6.4) in this text. That density can in turn be used to find a theoretical mean for s. As it turns out, although $Es^2 = \sigma^2$, the theoretical mean of s is not quite σ, but rather a multiple of σ (for a given sample size m). The constant of proportionality is typically called c_4, and in symbols,

$$Es = c_4 \sigma \quad (7.75)$$

or equivalently,

$$\sigma = \frac{Es}{c_4} \quad (7.76)$$

It is possible to write out an explicit expression for c_4, namely

$$c_4 = \sqrt{\frac{2}{m-1}} \left(\frac{\Gamma\left(\frac{m}{2}\right)}{\Gamma\left(\frac{m-1}{2}\right)} \right)$$

Values of c_4 for various m are given in Table B.2. From that table, it is easy to see that as a function of m, c_4 increases from about .8 when $m = 2$ to essentially 1 for large m.

The practical use made of the theoretical statements (7.75) and (7.76) is to think that the sample average of the sample standard deviations

Average sample standard deviation

$$\bar{s} = \frac{1}{r} \sum_{i=1}^{r} s_i \qquad (7.77)$$

can be expected to approximate the theoretical mean (sample) standard deviation Es, so that (from statement (7.76)) a plausible estimator of σ becomes

A standard deviation-based estimator of σ

$$\hat{\sigma} = \frac{\bar{s}}{c_4} \qquad (7.78)$$

(It is worth remarking that $\bar{s}$ is not the same as s_P, even when all sample sizes are the same. s_P is derived by averaging sample variances and then taking a square root. $\bar{s}$ comes from taking the square roots of the sample variances and then averaging. In general, these two orders of operation do not produce the same results.)

In any case, commonly used retrospective control limits for $\bar{x}$ are obtained by substituting $\bar{\bar{x}}$ given in formula (7.67) for μ and either of the estimates of σ given in displays (7.74) or (7.78) for σ in the formulas (7.70). Further, an "as past data" *center line* for an $\bar{x}$ chart is typically set at $\bar{\bar{x}}$.

Example 9
(continued)

Consider retrospective $\bar{x}$ control charting for the ream cutter data. Using the column totals given in Table 7.17, one finds from formulas (7.67), (7.73), and (7.77) that

$$\bar{\bar{x}} = \frac{183.4}{22} = 8.3$$

$$\bar{R} = \frac{109}{22} = 4.95$$

$$\bar{s} = \frac{44.41}{22} = 2.019$$

Then, consulting Table B.2 with a sample size of $m = 5$, $d_2 = 2.326$, so an estimate of σ based on $\bar{R}$ is (from expression (7.74))

$$\frac{\bar{R}}{d_2} = \frac{4.95}{2.326} = 2.13$$

Also, Table B.2 shows that for a sample size of $m = 5$, $c_4 = .9400$, so an estimate of σ based on $\bar{s}$ is (from expression (7.78))

$$\frac{\bar{s}}{c_4} = \frac{2.019}{.94} = 2.15$$

(Note that beginning from the standard deviations in Table 7.17, $s_P = 2.19$, and clearly $s_P \neq \bar{s}$.)

Using (for example) statistic (7.74), one is thus led to substitute 8.3 for μ and 2.13 for σ in "standards given" formulas (7.70) to obtain the retrospective limits

$$LCL_{\bar{x}} = 8.3 - 3\frac{2.13}{\sqrt{5}} = 5.44 \quad \text{and} \quad UCL_{\bar{x}} = 8.3 + 3\frac{2.13}{\sqrt{5}} = 11.16$$

Figure 7.16 shows an "as past data" Shewhart $\bar{x}$ control chart for the ream cutter data, using limits based on $\overline{R}$.

Notice the contrast between the pictures of the ream cutter performance given in Figures 7.15 and 7.16. Figure 7.15 shows clearly that process parameters are not at their standard values, but Figure 7.16 shows that it is perhaps plausible to think of the data in Table 7.16 as coming from *some* stable data-generating mechanism. The observed $\bar{x}$'s hover nicely (indeed—as will be argued at the end of the next section—perhaps too nicely) about a central value, showing no "out of control" points or obvious trends. That hypothesis (7.66) is at least approximately true is believable on the basis of Figure 7.16.

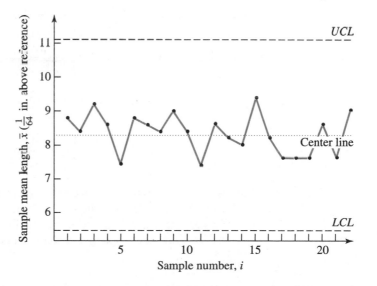

Figure 7.16   Retrospective Shewhart $\bar{x}$ control chart for cut sheet lengths

Several comments should be made before turning to a discussion of other Shewhart control charts for measurements. First, note that what is represented on an $\bar{x}$ chart is behavior (both expected and observed) of sample means, *not* individual measurements. It is unfortunately all too common to see engineering specifications (which refer to individual measurements) marked on $\bar{x}$ control charts either in place of, or in addition to, proper control limits. But how *sample means* compare to specifications for *individual measurements* tells nothing about either the stability of the process as represented in the means or the acceptability of individual measurements according to the stated engineering requirements. It is simply bad practice to mix (or mix up) control limits and specifications.

Control limits for $\bar{x}$ versus specifications for x

A second comment has to do with the fairly arbitrary choice of 3-sigma control limits in formulas (7.70). A legitimate question is, "Why not 2-sigma or 2.5-sigma or 3.09-sigma limits?" There is no completely convincing theoretical answer to this question. Indeed, arguments in favor of other multiples than 3 for use in formulas (7.70) are heard from time to time. But the forces of historical precedent and many years of successful application combine to make the use of 3-sigma limits nearly universal.

As a final point regarding $\bar{x}$ charts, the basic "standards given" formulas for control limits (7.70) are sometimes combined with formula (7.74) or (7.78) for estimating σ, and $\bar{\bar{x}}$ is put in place of μ to obtain formulas for retrospective control limits for $\bar{x}$. For example, using the estimate of σ in display (7.74), one obtains the formulas

$$LCL_{\bar{x}} = \bar{\bar{x}} - 3\frac{\overline{R}}{d_2\sqrt{m}} \quad \text{and} \quad UCL_{\bar{x}} = \bar{\bar{x}} + 3\frac{\overline{R}}{d_2\sqrt{m}} \tag{7.79}$$

In fact, it is standard practice to use the abbreviation

$$A_2 = \frac{3}{d_2\sqrt{m}}$$

and rewrite the limits in formulas (7.79) as

Range-based retrospective control limits for $\bar{x}$

$$\boxed{LCL_{\bar{x}} = \bar{\bar{x}} - A_2\overline{R} \quad \text{and} \quad UCL_{\bar{x}} = \bar{\bar{x}} + A_2\overline{R}} \tag{7.80}$$

Values of A_2 are given along with the other control chart constants in Table B.2. It is worthwhile to verify that the use of formulas (7.80) in the context of Example 9 produces exactly the retrospective control limits for $\bar{x}$ found earlier.

The version of retrospective $\bar{x}$ chart limits related to the estimate of σ in display (7.78) is

$$LCL_{\bar{x}} = \bar{\bar{x}} - 3\frac{\bar{s}}{c_4\sqrt{m}} \quad \text{and} \quad UCL_{\bar{x}} = \bar{\bar{x}} + 3\frac{\bar{s}}{c_4\sqrt{m}} \tag{7.81}$$

It is also standard practice to use the abbreviation

$$A_3 = \frac{3}{c_4\sqrt{m}}$$

and rewrite the limits in display (7.81) as

Standard deviation-based retrospective control limits for $\bar{x}$

$$\boxed{LCL_{\bar{x}} = \bar{\bar{x}} - A_3\bar{s} \quad \text{and} \quad UCL_{\bar{x}} = \bar{\bar{x}} + A_3\bar{s}}$$

(7.82)

Values of A_3 are given in Table B.2.

7.5.4   Control Charts for Ranges

The $\bar{x}$ control chart is aimed primarily at monitoring the constancy of the average process response, μ, over time. It deals only indirectly with the process short-term variation σ. (If σ increases beyond a standard value, it will produce $\bar{x}_i$ more variable than expected and eventually trigger an "out of control" point. But such a possible change in σ is detected most effectively by directly monitoring the spread of samples.) Thus, in applications, $\bar{x}$ charts are almost always accompanied by companion charts intended to monitor σ.

The conceptually simplest and most common Shewhart control charts for monitoring the process standard deviation are the **R charts**, the charts for sample ranges. In their "standards given" version, they are based again on the fact that it is possible to find a probability density for R based on m iid normal (μ, σ^2) random variables. Using this density, not only is it possible to show that $ER = d_2\sigma$ but the standard deviation of the probability distribution can be found as well. It turns out (for a given m) to be proportional to σ. The constant of proportionality is called d_3 and is tabled for various m in Table B.2. That is, for R based on m iid normal observations,

$$\sqrt{\text{Var } R} = d_3\sigma$$

(7.83)

Although the information about the theoretical distribution of R provided by formulas (7.71) and (7.83) is somewhat sketchy, it is enough to suggest possible "standards given" 3-sigma (of R) control limits for R. A plausible *center line* for a "standards given" R chart is at $ER = d_2\sigma$, and (using formula (7.83)) control limits are

$$LCL_R = ER - 3\sqrt{\text{Var } R} = d_2\sigma - 3d_3\sigma = (d_2 - 3d_3)\sigma$$

(7.84)

$$UCL_R = ER + 3\sqrt{\text{Var } R} = (d_2 + 3d_3)\sigma$$

(7.85)

The limit indicated in formula (7.84) turns out to be negative for $m \leq 6$. For those sample sizes, since ranges are nonnegative, no lower control limit is used. Formulas

(7.84) and (7.85) are typically simplified by the introduction of yet more notation. That is, standard quality control usage is to let

$$D_1 = (d_2 - 3d_3) \quad \text{and} \quad D_2 = (d_2 + 3d_3)$$

and rewrite formulas (7.84) and (7.85) as

"Standards given"
control limits for R

$$LCL_R = D_1\sigma \quad \text{and} \quad UCL_R = D_2\sigma \qquad\qquad (7.86)$$

Like the other control chart constants, D_1 and D_2 appear in Table B.2. Note that for $m \le 6$, there is no tabled value for D_1, as no lower limit is in order.

Example 9
(continued)

Consider a "standards given" control chart analysis for the sheet length ranges given in Table 7.17, using a standard $\sigma = 1.9$ ($\frac{1}{64}$ in.). Since samples of size $m = 5$ are involved, Table B.2 shows that $d_2 = 2.326$ and $D_2 = 4.918$ are appropriate for establishing a "standards given" control chart for R. The center line should be drawn at

$$d_2\sigma = 2.326(1.9) = 4.4$$

and the upper control limit should be set at

$$D_2\sigma = 4.918(1.9) = 9.3$$

(Since $m \le 6$, no lower control limit will be used.) Figure 7.17 shows a "standards given" control chart for ranges of the sheet lengths. It is clear from the figure that

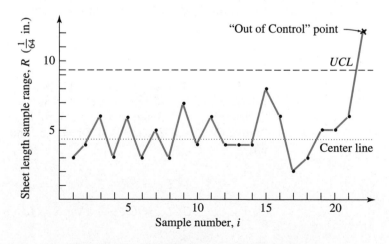

Figure 7.17   "Standards given" Shewhart *R* chart for cut sheet lengths

for the most part, a constant process standard deviation of $\sigma = 1.9$ is plausible, except for the clear indication to the contrary at sample 22. The 22nd observed range, $R = 12$, is simply larger than expected based on a sample of size $m = 5$ from a normal distribution with $\sigma = 1.9$. In practice, it would be appropriate to undertake a physical search for the cause of the apparent increase in process variability associated with the last sample taken.

As was the case for $\bar{x}$ charts, combination of formulas for the estimation of (supposedly constant) process parameters with the "standards given" limits (7.86) produces retrospective control limits for R charts. For example, basing an estimate of σ on $\overline{R}$ as in display (7.74), leads (not too surprisingly) to a retrospective *center line* for R at $d_2(\overline{R}/d_2) = \overline{R}$ and retrospective control limits

$$LCL_R = \frac{D_1\overline{R}}{d_2} \quad \text{and} \quad UCL_R = \frac{D_2\overline{R}}{d_2} \tag{7.87}$$

The abbreviations

$$D_3 = \frac{D_1}{d_2} \quad \text{and} \quad D_4 = \frac{D_2}{d_2}$$

are commonly used, and limits (7.87) are written as

Retrospective control limits for R

$$LCL_R = D_3\overline{R} \quad \text{and} \quad UCL_R = D_4\overline{R} \tag{7.88}$$

Values of the constants D_3 and D_4 are found in Table B.2.

Example 9
(continued)

For the ream cutter data, $\overline{R} = \frac{109}{22}$, so retrospective control limits for ranges of the type (7.88) put a center line at

$$\overline{R} = 4.95$$

and since for $m = 5$, $D_4 = 2.114$,

$$UCL_R = 2.114\left(\frac{109}{22}\right) = 10.5$$

Look again at Figure 7.17 and note that the use of these retrospective limits (instead of the $\sigma = 1.9$ "standards given" limits of Figure 7.17) does not materially alter the appearance of the plot. The range for sample 22 still plots above the upper control limit. It is not plausible that a single σ stands behind all of the 22

Example 9
(continued)

plotted ranges (not even $\sigma \approx \overline{R}/d_2 = 2.13$). It is pretty clear that a different physical mechanism must have been acting at sample 22 than was operative earlier.

For pedagogical reasons, $\bar{x}$ charts were considered first before turning to charts aimed at monitoring σ. In terms of order of attention in an application, however, R (or s) charts are traditionally (and correctly) given first priority. They deal directly with the baseline component of process variation. Thus (so conventional wisdom goes), if they show lack of stability, there is little reason to go on to considering the behavior of means (which deals primarily with the long-term component of process variation) until appropriate physical changes bring the ranges (or standard deviations) to the place of repeatability.

7.5.5   Control Charts for Standard Deviations

Less common but nevertheless important alternatives to range charts are control charts for standard deviations, s. In their "standards given" version, s **charts** are based on the fact that it is possible to find both a mean and variance for s calculated from m iid normal (μ, σ^2) random variables. We have already used the fact that $Es = c_4\sigma$. And it turns out that

$$\sqrt{\operatorname{Var} s} = \sqrt{1 - c_4^2}\,\sigma \tag{7.89}$$

Then formulas (7.75) and (7.89) taken together yield "standards given" 3-sigma control limits for s. That is, with a *center line* at $c_4\sigma$, one employs the limits

$$LCL_s = c_4\sigma - 3\sqrt{1 - c_4^2}\,\sigma = \left(c_4 - 3\sqrt{1 - c_4^2}\right)\sigma$$
$$UCL_s = c_4\sigma + 3\sqrt{1 - c_4^2}\,\sigma = \left(c_4 + 3\sqrt{1 - c_4^2}\right)\sigma$$

Standard notation is to let

$$B_5 = \left(c_4 - 3\sqrt{1 - c_4^2}\right) \quad \text{and} \quad B_6 = \left(c_4 + 3\sqrt{1 - c_4^2}\right)$$

so, ultimately, "standards given" control limits for s become

"Standards given" control limits for s

$$\boxed{LCL_s = B_5\sigma \quad \text{and} \quad UCL_s = B_6\sigma} \tag{7.90}$$

As expected, the constants B_5 and B_6 are tabled in Table B.2. For $m \leq 5$, $c_4 - 3\sqrt{1 - c_4^2}$ turns out to be negative, so no value is shown in Table B.2 for B_5, and no lower control limit for s is typically used for such sample sizes.

Example 9
(*continued*)

Returning once more to the ream cutter example of Shervheim and Snider, consider the monitoring of σ through the use of sample standard deviations rather than ranges, based on a standard of $\sigma = 1.9$ ($\frac{1}{64}$ in.). Table B.2 with sample size $m = 5$ once again gives $c_4 = .9400$ and also shows that $B_6 = 1.964$. So an s chart for the data of Table 7.16 has a center line at

$$c_4\sigma = (.94)(1.9) = 1.79$$

and an upper control limit at

$$UCL_s = B_6\sigma = 1.964(1.9) = 3.73$$

and, since the sample size is only 5, no lower control limit.

Figure 7.18 is a "standards given" Shewhart s chart for the s values given in Table 7.17. The story told by Figure 7.18 is essentially identical to that conveyed by the range chart in Figure 7.17. Only at sample 22 does the hypothesis that $\sigma = 1.9$ become untenable, and the need for physical intervention is indicated there.

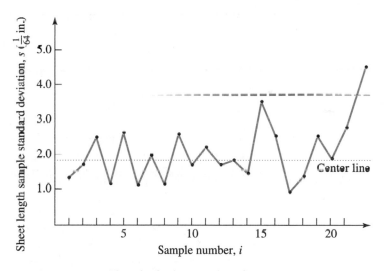

Figure 7.18 "Standards given" s chart for cut sheet lengths

As was the case for $\bar{x}$ and R charts, retrospective control limits for s can be had by replacing the parameter σ in the "standards given" limits (7.90) with any appropriate estimate. The most common way of proceeding is to employ the estimator $\bar{s}/c_4$ and thus end up with a retrospective *center line* for an s chart at $c_4(\bar{s}/c_4) = \bar{s}$ and retrospective control limits

$$LCL_s = \frac{B_5\bar{s}}{c_4} \quad \text{and} \quad UCL_s = \frac{B_6\bar{s}}{c_4} \qquad \textbf{(7.91)}$$

And using the abbreviations

$$B_3 = \frac{B_5}{c_4} \quad \text{and} \quad B_4 = \frac{B_6}{c_4}$$

the retrospective limits (7.91) are written as

Retrospective control limits for s

$$LCL_s = B_3 \bar{s} \quad \text{and} \quad UCL_s = B_4 \bar{s} \qquad (7.92)$$

Values of B_3 and B_4 are given in Table B.2.

Example 9 *(continued)*

For the ream cutter data, $\bar{s} = \frac{44.41}{22} = 2.02$, so retrospective control limits for standard deviations of the type (7.92) put a center line at

$$\bar{s} = 2.02$$

and, since $B_4 = 2.089$ for $m = 5$,

$$UCL_s = 2.089 \left(\frac{44.41}{22} \right) = 4.22$$

Look again at Figure 7.18 and verify that the use of these retrospective limits (instead of the $\sigma = 1.9$ "standards given" limits) wouldn't much change the appearance of the plot. As was the case for the retrospective R chart analysis, these retrospective s chart limits still put sample 22 in a class by itself, suggesting that a different physical mechanism produced it than that which led to the other 21 samples.

Ranges are easier to calculate "by hand" than standard deviations and are easier to explain as well. As a result, R charts are more popular than s charts. In fact, R charts are so common that the phrase "$\bar{x}$ and R charts" is often spoken in quality control circles in such a way that the $\bar{x}/R$ pair is almost implied to be a single inseparable entity. However, when computational problems and conceptual understanding are not issues, s charts are preferable to R charts because of their superior sensitivity to changes in σ.

A useful final observation about the s chart idea is that for r-sample statistical engineering studies where all sample sizes are the same, the "as past data" control limits in display (7.92) can provide some rough help in the model-checking activities of Section 7.1 (in reference to the "single variance" assumption of the one-way model). $B_3 \bar{s}$ and $B_4 \bar{s}$ can be treated as rough limits on the variation in sample standard deviations deemed to be consistent with the one-way model's single variance assumption.

Example 10
(*Example 1 revisited*)

s Chart Control Limits and the "Equal Variances" Assumption
in the Concrete Strength Study

In the concrete compressive strength study of Armstrong, Babb, and Campen, the $r = 8$ sample standard deviations based on samples of size $m = 3$ given in Table 7.3 (page 450) have $\bar{s} = 534.8$ psi. Then for $m = 3$, $B_4 = 2.568$, and so

$$B_4\bar{s} = 2.568(534.8) = 1,373 \text{ psi}$$

The largest of the eight values s_i in Table 7.3 is 965.6, and there are thus no "out of control" standard deviations. So as in Section 7.1, no strong evidence against the relevance of the "single variance" model assumption is discovered here.

7.5.6 Control Charts for Measurements and Industrial Process Improvement

The $\bar{x}$ and R (or $\bar{x}$ and s) control chart combination is an important engineering tool for the improvement of manufacturing processes. U.S. companies have trained literally hundreds of thousands of workers in the making of Shewhart $\bar{x}$ and R charts over the past few years, hoping for help in meeting the challenge of international competition. The record of success produced by this training effort is mixed. It is thus worth pausing briefly to reflect on what aid the tools of this section can and cannot rationally be expected to provide in the effort to improve industrial processes.

Out-of-control signals must produce action

In the first place, warnings of assignable variation provided by Shewhart control charts are helpful in reducing the variation of an industrial process only to the extent that they are acted on in a timely and competent fashion. If "out of control" signals don't lead to appropriate physical investigation and action to eliminate assignable causes, they contribute nothing toward improved process behavior. If workers collect data to be archived away on $\bar{x}$ and R chart forms and do not have the authority, skills, or motivation to intervene intelligently when excess process variation is indicated, they are engaged in a futile activity.

Control charts can prevent over-adjustment

Control charts can signal the need for process intervention. But perhaps nearly as important is the fact that they also tell a user when *not* to be alarmed at observed variation and give in to the temptation to adjust a stable process. This is the other side of the intervention coin. Inadvisably adjusting an industrial process that is subject only to common or random causes degrades its behavior rather than improves it. Rational use of Shewhart control charts can help prevent this possibility.

Control charts help maintain current process best performance

It is also important to say that even when properly made and acted on, Shewhart control charts can do only so much towards the improvement of industrial processes. They can be a tool for helping to reduce variation to the minimum possible for a given system configuration (in terms of equipment, methods of operation, etc.). But once that minimum has been reached, all that Shewhart charting does is to help maintain that configuration's best performance—to maintain the "baseline variation only" situation corresponding to the status quo way of doing things.

Control charts are not directly tools for innovation

In a modern world economy, however, companies cannot hope to be leaders in their industries by being content simply to maintain stable, status quo methods of operation. Instead, ways must be found for improving beyond today's methods for tomorrow. This requires thought and, often, engineering experimentation. The philosophies and methods of experimental design and engineering data collection and analysis discussed in this book have an important role in that search for improvement beyond today's best industrial methodology. But the particular role of control charting in such efforts is only indirect. By using control charts and bringing a current process to stability, a basis or foundation for improvement through experimentation and reconfiguration is provided. Indeed, it can be argued fairly convincingly that unless an existing process is repeatable, there is no sensible way of evaluating the impact of experimental changes made to it, trying to find tomorrow's improved version of the process. It is important to realize, however, that the Shewhart control charts provide only the foundation rather than the necessary subject matter expertise or statistical tools needed to guide the experimental search for improved ways of doing things.

Section 5 Exercises

1. The following are some data taken from a larger set in *Statistical Quality Control* by Grant and Leavenworth, giving the drained weights (in ounces) of contents of size No. $2\frac{1}{2}$ cans of standard grade tomatoes in puree. Twenty samples of three cans taken from a canning process at regular intervals are represented.

Sample	x_1	x_2	x_3	Sample	x_1	x_2	x_3
1	22.0	22.5	22.5	11	20.0	19.5	21.0
2	20.5	22.5	22.5	12	19.0	21.0	21.0
3	20.0	20.5	23.0	13	19.5	20.5	21.0
4	21.0	22.0	22.0	14	20.0	21.5	24.0
5	22.5	19.5	22.5	15	22.5	19.5	21.0
6	23.0	23.5	21.0	16	21.5	20.5	22.0
7	19.0	20.0	22.0	17	19.0	21.5	23.0
8	21.5	20.5	19.0	18	21.0	20.5	19.5
9	21.0	22.5	20.0	19	20.0	23.5	24.0
10	21.5	23.0	22.0	20	22.0	20.5	21.0

(a) Suppose that standard values for the process mean and standard deviation of drained weights (μ and σ) in this canning plant are 21.0 oz and 1.0 oz, respectively. Make and interpret "standards given" $\bar{x}$ and R charts based on these samples. What do these charts indicate about the behavior of the filling process over the time period represented by these data?

(b) As an alternative to the "standards given" range chart made in part (a), make a "standards given" s chart based on the 20 samples. How does its appearance compare to that of the R chart?

Now suppose that no standard values for μ and σ have been provided.

(c) Find one estimate of σ for the filling process based on the average of the 20 sample ranges, $\bar{R}$, and another based on the average of 20 sample standard deviations, $\bar{s}$. How do these compare to the pooled sample standard deviation (of Section 7.1), s_P, here?

(d) Use $\bar{\bar{x}}$ and your estimate of σ based on $\bar{R}$ and make retrospective control charts for $\bar{x}$ and R.

What do these indicate about the stability of the filling process over the time period represented by these data?

(e) Use $\bar{\bar{x}}$ and your estimate of σ based on $\bar{s}$ and make retrospective control charts for $\bar{x}$ and s. How do these compare in appearance to the retrospective charts for process mean and variability made in part (d)?

2. A manufacturer of U-bolts collects data on the thread lengths of the bolts that it produces. Nineteen samples of five consecutive bolts gave the thread lengths indicated the accompanying table (in .001 in. above nominal).

Sample	Thread Lengths	$\bar{x}$	R	s
1	11, 14, 14, 10, 8	11.4	6	2.61
2	14, 10, 11, 10, 11	11.2	4	1.64
3	8, 13, 14, 13, 10	11.6	6	2.51
4	11, 8, 13, 11, 13	11.2	5	2.05
5	13, 10, 11, 11, 11	11.2	3	1.10
6	11, 10, 10, 11, 13	11.0	3	1.22
7	8, 6, 11, 11, 11	9.4	5	2.30
8	10, 11, 10, 14, 10	11.0	4	1.73
9	11, 8, 11, 8, 10	9.6	3	1.52
10	6, 6, 11, 13, 11	9.4	7	3.21
11	11, 14, 13, 8, 11	11.4	6	2.30
12	8, 11, 10, 11, 14	10.8	6	2.17
13	11, 11, 13, 8, 13	11.2	5	2.05
14	11, 8, 11, 11, 11	10.4	3	1.34
15	11, 11, 13, 11, 11	11.4	2	.89
16	14, 13, 13, 13, 14	13.4	1	.55
17	14, 13, 14, 13, 11	13.0	3	1.22
18	13, 11, 11, 11, 13	11.8	2	1.10
19	14, 11, 11, 11, 13	12.0	3	1.41

$$\sum \bar{x} = 212.4 \quad \sum R = 77 \quad \sum s = 32.92$$

(a) Compute two different estimates of the process short-term standard deviation of thread length, one based on the sample ranges and one based on the sample standard deviations.

(b) Use your estimate from (a) based on sample standard deviations and compute control limits for the sample ranges R, and then compute control limits for the sample standard deviations s. Applying these to the R and s values, what is suggested about the threading process?

(c) Using a center line at $\bar{\bar{x}}$, and your estimate of σ based on the sample standard deviations, compute control limits for the sample means $\bar{x}$. Applying these to the $\bar{x}$ values here, what is suggested about the threading process?

(d) A check of the control chart form from which these data were taken shows that the coil of the heavy wire from which these bolts are made was changed just before samples 1, 9, and 16 were taken. What insight, if any, does this information provide into the possible origins of any patterns you see in the data?

(e) Suppose that a customer will purchase bolts of the type represented in the data only if essentially all bolts received can be guaranteed to have thread lengths within .01 in. of nominal. Does it appear that with proper process monitoring and adjustment, the equipment and manufacturing practices in use at this company will be able to produce only bolts meeting these standards? Explain in quantitative terms. If the equipment was not adequate to meet such requirements, name two options that might be taken and their practical pros and cons.

3. State briefly the practical goals of control charting and action on "out of control" signals produced by the charts.

4. Why might it well be argued that the name *control* chart invites confusion?

5. What must an engineering application of control charting involve beyond the simple naming of points plotting out of control if it is to be practically effective?

6. Explain briefly how a Shewhart $\bar{x}$ chart can help reduce variation in, say, a widget diameter, first by signaling the need for process intervention/ adjustment and then also by preventing adjustments when no "out of control" signal is given.

7.6 Shewhart Control Charts for Qualitative and Count Data

The previous section discussed Shewhart $\bar{x}$, R, and s control charts, treating them as tools for studying the stability of a system over time. This section focuses on how the Shewhart control charting idea can be applied to attributes data (i.e., counts).

The discussion begins with p charts. Next u charts and their specialization to the case of a constant-size inspection unit, the c charts, are introduced. Finally, consideration is given to a number of common nonrandom patterns that can appear on both variables control charts and attributes control charts. Possible physical causes for them and some formal rules that are often recommended for automating their recognition are discussed.

7.6.1 p Charts

This text has consistently indicated that measurements are generally preferable to attributes data. But in some situations, the only available information on the stability of a process takes the form of qualitative or count data. Consideration of the topic of control charting in such situations will begin here with **p charts** for cases where what is available for plotting are sample fractions, $\hat{p}_i$. The most common use of this is where $\hat{p}_i$ is the fraction of a sample of n_i items that is nonconforming according to some engineering standard or specification. So this section will use the "fraction nonconforming" language, in spite of the fact that $\hat{p}_i$ can be the sample fraction having any attribute of interest (desirable, undesirable, or indifferent).

The probability facts supporting control charting for the fraction nonconforming are exactly those used in Section 6.5 to develop inference methods based on $\hat{p}$. That is, if a process is stable over time, each $n_i \hat{p}_i$ is usefully modeled as binomial (n_i, p), where p is a constant likelihood that any sampled item is nonconforming. (This section will explicitly allow for sample sizes n_i varying in time. Charts for measurements are almost always based on fairly small but constant sample sizes. But charts for attributes data typically involve larger sample sizes that sometimes vary.)

As in Section 6.5, a binomial model for $n_i \hat{p}_i$ leads immediately to

$$E\hat{p}_i = p \tag{7.93}$$

and

$$\sqrt{\operatorname{Var}\hat{p}_i} = \sqrt{\frac{p(1-p)}{n_i}} \tag{7.94}$$

But then formulas (7.93) and (7.94) suggest obvious "standards given" 3-sigma control limits for the sample "fraction nonconforming" $\hat{p}_i$. That is, if p is a standard likelihood that any single item is nonconforming, then a "standards given" p chart has a *center line* at p and control limits

"Standards given" p
chart control limits

$$LCL_{\hat{p}_i} = p - 3\sqrt{\frac{p(1-p)}{n_i}}$$ (7.95)

$$UCL_{\hat{p}_i} = p + 3\sqrt{\frac{p(1-p)}{n_i}}$$ (7.96)

In the event that formula (7.95) produces a negative value, no lower control limit is used.

Example 11

p Chart Monitoring of a Pelletizing Process

Kaminski, Rasavahn, Smith, and Weitekamper worked on the same pelletizing process already used as an example several times in this book. (See Examples 2 (Chapter 1), 14 (Chapter 3), 4 (Chapter 5), and 18 (Chapter 6).) Extensive data collection on two different days led the students to establish $p = .61$ as a standard rate of nonconforming tablets produced by the process, when run under a shop standard operating regimen. On a third day, the students took $r = 25$ samples of $n_1 = n_2 = \cdots = n_{25} = m = 30$ consecutive pellets at intervals as they came off the machine and plotted sample fractions nonconforming $\hat{p}_i$, on a "standards given" p chart made with $p = .61$. Their data are given in Table 7.18.

For samples of size $n_i = m = 30$, 3-sigma "standards given" p chart control limits are, from formulas (7.95) and (7.96),

$$LCL_{\hat{p}_i} = .61 - 3\sqrt{\frac{(.61)(1-.61)}{30}} = .34$$

$$UCL_{\hat{p}_i} = .61 + 3\sqrt{\frac{(.61)(1-.61)}{30}} = .88$$

and a center line at .61 is appropriate. Figure 7.19 is a "standards given" p chart for the data of Table 7.18.

Four $\hat{p}_i$ values plot below the lower control limit in Figure 7.19, and the $\hat{p}_i$ values run consistently below the chart's center line. These facts make untenable the hypothesis that the pelletizing process was stable at the standard value of 61% nonconforming on the day these data were gathered. In this example, points plotting "out of control" on the low side are an indication of process *improvement*. They nevertheless represent a circumstance warranting physical attention to determine the physical cause for the reduced fraction defective and possibly to learn how to make the improvement permanent.

Example 11
(*continued*)

Table 7.18
Numbers and Fractions of Nonconforming
Pellets in 25 Samples of Size 30

i, Sample	$n_i \hat{p}_i$, Number Nonconforming	$\hat{p}_i$
1	13	.43
2	12	.40
3	9	.30
4	15	.50
5	17	.57
6	13	.43
7	20	.67
8	18	.60
9	18	.60
10	16	.53
11	15	.50
12	17	.57
13	15	.50
14	20	.67
15	10	.33
16	12	.40
17	17	.57
18	14	.47
19	16	.53
20	10	.33
21	14	.47
22	13	.43
23	17	.57
24	10	.33
25	12	.40

$$\sum n_i \hat{p}_i = 363$$

To make retrospective limits for a p chart, one must settle on a method of estimating the (supposedly constant) process parameter p. Here the pooling idea introduced in the two-sample context of Section 6.5 can be used. That is, as a direct extension of formula (6.71) of Section 6.5, let

Pooled estimator of a common p

$$\hat{p} = \frac{n_1 \hat{p}_1 + n_2 \hat{p}_2 + \cdots + n_r \hat{p}_r}{n_1 + n_2 + \cdots + n_r} \qquad (7.97)$$

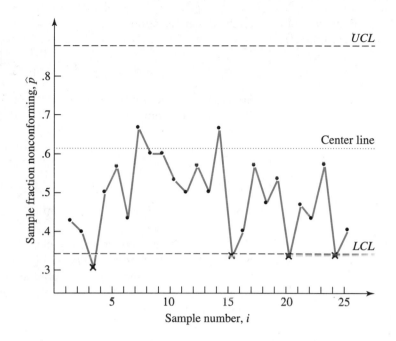

Figure 7.19   "Standards given" p chart for nonconforming pellets

($\hat{p}$ is the total number nonconforming divided by the total number inspected. When sample sizes vary, it is a weighted average of the $\hat{p}_i$.)

With $\hat{p}$ as in formula (7.97), an "as past data" Shewhart p chart has a *center line* at $\hat{p}$ and

Retrospective p chart control limits

$$LCL_{\hat{p}_i} = \hat{p} - 3\sqrt{\frac{\hat{p}(1-\hat{p})}{n_i}} \tag{7.98}$$

$$UCL_{\hat{p}_i} = \hat{p} + 3\sqrt{\frac{\hat{p}(1-\hat{p})}{n_i}} \tag{7.99}$$

As in the "standards given" context, when formula (7.98) produces a negative value, no lower control limit is used for $\hat{p}_i$.

Example 11
(*continued*)

In the pelletizing case, the total number nonconforming in the samples was $\sum n_i \hat{p}_i = 363$. Then, since $mr = 30(25) = 750$ pellets were actually inspected

Example 11
(*continued*)

on the day in question,

$$\hat{p} = \frac{363}{750} = .484$$

So a retrospective 3-sigma p chart for the data of Table 7.18 has a center line at $\hat{p} = .484$ and, from formulas (7.98) and (7.99),

$$LCL_{\hat{p}_i} = .484 - 3\sqrt{\frac{(.484)(1 - .484)}{30}} = .21$$

$$UCL_{\hat{p}_i} = .484 + 3\sqrt{\frac{(.484)(1 - .484)}{30}} = .76$$

Figure 7.20 is a retrospective p chart for the situation of Kaminski et al. All points plot within control limits on Figure 7.20. So although it is not tenable that the pelletizing process was stable at $p = .61$ over the study period, it is completely plausible that it was stable at some value of p (and $\hat{p} = .484$ is a sensible guess for that value).

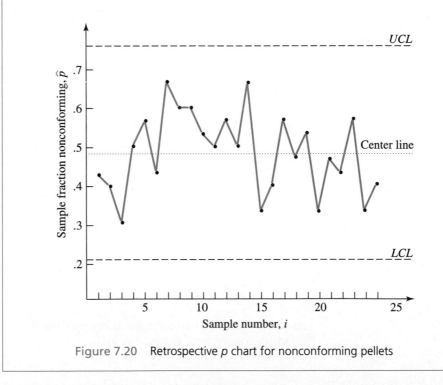

Figure 7.20   Retrospective p chart for nonconforming pellets

Because of the inherent limitations of categorical data in engineering contexts, little more will be said in this book about formal inference based on sample fractions beyond what is in Section 6.5. For example, formal significance tests of equality of r proportions, parallel to the tests of equality of r means presented in Section 7.4, won't be discussed. However, the retrospective p chart can be interpreted as a rough graphical tool for judging how sensible the hypothesis H_0: $p_1 = p_2 = \cdots = p_r$ appears.

7.6.2 u Charts

Section 3.4 introduced the notation $\hat{u}$ for the ratio of the number of occurrences of a phenomenon of interest to the total number of inspection units or items sampled in contexts where there may be multiple occurrences on a given item or inspection unit. The most common application of **u charts** based on such ratios is that of nonconformance to some engineering standard or specification. This section will use the terminology of "nonconformances per unit" in spite of the fact that $\hat{u}$ can be the sample occurrence rate for any type of phenomenon (desirable, undesirable, or indifferent).

The theoretical basis for control charting based on nonconformances per unit is found in the Poisson distributions of Section 5.1. That is, suppose that for some specified inspection unit or unit of process output of a given size, a physically stable process has an associated mean nonconformances per unit of λ and

X_i = the number of nonconformances observed **on k_i units** inspected at time i

Then a reasonable model for X_i is often the Poisson distribution with mean $k_i\lambda$. The material in Section 5.1 then says that both $EX_i = k_i\lambda$ and $\text{Var } X_i = k_i\lambda$.

But notice that if $\hat{u}_i$ is the sample nonconformances per unit observed at period i,

Rate plotted on a u chart

$$\hat{u}_i = \frac{X_i}{k_i}$$

so Proposition 1 in Chapter 5 (page 307) can be applied to produce a mean and standard deviation for $\hat{u}_i$. That is,

$$E\hat{u}_i = E\frac{X_i}{k_i} = \frac{1}{k_i}EX_i = \frac{1}{k_i}(k_i\lambda) = \lambda$$

$$\text{Var } \hat{u}_i = \text{Var } \frac{X_i}{k_i} = \frac{1}{k_i^2}\text{Var } X_i = \frac{1}{k_i^2}(k_i\lambda) = \frac{\lambda}{k_i}$$

(7.100)

so

$$\sqrt{\text{Var } \hat{u}_i} = \sqrt{\frac{\lambda}{k_i}}$$

(7.101)

The relationships (7.100) and (7.101) then motivate "standards given" 3-sigma control limits for $\hat{u}_i$. That is, if λ is a standard mean nonconformances per unit, then a "standards given" u chart has a *center line* at λ and

"Standards given"
u chart control
limits

$$LCL_{\hat{u}_i} = \lambda - 3\sqrt{\frac{\lambda}{k_i}} \qquad\qquad (7.102)$$

$$UCL_{\hat{u}_i} = \lambda + 3\sqrt{\frac{\lambda}{k_i}} \qquad\qquad (7.103)$$

The difference in formula (7.102) can turn out negative. When it does, no lower control limit is used.

Another matter of notation must be discussed at this point. λ is the symbol commonly used (as in Section 5.1) for a Poisson mean, and this fact is the basis for the usage here. However, it is more common in statistical quality control circles to use c or even c' for a standard mean nonconformances per unit. In fact, the case of the u chart where all k_i are 1 is usually referred to as a **c chart**. The λ notation used here represents the path of least confusion through this notational conflict and thus c or c' will not be used in this text. However, be aware that at least in the quality control world, there is a more popular alternative to the present λ convention.

When the limits (7.102) and (7.103) are used with nonconformances per unit data, one is essentially checking whether the prespecified λ is a plausible description of a physical process at each time period covered by the data. Often, however, there is no obvious standard occurrence rate λ, and u charting is to be done retrospectively. The question is then whether or not it is plausible that some (single) λ describes the process over all time periods covered by the data. What is needed in order to produce retrospective control limits for such cases is a way to use the $\hat{u}_i$ to make a single estimate of a supposedly constant λ. This text's approach to this problem is to make an estimate exactly analogous to the pooled estimate of p in formula (7.97). That is, let

Pooled estimator
of a common λ

$$\hat{\lambda} = \frac{k_1\hat{u}_1 + k_2\hat{u}_2 + \cdots + k_r\hat{u}_r}{k_1 + k_2 + \cdots + k_r} \qquad\qquad (7.104)$$

$\hat{\lambda}$ is the total number of nonconformances observed divided by the total number of units inspected. Then combining formula (7.104) with limits (7.102) and (7.103), a retrospective 3-sigma u chart has a *center line* at $\hat{\lambda}$ and

Retrospective u
chart control limits

$$LCL_{\hat{u}_i} = \hat{\lambda} - 3\sqrt{\frac{\hat{\lambda}}{k_i}} \qquad\qquad (7.105)$$

$$UCL_{\hat{u}_i} = \hat{\lambda} + 3\sqrt{\frac{\hat{\lambda}}{k_i}} \qquad \textbf{(7.106)}$$

As the reader might by now expect, when formula (7.105) gives a negative value, no lower control limit is employed.

Example 12

(Example 13, Chapter 3, revisited—see page 110)

---

u Chart Monitoring of the Defects per Truck Found at Final Assembly

In his book *Statistical Quality Control Methods*, I. W. Burr discusses the use of *u* charts to monitor the performance of an assembly process at a station in a truck assembly plant. Part of Burr's data were given earlier in Table 3.19. Table 7.19 gives a (partially overlapping) $r = 30$ production days' worth of Burr's data. (The values were extrapolated from Burr's figures and the fact that truck production through sample 13 was 95 trucks/day and was 130 trucks/day thereafter. Burr gives only $\hat{u}_i$ values, production rates, and the fact that all trucks produced were inspected.)

Consider the problem of control charting for these data. Since Burr gave no figure λ for the plant's standard errors per truck, this problem will be approached as one of making a retrospective *u* chart. Using formula (7.104), and the column totals from Table 7.19,

$$\hat{\lambda} = \frac{\sum X_i}{\sum k_i} = \frac{6{,}078}{3{,}445} = 1.764$$

So an "as past data" *u* chart will have a center line at 1.764 errors/truck. From formulas (7.105) and (7.106), for the first 13 days (where each k_i was 95),

$$LCL_{\hat{u}_i} = 1.764 - 3\sqrt{\frac{1.764}{95}} = 1.355 \text{ errors/truck}$$

$$UCL_{\hat{u}_i} = 1.764 + 3\sqrt{\frac{1.764}{95}} = 2.173 \text{ errors/truck}$$

On the other hand, for the last 17 days (during which 130 trucks were produced each day),

$$LCL_{\hat{u}_i} = 1.764 - 3\sqrt{\frac{1.764}{130}} = 1.415 \text{ errors/truck}$$

$$UCL_{\hat{u}_i} = 1.764 + 3\sqrt{\frac{1.764}{130}} = 2.113 \text{ errors/truck}$$

Example 12
(*continued*)

Table 7.19
Numbers and Rates of Nonconformances for a Truck Assembly Process

i, Sample	Date	k_i, Trucks Produced	$X_i = k_i \hat{u}_i$, Errors Found	$\hat{u}_i$, Errors/Truck
1	11/4	95	114	1.20
2	11/5	95	142	1.50
3	11/6	95	146	1.54
4	11/7	95	257	2.70
5	11/8	95	185	1.95
6	11/11	95	228	2.40
7	11/12	95	327	3.44
8	11/13	95	269	2.83
9	11/14	95	167	1.76
10	11/15	95	190	2.00
11	11/18	95	199	2.09
12	11/19	95	180	1.89
13	11/20	95	171	1.80
14	11/21	130	163	1.25
15	11/22	130	205	1.58
16	11/25	130	292	2.25
17	11/26	130	325	2.50
18	11/27	130	267	2.05
19	11/29	130	190	1.46
20	12/2	130	200	1.54
21	12/3	130	185	1.42
22	12/4	130	204	1.57
23	12/5	130	182	1.40
24	12/6	130	196	1.51
25	12/9	130	140	1.08
26	12/10	130	165	1.27
27	12/11	130	153	1.18
28	12/12	130	181	1.39
29	12/13	130	185	1.42
30	12/16	130	270	2.08

$$\sum k_i = 3,445 \qquad \sum X_i = 6,078$$

Notice that since k_i appears in the denominator of the plus-or-minus part of control limit formulas (7.102), (7.103), (7.105), and (7.106), the larger the inspection effort at a given time period, the tighter the corresponding control limits. This is perfectly logical. A bigger "sample size" at a given period ought to make the corresponding $\hat{u}_i$ a more reliable indicator of λ, so less variation of $\hat{u}_i$'s about a standard or estimated common value is tolerated.

Figure 7.21 is a retrospective u chart for the data of Table 7.19. The figure shows that the data-generating process can in no way be thought of as stable or subject to only random causes. There is too much variation in the $\hat{u}_i$ to be explainable as due only to small unidentifiable causes. Some of the variation can probably be thought of in terms of a general downward trend, perhaps associated with workers gaining job skills. But even accounting for that, there is substantial erratic fluctuation of the $\hat{u}_i$—which couldn't fit between control limits no matter where they might be centered. These data simply represent a real engineering process that, according to accepted standards, is not repeatable enough to allow (without appropriate sleuthing and elimination of large causes of variation) anything but "one day at a time" inferences about its behavior.

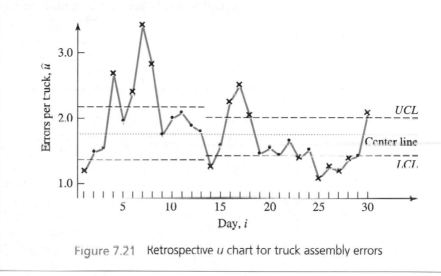

Figure 7.21 Retrospective u chart for truck assembly errors

This book has had little to say about formal inference from data with an underlying Poisson distribution. But retrospective u charts like the one in Example 12 can be thought of as rough graphical tests of the hypothesis $H_0: \lambda_1 = \lambda_2 = \cdots = \lambda_r$ for Poisson-distributed $X_i = k_i \hat{u}_i$.

7.6.3 Common Control Chart Patterns and Special Checks

Shewhart control charts (both those for measurements and those for attributes data) are useful for reasons beyond the fact that they supply semiformal information of a hypothesis-testing type. Much important qualitative information is also carried by **patterns** that can sometimes be seen in the charts' simple plots. Section 3.3 included some comments about engineering information carried in plots of summary statistics against time. Shewhart charts are such plots augmented with control limits. It is thus appropriate to amplify and extend those comments somewhat, in light of the extra element provided by the control limits.

What is expected
if a process is stable?

Before discussing interesting possible departures from the norm, it should probably be explicitly stated how a 3-sigma control chart is expected to look if a process is physically stable. One expects (tacitly assuming the distribution of the plotted statistic to be mound-shaped) that

1. most plotted points will lie in the middle, (say, the middle $\frac{2}{3}$) of the region delineated by the control limits around the center line,

2. a few (say, on the order of 1 in 20) points will lie outside this region but inside the control limits,

3. essentially no points will lie outside the control limits, and

4. there will be no obvious trends in time for any sizable part of the chart.

That is, one expects to see a random-scatter/white-noise plot that fills, but essentially remains within, the region bounded by the control limits. When something else is seen, even if no points plot outside the control limits, there is reason to consider the possibility that something in addition to chance causes is active in the data-generating mechanism.

Cyclical patterns
on a control chart

Cyclical (repeated "up, then back down again") **patterns** sometimes show up on Shewhart control charts. Such behavior is not characteristic of plots resulting from a stable-process data-generating mechanism. When it occurs, the alert engineer will look for identifiable physical causes of variation whose effects would come and go on about the same schedule as the ups and downs seen on the chart. Sometimes cyclical patterns are associated with daily or seasonal variables like ambient temperature effects, which may be largely beyond a user's control. But at other times, they have to do with things like different (rotating) operators' slightly different methods of machine operation, which can be mostly eliminated via standardization, training, and awareness.

Too much variation
on a control chart

Again, the expectation is that points plotted on a Shewhart control chart should (over time) pretty much fill up but rarely plot outside the region delineated by control limits. This can be violated in two different ways, both of which suggest the need for engineering attention. In the first place, more variation than expected (like that evident on Figure 7.21), which produces multiple points outside the control limits, is often termed **instability**. And (after eliminating the possibility of a blunder in calculations) it is nearly airtight evidence of one or more unregulated process variables having effects so large that they must be regulated. Such erratic behavior can sometimes be traced to material or components from several different suppliers having somewhat different physical properties and entering a production line in a mixed or haphazard order. Also, ill-advised operators may overadjust equipment (without any basis in control charting). This can take a fairly stable process and make it unstable.

Too little variation
on a control chart

Less variation than expected on a Shewhart chart presents an interesting puzzle. Look again at Figure 7.16 on page 507 and reflect on the fact that the plotted $\bar{x}$'s on that chart hug the center line. They don't come close to filling up the region between the control limits. The reader's first reaction to this might well be, "So what? Isn't small variation good?" Small variation is indeed a virtue, but when points on a control chart hug the center line, what one has is *unbelievably* small variation, which may conceal a blunder in calculation or (almost paradoxically) unnecessarily large but nonrandom variation.

In the first place, the simplest possible explanation of a plot like Figure 7.16 is that the process short-term variation, σ, has been overestimated—either because a standard σ is not applicable or because of some blunder in calculation or logic. Notice that using a value for σ that is bigger than what is really called for when making the limits

$$LCL_{\bar{x}} = \mu - 3\frac{\sigma}{\sqrt{m}} \quad \text{and} \quad UCL_{\bar{x}} = \mu + 3\frac{\sigma}{\sqrt{m}}$$

will spread the control limits too wide and produce an $\bar{x}$ chart that is insensitive to changes in μ. So this possibility should not be taken lightly.

Systematic differences and too little variation on a control chart/ stratification

A more subtle possible source of unbelievably small variation on a Shewhart chart has to do with the (usually unwitting) mixing of several consistently different streams of observations in the calculation of a single statistic that is naively thought to be representing only one stream of observations. This can happen when data are being taken from a production stream where multiple heads or cavities on a machine (or various channels of another type of multiple-channel process) are represented in a regular order in the stream. For example, items machined on heads 1, 2, and 3 of a machine might show up downstream in a production process in the order 1, 2, 3, 1, 2, 3, 1, 2, 3, etc. Then, if there is more difference between the different types of observations than there is within a given type, values of a single statistic calculated using observations of several types can be remarkably (excessively) consistent.

Consider, for example, the possibility that a five-head machine has heads that are detectably/consistently different. Suppose four of the five are perfectly adjusted and always produce conforming items and the fifth is severely misadjusted and always produces nonconforming items. Although 20% of the items produced are nonconforming, a binomial distribution model with $p = .2$ will typically overpredict the variation that will be seen in $n_i \hat{p}_i$ for samples of items from this process. Indeed, samples of size $m = 5$ of consecutive items coming off this machine will have $\hat{p}_i = .2$, always. Clearly, no $\hat{p}_i$'s would approach p chart control limits.

Or in a measurement data context, with the same hypothetical five-head machine, consider the possibility that four of the five heads always produce a part dimension at the target of 8 in. (plus or minus, say, .01 in.), whereas the fifth head is grossly misadjusted, always producing the dimension at 9 in. (plus or minus .01 in.). Then, in this exaggerated example, naive mixing together of the output of all five heads will produce ranges unbelievably stable at about 1 in. and sample means (of five consecutive pieces) unbelievably stable at about 8.2 in. But the super-stability is not a cause for rejoicing. Rather it is a cause for thought and investigation that could well lead to the physical elimination of the differences between the various mechanisms producing the data—in this case, the fixing of the faulty head.

The possibility of unnatural consistency on a Shewhart chart, brought on by more or less systematic sampling of detectably different data streams, is often called **stratification** in quality control circles. Although there is presently no way of verifying this suspicion, some form of stratification may have been at work in the production of the ream cutter data of Shervheim and Snider and the $\bar{x}$ chart in Figure 7.16. For example, multiple blades set at not quite equal angles on a roller that cuts sheets (as sketched in Figure 7.22) could produce consistently different consecutive sheet lengths and unbelievably stable $\bar{x}$'s. Or even with only a single blade on the cutter roller, regular patterns in material tension, brought on by slight eccentricities of feeder rollers, could also produce consistent patterns in consecutive sheet lengths and thus too much stability on the $\bar{x}$ chart.

Changes in level
Other nonrandom patterns sometimes appearing on control charts include both gradual and more sudden **changes in level** and unabated **trends** up or down. Gradual changes in level can sometimes be traced to machine warm-up phenomena, slow changeovers in a raw material source, or introduction of operator training. And phenomena like tool wear and machine degradation over time will typically produce patterns of plotted points moving in a single direction until there is some sort of human intervention.

Bunching
The terms **grouping** and **bunching** are used to describe irregular patterns on control charts where plotted points tend to come in sets of similar values but where the pattern is neither regular/repeatable enough to be termed cyclical nor consistent enough in one direction to merit the use of the term *trend*. Such grouping can be brought about (for example) by calibration changes in a measuring instrument and, in machining processes, by fixture changes.

Runs
Finally, **runs** of many consecutive points on one side of a center line are sometimes seen on control charts. Figure 7.15, the "standards given" $\bar{x}$ chart for the sheet-length data on page 502, is an extreme example of a chart exhibiting a run. On "standards given" charts, runs (even when not accompanied by points plotting outside control limits) tend to discredit the chart's center line value as a plausible median for the distribution of the plotted statistic. On $\bar{x}$ charts, that translates to a discrediting of the target process mean as the value of the true process mean, thus indicating that the process is misaimed. (In the sheet-length situation of Figure 7.15, average sheet length is clearly below the target length.) And on a p or u chart, it

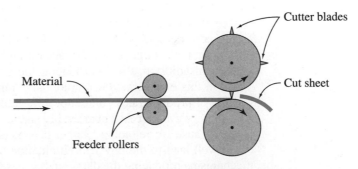

Figure 7.22   Schematic of a roller cutter

indicates the inappropriateness of the supposedly standard rate of nonconforming items or nonconformances. On retrospective control charts, runs on one side of the center line are usually matched by runs on the other side, and one of the earlier terms (cycles, trends, or grouping) can typically be applied in addition to the term *runs*.

In recognition of the fact that the elementary "wait for a point to plot outside of control limits" mode of using control charts is blind to the various interpretable patterns discussed here, a variety of **special checks** have been developed. To give the reader the flavor of these checks for unnatural patterns, two of the most famous sets are shown in Tables 7.20 and 7.21. Besides many other different sets appearing in quality control books, companies making serious use of control charts often develop their own collections of such rules. The two sets given here are included more to show what is possible than to advocate them in particular. The real bottom line of this discussion is simply that when used judiciously (overinterpretation of control chart patterns is a real temptation that also must be avoided), the qualitative information carried by patterns on Shewhart control charts can be an important engineering tool.

Table 7.20
Western Electric Alarm Rules (from the *AT&T Quality Control Handbook*)

- A single point outside 3-sigma limits
- 2 out of any 3 successive points outside 2-sigma limits on one side of the center line
- 4 out of any 5 successive points outside 1-sigma limits on one side of the center line
- 8 consecutive points on one side of the center line

Table 7.21
Alarm Rules of L. S. Nelson (from the *Journal of Quality Technology*)

- a single point outside 3-sigma limits
- 9 points in a row on one side of the center line
- 6 points in a row increasing or decreasing
- 14 points in a row alternating up and down
- 2 out of any 3 successive points outside 2-sigma limits on one side of the center line
- 4 out of any 5 successive points outside 1-sigma limits on one side of the center line
- 15 points in a row inside 1-sigma limits
- 8 points in a row with none inside 1-sigma limits

Section 6 Exercises

1. The accompanying data are some taken from *Statistical Quality Control Methods* by I. W. Burr, giving the numbers of beverage cans found to be defective in periodic samples of 312 cans at a bottling facility.

Sample	Defectives	Sample	Defectives
1	6	11	7
2	7	12	7
3	5	13	6
4	7	14	6
5	5	15	6
6	5	16	6
7	4	17	23
8	5	18	10
9	12	19	8
10	6	20	5

(a) Suppose that the company standard for the fraction of cans defective is that $p = .02$ of the cans be defective on average. Use this value and make a "standards given" p chart based on these data. Does it appear that the process fraction defective was stable at the $p = .02$ value over the period represented by these data?

(b) Make a retrospective p chart for these data. What does this chart indicate about the stability of the canning process?

2. The accompanying table lists some data on outlet leaks found in the first assembling of two radiator parts, again taken from Burr's *Statistical Quality Control Methods*. Each radiator may have several leaks.

Date	Number Tested	Leaks
6/3	39	14
6/4	45	4
6/5	46	5
6/6	48	13
6/7	40	6
6/10	58	2

Date	Number Tested	Leaks
6/11	50	4
6/12	50	11
6/13	50	8
6/14	50	10
6/17	32	3
6/18	50	11
6/19	33	1
6/20	50	3
6/24	50	6
6/25	50	8
6/26	50	5
6/27	50	2

(There were 841 radiators tested and a total of 116 leaks detected.) Make a retrospective u chart based on these data. What does it indicate about the stability of the assembly process?

3. In a particular defects/unit context, the number of standard size units inspected at a given opportunity varies. With

X_i = the number of defects found on sample i

k_i = the number of units inspected at time i

$\hat{u}_i = X_i/k_i$

the following were obtained at eight consecutive periods:

i	1	2	3	4	5	6	7	8
k_i	1	2	1	3	2	1	1	3
$\hat{u}_i$	0	1.5	0	.67	2	0	0	.33

(a) What do these values suggest about the stability of the process?

(b) Suppose that from now on, k_i is going to be held constant and that standard quality will be defined as a mean of 1.2 defects per unit. Compare 3-sigma Shewhart c charts based on

$k_i = 1$ and on $k_i = 2$ in terms of the probabilities that a given sample produces an "out of control" signal if
(i) the actual defect rate is standard.
(ii) the actual defect rate is twice standard.

4. Successive samples of carriage bolts are checked for length using "a go–no go" gauge. The results from ten successive samples are as follows:

Sample	1	2	3	4	5	6	7	8	9	10
Sample Size	30	20	40	30	20	20	30	20	20	20
Nonconforming	2	1	5	1	2	1	3	0	1	2

What do these values indicate about the stability of the bolt cutting process?

5. Why is it essential to have an operational definition of a nonconformance to make effective practical use of a Shewhart c chart?

6. Explain why too little variation appearing on a Shewhart control chart need not be a good sign.

Chapter 7 Exercises ....

1. Hoffman, Jabaay, and Leuer did a study of pencil lead strength. They loaded pieces of lead of the same diameter (supported on two ends) in their centers and recorded the forces at which they failed. Part of their data are given here (in grams of load applied at failure).

4H lead	H lead	B lead
56.7, 63.8, 56.7	99.2, 99.2, 92.1	56.7, 63.8, 70.9
63.8, 49.6	106.0, 99.2	63.8, 70.9

(a) In applying the methods of this chapter in the analysis of these data, what model assumptions must be made? Make three normal plots of these samples on the same set of axes and also make a normal plot of residuals for the one-way model as means of investigating the reasonableness of these assumptions. Comment on the plots.

(b) Compute a pooled estimate of variance based on these three samples. What is the corresponding value of s_p?

(c) Use the value of s_p that you calculated in (b) and make (individual) 95% two-sided confidence intervals for each of the three mean lead strengths, μ_{4H}, μ_H, and μ_B.

(d) Use s_p and make (individual) 95% two-sided confidence intervals for each of the three differences in mean lead strengths, $\mu_{4H} - \mu_H$, $\mu_{4H} - \mu_B$, and $\mu_H - \mu_B$.

(e) Suppose that for some reason it is desirable to compare the mean strength of B lead to the average of the mean strengths of 4H and H leads. Give a 95% two-sided confidence interval for the quantity $\frac{1}{2}(\mu_{4H} + \mu_H) - \mu_B$.

(f) Use the P-R method of simultaneous confidence intervals and make simultaneous 95% two-sided confidence intervals for the three mean strengths, μ_{4H}, μ_H, and μ_B. How do the lengths of these intervals compare to the lengths of the intervals you found in part (c)? Why is it sensible that the lengths should be related in this way?

(g) Use the Tukey method of simultaneous confidence intervals and make simultaneous 95% two-sided confidence intervals for the three differences in mean lead strengths, $\mu_{4H} - \mu_H$, $\mu_{4H} - \mu_B$, and $\mu_H - \mu_B$. How do the lengths of these intervals compare to the lengths of the intervals you found in part (d)?

(h) Use the one-way ANOVA test statistic and assess the strength of the evidence against $H_0: \mu_{4H} = \mu_H = \mu_B$ in favor of H_a: not H_0. Show the whole five-step format.

(i) Make the ANOVA table corresponding to the significance test you carried out in part (h).

(j) As a means of checking your work for parts (h) and (i) of this problem, use a statistical package to produce the required ANOVA table, F statistic, and p-value.

2. Allan, Robbins, and Wyckoff worked with a machine shop that employs a CNC (computer numerically controlled) lathe in the manufacture of a part for a heavy equipment maker. Some summary statistics for measurements of a particular diameter on the part for 20 hourly samples of $m = 4$ parts turned on the lathe are given here. (The means are in 10^{-4} in. above 1.1800 in. and the ranges are in 10^{-4} in.)

Sample	1	2	3	4	5
$\bar{x}$	9.25	8.50	9.50	6.25	5.25
R	1	2	2	8	7

Sample	6	7	8	9	10
$\bar{x}$	5.25	5.75	19.50	10.0	9.50
R	5	5	1	3	1

Sample	11	12	13	14	15
$\bar{x}$	9.50	9.75	12.25	12.75	14.50
R	6	1	9	2	7

Sample	16	17	18	19	20
$\bar{x}$	8.00	10.0	10.25	8.75	10.0
R	3	0	1	3	0

(a) The midspecification for the diameter in question was 1.1809 in. Suppose that a standard σ for diameters turned on this machine is 2.5×10^{-4} in. Use these two values and find "standards given" control limits for $\bar{x}$ and R. Make both $\bar{x}$ and R charts using these and comment on what the charts indicate about the turning process.

(b) In contrast to part (a) where standards were furnished, compute retrospective or "as past data" control limits for both $\bar{x}$ and R. Make both $\bar{x}$ and R charts using these and comment

on what the charts indicate about the turning process.

(c) If you were to judge the sample ranges to be stable, it would then make sense to use $\bar{R}$ to develop an estimate of the turning process short-term standard deviation σ. Find such an estimate.

(d) The engineering specifications for the turned diameter are (still in .0001 in. above 1.1800 in.) from 4 to 14. Supposing that the average diameter could be kept on target (at the mid-specification), does your estimate of σ from part (c) suggest that the turning process would then be *capable* of producing most diameters in these specifications? Explain.

3. Becker, Francis, and Nazarudin conducted a study of the effectiveness of commercial clothes dryers in removing water from different types of fabric. The following are some summary statistics from a part of their study, where a garment made of one of $r = 3$ different blends was wetted and dried for 10 minutes in a particular dryer and the (water) weight loss (in grams) measured. Each of the three different garments was tested three times.

100% Cotton	Cotton/Polyester	Cotton/Acrylic
$n_1 = 3$	$n_2 = 3$	$n_3 = 3$
$\bar{y}_1 = 85.0$ g	$\bar{y}_2 = 348.3$ g	$\bar{y}_3 = 258.3$ g
$s_1 = 25.0$ g	$s_2 = 88.1$ g	$s_3 = 63.3$ g

(a) What restrictions/model assumptions are required in order to do formal inference based on the data summarized here (if information on the baseline variability involved is pooled and the formulas of this chapter are used)? Assume that those model assumptions are a sensible description of this situation.

(b) Find s_P and the associated degrees of freedom.

(c) What does s_P measure?

(d) Give a 90% lower confidence bound for the mean amount of water that can be removed from the cotton garment by this dryer in a 10-minute period.

(e) Give a 90% two-sided confidence interval for comparing the means for the two blended garments.

(f) Suppose that all pairs of fabric means are to be compared using intervals of the form $\bar{y}_i - \bar{y}_{i'} \pm \Delta$ and that simultaneous 95% confidence is desired. Find Δ.

(g) A partially completed ANOVA table for testing $H_0: \mu_1 = \mu_2 = \mu_3$ follows. Finish filling in the table then find a p-value for a significance test of this hypothesis.

ANOVA Table

Source	SS	df	MS	F
	24,181			
	132,247			

4. The article "Behavior of Rubber-Based Elastomeric Construction Adhesive in Wood Joints" by P. Pellicane (*Journal of Testing and Evaluation*, 1990) compared the performance of $r = 8$ different commercially available construction adhesives. $m = 8$ joints glued with each glue were tested for strength, giving results summarized as follows (the units are kN):

Glue (i)	1	2	3	4	5	6	7	8
$\bar{y}_i$	1821	1968	1439	616	1354	1424	1694	1669
s_i	214	435	243	205	135	191	225	551

(a) Temporarily considering only the test results for glue 1, give a 95% lower tolerance bound for the strengths of 99% of joints made with glue 1.

(b) Still considering only the test results for glue 1, give a 95% lower confidence bound for the mean strength of joints made with glue 1.

(c) Now considering only the test results for glues 1 and 2, assess the strength of the evidence against the possibility that glues 1 and 2 produce joints with the same mean strength. Show the whole five-step significance-testing format.

(d) What model assumptions stand behind the formulas you used in parts (a) and (b)? In part (c)?

For the following questions, consider test results from all eight glues when making your analyses.

(e) Find a pooled sample standard deviation and give its degrees of freedom.

(f) Repeat parts (a) and (b) using the pooled standard deviation instead of only s_1. What extra model assumption is required to do this (beyond what was used in parts (a) and (b))?

(g) Find the value of an F statistic for testing $H_0: \mu_1 = \mu_2 = \cdots = \mu_8$ and give its degrees of freedom. (*Hint:* These data are balanced. You ought to be able to use the $\bar{y}$'s and the sample variance routine on your calculator to help get the numerator for this statistic.)

(h) Simultaneous 95% two-sided confidence limits for the mean strengths for the eight glues are of the form $\bar{y}_i \pm \Delta$ for an appropriate number Δ. Find Δ.

(i) Simultaneous 95% two-sided confidence limits for all differences in mean strengths for the eight glues are of the form $\bar{y}_i - \bar{y}_{i'} \pm \Delta$ for a number Δ. Find Δ.

5. Example 7 in Chapter 4 treats some data collected by Kotlers, MacFarland, and Tomlinson while studying strength properties of wood joints. Part of those data (stress at failure values in units of psi for four out of the original nine wood/joint type combinations) are reproduced here, along with $\bar{y}$ and s for each of the four samples represented:

		Wood Type	
		Pine	Oak
Joint Type	Butt	829	1169
		596	
		$\bar{y} = 712.5$	$\bar{y} = 1169$
		$s = 164.8$	
	Lap	1000	1295
		859	1561
		$\bar{y} = 929.5$	$\bar{y} = 1428.0$
		$s = 99.7$	$s = 188.1$

(a) Treating pine/butt joints alone, give a 95% two-sided confidence interval for mean strength for such joints. (Here, base your interval on only the pine/butt data.)

(b) Treating only lap joints, how strong is the evidence shown here of a difference in mean joint strength between pine and oak woods? (Here use only the pine/lap and oak/lap data.) Use the five-step format.

(c) Give a 90% two-sided confidence interval for comparing the strength standard deviations for pine/lap and oak/lap joints.

Consider all four samples in the following questions.

(d) Assuming that all four wood type/joint type conditions are thought to have approximately the same associated variability in joint strength, give an estimate of this supposedly common standard deviation.

(e) It is possible to compute simultaneous 95% lower (one-sided) confidence limits for mean joint strengths for all four wood type/joint type combinations. Give these (based on the P-R method).

(f) Suppose that you want to compare butt joint strength to lap joint strength and in fact want a 95% two-sided confidence interval for

$$\frac{1}{2}(\mu_{\text{pine/butt}} + \mu_{\text{oak/butt}}) - \frac{1}{2}(\mu_{\text{pine/lap}} + \mu_{\text{oak/lap}})$$

Give such a confidence interval, again making use of your answer to (d).

6. In an industrial application of Shewhart $\bar{x}$ and R control charts, 20 successive hourly samples of $m = 2$ high-precision metal parts were taken, and a particular diameter on the parts was measured. $\bar{x}$ and R values were calculated for each of the 20 samples, and these had

$$\bar{\bar{x}} = .35080 \text{ in.} \quad \text{and} \quad \overline{R} = .00019 \text{ in.}$$

(a) Give retrospective control limits that you would use in an analysis of the $\bar{x}$ and R values.

(b) The engineering specifications for the diameter being measured were .3500 in. $\pm$ .0020 in. Unfortunately, even practicing engineers

sometimes have difficulty distinguishing in their thinking and speech between specifications and control limits. Briefly (but carefully) discuss the difference in meaning between the control limits for $\bar{x}$ found in part (a) and these engineering specifications. (To what quantities do the two apply? What are the different purposes for the two? Where do the two come from? And so on.)

7. Here are some summary statistics produced by Davies and Sehili for ten samples of $m = 4$ pin head diameters formed on a type of electrical component. The sampled components were groups of consecutive items taken from the output of a machine approximately once every ten minutes. The units are .001 in.

Sample	$\bar{x}$	R	s	Sample	$\bar{x}$	R	s
1	31.50	3	1.29	6	33.00	3	1.41
2	30.75	2	.96	7	33.00	2	.82
3	29.75	3	1.26	8	33.00	4	1.63
4	30.50	3	1.29	9	34.00	2	.82
5	32.00	0	0	10	26.00	0	0

Some summaries for the statistics are

$$\sum \bar{x} = 313.5 \quad \sum R = 22 \quad \text{and} \quad \sum s = 9.48$$

(a) Assuming that the basic short-term variability of the mechanism producing pin head diameters is constant, it makes sense to try to quantify it in terms of a standard deviation σ. Various estimates of that σ are possible. Give three such possible estimates based on $\overline{R}$, $\bar{s}$, and s_P.

(b) Using each of your estimates from (a), give retrospective control limits for both $\bar{x}$ and R.

(c) Compare the $\bar{x}$'s and R's given above to your control limits from (b) based on $\overline{R}$. Are there any points that would plot outside control limits on a Shewhart $\bar{x}$ chart? On a Shewhart R chart?

(d) For the company manufacturing these parts, what are the practical implications of your analysis in parts (b) and (c)?

8. Dunnwald, Post, and Kilcoin studied the viscosities of various weights of various brands of motor oil. Some summary statistics for part of their data are given here. Summarized are $m = 10$ measurements of the viscosities of each of $r = 4$ different weights of Brand M motor oil at room temperature. Units are seconds required for a ball to drop a particular distance through the oil.

10W30	SAE 30	10W40	20W50
$\bar{y} = 1.385$	$\bar{y}_2 = 2.066$	$\bar{y}_3 = 1.414$	$\bar{y}_4 = 4.498$
$s_1 = .091$	$s_2 = .097$	$s_3 = .150$	$s_4 = .204$

(a) Find the pooled sample standard deviation here. What are the associated degrees of freedom?

(b) If the P-R method is used to find simultaneous 95% two-sided confidence intervals for all four mean viscosities, the intervals produced are of the form $\bar{y}_i \pm \Delta$, for Δ an appropriate number. Find Δ.

(c) If the Tukey method is used to find simultaneous 95% two-sided confidence intervals for all differences in mean viscosities, the intervals produced are of the form $\bar{y}_i - \bar{y}_{i'} \pm \Delta$, for Δ an appropriate number. Find Δ.

(d) Carry out an ANOVA test of the hypothesis that the four oil weights have the same mean viscosity.

9. Because of modern business pressures, it is not uncommon for standards for fractions nonconforming to be in the range of 10^{-4} to 10^{-6}.

(a) What are "standards given" 3-sigma control limits for a p chart with standard fraction nonconforming 10^{-4} and sample size 100?

(b) If p becomes twice the standard value (of 10^{-4}), what is the probability that the scheme from (a) detects this state of affairs at the first subsequent sample? (Use your answer to (a) and the binomial distribution for $n = 100$ and $p = 2 \times 10^{-4}$.)

(c) What does (b) suggest about the feasibility of doing process monitoring for very small fractions defective based on attributes data?

10. Suppose that a company standard for the mean number of visual imperfections on a square foot of plastic sheet is $\lambda = .04$.

(a) Give upper control limits for the number of imperfections found on pieces of material .5 ft × .5 ft and then 5 ft × 5 ft.

(b) What would you tell a worker who, instead of inspecting a 10 ft × 10 ft specimen of the plastic (counting total imperfections on the whole), wants to inspect only a 1 ft × 1 ft specimen and multiply the observed count of imperfections by 100?

11. Bailey, Goodman, and Scott worked on a process for attaching metal connectors to the ends of hydraulic hoses. One part of that process involved grinding rubber off the ends of the hoses. The amount of rubber removed is termed the skive length. The values in the accompanying table are skive length means and standard deviations for 20 samples of five consecutive hoses ground on one grinder. Skive length is expressed in .001 in. above the target length.

Sample	$\bar{x}$	s	Sample	$\bar{x}$	s
1	−.4	5.27	11	−2.2	5.50
2	0.0	4.47	12	−5.2	2.86
3	−1.4	3.29	13	−.8	1.30
4	1.8	2.28	14	.8	2.68
5	1.4	1.14	15	−2.0	2.92
6	0.0	4.24	16	−.2	1.30
7	−.4	4.39	17	−6.6	2.30
8	1.4	4.51	18	−1.0	4.21
9	.2	4.32	19	−3.2	5.76
10	−3.2	2.05	20	−2.4	4.28
				−23.4	69.07

(a) What do these values indicate about the stability of the skiving process? Show appropriate work and explain fully.

(b) Give an estimate of the process short-term standard deviation based on the given values.

(c) If specifications on the skive length are ±.006 in. and, over short periods, skive length can be thought of as normally distributed, what does your answer to (b) indicate about the best possible fraction (for perfectly adjusted grinders) of skives in specifications? Give a number.

(d) Based on your answer to (b), give control limits for future control of skive length means and ranges for samples of size $m = 3$.

(e) Suppose that hoses from all grinders used during a given shift are all dumped into a common bin. If upon sampling, say, 20 hoses from this bin at the end of a shift, the 20 measured skive lengths have a standard deviation twice the size of your answer to (b), what possible explanations come to mind for this?

(f) Suppose current policy is to sample five consecutive hoses once an hour for each grinder. An alternative possibility is to sample one hose every 12 minutes for each grinder.

(i) Briefly discuss practical trade-offs that you see between the two possible sampling methods.

(ii) If in fact the new sampling scheme were adopted, would you recommend treating the five hoses from each hour as a sample of size 5 and doing $\bar{x}$ and R charting with $m = 5$? Explain.

12. Two different types of nonconformance can appear on widgets manufactured by Company V. Counts of these on ten widgets produced one per hour are given here.

Widget	1	2	3	4	5	6	7	8	9	10
Type A Defects	4	2	1	2	2	2	0	2	1	0
Type B Defects	0	2	2	4	2	4	3	3	7	2
Total Defects	4	4	3	6	4	6	3	5	8	2

(a) Considering first total nonconformances, is there evidence here of process instability? Show appropriate work.

(b) What statistical indicators might you expect to observe in data like these if in fact type A and B defects have a common cause mechanism?

(c) **(Charts for Demerits)** For the sake of example, suppose that type A defects are judged twice as important as type B defects. One

might then consider charting

$$X = \text{demerits}$$
$$= 2(\text{number of A defects})$$
$$+ (\text{number of B defects})$$

If one can model (number of A defects) and (number of B defects) as independent Poisson random variables, it is relatively easy to come up with sensible control limits. (Remember that the variance of a sum of independent random variables is the sum of the variances.)

(i) If the mean number of A defects per widget is λ_1 and the mean number of B defects per widget is λ_2, what are the mean and variance for X? Use your answers to give "standards given" control limits for X.

(ii) In light of your answer to (i), what numerical limits for X would you use to analyze these values "as past data"?

13. **(Variables Versus Attributes Control Charting)** Suppose that a dimension of parts produced on a certain machine over a short period can be thought of as normally distributed with some mean μ and standard deviation $\sigma = .005$ in. Suppose further that values of this dimension more than .0098 in. from the 1.000 in. nominal value are considered nonconforming. Finally, suppose that hourly samples of ten of these parts are to be taken.

(a) If μ is exactly on target (i.e., $\mu = 1.000$ in.), about what fraction of parts will be nonconforming? Is it possible for the fraction nonconforming ever to be any less than this figure?

(b) One could use a p chart based on $m = 10$ to monitor process performance in this situation. What would be "standards given" 3-sigma control limits for the p chart, using your answer from part (a) as the standard value of p?

(c) What is the probability that a particular sample of $m = 10$ parts will produce an "out of control" signal on the chart from (b) if μ remains at its standard value of $\mu = 1.000$ in.? How does this compare to the same probability for a 3-sigma $\bar{x}$ chart for $m = 10$ set up with a center line at 1.000? (For the p chart, use a binomial probability calculation. For the $\bar{x}$ chart, use the facts that $\mu_{\bar{x}} = \mu$ and $\sigma_{\bar{x}} = \sigma/\sqrt{m}$.)

(d) Compare the probability that a particular sample of $m = 10$ parts will produce an "out of control" signal on the p chart from (b) to the probability that the sample will produce an "out of control" signal on the ($m = 10$) 3-sigma $\bar{x}$ chart first mentioned in (c), supposing that in fact $\mu = 1.005$ in. What moral is told by your calculations here and in part (c)?

14. The article "How to Use Statistics Effectively in a Pseudo-Job Shop" by G. Fellers (*Quality Engineering*, 1990) discusses some applications of statistical methods in the manufacture of corrugated cardboard boxes. One part of the article concerns the analysis of a variable called box "skew," which quantifies how far from being perfectly square boxes are. This response variable, which will here be called y, is measured in units of $\frac{1}{32}$ in. $r = 24$ customer orders (each requiring a different machine setup) were studied, and from each, the skews, y, of five randomly selected boxes were measured. A partial ANOVA table made in summary of the data follows.

ANOVA Table

Source	SS	df	MS	F
Order (setup)	1052.39			
Error				
Total	1405.59	119		

(a) Complete the ANOVA table.

(b) In a given day, hundreds of different orders are run in this plant. This situation is one in which a random effects analysis is most natural. Explain why.

(c) Find estimates of σ and σ_τ. What, in the context of this situation, do these two estimates measure?

(d) Find and interpret a two-sided 90% confidence interval for σ and then the ratio σ_τ/σ.

(e) If there is variability in skew, customers must continually adjust automatic folding and packaging equipment in order to prevent machine jam-ups. Such variability is therefore highly undesirable for the box manufacturer, who wishes to please customers. What does your analysis from (c) and (d) indicate about how the manufacturer should proceed in any attempts to reduce variability in skew? (What is the big component of variance, and what kind of actions might be taken to reduce it? For example, is there a need for the immediate purchase of new high-precision manufacturing equipment?)

15. The article "High Tech, High Touch" by J. Ryan (*Quality Progress*, 1987) discusses the quality enhancement processes used by Martin Marietta in the production of the space shuttle external (liquid oxygen) fuel tanks. It includes a graph giving counts of major hardware nonconformances for each of 41 tanks produced. The accompanying data (see next page) are approximate counts read from that graph for the last 35 tanks. (The first 6 tanks were of a different design than the others and are therefore not included here.)

(a) Make a retrospective c chart for these data. Is there evidence of real quality improvement in this series of counts of nonconformances? Explain.

(b) Consider only the last 17 tanks. Does it appear that quality was stable over the production period represented by these tanks? (Make another retrospective c chart.)

(c) It is possible that some of the figures read from the graph in the original article may differ from the real figures by as much as, say, 15 nonconformances. Would this measurement error account for the apparent lack of stability you found in (a) or (b) above? Explain.

Tank	Nonconformances	Tank	Nonconformances
1	537	19	157
2	463	20	120
3	417	21	148
4	370	22	65
5	333	23	130
6	241	24	111
7	194	25	65
8	185	26	74
9	204	27	65
10	185	28	148
11	167	29	74
12	157	30	65
13	139	31	139
14	130	32	213
15	130	33	222
16	267	34	93
17	102	35	194
18	130		

	Day 1		Day 2
Sample	Nonconforming	Sample	Nonconforming
1	16	1	14
2	18	2	20
3	17	3	17
4	18	4	13
5	22	5	12
6	14	6	12
7	16	7	14
8	18	8	15
9	18	9	19
10	19	10	21
11	20	11	18
12	25	12	14
13	14	13	13
14	13	14	9
15	23	15	16
16	13	16	16
17	23	17	15
18	15	18	11
19	14	19	17
20	23	20	8
21	17	21	16
22	20	22	13
23	16	23	16
24	19	24	15
25	22	25	13

16. Kaminski, Rasavahn, Smith, and Weitekamper worked with the same pelletizing machine referred to in Examples 2 (Chapter 1), 14 (Chapter 3), and 18 (Chapter 6). They collected process monitoring data on several different days of operation. The accompanying table shows counts of nonconforming pellets in periodic samples of size $m = 30$ from two different days. (The pelletizing on day 1 was done with 100% fresh material, and on the second day, a mixture of fresh and reground materials was used.)

(a) Make a retrospective p chart for the day 1 data. Is there evidence of process instability in the day 1 data? Explain.

(b) Treating the day 1 data as a single sample of size 750 from the day's production of pellets, give a 90% two-sided confidence interval for the fraction nonconforming produced on the day in question.

(c) In light of your answers to parts (a) and (b), explain why a process being in control or stable does not necessarily mean that it is producing a satisfactory fraction of conforming product.

(d) Repeat parts (a) and (b) for the day 2 data.

(e) Try making a single retrospective control chart for the two days taken together. Do points plot out of control on this single chart? Explain why this does or does not contradict the results of parts (a), (b), and (d).

(f) Treating the data from days 1 and 2 as two samples of size 750 from the respective days' production of pellets, give a two-sided 98% confidence interval for the difference in fractions of nonconforming pellets produced on the two days.

17. Eastman, Frye, and Schnepf counted defective plastic bags in 15 consecutive groups of 250 coming off a converting machine immediately after a changeover to a new roll of plastic. Their counts are as follows:

Sample	Nonconforming	Sample	Nonconforming
1	147	9	0
2	93	10	0
3	41	11	0
4	0	12	0
5	18	13	0
6	0	14	0
7	31	15	0
8	22		

Is it plausible that these data came from a physically stable process, or is it clear that there is some kind of start-up phenomenon involved here? Make and interpret an appropriate control chart to support your answer.

18. Sinnott, Thomas, and White compared several properties of five different brands of 10W30 motor oil. In one part of their study, they measured the boiling points of the oils, $m = 3$ measurements for each of the $r = 5$ oils follow. (Units are degrees F.)

Brand C	Brand H	Brand W	Brand Q	Brand P
378	357	321	353	390
386	365	303	349	378
388	361	306	353	381

(a) Compute and make a normal plot for the residuals for the one-way model. What does the plot indicate about the appropriateness of the one-way model assumptions?

(b) Using the five samples, find s_p, the pooled estimate of σ. What does this value measure? Give a two-sided 90% confidence interval for σ based on s_p.

(c) Individual two-sided confidence intervals for the five different means here would be of the form $\bar{y}_i \pm \Delta$, for an appropriate number Δ. If 90% individual confidence is desired, what value of Δ should be used?

(d) Individual two-sided confidence intervals for the differences in the five different means would be of the form $\bar{y}_i - \bar{y}_{i'} \pm \Delta$, for a number Δ. If 90% individual confidence is desired, what value of Δ should be used here?

(e) Using the P-R method, what Δ would be used to make two-sided intervals of the form $\bar{y}_i \pm \Delta$ for all five mean boiling points, possessing simultaneous 95% confidence?

(f) Using the Tukey method, what Δ would be used to make two-sided intervals of the form $\bar{y}_i - \bar{y}_{i'} \pm \Delta$ for all differences in the five mean boiling points, possessing simultaneous 99% confidence?

(g) Make an ANOVA table for these data. Then use the calculations to find both R^2 for the one-way model and also the observed level of significance for an F test of the null hypothesis that all five oils have the same mean boiling point.

(h) It is likely that the measurements represented here were all made on a single can of each brand of oil. (The students' report was not explicit about this point.) If so, the formal inferences made here are really most honestly thought of as applying to the five particular cans used in the study. Discuss why the inferences would not necessarily extend to all cans of the brands included in the study and describe the conditions under which you might be willing to make such an extension. Is the situation different if, for example, each of the measurements comes from a different can of oil, taken from different shipping lots? Explain.

19. Baik, Johnson, and Umthun worked with a small metal fabrication company on monitoring the performance of a process for cutting metal rods. Specifications for the lengths of these rods were 33.69 in. $\pm$ .03 in. Measured lengths of rods in 15 samples of $m = 4$ rods, made over a period of two

days, are shown in the accompanying table. (The data are recorded in inches above the target value of 33.69, and the first five samples were made on day 1, while the remainder were made on day 2.)

Sample	Rod Lengths	$\bar{x}$	R	s
1	.0075, .0100			
	.0135, .0135	.01113	.0060	.00293
2	−.0085, .0035			
	−.0180, .0010	−.00550	.0215	.00981
3	.0085, .0000			
	.0100, .0020	.00513	.0100	.00487
4	.0005, −.0005			
	.0145, .0170	.00788	.0175	.00916
5	.0130, .0035			
	.0120, .0070	.00888	.0095	.00444
6	−.0115, −.0110			
	−.0085, −.0105	−.01038	.0030	.00131
7	−.0080, −.0070			
	−.0060, −.0045	−.00638	.0035	.00149
8	−.0095, −.0100			
	−.0130, −.0165	−.01225	.0070	.00323
9	.0090, .0125			
	.0125, .0080	.01050	.0045	.00235
10	−.0105, −.0100			
	−.0150, −.0075	−.01075	.0075	.00312
11	.0115, .0150			
	.0175, .0180	.01550	.0065	.00297
12	.0020, .0005			
	.0010, .0010	.00113	.0015	.00063
13	−.0010, −.0025			
	−.0020, −.0030	−.00213	.0020	.00085
14	−.0020, .0015			
	.0025, .0025	.00113	.0045	.00214
15	−.0010, −.0015			
	−.0020, −.0045	−.00225	.0035	.00155

$$\bar{\bar{x}} = .00078 \qquad \overline{R} = .0072 \qquad \bar{s} = .00339$$

(a) Find a retrospective center line and control limits for all 15 sample ranges. Apply them to the ranges and say what is indicated about the rod cutting process.

(b) Repeat part (a) for the sample standard deviations rather than ranges.

The initial five samples were taken while the operators were first learning to cut these particular rods. Suppose that it therefore makes sense to look separately at the last ten samples. These samples have $\bar{\bar{x}} = -.00159$, $\overline{R} = .00435$, and $\bar{s} = .001964$.

(c) Both the ranges and standard deviations of the last ten samples look reasonably stable. What about the last ten $\bar{x}$'s? (Compute control limits for the last ten $\bar{x}$'s, based on either $\overline{R}$ or $\bar{s}$, and say what is indicated about the rod cutting process.)

As a matter of fact, the cutting process worked as follows. Rods were welded together at one end in bundles of 80, and the whole bundle cut at once. The four measurements in each sample came from a single bundle. (There are 15 bundles represented.)

(d) How does this explanation help you understand the origin of patterns discovered in the data in parts (a) through (c)?

(e) Find an estimate of the "process short-term σ" for the last ten samples. What is it really measuring in the present context?

(f) Use your estimate from (e) and, assuming that lengths of rods from a single bundle are approximately normally distributed, compute an estimate of the fraction of lengths in a bundle that are in specifications, if in fact $\mu = 33.69$ in.

(g) Simply pooling together the last ten samples (making a single sample of size 40) and computing the sample standard deviation gives the value $s = .00898$. This is much larger than any s recorded for one of the samples and should be much larger than your value from (e). What is the origin of this difference in magnitude?

20. Consider the last ten samples from Exercise 19. Upon considering the physical circumstances that produced the data, it becomes sensible to replace the control chart analysis done there with a random effects analysis simply meant to quantify

the within- and between-bundle variance components.

(a) Make an ANOVA table for these ten samples of size 4. Based on the mean squares, find estimates of σ, the standard deviation of lengths for a given bundle, and σ_τ, the standard deviation of bundle mean lengths.

(b) Find and interpret a two-sided 90% confidence interval for the ratio σ_τ/σ.

(c) What is the principal origin of variability in the lengths of rods produced by this cutting method? (Is it variability of lengths within bundles or differences between bundles?)

21. The following data appear in the text *Quality Control and Industrial Statistics* by A. J. Duncan. They represent the numbers of disabling injuries suffered and millions of man-hours worked at a large corporation in 12 consecutive months.

Month	1	2	3	4	5	6
Injuries	11	4	5	8	4	4
10^6 man-hr	.175	.178	.175	.180	.183	.198

Month	7	8	9	10	11	12
Injuries	9	12	2	6	6	7
10^6 man-hr	.210	.212	.210	.211	.195	.200

(a) Temporarily assuming the injury rate per man-hour to be stable over the period studied, find a sensible estimate of the mean injuries per 10^6 man-hours.

(b) Based on your figure from (a), find "control limits" for the observed rates in each of the 12 months. Do these data appear to be consistent with a "stable system" view of the corporation's injury production mechanisms? Or are there months that are clearly distinguishable from the others in terms of accident rates?

22. Eder, Williams, and Bruster studied the force (applied to the cutting arm handle) required to cut various types of paper in a standard paper trimmer. The students used stacks of five sheets of four different types of paper and recorded the forces needed to move the cutter arm (and thus cut the

paper). The data that follow (the units are ounces) are for $m = 3$ trials with each of the four paper types and also for a "baseline" condition where no paper was loaded into the trimmer.

No Paper	Newsprint	Construction	Computer	Magazine
24, 25, 31	61, 51, 52	72, 70, 77	59, 59, 70	54, 59, 61

(a) If the methods of this chapter are applied in the analysis of these data, what model assumptions must be made? With small sample sizes such as those here, only fairly crude checks on the appropriateness of the assumptions are possible. One possibility is to compute residuals and normal-plot them. Do this and comment on the appearance of the plot.

(b) Compute a pooled estimate of the standard deviation based on these five samples. What is s_P supposed to be measuring in the present situation?

(c) Use the value of s_P and make (individual) 95% two-sided confidence intervals for each of the five mean force requirements $\mu_{\text{No paper}}$, $\mu_{\text{Newsprint}}$, $\mu_{\text{Construction}}$, μ_{Computer}, and μ_{Magazine}.

(d) Individual confidence intervals for the differences between particular pairs of mean force requirements are of the form $\bar{y}_i - \bar{y}_{i'} \pm \Delta$, for an appropriate value of Δ. Use s_P and find Δ if individual 95% two-sided intervals are desired.

(e) Suppose that it is desirable to compare the "no paper" force requirement to the average of the force requirements for the various papers. Give a 95% two-sided confidence interval for the quantity $\mu_{\text{No paper}} - \frac{1}{4}(\mu_{\text{Newsprint}} + \mu_{\text{Construction}} + \mu_{\text{Computer}} + \mu_{\text{Magazine}})$.

(f) Use the P-R method of simultaneous confidence intervals and make simultaneous 95% two-sided confidence intervals for the five mean force requirements. How do the lengths of these intervals compare to the lengths of the intervals you found in part (c)? Why is it sensible that the lengths should be related in this way?

(g) Simultaneous confidence intervals for the differences between all pairs of mean force requirements are of the form $\bar{y}_i - \bar{y}_{i'} \pm \Delta$, for an appropriate value of Δ. Use s_P and find Δ if Tukey simultaneous 95% two-sided intervals are desired. How does this value compare to the value you found in part (d)?

(h) Use the one-way ANOVA test statistic and assess the strength of the students' evidence against $H_0: \mu_{\text{No paper}} = \mu_{\text{Newsprint}} = \mu_{\text{Construction}} = \mu_{\text{Computer}} = \mu_{\text{Magazine}}$ in favor of H_a: not H_0. Show the whole five-step format.

(i) Make the ANOVA table corresponding to the significance test you carried out in part (h).

23. Duffy, Marks, and O'Keefe did some testing of the 28-day compressive strengths of 3 in. × 6 in. concrete cylinders. In part of their study, concrete specimens made with a .50 water/cement ratio and different percentages of entrained air were cured in a moisture room and subsequently strength tested. $m = 4$ specimens of each type produced the measured strengths (in 10^3 psi) summarized as follows:

3% Air	6% Air	10% Air
$\bar{y}_1 = 5.3675$	$\bar{y}_2 = 4.9900$	$\bar{y}_3 = 2.9250$
$s_1 = .1638$	$s_2 = .1203$	$s_3 = .2626$

(a) Find the pooled sample standard deviation and its associated degrees of freedom.
Use your answer to part (a) throughout the rest of this problem.

(b) Give a 99% lower confidence bound for the mean strength of 3% air specimens.

(c) Give a 99% two-sided confidence interval for comparing the mean strengths of 3% air and 10% air specimens.

(d) Suppose that mean strengths of specimens for all pairs of levels of entrained air are to be compared using intervals of the form $\bar{y}_i - \bar{y}_{i'} \pm \Delta$. Find Δ for Tukey simultaneous 99% two-sided confidence limits.

(e) A partially completed ANOVA table for testing $H_0: \mu_1 = \mu_2 = \mu_3$ follows. Finish filling in the table, then find a p-value for an F test

of this hypothesis.

ANOVA Table				
Source	SS	df	MS	F
Total	14.1608			

24. Davis, Martin, and Poppinga used a ytterbium argon gas laser to make some cuts in stainless steel-316. Using 95 mJ/pulse and 20 Hz settings on the laser and a 15.5 mm distance to the steel specimens (set at a 45°angle to the laser beam), the students made cuts in specimens using 100, 500, and 1,000 pulses. (Although this is not absolutely clear from the students' report, it seems that four specimens were cut using each number of pulses.) The depths of cut the students measured were then as follows:

100 Pulses	500 Pulses
7.4, 8.6, 5.6, 8.0	24.2, 29.5, 26.5, 23.8

1000 Pulses
33.4, 37.5, 35.9, 34.8

(a) If the methods of this chapter are applied in the analysis of these three samples, what model assumptions must be made? Compute residuals and normal plot them as something of a check on the reasonableness of these assumptions. Comment on the appearance of the plot.

(b) Compute a pooled estimate of the standard deviation based on these three samples. What is s_P supposed to be measuring in the present situation?

(c) Make (individual) 95% two-sided confidence intervals for each of the three mean depths of cut, μ_{100}, μ_{500}, and μ_{1000}.

(d) Confidence intervals for the differences between particular pairs of mean depths of cut are of the form $\bar{y}_i - \bar{y}_{i'} \pm \Delta$, for a number Δ.

Find Δ if individual 95% two-sided intervals are desired.

(e) Suppose that it is desirable to compare the per pulse change in average depth of cut between 100 pulses and 500 pulses to the per pulse change in average depth of cut between 500 pulses and 1,000 pulses. Give a 90% two-sided confidence interval for the quantity

$$\frac{1}{400}\left(\mu_{500} - \mu_{100}\right) - \frac{1}{500}\left(\mu_{1000} - \mu_{500}\right)$$

(You will need to write this out as a linear combination of the three means before applying any formulas from Section 7.2.) Based on this interval, does it appear plausible that the depth of cut changes linearly in the number of pulses over the range from 100 to 1,000 pulses? Explain.

(f) Use the P-R method of simultaneous confidence intervals and make simultaneous 95% two-sided confidence intervals for the three mean depths of cut. How do the lengths of these intervals compare to the lengths of the intervals you found in part (c)? Why is it sensible that the lengths should be related in this way?

(g) Simultaneous confidence intervals for the differences between all pairs of mean depths of cut are of the form $\bar{y}_i - \bar{y}_{i'} \pm \Delta$, for a number Δ. Find Δ if Tukey simultaneous 95% two-sided intervals are desired. How does this value compare to the one you found in part (d)?

(h) Use the one-way ANOVA test statistic and assess the strength of the evidence against $H_0: \mu_1 = \mu_2 = \mu_3$. Show the whole five-step format.

(i) Make the ANOVA table corresponding to the significance test you carried out in part (h).

25. Anderson, Panchula, and Patrick tested several designs of "paper helicopters" for flight times when dropped from a point approximately 8 feet above the ground. Four different helicopters were made and tested for each design. Some summary statistics for the tests on four particular designs are given next. (The units are seconds.)

Design #1	Design #2	Design #3	Design #4
$n_1 = 4$	$n_2 = 4$	$n_3 = 4$	$n_4 = 4$
$\bar{y}_1 = 1.640$	$\bar{y}_2 = 2.545$	$\bar{y}_3 = 1.510$	$\bar{y}_4 = 2.600$
$s_1 = .096$	$s_2 = .426$	$s_3 = .174$	$s_4 = .168$

(a) Find a pooled estimate of σ in the one-way model. What does this quantity measure in the present context?

(b) Give 95% two-sided confidence limits for the mean flight time of helicopters of Design #1.

(c) P-R simultaneous two-sided 95% confidence limits for all mean flight times of the designs are of the form $\bar{y}_i \pm \Delta$. Find Δ.

(d) Give 95% two-sided confidence limits for the difference in mean flight times of helicopters of Designs #1 and #2.

(e) Tukey simultaneous two-sided 95% confidence limits for all differences in mean flight times of the designs are of the form $\bar{y}_t - \bar{y}_{t'} \pm \Delta$, for a number Δ. Find Δ.

(f) Based on your answer to part (e), do you believe that there are "statistically significant"/"statistically detectable" differences among these four designs in terms of mean flight times? Explain.

(g) Do a formal significance test of $H_0: \mu_1 = \mu_2 = \mu_3 = \mu_4$. Show the whole five-step format.

(h) As a matter of fact, the four designs considered here were Design #1, 2 in. wings and 1 in. body; Design #2, 4 in. wings and 1 in. body; Design #3, 2 in. wings and 3 in. body; Design #4, 4 in. wings and 3 in. body. So the quantity

$$\frac{1}{2}\left(\mu_1 + \mu_3\right) - \frac{1}{2}\left(\mu_2 + \mu_4\right)$$

is a measure of the effect of changing from 2 in. wings to 4 in. wings. Give 95% two-sided confidence limits for this quantity.

Inference for Full and Fractional Factorial Studies

Chapter 7 began this book's exposition of inference methods for multisample studies. The methods there neither require nor make use of any special structure relating the samples. They are both widely applicable and practically informative tools. But Chapter 4 illustrated on an informal or descriptive level the engineering importance of discovering, interpreting, and ultimately exploiting structure relating a response to one or more other variables. This chapter begins to provide inference methods to support these activities.

This chapter builds on the descriptive statistics material of Section 4.3 and the tools of Chapter 7 to provide methods for full and fractional factorial studies. It begins with a discussion of some inference methods for complete two-way factorials. Then complete p-way factorial inference is considered with special attention to the 2^p case. Then two successive sections describe what is possible in the way of factorial inference from well-chosen fractions of a 2^p factorial. First, half fractions are considered, and then $1/2^q$ fractions for $q > 1$.

8.1 Basic Inference in Two-Way Factorials with Some Replication

This section considers inference from complete two-way factorial data in cases where there is some replication—i.e., at least one of the sample sizes is larger than 1. It begins by pointing out that the material in Sections 7.1 through 7.4 can often be useful in sharpening the preliminary graphical analyses suggested in Section 4.3. Then there is a discussion of inference based on the fitted two-way factorial effects defined in Chapter 4. These are used to develop both individual and simultaneous confidence interval methods.

8.1.1   One-Way Methods in Two-Way Factorials

Example 1 revives a case used extensively in Section 4.3.

Example 1
(*Example 7, Chapter 4,
revisited—page 163*)

Joint Strengths for Three Different Joint Types in Three Different Woods

Consider again the wood joint strength study of Kotlers, MacFarland, and Tomlinson. Table 8.1 reorganizes the data given earlier in Table 4.11 into a 3×3 table showing the nine different samples of one or two joint strengths for all combinations of three woods and three joint types. The data in Table 8.1 have complete two-way factorial structure, and seven of the nine combinations represented in the table provide some replication.

Table 8.1
Joint Strengths for 3^2 Combinations of Joint Type and Wood

		Wood		
		1 (Pine)	2 (Oak)	3 (Walnut)
	1 (Butt)	829, 596	1169	1263, 1029
Joint	2 (Beveled)	1348, 1207	1518, 1927	2571, 2443
	3 (Lap)	1000, 859	1295, 1561	1489

The data in Table 8.1 constitute $r = 9$ samples of sizes 1 or 2. Provided the graphical and numerical checks of Section 7.1 reveal no obvious problems with the one-way model for joint strengths, all of the methods of Sections 7.2 through 7.4 can be brought to bear.

One way in which this is particularly helpful is in indicating the precision of estimated means on interaction plots. Section 4.3 discussed how near-parallelism on such plots leads to simple interpretations of two-way factorials. By marking either individual or simultaneous confidence limits as **error bars** around the sample means on an interaction plot, it is possible to get a rough idea of the detectability or statistical significance of any apparent lack of parallelism.

*Error bars
on interaction
plots*

Example 1
(*continued*)

The place to begin a formal analysis of the wood joint strength data is with consideration of the appropriateness of the one-way (normal distributions with a common variance) model for joint strength. Table 8.2 gives some summary statistics for the data of Table 8.1.

Residuals for the joint strength data are obtained by subtracting the sample means in Table 8.2 from the corresponding observations in Table 8.1. In this data set, the sample sizes are so small that the residuals will obviously be highly dependent. Those from samples of size 2 will be plus-and-minus a single number

Example 1
(*continued*)

Table 8.2
Sample Means and Standard Deviations for Nine Joint/Wood Combinations

		Wood		
		1 (Pine)	2 (Oak)	3 (Walnut)
	1 (Butt)	$\bar{y}_{11} = 712.5$ $s_{11} = 164.8$	$\bar{y}_{12} = 1,169$	$\bar{y}_{13} = 1,146$ $s_{13} = 165.5$
Joint	2 (Beveled)	$\bar{y}_{21} = 1,277.5$ $s_{21} = 99.7$	$\bar{y}_{22} = 1,722.5$ $s_{22} = 289.2$	$\bar{y}_{23} = 2,507$ $s_{23} = 90.5$
	3 (Lap)	$\bar{y}_{31} = 929.5$ $s_{31} = 99.7$	$\bar{y}_{32} = 1,428$ $s_{32} = 188.1$	$\bar{y}_{33} = 1489$

corresponding to that sample. Those from samples of size 1 will be zero. So there is reason to expect residual plots to show some effects of this dependence. Figure 8.1 is a normal plot of the 16 residuals, and its complete symmetry (with respect to the positive and negative residuals) is caused by this dependence.

Of course, the sample standard deviations in Table 8.2 vary somewhat, but the ratio between the largest and smallest (a factor of about 3) is in no way surprising based on these sample sizes of 2. (Even if only 2 rather than 7 sample variances were involved, since $9(= 3^2)$ is between the .75 and .9 quantiles of the $F_{1,1}$ distribution, the observed level of significance for testing the equality of the two underlying variances would exceed $.2 = 2(1 - .9)$.) And Figure 8.2, which is a plot of residuals versus sample means, suggests no trend in σ as a function of mean response, μ.

In sum, the very small sample sizes represented in Table 8.1 make definitive investigation of the appropriateness of the one-way normal model assumptions

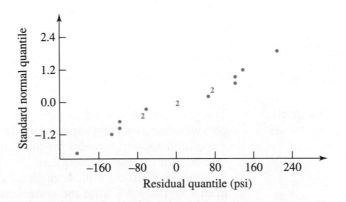

Figure 8.1   Normal plot of 16 residuals for the wood joint strength study

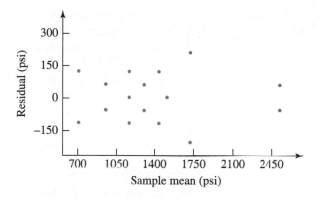

Figure 8.2   Plot of residuals versus sample means for the joint strength study

impossible. But the limited checks that are possible provide no indication of serious problems with operating under those restrictions.

Notice that for these data,

$$s_P^2 = \frac{(2-1)s_{11}^2 + (2-1)s_{13}^2 + (2-1)s_{21}^2 + \cdots + (2-1)s_{32}^2}{(2-1) + (2-1) + (2-1) + \cdots + (2-1)}$$

$$= \frac{1}{7}\left((164.8)^2 + (165.5)^2 + \cdots + (188.1)^2\right)$$

$$= 28,805 \ (\text{psi})^2$$

So

$$s_P = \sqrt{28,805} = 169.7 \ \text{psi}$$

where s_P has 7 associated degrees of freedom.

Then, for example, from formula (7.14) of Section 7.2, individual two-sided 99% confidence intervals for the combination mean strengths would have endpoints

$$\bar{y}_{ij} \pm 3.499(169.7)\frac{1}{\sqrt{n_{ij}}}$$

For the samples of size 1, this is

$$\bar{y}_{ij} \pm 593.9 \tag{8.1}$$

while for the samples of size 2, appropriate endpoints are

$$\bar{y}_{ij} \pm 419.9 \tag{8.2}$$

Example 1
(continued)

Figure 8.3 is an interaction plot (like Figure 4.22) enhanced with error bars made using limits (8.1) and (8.2). Notice, by the way, that the Bonferroni inequality puts the simultaneous confidence associated with all nine of the indicated intervals at a minimum of 91% $(.91 = 1 - 9(1 - .99))$.

The important message carried by Figure 8.3, not already present in Figure 4.22, is the relatively large imprecision associated with the sample means as estimates of long-run mean strengths. And that imprecision has implications regarding the statistical detectability of factorial effects. For example, by moving near the extremes on some error bars in Figure 8.3, one might find nine means within the indicated intervals such that their connecting line segments would exhibit parallelism. That is, the plot already suggests that the empirical interactions between Wood Type and Joint Type seen in these data may not be large enough to distinguish from background noise. Or if they are detectable, they may be only barely so.

The issues of whether the empirical differences between woods and between joint types are distinguishable from experimental variation are perhaps somewhat easier to call. There is consistency in the patterns "Walnut is stronger than oak is stronger than pine" and "Beveled is stronger than lap is stronger than butt." This, combined with differences at least approaching the size of indicated imprecisions,

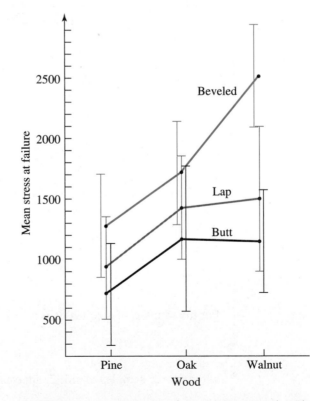

Figure 8.3 Interaction plot of mean joint strength with error bars based on individual 99% confidence intervals

suggests that firm statements about the main effects of Wood Type and Joint Type are likely possible.

The kind of analysis made thus far on the joint strength data is extremely important and illuminating. Our discussion will proceed to more complicated statistical methods for such problems. But these often amount primarily to a further refinement and quantification of the two-way factorial story already told graphically by a plot like Figure 8.3.

8.1.2 Two-Way Factorial Notation and Definitions of Effects

In order to discuss inference in two-way factorial studies, it is useful to modify the generic multisample notation used in Chapter 7. Consider combinations of factor A having I levels and factor B having J levels and use the triple subscript notation:

*Two-way
(triple subscript)
notation*

$$y_{ijk} = \text{the } k\text{th observation in the sample from the } i\text{th level of A}$$
$$\text{and } j\text{th plevel of B}$$

Then for $I \cdot J$ different samples corresponding to the possible combinations of a level of A with a level of B, let

$$n_{ij} = \text{the number of observations in the sample from the } i\text{th level of A}$$
$$\text{and } j\text{th level of B}$$

Use the notations $\bar{y}_{ij}$, $\bar{y}_{i\cdot}$, and $\bar{y}_{\cdot j}$ introduced in Section 4.3, and in the obvious way (actually already used in Example 1), let

$$s_{ij} = \text{the sample standard deviation of the } n_{ij} \text{ observations in the sample}$$
$$\text{from the } i\text{th level of A and the } j\text{th level of B}$$

This amounts to adding another subscript to the notation introduced in Chapter 7 in order to acknowledge the two-way structure. In Chapter 7, it was most natural to think of r samples as numbered $i = 1$ to r and laid out in a single row. Here it is appropriate to think of $r = I \cdot J$ samples laid out in the cells of a two-way table like Table 8.1 and named by their row number i and column number j.

In addition to using this notation for empirical quantities, it is also useful to modify the notation used in Chapter 7 for model parameters. That is, let

$$\mu_{ij} = \text{the underlying mean response corresponding to the } i\text{th level of A}$$
$$\text{and } j\text{th level of B}$$

The model assumptions that the $I \cdot J$ samples are roughly describable as independent samples from normal distributions with a common variance σ^2 can be written as

*Two-way model
statement*

$$y_{ijk} = \mu_{ij} + \epsilon_{ijk} \tag{8.3}$$

where the quantities $\epsilon_{111}, \ldots, \epsilon_{11n_{11}}, \epsilon_{121}, \ldots, \epsilon_{12n_{12}}, \ldots, \epsilon_{IJ1}, \ldots, \epsilon_{IJn_{IJ}}$ are independent normal $(0, \sigma^2)$ random variables. Equation (8.3) is sometimes called the **two-way (normal) model equation**. It is nothing but a rewrite of the basic one-way model equation of Chapter 7 in a notation that recognizes the special organization of $r = I \cdot J$ samples into rows and columns, as in Table 8.1.

The descriptive analysis of two-way factorials in Section 4.3 relied on computing row averages $\bar{y}_{i.}$ and column averages $\bar{y}_{.j}$ from the sample means $\bar{y}_{ij}$. These were then used to define fitted factorial effects. Analogous operations performed on the underlying or theoretical means μ_{ij} lead to appropriate definitions for theoretical factorial effects. That is, let

$$\mu_{i.} = \frac{1}{J} \sum_{j=1}^{J} \mu_{ij}$$

= the average underlying mean when factor A is at level i

$$\mu_{.j} = \frac{1}{I} \sum_{i=1}^{I} \mu_{ij}$$

= the average underlying mean when factor B is at level j

$$\mu_{..} = \frac{1}{IJ} \sum_{i, j} \mu_{ij}$$

= the grand average underlying mean

Figure 8.4 shows these as row, column, and grand averages of the μ_{ij}. (This is the theoretical counterpart of Figure 4.21.)

Then, following the pattern established in Definitions 5 and 6 in Chapter 4 for sample quantities, there are the following two definitions for theoretical quantities.

Definition 1	In a two-way complete factorial study with factors A and B, the **main effect of factor A at its ith level** is $$\alpha_i = \mu_{i.} - \mu_{..}$$ Similarly, the **main effect of factor B at its jth level** is $$\beta_j = \mu_{.j} - \mu_{..}$$

These main effects are measures of how (theoretical) mean responses change from row to row or from column to column in Figure 8.4. The fitted main effects of Section 4.3 can be thought of as empirical approximations to them. It is a

Factor B

	Level 1	Level 2		Level J	
Level 1	μ_{11}	μ_{11}	$\cdots$	μ_{1J}	$\mu_{1.}$
Level 2	μ_{21}	μ_{22}	$\cdots$	μ_{2J}	$\mu_{2.}$
Factor A $\vdots$	$\vdots$	$\vdots$		$\vdots$	$\vdots$
Level I	μ_{I1}	μ_{I2}	$\cdots$	μ_{IJ}	$\mu_{I.}$
	$\mu_{.1}$	$\mu_{.2}$	$\cdots$	$\mu_{.J}$	$\mu_{..}$

Figure 8.4   Underlying cell mean responses and their row, column, and grand averages

consequence of the form of Definition 1 that (like their empirical counterparts) main effects of a given factor sum to 0 over levels of that factor. That is, simple algebra shows that

$$\sum_{i=1}^{I} \alpha_i = 0 \quad \text{and} \quad \sum_{j=1}^{I} \beta_j = 0$$

Next is a definition of theoretical interactions.

Definition 2

In a two-way complete factorial study with factors A and B, the **interaction of factor A at its ith level and factor B at its jth level** is

$$\alpha\beta_{ij} = \mu_{ij} - (\mu_{..} + \alpha_i + \beta_j)$$

The interactions in a two-way set of underlying means μ_{ij} measure lack of parallelism on an interaction plot of the parameters μ_{ij}. They measure how much pattern there is in the theoretical means μ_{ij} that is not explainable in terms of the factors A and B acting individually. The fitted interactions of Section 4.3 are empirical approximations of these theoretical quantities. Small fitted interactions ab_{ij} indicate small underlying interactions $\alpha\beta_{ij}$ and thus make it justifiable to think of the two factors A and B as operating separately on the response variable.

Definition 2 has several simple algebraic consequences that are occasionally useful to know. One is that (like fitted interactions) interactions $\alpha\beta_{ij}$ sum to 0 over levels of either factor. That is, as defined,

$$\sum_{i=1}^{I} \alpha\beta_{ij} = \sum_{j=1}^{J} \alpha\beta_{ij} = 0$$

Another simple consequence is that upon adding $(\mu_{..} + \alpha_i + \beta_j)$ to both sides of the equation defining $\alpha\beta_{ij}$, one obtains a decomposition of each μ_{ij} into a grand mean plus an A main effect plus a B main effect plus an AB interaction:

$$\mu_{ij} = \mu_{..} + \alpha_i + \beta_j + \alpha\beta_{ij} \qquad (8.4)$$

The identity (8.4) is sometimes combined with the two-way model equation (8.3) to obtain the equivalent model equation

A second statement of the two-way model

$$y_{ijk} = \mu_{..} + \alpha_i + \beta_j + \alpha\beta_{ij} + \epsilon_{ijk} \qquad (8.5)$$

Here the factorial effects appear explicitly as going into the makeup of the observations. Although there are circumstances where representation (8.5) is essential, in most cases it is best to think of the two-way model assumptions in form (8.3) and just remember that the α_i, β_j, and $\alpha\beta_{ij}$ are simple functions of the $I \cdot J$ means μ_{ij}.

8.1.3 Individual Confidence Intervals for Factorial Effects

The primary new wrinkles in two-way factorial inference are

1. the drawing of inferences concerning the interactions and main effects, with

2. the possibility of finding A, B, or A and B "main effects only" models adequate to describe responses, and subsequently using such simplified descriptions in making predictions about system behavior.

Factorial effects are L's, fitted effects are corresponding L̂'s

The basis of inference for the α_i, β_j, and $\alpha\beta_{ij}$ is that they are linear combinations of the means μ_{ij}. (That is, for properly chosen "c's," the factorial effects are "L's" from Section 7.2.) And the fitted effects defined in Chapter 4's Definitions 5 and 6 are the corresponding linear combinations of the sample means $\bar{y}_{ij}$. (That is, the fitted factorial effects are the corresponding "L̂'s.")

Example 1
(continued)

To illustrate that the effects defined in Definitions 1 and 2 are linear combinations of the underlying means μ_{ij}, consider α_1 and $\alpha\beta_{23}$ in the wood joint strength study. First,

$$\alpha_1 = \mu_{1.} - \mu_{..}$$

$$= \frac{1}{3}(\mu_{11} + \mu_{12} + \mu_{13}) - \frac{1}{9}(\mu_{11} + \mu_{12} + \cdots + \mu_{32} + \mu_{33})$$

$$= \frac{2}{9}\mu_{11} + \frac{2}{9}\mu_{12} + \frac{2}{9}\mu_{13} - \frac{1}{9}\mu_{21} - \frac{1}{9}\mu_{22} - \frac{1}{9}\mu_{23} - \frac{1}{9}\mu_{31} - \frac{1}{9}\mu_{32} - \frac{1}{9}\mu_{33}$$

and a_1 is the corresponding linear combination of the $\bar{y}_{ij}$. Similarly,

$$\alpha\beta_{23} = \mu_{23} - (\mu_{..} + \alpha_2 + \beta_3)$$

$$= \mu_{23} - \left(\mu_{..} + (\mu_{2.} - \mu_{..}) + (\mu_{.3} - \mu_{..})\right)$$

$$= \mu_{23} - \mu_{2.} - \mu_{.3} + \mu_{..}$$

$$= \mu_{23} - \frac{1}{3}(\mu_{21} + \mu_{22} + \mu_{23}) - \frac{1}{3}(\mu_{13} + \mu_{23} + \mu_{33})$$

$$+ \frac{1}{9}(\mu_{11} + \mu_{12} + \cdots + \mu_{33})$$

$$= \frac{4}{9}\mu_{23} - \frac{2}{9}\mu_{21} - \frac{2}{9}\mu_{22} - \frac{2}{9}\mu_{13} - \frac{2}{9}\mu_{33} + \frac{1}{9}\mu_{11} + \frac{1}{9}\mu_{12}$$

$$+ \frac{1}{9}\mu_{31} + \frac{1}{9}\mu_{32}$$

and ab_{23} is the corresponding linear combination of the $\bar{y}_{ij}$.

Once one realizes that the factorial effects are simple linear combinations of the μ_{ij}, it is a small step to recognize that formula (7.20) of Section 7.2 can be applied to make confidence intervals for them. For example, the question of whether the lack of parallelism evident in Figure 8.3 is large enough to be statistically detectable can be approached by looking at confidence intervals for the $\alpha\beta_{ij}$. And quantitative comparisons between joint types can be based on confidence intervals for differences between the A main effects, $\alpha_i - \alpha_{i'} = \mu_{i.} - \mu_{i'.}$. And quantitative comparisons between woods can be based on differences between the B main effects, $\beta_j - \beta_{j'} = \mu_{.j} - \mu_{.j'}$.

The only obstacle to applying formula (7.20) of Section 7.2 to do inference for factorial effects is determining how the "$\sum c_i^2/n_i$" term appearing in the formula should look for quantities of interest. In the preceding example, a number of rather odd-looking coefficients c_{ij} appeared when writing out expressions for α_1 and $\alpha\beta_{23}$ in terms of the basic means μ_{ij}. However, it is possible to discover and write down general formulas for the sum $\sum c_{ij}^2/n_{ij}$ for some important functions of the factorial effects. Table 8.3 gives the relatively simple formulas for the balanced data case where all $n_{ij} = m$. The less pleasant general versions of the formulas are given in Table 8.4.

Table 8.3
Balanced Data Formulas to Use
with Limits (8.6)

L	$\hat{L}$	$\sum_{i,j} \dfrac{c_{ij}^2}{n_{ij}}$
$\alpha\beta_{ij}$	ab_{ij}	$\dfrac{(I-1)(J-1)}{mIJ}$
α_i	a_i	$\dfrac{I-1}{mIJ}$
$\alpha_i - \alpha_{i'}$	$a_i - a_{i'}$	$\dfrac{2}{mJ}$
β_j	b_j	$\dfrac{J-1}{mIJ}$
$\beta_j - \beta_{j'}$	$b_j - b_{j'}$	$\dfrac{2}{mI}$

Armed with Tables 8.3 and 8.4, the form of individual confidence intervals for any of the quantities $L = \alpha\beta_{ij}, \alpha_i, \beta_j, \alpha_i - \alpha_{i'}$, or $\beta_j - \beta_{j'}$ is obvious. In the formula for confidence interval endpoints

*Confidence limits
for a linear
combination of
two-way factorial
means*

$$\hat{L} \pm t s_{\mathrm{P}} \sqrt{\sum_{i,j} \frac{c_{ij}^2}{n_{ij}}} \tag{8.6}$$

1. s_{P} is computed by pooling the $I \cdot J$ sample variances in the usual way (arriving at an estimate with $n - r = n - IJ$ associated degrees of freedom),

2. the fitted effects from Section 4.3 are used to find $\hat{L}$,

3. an appropriate formula from Table 8.3 or 8.4 is chosen to give the quantity under the radical, and

4. t from Table B.4 is chosen according to a desired confidence and degrees of freedom $\nu = n - IJ$.

Table 8.4
General Formulas to use with Limits (8.6)

L	$\hat{L}$	$\sum_{i,j} \dfrac{c_{ij}^2}{n_{ij}}$
$\alpha\beta_{ij}$	ab_{ij}	$\left(\dfrac{1}{IJ}\right)^2 \left(\dfrac{(I-1)^2(J-1)^2}{n_{ij}} + (I-1)^2 \sum_{j'\neq j} \dfrac{1}{n_{ij'}} + (J-1)^2 \sum_{i'\neq i} \dfrac{1}{n_{i'j}} + \sum_{i'\neq i,\, j'\neq j} \dfrac{1}{n_{i'j'}} \right)$
α_i	a_i	$\left(\dfrac{1}{IJ}\right)^2 \left((I-1)^2 \sum_j \dfrac{1}{n_{ij}} + \sum_{i'\neq i,\, j} \dfrac{1}{n_{i'j}} \right)$
$\alpha_i - \alpha_{i'}$	$a_i - a_{i'}$	$\dfrac{1}{J^2} \left(\sum_j \dfrac{1}{n_{ij}} + \sum_j \dfrac{1}{n_{i'j}} \right)$
β_j	b_j	$\left(\dfrac{1}{IJ}\right)^2 \left((J-1)^2 \sum_i \dfrac{1}{n_{ij}} + \sum_{i,\, j'\neq j} \dfrac{1}{n_{ij'}} \right)$
$\beta_j - \beta_{j'}$	$b_j - b_{j'}$	$\dfrac{1}{I^2} \left(\sum_i \dfrac{1}{n_{ij}} + \sum_i \dfrac{1}{n_{ij'}} \right)$

Example 2

A Synthetic 3 × 3 Balanced Data Example

To illustrate how easy it is to do inference for factorial effects when complete two-way factorial data are balanced, consider a 3 × 3 factorial with $m = 2$ observations per cell. (This is the way that the wood joint strength study of Example 1 was planned. It was only circumstances beyond the control of the students that conspired to produce the unbalanced data of Table 8.1 through the loss of two specimens.) In this hypothetical situation, s_P has degrees of freedom $v = n - IJ = mIJ - IJ = 2 \cdot 3 \cdot 3 - 3 \cdot 3 = 9$. Definitions 5 and 6 in Chapter 4 show how to compute fitted main effects a_i and b_j and fitted interactions ab_{ij}.

To, for example, make a confidence interval for an interaction $\alpha\beta_{ij}$, consult the first row of Table 8.3 and compute

$$\sum_{i,j} \frac{c_{ij}^2}{n_{ij}} = \frac{(I-1)(J-1)}{mIJ} = \frac{2 \cdot 2}{2 \cdot 3 \cdot 3} = \frac{2}{9} \quad \text{and} \quad \sqrt{\sum_{i,j} \frac{c_{ij}^2}{n_{ij}}} = .4714$$

Example 2
(continued)

Then choosing t (as a quantile of the t_9 distribution) to produce the desired confidence level, equation (8.6) shows appropriate confidence limits to be

▶
$$ab_{ij} \pm ts_P(.4714)$$

As a second example of this methodology, consider the estimation of the difference in two factor B main effects, $L = \beta_j - \beta_{j'} = \mu_{.j} - \mu_{.j'}$. Consulting the last row of Table 8.3,

$$\sum_{i,j} \frac{c_{ij}^2}{n_{ij}} = \frac{2}{mI} = \frac{2}{2 \cdot 3} = \frac{1}{3} \quad \text{and} \quad \sqrt{\sum_{i,j} \frac{c_{ij}^2}{n_{ij}}} = .5774$$

Then again choosing t to produce the desired confidence level, equation (8.6) shows appropriate confidence limits to be

$$b_j - b_{j'} \pm ts_P(.5774)$$

that is,

▶
$$\bar{y}_{.j} - \bar{y}_{.j'} \pm ts_P(.5774)$$

Example 1
(continued)

Consider making formal inferences for the factorial effects in the (unbalanced) wood joint strength. Suppose that inferences are to be phrased in terms of two-sided 99% individual confidence intervals and begin by considering the interactions $\alpha\beta_{ij}$.

Despite the students' best efforts to the contrary, the sample sizes in Table 8.1 are not all the same. So one is forced to use formulas in Table 8.4 instead of the simpler ones in Table 8.3. Table 8.5 collects the sums of reciprocal sample sizes appearing in the first row of Table 8.4 for each of the nine combinations of $i = 1, 2, 3$ and $j = 1, 2, 3$.

For example, for the combination $i = 1$ and $j = 1$,

$$\frac{1}{n_{11}} = \frac{1}{2} = .5$$

$$\frac{1}{n_{12}} + \frac{1}{n_{13}} = \frac{1}{1} + \frac{1}{2} = 1.5$$

$$\frac{1}{n_{21}} + \frac{1}{n_{31}} = \frac{1}{2} + \frac{1}{2} = 1.0$$

$$\frac{1}{n_{22}} + \frac{1}{n_{23}} + \frac{1}{n_{32}} + \frac{1}{n_{33}} = \frac{1}{2} + \frac{1}{2} + \frac{1}{2} + \frac{1}{1} = 2.5$$

Table 8.5

Sums of Reciprocal Sample Sizes Needed in Making Confidence Intervals for Joint/Wood Interactions

i	j	$\dfrac{1}{n_{ij}}$	$\sum_{j' \neq j} \dfrac{1}{n_{ij'}}$	$\sum_{i' \neq i} \dfrac{1}{n_{i'j}}$	$\sum_{i' \neq i,\, j' \neq j} \dfrac{1}{n_{i'j'}}$
1	1	.5	1.5	1.0	2.5
1	2	1.0	1.0	1.0	2.5
1	3	.5	1.5	1.5	2.0
2	1	.5	1.0	1.0	3.0
2	2	.5	1.0	1.5	2.5
2	3	.5	1.0	1.5	2.5
3	1	.5	1.5	1.0	2.5
3	2	.5	1.5	1.5	2.0
3	3	1.0	1.0	1.0	2.5

The entries in Table 8.5 lead to values for $\sum c_{ij}^2/n_{ij}$ via the formula on the first row of Table 8.4. Then, since (from before) $s_{\mathsf{P}} = 169.7$ psi with 7 associated degrees of freedom, and since the .995 quantile of the t_7 distribution is 3.499, it is possible to calculate the plus-or-minus part of formula (8.6) in order to get two-sided 99% confidence intervals for the $\alpha\beta_{ij}$. In addition, remember that all nine fitted interactions were calculated in Section 4.3 and collected in Table 4.14 (page 170). Table 8.6 gives the $\sqrt{\sum c_{ij}^2/n_{ij}}$ values, the fitted interactions ab_{ij}, and the plus-or-minus part of two-sided 99% individual confidence intervals for the interactions $\alpha\beta_{ij}$.

To illustrate the calculations summarized in the third column of Table 8.6, consider the combination with $i = 1$ (butt joints) and $j = 1$ (pine wood). Since $I = 3$ and $J = 3$, the first row of Table 8.4 shows that for $L = \alpha\beta_{11}$

$$\sum \frac{c_{ij}^2}{n_{ij}} = \left(\frac{1}{3 \cdot 3} \right)^2 \left(\frac{2^2 \cdot 2^2}{2} + 2^2(1.5) + 2^2(1.0) + 2.5 \right) = .2531$$

from which

$$\sqrt{\sum \frac{c_{ij}^2}{n_{ij}}} = \sqrt{.2531} = .5031$$

Consider the practical implications of the calculations summarized in Table 8.6. All but one of the intervals centered at an ab_{ij} with a half width given in the last column of the table would cover 0. Only for $i = 2$ (beveled joints) and $j = 3$ (walnut wood) is the magnitude of the fitted interaction big enough to put its

Example 1
(*continued*)

Table 8.6
99% Individual Two-Sided Confidence Intervals for Joint Type/Wood Type Interactions

i	j	$\sqrt{\sum \dfrac{c_{ij}^2}{n_{ij}}}$	ab_{ij} (psi)	$ts_P \sqrt{\sum \dfrac{c_{ij}^2}{n_{ij}}}$ (psi)
1	1	.5031	105.83	298.7
1	2	.5720	95.67	339.6
1	3	.5212	−201.5	309.5
2	1	.4843	−155.67	287.6
2	2	.5031	−177.33	298.7
2	3	.5031	333.0	298.7
3	1	.5031	49.83	298.7
3	2	.5212	81.67	309.5
3	3	.5720	−131.5	339.6

associated confidence interval entirely to one side of 0. That is, most of the lack of parallelism seen in Figure 8.3 is potentially attributable to experimental variation. But that associated with beveled joints and walnut wood can be differentiated from background noise. This suggests that if mean joint strength differences on the order of 333 ± 299 psi are of engineering importance, it is not adequate to think of the factors Joint Type and Wood Type as operating separately on joint strength across all three levels of each factor. On the other hand, if attention was restricted to either butt and lap joints or to pine and oak woods, a "no detectable interactions" description of joint strength would perhaps be tenable.

To illustrate the use of formula (8.6) in making inferences about main effects on joint strength, consider comparing joint strengths for pine and oak woods. The rather extended analysis of interactions here and the character of Figure 8.3 suggest that the strength profiles of pine and oak across the three joint types are comparable. If this is so, estimation of $\beta_1 - \beta_2 = \mu_{.1} - \mu_{.2}$ amounts to more than the estimation of the difference in average (across joint types) mean strengths of pine and oak joints (pine minus oak). $\beta_1 - \beta_2$ is also the difference in mean strengths of pine and oak joints *for any of the three joint types individually*. It is thus a quantity of real interest.

Once again, since the data in Table 8.1 are not balanced, it is necessary to use the more complicated formula in Table 8.4 rather than the formula in Table 8.3 in making a confidence interval for $\beta_1 - \beta_2$. For $L = \beta_1 - \beta_2$, the last row of Table 8.4 gives

$$\sum_{i,j} \frac{c_{ij}^2}{n_{ij}} = \frac{1}{3^2}\left[\left(\frac{1}{2} + \frac{1}{2} + \frac{1}{2}\right) + \left(\frac{1}{1} + \frac{1}{2} + \frac{1}{2}\right)\right] = .3889$$

So, since from the fitted effects in Section 4.3

$$b_1 = -402.5 \text{ psi} \quad \text{and} \quad b_2 = 64.17 \text{ psi}$$

formula (8.6) shows that endpoints of a two-sided 99% confidence interval for $L = \beta_1 - \beta_2$ are

$$(-402.5 - 64.17) \pm 3.499(169.7)\sqrt{.3889}$$

that is,

$$-466.67 \pm 370.29$$

that is,

▶
$$-836.96 \text{ psi} \quad \text{and} \quad -96.38 \text{ psi}$$

This analysis establishes that the oak joints are on average from 96 psi to 837 psi stronger than comparable pine joints. This may seem a rather weak conclusion, given the apparent strong increase in sample mean strengths as one moves from pine to oak in Figure 8.3. But it is as strong a statement as is justified in the light of the large confidence requirement (99%) and the substantial imprecision in the students' data (related to the small sample sizes and a large pooled standard deviation, $s_P = 169.7 \text{ psi}$). If ± 370 psi precision for comparing pine and oak joint strength is not adequate for engineering purposes and large confidence is still desired, these calculations point to the need for more data in order to sharpen that comparison.

The computational unpleasantness of the previous discussion results from the fact that the data of Kotlers, MacFarland, and Tomlinson are unbalanced. Example 2 illustrated that with balanced data, "by hand" calculation is simple. Most statistical packages have routines that will eliminate the need for a user to grind through the most tedious of the computations just illustrated. Printout 1 is a MINITAB General Linear Model output for the wood strength study (which is part of Printout 6 of Chapter 4). The "Coef" values in that printout are (again) the fitted effects of Definitions 5 and 6 in Chapter 4. The "StDev" values are the quantities

$$s_P \sqrt{\sum_{i,j} \frac{c_{ij}^2}{n_{ij}}}$$

from formula (8.6) needed to make confidence limits for main effects and interactions. (The MINITAB printout lists this information for only $(I - 1)$ factor A main effects, $(J - 1)$ factor B main effects, and $(I - 1)(J - 1)$ A×B interactions. Renaming levels of the factors to change their alphabetical order will produce a different printout giving this information for the remaining main effects and interactions.)

Printout 1   Estimated Standard Deviations
             of Joint Strength Fitted Effects (*Example 1*)

```
General Linear Model


Factor     Type Levels Values
joint      fixed    3  beveled butt    lap
wood       fixed    3      oak pine walnut

Analysis of Variance for strength, using Adjusted SS for Tests
```

Source	DF	Seq SS	Adj SS	Adj MS	F	P
joint	2	2153879	1881650	940825	32.67	0.000
wood	2	1641095	1481377	740689	25.72	0.001
joint*wood	4	468408	468408	117102	4.07	0.052
Error	7	201614	201614	28802		
Total	15	4464996				

Term	Coef	StDev	T	P
Constant	1375.67	44.22	31.11	0.000
joint				
beveled	460.00	59.63	7.71	0.000
butt	-366.50	63.95	-5.73	0.001
wood				
oak	64.17	63.95	1.00	0.349
pine	-402.50	59.63	-6.75	0.000
joint* wood				
beveled oak	-177.33	85.38	-2.08	0.076
beveled pine	-155.67	82.20	-1.89	0.100
butt oak	95.67	97.07	0.99	0.357
butt pine	105.83	85.38	1.24	0.255

8.1.4   Tukey's Method for Comparing Main Effects (*Optional*)

Formula (8.6) is meant to guarantee *individual* confidence levels for intervals made using it. When interactions in a two-way factorial study are negligible, questions of practical engineering importance can usually be phrased in terms of comparing the various A or B main effects. It is then useful to have a method designed specifically to produce a *simultaneous* confidence level for the comparison of all pairs of A or B main effects. Tukey's method (discussed in Section 7.3) can be modified to produce simultaneous confidence intervals for all differences in α_i's or in β_j's. That is, two-sided simultaneous confidence intervals for all possible differences in A main effects $\alpha_i - \alpha_{i'} = \mu_{i.} - \mu_{i'.}$ can be made using endpoints

Tukey simultaneous confidence limits for all differences in A main effects

$$\bar{y}_{i.} - \bar{y}_{i'.} \pm \frac{q^*}{\sqrt{2}} s_P \frac{1}{J} \sqrt{\sum_j \frac{1}{n_{ij}} + \sum_j \frac{1}{n_{i'j}}} \qquad (8.7)$$

where q^* is taken from Tables B.9 using $\nu = n - IJ$ degrees of freedom, number of means to be compared I, and the .95 or .99 quantile figure (depending whether 95% or 99% simultaneous confidence is desired). Expression (8.7) amounts to the specialization of formula (8.6) to $L = \alpha_i - \alpha_{i'}$ with t replaced by $q^*/\sqrt{2}$. When all $n_{ij} = m$, formula (8.7) simplifies to

Balanced data Tukey simultaneous confidence limits for all differences in A main effects

$$\bar{y}_{i.} - \bar{y}_{i'.} \pm \frac{q^* s_{\mathrm{P}}}{\sqrt{Jm}} \qquad (8.8)$$

Corresponding to formulas (8.7) and (8.8) are formulas for simultaneous two-sided confidence limits for all possible differences in B main effects $\beta_j - \beta_{j'} = \mu_{.j} - \mu_{.j'}$—namely,

Tukey simultaneous confidence limits for all differences in B main effects

$$\bar{y}_{.j} - \bar{y}_{.j'} \pm \frac{q^*}{\sqrt{2}} s_{\mathrm{P}} \frac{1}{I} \sqrt{\sum_i \frac{1}{n_{ij}} + \sum_i \frac{1}{n_{ij'}}} \qquad (8.9)$$

and

Balanced data Tukey simultaneous confidence limits for all differences in B main effects

$$\bar{y}_{.j} - \bar{y}_{.j'} \pm \frac{q^* s_{\mathrm{P}}}{\sqrt{Im}} \qquad (8.10)$$

where q^* is taken from Tables B.9 using $\nu = n - IJ$ degrees of freedom and number of means to be compared J.

Example 3 | **A 3 × 2 Factorial Study of Ultimate Tensile Strength for Drilled Aluminum Strips**

Clubb and Goedken studied the effects on tensile strength of holes drilled in 6 in. by 2 in. 2024–T3 aluminum strips .0525 in. thick. A hole of diameter .149 in., .185 in., or .221 in. was centered either .5 in. or 1.0 in. from the edge (and 3.0 in. from each end) of 18 strips. Ultimate axial stress was then measured for each on an MTS machine. $m = 3$ tests were made for each of the 3×2 combinations of hole size and placement. Mean tensile strengths (in pounds) obtained in the study are given in Table 8.7. Some plotting with the original data (not given here) shows that (except for some hint that hole size 3 strengths were less variable than the others) the one-way normal model assumptions provide a plausible description of tensile strength. We will proceed to use the assumptions (8.3) in what follows.

Example 3
(*continued*)

Table 8.7
Sample Means for 3 × 2 Size/Placement Combinations

		B *Placement*	
		1 (.5 in. from Edge)	2 (1.0 in. from Edge)
A *Size*	1 (.149 in.)	$\bar{y}_{11} = 5635.3$ lb	$\bar{y}_{12} = 5730.3$ lb
	2 (.185 in.)	$\bar{y}_{21} = 5501.0$ lb	$\bar{y}_{22} = 5638.0$ lb
	3 (.221 in.)	$\bar{y}_{31} = 5456.3$ lb	$\bar{y}_{32} = 5602.7$ lb

Pooling the $3 \cdot 2 = 6$ sample variances in the usual way produced

$$s_P = 106.7 \text{ lb}$$

with $\nu = mIJ - IJ = 3 \cdot 3 \cdot 2 - 3 \cdot 2 = 12$ associated degrees of freedom. Then consider summarizing the experimental results graphically. Notice that the P-R method for making simultaneous two-sided 95% confidence intervals for $r = 6$ means based on $\nu = 12$ degrees of freedom is (from formula (7.28) of Section 7.3) to use endpoints

$$\bar{y}_{ij} \pm 3.095 \frac{106.7}{\sqrt{3}}$$

for estimating each μ_{ij}. ($k_2^* = 3.095$ was obtained from Table B.8A.) This is approximately

$$\bar{y}_{ij} \pm 191$$

Figure 8.5 is an interaction plot of the $3 \times 2 = 6$ sample mean tensile strengths enhanced with ± 191 lb error bars.

The lack of parallelism in Figure 8.5 is fairly small, both compared to the absolute size of the strengths being measured and also relative to the kind of uncertainty about the individual mean strengths indicated by the error bars. Letting factor A be size and factor B be placement, it is straightforward to use the methods of Section 4.3 to calculate

$$
\begin{array}{ll}
a_1 = 88.9 & b_1 = -63.1 \\
a_2 = -24.4 & b_2 = 63.1 \\
a_3 = -64.4 & \\
ab_{11} = 15.6 & ab_{12} = -15.6 \\
ab_{21} = -5.4 & ab_{22} = 5.4 \\
ab_{31} = -10.1 & ab_{32} = 10.1
\end{array}
$$

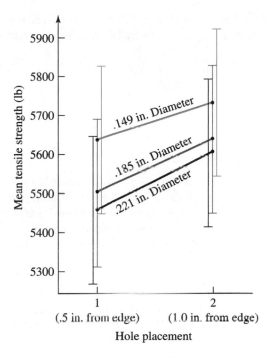

Figure 8.5   Interaction plot of aluminum strip sample means, enhanced with error bars based on 95% simultaneous confidence intervals

Then, since the data are balanced, one may use the formulas of Table 8.3 together with formula (8.6). So individual confidence intervals for the interactions $\alpha\beta_{ij}$ are of the form

$$ab_{ij} \pm t(106.7)\sqrt{\frac{(3-1)(2-1)}{3 \cdot 3 \cdot 2}}$$

that is,

$$ab_{ij} \pm t(35.6)$$

Clearly, for any sensible confidence level (producing t of at least 1), such intervals all cover 0. This confirms the lack of statistical detectability of the interactions already represented in Figure 8.5.

It thus seems sensible to proceed to consideration of the main effects in this tensile strength study. To illustrate the application of Tukey's method to factorial main effects, consider first simultaneous 95% two-sided confidence intervals for the three differences $\alpha_1 - \alpha_2$, $\alpha_1 - \alpha_3$, and $\alpha_2 - \alpha_3$. Applying formula (8.8)

Example 3
(*continued*)

with $\nu = 12$ degrees of freedom and $I = 3$ means to be compared, Table B.9A indicates that intervals with endpoints

$$\bar{y}_{i.} - \bar{y}_{i'.} \pm \frac{(3.77)(106.7)}{\sqrt{2 \cdot 3}}$$

that is,

$$\bar{y}_{i.} - \bar{y}_{i'.} \pm 164 \text{ lb}$$

are in order. No difference between the a_i's exceeds 164 lb. That is, if simultaneous 95% confidence is desired in the comparison of the hole size main effects, one must judge the students' data to be interesting—perhaps even suggestive of a decrease in strength with increased diameter—but nevertheless statistically inconclusive. To really pin down the impact of hole size on tensile strength, larger samples are needed.

To see that the Clubb and Goedken data do tell at least some story in a reasonably conclusive manner, finally consider the use of the last row of Table 8.3 with formula (8.6) to make a two-sided 95% confidence interval for $\beta_2 - \beta_1$, the difference in mean strengths for strips with centered holes as compared to ones with holes .5 in. from the strip edge. The desired interval has endpoints

$$b_2 - b_1 \pm t s_{\text{P}} \sqrt{\frac{2}{mI}}$$

that is,

$$63.1 - (-63.1) \pm 2.179(106.7) \sqrt{\frac{2}{3(3)}}$$

that is,

$$126.2 \pm 109.6$$

that is,

$$16.6 \text{ lb} \quad \text{and} \quad 235.8 \text{ lb}$$

Thus, although the students' data don't provide much precision, they *are* adequate to establish clearly the existence of some decrease in tensile strength as a hole is moved from the center of the strip towards its edge.

Formulas (8.7) through (8.10) are, mathematically speaking, perfectly valid providing only that the basic "equal variances, underlying normal distributions" model is a reasonable description of an engineering application. (Under the basic model (8.3), formulas (8.7) and (8.9) provide an actual simultaneous confidence at least as big as the nominal one, and when all $n_{ij} = m$, formulas (8.8) and (8.10) provide actual simultaneous confidence equal to the nominal one.) But in practical *terms*, the inferences they provide (and indeed the ones provided by formula (8.6) for individual differences in main effects) are not of much interest unless the interactions $\alpha\beta_{ij}$ have been judged to be negligible.

Nonnegligible interactions constitute a warning that the patterns of change in mean response, as one moves between levels of one factor, (say, B) are different for various levels of the other factor (say, A). That is, the pattern in the μ_{ij} is not a simple one generally describable in terms of the two factors acting separately. Rather than trying to understand the pattern in terms of main effects, something else must be done.

What if interactions are not negligible? As discussed in Section 4.4, sometimes a transformation can produce a response variable describable in terms of main effects only. At other times, restriction of attention to part of a factorial produces a study (of reduced scope) where it makes sense to think in terms of main effects. (In Example 1, consideration of only butt and lap joints gives an arena where "negligible interactions" may be a sensible description of joint strength.) Or it may be most natural to mentally separate an $I \times J$ factorial into $I(J)$ different $J(I)$ level studies on the effects of factor B(A) at different levels of A(B). (The 3×3 wood joint strength study in Example 1 might be thought of as three different studies, one for each joint type, of the effects of wood type on strength.) Or if none of these approaches to analyzing two-way factorial data with important interactions is attractive, it is always possible to ignore the two-way structure completely and treat the $I \cdot J$ samples as arising from simply $r = I \cdot J$ unstructured different conditions.

Section 1 Exercises

1. The accompanying table shows part of the data of Dimond and Dix, referred to in Examples 6 (Chapter 1) and 9 (Chapter 3). The values are the shear strengths (in lb) for $m = 3$ tests on joints of various combinations of Wood Type and Glue Type.

Wood	Glue	Joint Shear Strengths
pine	white	130, 127, 138
pine	carpenter's	195, 194, 189
pine	cascamite	195, 202, 207
fir	white	95, 119, 62
fir	carpenter's	137, 157, 145
fir	cascamite	152, 163, 155

(a) Make an interaction plot of the six combination means and enhance it with error bars derived

using the P-R method of making 95% simultaneous two-sided confidence intervals. (Plot mean strength versus glue type.)

(b) Compute the fitted main effects and interactions from the six combination sample means. Use these to make individual 95% confidence intervals for all of the main effects and interactions in this 2×3 factorial study. What do these indicate about the detectability of the various effects?

(c) Use Tukey's method for simultaneous comparison of main effects and give simultaneous 95% confidence intervals for all differences in Glue Type main effects.

2. B. Choi conducted a replicated full factorial study of the stopping properties of various types of bicycle tires on various riding surfaces. Three different Types of Tires were used on the bike, and

three different Pavement Conditions were used. For each Tire Type/Pavement Condition combination, $m = 6$ skid mark lengths were measured. The accompanying table shows some summary statistics for the study. (The units are cm.)

	Dry Concrete	Wet Concrete	Dirt
Smooth Tires	$\bar{y}_{11} = 359.8$ $s_{11} = 19.2$	$\bar{y}_{12} = 366.5$ $s_{12} = 26.4$	$\bar{y}_{13} = 393.0$ $s_{13} = 25.4$
Reverse Tread	$\bar{y}_{21} = 343.0$ $s_{21} = 15.5$	$\bar{y}_{22} = 356.7$ $s_{22} = 37.4$	$\bar{y}_{23} = 375.7$ $s_{23} = 39.9$
Treaded Tires	$\bar{y}_{31} = 384.8$ $s_{31} = 15.4$	$\bar{y}_{32} = 400.8$ $s_{32} = 60.8$	$\bar{y}_{33} = 402.5$ $s_{33} = 32.8$

(a) Compute s_P for Choi's data set. What is this supposed to be measuring?

(b) Make an interaction plot of the sample means similar to Figure 8.3. Use error bars for the means calculated from individual 95% two-sided confidence limits for the means. (Make use of your value of s_P from (a).)

(c) Based on your plot from (b), which factorial effects appear to be distinguishable from background noise? (Tire Type main effects? Pavement Condition main effects? Tire × Pavement interactions?)

(d) Compute all of the fitted factorial effects for Choi's data. (Find the a_i's, b_j's, and ab_{ij}'s defined in Section 4.3.)

(e) If one wishes to make individual 95% two-sided confidence intervals for the interactions $\alpha\beta_{ij}$, intervals of the form $ab_{ij} \pm \Delta$ are appropriate. Find Δ. Based on this value, are there statistically detectable interactions here? How does this conclusion compare with your more qualitative answer to part (c)?

(f) If one wishes to compare Tire Type main effects, confidence intervals for the differences $\alpha_i - \alpha_{i'}$ are in order. Find individual 95% two-sided confidence intervals for $\alpha_1 - \alpha_2$, $\alpha_1 - \alpha_3$, and $\alpha_2 - \alpha_3$. Based on these, are there statistically detectable differences in Tire Type main effects here? How does this conclusion compare with your answer to part (c)?

(g) Redo part (f), this time using (Tukey) simultaneous 95% two-sided confidence intervals.

8.2 *p*-Factor Studies with Two Levels for Each Factor

The previous section looked at inference for two-way factorial studies. This section presents methods of inference for complete *p*-way factorials, paying primary attention to those cases where each of *p* factors is represented at only two levels.

The discussion begins by again pointing out the relevance of the one-way methods of Chapter 7 to structured (in this case, *p*-way factorial) situations. Next, the *p*-way factorial normal model, definitions of effects in that model, and basic confidence interval methods for the effects are considered, paying particular attention to the 2^p case. Then attention is completely restricted to 2^p studies, and a further method for identifying detectable (2^p factorial) effects is presented. For balanced 2^p studies, there follows a review of the fitting of reduced models via the reverse Yates algorithm and the role of residuals in checking their efficacy. Finally, confidence interval methods based on simplified models in balanced 2^p studies are discussed.

8.2.1 One-Way Methods in *p*-Way Factorials

The place to begin the analysis of *p*-way factorial data is to recognize that fundamentally one is just working with several samples. Subject to the relevance of the model assumptions of Chapter 7, the inference methods of that chapter are available for use in analyzing the data.

Example 4

A 2^3 Factorial Study of Power Requirements in Metal Cutting

In *Fundamental Concepts in the Design of Experiments*, C. R. Hicks describes a study conducted by Purdue University engineering graduate student L. D. Miller on power requirements for cutting malleable iron using ceramic tooling. Miller studied the effects of the three factors

Factor A	Tool Type	(type 1 or type 2)
Factor B	Tool Bevel Angle	(15° or 30°)
Factor C	Type of Cut	(continuous or interrupted)

on the power required to make a cut on a lathe at a particular depth of cut, feed rate, and spindle speed. The response variable was the vertical deflection (in mm) of the indicator needle on a dynamometer (a measurement proportional to the horsepower required to make the particular cut). Miller's data are given in Table 8.8.

The most elementary view possible of the power requirement data in Table 8.8 is as $r = 8$ samples of size $m = 4$. Simple summary statistics for these $2^3 = 8$ samples are given in Table 8.9.

To the extent that the one-way normal model is an adequate description of this study, the methods of Chapter 7 are available for use in analyzing the data of Table 8.8. The reader is encouraged to verify that plotting of residuals (obtained by subtracting the $\bar{y}$ values in Table 8.9 from the corresponding raw data values of

Table 8.8
Dynamometer Readings for 2^3 Treatment Combinations in a Metal Cutting Study

Tool Type	Bevel Angle	Type of Cut	y, Dynamometer Reading (mm)
1	15°	continuous	29.0, 26.5, 30.5, 27.0
2	15°	continuous	28.0, 28.5, 28.0, 25.0
1	30°	continuous	28.5, 28.5, 30.0, 32.5
2	30°	continuous	29.5, 32.0, 29.0, 28.0
1	15°	interrupted	28.0, 25.0, 26.5, 26.5
2	15°	interrupted	24.5, 25.0, 28.0, 26.0
1	30°	interrupted	27.0, 29.0, 27.5, 27.5
2	30°	interrupted	27.5, 28.0, 27.0, 26.0

Example 4
(*continued*)

Table 8.9
Summary Statistics for 2^3 Samples of Dynamometer Readings in a Metal Cutting Study

Tool Type	Bevel Angle	Type of Cut	$\bar{y}$	s
1	15°	continuous	28.250	1.848
2	15°	continuous	27.375	1.601
1	30°	continuous	29.875	1.887
2	30°	continuous	29.625	1.702
1	15°	interrupted	26.500	1.225
2	15°	interrupted	25.875	1.548
1	30°	interrupted	27.750	0.866
2	30°	interrupted	27.125	0.854

Table 8.8) reveals only one slightly unpleasant feature of the power requirement data relative to the potential use of standard methods of inference. When plotted against levels of the Type of Cut variable, the residuals for interrupted cuts are shown to be on the whole somewhat smaller than those for continuous cuts. (This phenomenon is also obvious in retrospect from the sample standard deviations in Table 8.9. These are smaller for the second four samples than for the first four.) But the disparity in the sizes of the residuals is not huge. So although there may be some basis for suspecting improvement in power requirement consistency for interrupted cuts as opposed to continuous ones, the tractability of the one-way model and the kind of robustness arguments put forth at the end of Section 6.3 once again suggest that the standard model and methods be used. This is sensible, provided the resulting inferences are then treated as approximate and real-world "close calls" are not based on them.

The pooled sample variance here is

$$s_P^2 = \frac{(4-1)(1.848)^2 + (4-1)(1.601)^2 + \cdots + (4-1)(.854)^2}{(4-1) + (4-1) + \cdots + (4-1)} = 2.226$$

so

$$s_P = 1.492 \text{ mm}$$

with $\nu = n - r = 32 - 8 = 24$ associated degrees of freedom. Then, for example, the P-R method of simultaneous inference from Section 7.3 produces two-sided simultaneous 95% confidence intervals for mean dynamometer readings with endpoints

$$\bar{y}_{ijk} \pm 2.969 \frac{1.492}{\sqrt{4}}$$

that is,

$$\bar{y}_{ijk} \pm 2.21 \text{ mm}$$

(There is enough precision provided by the data to think of the sample means in Table 8.9 as roughly "all good to within 2.21 mm.") And the other methods of Sections 7.1 through 7.4 based on s_P might be used as well.

8.2.2 *p*-Way Factorial Notation, Definitions of Effects, and Related Confidence Interval Methods

Section 8.1 illustrated that standard notation in two-way factorials requires triple subscripts for naming observations. In a general *p*-way factorial, "$(p + 1)$-subscript" notation is required. As *p* grows, such notation quickly gets out of hand. As in Section 4.3 (on a descriptive level) the exposition here will explicitly develop only the general factorial notation for $p = 3$, leaving the reader to infer by analogy how things would have to go for $p = 4, 5$, etc. (When specializing to the 2^p situation later in this section, the special notation introduced in Section 4.3 makes it possible to treat even large-*p* situations fairly explicitly.)

Three factor (quadruple subscript) notation

Then for $p = 3$ factors A, B, and C having (respectively) *I*, *J*, and *K* levels, let

$$y_{ijkl} = \text{the } l\text{th observation in the sample from the } i\text{th level of A,}$$
$$j\text{th level of B, and } k\text{th level of C}$$

For the $I \cdot J \cdot K$ different samples corresponding to the possible combinations of a level of A with one of B and one of C, let

$$n_{ijk} = \text{the number of observations in the sample from the } i\text{th level of A,}$$
$$j\text{th level of B, and } k\text{th level of C}$$

$$\bar{y}_{ijk} = \text{the sample mean of the } n_{ijk} \text{ observations in the sample from the}$$
$$i\text{th level of A, } j\text{th level of B, and } k\text{th level of C}$$

$$s_{ijk} = \text{the sample standard deviation of the } n_{ijk} \text{ observations in the sample}$$
$$\text{from the } i\text{th level of A, } j\text{th level of B, and } k\text{th level of C}$$

and further continue the dot notations used in Section 4.3 for unweighted averages of the $\bar{y}_{ijk}$. In comparison to the notation of Chapter 7, this amounts to adding two subscripts in order to acknowledge the three-way structure in the samples.

The use of additional subscripts is helpful not only for naming empirical quantities but also for naming theoretical quantities. That is, with

$$\mu_{ijk} = \text{the underlying mean response corresponding to the}$$
$$i\text{th level of A, } j\text{th level of B, and } k\text{th level of C}$$

the standard one-way normal model assumptions can be rewritten as

Three-way model statement

$$y_{ijkl} = \mu_{ijk} + \epsilon_{ijkl} \qquad\qquad \textbf{(8.11)}$$

where the ϵ_{ijkl} terms are iid normal random variables with mean 0 and variance σ^2. Formula (8.11) could be called the **three-way (normal) model equation** because it recognizes the special organization of the $I \cdot J \cdot K$ samples according to combinations of levels of the three factors. But beyond this, it says no more or less than the one-way model equation from Section 7.1.

The initial objects of inference in three-way factorial analyses are linear combinations of theoretical means μ_{ijk}, analogous to the fitted effects of Section 4.3. Thus, it is necessary to carefully define the theoretical or underlying main effects, 2-factor interactions, and 3-factor interactions for a three-way factorial study. In the definitions that follow, a dot appearing as a subscript will (as usual) be understood to indicate that an average has been taken over all levels of the factor corresponding to the dotted subscript. Consider first main effects. Parallel to Definition 7 in Chapter 4 (page 182) for fitted main effects is a definition of theoretical main effects.

Definition 3

In a three-way complete factorial study with factors A, B, and C, the **main effect of factor A at its *i*th level** is

$$\alpha_i = \mu_{i..} - \mu_{...}$$

the **main effect of factor B at its *j*th level** is

$$\beta_j = \mu_{.j.} - \mu_{...}$$

and the **main effect of factor C at its *k*th level** is

$$\gamma_k = \mu_{..k} - \mu_{...}$$

These main effects measure how (when averaged over all combinations of levels of the other factors) underlying mean responses change from level to level of the factor in question. Definition 3 has the algebraic consequences that

$$\sum_{i=1}^{I} \alpha_i = 0, \quad \sum_{j=1}^{J} \beta_j = 0, \quad \text{and} \quad \sum_{k=1}^{K} \gamma_k = 0$$

The theoretical counterpart of Definition 8 in Chapter 4 is a definition of theoretical 2-factor interactions.

Definition 4	In a three-way complete factorial study with factors A, B, and C, the **2-factor interaction of factor A at its *i*th level and factor B at its *j*th level** is $$\alpha\beta_{ij} = \mu_{ij.} - (\mu_{...} + \alpha_i + \beta_j)$$ the **2-factor interaction of A at its *i*th level and C at its *k*th level** is $$\alpha\gamma_{ik} = \mu_{i.k} - (\mu_{...} + \alpha_i + \gamma_k)$$ and the **2-factor interaction of B at its *j*th level and C at its *k*th level** is $$\beta\gamma_{jk} = \mu_{.jk} - (\mu_{...} + \beta_j + \gamma_k)$$

Like their empirical counterparts defined in Section 4.3, the 2-factor interactions in a three-way study are measures of lack of parallelism on two-way plots of means obtained by averaging out over levels of the "other" factor. And it is an algebraic consequence of the form of Definition 4 that

$$\sum_{i=1}^{I} \alpha\beta_{ij} = \sum_{j=1}^{J} \alpha\beta_{ij} = 0, \quad \sum_{i=1}^{I} \alpha\gamma_{ik} = \sum_{k=1}^{K} \alpha\gamma_{ik} = 0$$

and

$$\sum_{j=1}^{J} \beta\gamma_{jk} = \sum_{k=1}^{K} \beta\gamma_{jk} = 0$$

Finally, there is the matter of three-way interactions in a three-way factorial study. Direct analogy with the meaning of fitted three-way interactions given as Definition 9 in Chapter 4 (page 183) gives the following:

Definition 5	In a three-way complete factorial study with factors A, B, and C, the **3-factor interaction of factor A at its *i*th level, factor B at its *j*th level, and factor C at its *k*th level** is $$\alpha\beta\gamma_{ijk} = \mu_{ijk} - (\mu_{...} + \alpha_i + \beta_j + \gamma_k + \alpha\beta_{ij} + \alpha\gamma_{ik} + \beta\gamma_{jk})$$

Like their fitted counterparts, the (theoretical) 3-factor interactions are measures of patterns in the μ_{ijk} not describable in terms of the factors acting separately or in pairs. Or differently put, they measure how much what one would call the AB interactions at a single level of C change from level to level of C. And, like the fitted 3-factor interactions defined in Section 4.3, the theoretical 3-factor interactions defined here sum to 0 over levels of any one of the factors. That is,

$$\sum_{i=1}^{I} \alpha\beta\gamma_{ijk} = \sum_{j=1}^{J} \alpha\beta\gamma_{ijk} = \sum_{k=1}^{K} \alpha\beta\gamma_{ijk} = 0$$

Factorial effects are L's, fitted effects are corresponding $\hat{L}$'s   The fundamental fact that makes inference for the factorial effects defined in Definitions 3, 4, and 5 possible is that they are particular linear combinations of the means μ_{ijk} (*L*'s from Section 7.2). And the fitted effects from Section 4.3 are the corresponding linear combinations of the sample means $\bar{y}_{ijk}$ ($\hat{L}$'s from Section 7.2). So at least in theory, to make confidence intervals for the factorial effects, one needs only to figure out exactly what coefficients are applied to each of the means and use formula (7.20) of Section 7.2.

Example 5

Finding Coefficients on Means for a Factorial Effect in a Three-Way Factorial

Consider a hypothetical example in which A appears at $I = 2$ levels, B at $J = 2$ levels, and C at $K = 3$ levels. For the sake of illustration, consider how you would make a confidence interval for $\alpha\gamma_{23}$. By Definitions 3 and 4,

$$\alpha\gamma_{23} = \mu_{2.3} - (\mu_{...} + \alpha_2 + \gamma_3)$$

$$= \mu_{2.3} - (\mu_{2..} + \mu_{..3} - \mu_{...})$$

$$= \frac{1}{2}(\mu_{213} + \mu_{223}) - \frac{1}{6}(\mu_{211} + \mu_{221} + \mu_{212} + \mu_{222} + \mu_{213} + \mu_{223})$$

$$- \frac{1}{4}(\mu_{113} + \mu_{213} + \mu_{123} + \mu_{223}) + \frac{1}{12}(\mu_{111} + \mu_{211} + \cdots + \mu_{223})$$

$$= \frac{1}{6}\mu_{213} + \frac{1}{6}\mu_{223} - \frac{1}{12}\mu_{211} - \frac{1}{12}\mu_{221} - \frac{1}{12}\mu_{212} - \frac{1}{12}\mu_{222}$$

$$- \frac{1}{6}\mu_{113} - \frac{1}{6}\mu_{123} + \frac{1}{12}\mu_{111} + \frac{1}{12}\mu_{121} + \frac{1}{12}\mu_{112} + \frac{1}{12}\mu_{122}$$

so the "$\sum c_i^2 / n_i$" applicable to estimating $\alpha\gamma_{23}$ via formula (7.20) of Section 7.2 is

$$\sum \frac{c_{ijk}^2}{n_{ijk}} = \left(\frac{1}{6}\right)^2 \left(\frac{1}{n_{213}} + \frac{1}{n_{223}} + \frac{1}{n_{113}} + \frac{1}{n_{123}}\right)$$

$$+ \left(\frac{1}{12}\right)^2 \left(\frac{1}{n_{211}} + \frac{1}{n_{221}} + \frac{1}{n_{212}} + \frac{1}{n_{222}} + \frac{1}{n_{111}} + \frac{1}{n_{121}} + \frac{1}{n_{112}} + \frac{1}{n_{122}}\right)$$

and using this expression, endpoints for a confidence interval for $\alpha\gamma_{23}$ are

$$ac_{23} \pm ts_{\mathrm{P}} \sqrt{\sum \frac{c_{ijk}^2}{n_{ijk}}}$$

It is possible to work out (unpleasant) general formulas for the "$\sum c_i^2/n_i$" terms for factorial effects in arbitrary *p*-way factorials and implement them in computer software. It is not consistent with the purposes of this book to lay those out here. However, in the special case of 2^p factorials, there is no difficulty in describing how to make confidence intervals for the effects or in carrying out a fairly complete analysis of all of these "by hand" for *p* as large as even 4 or 5. This is because the 2^p case of the general *p*-way factorial structure allows three important simplifications.

Coefficients applied to means to produce 2^p factorial effects are all $\pm \frac{1}{2^p}$
First, for any factorial effect in a 2^p factorial, the coefficients "c_i" applied to the means to produce the effect are all $\pm \frac{1}{2^p}$. So the "$\sum c_i^2/n_i$" term needed to make a confidence interval for any effect in a 2^p factorial is

$$\left(\pm \frac{1}{2^p}\right)^2 \left(\frac{1}{n_{(1)}} + \frac{1}{n_{\mathrm{a}}} + \frac{1}{n_{\mathrm{b}}} + \frac{1}{n_{\mathrm{ab}}} + \cdots\right)$$

where the subscripts (1), a, b, ab, etc. refer to the combination-naming convention for 2^p factorials introduced in Section 4.3.

So let *E* stand for a generic effect in a 2^p factorial (a particular kind of *L* from Section 7.2) and $\hat{E}$ be the corresponding fitted effect (the corresponding $\hat{L}$ from Section 7.2). Then endpoints of an individual two-sided confidence interval for *E* are

Individual confidence limits for an effect in a 2^p factorial

$$\hat{E} \pm ts_{\mathrm{P}} \frac{1}{2^p} \sqrt{\frac{1}{n_{(1)}} + \frac{1}{n_{\mathrm{a}}} + \frac{1}{n_{\mathrm{b}}} + \frac{1}{n_{\mathrm{ab}}} + \cdots} \qquad (8.12)$$

where the associated confidence is the probability that the *t* distribution with $v = n - r = n - 2^p$ degrees of freedom assigns to the interval between $-t$ and t. The usual device of using only one endpoint from formula (8.12) and halving the unconfidence produces a one-sided confidence interval for the effect. And in balanced-data situations where all sample sizes are equal to *m*, formula (8.12) can be written even more simply as

Balanced data confidence limits for an effect in a 2^p factorial

$$\hat{E} \pm t \frac{s_{\mathrm{P}}}{\sqrt{m2^p}} \qquad (8.13)$$

Estimating one 2^p effect of a given type is enough
There is a second simplification of the general *p*-way factorial situation afforded in the 2^p case. Because of the way factorial effects sum to 0 over levels of any factor

involved, estimating *one* effect of each type is sufficient to completely describe a 2^P factorial. For example, since in a 2^P factorial,

$$\alpha\beta_{11} = -\alpha\beta_{21} = -\alpha\beta_{12} = \alpha\beta_{22}$$

it is necessary to estimate only one AB interaction to have detailed what is known about 2-factor interactions of A and B. There is no need to labor in finding separate estimates of $\alpha\beta_{11}$, $\alpha\beta_{12}$, $\alpha\beta_{21}$, and $\alpha\beta_{22}$. Appropriate sign changes on an estimate of $\alpha\beta_{22}$ suffice to cover the matter.

The third important fact making analysis of 2^P factorial effects so tractable is the existence of the Yates algorithm. As demonstrated in Example 9 of Chapter 4, it is really quite simple to use the algorithm to mechanically generate one fitted effect of each type for a given 2^P data set: those effects corresponding to the high levels of all factors.

Example 4 (*continued*)	Consider again the metal working power requirement study. Agreeing to (arbitrarily) name tool type 2, the 30° tool bevel angle, and the interrupted cut type as the "high" levels of (respectively) factors A, B, and C, the eight combinations of the three factors are listed in Table 8.9 in Yates standard order. Taking the sample means from that table in the order listed, the Yates algorithm can be applied to produce the fitted effects for the high levels of all factors, as in Table 8.10. Recall that for the data of Table 8.8, $m = 4$ and $s_p = 1.492$ mm with $24(= 32 - 2^3)$ associated degrees of freedom. So one has (from formula (8.13)) that for (say) individual 90% confidence, the factorial effects in this example can be estimated with two-sided intervals having endpoints

$$\hat{E} \pm 1.711 \frac{1.492}{\sqrt{4 \cdot 2^3}}$$

Table 8.10
The Yates Algorithm Applied to the Means in Table 8.9

Combination	$\bar{y}$	Cycle 1	Cycle 2	Cycle 3	Cycle 3 ÷ 8
(1)	28.250	55.625	115.125	222.375	$27.7969 = \bar{y}_{...}$
a	27.375	59.500	107.250	−2.375	$-.2969 = a_2$
b	29.875	52.375	−1.125	6.375	$.7969 = b_2$
ab	29.625	54.875	−1.250	.625	$.0781 = ab_{22}$
c	26.500	−.875	3.875	−7.875	$-.9844 = c_2$
ac	25.875	−.250	2.500	−.125	$-.0156 = ac_{22}$
bc	27.750	−.625	.625	−1.375	$-.1719 = bc_{22}$
abc	27.125	−.625	0.000	−.625	$-.0781 = abc_{222}$

that is,

$$\hat{E} \pm .45$$

Then, comparing the fitted effects in the last column of Table 8.10 to the $\pm.45$ value, note that only the main effects of Tool Bevel Angle (factor B) and Type of Cut (factor C) are statistically detectable. And for example, it appears that running the machining process at the high level of factor B (the $30°$ bevel angle) produces a dynamometer reading that is on average between approximately

$$2(.80 - .45) = .7 \text{ mm} \quad \text{and} \quad 2(.80 + .45) = 2.5 \text{ mm}$$

The difference between main effects at high and low levels of a factor is twice the effect

higher than when the process is run at the low level of factor B (the $15°$ bevel angle). (The difference between B main effects at the high and low levels of B is $\beta_2 - \beta_1 = \beta_2 - (-\beta_2) = 2\beta_2$, hence the multiplication by 2 of the endpoints of the confidence interval for β_2.)

8.2.3 2^p Studies Without Replication and the Normal-Plotting of Fitted Effects

The use of formula (8.12) or (8.13) in judging the detectability of 2^p factorial effects is an extremely practical and effective method. But it depends for its applicability on there being replication somewhere in the data set. One must have a pooled sample standard deviation s_p. Unfortunately, it is not uncommon that poorly informed people do unreplicated 2^p factorial studies. Although such studies should be avoided whenever possible, various methods of analysis have been suggested for them. The most popular one follows from a very clever line of reasoning due originally to Cuthbert Daniel.

Daniel's idea was to invoke a principle of **effect sparsity**. He reasoned that in many real engineering systems, the effects of only a relatively few factors are the primary determiners of system mean response. Thus, in terms of the 2^p factorial effects used here, a relatively few of $\alpha_2, \beta_2, \alpha\beta_{22}, \gamma_2, \alpha\gamma_{22}, \ldots$, etc., often dominate the rest (are much larger in absolute value than the majority). In turn, this would imply that often among the fitted effects $a_2, b_2, ab_{22}, c_2, ac_{22}, \ldots$, etc., there are a few with sizable means, and the others have means that are (relatively speaking) near 0. Daniel's idea for identifying those cases where a few effects dominate the rest was to normal-plot the fitted effects for the "all high treatment" combination (obtained, for example, by use of the Yates algorithm). When a few plot in positions much more extreme than would be predicted from putting a line through the majority of the points, they are identified as the likely principal determiners of system behavior. (Actually, Daniel originally proposed making a **half normal plot** of the absolute values of the fitted effects. This was to eliminate any visual effect of the somewhat arbitrary naming of one level of each factor as the high level. For several reasons, among them simplicity, this presentation will use the full normal plot modification of Daniel's method. The idea of half normal plotting is considered further in Chapter Exercise 9.)

Example 6
(*Example 12, Chapter 4, revisited—page 195*)

Identifying Detectable Effects in an Unreplicated 2^4 Factorial Drill Advance Rate Study

Section 4.4 discussed an example of an unreplicated 2^4 factorial experiment taken from Daniel's *Applications of Statistics to Industrial Experimentation*. There the effects of the four factors

Factor A   Load

Factor B   Flow Rate

Factor C   Rotational Speed

Factor D   Type of Mud

on the logarithm of an advance rate of a small stone drill were considered. (The raw data are in Table 4.24.) The Yates algorithm applied to the $16 = 2^4$ observed log advance rates produced the following fitted effects:

$$\bar{y}_{...} = 1.5977$$
$$a_2 = .0650 \quad b_2 = .2900 \quad c_2 = .5772 \quad d_2 = .1633$$
$$ab_{22} = -.0172 \quad ac_{22} = .0052 \quad ad_{22} = .0334$$
$$bc_{22} = -.0251 \quad bd_{22} = -.0075 \quad cd_{22} = .0491$$
$$abc_{222} = .0052 \quad abd_{222} = .0261 \quad acd_{222} = .0266$$
$$bcd_{222} = -.0173 \quad abcd_{2222} = .0193$$

Figure 8.6 is a normal plot of the 15 fitted effects a_2 through $abcd_{2222}$.

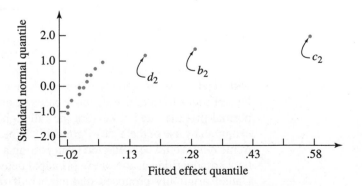

Figure 8.6 Normal plot of the fitted effects for Daniel's drill advance rate study

Applying Daniel's reasoning, it is obvious that the points corresponding to the C, B, and D main effects plot off any sensible line established through the bulk of the plotted points. So it becomes natural to think that these main effects are detectably larger than the other effects, and therefore distinguishable from experimental error even if the others are not. Thus, it seems that drill behavior is potentially describable in terms of the (separate) action of the factors Rotational Speed, Flow Rate, and Mud Type.

The plotted fitted effects concern the natural logarithm of advance rate. So the fact that $c_2 = .5772$ says that changing from the low level of rotational speed to the high level produces roughly an increase of $2(.5772) \approx 1.15$ in the natural log of the advance rate—i.e., increases the advance rate by a factor of $e^{1.15} \approx 3.2$.

Interpreting a normal plot of fitted effects

Example 6 is one in which the normal plotting clearly identifies a few effects as larger than the others. However, a normal plot of fitted effects sometimes has a fairly straight-line appearance. When this happens, the message is that the fitted effects are potentially explainable as resulting from background variation. And it is risky to make real-world engineering decisions based on fitted effects that haven't been definitively established as representing consistent system reactions to changes in level of the corresponding factors. A linear normal plot of fitted effects from an unreplicated 2^p study says that more data are needed.

This normal-plotting device has been introduced primarily as a tool for analyzing data lacking any replication. However, the method is useful even in cases where there is some replication and s_P can therefore be calculated and formula (8.12) or (8.13) used to judge the detectability of the various factorial effects. Some practice making and using such plots will show that the process often amounts to a helpful kind of "data fondling." Many times, a bit of thought makes it possible to trace an unusual pattern on such a plot back to a previously unnoticed peculiarity in the data.

As an example, consider what a normal plot of fitted effects would point out about the following eight hypothetical sample means.

$$\bar{y}_{(1)} = 95 \qquad \bar{y}_c = 145$$
$$\bar{y}_a = 101 \qquad \bar{y}_{ac} = 103$$
$$\bar{y}_b = 106 \qquad \bar{y}_{bc} = 107$$
$$\bar{y}_{ab} = 106 \qquad \bar{y}_{abc} = 97$$

This is an exaggerated example of a phenomenon that sometimes occurs less blatantly in practice. $2^p - 1$ of the sample means are more or less comparable, while one of the means is clearly different. When this occurs (unless the unusual mean corresponds to the "all high treatment" combination), a normal plot of fitted effects roughly like the one in Figure 8.7 will follow. About half the fitted effects will be large positively and the other half large negatively. (When the unusual mean is the one corresponding to the "all high" combination, the fitted effects will all have the same sign.)

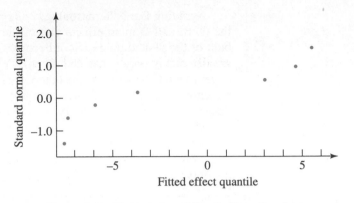

Figure 8.7   Normal plot of fitted effects for eight hypothetical means

8.2.4   Fitting and Checking Simplified Models in Balanced 2^p Factorial Studies and a Corresponding Variance Estimate (*Optional*)

When beginning the analysis of a 2^p factorial, one hopes that a simplified p-way model involving only a few main effects and/or low-order interactions will be adequate to describe it. Analyses based on formulas (8.12) or (8.13) or normal-plotting are ways of identifying such potential descriptions of special p-way structure. Once a potential simplification of the 2^p analog of model (8.11) has been identified, it is often of interest to go beyond that identification to

1. the fitting and checking (residual analysis) of the simplified model, and even to

2. the making of formal inferences under the restricted/simplified model assumptions.

When a 2^p factorial data set is *balanced*, the model fitting, checking, and subsequent interval-oriented inference is straightforward.

With balanced 2^p factorial data, producing least squares fitted values is no more difficult than adding together (with appropriate signs) desired fitted effects and the grand sample mean. Or equivalently and more efficiently, the reverse Yates algorithm can be used.

Example 4
(*continued*)

In the power requirement study and the data of Table 8.8, only the B and C main effects seem detectably nonzero. So it is reasonable to think of the simplified version of model (8.11),

$$y_{ijkl} = \mu_{...} + \beta_j + \gamma_k + \epsilon_{ijkl} \qquad \textbf{(8.14)}$$

for possible use in describing dynamometer readings. From Table 8.10, the fitted version of $\mu_{...}$ is $\bar{y}_{...} = 27.7969$, the fitted version of β_2 is $b_2 = .7969$, and the fitted version of γ_2 is $c_2 = -.9844$. Then, simply adding together appropriate signed versions of the fitted effects, for the four possible combinations of j and k, produces the corresponding fitted responses in Table 8.11. So for example, as long as the 15° bevel angle (low level of B) and a continuous cut (low level of C) are being considered, a fitted dynamometer reading of about 27.98 is appropriate under the simplified model (8.14).

Table 8.11
Fitted Responses for a "B and C Main Effects Only"
Description of Power Requirement

j	k	b_j	c_k	$\hat{y} = \bar{y}_{...} + b_j + c_k$
1	1	−.7969	.9844	27.9844
2	1	.7969	.9844	29.5782
1	2	−.7969	−.9844	26.0156
2	2	.7969	−.9844	27.6094

Example 6
(continued)

Having identified the C, B, and D main effects as detectably larger than the A main effect or any of the interactions in the drill advance rate study, it is natural to consider fitting the model

$$y_{ijkl} = \mu_{...} + \beta_j + \gamma_k + \delta_l + \epsilon_{ijkl} \tag{8.15}$$

to the logarithms of the unreplicated 2^4 factorial data of Table 4.24. (Note that even though $p = 4$ factors are involved here, five subscripts are not required, since a subscript is not needed to differentiate between multiple members of the 2^4 different samples in this unreplicated context. y_{ijkl} is the single observation at the ith level of A, jth level of B, kth level of C, and lth level of D.) Since the drill advance rate data are balanced (all sample sizes are $m = 1$), the fitted effects given earlier (calculated without reference to the simplified model) serve as fitted effects under model (8.15). And fitted responses under model (8.15) are obtainable by simple addition and subtraction using those.

Since there are eight different combinations of j, k, and l, eight different linear combinations of $\bar{y}_{...}$, b_2, c_2, and d_2 are required. While these could be treated one at a time, it is more efficient to generate them all at once using the reverse Yates algorithm (from Section 4.3) as in Table 8.12. From Table 8.12 it is evident, for example, that the fitted mean responses for combinations bcd and abcd ($\hat{y}_{bcd}$ and $\hat{y}_{abcd}$) are both 2.6282.

Example 6
(*continued*)

Table 8.12
The Reverse Yates Algorithm Used to Fit the "B, C, and D Main Effects" Model to Daniel's Data

Fitted Effect	Value	Cycle 1	Cycle 2	Cycle 3	Cycle 4 ($\hat{y}$)
$abcd_{2222}$	0	0	0	.1633	2.6282
bcd_{222}	0	0	.1633	2.4649	2.6282
acd_{222}	0	0	.5772	.1633	2.0482
cd_{22}	0	.1633	1.8877	2.4649	2.0482
abd_{222}	0	0	0	.1633	1.4738
bd_{22}	0	.5772	.1633	1.8849	1.4738
ad_{22}	0	.2900	.5772	.1633	.8938
d_2	.1633	1.5977	1.8877	1.8849	.8938
abc_{222}	0	0	0	.1633	2.3016
bc_{22}	0	0	.1633	1.3105	2.3016
ac_{22}	0	0	.5772	.1633	1.7216
c_2	.5772	.1633	1.3077	1.3105	1.7216
ab_{22}	0	0	0	.1633	1.1472
b_2	.2900	.5772	.1633	.7305	1.1472
a_2	0	.2900	.5772	.1633	.5672
$\bar{y}_{...}$	1.5977	1.5977	1.3077	.7305	.5672

Fitted means derived as in these examples lead in the usual way to residuals, R^2 values, and plots for checking on the reasonableness of simplified versions of the general 2^p version of model (8.11). In addition, corresponding to simplified or *reduced* models like (8.14) or (8.15), there are what will here be called **few-effects** s^2 values. When $m > 1$, these can be compared to s_P^2 as another means of investigating the reasonableness of the corresponding models.

Definition 6

In a balanced complete 2^p factorial study, if a reduced or simplified model involving u different effects (including the grand mean) has corresponding fitted values $\hat{y}$ and thus residuals $y - \hat{y}$, the quantity

$$s_{\text{FE}}^2 = \frac{1}{m2^p - u} \sum (y - \hat{y})^2 \tag{8.16}$$

will be called a **few-effects sample variance**. Associated with it are $\nu = m2^p - u$ degrees of freedom.

The quantity (8.16) represents an estimator of the basic background variance whenever the corresponding simplified/reduced/few-effects model is an adequate description of the study. When it is not, s_{FE} will tend to overestimate σ. So comparing s_{FE} to s_{P} is a way of investigating the appropriateness of that description.

It is not obvious at this point, but there is a helpful alternative way to calculate the value of s_{FE}^2 given in formula (8.16). It turns out that

An alternative formula for a few effects sample variance

$$s_{\mathrm{FE}}^2 = \frac{1}{m2^p - u}\left[SSTot - m2^p \sum \hat{E}^2\right] \tag{8.17}$$

where the sum is over the squares of the $u - 1$ fitted effects corresponding to those main effects and interactions appearing in the reduced model equation, and (as always) $SSTot = \sum(y - \bar{y})^2 = (n - 1)s^2$.

Example 4
(continued)

Residuals for the power requirement data based on the full model (8.11) are obtained by subtracting sample means in Table 8.9 from observations in Table 8.8. Under the reduced model (8.14), however, the fitted values in Table 8.11 are appropriate for producing residuals. The fitted means and residuals for a "B and C main effects only" description of this 2^3 data set are given in Table 8.13. Figure 8.8 is a normal plot of these residuals, and Figure 8.9 is a plot of the residuals against the fitted values.

If there is anything remarkable in these plots, it is that Figure 8.9 contains a hint that smaller mean response has associated with it smaller response variability. In fact, looking back at Table 8.13, it is easy to see that the two smallest fitted means correspond to the high level of C (i.e., interrupted cuts). That is, the hint of change in response variation shown in Figure 8.9 is the same phenomenon related

Table 8.13
Residuals for the "B and C Main Effects Only" Model of Power Requirement

Combination	$\hat{y}$	Residuals $(y - \hat{y})$
(1)	27.9844	1.0156, 1.4844, 2.5156, −.9844
a	27.9844	.0156, .5156, .0156, −2.9844
b	29.5782	−1.0782, −1.0782, .4218, 2.9218
ab	29.5782	−.0782, 2.4218, −.5782, −1.5782
c	26.0156	1.9844, −1.0156, .4844, .4844
ac	26.0156	−1.5156, −1.0156, 1.9844, −.0156
bc	27.6094	−.6094, 1.3906, −.1094, −.1094
abc	27.6094	−.1094, .3906, −.6094, −1.6094

Example 4
(*continued*)

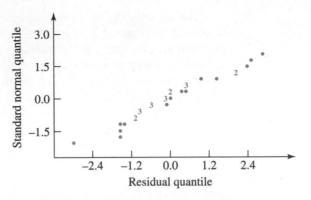

Figure 8.8 Normal plot of residuals for the power requirement study (B and C main effects only)

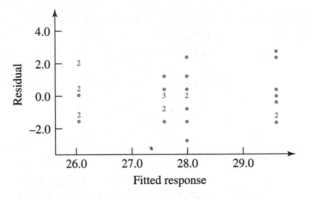

Figure 8.9 Plot of residuals versus fitted power requirements (B and C main effects only)

to cut type that was discussed when these data were first introduced. It appears that power requirements for interrupted cuts may be slightly more consistent than for continuous cuts. But on the whole, there is little in the two figures to invalidate model (8.14) as at least a rough-and-ready description of the mechanism behind the data of Table 8.8.

For the power requirement data,

$$SSTot = (n-1)s^2 = 108.93$$

Then, since $s_P^2 = 2.226$, the one-way ANOVA identity (7.49, 7.50, or 7.51) of Section 7.4 says that

$$SSTr = SSTot - SSE = 108.93 - 24(2.226) = 55.51$$

so R^2 corresponding to the general or "full" model (8.11) is (as in equations (7.52) or (7.53))

$$R^2 = \frac{SStr}{SStot} = \frac{55.51}{108.93} = .51$$

On the other hand, it is possible to verify that for the simplified model (8.14), squaring and summing the residuals in Table 8.13 gives

$$SSE = \sum (y - \hat{y})^2 = 57.60$$

(Recall Definition 6 in Chapter 7 for SSE.) So for the "B and C main effects only" description of dynamometer readings,

$$R^2 = \frac{SStot - SSE}{SStot} = \frac{108.93 - 57.60}{108.93} = .47$$

Thus, although at best only about 51% of the raw variation in dynamometer readings will be accounted for, fitting the simple model (8.14) will account for nearly all of that potentially assignable variation. So from this point of view as well, model (8.14) seems attractive as a description of power requirement.

Note that formulas (8.16) and (8.17) imply that for balanced 2^p factorial data, fitting reduced models gives

$$\sum (y - \hat{y})^2 = SSTot - m2^p \sum \hat{E}^2$$

So it is not surprising that using the $b_2 = .7969$ and $c_2 = -.9844$ figures from before,

$$SSTot - m2^p \sum \hat{E}^2 = 108.93 - 4 \cdot 2^3 \cdot \left((.7969)^2 + (-.9844)^2 \right)$$
$$= 108.93 - 51.33$$
$$= 57.60$$

which is the value of $\sum (y - \hat{y})^2$ just used in finding R^2 for the reduced model. From formula (8.16) or (8.17), it is then clear that (corresponding to reduced model (8.14))

$$s^2_{\text{FE}} = \frac{1}{4 \cdot 2^3 - 3} (57.60) = 1.986$$

so

$$s_{\text{FE}} = \sqrt{1.986} = 1.409 \text{ mm}$$

Example 4
(continued)

which agrees closely with $s_P = 1.492$. Once again on this account, description (8.14) seems quite workable.

Example 6
(continued)

Table 8.14 contains the log advance rates, fitted values, and residuals for Daniel's unreplicated 2^4 example. (The raw data were given in Table 4.24, and it is the few-effects model (8.15) that is under consideration.)

The reader can verify by plotting that the residuals in Table 8.14 are not in any way remarkable. Further, it is possible to check that

$$SSTot = \sum(y - \bar{y})^2 = 7.2774$$

and

$$SSE = \sum(y - \hat{y})^2 = .1736$$

So (as indicated earlier in Example 12 in Chapter 4) for the use of model (8.15),

$$R^2 = \frac{SStot - SSE}{SStot} = \frac{7.2774 - .1736}{7.2774} = .976$$

Table 8.14
Responses, Fitted Values, and Residuals for the "B, C, and D Main Effects" Model and Daniel's Drill Advance Rate Data

Combination	y, ln(advance rate)	$\hat{y}$	$e = y - \hat{y}$
(1)	.5188	.5672	−.0484
a	.6831	.5672	.1159
b	1.1878	1.1472	.0406
ab	1.2355	1.1472	.0883
c	1.6054	1.7216	−.1162
ac	1.7405	1.7216	.0189
bc	2.2996	2.3016	−.0020
abc	2.2050	2.3016	−.0966
d	.7275	.8938	−.1663
ad	.8920	.8938	−.0018
bd	1.4085	1.4738	−.0653
abd	1.5107	1.4738	.0369
cd	2.0503	2.0482	.0021
acd	2.2439	2.0482	.1957
bcd	2.4639	2.6282	−.1643
abcd	2.7912	2.6282	.1630

Since there is no replication in this data set, fitting the 4-factor version of the general model (8.11) would give a perfect fit, R^2 equal to 1.000, all residuals equal to 0, and no value of s_P^2. Thus, there is really nothing to judge $R^2 = .976$ against in relative terms. But even in absolute terms it appears that the "B, C, and D main effects only" model for log advance rate fits the data well.

An estimate of the variability of log advance rates for a fixed combination of factor levels derived under the assumptions of model (8.15), is (from formula (8.16))

$$s_{FE} = \sqrt{\frac{1}{1 \cdot 2^4 - 4}(.1736)} = .120$$

As noted, there's no s_P to compare this to, but it is at least consistent with the kind of variation in y seen in Table 8.14 when responses are compared for pairs of combinations that (like combinations b and ab) differ only in level of the factor A.

8.2.5 Confidence Intervals for Balanced 2^p Studies under Few-Effects Models (*Optional*)

Since the basic *p*-way factorial model is just a rewritten version of the one-way normal model from Chapter 7, the confidence interval methods of that chapter can all see application in *p*-way factorial studies. But when a simplified/few-effects model is appropriate, sharper real-world engineering conclusions can usually be had by using methods based on the simplified model than by applying the general methods of Chapter 7. And for balanced 2^p studies, it is possible to write down simple, explicit formulas for several useful forms of interval-oriented inference.

As a first example of what is possible under a few-effects model in a balanced 2^p factorial study, consider the estimation of a particular mean response. For balanced data, the 2^p fitted effects (including the grand mean) that come out of the Yates algorithm are independent normal variables with means equal to the corresponding underlying effects and variances $\sigma^2/m2^p$. So, if a simplified version of model (8.11) involving u effects (including the overall mean) is appropriate, a fitted response $\hat{y}$ has mean equal to the corresponding underlying mean, and

$$\text{Var } \hat{y} = u\frac{\sigma^2}{m2^p}$$

It should then be plausible that under a few-effects model in a balanced 2^p factorial study, a two-sided interval with endpoints

Balanced data individual confidence limits for a mean repsonse under a simplified model

$$\hat{y} \pm ts_{FE}\sqrt{\frac{u}{m2^p}}$$

(8.18)

may be used as an individual confidence interval for the corresponding mean response. The associated confidence is the probability that the t distribution with $\nu = m2^p - u$ degrees of freedom assigns to the interval between $-t$ and t. And a one-sided confidence interval for the mean response can be obtained in the usual way, by employing only one of the endpoints indicated in formula (8.18) and appropriately adjusting the confidence level.

Example 4
(continued)

Consider estimating the mean dynamometer reading corresponding to a $15°$ bevel angle and interrupted cut using the "B and C main effects only" description of Miller's power requirement study. (These are the conditions that appear to produce the smallest mean power requirement.) Using (for example) 95% confidence, a fitted value of 26.02 from Table 8.11, and $s_{\text{FE}} = 1.409$ mm possessing $\nu = 4 \cdot 2^3 - 3 = 29$ associated degrees of freedom in formula (8.18), leads to a two-sided interval with endpoints

$$26.02 \pm 2.045(1.409)\sqrt{\frac{3}{4 \cdot 2^3}}$$

that is, endpoints

$$26.02 \text{ mm} \pm .88 \text{ mm} \tag{8.19}$$

that is,

$$25.14 \text{ mm} \quad \text{and} \quad 26.90 \text{ mm}$$

In contrast to this interval, consider what the method of Section 7.2 provides for a 95% confidence interval for the mean reading for tool type 1, a $15°$ bevel angle, and interrupted cuts. Since $s_{\text{P}} = 1.492$ with $\nu = 24$ associated degrees of freedom, and (from Table 8.9) $\bar{y}_{\text{c}} = 26.50$, formula (7.14) of Section 7.2 produces a two-sided confidence interval for μ_{c} with endpoints

$$26.50 \pm 2.064(1.492)\frac{1}{\sqrt{4}}$$

that is,

$$26.50 \text{ mm} \pm 1.54 \text{ mm} \tag{8.20}$$

A major practical difference between intervals (8.19) and (8.20) is the apparent increase in precision provided by interval (8.19), due in numerical terms primarily to the "extra" $\sqrt{3/8}$ factor present in the first plus-or-minus calculation but not in the second. However, it must be remembered that the extra precision is bought at the price of the use of model (8.14) and the consequent use of all observations in the generation of $\hat{y}_c$ (rather than only the observations from the single sample corresponding to combination c).

A second balanced-data confidence interval method based on a few-effects simplification of the general 2^p model is that for estimating the effects included in the model. It comes about by replacing s_P in formula (8.13) with s_{FE} and appropriately adjusting the degrees of freedom associated with the t quantile. That is, under a few-effects model in a 2^p study with balanced data, a two-sided individual confidence interval for an effect included in the model is

Balanced data individual confidence limits for a 2^p effect under a simplified model

$$\hat{E} \pm t\frac{s_{FE}}{\sqrt{m2^p}} \tag{8.21}$$

where $\hat{E}$ is the corresponding fitted effect and the confidence associated with the interval is the probability that the t distribution with $\nu = m2^p - u$ degrees of freedom assigns to the interval between $-t$ and t. One-sided intervals are made from formula (8.21) in the usual way.

Unlike formula (8.13), formula (8.21) can be used in studies where $m = 1$. This makes it possible to attach precision figures to estimated effects in unreplicated factorial studies, provided one is willing to base them on a reduced or simplified model.

Example 6
(*continued*)

Consider again Daniel's drill advance rate study and, for example, the effect of the high level of rotational speed on the natural logarithm of advance rate. Under the "B, C, and D main effects only" description of log advance rate, $s_{FE} = .120$ with $\nu = 1 \cdot 2^4 - 4 = 12$ associated degrees of freedom. Also, $c_2 = .5772$. Then (for example) using a 95% confidence level, from formula (8.21), a two-sided interval for γ_2 under the simplified model has endpoints

$$.5772 + 2.179\frac{.120}{\sqrt{1 \cdot 2^4}}$$

that is,

$$.5772 \pm .0654$$

that is,

$$.5118 \quad \text{and} \quad .6426$$

Example 6
(*continued*)

This in turn translates (via multiplication by 2, since $\gamma_2 - \gamma_1 = 2\gamma_2$) to an increase of between

$$1.0236 \quad \text{and} \quad 1.2852$$

in average log advance rate as one moves from the low level of rotational speed to the high level. And upon exponentiation, a multiplication of median advance rate by a factor between

$$2.78 \quad \text{and} \quad 3.62$$

is indicated as one moves between levels of rotational speed. (A normal mean is also the distribution's median, and under a transformation the median of the transformed values is the transformation applied to the median. So the inference about the mean logged rate can be translated to one about the median rate. However, since the mean of transformed values is not in general the transformed mean, the interval obtained by exponentiation unfortunately does not apply to the mean advance rate.)

There are other ways to use the reduced model ideas discussed here. For example, a simplified model for responses can be used to produce prediction and tolerance intervals for individuals. Section 8.3 of Vardeman's *Statistics for Engineering Problem Solving* is one place to find an exposition of these additional methods.

Section 2 Exercises ....

1. Consider again the situation of Exercise 2 of Section 4.3.
   (a) For the logged responses, make individual 95% confidence intervals for the effects corresponding to the high levels of all three factors. Which effects are statistically detectable?
   (b) Fit an appropriate few-effects model suggested by your work in (a) to these data. Compare the corresponding value of s_{FE} to the value of s_P.
   (c) Compare a two-sided individual 95% confidence interval for the mean (logged) response for combination (1) made using the fitted few-effects model to one based on the methods of Section 7.2.

2. Chapter Exercise 9 in Chapter 4 concerns the making of Dual In-line Packages and the number of pullouts produced on such devices under 2^4 different combinations of manufacturing conditions. Return to that exercise, and if you have not already

done so, use the Yates algorithm and compute fitted 2^4 factorial effects for the data set.
   (a) Use normal-plotting to identify statistically detectable effects here.
   (b) Based on your analysis from (a), postulate a possible few-effects model for this situation. Use the reverse Yates algorithm to fit such a model to these data. Use the fitted values to compute residuals. Normal-plot these and plot them against levels of each of the four factors, looking for obvious problems with the model.
   (c) Based on your few-effects model, make a recommendation for the future making of these devices. Give a 95% two-sided confidence interval (based on the few-effects model) for the mean pullouts you expect to experience if your advice is followed.

3. A classic unreplicated 2^4 factorial study, used as an example in *Experimental Statistics* (NBS Handbook

#91) by M. G. Natrella, concerns flame tests of fire-retardant treatments for cloth. The factors and levels used in the study were

A   Fabric Tested   sateen ($-$) vs. monk's cloth ($+$)

B   Treatment   X ($-$) vs. Y ($+$)

C   Laundering   before ($-$) vs. after ($+$)
    Condition

D   Direction of Test   warp ($-$) vs. fill ($+$)

The response variable, y, is the inches burned on a standard-size sample in the flame test. The data reported by Natrella follow:

Combination	y	Combination	y
(1)	4.2	d	4.0
a	3.1	ad	3.0
b	4.5	bd	5.0
ab	2.9	abd	2.5
c	3.9	cd	4.0
ac	2.8	acd	2.5
bc	4.6	bcd	5.0
abc	3.2	abcd	2.3

(a) Use the (four-cycle) Yates algorithm and compute the fitted 2^4 factorial effects for the study.

(b) Make either a normal plot or a half normal plot using the fitted effects from part (a). What subject-matter interpretation of the data is suggested by the plot? (See Chapter Exercise 9 regarding half normal-plotting.)

(c) Natrella's original analysis of these data produced the conclusion that both the A main effects and the AB two-factor interactions are statistically detectable and of practical importance. We (based on a plot like the one asked for in (b)) are inclined to doubt that the data are really adequate to detect the AB interaction. But for the sake of example, temporarily accept the conclusion of Natrella's analysis. What does it say in practical terms about the fire-retardant treating of cloth? (How would you explain the results to a clothing manufacturer?)

8.3 Standard Fractions of Two-Level Factorials, Part I: $\frac{1}{2}$ Fractions

The notion of a fractional factorial data structure was first introduced in Section 1.2. But as yet, this text has done little to indicate either how such a structure might be chosen or how analysis of fractional factorial data might proceed. The delay is a reflection of the subtle nature of these topics rather than any lack of importance. Indeed, fractional factorial experimentation and analysis is one of the most important tools in the modern engineer's kit. This is especially true where many factors potentially affect a response and there is little a priori knowledge about the relative impacts of these factors.

This section and the next treat the (standard) 2^{p-q} fractional factorials—the class of fractional factorials for which advantageous methods of data collection and analysis can be presented most easily and completely. These structures, involving $\frac{1}{2^q}$ of all possible combinations of levels of p two-level factors, are among the most useful fractional factorial designs for application in engineering experimentation. In addition, they clearly illustrate the general issues that arise any time only a fraction of a complete factorial set of factor-level combinations can be included in a multifactor study.

This section begins with some general qualitative remarks about fractional factorial experimentation. The standard $\frac{1}{2}$ fractions of 2^p studies (the 2^{p-1} fractional factorials) are then discussed in detail. The section covers in turn (1) the proper choice of such fractions, (2) the resultant aliasing or confounding patterns, and (3) corresponding methods of data analysis. The section closes with a few remarks about qualitative issues, addressed to the practical use of 2^{p-1} designs.

8.3.1 General Observations about Fractional Factorial Studies

In many of the physical systems engineers work on, there are many factors potentially affecting a response y. In such cases, even when the number of levels considered for each factor is only two, there are a huge number of different combinations of levels of the factors to consider. For instance, if $p = 10$ factors are considered, even when limiting attention to only two levels of each factor, at least $2^{10} = 1,024$ data points must be collected in order to complete a full factorial study. In most engineering contexts, restrictions on time and other resources would make a study of that size infeasible. One could try to guess which few factors are most important in determining the response and do a smaller complete factorial study on those factors (holding the levels of the remaining factors fixed). But there is obviously a risk of guessing wrong and therefore failing to discover the real pattern of how factors affect the response.

A superior alternative is to conduct the investigation in at least two stages. A relatively small screening study (or several of them), intended to identify those factors most likely influencing the response, can be done first. This can be followed up with a more detailed study (or studies) in those variables. It is in the initial screening phase of such a program that fractions of 2^p studies are most appropriate. Tools such as full factorials are appropriate for the later stage (or stages) of study.

Once the reality of resource limitations leads to consideration of fractional factorial experimentation, several qualitative points become clear. For one, there is no way to learn as much from a fraction of a full factorial study as from the full factorial itself. (There is no Santa Claus who for the price of eight observations will give as much information as can be obtained from 16.) Fractional factorial experiments inevitably leave some ambiguity in the interpretation of their results. Through careful planning of exactly which fraction of a full factorial to use, the object is to hold the ambiguity to a minimum and to make sure it is of a type that is most tolerable. Not all fractions of a given size from a particular full factorial study have the same potential for producing useful information.

Example 7

Choosing Half of a 2^2 Factorial Study

As a completely artificial but instructive example of the preceding points, suppose that two factors A and B each have two levels (low and high) and that instead of conducting a full 2^2 factorial study, data at only $\frac{1}{2}$ of the four possible combinations will be collected

$$(1), a, b, \text{ and } ab$$

If (1) is chosen as one of the two combinations to be studied, two of the three possible choices of the other combination can easily be eliminated from consideration. The possibility of studying the combinations

$$(1) \text{ and } a$$

is no good, since in both cases the factor B is held at its low level. Therefore, no information at all would be obtained on B's impact on the response. Similarly, the possibility of studying the combinations

$$(1) \text{ and } b$$

can be eliminated, since no information would be obtained on factor A's impact on the response. So that leaves only the set of combinations

$$(1) \text{ and } ab$$

as a $\frac{1}{2}$ fraction of the full 2^2 factorial that is at all sensible (if combination (1) is to be included). Similar reasoning eliminates all other pairs of combinations from potential use except the pair

$$a \text{ and } b$$

But now notice that any experiment that includes only combinations

$$(1) \text{ and } ab$$

or combinations

$$a \text{ and } b$$

must inevitably produce somewhat ambiguous results. Since one moves from combination (1) to combination ab (or from a to b) by changing levels of *both* factors, if a large difference in response is observed, it will not be clear whether the difference is due to A or due to B.

At least in qualitative terms, such is the nature of all fractional factorial studies. Although very poor choices of experimental combinations may be avoided, some level of ambiguity must be accepted as the price for not conducting a full factorial.

Example 8

Half of a Hypothetical 2^3 Factorial

As a second hypothetical but instructive example of the issues that must be dealt with in fractional factorial experimentation, consider a system whose behavior is governed principally by the levels of three factors: A, B, and C. (For the sake of concreteness, suppose that A is a temperature, B is a pressure, and C is a catalyst type, and that the effects of these on the yield y of a chemical process are under consideration.) Suppose further that in a 2^3 study of this system, the factorial effects on an underlying mean response μ are given by

$$\mu_{...} = 10, \qquad \alpha_2 = 3, \qquad \beta_2 = 1, \qquad \gamma_2 = 2,$$
$$\alpha\beta_{22} = 2, \qquad \alpha\gamma_{22} = 0, \qquad \beta\gamma_{22} = 0, \qquad \alpha\beta\gamma_{222} = 0$$

Either through the use of the reverse Yates algorithm or otherwise, it is possible to verify that corresponding to these effects are then the eight combination means

$$\mu_{(1)} = 6, \qquad \mu_a = 8, \qquad \mu_b = 4, \qquad \mu_{ab} = 14,$$
$$\mu_c = 10, \qquad \mu_{ac} = 12, \qquad \mu_{bc} = 8, \qquad \mu_{abc} = 18$$

Now imagine that for some reason, only four of the eight combinations of levels of A, B, and C will be included in a study of this system, namely the combinations

a, b, c, and abc

Suppose further that the background noise is negligible, so that observations for a given treatment combination are essentially equal to the corresponding underlying mean. Then one essentially knows the values of

$$\mu_a = 8, \qquad \mu_b = 4, \qquad \mu_c = 10, \qquad \mu_{abc} = 18$$

Figure 8.10 shows the complete set of eight combination means laid out on a cube plot, with the four observed means circled.

As a sidelight, note the admirable symmetry possessed by the four circled corners on Figure 8.10. Each face of the cube has two circled corners (both levels of all factors appear twice in the choice of treatment combinations). Each edge has one circled corner (each combination of all pairs of factors appears once). And collapsing the cube in any one of the three possible directions (left to right, top to bottom, or front to back) gives a full factorial set of four combinations. (Ignoring the level of any one of A, B, or C in the four combinations a, b, c, and abc gives a full factorial in the other two factors.)

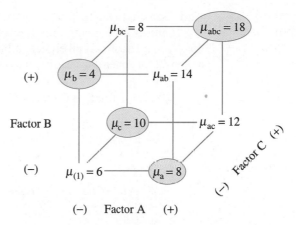

Figure 8.10 2^3 hypothetical means, with four known means circled

Now consider what an engineer possessing only the values of μ_a, μ_b, μ_c, and μ_{abc} might be led to conclude about the system. In particular, begin with the matter of evaluating an A main effect. Definition 3 says that

$$\alpha_2 = \mu_{2..} - \mu_{...}$$

$$= \left(\begin{array}{c} \text{the average of all four mean} \\ \text{responses where A is at its} \\ \text{second or high level} \end{array} \right) \left(\begin{array}{c} \text{the grand average of all} \\ \text{eight mean responses} \end{array} \right)$$

which can be thought of as the right-face average minus the grand average for the cube in Figure 8.10. Armed only with the four means μ_a, μ_b, μ_c, and μ_{abc} (the four circled corners on Figure 8.10), it is not possible to compute α_2. But what might be done is to make a calculation similar to the one that produces α_2 using only the available means. That is,

$$\alpha_2^* = \text{a "} \tfrac{1}{2} \text{ fraction A main effect"}$$

$$= \left(\begin{array}{c} \text{the average of the available} \\ \text{two means where A is at its} \\ \text{high level} \end{array} \right) - \left(\begin{array}{c} \text{the grand average of the} \\ \text{available four means} \end{array} \right)$$

$$= \frac{1}{2}(\mu_a + \mu_{abc}) - \frac{1}{4}(\mu_a + \mu_b + \mu_c + \mu_{abc})$$

$$= \frac{1}{2}(8 + 18) - \frac{1}{4}(8 + 4 + 10 + 18)$$

$$= 13 - 10$$

$$= 3$$

Example 8
(continued)

And, amazingly enough, $\alpha_2^* = \alpha_2$ here.

It appears that using only four combinations, as much can be learned about the A main effect as if all eight combination means were in hand! This is too good to be true in general, as is illustrated by a parallel calculation for a C main effect.

$$\gamma_2^* = \text{a "}\tfrac{1}{2}\text{ fraction C main effect"}$$

$$= \left(\begin{array}{c} \text{the average of the two} \\ \text{available means where} \\ \text{C is at its high level} \end{array} \right) - \left(\begin{array}{c} \text{the grand average of the} \\ \text{four available means} \end{array} \right)$$

$$= \frac{1}{2}(\mu_c + \mu_{abc}) - \frac{1}{4}(\mu_a + \mu_b + \mu_c + \mu_{abc})$$

$$= 4$$

while this hypothetical example began with $\gamma_2 = 2$. Here, the $\frac{1}{2}$ fraction calculation gives something quite different from the full factorial calculation.

The key to understanding how one can apparently get something for nothing in the case of the A main effects in this example, but cannot do so in the case of the C main effects, is to know that (in general) for this $\frac{1}{2}$ fraction,

$$\alpha_2^* = \alpha_2 + \beta\gamma_{22}$$

and

$$\gamma_2^* = \gamma_2 + \alpha\beta_{22}$$

Since this numerical example began with $\beta\gamma_{22} = 0$, one is "fortunate"—it turns out numerically that $\alpha_2^* = \alpha_2$. On the other hand, since $\alpha\beta_{22} = 2 \neq 0$, one is "unfortunate"—it turns out numerically that $\gamma_2^* = \gamma_2 + 2 \neq \gamma_2$.

Relationships like these for α_2^* and γ_2^* hold for all $\frac{1}{2}$ fraction versions of the full factorial effects. These relationships detail the nature of the ambiguity inherent in the use of the $\frac{1}{2}$ fraction of the full 2^3 factorial set of combinations. Essentially, based on data from four out of eight possible combinations, one will be unable to distinguish between certain pairs of effects, such as the A main effect and BC 2-factor interaction pair here.

8.3.2   Choice of Standard $\frac{1}{2}$ Fractions of 2^p Studies

The use of standard 2^{p-q} fractional factorial data structures depends on having answers for the following three basic questions:

Three fundamental issues in the use of a fractional factorial

1. How is $\frac{1}{2^q}$ of 2^p possible combinations of factor levels to include in a study rationally chosen?

2. How is the pattern of ambiguities implied by a given choice of 2^{p-q} combinations determined?

3. How is data analysis done for a particular choice of 2^{p-q} combinations?

These questions will be answered in this section for the case of $\frac{1}{2}$ fractions (2^{p-1} fractional factorials) and for general q in the next section.

Prescription for   In order to arrive at what is in some sense a best possible choice of $\frac{1}{2}$ of 2^p
a best half fraction   combinations of levels of p factors, do the following. For the first $p - 1$ factors,
of a 2^p factorial   write out all 2^{p-1} possible combinations of these factors. By multiplying plus and minus signs (thinking of multiplying plus and minus 1's) corresponding to levels of the first factors, then arrive at a set of plus and minus signs that can be used to prescribe how to choose levels for the last factor (to be used in combination with the indicated levels of the first $p - 1$ factors).

Example 9

A 2^{5-1} Chemical Process Experiment

In his article "Experimenting with a Large Number of Variables" (*ASQC Technical Supplement Experiments in Industry*, 1985), R. Snee discusses a successful 2^{5-1} experiment on a chemical process, where the response of interest, y, was a coded color index of the product. The factors studied and their levels are as in Table 8.15.

The standard recommendation for choosing a $\frac{1}{2}$ fraction was followed in Snee's study. Table 8.16 shows an appropriate set of 16 lines of plus and minus signs for generating the $\frac{1}{2} \cdot 32 = 16$ combinations included in Snee's study. The first four columns of this table specify levels of factors A, B, C, and D for the $16 = 2^4$ possible combinations of levels of these factors (written in Yates standard order). (The first line, for example, indicates the low level of all of these first four factors.) The last column of this table is obtained by multiplying the first four plus or minus signs (plus or minus 1's) in a given row. It is this last column that can be used to determine how to choose a level of factor E for use when the factors A through D are at the levels indicated in the first four columns.

Table 8.15
Five Chemical Process Variables and Their Experimental Levels

Factor	Process Variable	Factor Levels
A	Solvent/Reactant	low ($-$) vs. high ($+$)
B	Catalyst/Reactant	.025 ($-$) vs. .035 ($+$)
C	Temperature	150°C ($-$) vs. 160°C ($+$)
D	Reactant Purity	92% ($-$) vs. 96% ($+$)
E	pH of Reactant	8.0 ($-$) vs. 8.7 ($+$)

Example 9
(continued)

Table 8.16
Signs for Specifying a Standard 2^{5-1}
Fractional Factorial

A	B	C	D	ABCD Product
−	−	−	−	+
+	−	−	−	−
−	+	−	−	−
+	+	−	−	+
−	−	+	−	−
+	−	+	−	+
−	+	+	−	+
+	+	+	−	−
−	−	−	+	−
+	−	−	+	+
−	+	−	+	+
+	+	−	+	−
−	−	+	+	+
+	−	+	+	−
−	+	+	+	−
+	+	+	+	+

In Snee's study, the signs in the ABCD Product column were used without modification to specify levels of E. The corresponding treatment combination names (written in the same order as in Table 8.16) and the data reported by Snee are given in Table 8.17. Notice that the 16 combinations listed in Table 8.17 are $\frac{1}{2}$ of the $2^5 = 32$ possible combinations of levels of these five factors. (They are those 16 that have an odd number of factors appearing at their high levels).

Example 10

A 2^{5-1} Agricultural Engineering Study

The article "An Application of Fractional Factorial Experimental Designs" by Mary Kilgo (*Quality Engineering*, 1988) provides an interesting complement to the previous example. In one part of an agricultural engineering study concerned with the use of carbon dioxide at very high pressures to extract oil from peanuts, the effects of five factors on a percent yield variable y were studied in a 2^{5-1} fractional factorial experiment. The five factors and their levels (as named in Kilgo's article) are given in Table 8.18.

Interestingly enough, rather than studying the 16 combinations obtainable using the final column of Table 8.16 directly, Kilgo *switched* all of the signs in the ABCD product column before assigning levels of E. This leads to the use of "the other" 16 out of 32 possible combinations (those having an even number of

Table 8.17

16 Combinations and Observed Color Indices in Snee's 2^{5-1} Study (*Example 9*)

Combination	Color Index, y
e	−.63
a	2.51
b	−2.68
abe	1.66
c	2.06
ace	1.22
bce	−2.09
abc	1.93
d	6.79
ade	6.47
bde	3.45
abd	5.68
cde	5.22
acd	9.38
bcd	4.30
abcde	4.05

Table 8.18

Five Peanut Processing Variables and Their Experimental Levels

Factor	Process Variable	Factor Levels
A	Pressure	415 bars (−) vs. 550 bars (+)
B	Temperature	25°C (−) vs. 95°C (+)
C	Peanut Moisture	5% (−) vs. 15% (+)
D	Flow Rate	40 1/min (−) vs. 60 1/min (+)
E	Average Particle Size	1.28 mm (−) vs. 4.05 mm (+)

factors appearing at their high levels). The 16 combinations studied and corresponding responses reported by Kilgo are given in Table 8.19 in the same order for factors A through D as in Table 8.16.

The difference between the combinations listed in Tables 8.17 and 8.19 deserves some thought. As Kilgo named the factor levels, the two lists of combinations are quite different. But verify that if she had made the slightly less natural but nevertheless permissible choice to call the 4.05 mm level of factor E the low (−) level

Table 8.19
16 Combinations and Observed Yields in Kilgo's 2^{5-1} Study (*Example 10*)

Combination	Yield, y (%)
(1)	63
ae	21
be	36
ab	99
ce	24
ac	66
bc	71
abce	54
de	23
ad	74
bd	80
abde	33
cd	63
acde	21
bcde	44
abcd	96

and the 1.28 mm level the high (+) level, the names of the physical combinations actually studied would be exactly those in Table 8.17 rather than those in Table 8.19.

The point here is that due to the rather arbitrary nature of how one chooses to name high and low levels of two factors, the names of different physical combinations are themselves to some extent arbitrary. In choosing fractional factorials, one chooses some particular naming convention and then has the freedom to choose levels of the last factor (or factors for $q > 1$ cases) by either using the product column(s) directly or after switching signs. The decision whether or not to switch signs *does affect* exactly which physical combinations will be run and thus how the data should be interpreted in the subject-matter context. But generally, the different possible choices (to switch or not switch signs) are a priori equally attractive. For systems that happen to have relatively simple structure, all possible results of these arbitrary choices typically lead to similar engineering conclusions. When systems turn out to have complicated structures, the whole notion of fractional factorial experimentation loses its appeal. Different arbitrary choices lead to different perceptions of system behavior, none of which (usually) correctly portrays the complicated real situation.

Fractional factorials fully reveal system structure only for simple cases

8.3.3 Aliasing in the Standard $\frac{1}{2}$ Fractions

Once a $\frac{1}{2}$ fraction of a 2^p study is chosen, the next issue is determining the nature of the ambiguities that must arise from its use. For 2^{p-1} data structures of the type described here, one can begin with a kind of statement of how the fractional factorial plan was derived and through a **system of formal multiplication** arrive at an understanding of which (full) factorial effects cannot be separated on the basis of the fractional factorial data. Some terminology is given next, in the form of a definition.

Definition 7	When it is only possible to estimate the sum (or difference) of two or more (full) factorial effects on the basis of data from a fractional factorial, those effects are said to be **aliased** or **confounded** and are sometimes called **aliases**. In this text, the phrase **alias structure** of a fractional factorial plan will mean a complete specification of all sets of aliased effects.

As an example of the use of this terminology, return to Example 8. There, it is possible only to estimate $\alpha_2 + \beta\gamma_{22}$, not either of α_2 or $\beta\gamma_{22}$ individually. So the A main effect is confounded with (or aliased with) the BC 2-factor interaction.

The way the system of formal multiplication works for detailing the alias structure of one of the recommended 2^{p-1} factorials is as follows. One begins by writing

Generator for a standard half fraction of a 2^p factorial

$$\left(\begin{array}{c} \text{the name of the} \\ \text{last factor} \end{array} \right) \leftrightarrow \pm \left(\begin{array}{c} \text{the product of names of} \\ \text{the first } p - 1 \text{ factors} \end{array} \right) \qquad \textbf{(8.22)}$$

where the plus or minus sign is determined by whether the signs were left alone or switched in the specification of levels of the last factor. The double arrow in expression (8.22) will be read as "is aliased with." And since expression (8.22) really says how the fractional factorial under consideration was chosen, expression (8.22) will be called the plan's **generator**. The generator (8.22) for a 2^{p-1} plan says that the (high level) main effect of the last factor will be aliased with plus or minus the (all factors at their high levels) $p - 1$ factor interaction of the first $p - 1$ factors.

Example 9 *(continued)*	In Snee's 2^{5-1} study, the generator

$$E \leftrightarrow ABCD$$

was used. Therefore the (high level) E main effect is aliased with the (all high levels) ABCD 4-factor interaction. That is, only $\epsilon_2 + \alpha\beta\gamma\delta_{2222}$ can be estimated based on the $\frac{1}{2}$ fraction data, not either of its summands individually.

Example 10 *(continued)*	In Kilgo's 2^{5-1} study, the generator

$$E \leftrightarrow -ABCD$$

Example 10
(*continued*)

was used. The (high level) E main effect is aliased with minus the (all high levels) ABCD 4-factor interaction. That is, only $\epsilon_2 - \alpha\beta\gamma\delta_{2222}$ can be estimated based on the $\frac{1}{2}$ fraction data, not either of the terms individually.

Conventions for the system of formal multiplication

The entire alias structure for a $\frac{1}{2}$ fraction follows from the generator (8.22) by multiplying both sides of the expression by various factor names, using two special conventions. These are that any letter multiplied by itself produces the symbol "I" and that any letter multiplied by "I" is that letter again. Applying the first of these conventions to expression (8.22), both sides of the expression may be multiplied by the name of the last factor to produce the relation

Defining relation for a standard half fraction of a 2^p factorial

$$\boxed{\text{I} \leftrightarrow \pm \text{ the product of names of all } p \text{ factors}} \qquad (8.23)$$

Expression (8.23) means that the grand mean is aliased with plus or minus the (all factors at their high level) p-factor interaction. There is further special terminology for an expression like that in display (8.23).

Definition 8

The list of all aliases of the grand mean for a 2^{p-q} fractional factorial is called the **defining relation** for the design.

By first translating a generator (or generators in the case of $q > 1$) into a defining relation and then multiplying through the defining relation by a product of letters corresponding to an effect of interest, one can identify all aliases of that effect.

Example 9
(*continued*)

In Snee's 2^{5-1} experiment, the generator was

$$\text{E} \leftrightarrow \text{ABCD}$$

When multiplied through by E, this gives the experiment's defining relation

$$\text{I} \leftrightarrow \text{ABCDE} \qquad (8.24)$$

which indicates that the grand mean $\mu_{\dots}$ is aliased with the 5-factor interaction $\alpha\beta\gamma\delta\epsilon_{22222}$. Then, for example, multiplying through defining relation (8.24) by the product AC produces the relationship

$$\text{AC} \leftrightarrow \text{BDE}$$

Thus, the AC 2-factor interaction is aliased with the BDE 3-factor interaction. In fact, the entire alias structure for the Snee study can be summarized in terms of the aliasing of 16 different pairs of effects. These are indicated in Table 8.20, which was developed by using the defining relation (8.24) to find successively (in Yates order) the aliases of all effects involving only factors A, B, C, and D. Table 8.20 shows that main effects are confounded with 4-factor interactions and 2-factor interactions with 3-factor interactions. This degree of ambiguity is as mild as is possible in a 2^{5-1} study.

Table 8.20
The Complete Alias Structure for
Snee's 2^{5-1} Study

I $\leftrightarrow$ ABCDE	D $\leftrightarrow$ ABCE
A $\leftrightarrow$ BCDE	AD $\leftrightarrow$ BCE
B $\leftrightarrow$ ACDE	BD $\leftrightarrow$ ACE
AB $\leftrightarrow$ CDE	ABD $\leftrightarrow$ CE
C $\leftrightarrow$ ABDE	CD $\leftrightarrow$ ABE
AC $\leftrightarrow$ BDE	ACD $\leftrightarrow$ BE
BC $\leftrightarrow$ ADE	BCD $\leftrightarrow$ AE
ABC $\leftrightarrow$ DE	ABCD $\leftrightarrow$ E

Example 10
(*continued*)

In Kilgo's peanut oil extraction study, since the generator is E $\leftrightarrow$ −ABCD, the defining relation is I $\leftrightarrow$ −ABCDE, and the alias structure is that given in Table 8.20, except that a minus sign should be inserted on one side or the other of every row of the table. So, for example, $\alpha\beta_{22} - \gamma\delta\epsilon_{222}$ may be estimated based on Kilgo's data, but neither $\alpha\beta_{22}$ nor $\gamma\delta\epsilon_{222}$ separately.

8.3.4 Data Analysis for 2^{p-1} Fractional Factorials

Once the alias structure of a 2^{p-1} fractional factorial is understood, the question of how to analyze data from such a study has a simple answer.

1. Temporarily ignore the last factor and compute the estimated or fitted "effects."

2. Somehow judge the statistical significance and apparent real importance of the "effects" computed for the complete factorial in $p - 1$ two-level factors. (Where some replication is available, the judging of statistical significance can be done through the use of confidence intervals. Where all 2^{p-1} samples are of size 1, the device of normal-plotting fitted "effects" is standard.)

3. Finally, seek a plausible simple interpretation of the important fitted "effects," recognizing that they are estimates not of the effects in the first $p - 1$ factors alone, but of *those effects plus their aliases*.

Example 9
(*continued*)

Consider the analysis of Snee's data, listed in Table 8.17 in Yates standard order for factors A, B, C, and D (ignoring the existence of factor E). Then, according to the prescription for analysis just given, the first step is to use the Yates algorithm (for four factors) on the data. These calculations are summarized in Table 8.21.

Each entry in the final column of Table 8.21 gives the name of the effect that the corresponding numerical value in the "Cycle $4 \div 16$" column would be estimating if factor E weren't present, *plus* the alias of that effect. The numbers in the next-to-last column must be interpreted in light of the fact that they are estimating sums of 2^5 factorial effects.

Since there is no replication indicated in Table 8.17, only normal-plotting fitted (sums of) effects is available to identify those that are distinguishable from noise. Figure 8.11 is a normal plot of the last 15 entries of the Cycle $4 \div 16$ column of Table 8.21. (Since in most contexts one is a priori willing to grant that the overall mean response is other than 0, the estimate of it plus its alias(es) is rarely included in such a plot.)

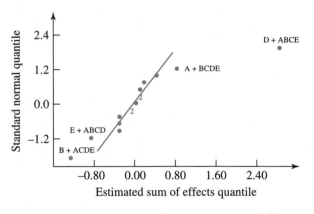

Figure 8.11 Normal plot of estimated sums of effects in Snee's 2^{5-1} study

Depending upon how the line is drawn through the small estimated (sums of) effects in Figure 8.11, the estimates corresponding to D + ABCE, and possibly B + ACDE, E + ABCD, and A + BCDE as well, are seen to be distinguishable in magnitude from the others. (The line in Figure 8.11 has been drawn in keeping with the view that there are four statistically detectable sums of effects, primarily because a half normal plot of the absolute values of the estimates—not included here—supports that view.) If one adopts the view that there are indeed four detectable (sums of) effects indicated by Figure 8.11, it is clear that the simplest possible interpretation of this outcome is that the four large estimates are each reflecting primarily the corresponding main effects (and not the aliased 4-factor

Table 8.21
The Yates Algorithm for a 2^4 Factorial Applied to Snee's 2^{5-1} Data

y	Cycle 1	Cycle 2	Cycle 3	Cycle 4	Cycle 4 ÷ 16	Sum Estimated
−.63	1.88	−2.46	.66	46.00	2.875	$\mu_{.....} + \alpha\beta\gamma\delta\epsilon_{22222}$
2.51	−4.34	3.12	45.34	13.16	.823	$\alpha_2 + \beta\gamma\delta\epsilon_{2222}$
−2.68	3.28	22.39	7.34	−20.04	−1.253	$\beta_2 + \alpha\gamma\delta\epsilon_{2222}$
−1.66	−.16	22.95	5.82	.88	.055	$\alpha\beta_{22} + \gamma\delta\epsilon_{222}$
2.06	13.26	4.16	−9.66	6.14	.384	$\gamma_2 + \alpha\beta\delta\epsilon_{2222}$
1.22	9.13	3.18	−10.38	1.02	.064	$\alpha\gamma_{22} + \beta\delta\epsilon_{222}$
−2.09	14.60	1.91	2.74	.66	.041	$\beta\gamma_{22} + \alpha\delta\epsilon_{222}$
1.93	8.35	3.91	−1.86	.02	.001	$\alpha\beta\gamma_{222} + \delta\epsilon_{22}$
6.79	3.14	−6.22	5.58	44.68	2.793	$\delta_2 + \alpha\beta\gamma\epsilon_{2222}$
6.47	1.02	−3.44	.56	−1.52	−.095	$\alpha\delta_{22} + \beta\gamma\epsilon_{222}$
3.45	−.84	−4.13	−.98	−.72	−.045	$\beta\delta_{22} + \alpha\gamma\epsilon_{222}$
5.68	4.02	−6.25	2.00	−4.60	−.288	$\alpha\beta\delta_{222} + \gamma\epsilon_{22}$
5.22	−.32	−2.12	2.78	−5.02	−.314	$\gamma\delta_{22} + \alpha\beta\epsilon_{222}$
9.38	2.23	4.86	−2.12	2.98	.186	$\alpha\gamma\delta_{222} + \beta\epsilon_{22}$
4.30	4.16	2.55	6.98	−4.90	.306	$\beta\gamma\delta_{222} + \alpha\epsilon_{22}$
4.05	−.25	−4.41	−6.96	−13.94	−.871	$\alpha\beta\gamma\delta_{2222} + \epsilon_2$

interactions). That is, a tentative (because of the incomplete nature of fractional factorial information) description of the chemical process is that D (reactant purity), B (catalyst/reactant), A (solvent/reactant), and E (pH of reactant) main effects are (in that order) the principal determinants of product color. Depending on the engineering objectives for product color index y, this tentative description of the system could have several possible interpretations. If large y were desirable, the high levels of A and D and low levels of B and E appear most attractive. If small y were desirable, the situation would be reversed. But in fact, Snee's study was done not to figure out how to maximize or minimize y, but rather to determine how to reduce variation in y. The engineering implications of the "D, B, A, and E main effects only" system description are thus to focus attention on the need to control variation first in level of factor D (reactant purity), then in level of factor B (catalyst/reactant), then in level of factor A (solvent/reactant), and finally in level of factor E (pH of reactant).

Example 10
(*continued*)

Verify that for Kilgo's data in Table 8.19, use of the (four-cycle) Yates algorithm on the data as listed (in standard order for factors A, B, C, and D, ignoring factor E) produces the estimated (differences of) effects given in Table 8.22.

Example 10
(*continued*)

Table 8.22
Estimated Differences of 2^5 Factorial Effects from Kilgo's 2^{5-1} Study

Value	Difference Estimated	Value	Difference Estimated
54.3	$\mu_{.....} - \alpha\beta\gamma\delta\epsilon_{22222}$	0.0	$\delta_2 - \alpha\beta\gamma\epsilon_{2222}$
3.8	$\alpha_2 - \beta\gamma\delta\epsilon_{2222}$	−2.0	$\alpha\delta_{22} - \beta\gamma\epsilon_{222}$
9.9	$\beta_2 - \alpha\gamma\delta\epsilon_{2222}$	−.9	$\beta\delta_{22} - \alpha\gamma\epsilon_{222}$
2.6	$\alpha\beta_{22} - \gamma\delta\epsilon_{222}$	−3.1	$\alpha\beta\delta_{222} - \gamma\epsilon_{22}$
.6	$\gamma_2 - \alpha\beta\delta\epsilon_{2222}$	1.1	$\gamma\delta_{22} - \alpha\beta\epsilon_{222}$
.6	$\alpha\gamma_{22} - \beta\delta\epsilon_{222}$	.1	$\alpha\gamma\delta_{222} - \beta\epsilon_{22}$
1.5	$\beta\gamma_{22} - \alpha\delta\epsilon_{222}$	3.5	$\beta\gamma\delta_{222} - \alpha\epsilon_{22}$
1.8	$\alpha\beta\gamma_{222} - \delta\epsilon_{22}$	22.3	$\alpha\beta\gamma\delta_{2222} - \epsilon_2$

The last 15 of these estimated differences are normal-plotted in Figure 8.12. It is evident from the figure that the two estimated (differences of) effects corresponding to

$$\beta_2 - \alpha\gamma\delta\epsilon_{2222} \quad \text{and} \quad \alpha\beta\gamma\delta_{2222} - \epsilon_2$$

are significantly larger than the other 13 estimates. The simplest possible interpretation of this outcome is that the two large estimates are each reflecting primarily the corresponding main effects (not the aliased 4-factor interactions). That is, a tentative description of the oil extraction process is that average particle size (factor E) and temperature (factor B), acting more or less separately, are the principle determinants of yield. This is an example where the ultimate engineering objective is to maximize response and the two large estimates are both positive. So, for best yield one would prefer the high level of B (95°C temperature) and

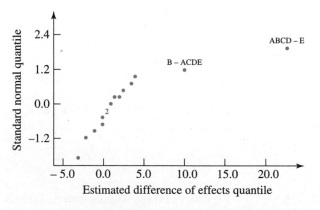

Figure 8.12 Normal plot of estimated differences of effects in Kilgo's 2^{5-1} study

low level of E (1.28 mm particle size). ($-\epsilon_2$ is apparently positive, and since $\epsilon_1 = -\epsilon_2$, the superiority of the low level of E is indicated.)

8.3.5 Some Additional Comments

The next section treats general $\frac{1}{2^q}$ fractions of 2^p factorials. But before closing this discussion of the special case of $q = 1$, several issues deserve comment. The first concerns the range of statistical methods that will be provided here for use with fractional factorials. The data analysis methods presented in this section and the next are confined to those for the identification of potential "few effects" descriptions of a p-factor situation. (For example, we do not go on to issues of inference under such a reduced model.) This stance is consistent with the fact that fractional factorials are primarily screening devices, useful for gaining some idea about which of many factors might be important. They are typically not suited (at least without additional data collection) to serve as the basis for detailed modeling of a response. The insights they provide must be seen as tentative and as steps along a path of learning about what factors influence a response.

A second matter regards the sense in which the $\frac{1}{2}$ fractions recommended here are the best ones possible. Other $\frac{1}{2}$ fractions could be developed (essentially by using a product column of signs derived from levels of fewer than all $p - 1$ of the first factors to assign levels of the last one). But the alias structures associated with those alternatives are less attractive than the ones encountered in this section. That is, here main effects have been aliased with $p - 1$ factor interactions, 2-factor interactions with $(p - 2)$-factor interactions, and so on. Any other $\frac{1}{2}$ fractions fundamentally different from the ones discussed here would have main effects aliased with interactions of $p - 2$ or less factors. They would thus be more likely to produce data incapable of separating important effects. The "l order effects aliased with $p - l$ order effects" structure of this section is simply the best one can do with a 2^{p-1} fractional factorial.

The last matter for discussion concerns what directions a follow-up investigation might take in order to resolve ambiguities left after a 2^{p-1} study is completed. Sometimes several different simple descriptions of system structure remain equally plausible after analysis of an initial $\frac{1}{2}$ fraction of a full factorial study. One approach to resolving these is to complete the factorial and "run the other $\frac{1}{2}$ fraction."

Example 11 | **A 2^{4-1} Fabric Tenacity Study Followed Up by a Second 2^{4-1} Study**

Researchers Johnson, Clapp, and Baqai, in "Understanding the Effect of Confounding in Design of Experiments: A Case Study in High Speed Weaving" (*Quality Engineering*, 1989), discuss a study done to evaluate the effects of four two-level factors on a measure of woven fabric tenacity. The factors that were studied are indicated in Table 8.23.

Example 11
(*continued*)

Table 8.23

Four Weaving Process Variables and Their Experimental Levels

Factor	Weaving Process Variable	Factor Levels
A	Side of Cloth (l. to r.)	nozzle side $(-)$ vs. opposite side $(+)$
B	Yarn Type	air spun $(-)$ vs. ring spun $(+)$
C	Pick Density	35 ppi $(-)$ vs. 50 ppi $(+)$
D	Air Pressure	30 psi $(-)$ vs. 45 psi $(+)$

Factor A reflects the left-to-right location on the fabric width from which a tested sample is taken. Factor C reflects a count of yarns per inch inserted in the cloth, top to bottom, during weaving. Factor D reflects the air pressure used to propel the yarn across the fabric width during weaving.

Initially, a replicated 2^{4-1} study was done using the generator D $\leftrightarrow$ ABC. $m = 5$ pieces of cloth were tested for each of the eight different factor-level combinations studied. The resulting mean fabric tenacities $\bar{y}$, expressed in terms of strength per unit linear density, are given in Table 8.24. Although it is not absolutely clear in the article, it also appears that pooling the eight s^2 values from the $\frac{1}{2}$ fraction gave $s_\mathrm{p} \approx 1.16$.

Apply the (three-cycle) Yates algorithm to the means listed in Table 8.24 (in the order given) and verify that the estimated sums of effects corresponding to the means in Table 8.24 are those given in Table 8.25.

Temporarily ignoring the existence of factor D, confidence intervals based on these estimates can be made using the $m = 5$ and $p = 3$ version of formula (8.13) from Section 8.2. That is, using 95% two-sided individual confidence intervals, since $\nu = 8(5 - 1) = 32$ degrees of freedom are associated with s_p, a precision of roughly

$$\pm\frac{(2.04)(1.16)}{\sqrt{5 \cdot 8}} = \pm.375$$

should be associated with each of the estimates in Table 8.25. By this standard, the estimates corresponding to the A + BCD, AB + CD, C + ABD, and BC + AD

Table 8.24

Eight Sample Means from a 2^{4-1} Fabric Tenacity Experiment

Combination	$\bar{y}$ (g/den.)	Combination	$\bar{y}$ (g/den.)
(1)	24.50	cd	25.68
ad	22.05	ac	24.51
bd	24.52	bc	24.68
ab	25.00	abcd	24.23

Table 8.25
Estimated Sums of 2^4 Effects in a 2^{4-1}
Fabric Weaving Experiment

Estimate	Sum of Effects Estimated
24.396	$\mu_{....} + \alpha\beta\gamma\delta_{2222}$
−.449	$\alpha_2 + \beta\gamma\delta_{222}$
.211	$\beta_2 + \alpha\gamma\delta_{222}$
.456	$\alpha\beta_{22} + \gamma\delta_{22}$
.379	$\gamma_2 + \alpha\beta\delta_{222}$
.044	$\alpha\gamma_{22} + \beta\delta_{22}$
−.531	$\beta\gamma_{22} + \alpha\delta_{22}$
−.276	$\alpha\beta\gamma_{222} + \delta_2$

sums are statistically significant. Two reasonably plausible and equally simple tentative interpretations of this outcome are

1. There are detectable A and C main effects and detectable 2-factor interactions of A with B and D.

2. There are detectable A and C main effects and detectable 2-factor interactions of C with B and D.

(For that matter, there are others that you may well find as plausible as these two.)

In any case, the ambiguities left by the collection of the data summarized in Table 8.24 were unacceptable. To remedy the situation, the authors subsequently completed the 2^4 factorial study by collecting data from the other eight combinations defined by the generator $D \leftrightarrow -ABC$. The means they obtained are given in Table 8.26.

One should honestly consider (and hopefully eliminate) the possibility that there is a systematic difference between the values in Table 8.24 and in Table 8.26 as a result of some unknown factor or factors that changed in the time lapse between the collection of the first block of observations and the second block. If

Table 8.26
Eight More Sample Means from a Second 2^{4-1}
Fabric Tenacity Study

Combination	$\bar{y}$	Combination	$\bar{y}$
d	23.73	c	24.63
a	23.55	acd	25.78
b	25.98	bcd	24.10
abd	23.64	abc	23.93

Example 11
(*continued*)

that possibility can be eliminated, it would make sense to put together the two data sets, treat them as a single full 2^4 factorial data set, and employ the methods of Section 8.2 in their analysis. (Some repetition of a combination or combinations included in the first study phase—e.g., the center point of the design—would have been advisable to allow at least a cursory check on the possibility of a systematic block effect.)

Johnson, Clapp, and Baqai don't say explicitly what sample sizes were used to produce the $\bar{y}$'s in Table 8.26. (Presumably, $m = 5$ was used.) Nor do they give a value for s_p based on all 2^4 samples, so it is not possible to give a complete analysis of the full factorial data à la Section 8.2. But it is possible to note what results from the use of the Yates algorithm with the full factorial set of $\bar{y}$'s. This is summarized in Table 8.27.

Table 8.27
Fitted Effects from the Full 2^4 Factorial Fabric Tenacity Study

Effect	Estimate	Effect	Estimate
$\mu_{\ldots}$	$\bar{y}_{\ldots} = 24.407$	δ_2	$d_2 = -.191$
α_2	$a_2 = -.321$	$\alpha\delta_{22}$	$ad_{22} = .029$
β_2	$b_2 = .103$	$\beta\delta_{22}$	$bd_{22} = -.197$
$\alpha\beta_{22}$	$ab_{22} = .011$	$\alpha\beta\delta_{222}$	$abd_{222} = .093$
γ_2	$c_2 = .286$	$\gamma\delta_{22}$	$cd_{22} = .446$
$\alpha\gamma_{22}$	$ac_{22} = .241$	$\alpha\gamma\delta_{222}$	$acd_{222} = .108$
$\beta\gamma_{22}$	$bc_{22} = -.561$	$\beta\gamma\delta_{222}$	$bcd_{222} = -.128$
$\alpha\beta\gamma_{222}$	$abc_{222} = -.086$	$\alpha\beta\gamma\delta_{2222}$	$abcd_{2222} = -.011$

The statistical significance of the entries of Table 8.27 will not be judged here. But note that the picture of fabric tenacity given by the fitted effects in this table is somewhat more complicated than either of the tentative descriptions derived from the original 2^{4-1} study. The fitted effects, listed in order of decreasing absolute value, are

$$BC, CD, A, C, AC, BD, D, \ldots, \text{etc.}$$

Although tentative description (2) (page 609) accounts for the first four of these, the A and C main effects indicated in Table 8.27 are not really as large as one might have guessed looking only at Table 8.25. Further, the AC 2-factor interaction appears from Table 8.27 to be nearly as large as the C main effect. This is obscured in the original 2^{4-1} fractional factorial because the AC 2-factor interaction is aliased with an apparently fairly large BD 2-factor interaction *of opposite sign*.

> Ultimately, this example is one of a fairly complicated system of effects. It admirably illustrates the difficulties and even errors of interpretation that can arise when only fractional factorial data are available for use in studying such systems.

In conclusion, it should be said that when a 2^{p-1} fractional factorial seems to leave only very mild ambiguities of interpretation, it can be possible to resolve those with the use of only a *few* additional data points (rather than requiring the addition of the entire other $\frac{1}{2}$ fraction of combinations). But this is a more advanced topic than is sensibly discussed here. The interested reader can refer to Chapter 14 of Daniel's *Applications of Statistics to Industrial Experimentation* for an illuminating discussion of this matter.

Section 3 Exercises

1. In a 2^{5-1} study with defining relation I $\leftrightarrow$ ABCDE, it is possible for both the A main effect and the BCDE 4-factor interaction to be of large magnitude but for both of them to go undetected. How might this quite easily happen?

2. The paper "How to Optimize and Control the Wire Binding Process: Part I" by Scheaffer and Levine (*Solid State Technology*, November 1990) contains the results of a 2^{5-1} fractional factorial experiment with additional repeated center point, run in an effort to determine how to improve the operation of a K&S Model 1484 XQ wire bonder. The generator E $\leftrightarrow$ ABCD was used in setting up the 2^{5-1} part of the experiment involving the factors and levels indicated in the accompanying table.

Factor A	Constant Velocity	.6 in./sec ($-$) vs. 1.2 in./sec ($+$)
Factor B	Temperature	150°C ($-$) vs. 200°C ($+$)
Factor C	Bond Force	80 g ($-$) vs. 120 g ($+$)
Factor D	Ultrasonic Power	120 mW ($-$) vs. 200 mW ($+$)
Factor E	Bond Time	10 ms ($-$) vs. 20 ms ($+$)

The response variable, y, was a force (in grams) required to pull wire bonds made on the machine under a particular combination of levels of the factors. (Each y was actually an average of the pull forces required on a 30 lead test sample.) The responses from the 2^{5-1} part of the study were as follows:

Combination	y	Combination	y
e	8.5	d	5.8
a	7.9	ade	8.0
b	7.7	bde	7.8
abe	8.7	abd	8.7
c	9.0	cde	6.9
ace	9.2	acd	8.5
bce	8.6	bcd	8.6
abc	9.5	abcde	8.3

In addition, three runs were made at a constant velocity of .9 in./sec, a temperature of 175°C, a bond force of 100 g, a power of 160 mW, and a bond time of 15 ms. These produced y values of 8.1, 8.6, and 8.1.

(a) Place the 16 observations from the 2^{5-1} part of the experiment in Yates standard order as regards levels of factors A through D. Use the four-cycle Yates algorithm to compute fitted sums of 2^5 effects. Identify what sum of effects each of these estimates. (For example, the first estimates $\mu_{\ldots} + \alpha\beta\gamma\delta\epsilon_{22222}$.)

(b) The three center points can be thought of as providing a pooled sample variance here. You may verify that $s_p = .29$. If one then wishes to make confidence intervals for the sums of effects, it is possible to use the $m = 1$, $p = 4$, and

$\nu = 2$ version of formula (8.13) of Section 8.2. What is the plus-or-minus value that comes from this program, for individual 95% two-sided confidence intervals? Using this value, which of the fitted sums of effects would you judge to be statistically detectable? Does this list suggest to you any particularly simple/intuitive description of how bond strength depends on the levels of the five factors?

(c) Based on your analysis from (b), if you had to guess what levels of the factors A, C, and D should be used for high bond strength, what would you recommend? If the CE + ABD fitted sum reflects primarily the CE 2-factor interaction, what level of E then seems best? Which of the combinations actually observed had these levels of factors A, C, D, and E? How does its response compare to the others?

3. Return to the fire retardant flame test study of Exercise 3 of Section 8.2. The original study, summarized in that exercise, was a full 2^4 factorial study.

(a) If you have not done so previously, use the (four-cycle) Yates algorithm and compute the fitted 2^4 factorial effects for the study. Normal-plot these. What subject-matter interpretation of the data is suggested by the normal plot?

Now suppose that instead of a full factorial study, only the $\frac{1}{2}$ fraction with generator D ↔ ABC had been conducted.

(b) Which 8 of the 16 treatment combinations would have been run? List these combinations in Yates standard order as regards factors A, B, and C and use the (three-cycle) Yates algorithm to compute the 8 estimated sums of effects that it is possible to derive from these 8 treatment combinations. Verify that each of these 8 estimates is the sum of two of your fitted effects from part (a). (For example, you should find that the first estimated sum here is $\bar{y}_{....} + abcd_{2222}$ from part (a).)

(c) Normal-plot the last 7 of the estimated sums from (b). Interpret this plot. If you had only the data from this 2^{4-1} fractional factorial, would your subject-matter conclusions be the same as those reached in part (a), based on the full 2^4 data set?

8.4 Standard Fractions of Two-Level Factorials Part II: General 2^{p-q} Studies

Section 8.3 began the study of fractional factorials with the $\frac{1}{2}$ fractions of 2^p factorials, considering in turn the issues of (1) choice, (2) determination of the corresponding alias structure, and (3) data analysis. The approaches used to treat 2^{p-1} studies extend naturally to the smaller $\frac{1}{2^q}$ fractions of 2^p factorials for $q > 1$.

This section first shows how the ideas of Section 8.3 are generalized to cover the general 2^{p-q} situation. Then it considers the notion of design resolution and its implications for comparing alternative possible 2^{p-q} plans. Next an introduction is given to how the 2^{p-q} ideas can be employed where a blocking variable (or variables) dictate the use of a number of blocks equal to a power of 2. The section concludes with some comments regarding wise use of this 2^{p-q} material.

8.4.1 Using 2^{p-q} Fractional Factorials

The recommended method of choosing a $\frac{1}{2}$ fraction of a 2^p factorial uses a column of signs developed as products of plus and minus signs for *all* of the first $p-1$ factors. The key to understanding how the ideas of the previous section generalize to $\frac{1}{4}$, $\frac{1}{8}$, $\frac{1}{16}$, etc. fractions of 2^p studies is to realize that there are several possible similar columns that could be developed using only *some* of the first $p - 1$ factors. When moving from $\frac{1}{2}$ fractions to $\frac{1}{2^q}$ fractions of 2^p factorials, one makes use of such columns in assigning levels of the last q factors and then develops and uses an alias structure consistent with the choice of columns.

For example, first consider the situation for cases where $p - q = 3$—that is, where $2^3 = 8$ different combinations of levels of p two-level factors are going to be included in a study. A table of signs specifying all eight possible combinations of levels of the first three factors A, B, and C, with four additional columns made up as the possible products of the first three columns, is given in Table 8.28.

Choosing a 2^{p-q} fractional factorial with $p - q = 3$

The final column of Table 8.28 can be used to choose levels of factor D for a best possible 2^{4-1} fractional factorial study. But it is also true that two or more of the product columns in Table 8.28 can be used to choose levels of several additional factors (beyond the first three). If this is done, one winds up with a fractional factorial that can be understood in the same ways it is possible to make sense of the standard 2^{p-1} data structures discussed in Section 8.3.

Table 8.28
Signs for Specifying all Eight Combinations of Three Two-Level Factors and Four Sets of Products of Those Signs

A	B	C	AB Product	AC Product	BC Product	ABC Product
−	−	−	+	+	+	−
+	−	−	−	−	+	+
−	+	−	−	+	−	+
+	+	−	+	−	−	−
−	−	+	+	−	−	+
+	−	+	−	+	−	−
−	+	+	−	−	+	−
+	+	+	+	+	+	+

Example 12

A 2^{6-3} Propellant Slurry Study

The text *Probability and Statistics for Engineers and Scientists*, by Walpole and Myers, contains an interesting 2^{6-3} fractional factorial data set taken originally from the *Proceedings of the 10th Conference on the Design of Experiments in Army Research, Development and Testing* (ARO-D Report 65-3). The study investigated the effects of six two-level factors on X-ray intensity ratios associated with a particular component of propellant mixtures in X-ray fluorescent analyses of propellant slurry. Factors A, B, C, and D represent the concentrations (at low and high levels) of four propellant components. Factors E and F represent the weights (also at low and high levels) of fine and coarse particles present.

Example 12
(continued)

Eight different combinations of levels of factors A, B, C, D, E, and F were each tested twice for X-ray intensity ratio, y. The eight combinations actually included in the study can be thought of as follows. Using the columns of Table 8.28, levels of factor D were chosen using the signs in the ABC product column directly; levels of factor E were chosen by reversing the signs in the BC product column; and levels of factor F were chosen by reversing the signs of the AC product column. Verify that such a prescription implies that the eight combinations included in the study (written down in Yates order for factors A, B, and C) were as displayed in Table 8.29. The eight combinations indicated in Table 8.29 are, of course, $\frac{1}{8}$ of the 64 different possible combinations of levels of the six factors.

Table 8.29
Combinations Included in the 2^{6-3} Propellant Slurry Study

A	B	C	F	E	D	Combination Name
−	−	−	−	−	−	(1)
+	−	−	+	−	+	adf
−	+	−	−	+	+	bde
+	+	−	+	+	−	abef
−	−	+	+	+	+	cdef
+	−	+	−	+	−	ace
−	+	+	+	−	−	bcf
+	+	+	−	−	+	abcd

The development of 2^{p-q} fractional factorials has been illustrated with eight-combination (i.e., $p - q = 3$) plans. But it should be obvious that there are 16-row, 32-row, 64-row, . . . , etc. versions of Table 8.28. Using any of these, one can assign levels for the last q factors according to signs in product columns and end up with a $\frac{1}{2^q}$ fraction of a full 2^p factorial plan. When this is done, the 2^p factorial effects are aliased in 2^{p-q} groups of 2^q effects each. The determination of this alias structure can be made by using q **generators** to develop a defining relation for the fractional factorial. A general definition of the notion of generators for a 2^{p-q} fractional factorial is next.

Determining the alias structure of a 2^{p-q} factorial

Definition 9

When a 2^{p-q} fractional factorial comes about by assigning levels of each of the "last" q factors based on a different column of products of signs for the "first" $p - q$ factors, the q different relationships

$$\left(\begin{array}{c} \text{the name of an} \\ \text{additional factor} \end{array} \right) \leftrightarrow \pm \left(\begin{array}{c} \text{a product of names of some} \\ \text{of the first } p - q \text{ factors} \end{array} \right)$$

corresponding to how the combinations are chosen are called **generators** of the plan.

Each generator can be translated into a statement with I on the left side and then taken individually, multiplied in pairs, multiplied in triples, and so on until the whole **defining relation** is developed. (See again Definition 8, page 602, for the meaning of this term.) In doing so, use can be made of the convention that minus any letter times minus that letter is I.

Example 12
(*continued*)

In the Army propellant example, the $q = 3$ generators that led to the combinations in Table 8.29 were

$$D \leftrightarrow ABC$$
$$E \leftrightarrow -BC$$
$$F \leftrightarrow -AC$$

Multiplying through by the left sides of these, one obtains the three relationships

$$I \leftrightarrow ABCD \qquad (8.25)$$
$$I \leftrightarrow -BCE \qquad (8.26)$$
$$I \leftrightarrow -ACF \qquad (8.27)$$

But in light of the conventions of formal multiplication, if I $\leftrightarrow$ ABCD and I $\leftrightarrow$ −BCE, it should also be the case that

$$I \leftrightarrow (ABCD) \cdot (-BCE)$$

that is,

$$I \leftrightarrow -ADE$$

Similarly, using relationships (8.25) and (8.27), one obtains

$$I \leftrightarrow -BDF$$

using relationships (8.26) and (8.27), one obtains

$$I \leftrightarrow ABEF$$

Example 12
(*continued*)

and finally, using all three relationships (8.25), (8.26), and (8.27), one has

$$I \leftrightarrow CDEF$$

Combining all of this, the complete defining relation for this 2^{6-3} study is

$$I \leftrightarrow ABCD \leftrightarrow -BCE \leftrightarrow -ACF \leftrightarrow$$
$$-ADE \leftrightarrow -BDF \leftrightarrow ABEF \leftrightarrow CDEF \tag{8.28}$$

Defining relation (8.28) is rather formidable, but it tells the whole truth about what can be learned based on the $\frac{1}{8}$ of 64 possible combinations of six two-level factors. Relation (8.28) specifies all effects that will be aliased with the grand mean. Appropriately multiplying through expression (8.28) gives all aliases of any effect of interest. For example, multiplying through relation (8.28) by A gives

$$A \leftrightarrow BCD \leftrightarrow -ABCE \leftrightarrow -CF \leftrightarrow -DE \leftrightarrow -ABDF \leftrightarrow BEF \leftrightarrow ACDEF$$

and for example, the (high level) A main effect will be indistinguishable from minus the (all high levels) CF 2-factor interaction.

Data analysis for a 2^{p-q} study

With a 2^{p-q} fractional factorial's defining relation in hand, the analysis of data proceeds exactly as indicated earlier for $\frac{1}{2}$ fractions. It is necessary to

1. compute estimates of (sums and differences of) effects ignoring the last q factors,

2. judge their statistical detectability using confidence interval or normal plotting methods, and then

3. seek a plausible tentative interpretation of the important estimates in light of the alias structure.

Example 12
(*continued*)

In the Army propellant study, $m = 2$ trials for each of the 2^{6-3} combinations listed in Table 8.29 gave $s_P^2 = .02005$ and the sample averages listed in Table 8.30.

Temporarily ignoring all but the ("first") three factors A, B, and C (since the levels of D, E and F were derived or generated from the levels of A, B and C), the (three-cycle) Yates algorithm can be used on the sample means, as shown in Table 8.31. Remember that the estimates in the next-to-last column of Table 8.31 must be interpreted in light of the alias structure for the original experimental plan. So for example, since (both from the original generators and

Table 8.30
Eight Sample Means from the 2^{6-3} Propellant Slurry Study

Combination	$\bar{y}$	Combination	$\bar{y}$
(1)	1.1214	cdef	.9285
adf	1.0712	ace	1.1635
bde	.9415	bcf	.9561
abef	1.1240	abcd	.9039

from relation (8.28)) one knows that D $\leftrightarrow$ ABC, the $-.0650$ value on the last line of Table 8.31 is estimating

$$\alpha\beta\gamma_{222} + \delta_2 \pm \text{(six other effects)}$$

So if one were expecting a large main effect of factor D, one would expect it to be evident in the $-.0650$ value.

Since a value of s_P is available here, there is no need to resort to normal-plotting to judge the statistical detectability of the values coming out of the Yates algorithm. Instead (still temporarily calculating as if only the first three factors were present) one can make confidence intervals based on the estimates, by employing the $\nu = 8 = 16 - 8$, $m = 2$, and $p = 3$ version of formula (8.13) from Section 8.2. That is, using 95% two-sided individual confidence intervals, a precision of

$$\pm 2.306 \frac{\sqrt{.02005}}{\sqrt{2 \cdot 2^3}} = \pm.0817$$

should be attached to each of the estimates in Table 8.31. By this standard, none of the estimates from the propellant study are clearly different from 0. For

Table 8.31
The Yates Algorithm for a 2^3 Factorial Applied to the 2^{6-3} Propellant Data

$\bar{y}$	Cycle 1	Cycle 2	Cycle 3	Cycle 3 ÷ 8	Sum Estimated
1.1214	2.1926	4.2581	8.2101	1.0263	$\mu_{....}$ + aliases
1.0712	2.0655	3.9520	.3151	.0394	α_2 + aliases
.9415	2.0920	.1323	$-.3591$	$-.0449$	β_2 + aliases
1.1240	1.8600	.1828	$-.0545$	$-.0068$	$\alpha\beta_{22}$ + aliases
.9285	$-.0502$	$-.1271$	$-.3061$	$-.0383$	γ_2 + aliases
1.1635	.1825	$-.2320$	.0505	.0063	$\alpha\gamma_{22}$ + aliases
.9561	.2350	.2327	$-.1049$	$-.0131$	$\beta\gamma_{22}$ + aliases
.9039	$-.0522$	$-.2872$	$-.5199$	$-.0650$	$\alpha\beta\gamma_{222}$ + aliases

Example 12
(*continued*)

engineering purposes, the bottom line is that more data are needed before even the most tentative conclusions about system behavior should be made.

Example 13

A 2^{5-2} Catalyst Development Experiment

Hansen and Best, in their paper "How to Pick a Winner" (presented at the 1986 annual meeting of the American Statistical Association), described several industrial experiments conducted in a research program aimed at the development of an effective catalyst for producing ethyleneamines by the amination of monoethanolamine. One of these was a partially replicated 2^{5-2} fractional factorial study in which the response variable, y, was percent water produced during the reaction period. The five two-level experimental factors were as in Table 8.32. (The T-372 support was an alpha-alumina support and the T-869 support was a silica alumina support.)

The fractional factorial described by Hansen and Best has ($q = 2$) generators D $\leftrightarrow$ ABC and E $\leftrightarrow$ BC. The resulting defining relation (involving $2^2 = 4$ strings of letters) is then

$$I \leftrightarrow ABCD \leftrightarrow BCE \leftrightarrow ADE$$

where the fact that the ADE 3-factor interaction is aliased with the grand mean can be seen by multiplying together ABCD and BCE, which (from the generators) themselves represent effects aliased with the grand mean. Here one sees that effects will be aliased together in eight groups of four.

The data reported by Hansen and Best, and some corresponding summary statistics, are given in Table 8.33. The pooled sample variance derived from the values in Table 8.33 is

$$s_P^2 = \frac{(3-1)(2.543) + (2-1)(2.163) + (2-1)(.238)}{(3-1) + (2-1) + (2-1)}$$

$$= 1.872$$

Table 8.32
Five Catalysis Variables and Their Experimental Levels

Factor	Process Variable	Levels
A	Ni/Re Ratio	2/1 ($-$) vs. 20/1 ($+$)
B	Precipitant	$(NH_4)_2CO_3$ ($-$) vs. none ($+$)
C	Calcining Temperature	300°C ($-$) vs. 500°C ($+$)
D	Reduction Temperature	300°C ($-$) vs. 500°C ($+$)
E	Support Used	T-372 ($-$) vs. T-869 ($+$)

Table 8.33
Data from a 2^{5-2} Catalyst Study and Corresponding Sample Means and Variances

Combination	% Water Produced, y	$\bar{y}$	s^2
e	8.70, 11.60, 9.00	9.7670	2.543
ade	26.80	26.800	—
bd	24.88	24.880	—
ab	33.15	33.150	—
cd	28.90, 30.980	29.940	2.163
ac	30.20	30.200	—
bce	8.00, 8.69	8.345	.238
abcde	29.30	29.300	—

with $\nu = (3 - 1) + (2 - 1) + (2 - 1) = 4$ associated degrees of freedom. The corresponding pooled sample standard deviation is

$$\sqrt{s_P^2} = \sqrt{1.872} = 1.368$$

So temporarily ignoring the existence of factors D and E, it is possible to use the $p = 3$ version of formula (8.12) to derive precisions to attach to the estimates (of sums of 2^5 factorial effects) that result from the use of the Yates algorithm on the sample means in Table 8.33. That is, for 95% two-sided individual confidence intervals, precisions of

$$\pm 2.776(1.368)\frac{1}{2^3}\sqrt{\frac{1}{3} + \frac{1}{1} + \frac{1}{1} + \frac{1}{1} + \frac{1}{2} + \frac{1}{1} + \frac{1}{2} + \frac{1}{1}}$$

that is,

$$\pm 1.195\% \text{ water}$$

can be attached to the estimates.

The reader can verify that the (three-cycle) Yates algorithm applied to the means in Table 8.33 gives the estimates in Table 8.34. Identifying those estimates in Table 8.34 whose magnitudes make them statistically detectable according to a criterion of ± 1.195, there are (in order of decreasing magnitude)

$$\alpha_2 + \beta\gamma\delta_{222} + \alpha\beta\gamma\epsilon_{2222} + \delta\epsilon_{22} \text{ estimated as } 5.815$$
$$\beta\gamma_{22} + \alpha\delta_{22} + \epsilon_2 + \alpha\beta\gamma\delta\epsilon_{22222} \text{ estimated as } -5.495$$
$$\alpha\beta\gamma_{222} + \delta_2 + \alpha\epsilon_{22} + \beta\gamma\delta\epsilon_{2222} \text{ estimated as } 3.682$$
$$\alpha\beta_{22} + \gamma\delta_{22} + \alpha\gamma\epsilon_{222} + \beta\delta\epsilon_{222} \text{ estimated as } 1.492$$

Example 13
(*continued*)

Table 8.34
Estimates of Sums of Effects
for the Catalyst Study

Sum of Effects Estimated	Estimate
grand mean + aliases	24.048
A + aliases	5.815
B + aliases	−.129
AB + aliases	1.492
C + aliases	.399
AC + aliases	−.511
BC + aliases	−5.495
ABC + aliases	3.682

The simplest possible tentative interpretation of the first two of these results is that the A and E main effects are large enough to see above the background variation. What to make of the third, given the first two, is not so clear. The large 3.682 estimate can equally simply be tentatively attributed to a D main effect or to an AE 2-factor interaction. (Interestingly, Hansen and Best reported that subsequent experimentation was done with the purpose of determining the importance of the D main effect, and indeed, the importance of this factor in determining y was established.)

Exactly what to make of the fourth statistically significant estimate is even less clear. It is therefore comforting that, although big enough to be detectable, it is less than half the size of the third largest estimate. In the particular real situation, the authors seem to have found an "A, E, and D main effects only" description of y useful in subsequent work with the chemical system.

The reader may have noticed that the possibilities discussed in the previous example do not even exhaust the plausible interpretations of the fact that three estimated sums of effects are especially large. For example, "large DE 2-factor interactions and large D and E main effects" is yet another alternative possibility. This ambiguity serves to again emphasize the tentative nature of conclusions that can be drawn on the basis of small fractions of full factorials. And it also underlines the absolute necessity of subject-matter expertise and follow-up study in sorting out the possibilities in a real problem. There is simply no synthetic way to tell which of various simple alternative explanations suggested by a fractional factorial analysis is the right one.

8.4.2   Design Resolution

The results of five different real applications of 2^{p-q} plans have been discussed in Examples 9, 10, 11, 12, and 13. From them, it should be clear how important it is to

Good choice of a fractional factorial

have the simplest alias structure possible when it comes time to interpret the results of a fractional factorial study. The object is to have low-order effects (like main effects and 2-factor interactions) aliased not with other low-order effects, but rather only with high-order effects (many-factor interactions). It is the defining relation that governs how the 2^p factorial effects are divided up into groups of aliases. If there are only long products of factor names appearing in the defining relation, low-order effects are aliased only with high-order effects. On the other hand, if there are short products of factor names appearing, there will be low-order effects aliased with other low-order effects. As a kind of measure of quality of a 2^{p-q} plan, it is thus common to adopt the following notion of **design resolution**.

Definition 10

> The **resolution** of a 2^{p-q} fractional factorial plan is the number of letters in the shortest product appearing in its defining relation.

In general, when contemplating the use of a 2^{p-q} design, one wants the largest resolution possible for a given investment in 2^{p-q} combinations. Not all choices of generators give the same resolution. In Section 8.3, the prescription given for the $\frac{1}{2}$ fractions was intended to give 2^{p-1} fractional factorials of resolution p (the largest resolution possible). For general 2^{p-q} studies, one must be a bit careful in choosing generators. What seems like the most obvious choice need not be the best in terms of resolution.

Example 14

Resolution 4 in a 2^{6-2} Study

Consider planning a 2^{6-2} study—that is, a study including 16 out of 64 possible combinations of levels of factors A, B, C, D, E, and F. A rather natural choice of two generators for such a study is

$$E \leftrightarrow ABCD$$
$$F \leftrightarrow ABC$$

The corresponding defining relation is

$$I \leftrightarrow ABCDE \leftrightarrow ABCF \leftrightarrow DEF$$

The resulting design is of resolution 3, and there are some main effects aliased with (only) 2-factor interactions.

On the other hand, the perhaps slightly less natural choice of generators

$$E \leftrightarrow BCD$$
$$F \leftrightarrow ABC$$

Example 14
(*continued*)

has defining relation

$$I \leftrightarrow BCDE \leftrightarrow ABCF \leftrightarrow ADEF$$

and is of resolution 4. No main effect is aliased with any interaction of order less than 3. This second choice is better than the first in terms of resolution.

Table 8.35 indicates what is possible in terms of resolution for various numbers of factors and combinations for a 2^{p-q} fractional factorial. The table was derived from a more detailed one on page 410 of *Statistics for Experimenters* by Box, Hunter, and Hunter, which gives not only the best resolutions possible but also generators for designs achieving those resolutions. The more limited information in Table 8.35 is sufficient for most purposes. Once one is sure what is possible, it is usually relatively painless to do the trial-and-error work needed to produce a plan of highest possible resolution. And it is probably worth doing as an exercise, to help one consider the pros and cons of various choices of generators for a given set of real factors.

Table 8.35 has no entries in the "8 combinations" row for more than 7 factors. If the table were extended beyond 11 factors, there would be no entries in the "16 samples" row beyond 15 factors, no entries in the "32 samples" row beyond 31 factors, etc. The reason for this should be obvious. For 8 combinations, there are only 7 columns total to use in Table 8.28. Corresponding tables for 16 combinations would have only 15 columns total, for 32 combinations only 31 columns total, etc.

As they have been described here, 2^{p-q} fractional factorials can be used to study at most $2^t - 1$ factors in 2^t samples. The cases of 7 factors in 8 combinations, 15 factors in 16 combinations, 31 factors in 32 combinations, etc. represent a kind of extreme situation where a maximum number of factors is studied (at the price of creating a worst possible alias structure) in a given number of combinations. For the case of $p = 7$ factors in 8 combinations, effects are aliased in $2^{7-4} = 8$ groups of $2^4 = 16$; for the case of $p = 15$ factors in 16 combinations, the effects are aliased in $2^{15-11} = 16$ groups of $2^{11} = 2,048$; etc. These extreme cases of $2^t - 1$ factors in 2^t combinations are sometimes called **saturated fractional factorials**. They have very complicated alias structures and can support only the most tentative of conclusions.

Table 8.35
Best Resolutions Possible for Various Numbers of Combinations in a 2^{p-q} Study

		Number of Factors (p)							
		4	5	6	7	8	9	10	11
	8	4	3	3	3	—	—	—	—
Number of	16		5	4	4	4	3	3	3
Combinations (2^{p-q})	32			6	4	4	4	4	4
	64				7	5	4	4	4
	128					8	6	5	5

Example 15

A 16-Run 15-Factor Process Development Study

The article "What Every Technologist Should Know About Experimental Design" by C. Hendrix (*Chemtech*, 1979) includes the results from an unreplicated 16-run (saturated) 15-factor experiment. The response, y, was a measure of cold crack resistance for an industrial product. Experimental factors and levels were as listed in Table 8.36.

Table 8.36
15 Process Variables and Their Experimental Levels

Factor	Process Variable	Levels
A	Coating Roll Temperature	115° (−) vs. 125° (+)
B	Solvent	Recycled (−) vs. Refined (+)
C	Polymer X-12 Preheat	No (−) vs. Yes (+)
D	Web Type	LX-14 (−) vs. LB-17 (+)
E	Coating Roll Tension	30 (−) vs. 40 (+)
F	Number of Chill Rolls	1 (−) vs. 2 (+)
G	Drying Roll Temperature	75° (−) vs. 80° (+)
H	Humidity of Air Feed to Dryer	75% (−) vs. 90% (+)
J	Feed Air to Dryer Preheat	No (−) vs. Yes (+)
K	Dibutylfutile in Formula	12% (−) vs. 15% (+)
L	Surfactant in Formula	.5% (−) vs. 1% (+)
M	Dispersant in Formula	.1% (−) vs. 2% (+)
N	Wetting Agent in Formula	1.5% (−) vs. 2.5% (+)
O	Time Lapse Before Coating Web	10 min (−) vs. 30 min (+)
P	Mixer Agitation Speed	100 RPM (−) vs. 250 RPM (+)

The experimental plan used was defined by the $q = 11$ generators

E ↔ ABCD, F ↔ BCD, G ↔ ACD, H ↔ ABC, J ↔ ABD, K ↔ CD, L ↔ BD, M ↔ AD, N ↔ BC, O ↔ AC, and P ↔ AB

The combinations actually run and the cold crack resistances observed are given in Table 8.37.

Ignoring all factors but A, B, C, and D, the combinations listed in Table 8.37 are in Yates standard order and are therefore ready for use in finding estimates of sums of effects. Table 8.38 shows the results of using the (four-cycle) Yates algorithm on the 16 observations listed in Table 8.37. A normal plot of the last 15 of these estimates is shown in Figure 8.13. It is clear from the figure that the two corresponding to B + *aliases* and F + *aliases* are detectably larger than the rest.

Example 15
(*continued*)

Table 8.37
16 Experimental Combinations and Measured
Cold Crack Resistances

Combination	y	Combination	y
eklmnop	14.8	dfgjnop	17.8
aghjkln	16.3	adefhmn	18.9
bfhjkmo	23.5	bdeghlo	23.1
abefgkp	23.9	abdjlmp	21.8
cfghlmp	19.6	cdehjkp	16.6
acefjlo	18.6	acdgkmo	16.7
bcegjmn	22.3	bcdfkln	23.5
abchnop	22.2	abcdefghjklmnop	24.9

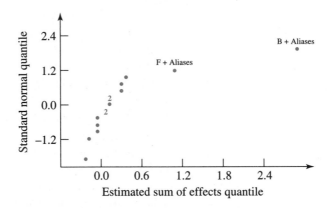

Figure 8.13   Normal plot of estimated sums of effects in
the 2^{15-11} process development study

It is not feasible to write out the whole defining relation for this 2^{15-11} study. Effects are aliased in $2^{p-q} = 2^{15-11} = 16$ groups of $2^q = 2^{11} = 2,048$. In particular (though it would certainly be convenient if the 2.87 estimate in Table 8.38 could be thought of as essentially representing β_2), β_2 has 2,047 aliases, some of them as simple as 2-factor interactions. By the same token, it would certainly be convenient if the small estimates in Table 8.38 were indicating that all summands of the sums of effects they represent were small. But the possibility of cancellation in the summation must not be overlooked.

The point is that only the most tentative description of this system should be drawn from even this very simple "two large estimates" outcome. The data in Table 8.37 hint at the primary importance of factors B and F in determining cold crack resistance, but the case is hardly airtight. There is a suggestion of a direction for further experimentation and discussion with process experts but certainly no detailed map of the countryside where one is going.

Table 8.38
Estimates of Sums of Effects for the 2^{15-11}
Process Development Study

Sum of Effects Represented	Estimate
grand mean + aliases	20.28
A + aliases	.13
B + aliases	2.87
P + aliases (including AB)	−.08
C + aliases	.27
O + aliases (including AC)	−.08
N + aliases (including BC)	.19
H + aliases (including ABC)	.36
D + aliases	.13
M + aliases (including AD)	.03
L + aliases (including BD)	.04
J + aliases (including ABD)	−.06
K + aliases (including CD)	−.26
G + aliases (including ACD)	.29
F + aliases (including BCD)	1.06
E + aliases (including ABCD)	.11

One thing that can be said fairly conclusively on the basis of this study is that the analysis points out what is in retrospect obvious in Table 8.37. Consistent with the "B + *aliases* and F + *aliases* sums are positive and large" story told in Figure 8.13, the largest four values of y listed in Table 8.37 correspond to combinations where both B and F are at their high levels.

8.4.3 Two-Level Factorials and Fractional Factorials in Blocks (*Optional*)

A somewhat specialized but occasionally useful adaptation of the 2^{p-q} material presented here has to do with the analysis of full or fractional two-level factorial studies run in complete or incomplete blocks. When the number of blocks under consideration is itself a power of 2, clever use of the methods developed in this chapter can guide the choice of which combinations to place in incomplete blocks, as well as the analysis of data from both incomplete and complete block studies.

The basic idea used is to formally represent one 2^t-level factor "Blocks" as t "extra" two-level factors. One lets combinations of levels of these extra factors define the blocks into which combinations of levels of the factors of interest are placed. In data analysis, effects involving only the extra factors as Block main effects and effects involving both the extra factors and the factors of interest are recognized as Block×Treatment interactions. In carrying out this program, it is fairly common (though not necessarily safe) to operate as if the Block×Treatment interactions were all negligible. How choice and analysis of blocked 2^{p-q} studies proceed will be illustrated with a series of three examples that are variations on Example 11.

Example 16
(*Example 11 revisited*)

A 2^4 Fabric Tenacity Study Run in Two Blocks

In the weaving study of Johnson, Clapp, and Baqai, four factors—A, B, C, and D—were studied. The discussion in Section 8.3 described how the authors originally ran a replicated 2^{4-1} fractional factorial with defining relation I ↔ ABCD. This was followed up later with a second 2^{4-1} fractional factorial having defining relation I ↔ −ABCD, thus completing the full 2^4 factorial. However, since the study of the two $\frac{1}{2}$ fractions was separated in time, it is sensible to think of the two parts of the study as different *blocks*—that is, to think of a fifth two-level factor (say, E) representing the time of observation.

How then to use the formal multiplication idea to understand the alias structure? Notice that there are 16 different samples and five factors for consideration. This suggests that somehow (at least in formal terms) this situation might be thought of as a 2^{5-1} data structure. Further, the two formal expressions

$$I \leftrightarrow ABCD \tag{8.29}$$

$$I \leftrightarrow -ABCD \tag{8.30}$$

define the two sets of 8 out of 16 ABCD combinations actually run. These result from a formal expression like

$$I \leftrightarrow ABCDE \tag{8.31}$$

where E can be thought of as contributing either the plus or the minus signs in expressions (8.29) and (8.30). If one calls block 1 (the first set of 8 samples) the high level of E, expression (8.31) leads to exactly the I ↔ ABCD $\frac{1}{2}$-fraction of 2^4 combinations of A, B, C, and D for use as block 1. And the I ↔ −ABCD $\frac{1}{2}$-fraction for use as block 2. This can be seen in Table 8.39.

With factor E designating block number, the two columns of Table 8.39 taken together designate the I ↔ ABCDE $\frac{1}{2}$-fraction of 2^5 A, B, C, D, and E combinations. And (ignoring the e) the first column of Table 8.39 designates the I ↔ ABCD $\frac{1}{2}$-fraction of 2^4 A, B, C, and D combinations, while the second designates the I ↔ −ABCD $\frac{1}{2}$-fraction of 2^4 A, B, C, and D combinations.

Once it is clear that the Johnson, Clapp, and Baqai study can be thought of in terms of expression (8.31) with the two-level blocking factor E, it is also clear how any *block effects* will show up during data analysis. One temporarily ignores the blocks and uses the Yates algorithm to compute fitted 2^4 factorial effects. It is then necessary to remember, for example, that the fitted ABCD 4-factor interaction reflects not only $\alpha\beta\gamma\delta_{2222}$ but any block main effects as well.

Table 8.39
A 2^{5-1} Fractional Factorial or
a 2^4 Factorial in Two Blocks

Block 1	Block 2
e	a
abe	b
ace	c
bce	abc
ade	d
bde	abd
cde	acd
abcde	bcd

And for example, any 2-factor interaction of A and blocks will be reflected in the fitted BCD 3-factor interaction. Of course, if all interactions with blocks are negligible, all fitted effects except that for the ABCD 4-factor interaction would indeed represent the appropriate 2^4 factorial effects.

Example 17

A 2^4 Factorial Run in Four Blocks

For the sake of illustration, suppose that Johnson, Clapp, and Daqai had a priori planned to conduct a full 2^4 factorial set of ABCD combinations in four incomplete blocks (of four combinations each). Consider how those blocks might have been chosen and how subsequent data analysis would have proceeded.

The one four-level factor Blocks can here be thought of in terms of the combinations of two extra two-level factors, which can be designated as E and F. In order to accommodate the original four factors and these two additional ones in 16 ABCDEF combinations, one must choose a 2^{6-2} design by specifying two generators. The choices

$$E \leftrightarrow BCD \tag{8.32}$$

$$F \leftrightarrow ABC \tag{8.33}$$

leading to the defining relation

$$I \leftrightarrow BCDE \leftrightarrow ABCF \leftrightarrow ADEF \tag{8.34}$$

will be used here. Table 8.40 indicates the 16 combinations of levels of factors A through F prescribed by the generators (8.32) and (8.33).

Example 17
(*continued*)

The four different combinations of levels of E and F ((1), e, f, and ef) can be thought as designating in which block a given ABCD combination should appear. So generators (8.32) and (8.33) prescribe the division of the full 2^4 factorial (in the factors A through D) into the blocks indicated in Table 8.40 and Table 8.41.

As always, the defining relation (given here in display (8.34)) describes how effects are aliased. Table 8.42 indicates the aliases of each of the 2^4 factorial effects, obtained by multiplying through relation (8.34) by the various combinations of the letters A, B, C, and D. Notice from Table 8.42 that the BCD and ABC 3-factor interactions are aliased with block main effects. So is the AD 2-factor

Table 8.40
16 Combinations of Levels of A through F

A	B	C	D	E	F	Block Prescribed by Levels of E and F
−	−	−	−	−	−	1
+	−	−	−	−	+	3
−	+	−	−	+	+	4
+	+	−	−	+	−	2
−	−	+	−	+	+	4
+	−	+	−	+	−	2
−	+	+	−	−	−	1
+	+	+	−	−	+	3
−	−	−	+	+	−	2
+	−	−	+	+	+	4
−	+	−	+	−	+	3
+	+	−	+	−	−	1
−	−	+	+	−	+	3
+	−	+	+	−	−	1
−	+	+	+	+	−	2
+	+	+	+	+	+	4

Table 8.41
A 2^4 Factorial in Four Blocks
(from a 2^{6-2} Fractional Factorial)

Block 1	Block 2	Block 3	Block 4
(1)	ab	a	b
bc	ac	abc	c
abd	d	bd	ad
acd	bcd	cd	abcd

Table 8.42
Aliases of the 2^4 Factorial Effects
When Run in Four Blocks Prescribed
by Generators (8.32) and (8.33)

$$I \leftrightarrow BCDE \leftrightarrow ABCF \leftrightarrow ADEF$$
$$A \leftrightarrow ABCDE \leftrightarrow BCF \leftrightarrow DEF$$
$$B \leftrightarrow CDE \leftrightarrow ACF \leftrightarrow ABDEF$$
$$AB \leftrightarrow ACDE \leftrightarrow CF \leftrightarrow BDEF$$
$$C \leftrightarrow BDE \leftrightarrow ABF \leftrightarrow ACDEF$$
$$AC \leftrightarrow ABDE \leftrightarrow BF \leftrightarrow CDEF$$
$$BC \leftrightarrow DE \leftrightarrow AF \leftrightarrow ABCDEF$$
$$ABC \leftrightarrow ADE \leftrightarrow F \leftrightarrow BCDEF$$
$$D \leftrightarrow BCE \leftrightarrow ABCDF \leftrightarrow AEF$$
$$AD \leftrightarrow ABCE \leftrightarrow BCDF \leftrightarrow EF$$
$$BD \leftrightarrow CE \leftrightarrow ACDF \leftrightarrow ABEF$$
$$ABD \leftrightarrow ACE \leftrightarrow CDF \leftrightarrow BEF$$
$$CD \leftrightarrow BE \leftrightarrow ABDF \leftrightarrow ACEF$$
$$ACD \leftrightarrow ABE \leftrightarrow BDF \leftrightarrow CEF$$
$$BCD \leftrightarrow E \leftrightarrow ADF \leftrightarrow ABCEF$$
$$ABCD \leftrightarrow AE \leftrightarrow DF \leftrightarrow BCEF$$

interaction, since one of its aliases is EF, which involves only the two-level extra factors E and F used to represent the four-level factor Blocks. On the other hand, if interactions with Blocks are negligible, it is *only* these three of the 2^4 factorial effects that are aliased with other possibly nonnegligible effects. (For any other of the 2^4 factorial effects, each alias involves letters both from the group A, B, C, and D and also from the group E and F—and is therefore some kind of Block $\times$ Treatment interaction.)

Analysis of data from a plan like that in Table 8.41 would proceed as indicated repeatedly in this chapter. The Yates algorithm applied to sample means listed in Yates standard order for factors A, B, C, and D produces estimates that are interpreted in light of the alias structure laid out in Table 8.42.

Example 18

A 2^{4-1} Fractional Factorial Run in Four Blocks

As a final variant on the 4-factor weaving example, consider how the original $\frac{1}{2}$ fraction of the 2^4 factorial might itself have been run in four incomplete blocks of two combinations. (Imagine that for some reason, only two combinations could be prepared on any single day and that there was some fear of Day effects related to environmental changes, instrument drift, etc.)

Example 18
(*continued*)

Only eight combinations are to be chosen. In doing so, one needs to account for the four experimental factors A, B, C, and D and two extras E and F, which can be used to represent the four-level factor Blocks. Starting with the first three experimental factors A, B, and C (three of them because $2^3 = 8$), one needs to choose three generators. The original 2^{4-1} study had generator

$$D \leftrightarrow ABC$$

so it is natural to begin there. For the sake of example, consider also the generators

$$E \leftrightarrow BC$$

$$F \leftrightarrow AC$$

These give the defining relation

$$I \leftrightarrow ABCD \leftrightarrow BCE \leftrightarrow ACF \leftrightarrow ADE \leftrightarrow BDF \leftrightarrow ABEF \leftrightarrow CDEF \quad \textbf{(8.35)}$$

and the prescribed set of combinations listed in Table 8.43. (The four different combinations of levels of E and F ((1), e, f, and ef) designate in which block a given ABCD combination from the $\frac{1}{2}$ fraction should appear.)

Table 8.43
A 2^{6-3} Fractional Factorial or a 2^{4-1} Fractional Factorial in Four Blocks

Block 1	Block 2	Block 3	Block 4
ab	ade	bdf	ef
cd	bce	acf	abcdef

Some experimenting with relation (8.35) will show that *all* 2-factor interactions of the four original experimental factors A, B, C, and D are aliased not only with other 2-factor interactions of experimental factors but also with Block main effects. Thus, any systematic block-to-block changes would further confuse one's perception of 2-factor interactions of the experimental factors. But at least the main effects of A, B, C, and D are not aliased with Block main effects.

Examples 16 through 18 all treat situations where blocks are *incomplete*—in the sense that they don't each contain every combination of the experimental factors studied. **Complete block plans** with 2^t blocks can also be developed and analyzed through the use of t "extra" two-level factors to represent the single (2^t-level) factor Blocks. The path to be followed is by now worn enough through use in this chapter that further examples will not be included. But the reader should have no trouble figuring out, for example, how to analyze a full 2^4 factorial that is run completely once in each of two blocks, or even how to analyze a standard 2^{4-1} fractional factorial that is run completely once in each of four blocks.

8.4.4 Some Additional Comments

This 2^{p-q} fractional factorial material is fascinating, and extremely useful when used with a proper understanding of both its power and its limitations. However, an engineer who tries to use it in a cookbook fashion will usually wind up frustrated and disillusioned. The implications of aliasing must be thoroughly understood for successful use of the material. And a clear understanding of these implications will work to keep the engineer from routinely trying to study many factors based on very small amounts of data in a one-shot experimentation mode.

Engineers newly introduced to fractional factorial experimentation sometimes try to routinely draw final engineering conclusions about multifactor systems based on as few as eight data points. The folly of such a method of operation should be apparent. Economy of experimental effort involves not just collecting a small amount of data on a multifactor system, but rather collecting the minimum amount sufficient for a practically useful and reliable understanding of system behavior. Just a few expensive engineering errors, traceable to naive and overzealous use of fractional factorial experimentation, will easily negate any supposed savings generated by overly frugal data collection.

Choice of experiment size   Although several 8-combination plans have been used as examples in this section, such designs are often too small to provide much information on the behavior of real engineering systems. Typically, 2^{p-q} studies with $p - q \geq 4$ are recommended as far more likely to lead to a satisfactory understanding of system behavior.

It has been said several times that when intelligently used as factor-screening tools, 2^{p-q} fractional factorial studies will usually be followed up with more complete experimentation, such as a larger fraction or a complete factorial (often in a reduced set of factors). It is also true that techniques exist for choosing a relatively small second fraction in such a way as to resolve certain particular types of ambiguities of interpretation that can remain after the analysis of an initial fractional factorial. The interested reader can refer to Section 12.5 of *Statistics for Experimenters* by Box, Hunter, and Hunter for discussions of how to choose an additional fraction to "dealias" a particular main effect and all its associated interactions or to "dealias" all main effects.

Section 4 Exercises

1. What are the advantages and disadvantages of fractional factorial experimentation in comparison to factorial experimentation?

2. Under what circumstances can one hope to be successful experimenting with (say) 12 factors in (say) 16 experimental runs (i.e., based on 16 data points)?

3. What is the principle of "sparsity of effects" and how can it be used in the analysis of unreplicated 2^p and 2^{p-q} experiments?

4. In a 7-factor study, only 32 different combinations of levels of (two-level factors) A, B, C, D, E, F, and

G will be included, at least initially. The genera-
tors F $\leftrightarrow$ ABCD and G $\leftrightarrow$ ABCE will be used to
choose the 32 combinations to include in the study.

(a) Write out the whole defining relation for the
experiment that is contemplated here.

(b) Based on your answer to part (a), what effects
will be aliased with the C main effect in the
experiment that is being planned?

(c) When running the experiment, what levels of
factors F and G are used when all of A, B, C,
D, and E are at their low levels? What levels
of factors F and G are used when A, B, and C
are at their high levels and D and E are at their
low levels?

(d) Suppose that after listing the data (observed
y's) in Yates standard order as regards factors
A, B, C, D, and E, you use the Yates algorithm to compute 32 fitted sums of effects.
Suppose further that the fitted values appearing on the A + aliases, ABCD + aliases, and
BCD + aliases rows of the Yates computations
are the only ones judged to be of both statistical significance and practical importance.
What is the simplest possible interpretation of
this result?

5. In a 2^{5-2} study, where four sample sizes are 1 and
four sample sizes are 2, $s_p = 5$. If 90% two-sided
confidence limits are going to be used to judge
the statistical detectability of sums of effects, what
plus-or-minus value will be used?

6. Consider planning, executing, and analyzing the results of a 2^{6-2} fractional factorial experiment based
on the two generators E $\leftrightarrow$ ABC and F $\leftrightarrow$ BCD.

(a) Write out the defining relation (i.e., the whole
list of aliases of the grand mean) for such a
plan.

(b) When running the experiment, what levels of
factors E and F are used when all of A, B, C,
and D are at their low levels? When A is at
its high level but B, C, and D are at their low
levels?

(c) Suppose that $m = 3$ data points from each of
the 16 combinations of levels of factors (specified by the generators) give a value of $s_p \approx$
2.00. If individual 90% two-sided confidence
intervals are to be made to judge the statistical
significance of the estimated (sums of) effects,
what is the value of the plus-or-minus part of
each of those intervals?

Chapter 8 Exercises ....

1. Return to the situation of Chapter Exercise 4 of
Chapter 4. That exercise concerns some unreplicated 2^3 factorial data taken from a study of the
mechanical properties of a polymer. If you have
not already done so, use the Yates algorithm to
compute fitted 2^3 factorial effects for the data
given in that exercise. Then make a normal plot
of the seven fitted effects $a_2, b_2, \ldots, abc_{222}$ as a
means of judging the statistical detectability of
the various effects on impact strength. Interpret
this plot.

2. Chapter Exercise 5 in Chapter 4 concerns a 2^3
study of mechanical pencil lead strength done by
Timp and M-Sidek. Return to that exercise, and if
you have not already done so, use the Yates algorithm to compute fitted 2^3 effects for the logged
data.

(a) Compute s_p for the logged data. Individual
confidence intervals for the theoretical 2^3 effects are of the form $\hat{E} \pm \Delta$. Find Δ if 95%
individual two-sided intervals are of interest.

(b) Based on your value from part (a), which
of the factorial effects are statistically detectable? Considering only those effects that
are both statistically detectable and large
enough to have a material impact on the
breaking strength, interpret the results of the
students' experiment. (For example, if the A
main effect is judged to be both detectable
and of practical importance, what does moving from the .3 diameter to the .7 diameter do
to the breaking strength? Remember to translate back from the log scale when making
these interpretations.)

(c) Use the reverse Yates algorithm to produce fitted $\ln(y)$ values for a few-effects model corresponding to your answer to (b). Use the fitted values to compute residuals (still on the log scale). Normal-plot these and plot them against levels of each of the three factors and against the fitted values, looking for obvious problems with the few-effects model.

(d) Based on your few-effects model, give a 95% two-sided confidence interval for the mean $\ln(y)$ that would be produced by the abc treatment combination. By exponentiating the endpoints of this interval, give a 95% two-sided confidence interval for the median number of clips required to break a piece of lead under this set of conditions.

3. The following are the weights recorded by $I = 3$ different students when weighing the same nominally 5 g mass with $J = 2$ different scales $m = 2$ times apiece. (They are part of the much larger data set given in Chapter Exercise 5 of Chapter 3.)

	Scale 1	Scale 2
Student 1	5.03, 5.02	5.07, 5.09
Student 2	5.03, 5.01	5.02, 5.07
Student 3	5.06, 5.00	5.10, 5.08

Corresponding fitted factorial effects are: $a_1 = .00417$, $a_2 = -.01583$, $a_3 = .01167$, $b_1 = -.02333$, $b_2 = .02333$, $ab_{11} = -.00417$, $ab_{12} = .00417$, $ab_{21} = .01083$, $ab_{22} = .01083$, $ab_{31} = -.00667$, and $ab_{32} = .00667$. Further, a pooled standard deviation is $s_P = .02483$.

(a) To enhance an interaction plot of sample means with error bars derived from 95% two-sided individual confidence limits for the mean weights, what plus-or-minus value would be used to make those error bars? Make such a plot and discuss the likely statistical detectability of the interactions.

(b) Individual 95% two-sided confidence limits for the interactions $\alpha\beta_{ij}$ are of the form $ab_{ij} \pm \Delta$. Find Δ here. Based on this, are the interactions statistically detectable?

(c) Compare the Student main effects using individual 95% two-sided confidence intervals.

(d) Compare the Student main effects using simultaneous 95% two-sided confidence intervals.

4. The oil viscosity study of Dunnwald, Post, and Kilcoin (referred to in Chapter Exercise 8 of Chapter 7) was actually a 3×4 full factorial study. Some summary statistics for the entire data set are recorded in the accompanying tables. Summarized are $m = 10$ measurements of the viscosities of each of four different weights of three different brands of motor oil at room temperature. Units are seconds required for a ball to drop a particular distance through the oil.

	10W30	SAE 30
Brand M	$\bar{y}_{11} = 1.385$	$\bar{y}_{12} = 2.066$
	$s_{11} = .091$	$s_{12} = .097$
Brand C	$\bar{y}_{21} = 1.319$	$\bar{y}_{22} = 2.002$
	$s_{21} = .088$	$s_{22} = .089$
Brand H	$\bar{y}_{31} = 1.344$	$\bar{y}_{32} = 2.049$
	$s_{31} = .066$	$s_{32} = .089$

	10W40	20W50
Brand M	$\bar{y}_{13} = 1.414$	$\bar{y}_{14} = 4.498$
	$s_{13} = .150$	$s_{14} = .204$
Brand C	$\bar{y}_{23} = 1.415$	$\bar{y}_{24} = 4.662$
	$s_{23} = .115$	$s_{24} = .151$
Brand H	$\bar{y}_{33} = 1.544$	$\bar{y}_{34} = 4.549$
	$s_{33} = .068$	$s_{34} = .171$

(a) Find the pooled sample standard deviation here. What are the associated degrees of freedom?

(b) Make an interaction plot of sample means. Enhance this plot by adding error bars derived from 99% individual confidence intervals for the cell means. Does it appear that there are important and statistically detectable interactions here?

(c) If the Tukey method is used to find simultaneous 95% two-sided confidence intervals for all differences in Brand main effects, the intervals produced are of the form $\bar{y}_{i.} - \bar{y}_{i'.} \pm \Delta$. Find Δ.

(d) If the Tukey method is used to find simultaneous 95% two-sided confidence intervals for all differences in Weight main effects, the intervals produced are of the form $\bar{y}_{.j} - \bar{y}_{.j'} \pm \Delta$. Find Δ.

(e) Based on your answers to (c) and (d), would you say that there are statistically detectable Brand and/or Weight main effects on viscosity?

(f) We strongly suspect that the "$m = 10$" viscosity measurements made for each of the 12 Brand/Weight combinations were made on oil from a single quart of that type of oil. If this is the case, s_P, the baseline measure of variability, is measuring only the variability associated with experimental technique (not, for example, from quart to quart of a given type of oil). One might thus argue that the real-world inferences to be made, properly speaking, extend only to the particular quarts used in the study. Discuss how these interpretations (of s_P and the extent of statistically based inferences) would be different if in fact the students used different quarts of oil in producing the "$m = 10$" different viscosity measurements in each cell.

5. The article "Effect of Temperature on the Early-Age Properties of Type I, Type III and Type I/Fly Ash Concretes" by N. Gardner (*ACI Materials Journal*, 1990) contains summary statistics for a very large study of the properties of several concretes under a variety of curing conditions. The accompanying tables present some of the statistics from that paper. Given here are the sample means and standard deviations of 14-day compressive strengths for $m = 5$ specimens of Type I cement/fly ash concrete for all possible combinations of $I = 2$ water-cement ratios and $J = 4$ curing temperatures. The units are MPa.

	0°C	10°C
.55 Water/Cement Ratio	$\bar{y}_{11} = 28.99$	$\bar{y}_{12} = 30.24$
	$s_{11} = .91$	$s_{12} = 1.26$
.35 Water/Cement Ratio	$\bar{y}_{21} = 38.70$	$\bar{y}_{22} = 36.16$
	$s_{21} = .77$	$s_{22} = 1.92$

	20°C	30°C
.55 Water/Cement Ratio	$\bar{y}_{13} = 33.99$	$\bar{y}_{14} = 36.02$
	$s_{13} = 1.85$	$s_{14} = .93$
.35 Water/Cement Ratio	$\bar{y}_{23} = 40.18$	$\bar{y}_{24} = 42.36$
	$s_{23} = 2.86$	$s_{24} = 1.35$

(a) Find the pooled sample standard deviation here. What are the associated degrees of freedom?

(b) Make an interaction plot of sample means. Enhance this plot by adding error bars derived from simultaneous 95% confidence intervals for the cell means. Does it appear that there are important and statistically detectable interactions here? What practical implications would this have for a cold-climate civil engineer?

(c) Compute the fitted factorial effects from the eight sample means.

(d) If one wished to make individual 95% confidence intervals for the Ratio × Temperature interactions $\alpha\beta_{ij}$, these would be of the form $ab_{ij} \pm \Delta$, for an appropriate value of Δ. Find this Δ. Based on this value, do you judge any of the interactions to be statistically detectable?

6. The same article referred to in Exercise 5 reported summary statistics (similar to the ones for Type I cement/fly ash concrete) for the 14-day compressive strengths of Type III cement concrete. These are shown in the accompanying tables.

	0°C	10°C
.55 Water/Cement Ratio	$\bar{y}_{11} = 47.82$	$\bar{y}_{12} = 42.75$
	$s_{11} = 4.03$	$s_{12} = 2.96$
.35 Water/Cement Ratio	$\bar{y}_{21} = 42.14$	$\bar{y}_{22} = 36.72$
	$s_{21} = 2.64$	$s_{22} = 3.03$

	20°C	30°C
.55 Water/Cement Ratio	$\bar{y}_{13} = 42.38$	$\bar{y}_{14} = 43.45$
	$s_{13} = 2.62$	$s_{14} = 1.80$
.35 Water/Cement Ratio	$\bar{y}_{23} = 36.72$	$\bar{y}_{24} = 37.70$
	$s_{23} = 1.51$	$s_{24} = .89$

(a) Find the pooled sample standard deviation here. What are the associated degrees of freedom? ($m = 5$, as in Exercise 5.)

(b) Make an interaction plot of sample means useful for investigating the size of Ratio $\times$ Temperature interactions. Enhance this plot by adding error bars derived from simultaneous 95% confidence intervals for the cell means. Does it appear that there are important and statistically detectable interactions here? What practical implications would this have for a cold-climate civil engineer?

(c) Compute the fitted factorial effects from the eight sample means.

(d) If one wished to make individual 95% confidence intervals for the Ratio $\times$ Temperature interactions $\alpha\beta_{ij}$, these would be of the form $ab_{ij} \pm \Delta$, for an appropriate value of Δ. Find this Δ. Based on this value, do you judge any of the interactions to be statistically detectable?

(e) Give and interpret a 90% confidence interval for the difference in water/cement ratio main effects, $\alpha_2 - \alpha_1$. How would this be of practical use to a cold-climate civil engineer?

7. Suppose that in the context of Exercises 5 and 6, you judge that for the Type I cement/fly ash concrete there are important Ratio $\times$ Temperature interactions, but that for the Type III cement concrete there are not important Ratio $\times$ Temperature interactions. Taking the whole data set from both exercises together (both concrete types), would there be important (3-factor) Type $\times$ Ratio $\times$ Temperature interactions? Explain.

8. The ISU M.S. thesis, "An Accelerated Engine Test for Crankshaft and Bearing Compatibility," by P. Honan, discusses an industrial experiment run to investigate the effects of three factors on the wear of engine bearings. The factors and levels shown here were used in a 100-hour, 20-step engine probe test.

A	Crankshaft Material	cast nodular iron ($-$) vs. forged steel ($+$)
B	Bearing Material	aluminum ($-$) vs. copper/lead ($+$)
C	Debris Added to Oil	none ($-$) vs. 5.5 g SAE fine dust every 25 hours ($+$)

Two response variables were measured:

$$y_1 = \text{rod journal wear } (\mu m)$$
$$y_2 = \text{main journal wear } (\mu m)$$

The values of y_1 and y_2 reported by Honan are as follows.

Combination	y_1	y_2	Combination	y_1	y_2
(1)	2.7	5.6	c	3.1	3.2
a	.9	1.4	ac	18.6	27.3
b	3.0	7.1	bc	2.5	6.0
ab	1.1	1.6	abc	60.3	99.7

(a) Use the Yates algorithm and compute the fitted effects of the three experimental factors on both the rod and main bearing wear figures.

(b) Because there was no replication in this relatively expensive industrial experiment, there is no real option for judging the statistical significance of the 2^3 factorial effects except the use of normal-plotting. Make normal plots of the seven fitted effects, $a_2, b_2, \ldots, abc_{222}$ for both response variables. Do these identify one or two of the 2^3 factorial effects as clearly larger than the others? How hopeful are you that there is a simple, intuitively appealing few-effects description of the effects of factors A, B, and C on y_1 and y_2?

(c) Your normal plots from (b) ought to each have an interesting gap in the middle of the plot. Explain the origin of both that gap and the fact that all of your fitted effects should be positive, in terms of the relative magnitudes of the responses listed. (How, for example, does the response for combination abc enter into the calculation of the various fitted effects?)

(d) One simple way to describe the outcomes obtained in this study is as having *one very big response and one moderately big response*. Is there much chance that this pattern in y_1 and y_2 is in fact due only to random variation (i.e., that none of the factors have any effect here)? Make a normal plot of the raw y_1 values and one for the raw y_2 values to support your answer.

9. There is a certain degree of arbitrariness in the choice to use signs on the fitted effects corresponding to the "all high treatment" combination when normal-plotting fitted 2^p factorial effects. This can be eliminated by probability plotting the absolute values of the fitted effects and using not standard normal quantiles but rather quantiles for the distribution of the absolute value of a standard normal random variable. This notion is called *half normal-plotting the absolute fitted effects*, since the probability density of the absolute value of a standard normal variable looks like the right half of the standard normal density (multiplied by 2). The half normal quantiles are related to the standard normal quantiles via

$$Q(p) = Q_z \left(\frac{1+p}{2} \right)$$

and one interprets a half normal plot in essentially the same way that a normal plot is interpreted. That is, one thinks of the smaller plotted values as

establishing a pattern of random-looking variation and identifies any of the larger values plotting off a line on the plot established by the small values as detectably larger than the others.

(a) Redo part (a) of Exercise 2 of Section 8.2 using a half normal plot of the absolute values of the fitted effects. (Your ith plotted point will have a horizontal coordinate equal to the ith smallest absolute fitted effect and a vertical coordinate equal to the $p = \frac{i-.5}{15}$ half normal quantile.) Are the conclusions about the statistical detectability of effects here the same as those you reached in Exercise 2 of Section 8.2?

(b) Redo Exercise 1 here using a half normal plot of the absolute values of the fitted effects. (Your ith plotted point will have a horizontal coordinate equal to the ith smallest absolute fitted effect and a vertical coordinate equal to the $p = \frac{i-.5}{7}$ half normal quantile.) Are the conclusions about the statistical detectability of effects here the same as those you reached in Exercise 1?

10. The text *Engineering Statistics* by Hogg and Ledolter contains an account (due originally to R. Snee) of a partially replicated 2^3 factorial industrial experiment. Under investigation were the effects of the following factors and levels on the percentage impurity, y, in a chemical product:

A	Polymer Type	standard $(-)$ vs. new (but expensive) $(+)$
B	Polymer Concentration	.01% $(-)$ vs. .04% $(+)$
C	Amount of an Additive	2 lb $(-)$ vs. 12 lb $(+)$

The data that were obtained are as follows:

Combination	y (%)	Combination	y (%)
(1)	1.0	c	.9, .7
a	1.0, 1.2	ac	1.1
b	.2	bc	.2, .3
ab	.5	abc	.5

(a) Compute the fitted 2^3 factorial effects corresponding to the "all high treatment" combination.

(b) Compute the pooled sample standard deviation, s_P.

(c) Use your value of s_P from (b) and find the plus-or-minus part of 90% individual two-sided confidence limits for the 2^3 factorial effects.

(d) Based on your calculation in (c), which of the effects do you judge to be detectable in this 2^3 study?

(e) Write a paragraph or two for your engineering manager, summarizing the results of this experiment and making recommendations for the future running of this process. (Remember that you want low y and, all else being equal, low production cost.)

11. The article "Use of Factorial Designs in the Development of Lighting Products" by J. Scheesley (*Experiments in Industry: Design, Analysis and Interpretation of Results,* American Society for Quality Control, 1985) discusses a large industrial experiment intended to compare the use of two different types of lead wire in the manufacture of incandescent light bulbs under a variety of plant circumstances. The primary response variable in the study was

y = average number of leads missed per hour (because of misfeeds into automatic assembly equipment)

which was measured and recorded on the basis of eight-hour shifts. Consider here only part of the original data, which may be thought of as having replicated 2^4 factorial structure. That is, consider the following factors and levels:

A	Lead Type	standard $(-)$ vs. new $(+)$
B	Plant	1 $(-)$ vs. 2 $(+)$
C	Machine Type	standard $(-)$ vs. high speed $(+)$
D	Shift	1st $(-)$ vs. 2nd $(+)$

$m = 4$ values of y (each requiring an eight-hour shift to produce) for each combination of levels

of factors A, B, C, and D gave the accompanying $\bar{y}$ and s^2 values.

Combination	$\bar{y}$	s^2	Combination	$\bar{y}$	s^2
(1)	28.4	97.6	d	36.8	146.4
a	21.9	15.1	ad	19.2	24.8
b	20.2	5.1	bd	19.9	5.7
ab	14.3	61.1	abd	22.5	22.5
c	30.4	43.5	cd	25.5	53.4
ac	25.1	96.2	acd	21.5	56.6
bc	38.2	100.8	bcd	22.0	10.4
abc	12.8	23.6	abcd	22.5	123.8

(a) Compute the pooled sample standard deviation. What does it measure in the present context? (Variability in hour-to-hour missed lead counts? Variability in shift-to-shift missed lead per hour figures?)

(b) Use the Yates algorithm and compute the fitted 2^4 factorial effects.

(c) Which of the effects are statistically detectable here? (Use individual two-sided 98% confidence limits for the effects to make this determination.) Is there a simple interpretation of this set of effects?

(d) Would you be willing to say, on the basis of your analysis in (a) through (c), that the new lead type will provide an overall reduction in the number of missed leads? Explain.

(e) Would you be willing to say, on the basis of your analysis in (a) through (c), that a switch to the new lead type will provide a reduction in missed leads for every set of plant circumstances? Explain.

12. DeBlieck, Rohach, Topf, and Wilcox conducted a replicated 3×3 factorial study of the uniaxial force required to buckle household cans. A single brand of cola cans, a single brand of beer cans, and a single brand of soup cans were used in the study. The cans were prepared by bringing them to $0°C$, $22°C$, or $200°C$ before testing. The forces required to buckle each of $m = 3$ cans for the nine different Can Type/Temperature combinations follow.

Can Type	Temperature	Force Required, y (lb)
cola	0°C	174, 306, 192
cola	22°C	150, 188, 125
cola	200°C	200, 198, 204
beer	0°C	234, 246, 300
beer	22°C	204, 339, 254
beer	200°C	414, 200, 286
soup	0°C	570, 704, 632
soup	22°C	667, 593, 647
soup	200°C	600, 620, 596

(a) Make an interaction plot of the nine combination sample means. Enhance it with error bars derived using 98% individual two-sided confidence intervals.

(b) Compute the fitted main effects and interactions from the nine combination sample means. Use these to make individual 98% confidence intervals for all of the main effects and interactions in this 3×3 factorial study. What do these indicate about the detectability of the various effects?

(c) Use Tukey's method for simultaneous comparison of main effects and give simultaneous 99% confidence intervals for all differences in Can Type main effects. Then use the same method and give simultaneous 99% confidence intervals for all differences in Temperature main effects.

13. Consider again the 2^4 factorial data set in Chapter Exercise 20 of Chapter 4. (Paper airplane flight distances collected by K. Fellows were studied there.) As a means of making the evaluation of which of the fitted effects produced by the Yates algorithm appear to be detectable, normal-plot the fitted effects. Interpret the plot.

14. Boston, Franzen, and Hoefer conducted a 2×3 factorial study of the strengths of rubber bands. Two different brands of bands were studied. From both companies, bands of three different widths were used. For each Brand/Width combination, the strengths of $m = 5$ bands of length 18–20 cm were determined by loading the bands till fail-

ure. Some summary statistics from the study are presented in the accompanying table.

		Factor B *Width*		
		1 narrow (<2 mm)	2 medium (3.5 mm)	3 wide (5.5 mm)
Factor A *Brand*	1	$\bar{y}_{11} = 2.811$ kg $s_{11} = .0453$ kg	$\bar{y}_{12} = 4.164$ kg $s_{12} = .2490$ kg	$\bar{y}_{13} = 8.001$ kg $s_{13} = .8556$ kg
	2	$\bar{y}_{21} = 2.459$ kg $s_{21} = .4697$ kg	$\bar{y}_{22} = 4.111$ kg $s_{22} = .1030$ kg	$\bar{y}_{23} = 6.346$ kg $s_{23} = .1924$ kg

(a) Compute s_P for the rubber band strength data. What is this supposed to measure?

(b) Make an interaction plot of sample means. Use error bars for the means calculated from 95% two-sided individual confidence limits. (Make use of your value of s_P.)

(c) Based on your plot from (b), which factorial effects appear to be distinguishable from background noise? (Brand main effects? Width main effects? Brand × Width interactions?)

(d) Compute all of the fitted factorial effects for the strength data. (Find the a_i's, the b_j's, and the ab_{ij}'s defined in Section 4.3.)

(e) To find individual 95% confidence intervals for the interactions $\alpha\beta_{ij}$, intervals of the form

$ab_{ij} \pm \Delta$ are appropriate. Find Δ. Based on this value, are there statistically detectable interactions here? How does this conclusion compare with your more qualitative answer to part (c)?

(f) To compare Width main effects, confidence intervals for the differences $\beta_j - \beta_{j'}$ are in order. Find individual 95% two-sided confidence intervals for $\beta_1 - \beta_2$, $\beta_1 - \beta_3$, and $\beta_2 - \beta_3$. Based on these, are there statistically detectable Width main effects here? How does this compare with your answer to part (c)?

(g) Redo part (f), this time using simultaneous 95% two-sided confidence intervals.

15. In Section 8.3, you were advised to choose $\frac{1}{2}$ fractions of 2^p factorials by using the generator

last factor $\leftrightarrow$ product of all other factors

For example, this means that in choosing $\frac{1}{2}$ of 2^4 possible combinations of levels of factors A, B, C, and D, you were advised to use the generator D $\leftrightarrow$ ABC. There are other possibilities. For example, you could use the generator D $\leftrightarrow$ AB.

(a) Using this alternative plan (specified by D $\leftrightarrow$ AB), what eight different combinations of factor levels would be run? (Use the standard naming convention, listing for each of the eight sets of experimental conditions to be run those factors appearing at their high levels.)

(b) For the alternative plan specified by D $\leftrightarrow$ AB, list all eight pairs of effects of factors A, B, C, and D that would be aliased. (You may, if you wish, list eight sums of the effects $\mu_{....}, \alpha_2, \beta_2, \alpha\beta_{22}, \gamma_2, \ldots$ etc. that can be estimated.)

(c) Suppose that in an analysis of data from an experiment run according to the alternative plan (with D $\leftrightarrow$ AB), the Yates algorithm is used with $\bar{y}$'s listed according to Yates standard order for factors A, B, and C. Give four equally plausible interpretations of the eventuality that the first four lines of the Yates calculations produce large estimated sums of

effects (in comparison to the other four, for example).

(d) Why might it be well argued that the choice D $\leftrightarrow$ ABC is superior to the choice D $\leftrightarrow$ AB?

16. $p = 5$ factors A, B, C, D, and E are to be studied in a 2^{5-2} fractional factorial study. The two generators D $\leftrightarrow$ AB and E $\leftrightarrow$ AC are to be used in choosing the eight ABCDE combinations to be included in the study.

(a) Give the list of eight different combinations of levels of the factors that will be included in the study. (Use the convention of naming, for each sample, those factors that should be set at their high levels.)

(b) Give the list of all effects aliased with the A main effect if this experimental plan is adopted.

17. The following are eight sample means listed in Yates standard order (left to right), considering levels of three two-level factors A, B, and C:

$$70, 61, 72, 59, 68, 64, 69, 69$$

(a) Use the Yates algorithm here to compute eight estimates of effects from the sample means.

(b) Temporarily suppose that no value for s_p is available. Make a plot appropriate to identifying those estimates from (a) that are likely to represent something more than background noise. Based on the appearance of your plot, which if any of the estimated effects are clearly representing something more than background noise?

(c) As it turned out, $s_p = .9$, based on $m = 2$ observations at each of the eight different sets of conditions. Based on 95% individual two-sided confidence intervals for the underlying effects estimated from the eight $\bar{y}$'s, which estimated effects are clearly representing something other than background noise? (If confidence intervals $\hat{E} \pm \Delta$ were to be made, show the calculation of Δ and state which estimated effects are clearly representing more than noise.)

Still considering the eight sample means, henceforth suppose that by some criteria, only the estimates ending up on the first, second, and sixth lines of the Yates calculations are considered to be both statistically detectable and of practical importance.

(d) If in fact the eight $\bar{y}$'s came from a (4-factor) 2^{4-1} experiment with generator $D \leftrightarrow ABC$, how would one typically interpret the result that the first, second, and sixth lines of the Yates calculations (for means in standard order for factors A, B, and C) give statistically detectable and practically important values?

(e) If in fact the eight $\bar{y}$'s came from a (5-factor) 2^{5-2} experiment with generators $D \leftrightarrow ABC$ and $E \leftrightarrow AC$, how would one typically interpret the result that the first, second, and sixth lines of the Yates calculations (for means in standard order for factors A, B, and C) give statistically detectable and practically important values?

18. A production engineer who wishes to study six two-level factors in eight experimental runs decides to use the generators $D \leftrightarrow AB$, $E \leftrightarrow AC$, and $F \leftrightarrow BC$ in planning a 2^{6-3} fractional factorial experiment.

(a) What eight combinations of levels of the six factors will be run? (Name them using the usual convention of prescribing for each run which of the factors will appear at their high levels.)

(b) What seven other effects will be aliased with the A main effect in the engineer's study?

19. The article "Going Beyond Main-Effect Plots" by Kenett and Vogel (*Quality Progress*, 1991) outlines the results of a 2^{5-1} fractional factorial industrial experiment concerned with the improvement of the operation of a wave soldering machine. The effects of the five factors Conveyor Speed (A), Preheat Temperature (B), Solder Temperature (C), Conveyor Angle (D), and Flux Concentration (E) on the variable

$y =$ the number of faults per 100 solder joints (computed from inspection of 12 circuit boards)

were studied. (The actual levels of the factors employed were not given in the article.) The combinations studied and the values of y that resulted are given next.

Combination	y	Combination	y
(1)	.037	de	.351
a	.040	ade	.360
b	.014	bde	.329
ab	.042	abde	.173
ce	.063	cd	.372
ace	.100	acd	.184
bce	.067	bcd	.158
abce	.026	abcd	.131

Kenett and Vogel were apparently called in after the fact of experimentation to help analyze this nonstandard $\frac{1}{2}$ fraction of the full 2^5 factorial. The recommendations of Section 8.3 were not followed in choosing which 16 of the 32 possible combinations of levels of factors A through E to include in the wave soldering study. In fact, the generator $E \leftrightarrow -CD$ was apparently employed.

(a) Verify that the combinations listed above are in fact those prescribed by the relationship $E \leftrightarrow -CD$. (For example, with all of A through D at their low levels, note that the low level of E is indicated by multiplying minus signs for C and D by another minus sign. Thus, combination (1) is one of the 16 prescribed by the generator.)

(b) Write the defining relation for the experiment. What is the resolution of the design chosen by the authors? What resolution does the standard choice of $\frac{1}{2}$ fraction provide? Unless there were some unspecified extenuating circumstances that dictated the choice of $\frac{1}{2}$ fraction, why does it seem to be an unwise one?

(c) Write out the 16 different differences of effects that can be estimated based on the data given. (For example, one of these is $\mu_{....} - \gamma\delta\epsilon_{222}$, another is $\alpha_2 - \alpha\gamma\delta\epsilon_{2222}$, etc.)

(d) Notice that the combinations listed here are in Yates standard order as regards levels of factors A through D. Use the four-cycle Yates algorithm and find the fitted differences of effects. Normal-plot these and identify any statistically detectable differences. Notice that by virtue of the choice of $\frac{1}{2}$ fraction made by the engineers, the most obviously statistically significant difference is that of a main effect and a 2-factor interaction.

20. The article "Robust Design: A Cost-Effective Method for Improving Manufacturing Processes" by Kacker and Shoemaker (*AT&T Technical Journal*, 1986) discusses the use of a 2^{8-4} fractional factorial experiment in the improvement of the performance of a step in an integrated circuit fabrication process. The initial step in fabricating silicon wafers for IC devices is to grow an epitaxial layer of sufficient (and, ideally, uniform) thickness on polished wafers. The engineers involved in running this part of the production process considered the effects of eight factors (listed in the accompanying table) on the properties of the deposited epitaxial layer.

Factor A	Arsenic Flow Rate	55% (−) vs. 59% (+)
Factor B	Deposition Temperature	1210°C (−) vs. 1220°C (+)
Factor C	Code of Wafers	668G4 (−) vs. 678G4 (+)
Factor D	Susceptor Rotation	continuous (−) vs. oscillating (+)
Factor E	Deposition Time	high (−) vs. low (+)
Factor F	HC1 Etch Temperature	1180°C (−) vs. 1215°C (+)
Factor G	HC1 Flow Rate	10% (−) vs. 14% (+)
Factor H	Nozzle Position	2 (−) vs. 6 (+)

A batch of 14 wafers is processed at one time, and the experimenters measured thickness at five locations on each of the wafers processed during one experimental run. These $14 \times 5 = 70$ measurements from each run of the process were then reduced to two response variables:

y_1 = the mean of the 70 thickness measurements

y_2 = the logarithm of the variance of the 70 thickness measurements

y_2 is a measure of uniformity of the epitaxial thickness, and y_1 is (clearly) a measure of the magnitude of the thickness. The authors reported results from the experiment as shown in the accompanying table.

Combination	y_1 (μm)	y_2
(1)	14.821	−.4425
afgh	14.888	−1.1989
begh	14.037	−1.4307
abef	13.880	−.6505
cefh	14.165	−1.4230
aceg	13.860	−.4969
bcfg	14.757	−.3267
abch	14.921	−.6270
defg	13.972	−.3467
adeh	14.032	−.8563
bdfh	14.843	−.4369
abdg	14.415	−.3131
cdgh	14.878	−.6154
acdf	14.932	−.2292
bcde	13.907	−.1190
abcdefgh	13.914	−.8625

It is possible to verify that the combinations listed here come from the use of the four generators E $\leftrightarrow$ BCD, F $\leftrightarrow$ ACD, G $\leftrightarrow$ ABD, and H $\leftrightarrow$ ABC.

(a) Write out the whole defining relation for this experiment. (The grand mean will have 15 aliases.) What is the resolution of the design?

(b) Consider first the response y_2, the measure of uniformity of the epitaxial layer. Use the Yates algorithm and normal- and/or half normal-plotting (see Exercise 9) to identify statistically detectable fitted sums of effects. Suppose that only the two largest (in magnitude) of these are judged to be both statistically significant and of practical importance. What is suggested about how levels of the factors might henceforth be set in order to minimize y_2? From the limited description of the process above, does it appear that these settings require any extra manufacturing expense?

(c) Turn now to the response y_1. Again use the Yates algorithm and normal- and/or half normal-plotting to identify statistically detectable sums of effects. Which of the factors seems to be most important in determining the average epitaxial thickness? In fact, the target thickness for this deposition process was 14.5 μm. Does it appear that by appropriately choosing a level of this variable it may be possible to get the mean thickness on target? Explain. (As it turns out, the thought process outlined here allowed the engineers to significantly reduce the variability in epitaxial thickness while getting the mean on target, improving on previously standard process operating methods.)

21. Arndt, Cahill, and Hovey worked with a plastics manufacturer and experimented on an extrusion process. They conducted a 2^{6-2} fractional factorial study with some partial "replication" (the reason for the quote marks will be discussed later). The experimental factors in their study were as follows:

Factor A	Bulk Density, a measure of the weight per unit volume of the raw material used
Factor B	Moisture, the amount of water added to the raw material mix
Factor C	Crammer Current, the amperage supplied to the crammer-auger
Factor D	Extruder Screw Speed
Factor E	Front-End Temperature, a temperature controlled by heaters on the front end of the extruder
Factor F	Back-End Temperature, a temperature controlled by heaters on the back end of the extruder

Physically low and high levels of these factors were identified. Using the two generators E $\leftrightarrow$ AB and F $\leftrightarrow$ AC, 16 different combinations of levels of the factors were chosen for inclusion in a plant experiment, where the response of primary interest was the output of the extrusion process in terms of pounds of useful product per hour. A coded version of the data the students obtained is given in the accompanying table. (The data have been rescaled by subtracting a particular value and

dividing by another so as to disguise the original responses without destroying their basic structure. You may think of these values as output measured in numbers of some undisclosed units above an undisclosed baseline value.)

Combination	y
ef	13.99
a	6.76
bf	20.71
abe	11.11, 11.13
ce	19.61
acf	15.73
bc	23.45
abcef	20.00
def	24.94
ad	24.03, 25.03
bdf	24.97
abde	24.29
cde	24.94, 25.21
acdf	24.32, 24.48
bcd	30.00
abcdef	33.08

(a) The students who planned this experiment hadn't been exposed to the concept of design resolution. What does Table 8.35 indicate is the best possible resolution for a 2^{6-2} fractional factorial experiment? What is the resolution of the one that the students planned? Why would they have been better off with a different plan than the one specified by the generators E$\leftrightarrow$ AB and F $\leftrightarrow$ AC?

(b) Find a choice of generators E $\leftrightarrow$ (some product of letters A through D) and F $\leftrightarrow$ (some other product of letters A through D) that provides maximum resolution for a 2^{6-2} experiment.

(c) The combinations here are listed in Yates standard order as regards factors A through D. Compute $\bar{y}$'s and then use the (four-cycle) Yates algorithm and compute 16 estimated sums of 2^6 factorial effects.

(d) When the extrusion process is operating, many pieces of product can be produced in an hour, but the entire data collection process leading to the data here took over eight hours. (Note, for example, that changing temperatures on industrial equipment requires time for parts to heat up or cool down, changing formulas of raw material means that one must let one batch clear the system, etc.) The repeat observations above were obtained from *two consecutive pieces of product*, made minutes apart, without any change in the extruder setup in between their manufacture. With this in mind, discuss why a pooled standard deviation based on these four "samples of size 2" is quite likely to underrepresent the level of "baseline" variability in the output of this process under a fixed combination of levels of factors A through F. Argue that it would have been extremely valuable to have (for example) rerun one or more of the combinations tested early in the study again late in the study.

(e) Use the pooled sample standard deviation from the repeat observations and compute (using the $p = 4$ version of formula (8.12) in Section 8.2) the plus-or-minus part of 90% two-sided confidence limits for the 16 sums of effects estimated in part (c), acting as if the value of s_p were a legitimate estimate of background variability. Which sums of effects are statistically detectable by this standard? How do you interpret this in light of the information in part (d)?

(f) As an alternative to the analysis in part (e), make a normal plot of the last 15 of the 16 estimated sums of effects you computed in part (c). Which sums of effects appear to be statistically detectable? What is the simplest interpretation of your findings in the context of the industrial problem? (What has been learned about how to run the extruding process?)

(g) Briefly discuss where to go from here if it is your job to optimize the extrusion process (maximize y). What data would you collect next, and what would you be planning to do with them?

22. The article "The Successful Use of the Taguchi Method to Increase Manufacturing Process Capability" by S. Shina (*Quality Engineering*, 1991) discusses the use of a 2^{8-3} fractional factorial experiment to improve the operation of a wave soldering process for through-hole printed circuit boards. The experimental factors and levels studied were as shown in the accompanying table.

Factor A	Preheat Temperature	180° (−) vs. 220° (+)
Factor B	Solder Wave height	.250 (−) vs. .400 (+)
Factor C	Wave Temperature	490° (−) vs. 510° (+)
Factor D	Conveyor Angle	5.0 (−) vs. 6.1 (+)
Factor E	Flux Type	A857 (−) vs. K192 (+)
Factor F	Direction of Boards	0 (−) vs. 90 (+)
Factor G	Wave Width	2.25 (−) vs. 3.00 (+)
Factor H	Conveyor Speed	3.5 (−) vs. 6.0 (+)

The generators $F \leftrightarrow -CD$, $G \leftrightarrow -AD$, and $H \leftrightarrow -ABCD$ were used to pick 32 different combinations of levels of these factors to run. For each combination, four special test printed circuit boards were soldered, and the lead shorts per board, y_1, and touch shorts per board, y_2, were counted, giving the accompanying data. (The data here and on page 644 are exactly as given in the article, and we have no explanation for the fact that some of the numbers do not seem to have come from division of a raw count by 4.)

Combination	y_1	y_2
(1)	6.00	13.00
agh	10.00	26.00
bh	10.00	12.00
abg	8.50	14.00
cfh	1.50	18.75
acfg	.25	16.25
bcf	1.75	25.75
abcfgh	4.25	18.50
dfgh	6.50	6.50

(*continued*)

Combination	y_1	y_2
adf	.75	.00
bdfg	3.50	1.00
abdfh	3.25	6.50
cdg	6.00	7.25
acdh	9.50	11.25
bcdgh	6.25	10.00
abcd	6.75	12.50
e	20.00	29.25
aegh	16.50	31.25
beh	17.25	28.75
abeg	19.50	41.25
cefh	9.67	21.33
acefg	2.00	10.75
bcef	5.67	28.67
abcefgh	3.75	35.75
defgh	6.00	22.70
adef	7.30	25.70
bdefg	8.70	30.00
abdefh	9.00	29.70
cdeg	19.30	32.70
acdeh	26.70	25.70
bcdegh	17.70	45.30
abcde	10.30	37.00

(a) Verify that the 32 combinations of levels of the factors A through H listed here are those that are prescribed by the choice of generators. (For each combination of levels of the factors A through E, determine what levels of F, G, and H are prescribed by the generators and check that such a combination is listed.)

(b) Use the generators given here and write out the whole defining relation for this study. (You will end with I aliased with seven other strings of letters.) What is the resolution of the design used in this study? According to Table 8.35, what was possible in terms of resolution for a 2^{8-3} study? Could the engineers in charge here have done better at containing the ambiguity that unavoidably follows from use of a fractional factorial study?

(c) Note that the 32 combinations of the 8 factors above are listed in Yates standard order as regards Factors A through E (ignoring F, G, and H). By some means (using a statistical analysis package like MINITAB, implementing spreadsheet calculations, or doing the 5-cycle Yates algorithm "by hand") find the estimated sums of effects for the response y_1. Normal-plot the last 31 of these. You should find that the largest of these would be the CD 2-factor interaction, the E main effect, and the CDE 3-factor interaction if only 5 factors were involved (instead of 8). These are all positive and clearly larger in magnitude than the other estimates. If possible, give a simple interpretation of this in light of the alias structure specified by the defining relation you found in part (b).

(d) Now find and normal-plot the estimated sums of effects for the response y_2. (Normal-plot 31 estimates.) You should find the estimate corresponding to the E main effect plus aliases to be positive, larger in magnitude than the rest, and detectably nonzero.

(e) In light of your answers to (c) and (d), the signs of the fitted linear combinations of effects, and a desire to reduce both y_1 and y_2 to the minimum values possible, what combination of levels of the factors do you tentatively recommend here? Is the combination of levels that you see as promising one that is among the 32 tested? If it is not, how would you recommend proceeding in the real manufacturing scenario? (Would you, for example, order that any permanent process changes necessary to the use of your promising combination be adopted immediately?)

The original article reported a decrease in solder defects by nearly a factor of 10 in this process as a result of what was learned from this experiment.

23. In the situation of Exercise 22, the 32 different combinations of levels of factors A through H were run in the order listed. In fact, the first 16 runs were made by one shift of workers, and the last 16 were made by a second shift.

(a) In light of the material in Chapter 2 on experiment planning and the formal notion of confounding, what risk of a serious logical flaw did the engineers run in the execution of their experiment? (How would possible shift-to-shift differences show up in the data from an experiment run like this? One of the main things learned from the experiment was that factor E was very important. Did the engineers run the risk of clouding their view of this important fact?) Explain.

(b) Devise an alternative plan that could have been used to collect data in the situation of Exercise 22 without completely confounding the effects of Flux and Shift. Continue to use the 32 combinations of the original factors listed in Exercise 22, but give a better assignment of 16 of them to each shift. (*Hint:* Think of Shift as a ninth factor, pick a sensible generator, and use it to put half of the 32 combinations in each shift. There are a variety of possibilities here.)

(c) Discuss in qualitative terms how you would do data analysis if your suggestion in (b) were to be followed.

24. The article "Computer Control of a Butane Hydrogenolysis Reactor" by Tremblay and Wright (*The Canadian Journal of Chemical Engineering*, 1974) contains an interesting data set concerned with the effects of $p = 3$ process variables on the performance of a chemical reactor. The factors and their levels were as follows:

Factor A	Total Feed Flow (cc/sec at STP)	50 (−) vs. 180 (+)
Factor B	Reactor Wall Temperature (°F)	470 (−) vs. 520 (+)
Factor C	Feed Ratio (Hydrogen/Butane)	4 (−) vs. 8 (+)

The data had to be collected over a four-day period, and two combinations of the levels of factors A, B, and C above were run each day along with a center point—a data point with Total Feed Flow 115, Reactor Wall Temperature 495, and

Feed Ratio 6. The response variable was

$$y = \text{percent conversion of butane}$$

and the data in the accompanying table were collected.

Day	Feed Flow	Wall Temp.	Feed Ratio	Combination	y
1	115	495	6	—	78
1	50	470	4	(1)	99
1	180	520	8	abc	87
2	50	520	4	b	98
2	180	470	8	ac	18
2	115	495	6	—	87
3	50	520	8	bc	95
3	180	470	4	a	59
3	115	495	6	—	90
4	50	470	8	c	76
4	180	520	4	ab	92
4	115	495	6	—	89

(a) Suppose that to begin with, you ignore the fact that these data were collected over a period of four days and simply treat the data as a complete 2^3 factorial augmented with a repeated center point. Analyze these data using the methods of this chapter. (Compute s_p from four center points. Use the Yates algorithm and the eight corner points to compute fitted 2^3 factorial effects. Then judge the statistical significance of these using appropriate 95% two-sided confidence limits based on s_p.) Is any simple interpretation of the experimental results in terms of factorial effects obvious? According to the authors, there was the possibility of "process drift" during the period of experimentation. The one-per-day center points were added to the 2^3 factorial at least in part to provide some check on that possibility, and the allocation of two ABC combinations to each day was very carefully done in order to try to minimize the possible confounding introduced by any Day/Block effects. The rest of this problem considers analyses that might be performed on the experimenters' data in recognition of the possibility of process drift.

(b) Plot the four center points against the number of the day on which they were collected. What possibility is at least suggested by your plot? Would the plot be particularly troubling if your experience with this reactor told you that a standard deviation of around 5(%) was to be expected for values of y from consecutive runs of the reactor under fixed operating conditions on a given day? Would the plot be troubling if your experience with this reactor told you that a standard deviation of around 1(%) was to be expected for values of y from consecutive runs of the reactor under fixed operating conditions on a given day?

(c) The four-level factor Day can be formally thought of in terms of two extra two-level factors—say, D and E. Consider the choice of generators D $\leftrightarrow$ AB and E $\leftrightarrow$ BC for a 2^{5-2} fractional factorial. Verify that the eight combinations of levels of A through E prescribed by these generators divide the eight possible combinations of levels of A through C up into the four groups of two corresponding to the four days of experimentation. (To begin with, note that both A low, B low, C low and A high, B high, C high correspond to D high and E high. That is, the first level of Day can be thought of as the D high and E high combination.)

(d) The choice of generators in (c) produces the defining relation I $\leftrightarrow$ ABD $\leftrightarrow$ BCE $\leftrightarrow$ ACDE. Write out, on the basis of this defining relation, the list of eight groups of aliased 2^5 factorial effects. Any effect involving factors A, B, or C with either of the letters D (δ) or E (ϵ) in its name represents some kind of interaction with Days. Explain what it means for there to be no interactions with Days. Make out a list of eight smaller groups of aliased effects that are appropriate supposing that there are no interactions with Days.

(e) Allowing for the possibility of Day (Block) effects, it does not make sense to use the center points to compute s_P. However, one might

normal-plot (or half normal-plot) the fitted effects from (a). Do so. Interpret your plot, supposing that there were no interactions with Days in the reactor study. How do your conclusions differ (if at all) from those in (a)?

(f) One possible way of dealing with the possibility of Day effects in this particular study is to use the center point on each day as a sort of baseline and express each other response as a deviation from that baseline. (If on day i there is a Day effect γ_i, and on day i the mean response for any combination of levels of factors A through C is $\mu_{comb} + \gamma_i$, the mean of the difference $y_{comb} - y_{center}$ is $\mu_{comb} - \mu_{center}$; one can therefore hope to see 2^3 factorial effects uncontaminated by additive Day effects using such differences in place of the original responses.) For each of the four days, subtract the response at the center point from the other two responses and apply the Yates algorithm to the eight differences. Normal-plot the fitted effects on the (difference from the center point mean) response. Is there any substantial difference between the result of this analysis and that for the others suggested in this problem?

25. The article "Including Residual Analysis in Designed Experiments: Case Studies" by W. H. Collins and C. B. Collins (*Quality Engineering*, 1994) contains discussions of several machining experiments concerned with surface finish. Given here are the factors and levels studied in (part of) one of those experiments on a particular lathe.

	Factor	Levels
A	Speed	2500 RPM $(-)$ vs. 4500 RPM $(+)$
B	Feed	.003 in/rev $(-)$ vs. .009 in/rev $(+)$
C	Tool Condition	New $(-)$ vs. Used (after 250 parts) $(+)$

$m = 2$ parts were turned on the lathe for each of the 2^3 different combinations of levels of the 3 factors, and surface finish measurements, y, were made on these. (y is a measurement of the vertical distance traveled by a probe as it moves horizontally across a particular 1 inch section of the part.) Next are some summary statistics from the experiment.

Combination	$\bar{y}$	s	Combination	$\bar{y}$	s
(1)	33.0	0.0	c	35.5	6.4
a	45.5	7.8	ac	44.0	7.1
b	222.5	4.9	bc	216.5	6.4
ab	241.5	4.9	abc	216.5	0.7

(a) Find s_P and its degrees of freedom. What does this quantity intend to measure?

(b) 95% individual two-sided confidence limits for the mean surface finish measurement for a part turned under a given set of conditions are of the form $\bar{y}_{ijk} \pm \Delta$. Based on the value of s_P found above, find Δ.

(c) Would you say that the mean surface finish measurements for parts of types "(1)" and "a" are detectably different? Why or why not? (Show appropriate calculations.)

(d) 95% individual two-sided individual confidence limits for the 2^3 factorial effects in this study are of the form $\hat{E} \pm \Delta$. Find Δ.

(e) Compute the 2^3 factorial fitted effects for the "all high" combination (abc).

(f) Based on your answers to parts (d) and (e), which of the main effects and/or interactions do you judge to be statistically detectable? Explain.

(g) Give the practical implications of your answer to part (f). (How do you suggest running the lathe if small y and minimum machining cost are desirable?)

(h) Suppose you were to judge only the B main effect to be both statistically detectable and of practical importance in this study. What surface finish value would you then predict for a part made at a 2500 RPM speed and a .009 in/rev feed rate using a new tool?

26. Below are 2^4 factorial data for two response variables taken from the article "Chemical Vapor Deposition of Tungsten Step Coverage and Thickness Uniformity Experiments" by J. Chang (*Thin Solid Films*, 1992). The experiment concerned the blanket chemical vapor deposition of tungsten in the manufacture of integrated circuit chips. The factors studied were as follows:

A	Chamber Pressure	8 ($-$) vs. 9 ($+$)
B	H_2 Flow	500 ($-$) vs. 1000 ($+$)
C	SiH_4 Flow	15 ($-$) vs. 25 ($+$)
D	WF_6 Flow	50 ($-$) vs. 60 ($+$)

The pressure is measured in Torr and the flows are measured in standard cm^3/min. The response variable y_1 is the "percent step coverage," 100 times the ratio of tungsten film thickness at the top of the side wall to the bottom of the side wall (large is good). The response variable y_2 is an "average sheet resistance" (measured in $m\Omega$).

Combination	y_1	y_2	Combination	y_1	y_2
(1)	73	646	d	83	666
a	60	623	ad	80	597
b	77	714	bd	100	718
ab	90	643	abd	85	661
c	67	360	cd	77	304
ac	78	359	acd	90	309
bc	100	335	bcd	70	360
abc	77	318	abcd	75	318

(a) Make a normal plot of the 15 fitted effects $a_2, b_2, \ldots, abcd_{2222}$ as a means of judging the statistical detectability of the effects on the response, y_1. Interpret this plot and say what is indicated about producing good "percent step coverage."

(b) Repeat part (a) for the response variable y_2.

Now suppose that instead of a full factorial study, only the half fraction with defining relation $D \leftrightarrow ABC$ had been conducted.

(c) Which 8 of the 16 treatment combinations would have been run? List these combinations in Yates standard order as regards factors A, B, and C and use the (3-cycle Yates algorithm) to compute the 8 estimated sums of effects that it is possible to derive from these 8 treatment combinations for response y_2. Verify that each of these 8 estimates is the sum of two of your fitted effects from part (b). (For example, you should find that the first estimated sum here is $\bar{y}_{....} + abcd_{2222}$ from part (b).)

(d) Normal-plot the last 7 of the estimated sums from (c). Interpret this plot. If you had only the data from this 2^{4-1} fractional factorial, would your subject-matter conclusions be the same as those reached in part (b), based on the full 2^4 data set?

27. An engineer wishes to study seven experimental factors, A, B, C, D, E, F and G, each at 2 levels, using only 16 combinations of factor levels. He plans initially to use generators $E \leftrightarrow ABCD$, $F \leftrightarrow ABC$, and $G \leftrightarrow BCD$.

(a) With this initial choice of generators, what 16 combinations of levels of the seven factors will be run?

(b) In a 2^{7-3} fractional factorial, each effect is aliased with 7 other effects. Starting from the engineer's choice of generators, find the defining relation for his study. (You will need not only to consider products of pairs but also a product of a triple.)

(c) An alternative choice of generators is $E \leftrightarrow ABC$, $F \leftrightarrow BCD$, $G \leftrightarrow ABD$. This choice yields the defining relation

$$I \leftrightarrow ABCE \leftrightarrow BCDF \leftrightarrow ABDG$$

$$\leftrightarrow ADEF \leftrightarrow CDEG \leftrightarrow ACFG \leftrightarrow BEFG$$

Which is preferable, the defining relation in part (b), or the one here? Why?

28. The article "Establishing Optimum Process Levels of Suspending Agents for a Suspension Product" by A. Gupta (*Quality Engineering*, 1997–1998) discussed an unreplicated fractional factorial experiment. The experimental factors and their levels in the study were:

A	Method of Preparation	Usual (−) vs. Modified (+)
B	Sugar Content	50% (−) vs. 60% (+)
C	Antibiotic Level	8% (−) vs. 16% (+)
D	Aerosol	.4% (−) vs. .6% (+)
E	CMC	.2% (−) vs. .4% (+)

The response variable was

$$y = \text{separated clear volume (\%)}$$
$$\text{for a suspension of antibiotic after 45 days}$$

and the manufacturer hoped to find a way to make y small. The experimenters failed to follow the recommendation in Section 8.3 for choosing a best half fraction of the factorial and used the generator $E \leftrightarrow ABC$ (instead of the better one $E \leftrightarrow ABCD$).

(a) In what sense was the experimental plan used in the study inferior to the one prescribed in Section 8.3? (How is the one from Section 8.3 "better"?)

The Yates algorithm applied to the 16 responses given in the paper produced the 16 fitted sums of effects:

mean + *alias* = 37.563	D + *alias* = −7.437
A + *alias* = .187	AD + *alias* = .937
B + *alias* = 2.437	BD + *alias* = .678
AB + *alias* = .312	ABD + *alias* = .812
C + *alias* = −1.062	CD + *alias* = 1.438
AC + *alias* = .312	ACD + *alias* = .062
BC + *alias* = −1.187	BCD + *alias* = .062
ABC + *alias* = −2.063	ABCD + *alias* = −.062

(b) Make a normal plot of the last 15 of these fitted sums.

(c) If you had to guess (based on the results of this experiment) the order of the magnitudes of the five main effects (A, B, C, D and E) from smallest to largest, what would you guess? Explain.

(d) Based on the normal plot in (b), which sums of effects do you judge to be statistically detectable? Explain.

(e) Based on your answers to (c) and (d), how do you suggest that suspensions of this antibiotic be made in order to produce small y? What mean y do you predict if your recommendations are followed?

(f) Actually, the company that ran this study planned to make suspensions using both high and low levels of antibiotic (factor C). Does your answer to (d) suggest that the company needs to use different product formulations for the two levels of antibiotic? Explain.

29. The paper "Achieving a Target Value for a Manufacturing Process," by Eibl, Kess, and Pukelsheim (*Journal of Quality Technology*, 1992) describes a series of experiments intended to guide the adjustment of a paint coating process. The first of these was a 2^{6-3} fractional factorial study. The experimental factors studied were as follows (exact levels of these factors are not given in the paper, presumably due to corporate security considerations):

A	Tube Height	low ($-$) vs. high ($+$)
B	Tube Width	low ($-$) vs. high ($+$)
C	Paint Viscosity	low ($-$) vs. high ($+$)
D	Belt Speed	low ($-$) vs. high ($+$)
E	Pump Pressure	low ($-$) vs. high ($+$)
F	Heating Temperature	low ($-$) vs. high ($+$)

The response variable was a paint coating thickness measurement, y, whose units are mm. $m = 4$ workpieces were painted and measured for each of the $r = 8$ combinations of levels of the factors studied. The $r = 8$ samples of size $m = 4$ produced a value of $s_p = .118$ mm.

(a) Suppose that you wish to attach a precision to one of the $r = 8$ sample means obtained in this study. This can be done using 95% two-sided confidence limits of the form $\bar{y} \pm \Delta$. Find Δ.

(b) Following are the mean thicknesses measured for the combinations studied, listed in Yates standard order as regards levels of factors A, B, and C. Use the Yates algorithm and find eight estimated (sums of) effects.

A	B	C	$\bar{y}$
$-$	$-$	$-$	.98
$+$	$-$	$-$	1.58
$-$	$+$	$-$	1.13
$+$	$+$	$-$	1.74
$-$	$-$	$+$	1.49
$+$	$-$	$+$	.84
$-$	$+$	$+$	2.18
$+$	$+$	$+$	1.45

(c) Two-sided confidence limits based on the estimated (sums of) effects calculated in part (b) are of the form $\hat{E} \pm \Delta$. Find Δ if (individual) 95% confidence is desired.

(d) Based on your answer to (c), list those estimates from part (b) that represent statistically detectable (sums of) effects.

In fact, the experimental plan used by the investigators had generators D $\leftrightarrow$ AC, E $\leftrightarrow$ BC, and F $\leftrightarrow$ ABC.

(e) Specify the combinations (of levels of the experimental factors A, B, C, D, E and F) that were included in the experiment.

(f) Write out the whole defining relation for this study. (You will need to consider here not only products of pairs but a product of a triple as well. The grand mean is aliased with seven other effects.)

(g) In light of your answers to part (d) and the aliasing pattern here, what is the simplest possible potential interpretation of the results of this experiment?

9

●●●●●●●●●●●●●●●●●●●●●●●

Regression Analysis—Inference for Curve- and Surface-Fitting

The two previous chapters began a study of inference methods for multisample studies by considering first those which make no explicit use of structure relating several samples and then discussing some directed at the analysis of factorial structure. The discussion in this chapter will primarily consider inference methods for multisample studies where factors involved are inherently quantitative and it is reasonable to believe that some approximate functional relationship holds between the values of the system/input/independent variables and observed system responses. That is, this chapter introduces and applies inference methods for the curve- and surface-fitting contexts discussed in Sections 4.1 and 4.2.

The chapter begins with a discussion of the simplest situation of this type—namely, where a response variable y is approximately linearly related to a single quantitative input variable x. In this specific context, it is possible to give explicit formulas and illustrate in concrete terms what is possible in the way of inference methods for surface-fitting analyses. The second section then treats the general problem of statistical inferences in multiple regression (curve- and surface-fitting) analyses. In the general case, it is not expedient to produce many computational formulas. So the exposition relies instead on summary measures commonly appearing on multiple regression printouts from statistical packages. A final section further illustrates the broad utility of the multiple regression methods by applying them to "response surface," and then factorial, analyses.

9.1 Inference Methods Related to the Least Squares Fitting of a Line (Simple Linear Regression)

This section considers inference methods that are applicable where a response y is approximately linearly related to an input/system variable x. It begins by introducing the (normal) simple linear regression model and discussing how to estimate response variance in this context. Next there is a look at standardized residuals. Then inference for the rate of change $(\Delta y / \Delta x)$ is considered, along with inference for the average response at a given x. There follows a discussion of prediction and tolerance intervals for responses at a given setting of x. Next is an exposition of ANOVA ideas in the present situation. The section then closes with an illustration of how statistical software expedites the calculations introduced in the section.

9.1.1 The Simple Linear Regression Model, Corresponding Variance Estimate, and Standardized Residuals

Chapter 7 introduced the one-way (equal variances, normal distributions) model as the most common probability basis of inference methods for multisample studies. It was represented in symbols as

$$y_{ij} = \mu_i + \epsilon_{ij} \tag{9.1}$$

where the means $\mu_1, \mu_2, \ldots, \mu_r$ were treated as r unrestricted parameters. In Chapter 8, it was convenient (for example) to rewrite equation (9.1) in two-way contexts as

$$y_{ijk} = \mu_{ij} + \epsilon_{ijk} \qquad (= \mu_{..} + \alpha_i + \beta_j + \alpha\beta_{ij} + \epsilon_{ijk}) \tag{9.2}$$

where the μ_{ij} are still unrestricted, and to consider restrictions/simplifications of model (9.2) such as

$$y_{ijk} = \mu_{..} + \alpha_i + \beta_j + \epsilon_{ijk} \tag{9.3}$$

Model (9.3) really differs from model (9.2) or (9.1) only in the fact that it postulates a special form or restriction for the means μ_{ij}. Expression (9.3) says that the means must satisfy a parallelism relationship.

Turning now to the matter of inference based on data pairs $(x_1, y_1), (x_2, y_2), \ldots, (x_n, y_n)$ exhibiting an approximately linear scatterplot, one once again proceeds by imposing a restriction on the one-way model (9.1). In words, the model assumptions will be that there are underlying normal distributions for the response y with a common variance σ^2 but means $\mu_{y|x}$ that change linearly in x. In symbols, it is typical to write that for $i = 1, 2, \ldots, n$,

$$y_i = \beta_0 + \beta_1 x_i + \epsilon_i \tag{9.4}$$

where the ϵ_i are (unobservable) iid normal $(0, \sigma^2)$ random variables, the x_i are known constants, and β_0, β_1, and σ^2 are unknown model parameters (fixed constants). Model (9.4) is commonly known as **the (normal) simple linear regression model**. If one thinks of the different values of x in an (x, y) data set as separating it into various samples of y's, expression (9.4) is the specialization of model (9.1) where the (previously unrestricted) means of y satisfy the linear relationship $\mu_{y|x} = \beta_0 + \beta_1 x$. Figure 9.1 is a pictorial representation of the "constant variance, normal, linear (in x) mean" model.

Inferences about quantities involving those x values represented in the data (like the mean response at a single x or the difference between mean responses at two different values of x) will typically be sharper when methods based on model (9.4) can be used in place of the general methods of Chapter 7. And to the extent that model (9.4) describes system behavior for values of x not included in the data, a model like (9.4) provides for inferences involving limited interpolation and extrapolation on x.

Section 4.1 contains an extensive discussion of the use of least squares in the fitting of the approximately linear relation

$$y \approx \beta_0 + \beta_1 x \tag{9.5}$$

to a set of (x, y) data. Rather than redoing that discussion, it is most sensible simply to observe that Section 4.1 can be thought of as an exposition of fitting and the use of residuals in model checking for the simple linear regression model (9.4). In

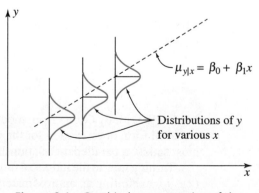

$\mu_{y|x} = \beta_0 + \beta_1 x$

Distributions of y
for various x

Figure 9.1 Graphical representation of the
simple linear regression model

particular, associated with the simple linear regression model are the estimates of β_1 and β_0

Estimator of β_1,
the slope

$$b_1 = \frac{\sum (x - \bar{x})(y - \bar{y})}{\sum (x - \bar{x})^2} \qquad (9.6)$$

and

Estimator of β_0,
the intercept

$$b_0 = \bar{y} - b_1 \bar{x} \qquad (9.7)$$

and corresponding fitted values

Fitted values for
simple linear
regression

$$\hat{y}_i = b_0 + b_1 x_i \qquad (9.8)$$

and residuals

Residuals for
simple linear
regression

$$e_i = y_i - \hat{y}_i \qquad (9.9)$$

Further, the residuals (9.9) can be used to make up an estimate of σ^2. As always, a sum of squared residuals is divided by an appropriate number of degrees of freedom. That is, there is the following definition of a (**simple linear regression or) line-fitting sample variance**.

Definition 1

For a set of data pairs $(x_1, y_1), (x_2, y_2), \ldots, (x_n, y_n)$ where least squares fitting of a line produces fitted values (9.8) and residuals (9.9),

$$s_{\mathrm{LF}}^2 = \frac{1}{n-2} \sum (y - \hat{y})^2 = \frac{1}{n-2} \sum e^2 \qquad (9.10)$$

will be called a **line-fitting sample variance**. Associated with it are $\nu = n - 2$ degrees of freedom and an estimated standard deviation of response, $s_{\mathrm{LF}} = \sqrt{s_{\mathrm{LF}}^2}$.

---

s_{LF}^2 estimates the level of basic background variation, σ^2, whenever the model (9.4) is an adequate description of the system under study. When it is not, s_{LF} will tend to overestimate σ. So comparing s_{LF} to s_{P} is another way of investigating the appropriateness of model (9.4). (s_{LF} much larger than s_{P} suggests the linear regression model is a poor one.)

Example 1

(Example 1, Chapter 4, revisited—page 124)

Inference in the Ceramic Powder Pressing Study

The main example in this section will be the pressure/density study of Benson, Locher, and Watkins (used extensively in Section 4.1 to illustrate the descriptive analysis of (x, y) data). Table 9.1 lists again those $n = 15$ data pairs (x, y) (first presented in Table 4.1) representing

$$x = \text{the pressure setting used (psi)}$$

$$y = \text{the density obtained (g/cc)}$$

in the dry pressing of a ceramic compound into cylinders, and Figure 9.2 is a scatterplot of the data.

Recall further from the calculation of R^2 in Example 1 of Chapter 4 that the data of Table 4.1 produce fitted values in Table 4.2 and then

$$\sum(y - \hat{y})^2 = .005153$$

So for the pressure/density data, one has (via formula (9.10)) that

$$s_{\text{LF}}^2 = \frac{1}{15 - 2}(.005153) = .000396 \text{ (g/cc)}^2$$

so

$$s_{\text{LF}} = \sqrt{.000396} = .0199 \text{ g/cc}$$

If one accepts the appropriateness of model (9.4) in this powder pressing example, for any fixed pressure the standard deviation of densities associated with many cylinders made at that pressure would be approximately .02 g/cc.

Table 9.1
Pressing Pressures and Resultant Specimen Densities

x, Pressure (psi)	y, Density (g/cc)	x, Pressure (psi)	y, Density (g/cc)
2,000	2.486	6,000	2.653
2,000	2.479	8,000	2.724
2,000	2.472	8,000	2.774
4,000	2.558	8,000	2.808
4,000	2.570	10,000	2.861
4,000	2.580	10,000	2.879
6,000	2.646	10,000	2.858
6,000	2.657		

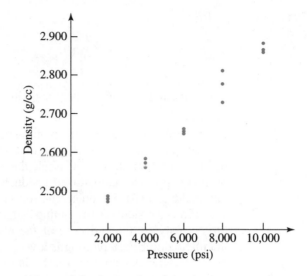

Figure 9.2   Scatterplot of density versus pressing pressure

Table 9.2
Sample Means and Standard Deviations of Densities for Five Different Pressing Pressures

x, Pressure (psi)	$\bar{y}$, Sample Mean	s, Sample Standard Deviation
2,000	2.479	.0070
4,000	2.569	.0110
6,000	2.652	.0056
8,000	2.769	.0423
10,000	2.866	.0114

The original data in this example can be thought of as organized into $r = 5$ separate samples of size $m = 3$, one for each of the pressures 2,000 psi, 4,000 psi, 6,000 psi, 8,000 psi, and 10,000 psi. It is instructive to consider what this thinking leads to for an alternative estimate of σ—namely, s_P. Table 9.2 gives $\bar{y}$ and s values for the five samples.

The sample standard deviations in Table 9.2 can be employed in the usual way to calculate s_P. That is, exactly as in Definition 1 of Chapter 7

$$s_\text{P}^2 = \frac{(3-1)(.0070)^2 + (3-1)(.0110)^2 + \cdots + (3-1)(.0114)^2}{(3-1) + (3-1) + \cdots + (3-1)}$$

$$= .000424 \ (\text{g/cc})^2$$

Example 1
(*continued*)

from which

$$s_P = \sqrt{s_P^2} = .0206 \text{ g/cc}$$

Comparing s_{LF} and s_P, there is no indication of poor fit carried by these values.

Section 4.1 includes some plotting of the residuals (9.9) for the pressure/density data (in particular, a normal plot that appears as Figure 4.7). Although the (raw) residuals (9.9) are most easily calculated, most commercially available regression programs provide standardized residuals as well as, or even in preference to, the raw residuals. (At this point, the reader should review the discussion concerning standardized residuals surrounding Definition 2 of Chapter 7.) In curve- and surface-fitting analyses, the variances of the residuals depend on the corresponding x's. Standardizing before plotting is a way to prevent mistaking a pattern on a residual plot that is explainable on the basis of these different variances for one that is indicative of problems with the basic model. Under model (9.4), for a given x with corresponding response y,

$$\text{Var}(y - \hat{y}) = \sigma^2 \left(1 - \frac{1}{n} - \frac{(x - \bar{x})^2}{\sum(x - \bar{x})^2} \right) \tag{9.11}$$

So using formula (9.11) and Definition 7.2, corresponding to the data pair (x_i, y_i) is the standardized residual for simple linear regression

Standardized residuals for simple linear regression

$$e_i^* = \frac{e_i}{s_{LF} \sqrt{1 - \dfrac{1}{n} - \dfrac{(x_i - \bar{x})^2}{\sum(x - \bar{x})^2}}} \tag{9.12}$$

The more sophisticated method of examining residuals under model (9.4) is thus to make plots of the values (9.12) instead of plotting the raw residuals (9.9).

Example 1
(*continued*)

Consider how the standardized residuals for the pressure/density data set are related to the raw residuals. Recalling that

$$\sum(x - \bar{x})^2 = 120,000,000$$

and that the x_i values in the original data included only the pressures 2,000 psi, 4,000 psi, 6,000 psi, 8,000 psi, and 10,000 psi, it is easy to obtain the necessary values of the radical in the denominator of expression (9.12). These are collected in Table 9.3.

Table 9.3
Calculations for Standardized Residuals
in the Pressure/Density Study

x	$\sqrt{1 - \dfrac{1}{15} - \dfrac{(x - 6{,}000)^2}{120{,}000{,}000}}$
2,000	.894
4,000	.949
6,000	.966
8,000	.949
10,000	.894

The entries in Table 9.3 show, for example, that one should expect residuals corresponding to $x = 6{,}000$ psi to be (on average) about $.966/.894 = 1.08$ times as large as residuals corresponding to $x = 10{,}000$ psi. Division of raw residuals by s_{LF} times the appropriate entry of the second column of Table 9.3 then puts them all on equal footing, so to speak. Table 9.4 shows both the raw residuals (taken from Table 4.5) and their standardized counterparts.

In the present case, since the values .894, .949, and .966 are roughly comparable, standardization via formula (9.12) doesn't materially affect conclusions about model adequacy. For example, Figures 9.3 and 9.4 are normal plots of (respectively) raw residuals and standardized residuals. For all intents and purposes, they are identical. So any conclusions (like those made in Section 4.1 based on Figure 4.7) about model adequacy supported by Figure 9.3 are equally supported by Figure 9.4, and vice versa.

In other situations, however (especially those where a data set contains a few very extreme x values), standardization can involve more widely varying denominators for formula (9.12) than those implied by Table 9.3 and thereby affect the results of a residual analysis.

Table 9.4
Residuals and Standardized Residuals for the Pressure/Density Study

x	e	Standardized Residual
2,000	.0137, .0067, −.0003	.77, .38, −.02
4,000	−.0117, .0003, .0103	−.62, .02, .55
6,000	−.0210, −.0100, −.0140	−1.09, −.52, −.73
8,000	−.0403, .0097, .0437	−2.13, .51, 2.31
10,000	−.0007, .0173, −.0037	−.04, .97, −.21

Example 1
(*continued*)

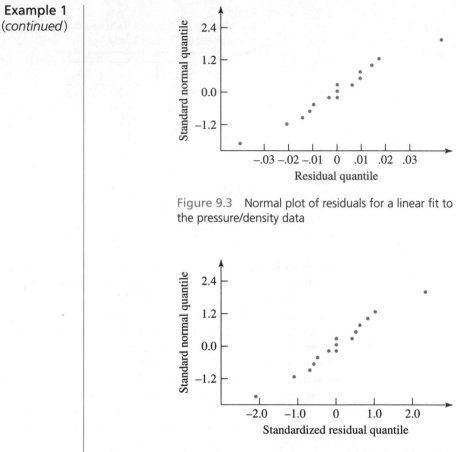

Figure 9.3 Normal plot of residuals for a linear fit to the pressure/density data

Figure 9.4 Normal plot of standardized residuals for a linear fit to the pressure/density data

9.1.2 Inference for the Slope Parameter

Especially in applications of the simple linear regression model (9.4) where x represents a variable that can be physically manipulated by the engineer, the slope parameter β_1 is of fundamental interest. It is the **rate of change of average response with respect to** x, and it governs the impact of a change in x on the system output. Inference for β_1 is fairly simple, because of the distributional properties that b_1 (the slope of the least squares line) inherits from the model. That is, under model (9.4), b_1 has a normal distribution with

$$Eb_1 = \beta_1$$

and

$$\operatorname{Var} b_1 = \frac{\sigma^2}{\sum (x - \bar{x})^2} \tag{9.13}$$

which in turn implies that

$$Z = \frac{\dfrac{b_1 - \beta_1}{\sigma}}{\sqrt{\sum (x - \bar{x})^2}}$$

is standard normal. In a manner similar to many of the arguments in Chapters 6 and 7, this motivates the fact that the quantity

$$T = \frac{\dfrac{b_1 - \beta_1}{s_{\text{LF}}}}{\sqrt{\sum (x - \bar{x})^2}} \tag{9.14}$$

has a t_{n-2} distribution. The standard arguments of Chapter 6 applied to expression (9.14) then show that

$$\text{H}_0 : \beta_1 = \# \tag{9.15}$$

can be tested using the test statistic

Test statistic for
$\text{H}_0 : \beta_1 = \#$

$$T = \frac{\dfrac{b_1 - \#}{s_{\text{LF}}}}{\sqrt{\sum (x - \bar{x})^2}} \tag{9.16}$$

and a t_{n-2} reference distribution. More importantly, under the simple linear regression model (9.4), a two-sided confidence interval for β_1 can be made using endpoints

Confidence limits
for the slope, β_1

$$b_1 \pm t \frac{s_{\text{LF}}}{\sqrt{\sum (x - \bar{x})^2}} \tag{9.17}$$

where the associated confidence is the probability assigned to the interval between $-t$ and t by the t_{n-2} distribution. A one-sided interval is made in the usual way, based on one endpoint from formula (9.17).

Example 1
(continued)

In the context of the powder pressing study, Section 4.1 showed that the slope of the least squares line through the pressure/density data is

$$b_1 = .0000486 \text{ (g/cc)/psi}$$

Example 1
(continued)

Then, for example, a 95% two-sided confidence interval for β_1 can be made using the .975 quantile of the t_{13} distribution in formula (9.17). That is, one can use endpoints

$$.000048\bar{6} \pm 2.160 \frac{.0199}{\sqrt{120,000,000}}$$

that is,

$$.000048\bar{6} \pm .0000039$$

that is,

$$.0000448 \text{ (g/cc)/psi} \quad \text{and} \quad .0000526 \text{ (g/cc)/psi}$$

A confidence interval like this one for β_1 can be translated into a confidence interval for a difference in mean responses for two different values of x. According to model (9.4), two different values of x differing by Δx have mean responses differing by $\beta_1 \Delta x$. One then simply multiplies endpoints of a confidence interval for β_1 by Δx to obtain a confidence interval for the difference in mean responses. For example, since $8,000 - 6,000 = 2,000$, the difference between mean densities at 8,000 psi and 6,000 psi levels has a 95% confidence interval with endpoints

$$2,000(.0000448) \text{ g/cc} \quad \text{and} \quad 2,000(.0000526) \text{ g/cc}$$

that is,

$$.0896 \text{ g/cc} \quad \text{and} \quad .1052 \text{ g/cc}$$

Considerations in the selection of x values

Formula (9.17) allows a kind of precision to be attached to the slope of the least squares line. It is useful to consider how that precision is related to study characteristics that are potentially under an investigator's control. Notice that both formulas (9.13) and (9.17) indicate that the larger $\sum(x - \bar{x})^2$ is (i.e., the more spread out the x_i values are), the more precision b_1 offers as an estimator of the underlying slope β_1. Thus, as far as the estimation of β_1 is concerned, in studies where x represents the value of a system variable under the control of an experimenter, he or she should choose settings of x with the largest possible sample variance. (In fact, if one has n observations to spend and can choose values of x anywhere in some interval $[a, b]$, taking $\frac{n}{2}$ of them at $x = a$ and $\frac{n}{2}$ at $x = b$ produces the best possible precision for estimating the slope β_1.)

However, this advice (to spread the x_i's out) must be taken with a grain of salt. The approximately linear relationship (9.4) may hold over only a limited range of possible x values. Choosing experimental values of x beyond the limits where it is reasonable to expect formula (9.4) to hold, hoping thereby to obtain a good estimate of slope, is of course nonsensical. And it is also important to recognize that precise estimation of β_1 under the assumptions of model (9.4) is not the only consideration when planning data collection. It is usually also important to be in a position to tell when the linear form of (9.4) is inappropriate. That dictates that data be collected at a number of different settings of x, not simply at the smallest and largest values possible.

9.1.3 Inference for the Mean System Response for a Particular Value of x

Chapters 7 and 8 repeatedly considered the problem of estimating the mean of y under a particular one (or combination) of the levels of the factor (or factors) of interest. In the present context, the analog is the problem of estimating the mean response for a fixed value of the system variable x,

$$\mu_{y|x} = \beta_0 + \beta_1 x \tag{9.18}$$

The natural data-based approximation of the mean in formula (9.18) is the corresponding y value taken from the least squares line. The notation

Estimator of
$\mu_{y|x} = \beta_0 + \beta_1 x$

$$\boxed{\hat{y} = b_0 + b_1 x} \tag{9.19}$$

will be used for this value on the least squares lines. (This is in spite of the fact that the value in formula (9.19) may not be a fitted value in the sense that the phrase has most often been used to this point. x need not be equal to any of $x_1, x_2, \ldots, x_n$ for both expressions (9.18) and (9.19) to make sense.) The simple linear regression model (9.4) leads to simple distributional properties for $\hat{y}$ that then produce inference methods for $\mu_{y|x}$.

Under model (9.4), $\hat{y}$ has a normal distribution with

$$E\hat{y} = \mu_{y|x} = \beta_0 + \beta_1 x$$

and

$$\text{Var}\,\hat{y} = \sigma^2 \left(\frac{1}{n} + \frac{(x - \bar{x})^2}{\sum (x - \bar{x})^2} \right) \tag{9.20}$$

(In expression (9.20), notation is being abused somewhat. The i subscripts and indices of summation in $\sum (x - \bar{x})^2$ have been suppressed. This summation runs over the n values x_i included in the original data set. On the other hand, in the $(x - \bar{x})^2$ term appearing as a numerator in expression (9.20), the x involved is not necessarily equal to any of $x_1, x_2, \ldots, x_n$. Rather, it is simply the value of the system variable at which the mean response is to be estimated.) Then

$$Z = \frac{\hat{y} - \mu_{y|x}}{\sigma \sqrt{\dfrac{1}{n} + \dfrac{(x - \bar{x})^2}{\sum(x - \bar{x})^2}}}$$

has a standard normal distribution. This in turn motivates the fact that

$$T = \frac{\hat{y} - \mu_{y|x}}{s_{LF} \sqrt{\dfrac{1}{n} + \dfrac{(x - \bar{x})^2}{\sum(x - \bar{x})^2}}} \qquad (9.21)$$

has a t_{n-2} distribution. The standard arguments of Chapter 6 applied to expression (9.21) then show that

$$H_0 : \mu_{y|x} = \# \qquad (9.22)$$

can be tested using the test statistic

Test statistic for
$H_0 : \mu_{y|x} = \#$

$$T = \frac{\hat{y} - \#}{s_{LF} \sqrt{\dfrac{1}{n} + \dfrac{(x - \bar{x})^2}{\sum(x - \bar{x})^2}}} \qquad (9.23)$$

and a t_{n-2} reference distribution. Further, under the simple linear regression model (9.4), a two-sided individual confidence interval for $\mu_{y|x}$ can be made using endpoints

Confidence limits
for the mean repsonse,
$\mu_{y|x} = \beta_0 + \beta_1 x$

$$\hat{y} \pm t s_{LF} \sqrt{\dfrac{1}{n} + \dfrac{(x - \bar{x})^2}{\sum(x - \bar{x})^2}} \qquad (9.24)$$

where the associated confidence is the probability assigned to the interval between $-t$ and t by the t_{n-2} distribution. A one-sided interval is made in the usual way based on one endpoint from formula (9.24).

Example 1
(*continued*)

Returning again to the pressure/density study, consider making individual 95% confidence intervals for the mean densities of cylinders produced first at 4,000 psi and then at 5,000 psi.

Treating first the 4,000 psi condition, the corresponding estimate of mean density is

$$\hat{y} = 2.375 + .000048\bar{6}(4,000) = 2.5697 \text{ g/cc}$$

Further, from formula (9.24) and the fact that the .975 quantile of the t_{13} distribution is 2.160, a precision of plus-or-minus

$$2.160(.0199)\sqrt{\frac{1}{15} + \frac{(4,000 - 6,000)^2}{120,000,000}} = .0136 \text{ g/cc}$$

can be attached to the 2.5697 g/cc figure. That is, endpoints of a two-sided 95% confidence interval for the mean density under the 4,000 psi condition are

▶   2.5561 g/cc   and   2.5833 g/cc

Under the $x = 5,000$ psi condition, the corresponding estimate of mean density is

$$\hat{y} = 2.375 + .000048\bar{6}(5,000) = 2.6183 \text{ g/cc}$$

Using formula (9.24), a precision of plus-or-minus

$$2.160(.0199)\sqrt{\frac{1}{15} + \frac{(5,000 - 6,000)^2}{120,000,000}} = .0118 \text{ g/cc}$$

can be attached to the 2.6183 g/cc figure. That is, endpoints of a two-sided 95% confidence interval for the mean density under the 5,000 psi condition are

▶   2.6065 g/cc   and   2.6301 g/cc

The reader should compare the plus-or-minus parts of the two confidence intervals found here. The interval for $x = 5,000$ psi is shorter and therefore more informative than the interval for $x = 4,000$ psi. The origin of this discrepancy should be clear, at least upon scrutiny of formula (9.24). For the students' data, $\bar{x} = 6,000$ psi. $x = 5,000$ psi is closer to $\bar{x}$ than is $x = 4,000$ psi, so the $(x - \bar{x})^2$ term (and thus the interval length) is smaller for $x = 5,000$ psi than for $x = 4,000$ psi.

The phenomenon noted in the preceding example—that the length of a confidence interval for $\mu_{y|x}$ increases as one moves away from $\bar{x}$—is an important one. And it has an intuitively plausible implication for the planning of experiments where an approximately linear relationship between y and x is expected, and x is under the investigator's control. If there is an interval of values of x over which one wants good precision in estimating mean responses, it is only sensible to center one's data collection efforts in that interval.

Inference for the intercept, β_0

Proper use of displays (9.22), (9.23), and (9.24) gives inference methods for the parameter β_0 in model (9.4). β_0 is the y intercept of the linear relationship (9.18). So by setting $x = 0$ in displays (9.22), (9.23), and (9.24), tests and confidence intervals for β_0 are obtained. However, unless $x = 0$ is a feasible value for the input variable and the region where the linear relationship (9.18) is a sensible description of physical reality includes $x = 0$, inference for β_0 alone is rarely of practical interest.

The confidence intervals represented by formula (9.24) carry individual associated confidence levels. Section 7.3 showed that it is possible (using the P-R method) to give simultaneous confidence intervals for r possibly different means, μ_i. This comes about essentially by appropriately increasing the t multiplier used in the plus-or-minus part of the formula for individual confidence limits. Here it is possible, by replacing t in formula (9.24) with a larger value, to give simultaneous confidence intervals for *all* means $\mu_{y|x}$. That is, under model (9.4), simultaneous two-sided confidence intervals for all mean responses $\mu_{y|x}$ can be made using respective endpoints

Simultaneous two-sided confidence limits for all means, $\mu_{y|x}$

$$(b_0 + b_1 x) \pm \sqrt{2f}\, s_{\text{LF}} \sqrt{\frac{1}{n} + \frac{(x - \bar{x})^2}{\sum (x - \bar{x})^2}} \qquad (9.25)$$

where for positive f, the associated simultaneous confidence is the $F_{2,n-2}$ probability assigned to the interval $(0, f)$.

Of course, the practical meaning of the phrase "for all means $\mu_{y|x}$" is more like "for all mean responses in an interval where the simple linear regression model (9.4) is a workable description of the relationship between x and y." As is always the case in curve- and surface-fitting situations, *extrapolation* outside of the range of x values where one has data (and even to some extent *interpolation* inside that range) is risky business. When it is done, it should be supported by subject-matter expertise to the effect that it is justifiable.

It may be somewhat difficult to grasp the meaning of a simultaneous confidence figure applicable to *all* possible intervals of the form (9.25). To this point, the confidence levels considered have been for finite sets of intervals. Probably the best way to understand the theoretically infinite set of intervals given by formula (9.25) is as defining a region in the (x, y)-plane thought likely to contain the line $\mu_{y|x} = \beta_0 + \beta_1 x$. Figure 9.5 is a sketch of a typical confidence region represented by formula (9.25). There is a region indicated about the least squares line whose vertical extent increases with distance from $\bar{x}$ and which has the stated confidence in covering the line describing the relationship between x and $\mu_{y|x}$.

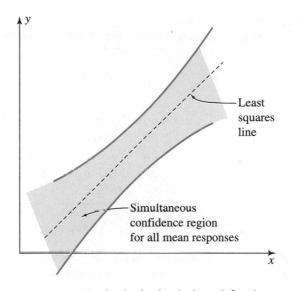

Figure 9.5   Region in the (x, y)-plane defined by simultaneous confidence intervals for all values of $\mu_{y|x}$

Example 1
(continued)

It is instructive to compare what the P-R method of Section 7.3 and formula (9.25) give for simultaneous 95% confidence intervals for mean cylinder densities produced under the five conditions actually used by the students in their study.

First, formula (7.28) of Section 7.3 shows that with $n - r = 15 - 5 = 10$ degrees of freedom for s_p and $r = 5$ conditions under study, 95% simultaneous two-sided confidence limits for all five mean densities are of the form

$$\bar{y}_i \pm 3.103 \frac{s_p}{\sqrt{n_i}}$$

which in the present context is

$$\bar{y}_i \pm 3.103 \frac{.0206}{\sqrt{3}}$$

that is,

$$\bar{y}_i \pm .0369 \text{ g/cc}$$

Then, since $v_1 = 2$ and $v_2 = 13$ degrees of freedom are involved in the use of formula (9.25), simultaneous limits of the form

$$\hat{y} \pm \sqrt{2(3.81)}\, s_{LF} \sqrt{\frac{1}{15} + \frac{(x - 6{,}000)^2}{120{,}000{,}000}}$$

Example 1
(*continued*)

Table 9.5

Simultaneous (and Individual) 95% Confidence Intervals for Mean Cylinder Densities

x, Pressure	$\mu_{y\|x}$ (P-R Method) Mean Density	$\mu_{y\|x}$ (from formula (9.25)) Mean Density	$\mu_{y\|x}$ (from formula (9.24)) Mean Density
2,000 psi	$2.4790 \pm .0369$ g/cc	$2.4723 \pm .0246$ g/cc	$2.4723 \pm .0136$ g/cc
4,000 psi	$2.5693 \pm .0369$ g/cc	$2.5697 \pm .0174$ g/cc	$2.5697 \pm .0118$ g/cc
6,000 psi	$2.6520 \pm .0369$ g/cc	$2.6670 \pm .0142$ g/cc	$2.6670 \pm .0111$ g/cc
8,000 psi	$2.7687 \pm .0369$ g/cc	$2.7643 \pm .0174$ g/cc	$2.7643 \pm .0118$ g/cc
10,000 psi	$2.8660 \pm .0369$ g/cc	$2.8617 \pm .0246$ g/cc	$2.8617 \pm .0136$ g/cc

are indicated. Table 9.5 shows the five intervals that result from the use of each of the two simultaneous confidence methods, together with individual intervals (9.24).

Two points are evident from Table 9.5. First, the intervals that result from formula (9.25) are somewhat wider than the corresponding individual intervals given by formula (9.24). But it is also clear that the use of the simple linear regression model assumptions in preference to the more general one-way assumptions of Chapter 7 can lead to shorter simultaneous confidence intervals and correspondingly sharper real-world engineering inferences.

9.1.4 Prediction and Tolerance Intervals (*Optional*)

Inference for $\mu_{y|x}$ is one kind of answer to the qualitative question, "If I hold the input variable x at some particular level, what can I expect in terms of a system response?" It is an answer in terms of *mean* or long-run average response. Sometimes an answer in terms of *individual responses* is of more practical use. And in such cases it is helpful to know that the simple linear regression model assumptions (9.4) lead to their own specialized formulas for prediction and tolerance intervals.

The basic fact that makes possible prediction intervals under assumptions (9.4) is that if y_{n+1} is one additional observation, coming from the distribution of responses corresponding to a particular x, and $\hat{y}$ is the corresponding fitted value at that x (based on the original n data pairs), then

$$T = \frac{y_{n+1} - \hat{y}}{s_{\text{LF}} \sqrt{1 + \dfrac{1}{n} + \dfrac{(x - \bar{x})^2}{\sum (x - \bar{x})^2}}}$$

has a t_{n-2} distribution. This fact leads in the usual way to the conclusion that under model (9.4) the two-sided interval with endpoints

Simple linear regression prediction limits for an additional y at a given x

$$\hat{y} \pm t s_{\text{LF}} \sqrt{1 + \frac{1}{n} + \frac{(x - \bar{x})^2}{\sum (x - \bar{x})^2}} \tag{9.26}$$

can be used as a prediction interval for an additional observation y at a particular value of the input variable x. The associated prediction confidence is the probability that the t_{n-2} distribution assigns to the interval between $-t$ and t. One-sided intervals are made in the usual way, by employing only one of the endpoints (9.26) and adjusting the confidence level appropriately.

It is possible not only to derive prediction interval formulas from the simple linear regression model assumptions but also to develop relatively simple formulas for approximate one-sided tolerance bounds. That is, the intervals

A one-sided tolerance interval for the y distribution at x

$$(\hat{y} - \tau s_{\text{LF}}, \infty) \tag{9.27}$$

and

Another one-sided tolerance interval for the y distribution at x

$$(-\infty, \hat{y} + \tau s_{\text{LF}}) \tag{9.28}$$

can be used as one-sided tolerance intervals for a fraction p of the underlying distribution of responses corresponding to a particular value of the system variable x, provided τ is appropriately chosen (depending upon the data, p, x, and the desired confidence level).

In order to write down a reasonably clean formula for τ, the notation

The ratio of $\sqrt{\text{Var}\,\hat{y}}$ to σ for simple linear regression

$$A = \sqrt{\frac{1}{n} + \frac{(x - \bar{x})^2}{\sum (x - \bar{x})^2}} \tag{9.29}$$

will be adopted for the multiplier that is used (e.g., in formula (9.24)) to go from an estimate of σ to an estimate of the standard deviation of $\hat{y}$. Then, for approximate

γ level confidence in locating a fraction p of the responses y at the x of interest, τ appropriate for use in interval (9.27) or (9.28) is

Multiplier to use in interval (9.27) or (9.28)

$$\tau = \frac{Q_z(p) + AQ_z(\gamma)\sqrt{1 + \dfrac{1}{2(n-2)}\left(\dfrac{Q_z^2(p)}{A^2} - Q_z^2(\gamma)\right)}}{1 - \dfrac{Q_z^2(\gamma)}{2(n-2)}} \qquad (9.30)$$

Example 1
(continued)

To illustrate the use of prediction and tolerance interval formulas in the simple linear regression context, consider a 90% lower prediction bound for a single additional density in powder pressing, if a pressure of 4,000 psi is employed. Then, additionally consider finding a 95% lower tolerance bound for 90% of many additional cylinder densities if that pressure is used.

Treating first the prediction problem, formula (9.26) shows that an appropriate prediction bound is

$$2.5697 - 1.350(.0199)\sqrt{1 + \frac{1}{15} + \frac{(4,000 - 6,000)^2}{120,000,000}} = 2.5796 - .0282$$

that is,

$$2.5514 \text{ g/cc}$$

If, rather than predicting a single additional density for $x = 4,000$ psi, it is of interest to locate 90% of additional densities corresponding to a 4,000 psi pressure, a tolerance bound is in order. First use formula (9.29) and find that

$$A = \sqrt{\frac{1}{15} + \frac{(4,000 - 6,000)^2}{120,000,000}} = .3162$$

Next, for 95% confidence, applying formula (9.30),

$$\tau = \frac{1.282 + (.3162)(1.645)\sqrt{1 + \dfrac{1}{2(15-2)}\left(\dfrac{(1.282)^2}{(.3162)^2} - (1.645)^2\right)}}{1 - \dfrac{(1.645)^2}{2(15-2)}} = 2.149$$

So finally, an approximately 95% lower tolerance bound for 90% of densities produced using a pressure of 4,000 psi is (via formula (9.27))

$$2.5697 - 2.149(.0199) = 2.5697 - .0428$$

that is,

$$2.5269 \text{ g/cc}$$

Cautions about prediction and tolerance intervals in regression   The fact that curve-fitting facilitates interpolation and extrapolation makes it imperative that care be taken in the interpretation of prediction and tolerance intervals. All of the warnings regarding the interpretation of prediction and tolerance intervals raised in Section 6.6 apply equally to the present situation. But the new element here (that formally, the intervals can be made for values of x where one has absolutely no data) requires additional caution. If one is to use formulas (9.26), (9.27), and (9.28) at a value of x not represented among $x_1, x_2, \ldots, x_n$, it must be plausible that model (9.4) not only describes system behavior at those x values where one has data, but at the additional value of x as well. And even when this is "plausible" the application of formulas (9.26), (9.27), and (9.28) to new values of x should be treated with a good dose of care. Should one's (unverified) judgment prove wrong, the nominal confidence level has unknown practical relevance.

9.1.5 Simple Linear Regression and ANOVA

Section 7.4 illustrates how, for unstructured studies, partition of the total sum of squares into interpretable pieces provides both (1) intuition and quantification regarding the origin of observed variation and also (2) the basis for an F test of "no differences between mean responses." It turns out that something similar is possible in simple linear regression contexts.

In the unstructured context of Section 7.4, it was useful to name the difference between *SSTot* and *SSE*. The corresponding convention for curve- and surface-fitting situations is stated next in definition form.

Definition 2   In curve- and surface-fitting analyses of multisample studies, the difference

$$SSR = SSTot - SSE$$

will be called the **regression sum of squares**.

It is not obvious, but the difference referred to in Definition 2 in general has the form of a sum of squares of appropriate quantities. In the present context of fitting a line by least squares,

$$SSR = \sum_{i=1}^{n} (\hat{y}_i - \bar{y})^2$$

Without using the particular terminology of Definition 2, this text has already made fairly extensive use of $SSR = SSTot - SSE$. A review of Definition 3 in Chapter 4 (page 130), and Definitions 4 and 6 in Chapter 7 (page 484) will show that in curve- and surface-fitting contexts,

The coefficient of determination for simple linear regression in sum of squares notation

$$R^2 = \frac{SSR}{SSTot} \tag{9.31}$$

That is, SSR is the numerator of the coefficient of determination defined first in Definition 3 (Chapter 4). It is commonly thought of as the part of the raw variability in y that is accounted for in the curve- or surface-fitting process.

SSR and SSE not only provide an appealing partition of $SSTot$ but also form the raw material for an F test of

$$H_0: \beta_1 = 0 \tag{9.32}$$

versus

$$H_a: \beta_1 \neq 0 \tag{9.33}$$

Under model (9.4), hypothesis (9.32) can be tested using the statistic

An F statistic for testing $H_0: \beta_1 = 0$

$$F = \frac{SSR/1}{s_{LF}^2} = \frac{SSR/1}{SSE/(n-2)} \tag{9.34}$$

and an $F_{1,n-2}$ reference distribution, where large observed values of the test statistic constitute evidence against H_0.

Earlier in this section, the general null hypothesis $H_0: \beta_1 = \#$ was tested using the t statistic (9.16). It is thus reasonable to consider the relationship of the F test indicated in displays (9.32), (9.33), and (9.34) to the earlier t test. The null hypothesis $H_0: \beta_1 = 0$ is a special form of hypothesis (9.15), $H_0: \beta_1 = \#$. It is the most frequently tested version of hypothesis (9.15) because it can (within limits) be interpreted as the null hypothesis that mean response doesn't depend on x. This is because when hypothesis (9.32) is true within the simple linear regression model (9.4), $\mu_{y|x} = \beta_0 + 0 \cdot x = \beta_0$, which doesn't depend on x. (Actually, a better interpretation of a test of hypothesis (9.32) is as a test of whether a linear term in

x adds significantly to one's ability to model the response y after accounting for an overall mean response.)

If one then considers testing hypotheses (9.32) and (9.33), it might appear that the # $= 0$ version of formula (9.16) and formula (9.34) represent two different testing methods. But they are equivalent. The statistic (9.34) turns out to be the square of the # $= 0$ version of statistic (9.16), and (two-sided) observed significance levels based on statistic (9.16) and the t_{n-2} distribution turn out to be the same as observed significance levels based on statistic (9.16) and the $F_{1,n-2}$ distribution. So, from one point of view, the F test specified here is redundant, given the earlier discussion. But it is introduced here because of its relationship to the ANOVA ideas of Section 7.4, and because it has an important natural generalization to more complex curve- and surface-fitting contexts. (This generalization is discussed in Section 9.2 and cannot be made equivalent to a t test.)

The partition of *SSTot* into its parts, *SSR* and *SSE*, and the calculation of the statistic (9.34) can be organized in ANOVA table format. Table 9.6 shows the general format that this book will use in the simple linear regression context.

Table 9.6
General Form of the ANOVA Table for Simple Linear Regression

ANOVA Table (for testing $H_0 : \beta_1 = 0$)

Source	SS	df	MS	F
Regression	*SSR*	1	*SSR*/1	*MSR/MSE*
Error	*SSE*	$n-2$	*SSE*/$(n-2)$	
Total	*SSTot*	$n-1$		

Example 1
(continued)

Recall again from the discussion of the pressure/density example in Section 4.1 that

$$SSTot = \sum(y - \bar{y})^2 = .289366$$

Also, from page 654 recall that

$$SSE = \sum(y - \hat{y})^2 = .005153$$

Thus,

$$SSR = SSTot - SSE = .289366 - .005153 = .284213$$

and the specific version of Table 9.6 for the present example is given as Table 9.7.

Example 1
(*continued*)

Then the observed level of significance for testing $H_0 : \beta_1 = 0$ is

$$P[\text{an } F_{1,13} \text{ random variable} > 717] < .001$$

and one has very strong evidence against the possibility that $\beta_1 = 0$. A linear term in Pressure is an important contributor to one's ability to describe the behavior of Cylinder Density. This is, of course, completely consistent with the earlier interval-oriented analysis that produced 95% confidence limits for β_1 of

$$.0000448 \text{ (g/cc)/psi} \quad \text{and} \quad .0000526 \text{ (g/cc)/psi}$$

that do not bracket 0.

The value of $R^2 = .9822$ (found first in Section 4.1) can also be easily derived, using the entries of Table 9.7 and the relationship (9.31).

Table 9.7
ANOVA Table for the Pressure/Density Data

| **ANOVA Table (for testing $H_0 : \beta_1 = 0$)** | | | | |
Source	SS	df	MS	F
Regression	.284213	1	.284213	717
Error	.005153	13	.000396	
Total	.289366	14		

9.1.6   Simple Linear Regression and Statistical Software

Many of the calculations needed for the methods of this section are made easier by statistical software packages. None of the methods of this section are so computationally intensive that they absolutely require the use of such software, but it is worthwhile to consider its use in the simple linear regression context. Learning where on a typical printout to find the various summary statistics corresponding to calculations made in this section helps in locating important summary statistics for the more complicated curve- and surface-fitting analyses of the next section. Printout 1 is from a MINITAB analysis of the pressure/density data.

Printout 1   Simple Linear Regression for the Pressure/Density Data (*Example 1*)

```
Regression Analysis

The regression equation is
density = 2.38 +0.000049 pressure
```

```
Predictor        Coef       StDev         T        P
Constant      2.37500     0.01206    197.01    0.000
pressure    0.00004867  0.00000182     26.78    0.000

S = 0.01991    R-Sq = 98.2%    R-Sq(adj) = 98.1%

Analysis of Variance

Source          DF         SS         MS        F       P
Regression       1     0.28421    0.28421   717.06   0.000
Residual Error  13     0.00515    0.00040
Total           14     0.28937

Obs    pressure   density       Fit   StDev Fit    Residual    St Resid
  1       2000    2.48600   2.47233     0.00890     0.01367        0.77
  2       2000    2.47900   2.47233     0.00890     0.00667        0.37
  3       2000    2.47200   2.47233     0.00890    -0.00033       -0.02
  4       4000    2.55800   2.56967     0.00630    -0.01167       -0.62
  5       4000    2.57000   2.56967     0.00630     0.00033        0.02
  6       4000    2.58000   2.56967     0.00630     0.01033        0.55
  7       6000    2.64600   2.66700     0.00514    -0.02100       -1.09
  8       6000    2.65700   2.66700     0.00514    -0.01000       -0.52
  9       6000    2.65300   2.66700     0.00514    -0.01400       -0.73
 10       8000    2.72400   2.76433     0.00630    -0.04033       -2.14R
 11       8000    2.77400   2.76433     0.00630     0.00967        0.51
 12       8000    2.80800   2.76433     0.00630     0.04367        2.31R
 13      10000    2.86100   2.86167     0.00890    -0.00067       -0.04
 14      10000    2.87900   2.86167     0.00890     0.01733        0.97
 15      10000    2.85800   2.86167     0.00890    -0.00367       -0.21
```

R denotes an observation with a large standardized residual

Predicted Values

```
   Fit   StDev Fit        95.0% CI            96.0% PI
2.61833    0.00545   ( 2.60655, 2.63011)   ( 2.57374, 2.66293)
```

Printout 1 is typical of summaries of regression analyses printed by commercially available statistical packages. The most basic piece of information on the printout is, of course, the fitted equation. Immediately below it is a table giving (to more significant digits) the estimated coefficients (b_0 and b_1), their estimated standard deviations, and the t ratios (appropriate for testing whether coefficients β are 0) made up as the quotients. The printout includes the values of s_{LF} and R^2 and an ANOVA table much like Table 9.7. For the several observed values of test statistics printed out (including the observed value of F from formula (9.34)), MINITAB gives observed levels of significance. The ANOVA table is followed by a table of values of y, fitted y,

$$\text{"StDev Fit"} = s_{LF}\sqrt{\frac{1}{n} + \frac{(x - \bar{x})^2}{\sum(x - \bar{x})^2}}$$

and residual, and standardized residual corresponding to the n data points. MINI-TAB's regression program has an option that allows one to request fitted values, confidence intervals for $\mu_{y|x}$, and prediction intervals for x values of interest, and Printout 1 finishes with this information for the value $x = 5,000$.

The reader is encouraged to compare the information on Printout 1 with the various results obtained in Example 1 and verify that everything on the printout (except the "adjusted R^2" value) is indeed familiar.

Section 1 Exercises ....

1. Return to the situation of Exercise 3 of Section 4.1 and the polymer molecular weight study of R. Harris.

   (a) Find s_{LF} for these data. What does this intend to measure in the context of the engineering problem?

   (b) Plot both residuals versus x and the standardized residuals versus x. How much difference is there in the appearance of these two plots?

   (c) Give a 90% two-sided confidence interval for the increase in mean average molecular weight that accompanies a 1°C increase in temperature here.

   (d) Give individual 90% two-sided confidence intervals for the mean average molecular weight at 212°C and also at 250°C.

   (e) Give simultaneous 90% two-sided confidence intervals for the two means indicated in part (d).

   (f) Give 90% lower prediction bounds for the next average molecular weight, first at 212°C and then at 250°C.

   (g) Give approximately 95% lower tolerance bounds for 90% of average molecular weights, first at 212°C and then at 250°C.

   (h) Make an ANOVA table for testing $H_0: \beta_1 = 0$ in the simple linear regression model. What is the p-value here for a two-sided test of this hypothesis?

2. Return to the situation of Chapter Exercise 1 of Chapter 4 and the concrete strength study of Nicholson and Bartle.

   (a) Find estimates of the parameters β_0, β_1, and σ in the simple linear regression model $y = \beta_0 + \beta_1 x + \epsilon$. How does your estimate of σ based on the simple linear regression model compare with the pooled sample standard deviation, s_P?

   (b) Compute residuals and standardized residuals. Plot both against x and $\hat{y}$ and normal-plot them. How much do the appearances of the plots of the standardized residuals differ from those of the raw residuals?

   (c) Make a 90% two-sided confidence interval for the increase in mean compressive strength that accompanies a .1 increase in the water/cement ratio. (This is $.1\beta_1$.)

   (d) Test the hypothesis that the mean compressive strength doesn't depend on the water/cement ratio. What is the p-value?

   (e) Make a 95% two-sided confidence interval for the mean strength of specimens with the water/cement ratio .5 (based on the simple linear regression model).

   (f) Make a 95% two-sided prediction interval for the strength of an additional specimen with the water/cement ratio .5 (based on the simple linear regression model).

   (g) Make an approximately 95% lower tolerance bound for the strengths of 90% of additional specimens with the water/cement ratio .5 (based on the simple linear regression model).

9.2 Inference Methods for General Least Squares Curve- and Surface-Fitting (Multiple Linear Regression)

The previous section presented formal inference methods available under the (normal) simple linear regression model. Confidence interval estimation, hypothesis testing, prediction and tolerance intervals, and ANOVA were all seen to have simple linear regression versions. This section makes a parallel study of more general curve- and surface-fitting contexts. First, the multiple linear regression model and its corresponding variance estimate and standardized residuals are introduced. Then, in turn, there are discussions of how multiple linear regression computer programs can (1) facilitate inference for rate of change parameters in the model, (2) make possible inference for the mean system response at a given combination of values for the input/system variables and the making of prediction and tolerance intervals, and (3) allow the use of ANOVA methods in multiple regression contexts.

9.2.1 The Multiple Linear Regression Model, Corresponding Variance Estimate, and Standardized Residuals

This section considers situations like those treated on a descriptive level in Section 4.2, where for k system variables $x_1, x_2, \ldots, x_k$ and a response y, an approximate relationship like

$$y \approx \beta_0 + \beta_1 x_1 + \beta_2 x_2 + \cdots + \beta_k x_k \tag{9.35}$$

holds. As in Section 4.2, the form (9.35) not only covers those circumstances where $x_1, x_2, \ldots, x_k$ all represent physically different variables but also describes contexts where some of the variables are functions of others. For example, the relationship

$$y \approx \beta_0 + \beta_1 x_1 + \beta_2 x_1^2$$

can be thought of as a $k = 2$ version of formula (9.35), where x_2 is a deterministic function of x_1, $x_2 = x_1^2$.

As in Section 4.2, a double subscript notation will be used for the values of the input variables. Thus, the problem considered is that of inference based on the data vectors $(x_{11}, x_{21}, \ldots, x_{k1}, y_1), (x_{12}, x_{22}, \ldots, x_{k2}, y_2), \ldots, (x_{1n}, x_{2n}, \ldots, x_{kn}, y_n)$. As always, a probability model is needed to support formal inferences for such data, and the one considered here is an appropriate specialization of the general one-way normal model of Section 7.1. That is, the standard assumptions of the multiple linear regression model are that there are underlying normal distributions for the response

y with a common variance σ^2 but means $\mu_{y|x_1,x_2,\ldots,x_k}$ that change linearly with each of $x_1, x_2, \ldots, x_k$. In symbols, it is typical to write that for $i = 1, 2, \ldots, n$,

The (normal) multiple linear regression model

$$y_i = \beta_0 + \beta_1 x_{1i} + \beta_2 x_{2i} + \cdots + \beta_k x_{ki} + \epsilon_i \tag{9.36}$$

where the ϵ_i are (unobservable) iid normal $(0, \sigma^2)$ random variables, the $x_{1i}, x_{2i}, \ldots, x_{ki}$ are known constants, and $\beta_0, \beta_1, \beta_2, \ldots, \beta_k$ and σ^2 are unknown model parameters (fixed constants). This is the specialization of the general one-way model

$$y_{ij} = \mu_i + \epsilon_{ij}$$

to the situation where the means $\mu_{y|x_1,x_2,\ldots,x_k}$ satisfy the relationship

$$\mu_{y|x_1,x_2,\ldots,x_k} = \beta_0 + \beta_1 x_1 + \beta_2 x_2 + \cdots + \beta_k x_k \tag{9.37}$$

If one thinks of formula (9.37) as defining a surface in $(k + 1)$-dimensional space, then the model equation (9.36) simply says that responses y differ from corresponding values on that surface by mean 0, variance σ^2 random noise. Figure 9.6 illustrates this point for the simple $k = 2$ case (where x_1 and x_2 are not functionally related).

Inferences about quantities involving those $(x_1, x_2, \ldots, x_k)$ combinations represented in the data, like the mean response at a single $(x_1, x_2, \ldots, x_k)$ or the difference between two such mean responses, will typically be sharper when methods based on model (9.36) can be used in place of the general methods of Chapter 7. And as was true for simple linear regression, to the extent that it is sensible to assume that model (9.36) describes system behavior for values of $x_1, x_2, \ldots, x_k$ not included

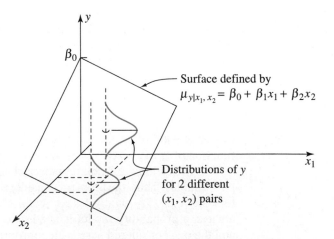

Figure 9.6 Graphical representation of the multiple linear regression model $y = \beta_0 + \beta_1 x_1 + \beta_2 x_2 + \epsilon$

in the data, it provides the basis for inferences involving limited interpolation and extrapolation on the system variables $x_1, x_2, \ldots, x_k$.

Estimators of the coefficients β in the multiple linear regression model

Section 4.2 contains a discussion of using statistical software in the least squares fitting of the approximate relationship (9.35) to a set of $(x_1, x_2, \ldots, x_k, y)$ data. That discussion can be thought of as covering the fitting and use of residuals in model checking for the multiple linear regression model (9.36). Section 4.2 did not produce explicit formulas for $b_0, b_1, b_2, \ldots, b_k$, the (least squares) estimates of $\beta_0, \beta_1, \beta_2, \ldots, \beta_k$. Instead it relied on the software to produce those estimates. Of course, once one has estimates of the β's, corresponding fitted values immediately become

Fitted values for the multiple linear regression model

$$\hat{y}_i = b_0 + b_1 x_{1i} + b_2 x_{2i} + \cdots + b_k x_{ki} \qquad (9.38)$$

with residuals

Residuals for the multiple linear regression model

$$e_i = y_i - \hat{y}_i \qquad (9.39)$$

The residuals (9.39) can be used to make up an estimate of σ^2. One divides a sum of squared residuals by an appropriate number of degrees of freedom. That is, one can make the following definition of a (**multiple linear regression or) surface-fitting sample variance.**

Definition 3

For a set of n data vectors $(x_{11}, x_{21}, \ldots, x_{k1}, y_1), (x_{12}, x_{22}, \ldots, x_{k2}, y_2), \ldots, (x_{1n}, x_{2n}, \ldots, x_{kn}, y_n)$ where least squares fitting produces fitted values given by formula (9.38) and residuals (9.39),

$$s_{\text{SF}}^2 = \frac{1}{n-k-1} \sum (y - \hat{y})^2 = \frac{1}{n-k-1} \sum e^2 \qquad (9.40)$$

will be called a **surface-fitting sample variance.** Associated with it are $\nu = n - k - 1$ degrees of freedom and an estimated standard deviation of response,

$$s_{\text{SF}} = \sqrt{s_{\text{SF}}^2}.$$

Compare Definitions 1 and 3 and notice that the $k = 1$ version of s_{SF}^2 is just s_{LF}^2 from simple linear regression. s_{SF} estimates the level of basic background variation, σ, whenever the model (9.36) is an adequate description of the system under study. When it is not, s_{SF} will tend to overestimate σ. So comparing s_{SF} to s_{P} is another way of investigating the appropriateness of that description. (s_{SF} much larger than s_{P} suggests that model (9.36) is a poor one.)

Example 2
(*Example 5, Chapter 4,
revisited—page 150*)

Inference in the Nitrogen Plant Study

The main example in this section will be the nitrogen plant data set given in Table 4.8. Recall that in the discussion of the example, with

$$x_1 = \text{a measure of air flow}$$

$$x_2 = \text{the cooling water inlet temperature}$$

$$y = \text{a measure of stack loss}$$

the fitted equation

$$\hat{y} = -15.409 - .069x_1 + .528x_2 + .007x_1^2$$

appeared to be a sensible data summary. Accordingly, consider the making of inferences based on the $k = 3$ version of model (9.36),

$$y_i = \beta_0 + \beta_1 x_{1i} + \beta_2 x_{2i} + \beta_3 x_{1i}^2 + \epsilon_i \qquad \textbf{(9.41)}$$

Printout 2 is from a MINITAB analysis of the data of Table 4.8. Among many other things, it gives the values of the residuals from the fitted version of formula (9.41) for all $n = 17$ data points. It is then possible to apply Definition 3 and produce a surface-fitting estimate of the parameter σ^2 in the model (9.41). That is,

$$s_{SF}^2 = \frac{1}{17 - 3 - 1} \left((.053)^2 + (-.125)^2 + \cdots + (.265)^2 + (2.343)^2 \right)$$

$$= 1.26$$

so a corresponding estimate of σ is

$$s_{SF} = \sqrt{1.26}$$

$$= 1.125$$

(The units of y—and therefore s_{SF}—are .1% of incoming ammonia escaping unabsorbed.)

In routine practice it is a waste to do even these calculations, since multiple regression programs typically output s_{SF} as part of their analysis. The reader should take time to locate the value $s_{SF} = 1.125$ on Printout 2. If one accepts the relevance of model (9.41), for fixed values of airflow and inlet temperature (and therefore airflow squared), the standard deviation associated with many days' stack losses produced under those conditions would then be expected to be approximately .1125%.

Printout 2   Multiple Linear Regression for the Stack Loss Data (*Example 2*)

```
Regression Analysis

The regression equation is
y = - 15.4 - 0.069 x1 + 0.528 x2 + 0.00682 x1**2

Predictor          Coef        StDev          T         P
Constant         -15.41        12.60      -1.22     0.243
x1              -0.0691        0.3984      -0.17     0.865
x2               0.5278        0.1501       3.52     0.004
x1**2           0.006818      0.003178      2.15     0.051

S = 1.125      R-Sq = 98.0%      R-Sq(adj) = 97.5%

Analysis of Variance

Source            DF           SS          MS         F         P
Regression         3        799.80      266.60     210.81     0.000
Residual Error    13         16.44        1.26
Total             16        816.24

Source            DF       Seq SS
x1                 1       775.48
x2                 1        18.49
x1**2              1         5.82

Obs       x1          y         Fit    StDev Fit    Residual    St. Resid
 1      80.0      37.000      36.947      1.121       0.053        0.57 X
 2      62.0      18.000      18.125      0.407      -0.125       -0.12
 3      62.0      18.000      18.653      0.462      -0.653       -0.64
 4      62.0      19.000      19.181      0.553      -0.181       -0.18
 5      62.0      20.000      19.181      0.553       0.819        0.84
 6      58.0      15.000      15.657      0.513      -0.657       -0.66
 7      50.0      14.000      13.018      0.475       0.982        0.96
 8      58.0      14.000      13.018      0.475       0.982        0.96
 9      58.0      13.000      12.490      0.595       0.510        0.53
10      58.0      11.000      13.018      0.475      -2.018       -1.98
11      58.0      12.000      13.546      0.378      -1.546       -1.46
12      50.0       8.000       7.680      0.493       0.320        0.32
13      50.0       7.000       7.680      0.493      -0.680       -0.67
14      50.0       8.000       8.208      0.499      -0.208       -0.21
15      50.0       8.000       8.208      0.499      -0.208       -0.21
16      50.0       9.000       8.735      0.548       0.265        0.27
17      56.0      15.000      12.657      0.298       2.343        2.16R
```

R denotes an observation with a large standardized residual
X denotes an observation whose X value gives it large influence.

Predicted Values

```
    Fit  StDev Fit        95.0% CI              95.0% PI
 15.544      0.383   ( 14.717,  16.372) ( 12.978,  18.111)
```

Example 2
(*continued*)

Among the 17 data points in Table 4.8, there are only 12 different airflow/inlet temperature combinations (and therefore 12 different (x_1, x_2, x_1^2) vectors). The original data can be thought of as organized into $r = 12$ separate samples, one for each different (x_1, x_2, x_1^2) vector and there is thus an estimate of σ that doesn't depend for its validity on the appropriateness of the assumption that $\mu_{y|x_1,x_2} = \beta_0 + \beta_1 x_1 + \beta_2 x_2 + \beta_3 x_1^2$. That is, s_P can be computed and compared it to s_{SF} as a check on the appropriateness of model (9.41). Table 9.8 organizes the calculation of that pooled estimate of σ.

Table 9.8
Twelve Sample Means and Four Sample Variances for the Stack Loss Data

x_1, Air Flow	x_2, Inlet Temperature	y, Stack Loss	$\bar{y}$	s^2
50	18	8, 7	7.5	.5
50	19	8, 8	8.0	0.0
50	20	9	9.0	—
56	20	15	15.0	—
58	17	13	13.0	—
58	18	14, 14, 11	13.0	3.0
58	19	12	12.0	—
58	23	15	15.0	—
62	22	18	18.0	—
62	23	18	18.0	—
62	24	19, 20	19.5	.5
80	27	37	37.0	—

Then

$$s_P^2 = \frac{1}{17 - 12} \left((2-1)(.5) + (2-1)(0.0) + (3-1)(3.0) + (2-1)(.5) \right)$$

$$= 1.40$$

so

$$s_P = \sqrt{s_P^2} = \sqrt{1.40} = 1.183$$

The fact that $s_{SF} = 1.125$ and $s_P = 1.183$ are in substantial agreement is consistent with the work in Example 5 of Chapter 4, which found the equation

$$\hat{y} = -15.409 - .069x_1 + .528x_2 + .007x_1^2$$

to be a good summarization of the nitrogen plant data.

s_{SF} is basic to all of formal statistical inference based on the multiple linear regression model. But before using it to make statistical intervals and do significance testing, note also that it is useful for producing standardized residuals for the multiple linear regression model. That is, it is possible to find positive constants $a_1, a_2, \ldots, a_n$ (which are each complicated functions of all of $x_{11}, x_{21}, \ldots, x_{k1}, x_{12}, x_{22}, \ldots, x_{k2}, \ldots, x_{1n}, x_{2n}, \ldots, x_{kn}$) such that the ith residual $e_i = y_i - \hat{y}_i$ has

$$\text{Var}(y_i - \hat{y}_i) = a_i \sigma^2$$

Then, recalling Definition 2 in Chapter 7 (page 458), corresponding to the data point $(x_{1i}, x_{2i}, \ldots, x_{ki}, y_i)$ is the standardized residual for multiple linear regression

Standardized residuals for multiple linear regression

$$e_i^* = \frac{e_i}{s_{SF}\sqrt{a_i}} \tag{9.42}$$

It is not possible to include here a simple formula for the a_i that are needed to compute standardized residuals. (They are of interest only as building blocks in formula (9.42) anyway.) But it is easy to read the standardized residuals (9.42) off a typical multiple regression printout and to plot them in the usual ways as means of checking the apparent appropriateness of a candidate version of model (9.36) fit to a set of n data points $(x_1, x_2, \ldots, x_k, y)$.

Example 2
(continued)

As an illustration of the use of standardized residuals, consider again Printout 2 on page 679. The annotations on that printout locate the columns of residuals and standardized residuals for model (9.41). Figure 9.7 depicts normal probability plots, first of the raw residuals and then of the standardized residuals.

There are only the most minor differences between the appearances of the two plots in Figure 9.7, suggesting that decisions concerning the appropriateness of model (9.41) based on raw residuals will not be much altered by the more sophisticated consideration of standardized residuals instead.

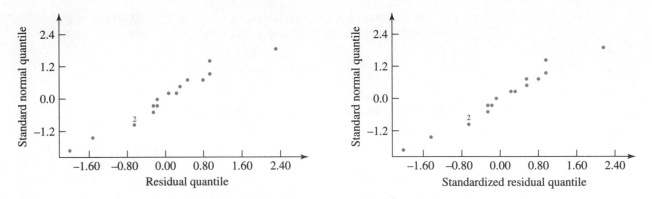

Figure 9.7 Normal plots of residuals and standardized residuals for the stack loss data (*Example 2*)

9.2.2 Inference for the Parameters $\beta_0, \beta_1, \beta_2, \ldots, \beta_k$

Section 9.1 considered inference for the slope parameter β_1 in simple linear regression, treating it as a rate of change (of average y as a function of x). In the multiple regression context, if $x_1, x_2, \ldots, x_k$ are all physically different system variables, the coefficients $\beta_1, \beta_2, \ldots, \beta_k$ can again be thought of as rates of change of average response with respect to $x_1, x_2, \ldots, x_k$, respectively. (They are partial derivatives of $\mu_{y|x_1,x_2,\ldots,x_k}$ with respect to the x's.) On the other hand, when some x's are functionally related to others (for instance, if $k = 2$ and $\mu_{y|x} = \beta_0 + \beta_1 x + \beta_2 x^2$), individual interpretation of the β's can be less straightforward. In any case, the β's do determine the nature of the surface represented by

$$\mu_{y|x_1,x_2,\ldots,x_k} = \beta_0 + \beta_1 x_1 + \beta_2 x_2 + \cdots + \beta_k x_k$$

and it is possible to do formal inference for $\beta_0, \beta_1, \ldots, \beta_k$ individually. In many instances, important physical interpretations can be found for such inferences. (For example, beginning with $\mu_{y|x} = \beta_0 + \beta_1 x + \beta_2 x^2$, an inference that β_2 is positive says that the mean response is concave up as a function of x and has a minimum value.)

The key to formal inference for the β's is that under model (9.36), there are positive constants $d_0, d_1, d_2, \ldots, d_k$ (which are each complicated functions of all of $x_{11}, \ldots, x_{k1}, x_{12}, \ldots, x_{k2} \ldots, x_{1n}, \ldots, x_{kn}$) such that the least squares coefficients $b_0, b_1, \ldots, b_k$ are normally distributed with

$$Eb_l = \beta_l$$

and

$$\mathrm{Var}\, b_l = d_l \sigma^2$$

This in turn makes it plausible that for $l = 0, 1, 2, \ldots, k$, the quantity

Estimated standard deviation of b_l

$$s_{\text{SF}}\sqrt{d_l} \tag{9.43}$$

is an estimate of the standard deviation of b_l and that

$$T = \frac{b_l - \beta_l}{s_{\text{SF}}\sqrt{d_l}} \tag{9.44}$$

has a t_{n-k-1} distribution.

There is no simple way to write down formulas for the constants d_l, but the estimated standard deviations of the coefficients, $s_{\text{SF}}\sqrt{d_l}$, are a typical part of the output from multiple linear regression programs.

The usual arguments of Chapter 6 applied to expression (9.44) then show that

$$\text{H}_0 : \beta_l - \# \tag{9.45}$$

can be tested using the test statistic

Test statistic for $H_0 : \beta_l = \#$

$$T = \frac{b_l - \#}{s_{\text{SF}}\sqrt{d_l}} \tag{9.46}$$

and a t_{n-k-1} reference distribution. More importantly, under the multiple linear regression model (9.36), a two-sided individual confidence interval for β_l can be made using endpoints

Confidence limits for β_l

$$b_l \pm t s_{\text{SF}}\sqrt{d_l} \tag{9.47}$$

where the associated confidence is the probability assigned to the interval between $-t$ and t by the t_{n-k-1} distribution. Appropriate use of only one of the endpoints (9.47) gives a one-sided interval for β_l.

Example 2
(continued)

Looking again at Printout 2 (see page 679), note that MINITAB's multiple regression output includes a table of estimated coefficients (b_l) and (estimated) standard deviations $(s_{\text{SF}}\sqrt{d_l})$. These are collected in Table 9.9.

Example 2
(*continued*)

Table 9.9
Fitted Coefficients and Estimates of Their Standard Deviations
for the Stack Loss Data

Estimated Coefficient	(Estimated) Standard Deviation of the Estimate
$b_0 = -15.41$	$s_{SF}\sqrt{d_0} = 12.60$
$b_1 = -.0691$	$s_{SF}\sqrt{d_1} = .3984$
$b_2 = .5278$	$s_{SF}\sqrt{d_2} = .1501$
$b_3 = .006818$	$s_{SF}\sqrt{d_3} = .003178$

Then since the upper .05 point of the t_{13} distribution is 1.771, from formula (9.47) a two-sided 90% confidence interval for β_2 in model (9.41) has endpoints

$$.5278 \pm 1.771(.1501)$$

that is,

$$.2620 \text{ (.1\% nitrogen loss/degree)} \quad \text{and} \quad .7936 \text{ (.1\% nitrogen loss/degree)}$$

This interval establishes that there is an increase in mean stack loss y with increased inlet temperature x_2 (the interval contains only positive values). It further gives a way of assessing the likely impact on y of various changes in x_2. For example, if x_1 (and therefore $x_3 = x_1^2$) is held constant but x_2 is increased by $2°$, one can anticipate an increase in mean stack loss of between

$$.5240 \text{ (.1\% nitrogen loss)} \quad \text{and} \quad 1.5873 \text{ (.1\% nitrogen loss)}$$

As a second example of the use of formula (9.47), note that a 90% two-sided confidence interval for β_3 has endpoints

$$.006818 \pm 1.771(.003178)$$

that is,

$$.0012 \quad \text{and} \quad .0124$$

β_3 controls the amount and direction of curvature (in the variable x_1) possessed by the surface specified by $\mu_{y|x_1,x_2} = \beta_0 + \beta_1 x_1 + \beta_2 x_2 + \beta_3 x_1^2$. Since the interval contains only positive values, it shows that at the 90% confidence level, there is some important concave-up curvature in the airflow variable needed to describe the stack loss variable. This is consistent with the picture of fitted mean response given previously in Figure 4.15 (see page 155).

However, check that if 95% confidence is used in the calculation of the two-sided interval for β_3, the resulting confidence interval contains values on both sides of 0. If this higher level of confidence is needed, the data in hand are not adequate to establish definitively the nature of any curvature in mean stack loss as a function of airflow. Any real curvature appears weak enough in comparison to the basic background variation that more data are needed to decide whether the surface is concave up, linear, or concave down in the variable x_1.

Very often multiple regression programs output not only the estimated standard deviations of fitted coefficients (9.43) but also the ratios

$$t = \frac{b_l}{s_{\text{SF}}\sqrt{d_l}}$$

and associated two-sided p-values for testing

$$H_0 : \beta_l = 0$$

Review Printout 2 and note that, for example, the two-sided p-value for testing $H_0 : \beta_3 = 0$ in model (9.41) is slightly larger than .05. This is completely consistent with the preceding discussion regarding the interpretation of interval estimates of β_3.

9.2.3 Inference for the Mean System Response for a Particular Set of Values for $x_1, x_2, \ldots, x_k$

Inference methods for the parameters $\beta_0, \beta_1, \ldots, \beta_k$ provide insight into the nature of the relationships between $x_1, x_2, \ldots, x_k$ and the mean response y. But other methods are needed to answer the important engineering question, "*What can be expected in terms of system response if I use a particular combination of levels of the system variables $x_1, x_2, \ldots, x_k$?*" An answer to this question will first be phrased in terms of inference methods for the mean system response $\mu_{y|x_1,x_2,\ldots,x_k}$.

In a manner similar to what was done in Section 9.1, the notation

Estimator of
$\mu_{y|x_1,x_2,\ldots,x_k}$

$$\hat{y} = b_0 + b_1 x_1 + b_2 x_2 + \cdots + b_k x_k \qquad (9.48)$$

will here be used for the value produced by the least squares equation when a particular set of numbers $x_1, x_2, \ldots, x_k$ is plugged into it. ($\hat{y}$ may not be a fitted value in the strict sense of the phrase, as the vector $(x_1, x_2, \ldots, x_k)$ may not match any data vector $(x_{1i}, x_{2i}, \ldots, x_{ki})$ used to produce the least squares coefficients $b_0, b_1, \ldots, b_k$.) As it turns out, the multiple linear regression model (9.36) leads to simple distributional properties for $\hat{y}$, which then produce inference methods for $\mu_{y|x_1,x_2,\ldots,x_k}$.

Under model (9.36), it is possible to find a positive constant A depending in a complicated way upon $x_1, x_2, \ldots, x_k$ and all of $x_{11}, \ldots, x_{k1}, x_{12}, \ldots, x_{k2}, \ldots, x_{1n}, \ldots, x_{kn}$ (the locations at which inference is desired and at which the original data points were collected) so that $\hat{y}$ has a normal distribution with

$$E\hat{y} = \mu_{y|x_1,x_2,\ldots,x_k} = \beta_0 + \beta_1 x_1 + \cdots + \beta_k x_k$$

and

$A = \sqrt{Var\hat{y}}/\sigma$

$$\text{Var } \hat{y} = \sigma^2 A^2 \tag{9.49}$$

In view of formula (9.49), it is thus plausible that

Estimated standard deviation of $\hat{y}$

$$s_{\text{SF}} \cdot A \tag{9.50}$$

can be used as an estimated standard deviation for $\hat{y}$ and that inference methods for the mean system response can be based on the fact that

$$T = \frac{\hat{y} - \mu_{y|x_1,x_2,\ldots,x_k}}{s_{\text{SF}} \cdot A}$$

has a t_{n-k-1} distribution. That is,

$$\text{H}_0: \mu_{y|x_1,x_2,\ldots,x_k} = \# \tag{9.51}$$

can be tested using the test statistic

Test statistic for
$\text{H}_0: \mu_{y|x_1,x_2,\ldots,x_k} = \#$

$$\boxed{T = \frac{\hat{y} - \#}{s_{\text{SF}} \cdot A}} \tag{9.52}$$

and a t_{n-k-1} reference distribution. Further, under the multiple linear regression model (9.36), a two-sided confidence interval for $\mu_{y|x_1,x_2,\ldots,x_k}$ can be made using endpoints

Confidence limits for the mean response
$\mu_{y|x_1,x_2,\ldots,x_k}$

$$\boxed{\hat{y} \pm t s_{\text{SF}} \cdot A} \tag{9.53}$$

where the associated confidence is the probability assigned to the interval between $-t$ and t by the t_{n-k-1} distribution. One-sided intervals based on formula (9.53) are made in the usual way.

Finding the factor A

    The practical obstacle to be overcome in the use of these methods is the computation of A. Although it is not possible to give a simple formula for A, most multiple regression programs provide A for $(x_1, x_2, \ldots, x_k)$ vectors of interest. MINITAB, for example, will fairly automatically produce values of $s_{\text{SF}} \cdot A$ corresponding to each data point $(x_{1i}, x_{2i}, \ldots, x_{ki}, y_i)$, labeled as (the estimated) **standard deviation (of the) fit**. And an option makes it possible to obtain similar information for *any* user-specified choice of $(x_1, x_2, \ldots, x_k)$. (Division of this by s_{SF} then produces A.)

Example 2
(continued)

Consider the problem of estimating the mean stack loss if the nitrogen plant of Example 5 in Chapter 4 is operated consistently with $x_1 = 58$ and $x_2 = 19$. (Notice that this means that $x_3 = x_1^2 = 3{,}364$ is involved.) Now the conditions $x_1 = 58$, $x_2 = 19$, and $x_3 = 3{,}364$ match perfectly those of data point number 11 on Printout 2 (see page 679). Thus, $\hat{y}$ and $s_{SF} \cdot A$ for these conditions may ▶ be read directly from the printout as 13.546 and .378, respectively. Then, for example, from formula (9.53), a 90% two-sided confidence interval for the mean stack loss corresponding to an airflow of 58 and water inlet temperature of 19 has endpoints

$$13.546 \pm 1.771(.378)$$

that is,

▶ 12.88 (.1% nitrogen loss)   and   14.22 (.1% nitrogen loss)

As a second illustration of the use of formula (9.53), suppose that setting plant operating conditions at an airflow of $x_1 = 60$ and a water inlet temperature of $x_2 = 20$ is contemplated and it is desireable to have an interval estimate for the mean stack loss implied by those conditions. Notice that the $x_1 = 60$, $x_2 = 20$, and $x_3 = x_1^2 = 3{,}600$ vector does not exactly match that of any of the $n = 17$ data points available. Therefore, some interpolation/extrapolation is required to make the desired interval. And it will not be possible to simply read appropriate values of $\hat{y}$ and $s_{SF} \cdot A$ off Printout 2 as related to one of the data points used to fit the equation.

Location of the point with coordinates $x_1 = 60$ and $x_2 = 20$ on a scatterplot of (x_1, x_2) values for the original $n = 17$ data points (like Figure 4.19) reveals that the candidate operating conditions are not wildly different from those used to develop the fitted equation. So there is hope that the use of formula (9.53) will provide an inference of some practical relevance. Accordingly, the coordinates $x_1 = 60$, $x_2 = 20$, and $x_3 = x_1^2 = 3{,}600$ were input into MINITAB and a "prediction" request made, resulting in the final section of Printout 2. Reading from that final section of the printout, $\hat{y} = 15.544$ and $s_{SF} \cdot A = .383$, so a 90% two-sided confidence interval for the mean stack loss has endpoints

$$15.544 \pm 1.771(.383)$$

that is,

▶ 14.86 (.1% nitrogen loss)   and   16.22 (.1% nitrogen loss)

(Of course, endpoints of a 95% interval can be read directly from the printout.)

Example 2
(continued)

It is impossible to overemphasize the fact that the preceding two intervals are dependent for their practical relevance on that of model (9.41) for not only those (x_1, x_2) pairs in the original data but (in the second case) also for the $x_1 = 60$ and $x_2 = 20$ set of conditions. Formulas like (9.53) always allow for imprecision due to statistical fluctuations/background noise in the data. They *do not*, however, allow for discrepancies related to the application of a model in a regime over which it is not appropriate. Formula (9.53) is an important and useful formula. But it should be used thoughtfully, with no expectation that it will magically do more than help quantify the precision provided by the data in the context of a particular set of model assumptions.

Lacking a simple explicit formula for A, it is difficult to be very concrete about how this quantity varies. In qualitative terms, it does change with the $(x_1, x_2, \ldots, x_k)$ vector under consideration. It is smallest when this vector is near the center of the cloud of points $(x_{1i}, x_{2i}, \ldots, x_{ki})$ in k-dimensional space corresponding to the n data points used to fit model (9.36). The fact that it can vary substantially is obvious from Printout 2. There for the nitrogen plant case, the estimated standard deviation of $\hat{y}$ given in display (9.50) varies from .298 to 1.121, indicating that A for data point 1 is about 3.8 times the size of A for data point 17 ($\frac{1.121}{.298} \approx 3.8$). That is, the precision with which a mean response is determined can vary widely over the region where it is sensible to use a fitted equation.

Formula (9.53) provides individual confidence intervals for mean responses. Simultaneous intervals are also easily obtained by a modification of formula (9.53) similar to the one provided for simple linear regression. That is, under the multiple linear regression model, simultaneous two-sided confidence intervals for all mean responses $\mu_{y|x_1, x_2, \ldots, x_k}$ can be made using respective endpoints

Simultaneous two-sided confidence limits for all mean repsonses

$$\mu_{y|x_1, x_2, \ldots, x_k}$$

$$\boxed{\hat{y} \pm \sqrt{(k+1)\, f}\, s_{SF} \cdot A} \qquad (9.54)$$

where for positive f, the associated confidence is the $F_{k+1, n-k-1}$ probability assigned to the interval $(0, f)$. Formula (9.54) is related to formula (9.53) through the replacement of the multiplier t by the (larger for a given nominal confidence) multiplier $\sqrt{(k+1)\, f}$. When it is applied only to $(x_1, x_2, \ldots, x_k)$ vectors found in the original n data points, formula (9.54) is an alternative to the P-R method of simultaneous intervals for means, appropriate to surface-fitting problems. When the multiple linear regression model is indeed appropriate, formula (9.54) will usually give shorter simultaneous intervals than the P-R method.

Example 2
(continued)

For making simultaneous 90% confidence intervals for the mean stack losses at the 12 different sets of plant conditions represented in the original data set, one can use formula (9.54) with $k = 3$, $f = 2.43$ (the .9 quantile of the $F_{4, 13}$ distribution) and the $\hat{y}$ and corresponding $s_{SF} \cdot A$ values appearing on Printout 2 (see page 679). For example, considering the $x_1 = 80$ and $x_2 = 27$ conditions of observation 1 on the printout, $s_{\text{SF}} \cdot A = 1.121$ and one of the simultaneous 90% confidence intervals associated with these conditions has endpoints

$$36.947 \pm \sqrt{(3+1)(2.43)(1.121)}$$

or

33.452 (.1% nitrogen loss)   and   40.442 (.1% nitrogen loss)

9.2.4 Prediction and Tolerance Intervals (*Optional*)

The second kind of answer that statistical theory can provide to the question, "What is to be expected in terms of system response if one uses a particular $(x_1, x_2, \ldots, x_k)$?", has to do with individual responses rather than mean responses. That is, the same factor A referred to in making confidence intervals for mean responses can be used to develop prediction and tolerance intervals for surface-fitting situations.

In the first place, under model (9.36), the two-sided interval with endpoints

Multiple regression prediction limits for an additional y at $(x_1, x_2, \ldots, x_k)$

$$\hat{y} \pm t s_{\text{SF}} \sqrt{1 + A^2} \qquad (9.55)$$

can be used as a prediction interval for an additional observation at a particular combination of levels of the variables $x_1, x_2, \ldots, x_k$. The associated prediction confidence is the probability that the t_{n-k-1} distribution assigns to the interval between $-t$ and t. One-sided intervals are made in the usual way, by employing only one of the endpoints (9.55) and adjusting the confidence level appropriately.

In order to use formula (9.55), $s_{\text{SF}} \cdot A$ and s_{SF} can be taken from a multiple regression printout and A obtained via division. Equivalently, it is possible to use a small amount of algebra to rewrite formula (9.55) as

An alternative formula for prediction limits

$$\hat{y} \pm t \sqrt{s_{\text{SF}}^2 + (s_{\text{SF}} \cdot A)^2} \qquad (9.56)$$

and substitute s_{SF} and $s_{\text{SF}} \cdot A$ directly into formula (9.56).

In order to find one-sided tolerance bounds in the surface-fitting context, begin with the value of A corresponding to a particular $(x_1, x_2, \ldots, x_k)$. If a confidence level of γ is desired in locating a fraction p of the underlying distribution of responses, compute

Multiplier to use in making tolerance intervals in multiple regression

$$\tau = \frac{Q_z(p) + A Q_z(\gamma) \sqrt{1 + \dfrac{1}{2(n-k-1)} \left(\dfrac{Q_z^2(p)}{A^2} - Q_z^2(\gamma) \right)}}{1 - \dfrac{Q_z^2(\gamma)}{2(n-k-1)}} \qquad (9.57)$$

Then, the interval

$$(\hat{y} - \tau s_{\text{SF}}, \infty)$$ **(9.58)**

or

$$(-\infty, \hat{y} + \tau s_{\text{SF}})$$ **(9.59)**

can be used as an approximately γ level one-sided tolerance interval for a fraction p of the underlying distribution of responses corresponding to $(x_1, x_2, \ldots, x_k)$.

Example 2
(continued)

Returning to the nitrogen plant example, consider first the calculation of a 90% lower prediction bound for a single additional stack loss y, if airflow of $x_1 = 58$ and water inlet temperature of $x_2 = 19$ are used. Then consider also a 95% lower tolerance bound for 90% of many additional stack loss values if the plant is run under those conditions.

Treating the prediction interval problem, recall that for $x_1 = 58$ and $x_2 = 19$, $\hat{y} = 13.546$ and $s_{\text{SF}} \cdot A = .378$. Since $s_{\text{SF}} = 1.125$ and the .9 quantile of the t_{13} distribution is 1.350, formula (9.56) shows that the desired 90% lower prediction bound for an additional stack loss under such plant operating conditions is

$$13.546 - 1.350\sqrt{(1.125)^2 + (.378)^2}$$

that is, approximately

$$11.94 \ (.1\% \text{ nitrogen loss})$$

To not predict a single additional stack loss, but rather to locate 90% of many additional stack losses with 95% confidence, expression (9.57) is the place to begin. Note that for $x_1 = 58$ and $x_2 = 19$,

$$A = .378/1.125 = .336$$

so, using expression (9.57),

$$\tau = \frac{1.282 + (.378)(1.645)\sqrt{1 + \dfrac{1}{2(17 - 3 - 1)}\left(\dfrac{(1.282)^2}{(.378)^2} - (1.645)^2\right)}}{1 - \dfrac{(1.645)^2}{2(17 - 3 - 1)}} = 2.234$$

So finally, a 95% lower tolerance bound for 90% of stack losses produced under operating conditions of $x_1 = 58$ and $x_2 = 19$ is, via display (9.58),

$$13.546 - 2.234(1.125) = 13.546 - 2.513$$

that is,

▶ $\qquad\qquad\qquad\qquad$ 11.033 (.1% nitrogen loss)

The warnings raised in the previous section concerning prediction and tolerance intervals in simple regression all apply equally to the present case of multiple regression. So do points similar to those made in Example 2 (page 688) in reference to confidence intervals for the mean system response. Although they are extremely useful engineering tools, statistical intervals are never any better than the models on which they are based.

9.2.5 Multiple Regression and ANOVA

Formal inference in curve- and surface-fitting contexts can (and typically should) be carried out primarily using interval-oriented methods. Nevertheless, testing and ANOVA methods do have their place. So the discussion now turns to the matter of what ANOVA ideas provide in multiple regression.

As always, $SSTot$ will stand for $\sum(y - \bar{y})^2$ and SSE for $\sum(y - \hat{y})^2$. Remember also that Definition 2 introduced the notation SSR for the difference $SSTot - SSE$. As remarked following Definition 2, the coefficient of determination can be written in terms of SSR and $SSTot$ as

$$R^2 = \frac{SSTot - SSE}{SSTot} = \frac{SSR}{SSTot}$$

Further, under model (9.36), these sums of squares ($SSTot$, SSE, and SSR) form the basis of an F test of the hypothesis

$$H_0 : \beta_1 = \beta_2 = \cdots = \beta_k = 0 \qquad\qquad\qquad \textbf{(9.60)}$$

versus

$$H_a : \text{not } H_0 \qquad\qquad\qquad \textbf{(9.61)}$$

Hypothesis (9.60) can be tested using the statistic

F statistic for testing
$H_0 : \beta_1 = \beta_2 = \cdots = \beta_k = 0$

$$F = \frac{SSR/k}{SSE/(n - k - 1)} \qquad\qquad\qquad \textbf{(9.62)}$$

and an $F_{k,n-k-1}$ reference distribution, where large observed values of the test statistic constitute evidence against H_0. (The denominator of statistic (9.62) is another way of writing s_{SF}^2.)

Hypothesis (9.60) in the context of the multiple linear regression model implies that the mean response doesn't depend on any of the process variables $x_1, x_2, \ldots, x_k$. That is, if all of β_1 through β_k are 0, model statement (9.36) reduces to

$$y_i = \beta_0 + \epsilon_i$$

Interpreting a test of
$H_0: \beta_1 = \beta_2 = \cdots = \beta_k = 0$

So a test of hypothesis (9.60) is often interpreted as a test of whether the mean response is related to any of the input variables under consideration. The calculations leading to statistic (9.62) are most often organized in a table quite similar to the one discussed in Section 9.1 for testing $H_0: \beta_1 = 0$ in simple linear regression. The general form of that table is given as Table 9.10.

Table 9.10
General Form of the ANOVA Table for Testing $H_0: \beta_1 = \beta_2 = \cdots = \beta_k = 0$ in Multiple Regression

ANOVA Table (for testing $H_0: \beta_1 = \beta_2 = \cdots = \beta_k = 0$)				
Source	SS	df	MS	F
Regression	SSR	k	SSR/k	MSR/MSE
Error	SSE	n − k − 1	SSE/(n − k − 1)	
Total	SSTot	n − 1		

Example 2
(continued)

Once again turning to the analysis of the nitrogen plant data under the model $y_i = \beta_0 + \beta_1 x_{1i} + \beta_2 x_{2i} + \beta_3 x_{1i}^2 + \epsilon_i$, consider testing $H_0: \beta_1 = \beta_2 = \beta_3 = 0$—that is, mean stack loss doesn't depend on airflow (or its square) or water inlet temperature. Printout 2 (see page 679) includes an ANOVA table for testing this hypothesis, which is essentially reproduced here as Table 9.11.

From Table 9.11, the observed value of the F statistic is 210.81, which is to be compared to $F_{3,13}$ quantiles in order to produce an observed level of significance. As indicated in Printout 2, the $F_{3,13}$ probability to the right of the value 210.81 is 0 (to three decimal places). This is definitive evidence that not all of β_1, β_2, and β_3 can be 0. Taken as a group, the variables x_1, x_2, and $x_3 = x_1^2$ definitely enhance one's ability to predict stack loss.

Table 9.11
ANOVA Table for Testing $H_0 : \beta_1 = \beta_2 = \beta_3 = 0$ for the Stack Loss Data

ANOVA Table (for testing $H_0 : \beta_1 = \beta_2 = \beta_3 = 0$)				
Source	SS	df	MS	F
Regression (on x_1, x_2, x_1^2)	799.80	3	266.60	210.81
Error	16.44	13	1.26	
Total	816.24	16		

Note also that the value of the coefficient of determination here can be calculated using sums of squares given in Table 9.11 as

$$R^2 = \frac{SSR}{SSTot} = \frac{799.80}{816.24} = .980$$

This is the value for R^2 advertised long ago in Example 5 in Chapter 4. Also, the error mean square, $MSE = 1.26$, is (as expected) exactly the value of s_{SF}^2 calculated earlier in this example.

It is a matter of simple algebra to verify that R^2 and the F statistic (9.62) are equivalent in the sense that

An expression for the F statistic (9.62) in terms of R^2

$$F = \frac{R^2/k}{(1 - R^2)/(n - k - 1)} \qquad (9.63)$$

so the F test of hypothesis (9.60) can be thought of in terms of attaching a p-value to the statistic R^2. This is a valuable development, but it should be remembered that it is R^2 (rather than F) that has the direct interpretation as a measure of what fraction of raw variability the fitted equation accounts for. F and its associated p-value take account of the sample size n in a way that R^2 doesn't. They really measure statistical detectability rather than variation accounted for. This means that an equation that accounts for a fraction of observed variation that is relatively small by most standards can produce a very impressive (small) p-value. If this point is not clear, try using formula (9.63) to find the p-value for a situation where $n = 1,000$, $k = 4$, and $R^2 = .1$.

From Section 4.2 on, R^2 values have been used in this book for informal comparisons of various potential summary equations for a single data set. It turns out that it is sometimes possible to attach p-values to such comparisons through the use of the corresponding regression sums of squares and *another* F test.

Suppose that there are two different regression models for describing a data set—the first of the usual form (9.36) for k input variables $x_1, x_2, \ldots, x_k$,

$$y_i = \beta_0 + \beta_1 x_{1i} + \beta_2 x_{2i} + \cdots + \beta_k x_{ki} + \epsilon_i$$

and the second being a specialization of the first where some p of the coefficients β (say, $\beta_{l_1}, \beta_{l_2}, \ldots, \beta_{l_p}$) are all 0 (i.e., a specialization not involving input variables $x_{l_1}, x_{l_2}, \ldots, x_{l_p}$). The first of these models will be called the **full regression model** and the second a **reduced regression model**. When one informally compares R^2 values for two such models, the comparison is essentially between SSR values, since the two R^2 values share the same denominator, $SSTot$. The two SSR values can be used to produce an observed level of significance for the comparison.

Under model the full model (9.36), the hypothesis

$$H_0: \beta_{l_1} = \beta_{l_2} = \cdots = \beta_{l_p} = 0 \tag{9.64}$$

(that the reduced model holds) can be tested against

$$H_a: \text{ not } H_0 \tag{9.65}$$

using the test statistic

F statistic for testing
$H_0: \beta_{l_1} = \cdots = \beta_{l_p} = 0$
in multiple regression

$$F = \frac{(SSR_f - SSR_r)/p}{SSE_f/(n - k - 1)} \tag{9.66}$$

and an $F_{p,n-k-1}$ reference distribution, where large observed values of the test statistic constitute evidence against H_0 in favor of H_a. In expression (9.66), the "f" and "r" subscripts refer to the *full* and *reduced* regressions. The calculation of statistic (9.66) can be facilitated by expanding the basic ANOVA table for the full model (Table 9.10). Table 9.12 shows one form this can take.

Table 9.12
Expanded ANOVA Table for Testing $H_0: \beta_{l_1} = \beta_{l_2} = \cdots = \beta_{l_p} = 0$ in Multiple Regression

ANOVA Table (for testing $H_0: \beta_{l_1} = \beta_{l_2} = \cdots = \beta_{l_p} = 0$)				
Source	SS	df	MS	F
Regression (full)	SSR_f	k		
Regression (reduced)	SSR_r	$k - p$		
Regression (full | reduced)	$SSR_f - SSR_r$	p	$(SSR_f - SSR_r)/p$	$MSR_{f|r}/MSE_f$
Error	SSE_f	$n - k - 1$	$SSE_f/(n - k - 1)$	
Total	$SSTot$	$n - 1$		

Example 2
(continued)

In the nitrogen plant example, consider the comparison of the two possible descriptions of stack loss

$$y \approx \beta_0 + \beta_1 x_1 \tag{9.67}$$

(stack loss is approximately a linear function of airflow only) and

$$y \approx \beta_0 + \beta_1 x_1 + \beta_2 x_2 + \beta_3 x_1^2 \tag{9.68}$$

(the description of stack loss that has been used throughout this section). Although a printout won't be included here to show it, it is a simple matter to verify that the fitting of expression (9.67) to the nitrogen plant data produces $SSR = 775.48$ and therefore $R^2 = .950$. Fitting expression (9.68), on the other hand, gives $SSR = 799.80$ and $R^2 = .980$. Since expression (9.67) is the specialization/reduction of expression (9.68) obtained by dropping the last $p = 2$ terms, the comparison of these two SSR (or R^2) values can be formalized with a p-value. A test of

$$H_0: \beta_2 = \beta_3 = 0$$

can be made in the (full) model (9.68). Table 9.13 organizes the calculation of the observed value of the statistic (9.66) for this problem. That is,

$$f = \frac{(799.80 - 775.48)/2}{16.44/13} = 9.7$$

When compared with tabled $F_{2,13}$ percentage points, the observed value of 9.7 is seen to produce a p-value between .01 and .001. There is strong evidence in the nitrogen plant data that an explanation of mean response in terms of expression (9.68) (pictured, for example, in Figure 4.15) is superior to one in terms of expression (9.67) (which could be pictured as a single linear mean response in x_1 for all x_2).

Table 9.13
ANOVA Table for Testing $H_0: \beta_2 = \beta_3 = 0$ in Model (9.68) for the Stack Loss Data

ANOVA Table (for testing $H_0: \beta_2 = \beta_3 = 0$)				
Source	SS	df	MS	F
Regression (x_1, x_2, x_1^2)	799.80	3		
Regression (x_1)	775.48	1		
Regression $(x_2, x_1^2 \mid x_1)$	24.32	2	12.16	9.7
Error (x_1, x_2, x_1^2)	16.44	13	1.26	
Total	816.24	16		

The F statistic (9.66) can be written in terms of R^2 values as

$$F = \frac{(R_f^2 - R_r^2)/p}{(1 - R_f^2)/(n - k - 1)} \tag{9.69}$$

Interpreting full and reduced R^2's and the F test

so that the test of hypothesis (9.64) is indeed a way of attaching a p-value to the comparison of two R^2's. However, just as was remarked earlier concerning the test of hypothesis (9.60), it is the R^2's themselves that indicate how much additional variation a full model accounts for over a reduced model. The observed F value or associated p-value measures the extent to which that increase is distinguishable from background noise.

p tests that single coefficients are 0 versus a test that p coefficients are all 0

To conclude this section, something needs to be said about the relationship between the tests of hypotheses (9.45) (with # = 0), mentioned earlier, and the tests of hypothesis (9.64) based on the F statistic (9.66). When $p = 1$ (the full model contains only one more term than the reduced model), observed levels of significance based on statistic (9.66) are in fact equal to two-sided observed levels of significance based on # = 0 versions of statistic (9.46). But for cases where $p \geq 2$, the tests of the hypotheses that individual β's are 0 (one at a time) are not an adequate substitute for the tests of hypothesis (9.64). For example, in the full model

$$y = \beta_0 + \beta_1 x_1 + \beta_2 x_2 + \beta_3 x_3 + \epsilon \tag{9.70}$$

testing

$$H_0: \beta_2 = 0 \tag{9.71}$$

and then testing

$$H_0: \beta_3 = 0 \tag{9.72}$$

need not be at all equivalent to making a single test of

$$H_0: \beta_2 = \beta_3 = 0 \tag{9.73}$$

This fact may at first seem paradoxical. But should the variables x_2 and x_3 be reasonably highly correlated in the data set, it is possible to get large p-values for tests of both hypothesis (9.71) and (9.72) and yet a tiny p-value for a test of hypothesis (9.73). The message carried by such an outcome is that (due to the fact that the variables x_2 and x_3 appear in the data set to be more or less equivalent) in the presence of x_1 and x_2, x_3 is not needed to model y. And in the presence of x_1 and x_3, x_2 is not needed to model y. But one or the other of the two variables x_2 and x_3 is needed to help model y even in the presence of x_1. So, the F test of hypothesis (9.64) is more than just a fancy version of several tests of hypotheses $H_0: \beta_l = 0$. It is an important addition to an engineer's curve- and surface-fitting tool kit.

Section 2 Exercises

1. Return to the situation of Chapter Exercise 2 of Chapter 4 and the carburetion study of Griffith and Tesdall. Consider an analysis of these data based on the model $y = \beta_0 + \beta_1 x + \beta_2 x^2 + \epsilon$.

   (a) Find s_{SF} for these data. What does this intend to measure in the context of the engineering problem?

   (b) Plot both residuals versus x and the standardized residuals versus x. How much difference is there in the appearance of these two plots?

   (c) Give 90% individual two-sided confidence intervals for each of β_0, β_1, and β_2.

   (d) Give individual 90% two-sided confidence intervals for the mean elapsed time with a carburetor jetting size of 70 and then with a jetting size of 76.

   (e) Give simultaneous 90% two-sided confidence intervals for the two means indicated in part (d).

   (f) Give 90% lower prediction bounds for an additional elapsed time with a carburetor jetting size of 70 and also with a jetting size of 76.

   (g) Give approximate 95% lower tolerance bounds for 90% of additional elapsed times, first with a carburetor jetting size of 70 and then with a jetting size of 76.

   (h) Make an ANOVA table for testing $H_0: \beta_1 = \beta_2 = 0$ in the model $y = \beta_0 + \beta_1 x + \beta_2 x^2 + \epsilon$. What is the meaning of this hypothesis in the context of the study and the quadratic model? What is the p-value?

   (i) Use a t statistic and test the null hypothesis $H_0: \beta_2 = 0$. What is the meaning of this hypothesis in the context of the study and the quadratic model?

2. Return to the situation of Exercise 2 of Section 4.2, and the chemithermomechanical pulp study of Miller, Shankar, and Peterson. Consider an analysis of the data there based on the model $y = \beta_0 + \beta_1 x_1 + \beta_2 x_2 + \epsilon$.

   (a) Find s_{SF}. What does this intend to measure in the context of the engineering problem?

   (b) Plot both residuals and standardized residuals versus x_1, x_2, and $\hat{y}$. How much difference is there in the appearance of these pairs of plots?

   (c) Give 90% individual two-sided confidence intervals for all of β_0, β_1, and β_2.

   (d) Give individual 90% two-sided confidence intervals for the mean specific surface area, first when $x_1 = 9.0$ and $x_2 = 60$ and then when $x_1 = 10.0$ and $x_2 = 70$.

   (e) Give simultaneous 90% two-sided confidence intervals for the two means indicated in part (d).

   (f) Give 90% lower prediction bounds for the next specific surface area, first when $x_1 = 9.0$ and $x_2 = 60$ and then when $x_1 = 10.0$ and $x_2 = 70$.

   (g) Give approximate 95% lower tolerance bounds for 90% of specific surface areas, first when $x_1 = 9.0$ and $x_2 = 60$ and then when $x_1 = 10.0$ and $x_2 = 70$.

   (h) Make an ANOVA table for testing $H_0: \beta_1 = \beta_2 = 0$ in the model $y = \beta_0 + \beta_1 x_1 + \beta_2 x_2 + \epsilon$. What is the p-value?

9.3 Application of Multiple Regression in Response Surface Problems and Factorial Analyses

The discussions in Sections 4.1, 4.2, 9.1, and 9.2 have, we hope, given you a growing appreciation of the wide utility of regression methods in engineering. The purpose of this final section is to further expand your range of experience with multiple regression by illustrating its usefulness in two additional contexts. First there is an illustration of how surface fitting is used in "response surface" (or response optimization) problems. Then there is a look at how regression has its applications even in factorial analyses.

9.3.1 Surface-Fitting and Response Surface Studies

Engineers are often called upon to address the following generic problem. A response or responses y are known to depend upon system variables $x_1, x_2, \ldots, x_k$. No simple physical theory is available for describing the dependence. Nevertheless, the variables $x_1, x_2, \ldots, x_k$ need adjustment to get good system behavior (as measured by the variables y). Multiple regression analysis and some specialized "response surface" considerations often prove effective in such problems.

Fitted linear and quadratic functions as empirical models
    For one thing, **linear and quadratic functions** of $x_1, x_2, \ldots, x_k$ are often useful empirical descriptions of a relationship between $x_1, x_2, \ldots, x_k$ and y. The material in Sections 4.2 and 9.2 directly addresses fitting and inference for a linear approximate relationship like

$$y \approx \beta_0 + \beta_1 x_1 + \beta_2 x_2 + \cdots + \beta_k x_k \tag{9.74}$$

Response surfaces specified by equation (9.74) are "planar" (see again Figure 9.6 in this regard). When such surfaces fail to capture the nature of dependence of y on $x_1, x_2, \ldots, x_k$ because of their "lack of curvature," quadratic approximate relationships often prove effective. The general version of a quadratic equation for y in k variables x has k linear terms, k quadratic terms, and cross product terms for all pairs of x variables. For example, the general 3-variable quadratic response surface is specified by

$$y \approx \beta_0 + \beta_1 x_1 + \beta_2 x_2 + \beta_3 x_3 + \beta_4 x_1^2 + \beta_5 x_2^2 + \beta_6 x_3^2 + \beta_7 x_1 x_2$$
$$+ \beta_8 x_1 x_3 + \beta_9 x_2 x_3 \tag{9.75}$$

Gathering adequate data
    One issue in using the k-variable version of quadratic function (9.75) is that of collecting adequate data to support the enterprise. 2^k factorial data are not sufficient. This is easy to see by considering the $k = 1$ case. Having data for only two different values of x_1, say $x_1 = 0$ and $x_1 = 1$, would not be adequate to support the fitting of

$$y \approx \beta_0 + \beta_1 x_1 + \beta_2 x_1^2 \tag{9.76}$$

There are, as an arbitrary example, many different versions of equation (9.76) with $y = 5$ for $x_1 = 0$ and $y = 7$ for $x_1 = 1$, including

$$y \approx 5 + 2x_1 + 0x_1^2$$
$$y \approx 5 - 8x_1 + 10x_1^2$$
$$y \approx 5 + 10x_1 - 8x_1^2$$

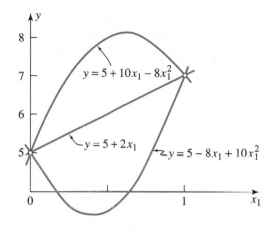

Figure 9.8   Plots of three different quadratic functions passing through the points $(x_1, y) = (0, 5)$ and $(x_1, y) = (1, 7)$

These three equations have plots with quite different shapes. The first is linear, the second is concave up with a minimum at $x_1 = .4$, and the third is concave down with a maximum at $x_1 = .625$. This is illustrated in Figure 9.8. The point is that data from at least three different x_1 values are needed in order to fit a one-variable quadratic equation.

What would happen if a regression program were used to fit equation (9.76) to a set of (x_1, y) data having only two different x_1 values in it? The program will typically refuse the user's request, perhaps fitting instead the simpler equation $y \approx \beta_0 + \beta_1 x_1$.

Exactly what *is* needed in the way of data in order to fit a k-variable quadratic equation is not easy to describe in elementary terms. 3^k factorial data are sufficient but for large k are really much more than are absolutely necessary. Statisticians have invested substantial effort in identifying patterns of $(x_1, x_2, \ldots, x_k)$ combinations that are both small (in terms of number of different combinations) and effective (in terms of facilitating precise estimation of the coefficients in a quadratic response function). See, for example, Section 7.2.2 of *Statistical Quality Assurance Methods for Engineers* by Vardeman and Jobe for a discussion of "central composite" plans often employed to gather data adequate to fit a quadratic. An early successful application of such a plan is described next.

Example 3

A Central Composite Study for Optimizing Bread Wrapper Seal Strength

The article "Sealing Strength of Wax-Polyethylene Blends" by Brown, Turner, and Smith (*Tappi*, 1958) contains an interesting central composite data set. The effects of the three process variables Seal Temperature, Cooling Bar Temperature, and % Polyethylene Additive on the seal strength y of a bread wrapper stock were studied. With the coding of the process variables indicated in Table 9.14, the data

Example 3
(*continued*)

Table 9.14

Coding of Three Process Variables in a Seal Strength Study

Factor		Variable	
A	Seal Temperature	$x_1 = \dfrac{t_1 - 255}{30}$	where t_1 is in °F
B	Cooling Bar Temperature	$x_2 = \dfrac{t_2 - 55}{9}$	where t_2 is in °F
C	Polyethylene Content	$x_3 = \dfrac{c - 1.1}{.6}$	where c is in %

Table 9.15

Seal Strengths Produced under 15 Different Sets of Process Conditions

x_1	x_2	x_3	Seal Strength, y (g/in.)
−1	−1	−1	6.6
1	−1	−1	6.9
−1	1	−1	7.9
1	1	−1	6.1
−1	−1	1	9.2
1	−1	1	6.8
−1	1	1	10.4
1	1	1	7.3
0	0	0	10.1
0	0	0	9.9
0	0	0	12.2
0	0	0	9.7
0	0	0	9.7
0	0	0	9.6
−1.682	0	0	9.8
1.682	0	0	5.0
0	−1.682	0	6.9
0	1.682	0	6.3
0	0	−1.682	4.0
0	0	1.682	8.6

in Table 9.15 were obtained. Notice that there are fewer than $3^3 = 27$ different (x_1, x_2, x_3) vectors in these data. (The central composite plan involves only 15 different combinations.)

If one fits a first-order (linear) model

$$y = \beta_0 + \beta_1 x_1 + \beta_2 x_2 + \beta_3 x_3 + \epsilon \qquad (9.77)$$

to the data points listed in Table 9.15, a coefficient of determination of only $R^2 = .38$ is obtained, along with $s_{SF} = 1.79$. The pooled sample standard deviation (coming from the six points with $x_1 = 0$, $x_2 = 0$, and $x_3 = 0$) is quite a bit smaller than s_{SF}—namely, $s_P = 1.00$. Between the small value of R^2 and the moderate difference between s_{SF} and s_P, there is already some indication that model (9.77) may be a poor description of the data. A residual analysis like those done in Section 4.2 would further confirm this.

On the other hand, fitting the expression (9.75) to the data in Table 9.15 produces the equation

$$\hat{y} = 10.165 - 1.104x_1 + .0872x_2 + 1.020x_3 - .7596x_1^2 - 1.042x_2^2$$

$$- 1.148x_3^2 - .3500x_1 x_2 - .5000x_1 x_3 + .1500x_2 x_3 \qquad (9.78)$$

with a coefficient of determination of $R^2 = .86$ and $s_{SF} = 1.09$. At least on the basis of the two measures R^2 and s_{SF}, this quadratic description of seal strength seems much superior to a first-order description.

Plots and
interpreting a
fitted quadratic   For small values of k, the interpretation of a fitted quadratic response function can be facilitated through the use of various plots. One possibility is to **plot $\hat{y}$ versus a particular system variable x**, with values of any other system variables held fixed. This was the method used in Figure 4.15 for the nitrogen plant data, in Figure 4.16 (see page 158) for the lift/drag ratio data of Burris, and in Figure 9.8 of this section for the hypothetical one-variable quadratics. (It is also worth noting that in light of the inference material presented in Section 9.2, one can enhance such plots of $\hat{y}$ by adding error bars based on confidence limits for the means $\mu_{y|x_1,x_2,...,x_k}$.)

A second kind of plot that can help in understanding a fitted quadratic function is the **contour plot**. A contour plot is essentially a topographic map. For a given pair of system variables (say x_1 and x_2) one can, for fixed values of all other input variables, sketch out the loci of points in the (x_1, x_2)-plane that produce several particular values of $\hat{y}$. Most statistical packages and engineering mathematics packages will make contour plots.

Example 3
(continued)   Figure 9.9 shows a series of five contour plots made using the fitted equation (9.78) for seal strength. These correspond to $x_3 = -2, -1, 0, 1$, and 2. The figure suggests that optimum predicted seal strength may be achievable for x_3 between 0 and 1, with x_1 between -2 and -1, and x_2 between 0 and 1.

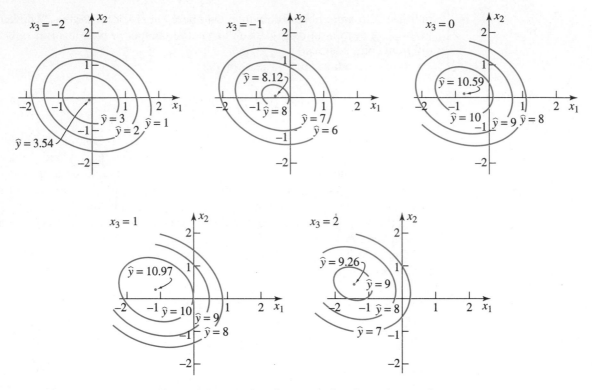

Figure 9.9 A series of contour plots for seal strength

Analytic interpretation of a fitted quadratic

Plotting is helpful in understanding a fitted quadratic primarily for small k. So it is important that there are also **analytical tools** that can be employed. To illustrate their character, consider the simple case of $k = 1$. The basic nature of the quadratic equation

$$\hat{y} = b_0 + b_1 x_1 + b_2 x_1^2$$

is governed by b_2. For $b_2 > 0$ it describes a parabola opening up. For $b_2 < 0$ it describes a parabola opening down. And for $b_2 = 0$ it describes a line. Provided $b_2 \neq 0$ the value

$$x_1 = -\frac{b_1}{2b_2}$$

produces the minimum ($b_2 > 0$) or maximum ($b_2 < 0$) value of $\hat{y}$. Something like this story is also true for $k > 1$.

It is necessary to use some matrix notation to say what happens for $k > 1$. Temporarily modify the way the b's are subscripted as follows. The meaning of b_0 will remain unchanged. b_1 through b_k will be the coefficients for the k system variables x_1 through x_k. b_{11} through b_{kk} will be the coefficients for the k squares x_1^2 through x_k^2. And for each $i \neq j$, b_{ij} will be the coefficient of the $x_i x_j$ cross product. One can define a $k \times 1$ vector $\boldsymbol{b}$ and a $k \times k$ matrix $\boldsymbol{B}$ as

Vector of linear coefficients and matrix of quadratic coefficients

$$\boldsymbol{b} = \begin{bmatrix} b_1 \\ b_2 \\ \vdots \\ b_k \end{bmatrix}$$

$$\boldsymbol{B} = \begin{bmatrix} b_{11} & \frac{1}{2}b_{12} & \cdots & \frac{1}{2}b_{1k} \\ \frac{1}{2}b_{12} & b_{22} & \cdots & \frac{1}{2}b_{2k} \\ \vdots & \vdots & & \vdots \\ \frac{1}{2}b_{1k} & \frac{1}{2}b_{2k} & \cdots & b_{kk} \end{bmatrix}$$

With

$$\boldsymbol{x} = \begin{bmatrix} x_1 \\ x_2 \\ \vdots \\ x_k \end{bmatrix}$$

Provided the matrix $\boldsymbol{B}$ is nonsingular, the corresponding k-variable quadratic then has a **stationary point** (i.e., a point at which first partial derivatives with respect to $x_1, x_2, \ldots, x_k$ are all 0) where

Location of a stationary point for a k-variable fitted quadratic

$$\boldsymbol{x} = -\tfrac{1}{2}\boldsymbol{B}^{-1}\boldsymbol{b} \tag{9.79}$$

And depending upon the nature of $\boldsymbol{B}$, the stationary point will be either **a minimum, a maximum, or a saddle point** of the fitted response. (Moving away from a saddle point in some directions produces an increase in $\hat{y}$, while moving away in other directions produces a decrease.)

It is the **eigenvalues** of $\boldsymbol{B}$ that are critical in determining the shape of the fitted quadratic surface. The eigenvalues of $\boldsymbol{B}$ are the k solutions of the equation (in λ)

Equation solved by the eigenvalues λ of the matrix $\boldsymbol{B}$

$$\det(\boldsymbol{B} - \lambda\boldsymbol{I}) = 0 \tag{9.80}$$

where $\boldsymbol{I}$ is the identity matrix. (Most statistical analysis packages and engineering mathematics packages will compute eigenvalues quite painlessly.)

When all solutions to equation (9.80) are positive, a fitted quadratic is bowl-shaped up and has a minimum at the point (9.79). When all solutions to equation

(9.80) are negative, a fitted quadratic is bowl-shaped down and has a maximum at the point (9.79). When some solutions to equation (9.80) are positive and some are negative, the fitted quadratic surface has neither a maximum nor minimum (unless one restricts attention to some bounded region of x vectors).

Printout 3   Analysis of the Fitted Quadratic for the Bread Wrapper Data (*Example 3*)

```
MTB > Read 3 3 M1.
DATA> -.7596 -.175 -.250
DATA> -.175 -1.042 .075
DATA> -.250 .075 -1.148
       3 rows read.
MTB > Read 3 1 M2.
DATA> -1.104
DATA> .0872
DATA> 1.020
       3 rows read.
MTB > Eigen M1 C1.
MTB > Print C1.

Data Display


C1
  -1.27090  -1.11680  -0.56190

MTB > Invert M1 M3.
MTB > Multiply M3 M2 M4.
MTB > Multiply M4 -.5 M5.
MTB > Print  M5.

Data Display


 Matrix M5

-1.01104
 0.26069
 0.68146
```

Example 3
(*continued*)

Printout 3 illustrates the use of MINITAB in the analytic investigation of the nature of the fitted surface (9.78) in the bread wrapper seal strength study. The printout shows the three eigenvalues of B to be negative. The fitted seal strength therefore has a maximum. This maximum is predicted to occur at the combination of values $x_1 = -1.01$, $x_2 = .26$, and $x_3 = .68$. (The MINITAB matrix functions used to make the printout are under the "Calc/Matrices" menu, and the display routine is under the "Manip/Display Data" menu.)

The discussion of response surface studies in this subsection isn't intended to be complete. Whole books, like, for example, Box and Draper's *Empirical Model-Building and Response Surfaces,* have been written on the subject. (Section 9.3 of Vardeman's *Statistics for Engineering Problem Solving* contains a more complete discussion than the present one, is still short of a book-length treatment.) We hope, however, this brief look at the topic suffices to indicate its importance to engineering practice.

9.3.2 Regression and Factorial Analyses

Many of the factorial inference methods discussed in this book are applicable only in balanced-data situations. For example, remember that the use of the reverse Yates algorithm to fit few-effects 2^p factorial models and the methods of interval-oriented inference for 2^p studies under few-effects models discussed in Section 8.2 are limited to balanced-data applications.

But by accident if not by design, an engineer will eventually face the analysis of unbalanced factorial data. Happily enough, this can be accomplished through use of the multiple regression formulas provided in Section 9.2. This subsection shows how factorial analyses can be thought of in multiple regression terms. It begins with a discussion of two-way factorial cases and then considers three-way (and higher) situations.

The basic multiple regression model equation used in Section 9.2,

$$y_i = \beta_0 + \beta_1 x_{1i} + \beta_2 x_{2i} + \cdots + \beta_k x_{ki} + \epsilon_i \tag{9.81}$$

looks deceptively simple. With proper choice of the inputs x, versions of it can be used in a wide variety of contexts, including factorial analyses. For purposes of illustration, consider the case of a complete two-way factorial study with $I = 3$ levels of factor A and $J = 3$ levels of factor B. In the usual two-way factorial notation introduced in Definitions 1 and 2 of Chapter 8, the basic constraints on the main effects and two-factor interactions are $\sum_i \alpha_i = 0$, $\sum_j \beta_j = 0$, and $\sum_i \alpha\beta_{ij} = \sum_j \alpha\beta_{ij} = 0$. These imply that the $I \cdot J = 3 \cdot 3 = 9$ different mean responses in such a study,

$$\mu_{ij} = \mu_{..} + \alpha_i + \beta_j + \alpha\beta_{ij} \tag{9.82}$$

can be written as displayed in Table 9.16.

At first glance, the advantage of writing out these mean responses in terms of only effects corresponding to the first 2 ($= I - 1$) levels of A and first 2 ($= J - 1$) levels of B is not obvious. But doing so expresses the 9 ($= I \cdot J$) different means in terms of only as many different parameters as there are means, and helps one find a regression-type analog of expression (9.82).

Notice first that $\mu_{..}$ appears in each mean response listed and therefore plays a role much like that of the intercept term β_0 in a regression model. Further, the two A main effects, α_1 and α_2, appear with positive signs when (respectively) $i = 1$

Table 9.16
Mean Responses in a 3^2 Factorial Study

i, Level of A	j, Level of B	Mean Response
1	1	$\mu_{..} + \alpha_1 + \beta_1 + \alpha\beta_{11}$
1	2	$\mu_{..} + \alpha_1 + \beta_2 + \alpha\beta_{12}$
1	3	$\mu_{..} + \alpha_1 - \beta_1 - \beta_2 - \alpha\beta_{11} - \alpha\beta_{12}$
2	1	$\mu_{..} + \alpha_2 + \beta_1 + \alpha\beta_{21}$
2	2	$\mu_{..} + \alpha_2 + \beta_2 + \alpha\beta_{22}$
2	3	$\mu_{..} + \alpha_2 - \beta_1 - \beta_2 - \alpha\beta_{21} - \alpha\beta_{22}$
3	1	$\mu_{..} - \alpha_1 - \alpha_2 + \beta_1 - \alpha\beta_{11} - \alpha\beta_{21}$
3	2	$\mu_{..} - \alpha_1 - \alpha_2 + \beta_2 - \alpha\beta_{12} - \alpha\beta_{22}$
3	3	$\mu_{..} - \alpha_1 - \alpha_2 - \beta_1 - \beta_2 + \alpha\beta_{11} + \alpha\beta_{12} + \alpha\beta_{21} + \alpha\beta_{22}$

or 2 but with negative signs when $i = 3 \ (= I)$. In a similar manner, the first two B main effects, β_1 and β_2, appear with positive signs when (respectively) $j = 1$ or 2 but with negative signs when $j = 3 \ (= J)$. If one thinks of the four A and B main effects used in Table 9.16 in terms of coefficients β in a regression model, it soon becomes clear how to invent "system variables" x to make the regression coefficients β appear with correct signs in the expressions for means μ_{ij}. That is, define four **dummy variables**

$$x_1^A = \begin{cases} 1 & \text{if the response } y \text{ is from level 1 of A} \\ -1 & \text{if the response } y \text{ is from level 3 of A} \\ 0 & \text{otherwise} \end{cases}$$

$$x_2^A = \begin{cases} 1 & \text{if the response } y \text{ is from level 2 of A} \\ -1 & \text{if the response } y \text{ is from level 3 of A} \\ 0 & \text{otherwise} \end{cases}$$

$$x_1^B = \begin{cases} 1 & \text{if the response } y \text{ is from level 1 of B} \\ -1 & \text{if the response } y \text{ is from level 3 of B} \\ 0 & \text{otherwise} \end{cases}$$

$$x_2^B = \begin{cases} 1 & \text{if the response } y \text{ is from level 2 of B} \\ -1 & \text{if the response } y \text{ is from level 3 of B} \\ 0 & \text{otherwise} \end{cases}$$

Then, making the correspondences indicated in Table 9.17, $\mu_{..} + \alpha_i + \beta_j$ can be written in regression notation as

$$\beta_0 + \beta_1 x_1^A + \beta_2 x_2^A + \beta_3 x_1^B + \beta_4 x_2^B$$

Table 9.17

Correspondences between Regression Coefficients and the Grand Mean and Main Effects in a 3^2 Factorial Study

Regression Coefficient	Corresponding 3×3 Factorial Effect
β_0	$\mu_{..}$
β_1	α_1
β_2	α_2
β_3	β_1
β_4	β_2

What is more, since the x's used here take only the values -1, 0, and 1, so also do their products. And taken in pairs (one x^A variable with one x^B variable), their products produce the correct (-1, 0, or 1) multipliers for the 2-factor interactions $\alpha\beta_{11}$, $\alpha\beta_{12}$, $\alpha\beta_{21}$, and $\alpha\beta_{22}$ appearing in Table 9.16. That is, if one thinks of the interactions $\alpha\beta_{ij}$ in terms of regression coefficients β, with the additional correspondences listed in Table 9.18, the entire expression (9.82) can be written in regression notation as

$$\mu_{y|x_1^A, x_2^A, x_1^B, x_2^B} = \beta_0 + \beta_1 x_1^A + \beta_2 x_2^A + \beta_3 x_1^B + \beta_4 x_2^B + \beta_5 x_1^A x_1^B$$
$$+ \beta_6 x_1^A x_2^B + \beta_7 x_2^A x_1^B + \beta_8 x_2^A x_2^B \qquad (9.83)$$

By rewriting the factorial-type expression (9.82) as a regression-type expression (9.83) it is then obvious how to fit few-effects models and do inference under those models even for unbalanced data. Nowhere in Section 9.2 was there any requirement that the data set be balanced. So the methods there can be used (employing properly constructed x variables and properly interpreting a corresponding regression printout) to fit reduced versions of model (9.83) and make confidence, prediction, and tolerance intervals under those reduced models.

Table 9.18

Correspondence between Regression Coefficients and Interactions in a 3^2 Factorial Study

Regression Coefficient	Corresponding 3×3 Factorial Effect
β_5	$\alpha\beta_{11}$
β_6	$\alpha\beta_{12}$
β_7	$\alpha\beta_{21}$
β_8	$\alpha\beta_{22}$

The general $I \times J$ two-way factorial version of this story is similar. One defines $I - 1$ factor A dummy variables $x_1^A, x_2^A, \ldots, x_{I-1}^A$ according to

I − 1 dummy variables for factor A

$$x_i^A = \begin{cases} 1 & \text{if the response } y \text{ is from level } i \text{ of A} \\ -1 & \text{if the response } y \text{ is from level } I \text{ of A} \\ 0 & \text{otherwise} \end{cases} \qquad (9.84)$$

and $J - 1$ factor B dummy variables $x_1^B, x_2^B, \ldots, x_{J-1}^B$ according to

J − 1 dummy variables for factor B

$$x_j^B = \begin{cases} 1 & \text{if the response } y \text{ is from level } j \text{ of B} \\ -1 & \text{if the response } y \text{ is from level } J \text{ of B} \\ 0 & \text{otherwise} \end{cases} \qquad (9.85)$$

Multiple regression and two-way factorial analyses

and uses a regression program to do the computations. Estimated regression coefficients of x_i^A or x_j^B variables alone are estimated main effects, while those for $x_i^A x_j^B$ cross products are estimated 2-factor interactions.

Example 4

(Examples 7, Chapter 4, and 1, Chapter 8, revisited—see pages 163, 547)

A Factorial Analysis of Unbalanced Wood Joint Strength Data Using a Regression Program

Consider again the wood joint strength study of Kotlers, MacFarland, and Tomlinson. The discussion in Section 8.1 showed that if only the wood types pine and oak are considered, a no-interaction description of joint strength for butt, beveled, and lap joints might be appropriate. The corresponding part of the (originally 3×3 factorial) data of Kotlers, MacFarland, and Tomlinson is given here in Table 9.19.

Table 9.19
Strengths of 11 Wood Joints

		B Wood Type	
		1 (Pine)	2 (Oak)
	1 (Butt)	829, 596	1169
A Joint Type	2 (Beveled)	1348, 1207	1518, 1927
	3 (Lap)	1000, 859	1295, 1561

Table 9.20
Joint Strength Data Prepared for a Factorial Analysis Using
a Regression Program

$i,$ Joint Type	$j,$ Wood Type	x_1^A	x_2^A	x_1^B	y
1	1	1	0	1	829, 596
1	2	1	0	−1	1169
2	1	0	1	1	1348, 1207
2	2	0	1	−1	1518, 1927
3	1	−1	−1	1	1000, 859
3	2	−1	−1	−1	1295, 1561

Notice that because these data are unbalanced (due to the unfortunate loss of one butt/oak response), it is not possible to fit a no-interaction model to these data by simply adding together fitted effects (defined in Section 4.3) or to use anything said in Chapter 8 to make inferences based on such a model. But it *is* possible to use the dummy variable regression approach based on formulas (9.84) and (9.85) to do so.

Consider the regression-data-set version of Table 9.19 given in Table 9.20. Printouts 4 and 5 show the results of fitting the two regression models

$$y = \beta_0 + \beta_1 x_1^A + \beta_2 x_2^A + \beta_3 x_1^B + \beta_4 x_1^A x_1^B + \beta_5 x_2^A x_1^B + \epsilon \qquad \textbf{(9.86)}$$

$$y = \beta_0 + \beta_1 x_1^A + \beta_2 x_2^A + \beta_3 x_1^B + \epsilon \qquad \textbf{(9.87)}$$

to the data of Table 9.20. Printout 4 corresponding to model (9.86) is the full model or $\mu_{ij} = \mu + \alpha_i + \beta_j + \alpha\beta_{ij}$ description of the data. For that regression run, the reader should verify the correspondences between fitted regression coefficients b and fitted effects (defined in Section 4.3), listed in Table 9.21. (For example,

Table 9.21
Correspondence between Fitted Regression Coefficients and Fitted Factorial
Effects for the Wood Joint Strength Data

Fitted Regression Coefficient	Value	Corresponding Fitted Effect
b_0	1206.5	$\bar{y}_{..}$
b_1	−265.75	a_1
b_2	293.50	a_2
b_3	−233.33	b_1
b_4	5.08	ab_{11}
b_5	10.83	ab_{21}

Example 4
(*continued*)

$\bar{y}_{..} = 1206.5$ and $\bar{y}_{1.} = 940.75$, so $a_1 = 940.75 - 1206.5 = -265.75$, which is the value of the fitted regression coefficient b_1.)

Model (9.86), like the two-way model (8.4) of Section 8.1, represents no restriction or simplification of the basic one-way model. So least squares estimates of parameters that are linear combinations of underlying means are simply the same linear combinations of sample means. Further, the fitted y values are (as expected) simply the sample means $\bar{y}_{ij}$.

Printout 5 corresponding to model (9.87) is the $\mu_{ij} = \mu_{..} + \alpha_i + \beta_j$ description of the data. The fitted regression coefficients b for model (9.87) are not equal to the (full-model) fitted factorial effects defined in Section 4.3. (The b's are least squares estimates of the underlying effects for the no-interaction model. When factorial data are unbalanced, these are not necessarily equal to the quantities defined in Section 4.3. For example, b_1 from Printout 5 is -264.48, which is the least squares estimate of α_1 in a no-interaction model but differs from $a_1 = -264.75$.) In a similar vein, the fitted responses are neither sample means nor sums of $\bar{y}_{..}$ plus the full-model fitted main effects defined in Section 4.3. (Of course, since the x variables take only values $-1, 0$, and 1, the fitted responses *are* sums and differences of the least squares estimates of the underlying parameters $\mu_{..}, \alpha_1, \alpha_2, \beta_1$ in the no-interaction model.)

Inference under model (9.86) is simply inference under the usual one-way normal model, and all of Sections 7.1 through 7.4 and 8.1 can be used. It is then reassuring that on Printout 4, $s_{SF} = s_P = 182.2$ and that (for example) for butt joints and pine wood (levels 1 of both A and B), the estimated standard deviation for $\hat{y} = \bar{y}_{11}$ is

$$128.9 = s_{SF} \cdot A = \frac{s_P}{\sqrt{n_{11}}} = \frac{182.2}{\sqrt{2}}$$

To illustrate how inference under a no-interaction model would proceed for the unbalanced 3×2 factorial joint strength data, consider making a 95% two-sided confidence interval for the mean strength of butt/pine joints and then a 90% lower prediction bound for the strength of a single joint of the same kind. Note that for data point 1 (a butt/pine observation) on Printout 5, $\hat{y} = 708.7$ and $s_{SF} \cdot A = 94.8$, where $s_{SF} = 154.7$ has seven associated degrees of freedom. So from formula (9.53) of Section 9.2 (page 686), two-sided 95% confidence limits for mean butt/pine joint strength are

$$708.7 \pm 2.365(94.8)$$

that is,

$$484.5 \text{ psi} \quad \text{and} \quad 932.9 \text{ psi}$$

Similarly, using formula (9.56) on page 689, a 90% lower prediction limit for a single additional butt/pine joint strength is

$$708.7 - 1.415\sqrt{(154.7)^2 + (94.8)^2} = 452.0 \text{ psi}$$

From these two calculations, it should be clear that other methods from Section 9.2 could be used here as well. The reader should have no trouble finding and using residuals and standardized residuals for the no-interaction model based on formulas (9.39) and (9.42), giving simultaneous confidence intervals for all six mean responses under the no-interaction model using formula (9.54) or giving one-sided tolerance bounds for certain joint/wood combinations under the no-interaction model using formula (9.58) or (9.59).

Printout 4 Multiple Regression Version of the With-Interactions Factorial Analysis of Joint Strength (*Example 4*)

```
Data Display


Row    xa1   xa2   xb1      y

 1      1     0     1      829
 2      1     0     1      596
 3      1     0    -1     1169
 4      0     1     1     1348
 5      0     1     1     1207
 6      0     1    -1     1518
 7      0     1    -1     1927
 8     -1    -1     1     1000
 9     -1    -1     1      859
10     -1    -1    -1     1295
11     -1    -1    -1     1561


Regression Analysis


The regression equation is
y = 1207 - 266 xa1 + 294 xa2 - 233 xb1 + 5.1 xa1*xb1 + 10.8 xa2*xb1

Predictor        Coef      StDev         T        P
Constant      1206.50      56.82     21.23    0.000
xa1           -265.75      85.91     -3.09    0.027
xa2            293.50      77.43      3.79    0.013
xb1           -233.33      56.82     -4.11    0.009
xa1*xb1          5.08      85.91      0.06    0.955
xa2*xb1         10.83      77.43      0.14    0.894

S = 182.2      R-Sq = 88.5%     R-Sq(adj) = 77.1%
```

```
Analysis of Variance

Source            DF          SS          MS        F        P
Regression         5     1283527      256705     7.73    0.021
Residual Error     5      166044       33209
Total             10     1449571

Source        DF     Seq SS
xa1            1     120144
xa2            1     577927
xb1            1     583908
xa1*xb1        1        897
xa2*xb1        1        650

Obs      xa1         y       Fit    StDev Fit    Residual    St Resid
  1     1.00     829.0     712.5      128.9       116.5        0.90
  2     1.00     596.0     712.5      128.9      -116.5       -0.90
  3     1.00    1169.0    1169.0      182.2        -0.0         * X
  4     0.00    1348.0    1277.5      128.9        70.5        0.55
  5     0.00    1207.0    1277.5      128.9       -70.5       -0.55
  6     0.00    1518.0    1722.5      128.9      -204.5       -1.59
  7     0.00    1927.0    1722.5      128.9       204.5        1.59
  8    -1.00    1000.0     929.5      128.9        70.5        0.55
  9    -1.00     859.0     929.5      128.9       -70.5       -0.55
 10    -1.00    1295.0    1428.0      128.9      -133.0       -1.03
 11    -1.00    1561.0    1428.0      128.9       133.0        1.03

X denotes an observation whose X value gives it large influence.
```

Printout 5   Multiple Regression Version of the No-Interactions Factorial Analysis
of Joint Strength (*Example 4*)

```
Regression Analysis


The regression equation is
y = 1207 - 264 xa1 + 293 xa2 - 234 xb1

Predictor        Coef       StDev          T        P
Constant      1207.14       47.38      25.48    0.000
xa1           -264.48       70.62      -3.74    0.007
xa2            292.86       65.11       4.50    0.003
xb1           -233.97       47.38      -4.94    0.002

S = 154.7      R-Sq = 88.4%     R-Sq(adj) = 83.5%

Analysis of Variance

Source            DF          SS          MS        F        P
Regression         3     1281980      427327    17.85    0.001
Residual Error     7      167591       23942
Total             10     1449571

Source        DF     Seq SS
xa1            1     120144
xa2            1     577927
xb1            1     583908
```

Obs	xa1	y	Fit	StDev Fit	Residual	St Resid
1	1.00	829.0	708.7	94.8	120.3	0.98
2	1.00	596.0	708.7	94.8	-112.7	-0.92
3	1.00	1169.0	1176.6	109.4	-7.6	-0.07
4	0.00	1348.0	1266.0	90.7	82.0	0.65
5	0.00	1207.0	1266.0	90.7	-59.0	-0.47
6	0.00	1518.0	1734.0	90.7	-216.0	-1.72
7	0.00	1927.0	1734.0	90.7	193.0	1.54
8	-1.00	1000.0	944.8	90.7	55.2	0.44
9	-1.00	859.0	944.8	90.7	-85.8	-0.68
10	-1.00	1295.0	1412.7	90.7	-117.7	-0.94
11	-1.00	1561.0	1412.7	90.7	148.3	1.18

The pattern of analysis set out for two-way factorials carries over quite naturally to three-way and higher factorials. To use a multiple regression program to fit and make inferences based on simplified versions of the p-way factorial model, proceed as follows. $I - 1$ dummy variables $x_1^A, x_2^A, \ldots, x_{I-1}^A$ are defined (as before) to carry information about I levels of factor A, $J - 1$ dummy variables $x_1^B, x_2^B, \ldots, x_{J-1}^B$ are defined (as before) to carry information about J levels of factor B, $K - 1$ dummy variables $x_1^C, x_2^C, \ldots, x_{K-1}^C$ are defined to carry information about K levels of factor C, $\ldots$, etc. Products of pairs of these, one each from the groups representing two different factors, carry information about 2-factor interactions of the factors. Products of triples of these, one each from the groups representing three different factors, carry information about 3-factor interactions of the factors. And so on.

Dummy variables for regression analysis of p-way factorials

When something short of the largest possible regression model is fitted to an unbalanced factorial data set, the estimated coefficients b that result are the least squares estimates of the underlying factorial effects *in the few-effects model*. (Usually, these differ somewhat from the (full-model) fitted effects defined in Section 4.3.) All of the regression machinery of Section 9.2 can be applied to create fitted values, residuals, and standardized residuals; to plot these to do model checking; to make confidence intervals for mean responses; and to create prediction and tolerance intervals.

When the regression with dummy variables approach is used as just described, the fitted coefficients b correspond to fitted effects for the levels 1 through $I - 1$, $J - 1$, $K - 1$, etc. of the factors. For two-level factorials, this means that the fitted coefficients are estimated factorial effects for the "all low" treatment combination. However, because of extensive use of the Yates algorithm in this text, you will probably think first in terms of the 2^p factorial effects for the "all high" treatment combination.

Two sensible courses of action then suggest themselves for the analysis of unbalanced 2^p factorial data. You can proceed exactly as just indicated, using dummy variables x_1^A, x_1^B, x_1^C, etc. and various products of the same, taking care to remember to interpret b's as "all low" fitted effects and subsequently to switch signs as appropriate to get "all high" fitted effects. The other possibility is to depart slightly from the program laid out for general p-way factorials in 2^p cases: Instead

Alternative choice of x variables for regression analysis of 2^p factorials

of using the variables x_1^A, x_1^B, x_1^C, etc. and their products when doing regression, one may use the variables

$$x_2^A = -x_1^A = \begin{cases} 1 & \text{if the response } y \text{ is from the high level of A} \\ -1 & \text{if the response } y \text{ is from the low level of A} \end{cases}$$

$$x_2^B = -x_1^B = \begin{cases} 1 & \text{if the response } y \text{ is from the high level of B} \\ -1 & \text{if the response } y \text{ is from the low level of B} \end{cases}$$

$$x_2^C = -x_1^C = \begin{cases} 1 & \text{if the response } y \text{ is from the high level of C} \\ -1 & \text{if the response } y \text{ is from the low level of C} \end{cases}$$

etc. and their products when doing regression. When the variables x_2^A, x_2^B, x_2^C, etc. are used, the fitted b's are the estimated "all high" 2^p factorial effects.

Example 5
(*Example 4, Chapter 8, revisited—page 569*)

A Factorial Analysis of Unbalanced 2^3 Power Requirement Data Using Regression

Return to the situation of the 2^3 metalworking power requirement study of Miller. The original data set (given in Table 8.8) is balanced, with the common sample size being $m = 4$. For the sake of illustrating how regression with dummy variables can be used in the analysis of unbalanced higher-way factorial data, consider artificially unbalancing Miller's data by supposing that the first data point appearing in Table 8.8 has gotten lost. The portion of Miller's data that will be used here is then given in Table 9.22.

Table 9.22
Dynamometer Readings for 2^3 Treatment Combinations

Tool Type	Bevel Angle	Type of Cut	y, Dynamometer Reading (mm)
1	15°	continuous	26.5, 30.5, 27.0
2	15°	continuous	28.0, 28.5, 28.0, 25.0
1	30°	continuous	28.5, 28.5, 30.0, 32.5
2	30°	continuous	29.5, 32.0, 29.0, 28.0
1	15°	interrupted	28.0, 25.0, 26.5, 26.5
2	15°	interrupted	24.5, 25.0, 28.0, 26.0
1	30°	interrupted	27.0, 29.0, 27.5, 27.5
2	30°	interrupted	27.5, 28.0, 27.0, 26.0

For this slightly altered data set, the Yates algorithm produces the fitted effects

$$a_2 = -.2656 \qquad ab_{22} = .0469 \qquad abc_{222} = -.0469$$
$$b_2 = .8281 \qquad ac_{22} = -.0469$$
$$c_2 = -.9531 \qquad bc_{22} = -.2031$$

and $s_P = 1.51$ with $\nu = 23$ associated degrees of freedom. Formula (8.12) (page 575) of Section 8.2 then shows that (say) two-sided 90% confidence intervals for effects have plus-and-minus parts

$$\pm 1.714(1.51)\frac{1}{2^3}\sqrt{\frac{7}{4} + \frac{1}{3}} = \pm .47$$

Just as in Example 4 in Chapter 8, where all $n = 32$ data points were used, one might thus judge only the B and C main effects to be clearly larger than background noise.

Printout 6 supports exactly these conclusions. This regression run was made using all seven of the terms

$$x_2^A, \ x_2^B, \ x_2^C, \ x_2^A x_2^B, \ x_2^A x_2^C, \ x_2^B x_2^C, \quad \text{and} \quad x_2^A x_2^B x_2^C$$

(i.e., using the *full model* in regression terminology and the *unrestricted* 2^3 *factorial model* in the terminology of Section 8.2). On Printout 6, one can identify the fitted regression coefficients b with the fitted factorial effects in the pairs indicated in Table 9.23.

Table 9.23

Correspondence Between Fitted Regression Coefficients and Fitted Factorial Effects for the Regression Run of Printout 6

Fitted Regression Coefficient	Fitted Factorial Effect
b_0	$\bar{y}_{...}$
b_1	a_2
b_2	b_2
b_3	c_2
b_4	ab_{22}
b_5	ac_{22}
b_6	bc_{22}
b_7	abc_{222}

Example 5
(continued)

Analysis of the data of Table 9.22 based on a full factorial model

$$y_{ijkl} = \mu_{\cdots} + \alpha_i + \beta_j + \gamma_k + \alpha\beta_{ij} + \alpha\gamma_{ik} + \beta\gamma_{jk} + \alpha\beta\gamma_{ijk} + \epsilon_{ijkl}$$

that is,

$$y_i = \beta_0 + \beta_1 x_{2i}^{A} + \beta_2 x_{2i}^{B} + \beta_3 x_{2i}^{C} + \beta_4 x_{2i}^{A} x_{2i}^{B} + \beta_5 x_{2i}^{A} x_{2i}^{C} + \beta_6 x_{2i}^{B} x_{2i}^{C}$$
$$+ \beta_7 x_{2i}^{A} x_{2i}^{B} x_{2i}^{C} + \epsilon_i$$

is a logical first step. Based on that step, it seems desirable to fit and draw inferences based on a "B and C main effects only" description of y. Since the data in Table 9.22 are unbalanced, the naive use of the reverse Yates algorithm with the (full-model) fitted effects will not produce appropriate fitted values. $\bar{y}_{\cdots}$, b_2, and c_2 are simply *not* the least squares estimates of $\mu_{\cdots}$, β_2, and γ_2 for the "B and C main effects only" model in this unbalanced data situation.

However, what can be done is to fit the reduced regression model

$$y_i = \beta_0 + \beta_2 x_{2i}^{B} + \beta_3 x_{2i}^{C} + \epsilon_i$$

to the data. Printout 7 represents the use of this technique. Locate on that printout the (reduced-model) estimates of the factorial effects $\mu_{\cdots}$, β_2, and γ_2 and note that they differ somewhat from $\bar{y}_{\cdots}$, b_2, and c_2 as defined in Section 4.3 and displayed on Printout 6. Note also that the four different possible fitted mean responses, along with their estimated standard deviations, are as given in Table 9.24.

The values in Table 9.24 can be used in the formulas of Section 9.2 to produce confidence intervals for the four mean responses, prediction intervals, tolerance intervals, and so on based on the "B and C main effects only" model. All of this can be done despite the fact that the data of Table 9.22 are unbalanced.

Table 9.24
Fitted Values and Their Estimated Standard Deviations for a "B and C Main Effects Only" Analysis of the Unbalanced Power Requirement Data

Bevel Angle	x_2^{B}	Type of Cut	x_2^{C}	$\hat{y}$	$s_{SF} \cdot A$
15°	−1	continuous	−1	27.88	.46
30°	1	continuous	−1	29.54	.44
15°	−1	interrupted	1	25.98	.44
30°	1	interrupted	1	27.64	.44

Printout 6 Multiple Regression Version of the With-Interactions Factorial Analysis of Power Requirement (*Example 5*)

Regression Analysis

The regression equation is
y = 27.8 - 0.266 xa2 + 0.828 xb2 - 0.953 xc2 + 0.047 xa*xb - 0.047 xa*xc
           - 0.203 xb*xc - 0.047 xa*xb*xc

Predictor	Coef	StDev	T	P
Constant	27.7656	0.2731	101.68	0.000
xa2	-0.2656	0.2731	-0.97	0.341
xb2	0.8281	0.2731	3.03	0.006
xc2	-0.9531	0.2731	-3.49	0.002
xa*xb	0.0469	0.2731	0.17	0.865
xa*xc	-0.0469	0.2731	-0.17	0.865
xb*xc	-0.2031	0.2731	-0.74	0.465
xa*xb*xc	-0.0469	0.2731	-0.17	0.865

S = 1.514   R-Sq = 51.0%   R-Sq(adj) = 36.0%

Analysis of Variance

Source	DF	SS	MS	F	P
Regression	7	54.748	7.821	3.41	0.012
Residual Error	23	52.687	2.291		
Total	30	107.435			

Source	DF	Seq SS
xa2	1	2.202
xb2	1	22.645
xc2	1	28.398
xa*xb	1	0.091
xa*xc	1	0.051
xb*xc	1	1.293
xa*xb*xc	1	0.068

Obs	xa2	y	Fit	StDev Fit	Residual	St Resid
1	-1.00	26.500	28.000	0.874	-1.500	-1.21
2	-1.00	30.500	28.000	0.874	2.500	2.02R
3	-1.00	27.000	28.000	0.874	-1.000	-0.81
4	1.00	28.000	27.375	0.757	0.625	0.48
5	1.00	28.500	27.375	0.757	1.125	0.86
6	1.00	28.000	27.375	0.757	0.625	0.48
7	1.00	25.000	27.375	0.757	-2.375	-1.81
8	-1.00	28.500	29.875	0.757	-1.375	-1.05
9	-1.00	28.500	29.875	0.757	-1.375	-1.05
10	-1.00	30.000	29.875	0.757	0.125	0.10
11	-1.00	32.500	29.875	0.757	2.625	2.00R
12	1.00	29.500	29.625	0.757	-0.125	-0.10
13	1.00	32.000	29.625	0.757	2.375	1.81
14	1.00	29.000	29.625	0.757	-0.625	-0.48
15	1.00	28.000	29.625	0.757	-1.625	-1.24
16	-1.00	28.000	26.500	0.757	1.500	1.14
17	-1.00	25.000	26.500	0.757	-1.500	-1.14
18	-1.00	26.500	26.500	0.757	-0.000	-0.00
19	-1.00	26.500	26.500	0.757	-0.000	-0.00
20	1.00	24.500	25.875	0.757	-1.375	-1.05
21	1.00	25.000	25.875	0.757	-0.875	-0.67
22	1.00	28.000	25.875	0.757	2.125	1.62
23	1.00	26.000	25.875	0.757	0.125	0.10

24	-1.00	27.000	27.750	0.757	-0.750	-0.57
25	-1.00	29.000	27.750	0.757	1.250	0.95
26	-1.00	27.500	27.750	0.757	-0.250	-0.19
27	-1.00	27.500	27.750	0.757	-0.250	-0.19
28	1.00	27.500	27.125	0.757	0.375	0.29
29	1.00	28.000	27.125	0.757	0.875	0.67
30	1.00	27.000	27.125	0.757	-0.125	-0.10
31	1.00	26.000	27.125	0.757	-1.125	-0.86

R denotes an observation with a large standardized residual

Printout 7 Multiple Regression Version of a "B and C Main Effects Only" Analysis of Power Requirement (*Example 5*)

Regression Analysis

The regression equation is
y = 27.8 + 0.832 xb2 - 0.949 xc2

Predictor	Coef	StDev	T	P
Constant	27.7619	0.2553	108.73	0.000
xb2	0.8319	0.2553	3.26	0.003
xc2	-0.9494	0.2553	-3.72	0.001

S = 1.420   R-Sq = 47.4%   R-Sq(adj) = 43.7%

Analysis of Variance

Source	DF	SS	MS	F	P
Regression	2	50.972	25.486	12.64	0.000
Residual Error	28	56.463	2.017		
Total	30	107.435			

Source	DF	Seq SS
xb2	1	23.093
xc2	1	27.879

Obs	xb2	y	Fit	StDev Fit	Residual	St Resid
1	-1.00	26.500	27.879	0.457	-1.379	-1.03
2	-1.00	30.500	27.879	0.457	2.621	1.95
3	-1.00	27.000	27.879	0.457	-0.879	-0.65
4	-1.00	28.000	27.879	0.457	0.121	0.09
5	-1.00	28.500	27.879	0.457	0.621	0.46
6	-1.00	28.000	27.879	0.457	0.121	0.09
7	-1.00	25.000	27.879	0.457	-2.879	-2.14R
8	1.00	28.500	29.543	0.437	-1.043	-0.77
9	1.00	28.500	29.543	0.437	-1.043	-0.77
10	1.00	30.000	29.543	0.437	0.457	0.34
11	1.00	32.500	29.543	0.437	2.957	2.19R
12	1.00	29.500	29.543	0.437	-0.043	-0.03
13	1.00	32.000	29.543	0.437	2.457	1.82
14	1.00	29.000	29.543	0.437	-0.543	-0.40
15	1.00	28.000	29.543	0.437	-1.543	-1.14
16	-1.00	28.000	25.981	0.437	2.019	1.49
17	-1.00	25.000	25.981	0.437	-0.981	-0.73
18	-1.00	26.500	25.981	0.437	0.519	0.38
19	-1.00	26.500	25.981	0.437	0.519	0.38
20	-1.00	24.500	25.981	0.437	-1.481	-1.10
21	-1.00	25.000	25.981	0.437	-0.981	-0.73
22	-1.00	28.000	25.981	0.437	2.019	1.49
23	-1.00	26.000	25.981	0.437	0.019	0.01

24	1.00	27.000	27.644	0.437	-0.644	-0.48
25	1.00	29.000	27.644	0.437	1.356	1.00
26	1.00	27.500	27.644	0.437	-0.144	-0.11
27	1.00	27.500	27.644	0.437	-0.144	-0.11
28	1.00	27.500	27.644	0.437	-0.144	-0.11
29	1.00	28.000	27.644	0.437	0.356	0.26
30	1.00	27.000	27.644	0.437	-0.644	-0.48
31	1.00	26.000	27.644	0.437	-1.644	-1.22

```
R denotes an observation with a large standardized residual
```

Example 5 has been treated as if the lack of balance in the data came about by misfortune. And the lack of balance in Example 4 *did* come about in such a way. But lack of balance in p-way factorial data can also be the result of careful planning. Consider, for example, a 2^4 factorial situation where the budget can support collection of 20 observations but not as many as 32. In such a case, complete replication of the 16 combinations of two levels of four factors in order to achieve balance is not possible. But it makes far more sense to replicate four of the 16 combinations (and thus be able to calculate s_p and honestly assess the size of background variation) than to achieve balance by using no replication. By now it should be obvious how to subsequently go about the analysis of the resulting partially replicated (and thus unbalanced) factorial data.

Section 3 Exercises

1. Flood and Shankwitz reported the results of a metallurgical engineering design project involving the tempering response of a certain grade of stainless steel. Slugs of this steel were preprocessed to reasonably uniform hardnesses, which were measured and recorded. The slugs were then tempered at various temperatures for various lengths of time. The hardnesses were then remeasured and the change in hardness computed. The data in the accompanying tables were obtained in this replicated 4 × 4 factorial study.

Time, x_1 (min)	Temperature, x_2 (°F)	Increase in Hardness, y
5	800	0, 0, −1
5	900	−3, −2, 1
5	1000	−1, −1, 0
5	1100	−4, 1, 3
50	800	3, 4, −1
50	900	−3, −1, 1
50	1000	−4, −1, −3
50	1100	−4, −4, −2

Time, x_1 (min)	Temperature, x_2 (°F)	Increase in Hardness, y
150	800	4, 2, −2
150	900	−1, −1, −2
150	1000	−4, −5, −7
150	1100	−7, −5, −8
500	800	1, −3, 0
500	900	−2, −8, −2
500	1000	−8, −7, −7
500	1100	−11, 9, 5

(a) Fit the quadratic model

$$y = \beta_0 + \beta_1 \ln(x_1) + \beta_2 x_2 + \beta_3 \left(\ln(x_1)\right)^2 + \beta_4 x_2^2 + \beta_5 x_2 \ln(x_1) + \epsilon$$

to these data. What fraction of the observed variability in hardness increase is accounted for in the fitting of the quadratic response surface? What is your estimate of the standard deviation of hardness changes that would be experienced at any fixed combination of time and temperature? How does this estimate compare with s_P? Does there appear to be enough difference between the two values to cast serious doubt on the appropriateness of the regression model?

(b) There was some concern on the project group's part that the 5-minute time was completely unlike the other times and should not be considered in the same analysis as the longer times. Temporarily delete the 12 slugs treated only 5 minutes from consideration, refit the quadratic model, and compare fitted values for the 36 slugs tempered longer than 5 minutes for this regression to those from part (a). How different are these two sets of values?

Henceforth consider the quadratic model fitted to all 48 data points.

(c) Make a contour plot showing how y varies with $\ln(x_1)$ and x_2. In particular, use it to identify the region of $\ln(x_1)$ and x_2 values where the tempering seems to provide an increase in hardness. Sketch the corresponding region in the (x_1, x_2)-plane.

(d) For the $x_1 = 50$ and $x_2 = 800$ set of conditions,
   (i) give a 95% two-sided confidence interval for the mean increase in hardness provided by tempering.
   (ii) give a 95% two-sided prediction interval for the increase in hardness produced by tempering an additional slug.

(iii) give an approximate 95% lower tolerance bound for the hardness increases of 90% of such slugs undergoing tempering.

2. Return to the situation of Chapter Exercise 10 of Chapter 8 and the chemical product impurity study. The analysis suggested in that exercise leads to the conclusion that only the A and B main effects are detectably nonzero. The data are unbalanced, so it is not possible to use the reverse Yates algorithm to fit the "A and B main effects only" model to the data.

(a) Use the dummy variable regression techniques to fit the "A and B main effects only" model. (You should be able to pattern what you do after Example 5.) How do A and B main effects estimated on the basis of this few-effects/simplified description of the pattern of response compare with what you obtained for fitted effects using the Yates algorithm?

(b) Compute and plot standardized residuals for the few-effects model. (Plot against levels of A, B, and C, against $\hat{y}$, and normal-plot them.) Do any of these plots indicate any problems with the few-effects model?

(c) How does s_{FE} (which you can read directly off your printout as s_{SF}) compare with s_P in this situation? Do the two values carry any strong suggestion of lack of fit?

Chapter 9 Exercises ....

1. Return to the situation of Chapter Exercise 3 of Chapter 4 and the grain growth study of Huda and Ralph. Consider an analysis of the researchers' data based on the model

$$y = \beta_0 + \beta_1 x_1 + + \beta_2 \ln(x_2) + \beta_3 x_1 \ln(x_2) + \epsilon$$

(a) Fit this model to the data given in Chapter 4. Based on this fit, what is your estimate of the standard deviation of grain size, y, associated with different specimens treated using a fixed temperature and time?

(b) Make a plot of the observed y's versus the corresponding $\ln(x_2)$'s. On this plot, sketch the linear fitted response functions ($\hat{y}$ versus $\ln(x_2)$) for $x_1 = 1443$, 1493, and 1543. Notice that the fit to the researchers' data is excellent. However, notice also that the model has four β's and was fit based on only nine data points. What possibility therefore needs to be kept in mind when making predictions based on this model?

(c) Make a 95% two-sided confidence interval for the mean y when a temperature of $x_1 = 1493°K$ and a time of $x_2 = 120$ minutes are used.

(d) Make a 95% two-sided prediction interval for an additional grain size, y, when a temperature of $x_1 = 1493°K$ and a time of $x_2 = 120$ minutes are used.

(e) Find a 95% two-sided confidence interval for the mean y when a temperature of $x_1 = 1500°K$ and a time of $x_2 = 500$ minutes are used. (This is not a set of conditions in the original data set. So you will need to inform your regression program of where you wish to predict.)

(f) What does the hypothesis $H_0: \beta_1 = \beta_2 = \beta_3 = 0$ mean in the context of this study and the model being used in this exercise? Find the p-value associated with an F test of this hypothesis.

(g) What does the hypothesis $H_0: \beta_3 = 0$ mean in the context of this study and the model being used in this exercise? Find the p-value associated with a two-sided t test of this hypothesis.

2. The article "Orthogonal Design for Process Optimization and its Application in Plasma Etching" by Yin and Jillie (*Solid State Technology*, 1987) discusses a 4-factor experiment intended to guide optimization of a nitride etch process on a single wafer plasma etcher. Data were collected at only nine out of $3^4 = 81$ possible combinations of three levels of each of the four factors (making up a so-called *orthogonal array*). The factors involved in the experimentation were the Power applied to the cathode x_1, the Pressure in the reaction chamber x_2, the spacing or Gap between the anode and the cathode x_3, and the Flow of the reactant gas C_2F_6, x_4. Three different responses were measured, an etch rate for SiN y_1, a uniformity for SiN y_2, and a selectivity of the process (for silicon nitride) between silicon nitride and polysilicon y_3. Eight of the nine different combinations were run once, while one combination was run three times. The researchers reported the data given in the accompanying table.

x_1 (W)	x_2 (mTorr)	x_3 (cm)	x_4 (sccm)	y_1 (Å/min)	y_2 (%)	y_3 (SiN/poly)
275	450	0.8	125	1075	2.7	1.63
275	500	1.0	160	633	4.9	1.37
275	550	1.2	200	406	4.6	1.10
300	450	1.0	200	860	3.4	1.58
300	500	1.2	125	561	4.6	1.26
275	450	0.8	125	1052	1.7	1.72
300	550	0.8	160	868	4.6	1.65
325	450	1.2	160	669	5.0	1.42
325	500	0.8	200	1138	2.9	1.69
325	550	1.0	125	749	5.6	1.54
275	450	0.8	125	1037	2.6	1.72

The data are listed in the order in which they were actually collected. Notice that the conditions under which the first, sixth, and eleventh data points were collected are the same—that is, there is some replication in this fractional factorial data set.

(a) The fact that the first, sixth, and last data points were collected under the same set of process conditions provides some check on the consistency of experimental results across time in this study. What else might (should) have been done in this study to try to make sure that time trends in an extraneous variable don't get confused with the effects of the experimental variables (in particular, the effect of x_1, as the experiment was run)? (Consider again the ideas of Section 2.3.)

(b) Fit a linear model in all of x_1, x_2, x_3, and x_4 to each of the three response variables. Notice that although such a model appears to provide a good fit to the y_3 data, the situations for y_1 and y_2 are not quite so appealing. (Compare s_{SF} to s_P for y_1 and note that R^2 for the second variable is relatively low, at least compared to what one can achieve for y_3.)

(c) In search of better-fitting equations for the y_2 (or y_1) data, one might consider fitting a full quadratic equation in x_1, x_2, x_3, and x_4 to the data. What happens when you attempt to do this using a regression package? (The problem is that the data given here are not adequate to distinguish between various possible quadratic response surfaces in four variables.)

(d) In light of the difficulty experienced in (c), a natural thing to do might be to try to fit quadratic surfaces involving only some of all possible second-order terms. Fit the two models for y_2 including (i) $x_1, x_2, x_3, x_4, x_1^2, x_2^2, x_3^2,$ and x_4^2 terms, and (ii) $x_1, x_2, x_4, x_1^2, x_2^2, x_4^2, x_1 x_2,$ and $x_2 x_4$ terms. How do these two fitted equations compare in terms of $\hat{y}_2$ values for (x_1, x_2, x_3, x_4) combinations in the data set? How do $\hat{y}_2$ values compare for the two fitted equations when $x_1 = 325, x_2 = 550, x_3 = 1.2,$ and $x_4 = 200$? (Notice that although this last combination is not in the data set, there are values of the individual variables in the data set matching these.) What is the practical engineering difficulty faced in a situation like this, where there is not enough data available to fit a full quadratic model but it doesn't seem that a model linear in the variables is an adequate description of the response?

Henceforth, confine attention to y_3 and consider an analysis based on a model linear in all of $x_1, x_2, x_3,$ and x_4.

(e) Give a 90% two-sided individual confidence interval for the increase in mean selectivity ratio that accompanies a 1 watt increase in power.

(f) What appear to be the optimal (large y_3) settings of the variables $x_1, x_2, x_3,$ and x_4 (within their respective ranges of experimentation)? Refer to the coefficients of your fitted equation from (b).

(g) Give a 90% two-sided confidence interval for the mean selectivity ratio at the combination of settings that you identified in (f). What cautions would you include in a report in which this interval is to appear? (Under what conditions is your calculated interval going to have real-world meaning?)

3. The article "How to Optimize and Control the Wire Bonding Process: Part II" by Scheaffer and Levine (*Solid State Technology*, 1991) discusses the use of a $k = 4$ factor central composite design in the improvement of the operation of the K&S 1484XQ

bonder. The effects of the variables Force, Ultrasonic Power, Temperature, and Time on the final ball bond shear strength were studied. The accompanying table gives data like those collected by the authors. (The original data were not given in the paper, but enough information was given to produce these simulated values that have structure like the original data.)

Force, x_1 (gm)	Power, x_2 (mw)	Temp., x_3 °C	Time, x_4 (ms)	Strength, y (gm)
30	60	175	15	26.2
40	60	175	15	26.3
30	90	175	15	39.8
40	90	175	15	39.7
30	60	225	15	38.6
40	60	225	15	35.5
30	90	225	15	48.8
40	90	225	15	37.8
30	60	175	25	26.6
40	60	175	25	23.4
30	90	175	25	38.6
40	90	175	25	52.1
30	60	225	25	39.5
40	60	225	25	32.3
30	90	225	25	43.0
40	90	225	25	56.0
25	75	200	20	35.2
45	75	200	20	46.9
35	45	200	20	22.7
35	105	200	20	58.7
35	75	150	20	34.5
35	75	250	20	44.0
35	75	200	10	35.7
35	75	200	30	41.8
35	75	200	20	36.5
35	75	200	20	37.6
35	75	200	20	40.3
35	75	200	20	46.0
35	75	200	20	27.8
35	75	200	20	40.3

(a) Fit both the full quadratic response surface and the simpler linear response surface to these data. On the basis of simple examination of the R^2 values, does it appear that the quadratic surface is enough better as a data summary to make it worthwhile to suffer the increased complexity that it brings with it? How do the s_{SF} values for the two fitted models compare to s_P computed from the final six data points listed here?

(b) Conduct a formal test (in the full quadratic model) of the hypothesis that the linear model $y = \beta_0 + \beta_1 x_1 + \beta_2 x_2 + \beta_3 x_3 + \beta_4 x_4 + \epsilon$ is an adequate description of the response. Does your p-value support your qualitative judgment from part (a)?

(c) In the linear model $y = \beta_0 + \beta_1 x_1 + \beta_2 x_2 + \beta_3 x_3 + \beta_4 x_4 + \epsilon$, give a 90% confidence interval for β_2. Interpret this interval in the context of the original engineering problem. (What is β_2 supposed to measure?) Would you expect the p-value from a test of $H_0: \beta_2 = 0$ to be large or to be small?

(d) Use the linear model and find an approximate 95% lower tolerance bound for 98% of bond shear strengths at the center point $x_1 = 35$, $x_2 = 15$, $x_3 = 200$, and $x_4 = 20$.

4. (**Testing for "Lack of Fit" to a Regression Model**) In curve- and surface-fitting problems where there is some replication, this text has used the informal comparison of s_{SF} (or s_{LF}) to s_P as a means of detecting poor fit of a regression model. It is actually possible to use these values to conduct a formal significance test for lack of fit. That is, under the one-way normal model of Chapter 7, it is possible to test

$$H_0: \mu_{y|x_1,x_2,\ldots,x_k} = \beta_0 + \beta_1 x_1 + \beta_2 x_2 + \cdots + \beta_k x_k$$

using the test statistic

$$F = \frac{\dfrac{(n-k-1)s_{SF}^2 - (n-r)s_P^2}{r-k-1}}{s_P^2}$$

and an $F_{r-k-1,n-r}$ reference distribution, where large values of F count as evidence against H_0. (If s_{SF} is much larger than s_P, the difference in the numerator of F will be large, producing a large sample value and a small observed level of significance.)

(a) It is not possible to use the lack of fit test in any of Exercise 3 of Section 4.1, Exercise 2 of Section 4.2, or Chapter Exercises 2 or 3 of Chapter 4. Why?

(b) For the situation of Exercise 2 of Section 9.1, conduct a formal test of lack of fit of the linear relationship $\mu_{y|x} = \beta_0 + \beta_1 x$ to the concrete strength data.

(c) For the situation of Exercise 1 of Section 9.3, conduct a formal test of lack of fit of the full quadratic relationship

$$\mu_{y|x_1,x_2} = \beta_0 + \beta_1 \ln(x_1) + \beta_2 x_2 + \beta_3 \left(\ln(x_1)\right)^2 + \beta_4 x_2^2 + \beta_5 x_2 \ln(x_1)$$

to the hardness increase data.

(d) For the situation of Chapter Exercise 3, conduct a formal test of lack of fit of the linear relationship

$$\mu_{y|x_1,x_2,x_3,x_4} = \beta_0 + \beta_1 x_1 + \beta_2 x_2 + \beta_3 x_3 + \beta_4 x_4$$

to the ball bond shear strength data.

5. Return to the situation of Chapter Exercises 18 and 19 of Chapter 4 and the ore refining study of S. Osoka. In that study, the object was to discover settings of the process variables x_1 and x_2 that would simultaneously maximize y_1 and minimize y_2.

(a) Fit full quadratic response functions for y_1 and y_2 to the data given in Chapter 4. Compute and plot standardized residuals for these two fitted equations. Comment on the appearance of these plots and what they indicate about the appropriateness of the fitted response surfaces.

(b) One useful rule of thumb in response surface studies (suggested by Box, Hunter, and Hunter in their book *Statistics for Experimenters*) is to check that for a fitted surface involving a total of l coefficients b (including b_0),

$$\max \hat{y} - \min \hat{y} > 4\sqrt{\frac{l \cdot s_{SF}^2}{n}}$$

before trying to make decisions based on its nature (bowl-shape up or down, saddle, etc.) or do even limited interpolation or extrapolation. This criterion is a comparison of the movement of the fitted surface across those n data points in hand, to four times an estimate of the root of the average variance associated with the n fitted values $\hat{y}$. If the criterion is not satisfied, the interpretation is that the fitted surface is so flat (relative to the precision with which it is determined) as to make it impossible to tell with any certainty the true nature of how mean response varies as a function of the system variables. Judge the usefulness of the surfaces fitted in part (a) against this criterion. Do the response surfaces appear to be determined adequately to support further analysis (involving optimization, for example)?

(c) Use the analytic method discussed in Section 9.3 to investigate the nature of the response surfaces fitted in part (a). According to the signs of the eigenvalues, what kinds of surfaces were fitted to y_1 and y_2, respectively?

(d) Make contour plots of the fitted y_1 and y_2 response surfaces from (a) on a single set of (x_1, x_2)-axes. Use these to help locate (at least approximately) a point (x_1, x_2) with maximum predicted y_1, subject to a constraint that predicted y_2 be no larger than 55.

(e) For the point identified in part (d), give 90% two-sided prediction intervals for the next values of y_1 and y_2 that would be produced by this refining process. Also give an approximate 95% lower tolerance bound for 90% of additional pyrite recoveries and an approximate 95% upper tolerance bound for 90% of additional kaolin recoveries at this combination of x_1 and x_2 settings.

6. Return to the concrete strength testing situation of Chapter Exercise 16 of Chapter 4.
   (a) Find estimates of the parameters β_0, β_1, and σ in the simple linear regression model $y = \beta_0 + \beta_1 x + \epsilon$.
   (b) Compute standardized residuals and plot them in the same ways that you were asked to plot the ordinary residuals in part (g) of the problem in Chapter 4. How much do the appearances of the new plots differ from the earlier ones?
   (c) Make a 95% two-sided confidence interval for the increase in mean compressive strength that accompanies a 5 psi increase in splitting tensile strength. (Note: This is $5\beta_1$.)
   (d) Make a 90% two-sided confidence interval for the mean strength of specimens with splitting tensile strength 300 psi (based on the simple linear regression model).
   (e) Make a 90% two-sided prediction interval for the strength of an additional specimen with splitting tensile strength 300 psi (based on the simple linear regression model).
   (f) Find an approximate 95% lower tolerance bound for the strengths of 90% of additional specimens with splitting tensile strength 300 psi (based on the simple linear regression model).

7. Wiltse, Blandin, and Schiesel experimented with a grain thresher built for an agricultural engineering design project. They ran efficiency tests on the cleaning chamber of the machine. This part of the machine sucks air through threshed material, drawing light (nonseed) material out an exhaust port, while the heavier seeds fall into a collection tray. Airflow is governed by the spacing of an air relief door. The following are the weights, y (in grams), of the portions of 14 gram samples of pure oat seeds run through the cleaning chamber that ended up in the collection tray. Four different door spacings x were used, and 20 trials were made at each door spacing.

.500 in. Spacing

12.00, 12.30, 12.45, 12.45, 12.50, 12.50, 12.50, 12.60, 12.65, 12.70, 12.70, 12.80, 12.90, 12.90, 13.00, 13.00, 13.00, 13.10, 13.20, 13.20

.875 in. Spacing

12.40, 12.80, 12.80, 12.90, 12.90, 12.90, 12.90, 13.00, 13.00, 13.00, 13.00, 13.20, 13.20, 13.20, 13.30, 13.40, 13.40, 13.45, 13.45, 13.70

1.000 in. Spacing

12.00, 12.80, 12.80, 12.90, 12.90, 13.00, 13.00, 13.00, 13.15, 13.20, 13.20, 13.30, 13.40, 13.40, 13.45, 13.50, 13.60, 13.60, 13.60, 13.70

1.250 in. Spacing

12.10, 12.20, 12.25, 12.25, 12.30, 12.30, 12.30, 12.40, 12.50, 12.50, 12.50, 12.60, 12.60, 12.85, 12.90, 12.90, 13.00, 13.10, 13.15, 13.25

Use the quadratic model $y = \beta_0 + \beta_1 x + \beta_2 x^2 + \epsilon$ and do the following.

(a) Find an estimate of σ in the model above. What is this supposed to measure? How does your estimate compare to s_P here? What does this comparison suggest to you?

(b) Use an F statistic and test the null hypothesis $H_0: \beta_1 = \beta_2 = 0$. (You may take values off a printout to do this but show the whole five-step significance-testing format.) What is the meaning of this hypothesis in the present context?

(c) Use a t statistic and test the null hypothesis $H_0: \beta_2 = 0$. (Again, you may take values off a printout to do this but show the whole five-step significance-testing format.) What is the meaning of this hypothesis in the present context?

(d) Give a 90% lower confidence bound for the mean weight of the part of such samples that would wind up in the collection tray using a 1.000 in. door spacing.

(e) Give a 90% lower prediction bound for the next weight of the part of such a sample that would wind up in the collection tray using a 1.000 in. door spacing.

(f) Give an approximate 95% lower tolerance for 90% of the weights of all such samples that would wind up in the collection tray using a 1.000 in. door spacing.

8. Return to the armor testing context of Chapter Exercise 21 of Chapter 4. In what follows, base your answers on the model $y = \beta_0 + \beta_1 x_1 + \beta_2 x_2 + \epsilon$.

(a) Based on this model, what is your estimate of the standard deviation of ballistic limit, y, associated with different specimens of a given thickness and Brinell hardness?

(b) Find and plot the standardized residuals. (Plot them versus x_1, versus x_2, and versus $\hat{y}$ and normal-plot them.) Comment on the appearance of your plots.

(c) Make 90% two-sided confidence intervals for β_1 and for β_2. Based on the second of these, what increase in mean ballistic limit would you expect to accompany a 20-unit increase in the Brinell hardness number?

(d) Make a 95% two-sided confidence interval for the mean ballistic limit when a thickness of $x_1 = 258$ (.001 in.) and a Brinell hardness of $x_2 = 391$ are involved.

(e) Make a 95% two-sided prediction interval for an additional ballistic limit when a thickness of $x_1 = 258$ (.001 in.) and a Brinell hardness of $x_2 = 391$ are involved.

(f) Find an approximate 95% lower tolerance bound for 98% of additional ballistic limits when a thickness of $x_1 = 258$ (.001 in.) and a Brinell hardness of $x_2 = 391$ are involved.

(g) Find a 95% two-sided confidence interval for the mean ballistic limit when a thickness of $x_1 = 260$ (.001 in.) and a Brinell hardness of $x_2 = 380$ are involved.

(h) What does the hypothesis $H_0: \beta_1 = \beta_2 = 0$ mean in the context of this study and the model being used in this exercise? Find the p-value associated with an F test of this hypothesis.

(i) What does the hypothesis $H_0: \beta_1 = 0$ mean in the context of this study and the model being used in this exercise? Find the p-value associated with a two-sided t test of this hypothesis.

9. Return to the PETN density/detonation velocity data of Chapter Exercise 23 of Chapter 4.

(a) Find estimates of the parameters β_0, β_1, and σ in the simple linear regression model $y = \beta_0 + \beta_1 x + \epsilon$. How does your estimate of σ compare to s_P? What does this comparison suggest about the reasonableness of the regression model for the data in hand?

(b) Compute standardized residuals and plot them in the same ways that you plotted the residuals in part (g) of Chapter Exercise 23 of Chapter 4. How much do the appearances of the new plots differ from the earlier ones?

(c) Make a 90% two-sided confidence interval for the increase in mean detonation velocity that accompanies a 1 g/cc increase in PETN density.

(d) Make a 90% two-sided confidence interval for the mean detonation velocity of charges with PETN density 0.65 g/cc.

(e) Make a 90% two-sided prediction interval for the next detonation velocity of a charge with PETN density 0.65 g/cc.

(f) Make an approximate 99% lower tolerance bound for the detonation velocities of 95% of charges having a PETN density of 0.65 g/cc.

10. Return to the thread stripping problem of Chapter Exercise 24 of Chapter 4.

(a) Find estimates of the parameters $\beta_0, \beta_1, \beta_2$, and σ in the model $y = \beta_0 + \beta_1 x + \beta_2 x^2 + \epsilon$. How does your estimate of σ compare to s_P? What does this comparison suggest about the reasonableness of the quadratic model for the data in hand? What is your estimate of σ supposed to be measuring?

(b) Use an F statistic and test the null hypothesis $H_0: \beta_1 = \beta_2 = 0$ for the quadratic model. (You may take values off a printout to help you do this but show the whole five-step sig-

nificance testing format.) What is the meaning of this hypothesis in the present context?

(c) Use a t statistic and test the hypothesis $H_0: \beta_2 = 0$ in the quadratic model. (Again, show the whole five-step significance testing format.) What is the meaning of this hypothesis in the present context?

(d) Give a 95% two-sided confidence interval for the mean torque at failure for a thread engagement of 40 (in the units of the problem) using the quadratic model.

(e) Give a 95% two-sided prediction interval for an additional torque at failure for a thread engagement of 40 using the quadratic model.

(f) Give an approximate 99% lower tolerance bound for 95% of torques at failure for studs having thread engagements of 40 using the quadratic model.

11. Return to the situation of Chapter Exercise 28 of Chapter 4 and the metal cutting experiment of Mielnick. Consider an analysis of the torque data based on the model $y_1' = \beta_0 + \beta_1 x_1' + \beta_2 x_2' + \epsilon$.

(a) Make a 90% two-sided confidence interval for the coefficient β_1.

(b) Make a 90% two-sided confidence interval for the mean log torque when a .318 in drill and a feed rate of .005 in./rev are used.

(c) Make a 95% two-sided prediction interval for an additional log torque when a .318 in drill and a feed rate of .005 in./rev are used. Exponentiate the endpoints of this interval to get a prediction interval for a raw torque under these conditions.

(d) Find a 95% two-sided confidence interval for the mean log torque for $x_1 = .300$ in and $x_2 = .010$ in./rev.

12. Return to Chapter Exercise 25 of Chapter 4 and the tire grip force study.

(a) Find estimates of the parameters β_0, β_1, and σ in the simple linear regression model $\ln(y) = \beta_0 + \beta_1 x + \epsilon$.

(b) Compute standardized residuals and plot them in the same ways you plotted the residuals in part (h) of Chapter Exercise 25 of

Chapter 4. How much do the appearances of the new plots differ from the earlier ones?

(c) Make a 90% two-sided confidence interval for the increase in mean log grip force that accompanies an increase in drag of 10% (e.g., from 30% drag to 40% drag). Note that this is $10\beta_1$.

(d) Make a 95% two-sided confidence interval for the mean log grip force of a tire of this type under 30% drag (based on the simple linear regression model).

(e) Make a 95% two-sided prediction interval for the raw grip force of another tire of this design under 30% drag. (*Hint:* Begin by making an interval for log grip force of such a tire.)

(f) Find an approximate 95% lower tolerance bound for the grip forces of 90% of tires of this design under 30% drag (based on the simple linear regression model for $\ln(y)$).

13. Consider again the asphalt permeability data of Woelfl, Wei, Faulstich, and Litwack given in Chapter Exercise 26 of Chapter 4. Use the quadratic model $y = \beta_0 + \beta_1 x + \beta_2 x^2 + \epsilon$ and do the following:

(a) Find an estimate of σ in the quadratic model. What is this supposed to measure? How does your estimate compare to s_P here? What does this comparison suggest to you?

(b) Use an F statistic and test the null hypothesis $H_0: \beta_1 = \beta_2 = 0$ for the quadratic model. (You may take values off a printout to help you do this, but show the whole five-step significance testing format.) What is the meaning of this hypothesis in the present context?

(c) Use a t statistic and test the null hypothesis $H_0: \beta_2 = 0$ in the quadratic model. Again, show the whole five-step significance testing format. What is the meaning of this hypothesis in the present context?

(d) Give a 90% two-sided confidence interval for the mean permeability of specimens of this type with a 6.5% asphalt content.

(e) Give a 90% two-sided prediction interval for the next permeability measured on a specimen of this type having a 6.5% asphalt content.

(f) Find an approximate 95% lower tolerance bound for the permeability of 90% of the specimens of this type having a 6.5% asphalt content.

14. Consider again the axial breaking strength data of Koh, Morden, and Ogbourne given in Chapter Exercise 27 of Chapter 4. At one point in that exercise, it is argued that perhaps the variable $x_3 = x_1^2/x_2$ is the principal determiner of axial breaking strength, y.

(a) Plot the 36 pairs (x_3, y) corresponding to the data given in Chapter 4. Note that a constant σ assumption is probably not a good one over the whole range of x_3's in the students' data.

In light of the point raised in part (a), for purposes of simple linear regression analysis, henceforth restrict attention to those 27 data pairs with $x_3 > .004$.

(b) Find estimates of the parameters β_0, β_1, and σ in the simple linear regression model $y = \beta_0 + \beta_1 x_3 + \epsilon$. How does your estimate of σ based on the simple linear regression model compare to s_P? What does this comparison suggest about the reasonableness of the regression model for the data in hand?

(c) Make a 98% two-sided confidence interval for the mean axial breaking strength of .250 in. dowels 8 in. in length based on the regression analysis. How does this interval compare with the use of formula (6.20) and the four measurements on dowels of this type contained in the data set?

(d) Make a 98% two-sided prediction interval for the axial breaking strength of a single additional .250 in. dowel 8 in. in length. Do the same if the dowel is only 6 in. in length.

(e) Make an approximate 95% lower tolerance bound for the breaking strengths of 98% of .250 in. dowels 8 in. in length.

A

More on Probability and Model Fitting

The introduction to probability theory in Chapter 5 was relatively brief. There are, of course, important engineering applications of probability that require more background in the subject. So this appendix gives a few more details and discusses some additional uses of the theory that are reasonably elementary, particularly in the contexts of reliability analysis and life data analysis.

The appendix begins by discussing the formal/axiomatic basis for the mathematics of probability and several of the most useful simple consequences of the basic axioms. It then applies those simple theorems of probability to the prediction of reliability for series, parallel, and combination series-parallel systems. A brief section treats principles of counting (permutations and combinations) that are sometimes useful in engineering applications of probability. There follows a section on special probability concepts used with life-length (or time-to-failure) variables. The appendix concludes with a discussion of maximum likelihood methods for model fitting and drawing inferences.

A.1   More Elementary Probability

Like any other mathematical theory or system, probability theory is built on a few basic definitions and some "rules of the game" called *axioms*. Logic is applied to determine what consequences (or theorems) follow from the definitions and axioms. These, in turn, can be helpful guides as an engineer seeks to understand and predict the behavior of physical systems that involve chance.

For the sake of logical completeness, this section gives the formal axiomatic basis for probability theory and states and then illustrates the use of some simple theorems that follow from this base. Conditional probability and the independence of events are then defined, and a simple theorem related to these concepts is stated and its use illustrated.

A.1.1 Basic Definitions and Axioms

As was illustrated informally in Chapter 5, the practical usefulness of probability theory is in assigning sensible likelihoods of occurrence to possible happenings in chance situations. The basic, irreducible, potential results in such a chance situation are called **outcomes** belonging to a **sample space**.

Definition 1	A single potential result of a chance situation is called an **outcome**. All outcomes of a chance situation taken together make up a **sample space** for the situation. A script capital S is often used to stand for a sample space.

Mathematically, outcomes are points in a universal set that is the sample space. And notions of simple set theory become relevant. For one thing, subsets of S containing more than one outcome can be of interest.

Definition 2	A collection of outcomes (a subset of S) is called an **event**. Capital letters near the beginning of the alphabet are sometimes used as symbols for events, as are English phrases describing the events.

Once one has defined events, the standard set-theoretic operations of complementation, union, and intersection can be applied to them. However, rather than using the typical "^{c}," "$\cup$," and "$\cap$" mathematical notation for these operations, it is common in probability theory to substitute the use of the words *not, or,* and *and,* respectively.

Definition 3	For event A and event B, subsets of some sample space S,

1. *notA* is an event consisting of all outcomes not belonging to A;

2. *AorB* is an event consisting of all outcomes belonging to one, the other, or both of the two events; and

3. *AandB* is an event consisting of all outcomes belonging simultaneously to the two events.

Example 1   A Redundant Inspection System for Detecting Metal Fatigue Cracks

Consider a redundant inspection system for the detection of fatigue cracks in metal specimens. Suppose the system involves the making of a fluorescent penetrant inspection (FPI) and also a (magnetic) eddy current inspection (ECI). When a

Example 1
(*continued*)

metal specimen is to be tested using this two-detector system, a potential sample space consists of four outcomes corresponding to the possible combinations of what can happen at each detector. That is, a possible sample space is specified in a kind of set notation as

$$\mathcal{S} = \{(\text{FPI signal and ECI signal}), (\text{no FPI signal and ECI signal}), \\ (\text{FPI signal and no ECI signal}), (\text{no FPI signal and no ECI signal})\} \quad \textbf{(A.1)}$$

and in tabular and pictorial forms as in Table A.1 and Figure A.1. Notice that Figure A.1 can be treated as a kind of Venn diagram—the big square standing for $\mathcal{S}$ and the four smaller squares making up $\mathcal{S}$ standing for events that each consist of one of the four different possible outcomes.

Using this four-outcome sample space to describe experience with a metal specimen, one can define several events of potential interest and illustrate the use of the notation described in Definition 3. That is, let

$$A = \{(\text{FPI signal and ECI signal}), (\text{FPI signal and no ECI signal})\} \quad \textbf{(A.2)}$$

$$B = \{(\text{FPI signal and ECI signal}), (\text{no FPI signal and ECI signal})\} \quad \textbf{(A.3)}$$

Table A.1
A List of the Possible Outcomes for Two Inspections

Possible Outcome	FPI Detection Signal?	ECI Detection Signal?
1	yes	yes
2	no	yes
3	yes	no
4	no	no

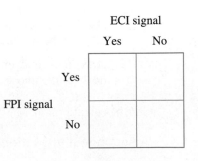

Figure A.1   Graphical representation of four outcomes of two inspections

Then in words,

$$A = \text{the FPI detector signals}$$
$$B = \text{the ECI detector signals}$$

Part 1 of Definition 3 means, for example, that using notations () and (A.2),

$$notA = \{(\text{no FPI signal and ECI signal}), (\text{no FPI signal and no ECI signal})\}$$
$$= \text{the FPI detector doesn't signal}$$

Part 2 of Definition 3 means, for example, that using notations (A.2) and (A.3),

$$AorB = \{(\text{FPI signal and ECI signal}), (\text{FPI signal and no ECI signal}),$$
$$(\text{no FPI signal and ECI signal})\}$$
$$= \text{at least one of the two detectors' signals}$$

And Part 3 of Definition 3 means that again using (A.2) and (A.3), one has

$$AandB = \{(\text{FPI signal and ECI signal})\}$$
$$= \text{both of the two detectors' signals}$$

$notA$, $AorB$, and $AandB$ are shown in Venn diagram fashion in Figure A.2.

Elementary set theory allows the possibility that a set can be empty—that is, have no elements. Such a concept is also needed in probability theory.

Definition 4

> The **empty event** is an event containing no outcomes. The symbol $\emptyset$ is typically used to stand for the empty event.

$\emptyset$ has the interpretation that none of the possible outcomes of a chance situation occur. The way in which $\emptyset$ is most useful in probability is in describing the relationship between two events that have no outcomes in common, and thus cannot both occur. There is special terminology for this eventuality (that $AandB = \emptyset$).

Definition 5

> If event A and event B have no outcomes in common (i.e., $AandB = \emptyset$), then the two events are called **disjoint** or **mutually exclusive**.

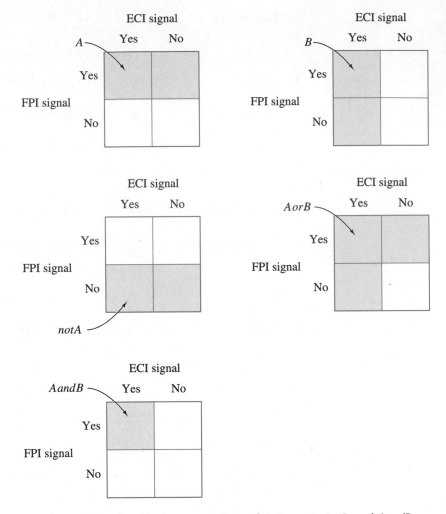

Figure A.2   Graphical representations of *A*, *B*, *notA*, *AorB*, and *AandB*

Example 1
(*continued*)

From Figure A.2 it is quite clear that, for example, the event *A* and the event *notA* are disjoint. And the event *AandB* and the event *not*(*AorB*), for example, are also mutually exclusive events.

Manipulation of events using complementation, union, intersection, etc. is necessary background, but it is hardly the ultimate goal of probability theory. The goal is assignment of likelihoods to events. In order to guarantee that such assignments are internally coherent, probabilists have devised what seem to be intuitively sensible axioms (or rules of operation) for probability models. Assignment of likelihoods in conformance to those rules guarantees that (at a minimum) the assignment is logically consistent. (Whether it is realistic or useful is a separate question.) The axioms of probability are laid out next.

Definition 6

A **system of probabilities** is an assignment of numbers (probabilities) $P[A]$ to events A in such a way that

1. for each event A, $0 \le P[A] \le 1$,

2. $P[\mathcal{S}] = 1$ and $P[\emptyset] = 0$, and

3. for mutually exclusive events $A_1, A_2, A_3, \ldots$,

$$P[A_1 \, or \, A_2 \, or \, A_3 \, or \ldots] = P[A_1] + P[A_2] + P[A_3] + \cdots$$

The relationships (1), (2), and (3) are the **axioms of probability theory**.

Definition 6 is meant to be in agreement with the ways that empirical relative frequencies behave. Axiom (1) says that, as in the case of relative frequencies, only probabilities between 0 and 1 make sense. Axiom (2) says that if one interprets a probability of 1 as sure occurrence and a probability of 0 as no chance of occurrence, it is certain that one of the outcomes in $\mathcal{S}$ will occur. Axiom (3) says that if an event can be made up of smaller nonoverlapping pieces, the probability assigned to that event must be equal to the sum of the probabilities assigned to the pieces.

Although it was not introduced in any formal way, the third axiom of probability was put to good use in Chapter 5. For example, when concluding that for a Poisson random variable X

$$P[2 \le X \le 5] = P[X = 2] + P[X = 3] + P[X = 4] + P[X = 5]$$
$$= f(2) + f(3) + f(4) + f(5)$$

one is really using the third axiom with

$$A_1 = \{X = 2\}$$
$$A_2 = \{X = 3\}$$
$$A_3 = \{X = 4\}$$
$$A_4 = \{X = 5\}$$

It is only in very simple situations that one would ever try to make use of Definition 6 by checking that an entire candidate set of probabilities satisfies the axioms of probability. It is more common to assign probabilities (totaling to 1) to individual outcomes and then simply declare that the third axiom of Definition 6 will be followed in making up any other probabilities. (This strategy guarantees that subsequent probability assignments will be logically consistent.)

Example 2

A System of Probabilities for Describing
a Single Inspection of a Metal Part

As an extremely simple illustration, consider the result of a single inspection of a metal part for fatigue cracks using fluoride penetrant technology. With a sample space

$$S = \{\text{crack signaled, crack not signaled}\}$$

there are only four events:

$$S$$
$$\{\text{crack signaled}\}$$
$$\{\text{no crack signaled}\}$$
$$\emptyset$$

An assignment of probabilities that can be seen to conform to Definition 6 is

$$P[S] = 1$$
$$P[\text{crack signaled}] = .3$$
$$P[\text{no crack signaled}] = .7$$
$$P[\emptyset] = 0$$

Since they conform to Definition 6, these values make up a mathematically valid system of probabilities. Whether or not they constitute a *realistic* or *useful* model is a separate question that can really be answered only on the basis of empirical evidence.

Example 1
(*continued*)

Returning to the situation of redundant inspection of metal parts using both fluoride penetrant and eddy current technologies, suppose that via extensive testing it is possible to verify that for cracks of depth .005 in., the following four values are sensible:

$$P[\text{FPI signal and ECI signal}] = .48 \tag{A.4}$$

$$P[\text{FPI signal and no ECI signal}] = .02 \tag{A.5}$$

$$P[\text{no FPI signal and ECI signal}] = .32 \tag{A.6}$$

$$P[\text{no FPI signal and no ECI signal}] = .18 \tag{A.7}$$

This assignment of probabilities to the basic outcomes in S is illustrated in Figure A.3. Since these four potential probabilities do total to 1, one can adopt them together with provision (3) of Definition 6 and have a mathematically consistent assignment. Then simple addition gives appropriate probabilities for all other events. For example, with event A and event B as defined earlier ($A =$ the FPI detector signals and $B =$ the ECI detector signals),

$$P[A] = P[\text{the FPI detector signals}]$$

$$= P[\text{FPI signal and ECI signal}] + P[\text{FPI signal and no ECI signal}]$$

$$= .48 + .02$$

$$= .50$$

And further,

$$P[A \text{ or } B] = P[\text{at least one of the two detectors signals}]$$

$$= P[\text{FPI signal and ECI signal}] + P[\text{FPI signal and no ECI signal}]$$

$$+ P[\text{no FPI signal and ECI signal}]$$

$$= .48 + .02 + .32$$

$$= .82$$

It is clear that to find the two values, one simply adds the numbers that appear in Figure A.3 in the regions that are shaded in Figure A.2 delimiting the events in question.

		ECI signal	
		Yes	No
	Yes	.48	.02
FPI signal			
	No	.32	.18

Figure A.3 An assignment of probabilities to four possible outcomes of two inspections

A.1.2   Simple Theorems of Probability Theory

The preceding discussion is typical of probability analyses, in that the probabilities for all possible events are not explicitly written down. Rather, probabilities for *some* events, together with logic and the basic rules of the game (the probability axioms), are used to deduce appropriate values for probabilities of *other* events that are of particular interest. This enterprise is often facilitated by the existence of a number of simple theorems. These are general statements that are logical consequences of the axioms in Definition 6 and thus govern the assigning of probabilities for all probability models.

One such simple theorem concerns the relationship between $P[A]$ and $P[not A]$.

Proposition 1	For any event A, $$P[not A] = 1 - P[A]$$

This fact is again one that was used freely in Chapter 5 without explicit reference. For example, in the context of independent, identical success-failure trials, the fact that the probability of at least one success (i.e., $P[X \geq 1]$ for a binomial random variable X) is 1 minus the probability of 0 successes (i.e., $1 - P[X = 0] = 1 - f(0)$) is really a consequence of Proposition 1.

Example 1 *(continued)*	Upon learning, via the addition of probabilities for individual outcomes given in displays (A.4) through (A.7), that the assignment $$P[A] = P[\text{the FPI detector signals}]$$ $$= .50$$ is appropriate, Proposition 1 immediately implies that $$P[not A] = P[\text{the FPI detector doesn't signal}]$$ $$= 1 - P[A]$$ $$= 1 - .50$$ $$= .50$$ is also appropriate. (Of course, if the point here weren't to illustrate the use of Proposition 1, this value could just as well have been gotten by adding .32 and .18.)

A second simple theorem of probability theory is a variation on axiom (3) of Definition 6 for two events that are not necessarily disjoint. It is sometimes called the **addition rule of probability**.

Proposition 2
(*The Addition Rule of Probability*)

For any two events, event A and event B,

$$P[A or B] = P[A] + P[B] - P[A and B] \qquad \text{(A.8)}$$

Note that when dealing with mutually exclusive events, the last term in equation (A.8) is $P[\emptyset] = 0$. Therefore, equation (A.8) simplifies to a two-event version of part (3) of Definition 6. When the event A and the event B are not mutually exclusive, the simple addition $P[A] + P[B]$ (so to speak) counts $P[A and B]$ twice, and the subtraction in equation (A.8) corrects for this in the computing of $P[A or B]$.

The practical usefulness of an equation like (A.8) is that when furnished with any three of the four terms appearing in it, the fourth can be gotten by using simple arithmetic.

Example 3

Describing the Dual Inspection of a Single Cracked Part

Suppose that two different inspectors, both using a fluoride penetrant inspection technique, are to inspect a metal part actually possessing a crack .007 in. deep. Suppose further that some relevant probabilities in this context are

$$P[\text{inspector 1 detects the crack}] = .50$$
$$P[\text{inspector 2 detects the crack}] = .45$$
$$P[\text{at least one inspector detects the crack}] = .55$$

Then using equation (A.8),

$$P[\text{at least one inspector detects the crack}] = P[\text{inspector 1 detects the crack}]$$
$$+ P[\text{inspector 2 detects the crack}] - P[\text{both inspectors detect the crack}]$$

Thus,

$$.55 = .50 + .45 - P[\text{both inspectors detect the crack}]$$

so

$$P[\text{both inspectors detect the crack}] = .40$$

Example 3
(*continued*)

Of course, the .40 value is only as good as the three others used to produce it. But it is at least logically consistent with the given probabilities, and if they have practical relevance, so does the .40 value.

A third simple theorem of probability concerns cases where the basic outcomes in a sample space are judged to be equally likely.

Proposition 3

If the outcomes in a finite sample space S all have the same probability, then for any event A,

$$P[A] = \frac{\text{the number of outcomes in } A}{\text{the number of outcomes in } S}$$

Proposition 3 shows that if one is clever or fortunate enough to be able to conceive of a sample space where an **equally likely outcomes** assignment of probabilities is sensible, the assessment of probabilities can be reduced to a simple counting problem. This fact is particularly useful in enumerative contexts (see again Definition 4 in Chapter 1 for this terminology) where one is drawing random samples from a finite population.

Example 4

Equally Likely Outcomes in a Random Sampling Scenario

Suppose that a storeroom holds, among other things, four integrated circuit chips of a particular type and that two of these are needed in the fabrication of a prototype of an advanced electronic instrument. Suppose further that one of these chips is defective. Consider assigning a probability that both of two chips selected on the first trip to the storeroom are good chips. One way to find such a value (there are others) is to use Proposition 3. Naming the three good chips G1, G2, and G3 and the single defective chip D, one can invent a sample space made up of ordered pairs, the first entry naming the first chip selected and the second entry naming the second chip selected. This is given in set notation as follows:

$$S = \{(G1, G2), (G1, G3), (G1, D), (G2, G1), (G2, G3), (G2, D), (G3, G1),$$
$$(G3, G2), (G3, D), (D, G1), (D, G2), (D, G3)\}$$

A pictorial representation of S is given in Figure A.4.

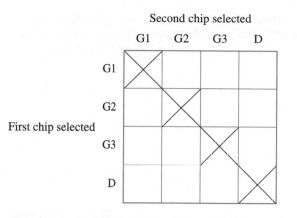

Figure A.4 Graphical representation of 12 possible outcomes when selecting two of four IC chips

Then, noting that the 12 outcomes in this sample space are reasonably thought of as equally likely and that 6 of them do not have D listed either first or second, Proposition 3 suggests the assessment

$$P[\text{two good chips}] = \frac{6}{12} = .50$$

A.1.3 Conditional Probability and the Independence of Events

Chapter 5 discussed the notion of independence for random variables. In that discussion, the idea of assigning probabilities for one variable conditional on the value of another was essential. The concept of conditional assignment of probabilities of *events* is spelled out next.

Definition 7

For event A and event B, provided event B has nonzero probability, the **conditional probability of A given B** is

$$P[A \mid B] = \frac{P[A\,and\,B]}{P[B]} \tag{A.9}$$

The ratio (A.9) ought to make reasonable intuitive sense. If, for example, $P[A\,and\,B] = .3$ and $P[B] = .5$, one might reason that "B occurs only 50% of the time, but of those times B occurs, A also occurs $\frac{.3}{.5} = 60\%$ of the time. So .6 is a sensible assessment of the likelihood of A knowing that indeed B occurs."

Example 4
(*continued*)

Return to the situation of selecting two integrated circuit chips at random from four residing in a storeroom, one of which is defective. Consider using expression (A.9) and evaluating

P[the second chip selected is defective | the first chip selected is good]

Simple counting in the 12-outcome sample space leads to the assignments

$$P\text{[the first chip selected is good]} = \frac{9}{12} = .75$$

$$P\text{[first chip selected is good and second is defective]} = \frac{3}{12} = .25$$

So using Definition 7,

$$P\text{[the second chip selected is defective | the first selected is good]} = \frac{\frac{3}{12}}{\frac{9}{12}} = \frac{1}{3}$$

Of the 9 equally likely outcomes in S for which the first chip selected is good, there are 3 for which the second chip selected is defective. If one thinks of the 9 outcomes for which the first chip selected is good as a kind of reduced sample space (brought about by the partial restriction that the first chip selected is good), then the $\frac{3}{9}$ figure above is a perfectly plausible value for the likelihood that the second chip is defective.

There are sometimes circumstances that make it obvious how a conditional probability ought to be assigned. For example, in the context of Example 4, one might argue that it is obvious that

$$P\text{[the second chip selected is defective | the first selected is good]} = \frac{1}{3}$$

because if the first is good, when the second is to be selected, the storeroom will contain three chips, one of which is defective.

When one does have a natural value for $P[A \mid B]$, the relationship between this and the probabilities $P[A and B]$ and $P[B]$ can sometimes be exploited to evaluate one or the other of them. This notion is important enough that the relationship (A.9) is often rewritten by multiplying both sides by the quantity $P[B]$ and calling the result the **multiplication rule of probability**.

Proposition 4
(*The Multiplication Rule of Probability*)

Provided $P[B] > 0$, so that $P[A \mid B]$ is defined,

$$P[A \, and \, B] = P[A \mid B] \cdot P[B] \qquad \textbf{(A.10)}$$

Example 5

The Multiplication Rule of Probability and a Probabilistic Risk Assessment

A probabilistic risk assessment of the solid rocket motor field joints used in space shuttles prior to the *Challenger* disaster was made in "Risk Analysis of the Space Shuttle: Pre-*Challenger* Prediction of Failure" (*Journal of the American Statistical Association*, 1989) by Dalal, Fowlkes, and Hoadley. They estimated that for each field joint (at 31° and 200 psi),

$$P[\text{primary O-ring erosion}] = .95$$
$$P[\text{primary O-ring blow-by} \mid \text{primary O-ring erosion}] = .29$$

Combining these two values according to rule (A.10), one then sees that the authors' assessment of the failure probability for each primary O-ring was

$$P[\text{primary O-ring erosion and blow-by}] = (.29)(.95) = .28$$

Typically, the numerical values of $P[A \mid B]$ and $P[A]$ are different. The difference can be thought of as reflecting the change in one's assessed likelihood of occurrence of A brought about by knowing that B's occurrence is certain. In cases where there is no difference, the terminology of **independence** is used.

Definition 8

If event A and event B are such that

$$P[A \mid B] = P[A]$$

they are said to be **independent**. Otherwise, they are called **dependent**.

Example 1
(*continued*)

Consider again the example of redundant fatigue crack inspection with probabilities given in Figure A.3. Since

$$P[\text{ECI signal}] = .80$$
$$P[\text{ECI signal} \mid \text{FPI signal}] = \frac{.48}{.50} = .96$$

the events {ECI signal} and {FPI signal} are dependent events.

Example 1
(*continued*)

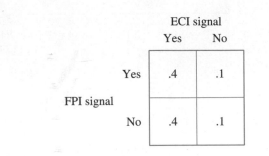

Figure A.5 A second assignment of probabilities to four possible outcomes of two inspections

Of course, different probabilities assigned to individual outcomes in this example can lead to the conclusion that the two events are independent. For example, the probabilities in Figure A.5 give

$$P[\text{ECI signal}] = .4 + .4 = .8$$

$$P[\text{ECI signal} \mid \text{FPI signal}] = \frac{.4}{.4 + .1} = .8$$

so with these probabilities, the two events would be independent.

The multiplication rule when A and B are independent

Independence is the mathematical formalization of the qualitative notion of *unrelatedness*. One way in which it is used in engineering applications is in conjunction with the multiplication rule. If one has values for $P[A]$ and $P[B]$ and judges the event A and the event B to be unrelated, then independence allows one to replace $P[A \mid B]$ with $P[A]$ in formula (A.10) and evaluate $P[AandB]$ as $P[A] \cdot P[B]$. (This fact was behind the scenes in Section 5.1 when sequences of independent identical success-failure trials and the binomial and geometric distributions were discussed.)

Example 5
(*continued*)

In their probabilistic risk assessment of the pre-*Challenger* space shuttle solid rocket motor field joints, Dalal, Fowlkes, and Hoadley arrived at the figure

$$P[\text{failure}] = .023$$

for a single field joint in a shuttle launch at 31°F. A shuttle's two solid rocket motors have a total of six such field joints, and it is perhaps plausible to think of their failures as independent events.

If a model of independence *is* adopted, it is possible to calculate as follows:

$$P[\text{joint 1 and joint 2 are both effective}] = P[\text{joint 1 is effective}] \times$$
$$P[\text{joint 2 is effective}]$$
$$= (1 - .023)(1 - .023)$$
$$= .95$$

And in fact, considering all six joints,

$$P[\text{at least one joint fails}] = 1 - P[\text{all 6 joints are effective}]$$
$$= 1 - (1 - .023)^6$$
$$= .13$$

Section 1 Exercises

1. Return to the situation of Chapter Exercise 30 of Chapter 5, where measured diameters of a turned metal part were coded as Green, Yellow, or Red, depending upon how close they were to a mid-specification. Suppose that the probabilities that a given diameter falls into the various zones are .6247 for the Green Zone, .3023 for the Yellow Zone, and .0730 for the Red Zone. Suppose further (as in the problem in Chapter 5) that the lathe turning the parts is checked once per hour according to the following rules: One diameter is measured, and if it is in the Green Zone, no further action is needed that hour. If it is in the Red Zone, the process is immediately stopped. If it is in the Yellow Zone, a second diameter is measured. If the second diameter is in the Green Zone, no further action is necessary, but if it is not, the process is stopped immediately. Suppose further that the lathe is physically stable, so that it makes sense to think of successive color codes as independent.
   (a) Show that the probability that the process is stopped in a given hour is .1865.
   (b) Given that the process is stopped, what is the conditional probability that the first diameter was in the Yellow Zone?

2. A bin of nuts is mixed, containing 30% $\frac{1}{2}$ in. nuts and 70% $\frac{9}{16}$ in. nuts. A bin of bolts has 40% $\frac{1}{2}$ in. bolts and 60% $\frac{9}{16}$ in. bolts. Suppose that one bolt and one nut are selected (independently and at random) from the two bins.
   (a) What is the probability that the nut and bolt match?
   (b) What is the conditional probability that the nut is a $\frac{9}{16}$ in. nut, given that the nut and bolt match?

3. A physics student is presented with six unmarked specimens of radioactive material. She knows that two are of substance A and four are of substance B. Further, she knows that when tested with a Geiger counter, substance A will produce an average of three counts per second, while substance B will produce an average of four counts per second. (Use Poisson models for the counts per time period.)
   (a) Suppose the student selects a sample at random and makes a one-second check of radioactivity. If one count is observed, how should the student assess the (conditional) probability that the specimen is of substance A?
   (b) Suppose the student selects a sample at random and makes a ten-second check of radioactivity.

If ten counts are observed, how should the student assess the (conditional) probability that the specimen is of substance A?

(c) Are your answers to (a) and (b) the same? How should this be understood?

4. At final inspection of certain integrated circuit chips, 20% of the chips are in fact defective. An automatic testing device does the final inspection. Its characteristics are such that 95% of good chips test as good. Also, 10% of the defective chips test as good.

   (a) What is the probability that the next chip is good and tests as good? *76%*

   (b) What is the probability that the next chip tests as good? *78%*

   (c) What is the (conditional) probability that the next chip that tests as good is in fact good? *.974*

5. In the process of producing piston rings, the rings are subjected to a first grind. Those rings whose thicknesses remain above an upper specification are reground. The history of the grinding process has been that on the first grind,

   60% of the rings meet specifications (and are done processing)

   25% of the rings are above the upper specification (and are reground)

   15% of the rings are below the lower specification (and are scrapped)

   The history has been that after the second grind,

   80% of the reground rings meet specifications

   20% of the reground rings are below the lower specification

   A ring enters the grinding process today.

   (a) Evaluate P[the ring is ground only once].

   (b) Evaluate P[the ring meets specifications].

   (c) Evaluate P[the ring is ground only once | the ring meets specifications].

   (d) Are the events {the ring is ground only once} and {the ring meets specifications} independent events? Explain.

(e) Describe any two mutually exclusive events in this situation.

6. A lot of machine parts is checked piece by piece for Brinell hardness and diameter, with the resulting counts as shown in the accompanying table. A single part is selected at random from this lot.

   (a) What is the probability that it is more than 1.005 in. in diameter?

   (b) What is the probability that it is more than 1.005 in. in diameter and has Brinell hardness of more than 210?

		Diameter		
		< 1.000 in.	1.000 to 1.005 in.	> 1.005 in.
Brinell Hardness	< 190	154	98	48
	190–210	94	307	99
	> 210	33	72	95 *1000*

   (c) What is the probability that it is more than 1.005 in. in diameter or has Brinell hardness of more than 210?

   (d) What is the conditional probability that it has a diameter over 1.005 in., given that its Brinell hardness is over 210?

   (e) Are the events {Brinell hardness over 210} and {diameter over 1.005 in.} independent? Explain.

   (f) Name any two mutually exclusive events in this situation.

7. Widgets produced in a factory can be classified as defective, marginal, or good. At present, a machine is producing about 5% defective, 15% marginal, and 80% good widgets. An engineer plans the following method of checking on the machine's adjustment: Two widgets will be sampled initially, and if either is defective, the machine will be immediately adjusted. If both are good, testing will cease without adjustment. If neither of these first two possibilities occurs, an additional three widgets will be sampled. If all three of these are good, or two are good and one is marginal, testing will

cease without machine adjustment. Otherwise, the machine will be adjusted.

(a) Evaluate P[only two widgets are sampled and no adjustment is made].
(b) Evaluate P[only two widgets are sampled].
(c) Evaluate P[no adjustment is made].
(d) Evaluate P[no adjustment is made | only two widgets are sampled].
(e) Are the events {only two widgets are sampled} and {no adjustment is made} independent events? Explain.
(f) Describe any two mutually exclusive events in this situation.

8. Glass vials of a certain type are conforming, blemished (but usable), or defective. Two large lots of these vials have the following compositions.

$3v$ Lot 1: 70% conforming, 20% blemished, and 10% defective 3

$1v$ Lot 2: 80% conforming, 10% blemished, and 10% defective

Lot 1 is three times the size of Lot 2 and these two lots have been mixed in a storeroom. Suppose that a vial from the storeroom is selected at random to use in a chemical analysis.

(a) What is the probability that the vial is from Lot 1 and not defective?
(b) What is the probability that the vial is blemished?
(c) What is the conditional probability that the vial is from Lot 1 given that it is blemished?

9. A digital communications system transmits information encoded as strings of 0's and 1's. As a means of reducing transmission errors, each digit in a message string is repeated twice. Hence the message string {0 1 1 0} would (ideally) be transmitted as {00 11 11 00} and if digits received in a given pair don't match, one can be sure that the pair has been corrupted in transmission. When each individual digit in a "doubled string" like {00 11 11 00} is transmitted, there is a probability p of transmission error. Further, whether or not a particular digit is correctly transferred is independent of whether any other one is correctly transferred.

Suppose first that the single pair {00} is transmitted.

(a) Find the probability that the pair is correctly received.
(b) Find the probability that what is received has obviously been corrupted.
(c) Find the conditional probability that the pair is correctly received given that it is not obviously corrupted.

Suppose now that the "doubled string" {00 00 11 11} is transmitted and that the string received is not obviously corrupted.

(d) What is then a reasonable assignment of the "chance" that the correct message string (namely {0 0 1 1}) is received? (*Hint:* Use your answer to part c).)

10. Figure A.6 is a Venn diagram with some probabilities of events marked on it. In addition to the values marked on the diagram, it is the case that, $P[B] = .4$ and $P[C \mid A] = .8$.

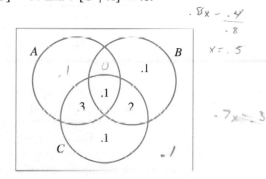

Figure A.6   Figure for Exercise 10

(a) Finish filling in the probabilities on the diagram. That is, evaluate the three probabilities $P[A \, and \, B \, and \, notC]$, $P[A \, and \, notB \, and \, notC]$ and $P[not(A \, or \, B \, or \, C)] = P[notA \, and \, notB \, and \, notC]$.
(b) Use the probabilities on the diagram (and your answers to (a)) and evaluate $P[A \, and \, B]$.
(c) Use the probabilities on the diagram and evaluate $P[B \mid C]$.
(d) Based on the information provided here, are the events B, C independent events? Explain.

A.2   Applications of Simple Probability to System Reliability Prediction

Sometimes engineering systems are made up of identifiable components or subsystems that operate reasonably autonomously and for which fairly accurate reliability information is available. In such cases, it is sometimes of interest to try to predict overall system reliability from the available component reliabilities. This section considers how the simple probability material from Section A.1 can be used to help do this for series, parallel, and combination series-parallel systems.

A.2.1   Series Systems

Definition 9

> A system consisting of components $C_1, C_2, C_3, \ldots, C_k$ is called a **series system** if its proper functioning requires the functioning of all k components.

Figure A.7 is a representation of a **series system** made up of $k = 3$ components. The interpretation to be attached to a diagram like Figure A.7 is that the system will function provided there is a path from point 1 to point 2 that crosses no failed component. (It is tempting, but *not* a good idea, to interpret a system diagram as a flow diagram or like an electrical circuit schematic. The flow diagram interpretation is often inappropriate because there need be no sequencing, time progression, communication, or other such relationship between components in a real series system. The circuit schematic notion often fails to be relevant, and even when it might seem to be, the independence assumptions typically used in arriving at a system reliability figure are of questionable practical appropriateness for electrical circuits.)

If it is sensible to model the functioning of the individual system components as *independent*, then the overall system reliability is easily deduced from component reliabilities via simple multiplication. For example, for a two-component series system,

$$P[\text{the system functions}] = P[C_1 \text{ functions and } C_2 \text{ functions}]$$
$$= P[C_2 \text{ functions} \mid C_1 \text{ functions}] \cdot P[C_1 \text{ functions}]$$
$$= P[C_2 \text{ functions}] \cdot P[C_1 \text{ functions}]$$

Figure A.7   Three-component series system

where the last step depends on the independence assumption. And in general, if the reliability of component C_i (i.e., $P[C_i$ functions]) is r_i, then assuming that the k components in a series system behave independently, the **(series) system reliability** (say, R_S), becomes

Series system reliability for independent components

$$R_S = r_1 \cdot r_2 \cdot r_3 \cdot \cdots \cdot r_k \qquad \text{(A.11)}$$

Example 6
(*Example 5 revisited*)

Space Shuttle Solid Rocket Motor Field Joints as a Series System

The probabilistic risk assessment of Dalal, Fowlkes, and Hoadley put the reliability (at 31°F) of pre-*Challenger* solid rocket motor field joints at .977 apiece. Since the proper functioning of six such joints is necessary for the safe operation of the solid rocket motors, assuming independence of the joints, the reliability of the system of joints is then

$$R_S = (.977)(.977)(.977)(.977)(.977)(.977) = .87$$

as in Example 5. (The .87 figure might well be considered optimistic with regard to the entire solid rocket motor system, as it doesn't take into account any potential problems other than those involving field joints.)

Since typically each r_i is less than 1.0, formula (A.11) shows (as intuitively it should) that system reliability decreases as components are added to a series system. And system reliability is no better (larger) than the worst (smallest) component reliability.

A.2.2 Parallel Systems

In contrast to series system structure is **parallel** system structure.

Definition 10

A system consisting of components $C_1, C_2, C_3, \ldots, C_k$ is called a **parallel system** if its proper functioning requires only the functioning of at least one component.

Figure A.8 is a representation of a parallel system made up of $k = 3$ components. This diagram is interpreted in a manner similar to Figure A.7 (i.e., the system will function provided there is a path from point 1 to point 2 that crosses no failed component).

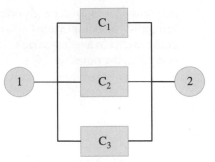

Figure A.8   Three-component parallel system

The fact that made it easy to develop formula (A.11) for the reliability of a series system is that for a series system to function, all components must function. The corresponding fact for a parallel system is that for a parallel system to *fail*, all components must *fail*. So if it is sensible to model the functioning of the individual components in a parallel system as independent, if r_i is the reliability of component i, and if R_P is the **(parallel) system reliability**,

$$1 - R_P = P[\text{the system fails}]$$

$$= P[\text{all components fail}]$$

$$= (1 - r_1)(1 - r_2)(1 - r_3) \cdots (1 - r_k)$$

Parallel system reliability for independent components

or

$$R_P = 1 - (1 - r_1)(1 - r_2)(1 - r_3) \cdots (1 - r_k) \qquad \textbf{(A.12)}$$

Example 7

Parallel Redundancy and Critical Safety Systems

The principle of parallel redundancy is often employed to improve the reliability of critical safety systems. For example, two physically separate automatic shutdown subsystems might be called for in the design of a nuclear power plant. The hope would be that in a rare overheating emergency, at least one of the two would successfully trigger reactor shutdown.

In such a case, if the shutdown subsystems are truly physically separate (so that independence could reasonably be used in a model of their emergency operation), relationship (A.12) might well describe the reliability of the overall safety system. And if, for example, each subsystem is 90% reliable, the overall reliability becomes

$$R_P = 1 - (.10)(.10) = 1 - .01 = .99$$

Expression (A.12) is perhaps a bit harder to absorb than expression (A.11). But the formula functions the way one would intuitively expect. System reliability increases as components are added to a parallel system and is no worse (smaller) than the best (largest) component reliability.

One useful type of calculation that is sometimes done using expression (A.12) is to determine how many equally reliable components of a given reliability r are needed in order to obtain a desired system reliability, R_P. Substitution of r for each r_i in formula (A.12) gives

$$R_P = 1 - (1 - r)^k$$

and this can be solved for an approximate number of components required, giving

Approximate number of components with individual reliability r needed to produce parallel system reliability R_P

$$k \approx \frac{\ln(1 - R_P)}{\ln(1 - r)} \qquad \text{(A.13)}$$

Using (for the sake of example) the values $r = .80$ and $R_P = .98$, expression (A.13) gives $k \approx 2.4$, so rounding *up* to an integer, 3 components of individual 80% reliability will be required to give a parallel system reliability of at least 98%.

A.2.3   Combination Series-Parallel Systems

Real engineering systems rarely have purely series or purely parallel structure. However, it is sometimes possible to conceive of system structure as a **combination** of these two basic types. When this is the case, formulas (A.11) and (A.12) can be used to analyze successively larger subsystems until finally an overall reliability prediction is obtained.

Example 8 | Predicting Reliability for a System with a Combination of Series and Parallel Structure

In order for an electronic mail message from individual A to reach individual B, the main computers at both A's site and B's site must be functioning, and at least one of three separate switching devices at a communications hub must be working. If the reliabilities for A's computer, B's computer, and each switching device are (respectively) .95, .99, and .90, a plausible figure for the reliability of the A-to-B electronic mail system can be determined as follows.

An appropriate system diagram is given in Figure A.9, with C_A, C_B, C_1, C_2, and C_3 standing (respectively) for the A site computer, the B site computer, and the three switching devices. Although this system is neither purely series nor purely parallel, mentally replacing components C_1, C_2, and C_3 with a single

Example 8
(*continued*)

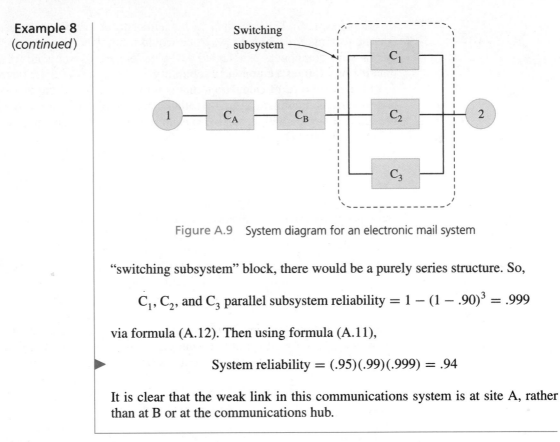

Figure A.9   System diagram for an electronic mail system

"switching subsystem" block, there would be a purely series structure. So,

$$C_1, C_2, \text{ and } C_3 \text{ parallel subsystem reliability} = 1 - (1 - .90)^3 = .999$$

via formula (A.12). Then using formula (A.11),

$$\text{System reliability} = (.95)(.99)(.999) = .94$$

It is clear that the weak link in this communications system is at site A, rather than at B or at the communications hub.

Section 2 Exercises

1. A series system is to consist of $k = 5$ independent components with comparable individual reliabilities. How reliable must each be if the system reliability is to be at least .999? Suppose that it is your job to guarantee components have this kind of individual reliability. Do you see any difficulty in empirically demonstrating this level of component performance? Explain.

2. A parallel system is to consist of k identical independent components. Design requirements are that system reliability be at least .99. Individual component reliability is thought to be at least .90. How large must k be?

3. A combination series-parallel system is to consist of $k = 3$ parallel subsystems, themselves in series.

Engineering design requirements are that the entire system have overall reliability at least .99. Two kinds of components are available. Type A components cost $8 apiece and have reliability .98. Type B components cost $5 apiece and have reliability .90.
   (a) If only type A components are used, what will be the minimum system cost? If only type B components are used, what will be the minimum system cost?
   (b) Find a system design meeting engineering requirements that uses some components of each type and is cheaper than the best option in part (a).

······················

A.3 Counting

Proposition 3 and Example 4 illustrate that using a model for a chance situation that consists of a finite sample space $\mathcal{S}$ with outcomes judged to be equally likely, the computation of probabilities for events of interest is conceptually a very simple matter. The number of outcomes in the event are simply counted up and divided by the total number of outcomes in the whole sample space. However, in most realistic applications of this simple idea, the process of writing down all outcomes in $\mathcal{S}$ and doing the counting involved would be most tedious indeed, and often completely impractical. Fortunately, there are some simple principles of counting that can often be applied to shortcut the process, allowing outcomes to be counted mentally. The purpose of this section is to present those counting techniques.

This section presents a multiplication principle of counting, the notion of permutations and how to count them, and the idea of combinations and how to count them, along with a few examples. This material is on the very fringe of what is appropriate for inclusion in this book. It is not statistics, nor even really probability, but rather a piece of discrete mathematics that has some engineering implications. It is included here for two reasons. First is the matter of tradition. Counting has traditionally been part of most elementary expositions of probability, because games of chance (cards, coins, and dice) are often assumed to be *fair* and thus describable in terms of sample spaces with equally likely outcomes. And for better or worse, games of chance have been a principal source of examples in elementary probability. A second and perhaps more appealing reason for including the material is that it does have engineering applications (regardless of whether they are central to the particular mission of this text). Ultimately, the reader should take this short section for what it is: a digression from the book's main story that can on occasion be quite helpful in engineering problems.

A.3.1 A Multiplication Principle, Permutations, and Combinations

The fundamental insight of this section is a multiplication principle that is simply stated but wide-ranging in its implications. To emphasize the principle, it will be stated in the form of a proposition.

Proposition 5
(*A Multiplication Principle*)

Suppose a complex action can be thought of as composed of r component actions, the first of which can be performed in n_1 different ways, the second of which can subsequently be performed in n_2 different ways, the third of which can subsequently be performed in n_3 different ways, etc. Then the total number of ways in which the complex action can be performed is

$$n = n_1 \cdot n_2 \cdot \cdots \cdot n_r$$

In graphical terms, this proposition is just a statement that a tree diagram that has n_1 first-level nodes, each of which leads to n_2 second-level nodes, and so on, must end up having a total of $n_1 \cdot n_2 \cdot \, \cdots \, \cdot n_r$ rth-level nodes.

Example 9

The Multiplication Principle and Counting the Number of Treatment Combinations in a Full Factorial

The familiar calculation of the number of different possible treatment combinations in a full factorial statistical study is an example of the use of Proposition 5. Consider a $3 \times 4 \times 2$ study in the factors A, B, and C. One may think of the process of writing down a combination of levels for A, B, and C as consisting of $r = 3$ component actions. There are $n_1 = 3$ different ways of choosing a level for A, subsequently $n_2 = 4$ different ways of choosing a level for B, and then subsequently $n_3 = 2$ different ways of choosing a level for C. There are thus

$$n_1 \cdot n_2 \cdot n_3 = 3 \cdot 4 \cdot 2 = 24$$

different ABC combinations.

Example 10

The Multiplication Principle and Counting the Number of Ways of Assigning 4 Out of 100 Pistons to Four Cylinders

Suppose that 4 out of a production run of 100 pistons are to be installed in a particular engine block. Consider the number of possible placements of (distinguishable) pistons into the four (distinguishable) cylinders. One may think of the installation process as composed of $r = 4$ component actions. There are $n_1 = 100$ different ways of choosing a piston for installation into cylinder 1, subsequently $n_2 = 99$ different ways of choosing a piston for installation into cylinder 2, subsequently $n_3 = 98$ different ways of choosing a piston for installation into cylinder 3, and finally, subsequently $n_4 = 97$ different ways of choosing a piston for installation into cylinder 4. There are thus

$$100 \cdot 99 \cdot 98 \cdot 97 = 94{,}109{,}400$$

different ways of placing 4 of 100 (distinguishable) pistons into four (distinguishable) cylinders. (Notice that the job of actually making a list of the different possibilities is not one that is practically doable.)

Example 10 is a generic type of enough importance that there is some special terminology and notation associated with it. The general problem is that of choosing an ordering for r out of n distinguishable objects, or equivalently, placing r out of n distinguishable objects into r distinguishable positions. The application of

Proposition 5 shows that the number of different ways in which this placement can be accomplished is

$$n(n-1)(n-2)\cdots(n-r+1) \tag{A.14}$$

since at each stage of sequentially placing objects into positions, there is one less object available for placement. The special terminology and notation for this are next.

Definition 11

If r out of n distinguishable objects are to be placed in an order 1 to r (or equivalently, placed into r distinguishable positions), each such potential arrangement is called a **permutation**. The number of such permutations possible will be symbolized as $P_{n,r}$, read **"the number of permutations of n things r at a time."**

In the notation of Definition 11, one has (from expression (A.14)) that

$$P_{n,r} = n(n-1)(n-2)\cdots(n-r+1)$$

that is,

Formula for the number of permutations of n things r at a time

$$P_{n,r} = \frac{n!}{(n-r)!} \tag{A.15}$$

Example 10
(*continued*)

In the special permutation notation, the number of different ways of installing the four pistons is

$$P_{100,4} = \frac{100!}{96!}$$

Example 11

Permutations and Counting the Number of Possible Circular Arrangements of 12 Turbine Blades

The permutation idea of Definition 11 can be used not only in straightforward ways, as in the previous example, but in slightly more subtle ways as well. To illustrate, consider a situation where 12 distinguishable turbine blades are to be placed into a central disk or hub at successive 30° angles, as sketched in Figure A.10. Notice that if one of the slots into which these blades fit is marked on the

Example 11
(*continued*)

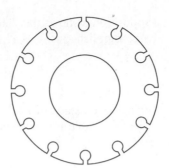

Figure A.10   Hub with 12
slots for blade installation

front face of the hub (and one therefore thinks of the blade positions as completely distinguishable), there are

$$P_{12,12} = 12 \cdot 11 \cdot 10 \cdot \cdots \cdot 2 \cdot 1$$

different possible arrangements of the blades.

But now also consider the problem of counting all possible (circular) arrangements of the 12 blades if relative position is taken into account but absolute position is not. (Moving each blade $30°$ counterclockwise after first installing them would not create an arrangement different from the first, with this understanding.) The permutation idea can be used here as well, thinking as follows. Placing blade 1 anywhere in the hub essentially establishes a point of reference and makes the remaining 11 positions distinguishable (relative to the point of reference). One then has 11 distinguishable blades to place in 11 distinguishable positions. Thus, there must be

$$P_{11,11} = 11 \cdot 10 \cdot 9 \cdot \cdots \cdot 2 \cdot 1$$

such circular arrangements of the 12 blades, where only relative position is considered.

A second generic counting problem is related to the permutation idea and is particularly relevant in describing simple random sampling. That is the problem of choosing an unordered collection of r out of n distinguishable objects. The special terminology and notation associated with this generic problem are as follows.

Definition 12

If an unordered collection of r out of n distinguishable objects is to be made, each such potential collection is called a **combination**. The number of such combinations possible will be symbolized as $\binom{n}{r}$, read "the number of combinations of n things r at a time."

There is in Definition 12 a slight conflict in terminology with other usage in this text. The "combination" in Definition 12 is not the same as the "treatment combination" terminology used in connection with multifactor statistical studies to describe a set of conditions under which a sample is taken. (The "treatment combination" terminology has been used in this very section in Example 9.) But this conflict rarely causes problems, since the intended meaning of the word *combination* is essentially always clear from context.

Appropriate use of Proposition 5 and formula (A.15) makes it possible to develop a formula for $\binom{n}{r}$ as follows. A permutation of r out of n distinguishable objects can be created through a two-step process. First a combination of r out of the n objects is selected and then those selected objects are placed in an order. This thinking suggests that $P_{n,r}$ can be written as

$$P_{n,r} = \binom{n}{r} \cdot P_{r,r}$$

But this means that

$$\frac{n!}{(n-r)!} = \binom{n}{r}\frac{r!}{0!}$$

that is,

Formula for the number of combinations of n things r at a time

$$\boxed{\binom{n}{r} = \frac{n!}{r!\,(n-r)!}}$$

(A.16)

The ratio in equation (A.16) ought to look familiar to readers who have studied Section 5.1. The multiplier of $p^x(1-p)^{n-x}$ in the binomial probability function is of the form $\binom{n}{x}$, counting up the number of ways of placing x successes in a series of n trials.

Example 12
(Example 10, Chapter 3, revisited—page 105)

Combinations and Counting the Numbers of Possible Samples of Cable Connectors with Prescribed Defect Class Compositions

In the cable connector inspection scenario of Delva, Lynch, and Stephany, 3,000 inspections of connectors produced 2,985 connectors classified as having no defects, 1 connector classified as having only minor defects, and 14 others classified as having moderately serious, serious, or very serious defects. Suppose that in an effort to audit the work of the inspectors, a sample of 100 of the 3,000 previously inspected connectors is to be reinspected.

Example 12
(*continued*)

Then notice that directly from expression (A.16), there are in fact

$$\binom{3000}{100} = \frac{3000!}{100!\,2900!}$$

different (unordered) possible samples for reinspection. Further, there are

$$\binom{2985}{100} = \frac{2985!}{100!\,2885!}$$

different possible samples of size 100 made up of only connectors judged to be defect-free. If (for some reason) the connectors to be reinspected were to be chosen as a simple random sample of the 3,000, the ratio

$$\frac{\binom{2985}{100}}{\binom{3000}{100}}$$

would then be a sensible figure to use for the probability that the sample is composed entirely of connectors initially judged to be defect-free.

It is instructive to take this example one step further and combine the use of Definition 12 and Proposition 5. So consider the problem of counting up the number of different samples containing 96 connectors initially judged defect-free, 1 judged to have only minor defects, and 3 judged to have moderately serious, serious, or very serious defects. To solve this problem, the creation of such a sample can be considered as a three-step process. In the first, 96 nominally defect-free connectors are chosen from 2,985. In the second, 1 connector nominally having minor defects only is chosen from 1. And finally, 3 connectors are chosen from the remaining 14. There are thus

$$\binom{2985}{96} \cdot \binom{1}{1} \cdot \binom{14}{3}$$

different possible samples of this rather specialized type.

Example 13

An Application of Counting Principles to the Calculation of a Probability in a Scenario of Equally Likely Outcomes

As a final example in this section, most of the ideas discussed here can be applied to the computation of a probability in another situation of equally likely outcomes where writing out a list of the possible outcomes is impractical.

Consider a hypothetical situation where 15 manufactured devices of a particular kind are to be sent 5 apiece to three different testing labs. Suppose further that 3 of the seemingly identical devices are defective. Consider the probability that each lab receives 1 defective device, if the assignment of devices to labs is done at random.

The total number of possible assignments of devices to labs can be computed by thinking first of choosing 5 of 15 to send to Lab A, then 5 of the remaining 10 to send to Lab B, then sending the remaining 5 to Lab C. There are thus

$$\binom{15}{5} \cdot \binom{10}{5} \cdot \binom{5}{5}$$

such possible assignments of devices to labs.

On the other hand, if each lab is to receive 1 defective device, there are $P_{3,3}$ ways to assign defective devices to labs and then subsequently $\binom{12}{4} \cdot \binom{8}{4} \cdot \binom{4}{4}$ possible ways of completing the three shipments. So ultimately, an appropriate probability assignment for the event that each lab receives 1 defective device is

$$\frac{P_{3,3} \cdot \binom{12}{4} \cdot \binom{8}{4} \cdot \binom{4}{4}}{\binom{15}{5} \cdot \binom{10}{5} \cdot \binom{5}{5}} = \frac{3 \cdot 2 \cdot 1 \cdot 12! \cdot 8! \cdot 5! \cdot 10! \cdot 5! \cdot 5!}{15! \cdot 10! \cdot 4! \cdot 8! \cdot 4! \cdot 4!}$$

$$= \frac{3 \cdot 2 \cdot 1 \cdot 5 \cdot 5 \cdot 5}{15 \cdot 14 \cdot 13}$$

$$= .27$$

Section 3 Exercises

1. A lot of 100 machine parts contains 10 with diameters that are out of specifications on the low side, 20 with diameters that are out of specifications on the high side, and 70 that are in specifications.
   (a) How many different possible samples of $n = 10$ of these parts are there?
   (b) How many different possible samples of size $n = 10$ are there that each contain exactly 1 part with diameter out of specifications on the low side, 2 parts with diameters out of specifications on the high side, and 7 parts with diameters that are in specifications?
   (c) Based on your answers to (a) and (b), what is the probability that a simple random sample of $n = 10$ of these contains exactly 1 part with diameter out of specifications on the low side, 2 parts with diameters out of specifications on

the high side, and 7 parts with diameters that are in specifications?

2. The lengths of bolts produced in a factory are checked with two "go–no go" gauges and the bolts sorted into piles of short, OK, and long bolts. Suppose that of the bolts produced, about 20% are short, 30% are long, and 50% are OK.
   (a) Find the probability that among the next ten bolts checked, the first three are too short, the next three are OK, and the last four are too long.
   (b) Find the probability that among the next ten bolts checked, there are three that are too short, three that are OK, and four that are too long. (*Hint:* In how many ways it is possible to choose three of the group to be short, three

to be OK, and four to be long? Then use your answer to (a).)

3. User names on a computer system consist of three letters A through Z, followed by two digits 0 through 9. (Letters and digits may appear more than once in a name.)

(a) How many user names of this type are there?

(b) Suppose that Joe has user name TPK66, but unfortunately he's forgotten it. Joe remembers only the format of the user names and that the letters K, P, and T appear in his name. If he picks a name at random from those consistent with his memory, what's the probability that he selects his own?

(c) If Joe in part (b) also remembers that his digits match, what's the probability that he selects his own user name?

4. A lot contains ten pH meters, three of which are miscalibrated. A technician selects these meters one at a time, at random without replacement, and checks their calibration.

(a) What is the probability that among the first four meters selected, exactly one is miscalibrated?

(b) What is the probability that the technician discovers his second miscalibrated meter when checking his fifth one?

5. A student decides to use the random digit function on her calculator to select a three-digit PIN number for use with her new ATM card. (Assume that all numbers 000 through 999 are then equally likely to be chosen.)

(a) What is the probability that her number uses only odd digits?

(b) What is the probability that all three digits in her number are different?

(c) What is the probability that her number uses three different digits and lists them in either ascending or descending order?

6. When ready to configure a PC order, a consumer must choose a Processor Chip, a MotherBoard, a Drive Controller and a Hard Drive. The choices are:

Processor Chip	Mother-Board	Drive Controller	Hard Drive
Fast New Generation	Premium	Premium	Premium
Slow New Generation	Standard	Standard	Standard
Fast Old Generation		Economy	Economy
Slow Old Generation			

(a) Suppose initially that all components are compatible with all components. How many different configurations are possible?

Suppose henceforth that:

(i) a Premium MotherBoard is needed to run a New Generation Processor,

(ii) a Premium MotherBoard is needed to run a Premium Drive Controller, and

(iii) a Premium Drive Controller is needed to run a Premium Hard Drive.

(b) How many permissible configurations are there with a Standard MotherBoard?

(c) How many permissible configurations are there total? Explain carefully.

A.4 Probabilistic Concepts Useful in Survival Analysis

Section A.2 is meant to provide only the most elementary insights into how probability might prove useful in the context of reliability modeling and prediction. The ideas discussed in that section are of an essentially "static" nature. They are most appropriate when considering the likelihood of a system performing adequately at a single point in time—for example, at its installation, or at the end of its warranty period.

Reliability engineers also concern themselves with matters possessing a more dynamic flavor, having to do with the modeling and prediction of life-length variables associated with engineering systems and their components. It is outside the intended scope of this text to provide anything like a serious introduction to the large body of methods available for probability modeling and subsequent formal inference for such variables. But what will be done here is to provide some material (supplementary to that found in Section 5.2) that is part of the everyday jargon and intellectual framework of reliability engineering. This section will consider several descriptions and constructs related to continuous random variables that, like system or component life lengths, take only positive values.

A.4.1 Survivorship and Force-of-Mortality Functions

In this section, T will stand for a continuous random variable taking only nonnegative values. The reader may think of T as the time till failure of an engineering component. In Section 5.2, continuous random variables X (or more properly, their distributions) were described through their probability densities $f(x)$ and cumulative probability functions $F(x)$. In the present context of lifetime random variables, there are several other, more popular ways of conveying the information carried by $f(t)$ or $F(t)$. Two of these devices are introduced next.

Definition 13	The **survivorship function** for a nonnegative random variable T is the function

$$S(t) = P[T > t] = 1 - F(t)$$

The survivorship function is also sometimes known as the **reliability function**. It specifies the probability that the component being described survives beyond time t.

Example 14	The Survivorship Function and Diesel Engine Fan Blades

Data on 70 diesel engines of a single type (given in Table 1.1 of Nelson's *Applied Life Data Analysis*) indicate that lifetimes in hours of a certain fan on such engines could be modeled using an exponential distribution with mean $\alpha \approx 27,800$. So from Definition 17 in Chapter 5, to describe a fan lifetime T, one could use the density

$$f(t) = \begin{cases} 0 & \text{if } t < 0 \\ \dfrac{1}{27,800}e^{-t/27,800} & \text{if } t > 0 \end{cases}$$

Example 14
(*continued*)

or the cumulative probability function

$$F(t) = \begin{cases} 0 & \text{if } t \leq 0 \\ 1 - e^{-t/27,800} & \text{if } t > 0 \end{cases}$$

or from Definition 13, the survivorship function

$$S(t) = \begin{cases} 1 & \text{if } t \leq 0 \\ e^{-t/27,800} & \text{if } t > 0 \end{cases}$$

The probability of a fan surviving at least 10,000 hours is then

$$S(10,000) = e^{-10,000/27,800} = .70$$

A second way of specifying the distribution of a life-length variable (unlike anything discussed in Section 5.2) is through a function giving a kind of "instantaneous rate of death of survivors."

Definition 14

The **force-of-mortality function** for a nonnegative continuous random variable T is, for $t > 0$, the function

$$h(t) = \frac{f(t)}{S(t)}$$

$h(t)$ is sometimes called the *hazard function* for T, but such usage tends to perpetuate unfortunate confusion with the *entirely different* concept of "hazard rate" for repairable systems. (The important difference between the two concepts is admirably explained in the paper "On the Foundations of Reliability" by W. A. Thompson (*Technometrics*, 1981) and in the book *Repairable Systems Reliability* by Ascher and Feingold.) This book will thus stick to the term *force of mortality*.

The force-of-mortality function can be thought of heuristically as

$$h(t) = \frac{f(t)}{S(t)} = \lim_{\Delta \to 0} \frac{P[t < T < t + \Delta]/\Delta}{P[t < T]} = \lim_{\Delta \to 0} \frac{P[t < T < t + \Delta \mid t < T]}{\Delta}$$

which is indeed a sort of "death rate of survivors at time t."

Example 14
(*continued*)

The force-of-mortality function for the diesel engine fan example is, for $t > 0$,

$$h(t) = \frac{f(t)}{S(t)} = \frac{\frac{1}{27,800} e^{-t/27,800}}{e^{-t/27,800}} = \frac{1}{27,800}$$

> The exponential (mean $\alpha = 27,800$) model for fan life implies a constant $\left(\frac{1}{27,800}\right)$ force of mortality.

Constant force of mortality is equivalent to exponential distribution

The property of the fan-life model shown in the previous example is characteristic of exponential distributions. That is, a distribution has constant force of mortality exactly when that distribution is exponential. So having a constant force of mortality is equivalent to possessing the *memoryless* property of the exponential distributions discussed in Section 5.2. If the lifetime of an engineering component is described using a constant force of mortality, there is no (mathematical) reason to replace such a component before it fails. The distribution of its remaining life from any point in time is the same as the distribution of the time till failure of a new component of the same type.

Potential probability models for lifetime random variables are often classified according to the nature of their force-of-mortality functions, and these classifications are taken into account when selecting models for reliability engineering applications. If $h(t)$ is increasing in t, the corresponding distribution is called an **increasing force-of-mortality (IFM) distribution**, and if $h(t)$ is decreasing in t, the corresponding distribution is called a **decreasing force-of-mortality (DFM) distribution**. The reliability engineering implications of an IFM distribution being appropriate for modeling the lifetimes of a particular type of component are often that (as a form of preventative maintenance) such components are retired from service once they reach a particular age, even if they have not failed.

Example 15 | The Weibull Distributions and Their Force-of-Mortality Functions

The Weibull family of distributions discussed in Section 5.2 is commonly used in reliability engineering contexts. Using formulas (5.26) and (5.27) of Section 5.2 for the Weibull cumulative probability function and probability density, the Weibull force-of-mortality function for shape parameter β and scale parameter α is, for $t > 0$

$$h(t) = \frac{f(t)}{S(t)} = \frac{f(t)}{1 - F(t)} = \frac{\frac{\beta}{\alpha^\beta} t^{\beta-1} e^{-(t/\alpha)^\beta}}{e^{-(t/\alpha)^\beta}} = \frac{\beta t^{\beta-1}}{\alpha^\beta}$$

For $\beta = 1$ (the exponential distribution case) this is constant. For $\beta < 1$, this is decreasing in t, and the Weibull distributions with $\beta < 1$ are DFM distributions. For $\beta > 1$, this is increasing in t, and the Weibull distributions with $\beta > 1$ are IFM distributions.

Example 16

Force-of-Mortality Function for a Uniform Distribution

As an artificial but instructive example, consider the use of a uniform distribution on the interval $(0, 1)$ as a life-length model. With

$$f(t) = \begin{cases} 1 & \text{if } 0 < t < 1 \\ 0 & \text{otherwise} \end{cases}$$

the survivorship function is

$$S(t) = \begin{cases} 1 & \text{if } t < 0 \\ 1 - t & \text{if } 0 \leq t < 1 \\ 0 & \text{if } 1 \leq t \end{cases}$$

so

$$h(t) = \frac{1}{1 - t} \qquad \text{if } 0 < t < 1$$

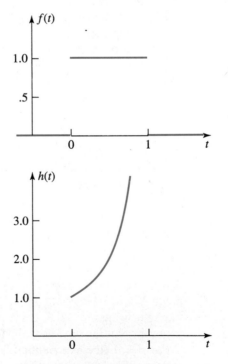

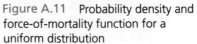

Figure A.11 Probability density and force-of-mortality function for a uniform distribution

Figure A.11 shows plots of both $f(t)$ and $h(t)$ for the uniform model. $h(t)$ is clearly increasing for $0 < t < 1$ (quite drastically so, in fact, as one approaches $t = 1$). And well it should be. Knowing that (according to the uniform model) life will certainly end by $t = 1$, nervousness about impending death should skyrocket as one nears $t = 1$.

Conventional wisdom in reliability engineering is that many kinds of manufactured devices have life distributions that ought to be described by force-of-mortality functions qualitatively similar to the hypothetical one sketched in Figure A.12.

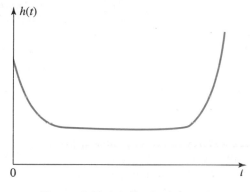

Figure A.12 A "bathtub curve" force-of-mortality function

The shape in Figure A.12 is often referred to as the **bathtub curve** shape. It includes an early region of decreasing force of mortality, a long central period of relatively constant force of mortality, and a late period of rapidly increasing force of mortality. Devices with lifetimes describable as in Figure A.12 are sometimes subjected to a **burn-in period** to eliminate the devices that will fail in the early period of decreasing force of mortality, and then sold with the recommendation that they be replaced before the onset of the late period of increasing force of mortality or **wear-out**. Although this story is intuitively appealing, the most tractable models for life length do not, in fact, have force-of-mortality functions with shapes like that in Figure A.12. For a further discussion of this matter and references to papers presenting models with bathtub-shaped force-of-mortality functions, refer to Chapter 2 of Nelson's *Applied Life Data Analysis*.

The functions $f(t)$, $F(t)$, $S(t)$, and $h(t)$ all carry the same information about a life distribution. They simply express it in different terms. Given one of them, the derivation of the others is (at least in theory) straightforward. Some of the relationships that exist among the four different characterizations are collected here for the reader's convenience. For $t > 0$,

Relationships between F(t), f(t), S(t), and h(t)

$$F(t) = \int_0^t f(x)\,dx$$

$$f(t) = \frac{d}{dt}F(t)$$

$$S(t) = 1 - F(t)$$

$$h(t) = \frac{f(t)}{S(t)}$$

$$S(t) = \exp\left(-\int_0^t h(x)\,dx\right)$$

$$f(t) = h(t)\exp\left(-\int_0^t h(x)\,dx\right)$$

Section 4 Exercises

1. An engineer begins a series of presentations to his corporate management with a working bulb in his slide projector and (an inferior-quality) Brand W replacement bulb in his briefcase. Suppose that the random variables

   X = the number of hours of service given by the bulb in the projector

   Y = the number of hours of service given by the spare bulb

   may be modeled as independent exponential random variables with respective means 15 and 5. The number of hours that the engineer may operate without disaster is $X + Y$.
   (a) Find the mean and standard deviation of $X + Y$ using Proposition 1 in Chapter 5.
   (b) Find, for $t > 0$, $P[X + Y \leq t]$.
   (c) Use your answer to (b) and find the probability density for $T = X + Y$.
   (d) Find the survivorship and force-of-mortality functions for T. What is the nature of the force-

   of-mortality function? Is it constant like those of X and Y?

2. A common modeling device in reliability applications is to assume that the (natural) logarithm of a lifetime variable, T, has a normal distribution. That is, one might suppose that for some parameters μ and σ, if $t > 0$

   $$F(t) = P[T \leq t] = \Phi\left(\frac{\ln t - \mu}{\sigma}\right)$$

   Consider the $\mu = 0$ and $\sigma = 1$ version of this.
   (a) Plot $F(t)$ versus t.
   (b) Plot $S(t)$ versus t.
   (c) Compute and plot $f(t)$ versus t.
   (d) Compute and plot $h(t)$ versus t.
   (e) Is this distribution for T an IFM distribution, a DFM distribution, or neither? What implication does your answer have for in-service replacement of devices possessing this lifetime distribution?

A.5 Maximum Likelihood Fitting of Probability Models and Related Inference Methods

The model-fitting and inference methods discussed in this text are, for the most part, methods for independent, *normally distributed* observations. This is in spite of the fact that there are strong hints in Chapter 5 and this appendix that other kinds of probability models often prove useful in engineering problem solving. (For example, binomial, geometric, Poisson, exponential, and Weibull distributions have been discussed, and parts of Sections A.1 and A.2 should suggest that probability models not even necessarily involving these standard distributions will often be helpful.) It thus seems wise to present at least a brief introduction to some principles of probability-model fitting and inference that can be applied more generally than to only scenarios involving independent, normal observations. This will be done to give at least the flavor of what is possible, as well as an idea of some kinds of things likely to be found in the engineering statistics literature.

This section considers the use of likelihood functions in the fitting of parametric probability models and in large-sample inference for the model parameters. It begins by discussing the idea of a likelihood function and maximum likelihood model fitting for discrete data. Similar discussions are then conducted for continuous and mixed data. Finally, there is a discussion of how for large samples, approximate confidence regions and tests can often be developed using likelihood functions.

A.5.1 Likelihood Functions for Discrete Data and Maximum Likelihood Model Fitting

To begin, consider scenarios where the outcome of a chance situation can be described in terms of a data vector of jointly discrete random variables (or a single discrete random variable) Y, whose probability function f depends on some (unknown) vector of parameters (or single parameter) Θ. To make the dependence of f on Θ explicit, this section will use the notation

$$f_\Theta(y)$$

for the (joint) probability function of Y.

Chapter 5 made heavy use of particular parametric probability functions, primarily thinking of them as functions of y. In this section, it will be very helpful to shift perspective. With data $Y = y$ in hand, think of

A discrete data likelihood function

$$\boxed{f_\Theta(y)}$$

(A.17)

or (often more conveniently) its natural logarithm

A discrete data log likelihood function

$$L(\Theta) = \ln\left(f_\Theta(y)\right) \tag{A.18}$$

as functions of Θ, specifying for various possible vectors of parameters "how likely" it would be to observe the particular data in hand. With this perspective, the function (A.17) is often called the **likelihood function** and function (A.18) the **log likelihood function** for the problem under consideration.

Example 17
(*Example 4, Chapter 5, revisited—page 235*)

The Log Likelihood Function for the Number Conforming in a Sample of Hexamine Pellets

In the pelletizing machine example used in Chapter 5 and earlier, it is possible to argue that under stable conditions,

$X =$ the number of conforming pellets produced in a batch of 100 pellets

might well be modeled using a binomial distribution with $n = 100$ and p some unknown parameter. The corresponding probability function is thus

$$f(x) = \begin{cases} \dfrac{100!}{x!\,(100-x)!}\, p^x(1-p)^{100-x} & x = 0, 1, \ldots, 100 \\ 0 & \text{otherwise} \end{cases}$$

Should $X = 66$ conforming pellets be observed in a batch, the material just introduced says that the function of p

$$L(p) = \ln(f(66)) = \ln(100!) - \ln(66!) - \ln(34!)$$
$$+ 66\ln(p) + 34\ln(1-p) \tag{A.19}$$

is an appropriate log likelihood function. Figure A.13 is a sketch of $L(p)$ for this problem. Notice that (in an intuitively appealing fashion) the value of p maximizing $L(p)$ is

$$\hat{p} = \frac{66}{100} = .66$$

That is, $p = .66$ makes the chance of observing the particular data in hand ($X = 66$) as large as possible.

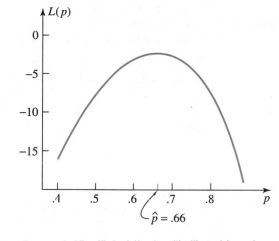

Figure A.13 Plot of the log likelihood function based on 66 conforming tablets out of 100

Example 18

The Log Likelihood Function for n Independent Poisson Observations

As a second, somewhat more abstract, example of the idea of a likelihood function, suppose that $X_1, X_2, \ldots, X_n$ are independent Poisson random variables, X_i with mean $k_i \lambda$ for $k_1, k_2, \ldots, k_n$ known constants, and λ an unknown parameter. Such a model might, for example, be appropriate in a quality monitoring context, where at time i, k_i standard-size units of product are inspected, X_i defects are observed, and λ is a constant mean defects per unit.

The joint probability function for $X_1, X_2, \ldots, X_n$ is

$$f(x_1, x_2, \ldots, x_n) = \begin{cases} \prod_{i=1}^{n} \dfrac{e^{-k_i \lambda}(k_i \lambda)^{x_i}}{x_i!} & \text{for each } x_i \text{ a nonnegative integer} \\ 0 & \text{otherwise} \end{cases}$$

The log likelihood function in the present context is thus

$$L(\lambda) = -\lambda \sum_{i=1}^{n} k_i + \sum_{i=1}^{n} x_i \ln(k_i) + \sum_{i=1}^{n} x_i \ln(\lambda) - \sum_{i=1}^{n} \ln(x_i!) \qquad \textbf{(A.20)}$$

The likelihood functions in Examples 17 and 18 are for individual (univariate) parameters. The next example involves two parameters.

Example 19

A Log Likelihood Function Based on Pre-*Challenger* Space Shuttle O-Ring Failure Data

Table A.2 contains pre-*Challenger* data on field joint primary O-ring failures on 23 (out of 24) space shuttle flights. (On one flight, the rocket motors were lost at sea, so no data are available.) The failure counts $x_1, x_2, \ldots, x_{23}$ are the numbers (out of 6 possible) of primary O-rings showing evidence of erosion or blow-by in postflight inspections of the solid rocket motors, and $t_1, t_2, \ldots, t_{23}$ are the corresponding *temperatures at the times of launch*.

Table A.2
Pre-*Challenger* Field Joint Primary O-Ring Failure Data

Flight Date	x, Number of Field Joint Primary O-Ring Incidents	t, Temperature at Launch (°F)
4/12/81	0	66
11/12/81	1	70
3/22/82	0	69
11/11/82	0	68
4/4/83	0	67
6/18/83	0	72
8/30/83	0	73
11/28/83	0	70
2/3/84	1	57
4/6/84	1	63
8/30/84	1	70
10/5/84	0	78
11/8/84	0	67
1/24/85	2	53
4/12/85	0	67
4/29/85	0	75
6/17/85	0	70
7/29/85	0	81
8/27/85	0	76
10/3/85	0	79
10/30/85	2	75
11/26/85	0	76
1/12/86	1	58

In "Risk Analysis of the Space Shuttle: Pre-*Challenger* Prediction of Failure" (*Journal of the American Statistical Association*, 1989), Dalal, Fowlkes, and Hoadley considered several analyses of the data in Table A.2 (and other pre-*Challenger* shuttle failure data). In one of their analyses of the data given here, Dalal, Fowlkes, and Hoadley used a likelihood approach based on the observations

$$y_i = \begin{cases} 1 & \text{if } x_i \geq 1 \\ 0 & \text{if } x_i = 0 \end{cases}$$

that indicate which flights experienced primary O-ring incidents. (They also considered a likelihood approach based on the counts x_i themselves. But here only the slightly simpler analysis based on the y_i's will be discussed.) The authors modeled $Y_1, Y_2, \ldots, Y_{23}$ as a priori independent variables and treated the probability of at least one O-ring incident on flight i,

$$p_i = P[Y_i = 1] = P[X_i \geq 1]$$

as a function of (temperature) t_i. The particular form of dependence of p_i on t_i used by the authors was a "linear (in t) log odds" form

$$\ln\left(\frac{p}{1-p}\right) = \alpha + \beta t \qquad \textbf{(A.21)}$$

for α and β some unknown parameters. Equation (A.21) can be solved for p to produce the function of t

$$p(t) = \frac{1}{1 + e^{-(\alpha + \beta t)}} \qquad \textbf{(A.22)}$$

From either equation (A.21) or (A.22), it is possible to see that if $\beta > 0$, the probability of at least one O-ring incident is increasing in t (low-temperature launches are best). On the other hand, if $\beta < 0$, p is decreasing in t (high-temperature launches are best).

The joint probability function for $Y_1, Y_2, \ldots, Y_{23}$ employed by Dalal, Fowlkes, and Hoadley was then

$$f(y_1, y_2, \ldots, y_{23}) = \begin{cases} \displaystyle\prod_{i=1}^{23} p(t_i)^{y_i} \left(1 - p(t_i)\right)^{1-y_i} & \text{for each } y_i = 0 \text{ or } 1 \\ 0 & \text{otherwise} \end{cases}$$

Example 19
(*continued*)

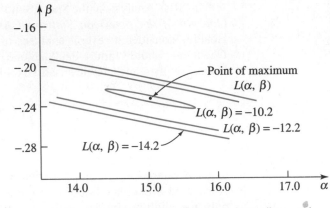

Figure A.14　Contour plot of the Dalal, Fowlkes, and Hoadley log likelihood function

The log likelihood function is then (using equations (A.21) and (A.22))

$$L(\alpha, \beta) = \sum_{i=1}^{23} y_i \ln\left(\frac{p(t_i)}{1 - p(t_i)}\right) + \sum_{i=1}^{23} \ln\left(1 - p(t_i)\right)$$

$$= \sum_{i=1}^{23} y_i(\alpha + \beta t_i) + \sum_{i=1}^{23} \ln\left(\frac{e^{-(\alpha+\beta t_i)}}{1 + e^{-(\alpha+\beta t_i)}}\right)$$

$$= 7\alpha + \beta(70 + 57 + 63 + 70 + 53 + 75 + 58)$$

$$+ \ln\left(\frac{e^{-(\alpha+66\beta)}}{1 + e^{-(\alpha+66\beta)}}\right) + \ln\left(\frac{e^{-(\alpha+70\beta)}}{1 + e^{-(\alpha+70\beta)}}\right)$$

$$+ \cdots + \ln\left(\frac{e^{-(\alpha+58\beta)}}{1 + e^{-(\alpha+58\beta)}}\right) \tag{A.23}$$

where the sum abbreviated in equation (A.23) is over all 23 t_i's. Figure A.14 is a contour plot of $L(\alpha, \beta)$ given in equation (A.23).

It is interesting (and sadly, of great engineering importance) that the region of (α, β) pairs making the data of Table A.2 most likely is in the $\beta < 0$ part of the (α, β)-plane—that is, where $p(t)$ is decreasing in t (i.e., increases as t falls). (Remember that the tragic *Challenger* launch was made at $t = 31°$.)

The binomial and Poisson examples of discrete-data likelihoods given thus far have arisen from situations that are most naturally thought of as intrinsically discrete. However, the details of how engineering data are collected sometimes lead to the production of essentially discrete data from intrinsically continuous variables. For example, consider a life test of some electrical components, where a technician begins a test by connecting 50 devices to a power source, goes away, and then returns every ten hours to note which devices are still functioning. The details of data collection produce only discrete data (which ten-hour period produces failure) from the intrinsically continuous life lengths of the 50 devices. The next example shows how the likelihood idea might be used in another situation where the underlying phenomenon is continuous.

Example 20

A Log Likelihood Function for a Crudely Gauged Normally Distributed Dimension of Five Machined Metal Parts

In many contexts where industrial process monitoring involves relatively stable processes and relatively crude gauging, intrinsically continuous product characteristics are measured and recorded as essentially discrete data. For example, Table A.3 gives values (in units of .0001 in. over nominal) of a critical dimension measured on a sample of $n = 5$ consecutive metal parts produced by a CNC lathe.

It might make sense to model underlying values of this critical dimension as normal, with some (unknown) mean μ and some (unknown) standard deviation σ, but nonetheless to want to explicitly recognize the discreteness of the recorded data. One way of doing so in this context is to think of the observed values as arising (after coding) from rounding normally distributed dimensions to the nearest integer. For a single metal part, this would mean that for any integer y,

$$
\begin{aligned}
P[\text{the value recorded is } y] &= P[\text{the actual dimension is between} \\
&\quad\quad y - .5 \text{ and } y + .5] \\
&= \Phi\left(\frac{y + .5 - \mu}{\sigma}\right) - \Phi\left(\frac{y - .5 - \mu}{\sigma}\right) \quad \textbf{(A.24)}
\end{aligned}
$$

Table A.3
Measurements of a Critical
Dimension on Five Metal Parts
Produced on a CNC Lathe

Part	Measured Dimension, y
1	4
2	3
3	3
4	2
5	3

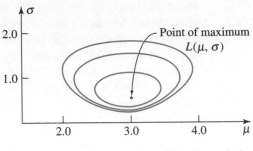

Figure A.15   Contour plot of the "rounded normal data" log likelihood for the data of Table A.3

So treating $n = 5$ consecutive recorded dimensions as independent, equation (A.24) leads to the joint probability function

$$f(y_1, y_2, \ldots, y_5) = \prod_{i=1}^{5} \left\{ \Phi\left(\frac{y_i + .5 - \mu}{\sigma} \right) - \Phi\left(\frac{y_i - .5 - \mu}{\sigma} \right) \right\}$$

and log likelihood function for the data in Table A.3

$$\left. \begin{aligned} L(\mu, \sigma) = \ & \ln\left(\Phi\left(\frac{2 + .5 - \mu}{\sigma} \right) - \Phi\left(\frac{2 - .5 - \mu}{\sigma} \right) \right) \\ & + 3\ln\left(\Phi\left(\frac{3 + .5 - \mu}{\sigma} \right) - \Phi\left(\frac{3 - .5 - \mu}{\sigma} \right) \right) \\ & + \ln\left(\Phi\left(\frac{4 + .5 - \mu}{\sigma} \right) - \Phi\left(\frac{4 - .5 - \mu}{\sigma} \right) \right) \end{aligned} \right\} \qquad \textbf{(A.25)}$$

Figure A.15 is a contour plot of $L(\mu, \sigma)$.

Consideration of a likelihood function $f_\Theta(y)$ or its log version $L(\Theta)$ can be thought of as a way of assessing how compatible various probability models indexed by Θ are with the data in hand, $Y = y$. Different parameter vectors Θ having the same value of $L(\Theta)$ can be viewed as equally compatible with data in hand. A value of Θ maximizing $L(\Theta)$ might then be considered to be as compatible with the observed data as is possible. This value is often termed the **maximum likelihood estimate** of the parameter vector Θ. Finding maximum likelihood estimates of parameters is a very common method of fitting probability models to data. In simple situations, calculus can sometimes be employed to see how to maximize $L(\Theta)$, but in most nonstandard situations, numerical or graphical methods are required.

Example 17
(*continued*)

In the pelletizing example, simple investigation of Figure A.13 shows

$$\hat{p} = \frac{66}{100}$$

to maximize $L(p)$ given in display (A.19) and thus to be the maximum likelihood estimate of p. The reader is encouraged to verify that by differentiating $L(p)$ with respect to p, setting the result equal to 0, and solving for p, this maximizing value can also be found analytically.

Example 18
(*continued*)

Differentiating the log likelihood (A.20) with respect to λ, one obtains

$$\frac{d}{d\lambda}L(\lambda) = -\sum_{i=1}^{n} k_i + \frac{1}{\lambda}\sum_{i=1}^{n} x_i$$

Setting this derivative equal to 0 and solving for λ produces

$$\lambda = \frac{\sum_{i=1}^{n} x_i}{\sum_{i=1}^{n} k_i} = \hat{u}$$

which is the total number of defects observed divided by the total number of units inspected. Since the second derivative of $L(\lambda)$ is easily seen to be negative for all λ, $\hat{u}$ is the unique maximizer of $L(\lambda)$—that is, the maximum likelihood estimate of λ.

Example 19
(*continued*)

Careful examination of contour plots like Figure A.14, or use of a numerical search method for the (α, β) pair maximizing $L(\alpha, \beta)$, produces maximum likelihood estimates

$$\hat{\alpha} = 15.043$$
$$\hat{\beta} = -.2322$$

based on the pre-*Challenger* data. Figure A.16 is a plot of $p(t)$ given in display (A.22) for these values of α and β. Notice the disconcerting fact that the corresponding estimate of $p(31)$ (the probability of at least one O-ring failure in a $31°$ launch) exceeds .99. ($t = 31$ is clearly a huge extrapolation away from any t values in Table A.2, but even so, this kind of analysis conducted before the *Challenger* launch could well have helped cast legitimate doubt on the advisability of a low-temperature launch.)

Example 19
(*continued*)

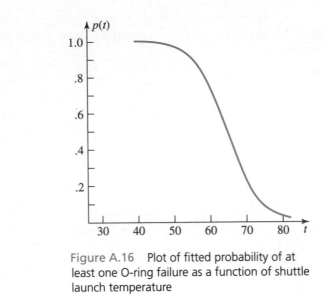

Figure A.16   Plot of fitted probability of at
least one O-ring failure as a function of shuttle
launch temperature

Example 20
(*continued*)

Examination of the contour plot in Figure A.15 shows maximum likelihood
estimates of μ and σ based on the rounded normal data model and the data in
Table A.3 to be approximately

$$\hat{\mu} = 3.0$$

$$\hat{\sigma} = .55$$

It is worth noting that for these data, $s = .71$, which is noticeably larger than
$\hat{\sigma}$. This illustrates a well-established piece of statistical folklore. It is fairly well
known that to ignore rounding of intrinsically continuous data will typically
have the effect of inappropriately inflating the apparent spread of the underlying
distribution.

A.5.2   Likelihood Functions for Continuous and Mixed Data and Maximum Likelihood Model Fitting

The likelihood function ideas discussed thus far depend on treating the Θ *probability*
of discrete data in hand, $Y = y$, as a function of Θ. When analyzing data using continuous distributions, a slight logical snag is therefore encountered: If a continuous
model is employed, the probability associated with observing any particular exact
realization y is always 0, for every Θ.

To understand how to employ likelihood methods in continuous models, it is then useful to consider the probability of observing a value of Y "near" y as a function of Θ. That is, suppose that

$$f_{\Theta}(y)$$

is a joint probability density for Y depending on an unknown parameter vector Θ. Then in rough terms, if Δ is a small positive number and $y = (y_1, y_2, \ldots, y_n)$,

$$P[\text{each } Y_i \text{ is within } \tfrac{\Delta}{2} \text{ of } y_i] \approx f_{\Theta}(y)\Delta^n \tag{A.26}$$

But in expression (A.26), Δ^n doesn't depend on Θ—that is, the approximate probability is proportional to the function of Θ, $f_{\Theta}(y)$. It is therefore plausible to use the joint density with data plugged in,

A continuous data likelihood function

$$\boxed{f_{\Theta}(y)} \tag{A.27}$$

as a likelihood function and to use its logarithm,

A continuous data log likelihood function

$$\boxed{L(\Theta) = \ln(f_{\Theta}(y))} \tag{A.28}$$

as a log likelihood for data modeled as jointly continuous. (Formulas (A.27) and (A.28) are formally identical to formulas (A.17) and (A.18), but they involve a different type of data.) Contemplation of formula (A.27) or (A.28) can be thought of as a way of assessing the consonance of different parameter vectors, Θ, with continuous data, y. And as for the discrete case, a vector Θ maximizing $L(\Theta)$ is often termed a *maximum likelihood estimate* of the parameter vector.

Example 21 | Maximum Likelihood Estimation Based on iid Exponential Data

The exponential distribution is a popular model for life-length variables. The following are hypothetical life lengths (in hours) for $n = 4$ nominally identical electrical components, which will be assumed to have been a priori adequately described as iid exponential variables with mean α,

$$75.4, \quad 39.4, \quad 3.7, \quad 4.5 \tag{A.29}$$

Example 21
(*continued*)

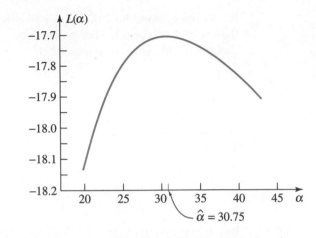

Figure A.17   Plot of a log likelihood based on four iid exponential observations

If Y_1, Y_2, Y_3, and Y_4 are iid exponential variables with means α, an appropriate joint probability density is

$$f(y) = \begin{cases} \displaystyle\prod_{i=1}^{4} \frac{1}{\alpha} e^{-y_i/\alpha} & \text{for each } y_i > 0 \\[2mm] 0 & \text{otherwise} \end{cases}$$

So with the data of display (A.29) in hand, the log likelihood function becomes

$$L(\alpha) = -4\ln(\alpha) - \frac{1}{\alpha}(75.4 + 39.4 + 3.7 + 4.5) \qquad \textbf{(A.30)}$$

It is easy to verify (using calculus and/or simply looking at the plot of $L(\alpha)$ in Figure A.17) that $L(\alpha)$ is maximized for

$$\hat{\alpha} = 30.75 = \frac{75.4 + 39.4 + 3.7 + 4.5}{4} = \bar{y}$$

This fact is a particular instance of the general result that the maximum likelihood estimate of an exponential mean is the sample average of the observations.

*Maximum
likelihood
and normal
observations*

Example 21 is fairly simple, in that only one parameter is involved and calculus can be used to find an explicit formula for the maximum likelihood estimator. The reader might be interested in working through the somewhat more complicated (two-parameter) situation involving n iid normal random variables with means μ and standard deviations σ. Two-variable calculus can be used to show that maximum likelihood estimates of the parameters based on observations $x_1, x_2, \ldots, x_n$ turn out to be, respectively,

$$\hat{\mu} = \bar{x}$$

$$\hat{\sigma} = \sqrt{\frac{n-1}{n}}\, s$$

The next example concerns an important continuous situation where no explicit formulas for maximum likelihood estimates seem to exist.

Example 22

> ## Maximum Likelihood Estimation Based on iid Weibull Steel Specimen Failure Times
>
> The data in Table A.4 are $n = 10$ ordered failure times for hardened steel specimens that were subjected to a particular rolling fatigue test. These data appeared originally in the paper of J. I. McCool, "Confidence Limits for Weibull Regression With Censored Data" (*IEEE Transactions on Reliability*, 1980). The Weibull probability plot of these data in Figure A.18 suggests the appropriateness of fitting a Weibull model to them (and indicates that β near 2 and α near .25 may be appropriate parameters for such a fitted model).
>
> Notice that the joint density function of $n = 10$ iid Weibull random variables $Y_1, Y_2, \ldots, Y_{10}$ with parameters α and β is
>
> $$f(y) = \begin{cases} \displaystyle\prod_{i=1}^{10} \frac{\beta}{\alpha^{\beta}} y_i^{\beta-1} e^{-(y_i/\alpha)^{\beta}} & \text{for each } y_i > 0 \\ \\ 0 & \text{otherwise} \end{cases}$$
>
> So using the data of Table A.4, the log likelihood
>
> $$L(\alpha, \beta) = 10\ln(\beta) - 10\beta\ln(\alpha) + (\beta-1)(\ln(.073) + \ln(.098) + \cdots + \ln(.456))$$
>
> $$- \frac{1}{\alpha^{\beta}}((.073)^{\beta} + (.098)^{\beta} + \cdots + (.456)^{\beta})$$
>
> $$= 10\ln(\beta) - 10\beta\ln(\alpha) - 16.267(\beta-1) - \frac{1}{\alpha^{\beta}}((.073)^{\beta} + (.098)^{\beta}$$
>
> $$+ \cdots + (.456)^{\beta})$$
>
> is indicated. Figure A.19 shows a contour plot of $L(\alpha, \beta)$ and indicates that maximum likelihood estimates of α and β are indeed in the vicinity of $\hat{\beta} = 2.0$ and $\hat{\alpha} = .26$.

Example 22
(continued)

Table A.4
Ten Ordered Failure Times of Steel Specimens

.073, .098, .117, .135, .175, .262, .270, .350, .386, .456

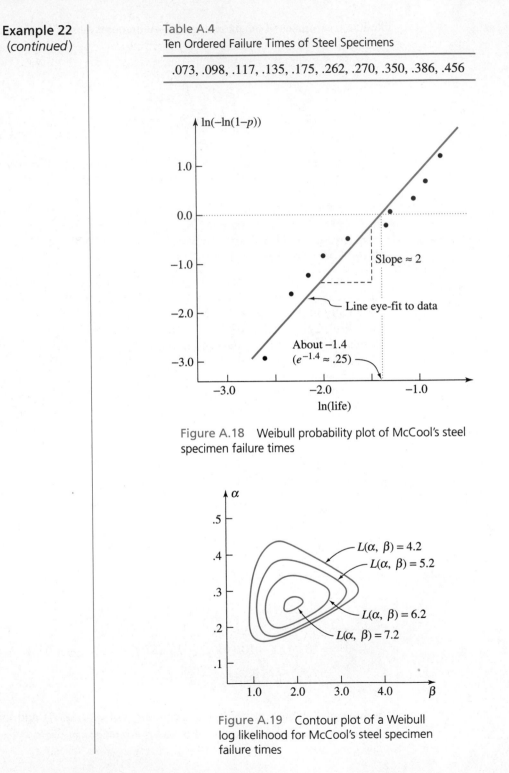

Figure A.18 Weibull probability plot of McCool's steel specimen failure times

Figure A.19 Contour plot of a Weibull log likelihood for McCool's steel specimen failure times

Analytical attempts to locate the maximum likelihood estimates for this kind of iid Weibull data situation are only partially fruitful. Setting partial derivatives of $L(\alpha, \beta)$ equal to 0, followed by some algebra, does lead to the two equations

$$\beta = \left(\frac{\sum y_i^{\beta} \ln(y_i)}{\sum y_i^{\beta}} - \frac{\sum \ln(y_i)}{n} \right)^{-1}$$

$$\alpha = \left(\frac{\sum y_i^{\beta}}{n} \right)^{1/\beta}$$

which maximum likelihood estimates must satisfy, but these must be solved numerically.

Discrete and continuous likelihood methods have thus far been discussed separately. However, particularly in life-data analysis contexts, statistical engineering studies occasionally yield data that are *mixed*—in the sense that some parts are discrete, while other parts are continuous. If it is sensible to think of the two parts as independent, a combination of things already said here can lead to an appropriate likelihood function and then, for example, to maximum likelihood parameter estimates.

That is, suppose that one has available discrete data, $Y_1 = y_1$, and continuous data, $Y_2 = y_2$, which can be thought of as independently generated—Y_1 from a discrete joint distribution with joint probability function

$$f_{\Theta}^{(1)}(y_1)$$

and Y_2 from a continuous joint distribution with joint probability density

$$f_{\Theta}^{(2)}(y_2)$$

Then a sensible likelihood function becomes

A mixed-data likelihood function

$$f_{\Theta}^{(1)}(y_1) \cdot f_{\Theta}^{(2)}(y_2) \qquad\qquad\text{(A.31)}$$

with corresponding log likelihood

A mixed-data log likelihood function

$$L(\Theta) = \ln\left(f_{\Theta}^{(1)}(y_1) \right) + \ln\left(f_{\Theta}^{(2)}(y_2) \right) \qquad\qquad\text{(A.32)}$$

Armed with equation (A.31) or (A.32), assessments of the compatibility of different parameter vectors Θ with the data in hand and maximum likelihood model fitting can proceed just as for purely discrete or purely continuous cases.

Example 23

Maximum Likelihood Estimation of a Mean Insulating Fluid Breakdown Time Using Censored Data

Table 2.1 of Nelson's *Applied Life Data Analysis* gives some data on times to breakdown (in seconds) of an insulating fluid at several different voltages. The results of $n = 12$ tests made at 30 kV are repeated below in Table A.5. The last two entries in Table A.5 mean that two tests were terminated at (respectively) 29,200 seconds and 86,100 seconds without failures having been observed. In common statistical jargon, these last two data values are *censored* (at the times 29,200 and 86,100, respectively).

Nelson remarks in his book that exponential distributions are often used to model life length for such fluids. Therefore, consider fitting an exponential distribution with mean α to the data of Table A.5. Notice that the first ten pieces of data in Table A.5 are continuous "exact" failure times, while the last two are essentially discrete pieces of information. Considering first the discrete part of the overall likelihood, the probability that two independent exponential variables exceed 29,200 and 86,100, respectively, is

$$f_\alpha^{(1)}(\mathbf{y}_1) = e^{-29,200/\alpha} \cdot e^{-86,100/\alpha}$$

Then considering the continuous part of the likelihood, the joint density of ten independent exponential variables with mean α is

$$f_\alpha^{(2)}(\mathbf{y}_2) = \begin{cases} \dfrac{1}{\alpha^{10}} e^{-\sum y_i/\alpha} & \text{for each } y_i > 0 \\ 0 & \text{otherwise} \end{cases}$$

Putting these two pieces together via equation (A.32), the log likelihood function appropriate here is

$$L(\alpha) = -10\ln(\alpha) - \frac{1}{\alpha}(50 + 134 + 187 + \cdots +$$

$$+\, 15{,}800 + 29{,}200 + 86{,}100)$$

$$= -10\ln\alpha - \frac{1}{\alpha}(144{,}673) \qquad \text{(A.33)}$$

Table A.5
12 Insulating Fluid Breakdown Times

50, 134, 187, 882, 1450, 1470, 2290, 2930, 4180, 15800, > 29200, > 86100

This function of α is easily seen via elementary calculus to be maximized at

$$\hat{\alpha} = \frac{144{,}673}{10} = 14{,}467.3 \text{ sec}$$

which has the intuitively appealing interpretation of the ratio of the total time on test to the number of failures observed during testing.

A.5.3 Likelihood-Based Large-Sample Inference Methods

One of the appealing things about the likelihood function idea is that in many situations, it is possible to base large-sample significance testing and confidence region methods on the likelihood function. Intuitively, it would seem that those parameter vectors Θ "most compatible" with the data in hand ought to form a sensible confidence set for Θ. And in significance-testing terms, if a hypothesized value of Θ (say, Θ_0) has a corresponding value of the likelihood function far smaller than the maximum possible, that circumstance ought to produce a small p-value— that is, strong evidence against $H_0: \Theta = \Theta_0$.

To make this thinking precise, let

The maximum of the log-likelihood function

$$L^* = \max_{\Theta} L(\Theta)$$

that is, L^* is the largest possible value of the log likelihood. (If $\hat{\Theta}$ is a maximum likelihood estimate of Θ, then $L^* = L(\hat{\Theta})$.) An intuitively appealing way to make a confidence set for the parameter vector Θ is to use the set of all Θ's with log likelihood not too far below L^*,

A likelihood-based confidence set for Θ

$$\left\{ \Theta \mid L(\Theta) > L^* - c \right\} \tag{A.34}$$

for an appropriate number c. And a plausible way of deriving a p-value for testing

$$H_0: \Theta = \Theta_0 \tag{A.35}$$

is by trying to identify a sensible probability distribution for

$$L^* - L(\Theta_0) \tag{A.36}$$

when H_0 holds, and using the upper-tail probability beyond an observed value of variable (A.36) as a p-value.

The practical gaps in this thinking are two: how to choose c in display (A.34) to get a desired confidence level and what kind of distribution to use to describe variable (A.36) under hypothesis (A.35). There are no general exact answers to these questions, but statistical theory does provide at least some indication of approximate answers that are often adequate for practical purposes when large samples are involved. That is, statistical theory suggests that in many large-sample situations, if Θ is of dimension k, choosing

Constant producing (large sample) approximate γ level confidence for $\{\Theta \mid L(\Theta) > L^* - c\}$

$$c = \tfrac{1}{2}U \tag{A.37}$$

for U the γ quantile of the χ_k^2 distribution, produces a confidence set (A.34) of confidence level roughly γ. And similar reasoning suggests that in many large-sample situations, if Θ is of dimension k, the hypothesis (A.35) can be tested using the test statistic

A test statistic for $H_0 : \Theta = \Theta_0$ with an approximately χ_k^2 reference distribution

$$2\left(L^* - L(\Theta_0)\right) \tag{A.38}$$

and a χ_k^2 approximate reference distribution, where large values of the test statistic (A.38) count as evidence against H_0.

Example 23
(continued)

Consider the problem of setting confidence limits on the mean time till breakdown of Nelson's insulating fluid tested at 30 kV. In this problem, Θ is $k = 1$-dimensional. So, for example, making use of the facts that the .9 quantile of the χ_1^2 distribution is 2.706 and that the maximum likelihood estimate of α is 14,467.3, displays (A.33), (A.34), and (A.37) suggest that those α with

$$L(\alpha) > -10\ln(14{,}467.3) - \frac{1}{14{,}467.3}(144{,}673) - \frac{1}{2}(2.706)$$

that is,

$$-10\ln(\alpha) - \frac{1}{\alpha}(144{,}673) > -107.15$$

form an approximate 90% confidence set for α. Figure A.20 shows a plot of the log likelihood (A.33) cut at the level -107.15 and the corresponding interval of α's. Numerical solution of the equation

$$-10\ln(\alpha) - \frac{1}{\alpha}(144{,}673) = -107.15$$

shows the interval for mean time till breakdown to extend from 8,963 sec to 25,572 sec.)

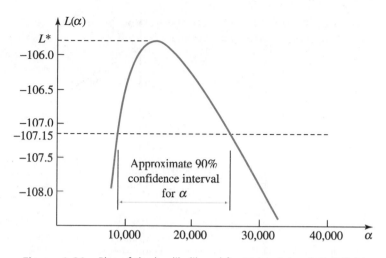

Figure A.20 Plot of the log likelihood for Nelson's insulating fluid breakdown time data and approximate confidence limits for α

The $n = 12$ pieces of data in Table A.5 do not constitute an especially large sample, so the 90% approximate confidence level associated with the interval (8,963, 25,572) should be treated as very approximate. But even so, this interval does give one some feeling about the precision with which α is known based on the data of Table A.5. There is clearly substantial uncertainty associated with the estimate $\hat{\alpha} = 14{,}467.3$.

Cautions concerning the large-sample likelihood-based inference methods

It is not a trivial matter to verify that the χ_k^2 approximations suggested here are adequate for a particular nonstandard probability model. In engineering situations where fairly exact confidence levels and/or p-values are critical, readers should seek genuinely expert statistical advice before placing too much faith in the χ_k^2 approximations. But for purposes of engineering problem solving requiring a rough, working quantification of uncertainty associated with parameter estimates, the use of the χ_k^2 approximation is certainly preferable to operating without any such quantification.

The insulating fluid example involved only a single parameter. As an example of a $k = 2$-parameter application, consider once again the space shuttle O-ring failure example.

Example 19
(continued)

Again use the log likelihood (A.23) and the fact that maximum likelihood estimates of α and β in equation (A.21) or (A.22) are $\hat{\alpha} = 15.043$ and $\hat{\beta} = -.2322$. These produce corresponding log likelihood -10.158. This, together with the

Example 19
(*continued*)

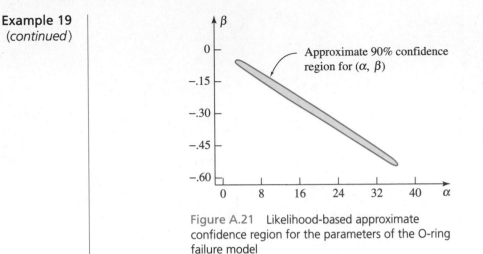

Figure A.21   Likelihood-based approximate confidence region for the parameters of the O-ring failure model

fact that the .9 quantile of the χ_2^2 distribution is 4.605, gives one (from displays (A.34) and (A.37)), that the set of (α, β) pairs with

$$L(\alpha, \beta) > -10.158 - \frac{1}{2}(4.605)$$

that is,

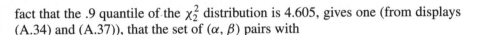

$$L(\alpha, \beta) > -12.4605$$

constitutes an approximate 90% confidence region for (α, β). This set of possible parameter vectors is shown in the plot in Figure A.21. Notice that one message conveyed by the contour plot is that β is pretty clearly negative. Low-temperature launches are more prone to O-ring failure than moderate- to high-temperature launches.

The approximate inference methods represented in displays (A.34) through (A.38) concern the entire parameter vector Θ in cases where it is multidimensional. It is reasonably common, however, to desire inferences only for particular parameters individually. (For example, in the case of the O-rings, it is the parameter β that determines whether $p(t)$ is increasing, constant, or decreasing in t, and for many purposes β is of primary interest.) It is thus worth mentioning that the likelihood ideas discussed here can be adapted to provide inference methods for a *part* of a parameter vector Θ of individual interest. An exposition of these adaptations will not be attempted here, but be aware of their existence. For details, refer to more complete expositions of likelihood methods (such as that in Meeker and Escobar's *Statistical Methods for Reliability Data* text).

APPENDIX B

Tables

Table B.1
Random Digits

12159	66144	05091	13446	45653	13684	66024	91410	51351	22772
30156	90519	95785	47544	66735	35754	11088	67310	19720	08379
59069	01722	53338	41942	65118	71236	01932	70343	25812	62275
54107	58081	82470	59407	13475	95872	16268	78436	39251	64247
99681	81295	06315	28212	45029	57701	96327	85436	33614	29070
27252	37875	53679	01889	35714	63534	63791	76342	47717	73684
93259	74585	11863	78985	03881	46567	93696	93521	54970	37601
84068	43759	75814	32261	12728	09636	22336	75629	01017	45503
68582	97054	28251	63787	57285	18854	35006	16343	51867	67979
60646	11298	19680	10087	66391	70853	24423	73007	74958	29020
97437	52922	80739	59178	50628	61017	51652	40915	94696	67843
58009	20681	98823	50979	01237	70152	13711	73916	87902	84759
77211	70110	93803	60135	22881	13423	30999	07104	27400	25414
54256	84591	65302	99257	92970	28924	36632	54044	91798	78018
36493	69330	94069	39544	14050	03476	25804	49350	92525	87941
87569	22661	55970	52623	35419	76660	42394	63210	62626	00581
22896	62237	39635	63725	10463	87944	92075	90914	30599	35671
02697	33230	64527	97210	41359	79399	13941	88378	68503	33609
20080	15652	37216	00679	02088	34138	13953	68939	05630	27653
20550	95151	60557	57449	77115	87372	02574	07851	22128	39189
72771	11672	67492	42904	64647	94354	45994	42538	54885	15983
38472	43379	76295	69406	96510	16529	83500	28590	49787	29822
24511	56510	72654	13277	45031	42235	96502	25567	23653	36707
01054	06674	58283	82831	97048	42983	06471	12350	49990	04809
94437	94907	95274	26487	60496	78222	43032	04276	70800	17378

(continued)

Table B.1
Random Digits (*continued*)

97842	69095	25982	03484	25173	05982	14624	31653	17170	92785
53047	13486	69712	33567	82313	87631	03197	02438	12374	40329
40770	47013	63306	48154	80970	87976	04939	21233	20572	31013
52733	66251	69661	58387	72096	21355	51659	19003	75556	33095
41749	46502	18378	83141	63920	85516	75743	66317	45428	45940
10271	85184	46468	38860	24039	80949	51211	35411	40470	16070
98791	48848	68129	51024	53044	55039	71290	26484	70682	56255
30196	09295	47685	56768	29285	06272	98789	47188	35063	24158
99373	64343	92433	06388	65713	35386	43370	19254	55014	98621
27768	27552	42156	23239	46823	91077	06306	17756	84459	92513
67791	35910	56921	51976	78475	15336	92544	82601	17996	72268
64018	44004	08136	56129	77024	82650	18163	29158	33935	94262
79715	33859	10835	94936	02857	87486	70613	41909	80667	52176
20190	40737	82688	07099	65255	52767	65930	45861	32575	93731
82421	01208	49762	66360	00231	87540	88302	62686	38456	25872

Table B.2
Control Chart Constants

m	d_2	d_3	c_4	A_2	A_3	B_3	B_4	B_5	B_6	D_1	D_2	D_3	D_4
2	1.128	0.853	0.7979	1.880	2.659		3.267		2.606		3.686		3.267
3	1.693	0.888	0.8862	1.023	1.954		2.568		2.276		4.358		2.575
4	2.059	0.880	0.9213	0.729	1.628		2.266		2.088		4.698		2.282
5	2.326	0.864	0.9400	0.577	1.427		2.089		1.964		4.918		2.114
6	2.534	0.848	0.9515	0.483	1.287	0.030	1.970	0.029	1.874		5.079		2.004
7	2.704	0.833	0.9594	0.419	1.182	0.118	1.882	0.113	1.806	0.205	5.204	0.076	1.924
8	2.847	0.820	0.9650	0.373	1.099	0.185	1.815	0.179	1.751	0.388	5.307	0.136	1.864
9	2.970	0.808	0.9693	0.337	1.032	0.239	1.761	0.232	1.707	0.547	5.394	0.184	1.816
10	3.078	0.797	0.9727	0.308	0.975	0.284	1.716	0.276	1.669	0.686	5.469	0.223	1.777
11	3.173	0.787	0.9754	0.285	0.927	0.321	1.679	0.313	1.637	0.811	5.535	0.256	1.744
12	3.258	0.778	0.9776	0.266	0.886	0.354	1.646	0.346	1.610	0.923	5.594	0.283	1.717
13	3.336	0.770	0.9794	0.249	0.850	0.382	1.618	0.374	1.585	1.025	5.647	0.307	1.693
14	3.407	0.763	0.9810	0.235	0.817	0.406	1.594	0.399	1.563	1.118	5.696	0.328	1.672
15	3.472	0.756	0.9823	0.223	0.789	0.428	1.572	0.421	1.544	1.203	5.740	0.347	1.653
20	3.735	0.729	0.9869	0.180	0.680	0.510	1.490	0.504	1.470	1.549	5.921	0.415	1.585
25	3.931	0.708	0.9896	0.153	0.606	0.565	1.435	0.559	1.420	1.806	6.056	0.459	1.541

This table was computed using Mathcad.

Table B.3
Standard Normal Cumulative Probabilities

$$\Phi(z) = \int_{-\infty}^{z} \frac{1}{\sqrt{2\pi}} \exp\left(-\frac{t^2}{2}\right) dt$$

z	.00	.01	.02	.03	.04	.05	.06	.07	.08	.09
−3.4	.0003	.0003	.0003	.0003	.0003	.0003	.0003	.0003	.0003	.0002
−3.3	.0005	.0005	.0005	.0004	.0004	.0004	.0004	.0004	.0004	.0003
−3.2	.0007	.0007	.0006	.0006	.0006	.0006	.0006	.0005	.0005	.0005
−3.1	.0010	.0009	.0009	.0009	.0008	.0008	.0008	.0008	.0007	.0007
−3.0	.0013	.0013	.0013	.0012	.0012	.0011	.0011	.0011	.0010	.0010
−2.9	.0019	.0018	.0018	.0017	.0016	.0016	.0015	.0015	.0014	.0014
−2.8	.0026	.0025	.0024	.0023	.0023	.0022	.0021	.0021	.0020	.0019
−2.7	.0035	.0034	.0033	.0032	.0031	.0030	.0029	.0028	.0027	.0026
−2.6	.0047	.0045	.0044	.0043	.0041	.0040	.0039	.0038	.0037	.0036
−2.5	.0062	.0060	.0059	.0057	.0055	.0054	.0052	.0051	.0049	.0048
−2.4	.0082	.0080	.0078	.0075	.0073	.0071	.0069	.0068	.0066	.0064
−2.3	.0107	.0104	.0102	.0099	.0096	.0094	.0091	.0089	.0087	.0084
−2.2	.0139	.0136	.0132	.0129	.0125	.0122	.0119	.0116	.0113	.0110
−2.1	.0179	.0174	.0170	.0166	.0162	.0158	.0154	.0150	.0146	.0143
−2.0	.0228	.0222	.0217	.0212	.0207	.0202	.0197	.0192	.0188	.0183
−1.9	.0287	.0281	.0274	.0268	.0262	.0256	.0250	.0244	.0239	.0233
−1.8	.0359	.0351	.0344	.0336	.0329	.0322	.0314	.0307	.0301	.0294
−1.7	.0446	.0436	.0427	.0418	.0409	.0401	.0392	.0384	.0375	.0367
−1.6	.0548	.0537	.0526	.0516	.0505	.0495	.0485	.0475	.0465	.0455
−1.5	.0668	.0655	.0643	.0630	.0618	.0606	.0594	.0582	.0571	.0559
−1.4	.0808	.0793	.0778	.0764	.0749	.0735	.0721	.0708	.0694	.0681
−1.3	.0968	.0951	.0934	.0918	.0901	.0885	.0869	.0853	.0838	.0823
−1.2	.1151	.1131	.1112	.1093	.1075	.1056	.1038	.1020	.1003	.0985
−1.1	.1357	.1335	.1314	.1292	.1271	.1251	.1230	.1210	.1190	.1170
−1.0	.1587	.1562	.1539	.1515	.1492	.1469	.1446	.1423	.1401	.1379
−0.9	.1841	.1814	.1788	.1762	.1736	.1711	.1685	.1660	.1635	.1611
−0.8	.2119	.2090	.2061	.2033	.2005	.1977	.1949	.1922	.1894	.1867
−0.7	.2420	.2389	.2358	.2327	.2297	.2266	.2236	.2206	.2177	.2148
−0.6	.2743	.2709	.2676	.2643	.2611	.2578	.2546	.2514	.2483	.2451
−0.5	.3085	.3050	.3015	.2981	.2946	.2912	.2877	.2843	.2810	.2776
−0.4	.3446	.3409	.3372	.3336	.3300	.3264	.3228	.3192	.3156	.3121
−0.3	.3821	.3783	.3745	.3707	.3669	.3632	.3594	.3557	.3520	.3483
−0.2	.4207	.4168	.4129	.4090	.4052	.4013	.3974	.3936	.3897	.3859
−0.1	.4602	.4562	.4522	.4483	.4443	.4404	.4364	.4325	.4286	.4247
−0.0	.5000	.4960	.4920	.4880	.4840	.4801	.4761	.4721	.4681	.4641

Table B.3
Standard Normal Cumulative Probabilities (*continued*)

z	.00	.01	.02	.03	.04	.05	.06	.07	.08	.09
0.0	.5000	.5040	.5080	.5120	.5160	.5199	.5239	.5279	.5319	.5359
0.1	.5398	.5438	.5478	.5517	.5557	.5596	.5636	.5675	.5714	.5753
0.2	.5793	.5832	.5871	.5910	.5948	.5987	.6026	.6064	.6103	.6141
0.3	.6179	.6217	.6255	.6293	.6331	.6368	.6406	.6443	.6480	.6517
0.4	.6554	.6591	.6628	.6664	.6700	.6736	.6772	.6808	.6844	.6879
0.5	.6915	.6950	.6985	.7019	.7054	.7088	.7123	.7157	.7190	.7224
0.6	.7257	.7291	.7324	.7357	.7389	.7422	.7454	.7486	.7517	.7549
0.7	.7580	.7611	.7642	.7673	.7704	.7734	.7764	.7794	.7823	.7852
0.8	.7881	.7910	.7939	.7967	.7995	.8023	.8051	.8078	.8106	.8133
0.9	.8159	.8186	.8212	.8238	.8264	.8289	.8315	.8340	.8365	.8389
1.0	.8413	.8438	.8461	.8485	.8508	.8531	.8554	.8577	.8599	.8621
1.1	.8643	.8665	.8686	.8708	.8729	.8749	.8770	.8790	.8810	.8830
1.2	.8849	.8869	.8888	.8907	.8925	.8944	.8962	.8980	.8997	.9015
1.3	.9032	.9049	.9066	.9082	.9099	.9115	.9131	.9147	.9162	.9177
1.4	.9192	.9207	.9222	.9236	.9251	.9265	.9279	.9292	.9306	.9319
1.5	.9332	.9345	.9357	.9370	.9382	.9394	.9406	.9418	.9429	.9441
1.6	.9452	.9463	.9474	.9484	.9495	.9505	.9515	.9525	.9535	.9545
1.7	.9554	.9564	.9573	.9582	.9591	.9599	.9608	.9616	.9625	.9633
1.8	.9641	.9649	.9656	.9664	.9671	.9678	.9686	.9693	.9699	.9706
1.9	.9713	.9719	.9726	.9732	.9738	.9744	.9750	.9756	.9761	.9767
2.0	.9773	.9778	.9783	.9788	.9793	.9798	.9803	.9808	.9812	.9817
2.1	.9821	.9826	.9830	.9834	.9838	.9842	.9846	.9850	.9854	.9857
2.2	.9861	.9864	.9868	.9871	.9875	.9878	.9881	.9884	.9887	.9890
2.3	.9893	.9896	.9898	.9901	.9904	.9906	.9909	.9911	.9913	.9916
2.4	.9918	.9920	.9922	.9925	.9927	.9929	.9931	.9932	.9934	.9936
2.5	.9938	.9940	.9941	.9943	.9945	.9946	.9948	.9949	.9951	.9952
2.6	.9953	.9955	.9956	.9957	.9959	.9960	.9961	.9962	.9963	.9964
2.7	.9965	.9966	.9967	.9968	.9969	.9970	.9971	.9972	.9973	.9974
2.8	.9974	.9975	.9976	.9977	.9977	.9978	.9979	.9979	.9980	.9981
2.9	.9981	.9982	.9983	.9983	.9984	.9984	.9985	.9985	.9986	.9986
3.0	.9987	.9987	.9987	.9988	.9988	.9989	.9989	.9989	.9990	.9990
3.1	.9990	.9991	.9991	.9991	.9992	.9992	.9992	.9992	.9993	.9993
3.2	.9993	.9993	.9994	.9994	.9994	.9994	.9994	.9995	.9995	.9995
3.3	.9995	.9995	.9996	.9996	.9996	.9996	.9996	.9996	.9996	.9997
3.4	.9997	.9997	.9997	.9997	.9997	.9997	.9997	.9997	.9997	.9998

This table was generated using MINITAB.

Table B.4
t Distribution Quantiles

ν	$Q(.9)$	$Q(.95)$	$Q(.975)$	$Q(.99)$	$Q(.995)$	$Q(.999)$	$Q(.9995)$
1	3.078	6.314	12.706	31.821	63.657	318.317	636.607
2	1.886	2.920	4.303	6.965	9.925	22.327	31.598
3	1.638	2.353	3.182	4.541	5.841	10.215	12.924
4	1.533	2.132	2.776	3.747	4.604	7.173	8.610
5	1.476	2.015	2.571	3.365	4.032	5.893	6.869
6	1.440	1.943	2.447	3.143	3.707	5.208	5.959
7	1.415	1.895	2.365	2.998	3.499	4.785	5.408
8	1.397	1.860	2.306	2.896	3.355	4.501	5.041
9	1.383	1.833	2.262	2.821	3.250	4.297	4.781
10	1.372	1.812	2.228	2.764	3.169	4.144	4.587
11	1.363	1.796	2.201	2.718	3.106	4.025	4.437
12	1.356	1.782	2.179	2.681	3.055	3.930	4.318
13	1.350	1.771	2.160	2.650	3.012	3.852	4.221
14	1.345	1.761	2.145	2.624	2.977	3.787	4.140
15	1.341	1.753	2.131	2.602	2.947	3.733	4.073
16	1.337	1.746	2.120	2.583	2.921	3.686	4.015
17	1.333	1.740	2.110	2.567	2.898	3.646	3.965
18	1.330	1.734	2.101	2.552	2.878	3.610	3.922
19	1.328	1.729	2.093	2.539	2.861	3.579	3.883
20	1.325	1.725	2.086	2.528	2.845	3.552	3.849
21	1.323	1.721	2.080	2.518	2.831	3.527	3.819
22	1.321	1.717	2.074	2.508	2.819	3.505	3.792
23	1.319	1.714	2.069	2.500	2.807	3.485	3.768
24	1.318	1.711	2.064	2.492	2.797	3.467	3.745
25	1.316	1.708	2.060	2.485	2.787	3.450	3.725
26	1.315	1.706	2.056	2.479	2.779	3.435	3.707
27	1.314	1.703	2.052	2.473	2.771	3.421	3.690
28	1.313	1.701	2.048	2.467	2.763	3.408	3.674
29	1.311	1.699	2.045	2.462	2.756	3.396	3.659
30	1.310	1.697	2.042	2.457	2.750	3.385	3.646
40	1.303	1.684	2.021	2.423	2.704	3.307	3.551
60	1.296	1.671	2.000	2.390	2.660	3.232	3.460
120	1.289	1.658	1.980	2.358	2.617	3.160	3.373
∞	1.282	1.645	1.960	2.326	2.576	3.090	3.291

This table was generated using MINITAB.

Table B.5
Chi-Square Distribution Quantiles

ν	$Q(.005)$	$Q(.01)$	$Q(.025)$	$Q(.05)$	$Q(.1)$	$Q(.9)$	$Q(.95)$	$Q(.975)$	$Q(.99)$	$Q(.995)$
1	0.000	0.000	0.001	0.004	0.016	2.706	3.841	5.024	6.635	7.879
2	0.010	0.020	0.051	0.103	0.211	4.605	5.991	7.378	9.210	10.597
3	0.072	0.115	0.216	0.352	0.584	6.251	7.815	9.348	11.345	12.838
4	0.207	0.297	0.484	0.711	1.064	7.779	9.488	11.143	13.277	14.860
5	0.412	0.554	0.831	1.145	1.610	9.236	11.070	12.833	15.086	16.750
6	0.676	0.872	1.237	1.635	2.204	10.645	12.592	14.449	16.812	18.548
7	0.989	1.239	1.690	2.167	2.833	12.017	14.067	16.013	18.475	20.278
8	1.344	1.646	2.180	2.733	3.490	13.362	15.507	17.535	20.090	21.955
9	1.735	2.088	2.700	3.325	4.168	14.684	16.919	19.023	21.666	23.589
10	2.156	2.558	3.247	3.940	4.865	15.987	18.307	20.483	23.209	25.188
11	2.603	3.053	3.816	4.575	5.578	17.275	19.675	21.920	24.725	26.757
12	3.074	3.571	4.404	5.226	6.304	18.549	21.026	23.337	26.217	28.300
13	3.565	4.107	5.009	5.892	7.042	19.812	22.362	24.736	27.688	29.819
14	4.075	4.660	5.629	6.571	7.790	21.064	23.685	26.119	29.141	31.319
15	4.601	5.229	6.262	7.261	8.547	22.307	24.996	27.488	30.578	32.801
16	5.142	5.812	6.908	7.962	9.312	23.542	26.296	28.845	32.000	34.267
17	5.697	6.408	7.564	8.672	10.085	24.769	27.587	30.191	33.409	35.718
18	6.265	7.015	8.231	9.390	10.865	25.989	28.869	31.526	34.805	37.156
19	6.844	7.633	8.907	10.117	11.651	27.204	30.143	32.852	36.191	38.582
20	7.434	8.260	9.591	10.851	12.443	28.412	31.410	34.170	37.566	39.997
21	8.034	8.897	10.283	11.591	13.240	29.615	32.671	35.479	38.932	41.401
22	8.643	9.542	10.982	12.338	14.041	30.813	33.924	36.781	40.290	42.796
23	9.260	10.196	11.689	13.091	14.848	32.007	35.172	38.076	41.638	44.181
24	9.886	10.856	12.401	13.848	15.659	33.196	36.415	39.364	42.980	45.559
25	10.520	11.524	13.120	14.611	16.473	34.382	37.653	40.647	44.314	46.928
26	11.160	12.198	13.844	15.379	17.292	35.563	38.885	41.923	45.642	48.290
27	11.808	12.879	14.573	16.151	18.114	36.741	40.113	43.195	46.963	49.645
28	12.461	13.565	15.308	16.928	18.939	37.916	41.337	44.461	48.278	50.994
29	13.121	14.256	16.047	17.708	19.768	39.087	42.557	45.722	49.588	52.336
30	13.787	14.953	16.791	18.493	20.599	40.256	43.773	46.979	50.892	53.672
31	14.458	15.655	17.539	19.281	21.434	41.422	44.985	48.232	52.192	55.003
32	15.134	16.362	18.291	20.072	22.271	42.585	46.194	49.480	53.486	56.328
33	15.815	17.074	19.047	20.867	23.110	43.745	47.400	50.725	54.775	57.648
34	16.501	17.789	19.806	21.664	23.952	44.903	48.602	51.966	56.061	58.964
35	17.192	18.509	20.569	22.465	24.797	46.059	49.802	53.204	57.342	60.275
36	17.887	19.233	21.336	23.269	25.643	47.212	50.998	54.437	58.619	61.581
37	18.586	19.960	22.106	24.075	26.492	48.364	52.192	55.668	59.893	62.885
38	19.289	20.691	22.878	24.884	27.343	49.513	53.384	56.896	61.163	64.183
39	19.996	21.426	23.654	25.695	28.196	50.660	54.572	58.120	62.429	65.477
40	20.707	22.164	24.433	26.509	29.051	51.805	55.759	59.342	63.691	66.767

This table was generated using MINITAB.

For $\nu > 40$, the approximation $Q(p) \approx \nu \left(1 - \dfrac{2}{9\nu} + Q_z(p)\sqrt{\dfrac{2}{9\nu}} \right)^3$ can be used.

Table B.6A
F Distribution .75 Quantiles

ν_2 (Denominator Degrees of Freedom)	\multicolumn{19}{c}{ν_1 (Numerator Degrees of Freedom)}
	1	2	3	4	5	6	7	8	9	10	12	15	20	24	30	40	60	120	∞
1	5.83	7.50	8.20	8.58	8.82	8.98	9.10	9.19	9.26	9.32	9.41	9.49	9.58	9.63	9.67	9.71	9.76	9.80	9.85
2	2.57	3.00	3.15	3.23	3.28	3.31	3.34	3.35	3.37	3.38	3.39	3.41	3.43	3.43	3.44	3.45	3.46	3.47	3.48
3	2.02	2.28	2.36	2.39	2.41	2.42	2.43	2.44	2.44	2.44	2.45	2.46	2.46	2.46	2.47	2.47	2.47	2.47	2.47
4	1.81	2.00	2.05	2.06	2.07	2.08	2.08	2.08	2.08	2.08	2.08	2.08	2.08	2.08	2.08	2.08	2.08	2.08	2.08
5	1.69	1.85	1.88	1.89	1.89	1.89	1.89	1.89	1.89	1.89	1.89	1.89	1.88	1.88	1.88	1.88	1.87	1.87	1.87
6	1.62	1.76	1.78	1.79	1.79	1.78	1.78	1.78	1.77	1.77	1.77	1.76	1.76	1.75	1.75	1.75	1.74	1.74	1.74
7	1.57	1.70	1.72	1.72	1.71	1.71	1.70	1.70	1.69	1.69	1.68	1.68	1.67	1.67	1.66	1.66	1.65	1.65	1.65
8	1.54	1.66	1.67	1.66	1.66	1.65	1.64	1.64	1.64	1.63	1.62	1.62	1.61	1.60	1.60	1.59	1.59	1.58	1.58
9	1.51	1.62	1.63	1.63	1.62	1.61	1.60	1.60	1.59	1.59	1.58	1.57	1.56	1.56	1.55	1.54	1.54	1.53	1.53
10	1.49	1.60	1.60	1.59	1.59	1.58	1.57	1.56	1.56	1.55	1.54	1.53	1.52	1.52	1.51	1.51	1.50	1.49	1.48
11	1.47	1.58	1.58	1.57	1.56	1.55	1.54	1.53	1.53	1.52	1.51	1.50	1.49	1.49	1.48	1.47	1.47	1.46	1.45
12	1.46	1.56	1.56	1.55	1.54	1.53	1.52	1.51	1.51	1.50	1.49	1.48	1.47	1.46	1.45	1.45	1.44	1.43	1.42
13	1.45	1.55	1.55	1.53	1.52	1.51	1.50	1.49	1.49	1.48	1.47	1.46	1.45	1.44	1.43	1.42	1.42	1.41	1.40
14	1.44	1.53	1.53	1.52	1.51	1.50	1.49	1.48	1.47	1.46	1.45	1.44	1.43	1.42	1.41	1.41	1.40	1.39	1.38
15	1.43	1.52	1.52	1.51	1.49	1.48	1.47	1.46	1.46	1.45	1.44	1.43	1.41	1.41	1.40	1.39	1.38	1.37	1.36
16	1.42	1.51	1.51	1.50	1.48	1.47	1.46	1.45	1.44	1.44	1.43	1.41	1.40	1.39	1.38	1.37	1.36	1.35	1.34
17	1.42	1.51	1.50	1.49	1.47	1.46	1.45	1.44	1.43	1.43	1.41	1.40	1.39	1.38	1.37	1.36	1.35	1.34	1.33
18	1.41	1.50	1.49	1.48	1.46	1.45	1.44	1.43	1.42	1.42	1.40	1.39	1.38	1.37	1.36	1.35	1.34	1.33	1.32
19	1.41	1.49	1.49	1.47	1.46	1.44	1.43	1.42	1.41	1.41	1.40	1.38	1.37	1.36	1.35	1.34	1.33	1.32	1.30
20	1.40	1.49	1.48	1.47	1.45	1.44	1.43	1.42	1.41	1.40	1.39	1.37	1.36	1.35	1.34	1.33	1.32	1.31	1.29
21	1.40	1.48	1.48	1.46	1.44	1.43	1.42	1.41	1.40	1.39	1.38	1.37	1.35	1.34	1.33	1.32	1.31	1.30	1.28
22	1.40	1.48	1.47	1.45	1.44	1.42	1.41	1.40	1.39	1.39	1.37	1.36	1.34	1.33	1.32	1.31	1.30	1.29	1.28
23	1.39	1.47	1.47	1.45	1.43	1.42	1.41	1.40	1.39	1.38	1.37	1.35	1.34	1.33	1.32	1.31	1.30	1.28	1.27
24	1.39	1.47	1.46	1.44	1.43	1.41	1.40	1.39	1.38	1.38	1.36	1.35	1.33	1.32	1.31	1.30	1.29	1.28	1.26
25	1.39	1.47	1.46	1.44	1.42	1.41	1.40	1.39	1.38	1.37	1.36	1.34	1.33	1.32	1.31	1.29	1.28	1.27	1.25
26	1.38	1.46	1.45	1.44	1.42	1.41	1.39	1.38	1.37	1.37	1.35	1.34	1.32	1.31	1.30	1.29	1.28	1.26	1.25
27	1.38	1.46	1.45	1.43	1.42	1.40	1.39	1.38	1.37	1.36	1.35	1.33	1.32	1.31	1.30	1.28	1.27	1.26	1.24
28	1.38	1.46	1.45	1.43	1.41	1.40	1.39	1.38	1.37	1.36	1.34	1.33	1.31	1.30	1.29	1.28	1.27	1.25	1.24
29	1.38	1.45	1.45	1.43	1.41	1.40	1.38	1.37	1.36	1.35	1.34	1.32	1.31	1.30	1.29	1.27	1.26	1.25	1.23
30	1.38	1.45	1.44	1.42	1.41	1.39	1.38	1.37	1.36	1.35	1.34	1.32	1.30	1.29	1.28	1.27	1.26	1.24	1.23
40	1.36	1.44	1.42	1.40	1.39	1.37	1.36	1.35	1.34	1.33	1.31	1.30	1.28	1.26	1.25	1.24	1.22	1.21	1.19
60	1.35	1.42	1.41	1.38	1.37	1.35	1.33	1.32	1.31	1.30	1.29	1.27	1.25	1.24	1.22	1.21	1.19	1.17	1.15
120	1.34	1.40	1.39	1.37	1.35	1.33	1.31	1.30	1.29	1.28	1.26	1.24	1.22	1.21	1.19	1.18	1.16	1.13	1.10
∞	1.32	1.39	1.37	1.35	1.33	1.31	1.29	1.28	1.27	1.25	1.24	1.22	1.19	1.18	1.16	1.14	1.12	1.08	1.00

This table was generated using MINITAB.

Table B.6B
F Distribution .90 Quantiles

ν_2 (Denominator Degrees of Freedom)	ν_1 (Numerator Degrees of Freedom)																		
	1	2	3	4	5	6	7	8	9	10	12	15	20	24	30	40	60	120	∞
1	39.86	49.50	53.59	55.84	57.24	58.20	58.90	59.44	59.85	60.20	60.70	61.22	61.74	62.00	62.27	62.53	62.79	63.05	63.33
2	8.53	9.00	9.16	9.24	9.29	9.33	9.35	9.37	9.38	9.39	9.41	9.42	9.44	9.45	9.46	9.47	9.47	9.48	9.49
3	5.54	5.46	5.39	5.34	5.31	5.28	5.27	5.25	5.24	5.23	5.22	5.20	5.18	5.18	5.17	5.16	5.15	5.14	5.13
4	4.54	4.32	4.19	4.11	4.05	4.01	3.98	3.95	3.94	3.92	3.90	3.87	3.84	3.83	3.82	3.80	3.79	3.78	3.76
5	4.06	3.78	3.62	3.52	3.45	3.40	3.37	3.34	3.32	3.30	3.27	3.24	3.21	3.19	3.17	3.16	3.14	3.12	3.10
6	3.78	3.46	3.29	3.18	3.11	3.05	3.01	2.98	2.96	2.94	2.90	2.87	2.84	2.82	2.80	2.78	2.76	2.74	2.72
7	3.59	3.26	3.07	2.96	2.88	2.83	2.78	2.75	2.72	2.70	2.67	2.63	2.59	2.58	2.56	2.54	2.51	2.49	2.47
8	3.46	3.11	2.92	2.81	2.73	2.67	2.62	2.59	2.56	2.54	2.50	2.46	2.42	2.40	2.38	2.36	2.34	2.32	2.29
9	3.36	3.01	2.81	2.69	2.61	2.55	2.51	2.47	2.44	2.42	2.38	2.34	2.30	2.28	2.25	2.23	2.21	2.18	2.16
10	3.28	2.92	2.73	2.61	2.52	2.46	2.41	2.38	2.35	2.32	2.28	2.24	2.20	2.18	2.16	2.13	2.11	2.08	2.06
11	3.23	2.86	2.66	2.54	2.45	2.39	2.34	2.30	2.27	2.25	2.21	2.17	2.12	2.10	2.08	2.05	2.03	2.00	1.97
12	3.18	2.81	2.61	2.48	2.39	2.33	2.28	2.24	2.21	2.19	2.15	2.10	2.06	2.04	2.01	1.99	1.96	1.93	1.90
13	3.14	2.76	2.56	2.43	2.35	2.28	2.23	2.20	2.16	2.14	2.10	2.05	2.01	1.98	1.96	1.93	1.90	1.88	1.85
14	3.10	2.73	2.52	2.39	2.31	2.24	2.19	2.15	2.12	2.10	2.05	2.01	1.96	1.94	1.91	1.89	1.86	1.83	1.80
15	3.07	2.70	2.49	2.36	2.27	2.21	2.16	2.12	2.09	2.06	2.02	1.97	1.92	1.90	1.87	1.85	1.82	1.79	1.76
16	3.05	2.67	2.46	2.33	2.24	2.18	2.13	2.09	2.06	2.03	1.99	1.94	1.89	1.87	1.84	1.81	1.78	1.75	1.72
17	3.03	2.64	2.44	2.31	2.22	2.15	2.10	2.06	2.03	2.00	1.96	1.91	1.86	1.84	1.81	1.78	1.75	1.72	1.69
18	3.01	2.62	2.42	2.29	2.20	2.13	2.08	2.04	2.00	1.98	1.93	1.89	1.84	1.81	1.78	1.75	1.72	1.69	1.66
19	2.99	2.61	2.40	2.27	2.18	2.11	2.06	2.02	1.98	1.96	1.91	1.86	1.81	1.79	1.76	1.73	1.70	1.67	1.63
20	2.97	2.59	2.38	2.25	2.16	2.09	2.04	2.00	1.96	1.94	1.89	1.84	1.79	1.77	1.74	1.71	1.68	1.64	1.61
21	2.96	2.57	2.36	2.23	2.14	2.08	2.02	1.98	1.95	1.92	1.87	1.83	1.78	1.75	1.72	1.69	1.66	1.62	1.59
22	2.95	2.56	2.35	2.22	2.13	2.06	2.01	1.97	1.93	1.90	1.86	1.81	1.76	1.73	1.70	1.67	1.64	1.60	1.57
23	2.94	2.55	2.34	2.21	2.11	2.05	1.99	1.95	1.92	1.89	1.84	1.80	1.74	1.72	1.69	1.66	1.62	1.59	1.55
24	2.93	2.54	2.33	2.19	2.10	2.04	1.98	1.94	1.91	1.88	1.83	1.78	1.73	1.70	1.67	1.64	1.61	1.57	1.53
25	2.92	2.53	2.32	2.18	2.09	2.02	1.97	1.93	1.89	1.87	1.82	1.77	1.72	1.69	1.66	1.63	1.59	1.56	1.52
26	2.91	2.52	2.31	2.17	2.08	2.01	1.96	1.92	1.88	1.86	1.81	1.76	1.71	1.68	1.65	1.61	1.58	1.54	1.50
27	2.90	2.51	2.30	2.17	2.07	2.00	1.95	1.91	1.87	1.85	1.80	1.75	1.70	1.67	1.64	1.60	1.57	1.53	1.49
28	2.89	2.50	2.29	2.16	2.06	2.00	1.94	1.90	1.87	1.84	1.79	1.74	1.69	1.66	1.63	1.59	1.56	1.52	1.48
29	2.89	2.50	2.28	2.15	2.06	1.99	1.93	1.89	1.86	1.83	1.78	1.73	1.68	1.65	1.62	1.58	1.55	1.51	1.47
30	2.88	2.49	2.28	2.14	2.05	1.98	1.93	1.88	1.85	1.82	1.77	1.72	1.67	1.64	1.61	1.57	1.54	1.50	1.46
40	2.84	2.44	2.23	2.09	2.00	1.93	1.87	1.83	1.79	1.76	1.71	1.66	1.61	1.57	1.54	1.51	1.47	1.42	1.38
60	2.79	2.39	2.18	2.04	1.95	1.87	1.82	1.77	1.74	1.71	1.66	1.60	1.54	1.51	1.48	1.44	1.40	1.35	1.29
120	2.75	2.35	2.13	1.99	1.90	1.82	1.77	1.72	1.68	1.65	1.60	1.55	1.48	1.45	1.41	1.37	1.32	1.26	1.19
∞	2.71	2.30	2.08	1.94	1.85	1.77	1.72	1.67	1.63	1.60	1.55	1.49	1.42	1.38	1.34	1.30	1.24	1.17	1.00

This table was generated using MINITAB.

v_2 (Denominator Degrees of Freedom)	v_1 (Numerator Degrees of Freedom)									
	1	2	3	4	5	6	7	8	9	10
1	161.44	199.50	215.69	224.57	230.16	233.98	236.78	238.89	240.55	241.89
2	18.51	19.00	19.16	19.25	19.30	19.33	19.35	19.37	19.39	19.40
3	10.13	9.55	9.28	9.12	9.01	8.94	8.89	8.85	8.81	8.79
4	7.71	6.94	6.59	6.39	6.26	6.16	6.09	6.04	6.00	5.96
5	6.61	5.79	5.41	5.19	5.05	4.95	4.88	4.82	4.77	4.74
6	5.99	5.14	4.76	4.53	4.39	4.28	4.21	4.15	4.10	4.06
7	5.59	4.74	4.35	4.12	3.97	3.87	3.79	3.73	3.68	3.64
8	5.32	4.46	4.07	3.84	3.69	3.58	3.50	3.44	3.39	3.35
9	5.12	4.26	3.86	3.63	3.48	3.37	3.29	3.23	3.18	3.14
10	4.96	4.10	3.71	3.48	3.33	3.22	3.14	3.07	3.02	2.98
11	4.84	3.98	3.59	3.36	3.20	3.09	3.01	2.95	2.90	2.85
12	4.75	3.89	3.49	3.26	3.11	3.00	2.91	2.85	2.80	2.75
13	4.67	3.81	3.41	3.18	3.03	2.92	2.83	2.77	2.71	2.67
14	4.60	3.74	3.34	3.11	2.96	2.85	2.76	2.70	2.65	2.60
15	4.54	3.68	3.29	3.06	2.90	2.79	2.71	2.64	2.59	2.54
16	4.49	3.63	3.24	3.01	2.85	2.74	2.66	2.59	2.54	2.49
17	4.45	3.59	3.20	2.96	2.81	2.70	2.61	2.55	2.49	2.45
18	4.41	3.55	3.16	2.93	2.77	2.66	2.58	2.51	2.46	2.41
19	4.38	3.52	3.13	2.90	2.74	2.63	2.54	2.48	2.42	2.38
20	4.35	3.49	3.10	2.87	2.71	2.60	2.51	2.45	2.39	2.35
21	4.32	3.47	3.07	2.84	2.68	2.57	2.49	2.42	2.37	2.32
22	4.30	3.44	3.05	2.82	2.66	2.55	2.46	2.40	2.34	2.30
23	4.28	3.42	3.03	2.80	2.64	2.53	2.44	2.37	2.32	2.27
24	4.26	3.40	3.01	2.78	2.62	2.51	2.42	2.36	2.30	2.25
25	4.24	3.39	2.99	2.76	2.60	2.49	2.40	2.34	2.28	2.24
26	4.23	3.37	2.98	2.74	2.59	2.47	2.39	2.32	2.27	2.22
27	4.21	3.35	2.96	2.73	2.57	2.46	2.37	2.31	2.25	2.20
28	4.20	3.34	2.95	2.71	2.56	2.45	2.36	2.29	2.24	2.19
29	4.18	3.33	2.93	2.70	2.55	2.43	2.35	2.28	2.22	2.18
30	4.17	3.32	2.92	2.69	2.53	2.42	2.33	2.27	2.21	2.16
40	4.08	3.23	2.84	2.61	2.45	2.34	2.25	2.18	2.12	2.08
60	4.00	3.15	2.76	2.53	2.37	2.25	2.17	2.10	2.04	1.99
120	3.92	3.07	2.68	2.45	2.29	2.18	2.09	2.02	1.96	1.91
∞	3.84	3.00	2.60	2.37	2.21	2.10	2.01	1.94	1.88	1.83

Table B.6C
F Distribution of .95 Quantiles (*continued*)

ν_2 (Denominator Degrees of Freedom)	ν_1 (Numerator Degrees of Freedom)								
	12	15	20	24	30	40	60	120	∞
1	243.91	245.97	248.02	249.04	250.07	251.13	252.18	253.27	254.31
2	19.41	19.43	19.45	19.45	19.46	19.47	19.48	19.49	19.50
3	8.74	8.70	8.66	8.64	8.62	8.59	8.57	8.55	8.53
4	5.91	5.86	5.80	5.77	5.75	5.72	5.69	5.66	5.63
5	4.68	4.62	4.56	4.53	4.50	4.46	4.43	4.40	4.36
6	4.00	3.94	3.87	3.84	3.81	3.77	3.74	3.70	3.67
7	3.57	3.51	3.44	3.41	3.38	3.34	3.30	3.27	3.23
8	3.28	3.22	3.15	3.12	3.08	3.04	3.01	2.97	2.93
9	3.07	3.01	2.94	2.90	2.86	2.83	2.79	2.75	2.71
10	2.91	2.85	2.77	2.74	2.70	2.66	2.62	2.58	2.54
11	2.79	2.72	2.65	2.61	2.57	2.53	2.49	2.45	2.40
12	2.69	2.62	2.54	2.51	2.47	2.43	2.38	2.34	2.30
13	2.60	2.53	2.46	2.42	2.38	2.34	2.30	2.25	2.21
14	2.53	2.46	2.39	2.35	2.31	2.27	2.22	2.18	2.13
15	2.48	2.40	2.33	2.29	2.25	2.20	2.16	2.11	2.07
16	2.42	2.35	2.28	2.24	2.19	2.15	2.11	2.06	2.01
17	2.38	2.31	2.23	2.19	2.15	2.10	2.06	2.01	1.96
18	2.34	2.27	2.19	2.15	2.11	2.06	2.02	1.97	1.92
19	2.31	2.23	2.16	2.11	2.07	2.03	1.98	1.93	1.88
20	2.28	2.20	2.12	2.08	2.04	1.99	1.95	1.90	1.84
21	2.25	2.18	2.10	2.05	2.01	1.96	1.92	1.87	1.81
22	2.23	2.15	2.07	2.03	1.98	1.94	1.89	1.84	1.78
23	2.20	2.13	2.05	2.01	1.96	1.91	1.86	1.81	1.76
24	2.18	2.11	2.03	1.98	1.94	1.89	1.84	1.79	1.73
25	2.16	2.09	2.01	1.96	1.92	1.87	1.82	1.77	1.71
26	2.15	2.07	1.99	1.95	1.90	1.85	1.80	1.75	1.69
27	2.13	2.06	1.97	1.93	1.88	1.84	1.79	1.73	1.67
28	2.12	2.04	1.96	1.91	1.87	1.82	1.77	1.71	1.65
29	2.10	2.03	1.94	1.90	1.85	1.81	1.75	1.70	1.64
30	2.09	2.01	1.93	1.89	1.84	1.79	1.74	1.68	1.62
40	2.00	1.92	1.84	1.79	1.74	1.69	1.64	1.58	1.51
60	1.92	1.84	1.75	1.70	1.65	1.59	1.53	1.47	1.39
120	1.83	1.75	1.66	1.61	1.55	1.50	1.43	1.35	1.25
∞	1.75	1.67	1.57	1.52	1.46	1.39	1.32	1.22	1.00

This table was generated using MINITAB.

Table B.6D
F Distribution .99 Quantiles

ν_2 (Denominator Degrees of Freedom)	ν_1 (Numerator Degrees of Freedom)									
	1	2	3	4	5	6	7	8	9	10
1	4052	4999	5403	5625	5764	5859	5929	5981	6023	6055
2	98.51	99.00	99.17	99.25	99.30	99.33	99.35	99.38	99.39	99.40
3	34.12	30.82	29.46	28.71	28.24	27.91	27.67	27.49	27.35	27.23
4	21.20	18.00	16.69	15.98	15.52	15.21	14.98	14.80	14.66	14.55
5	16.26	13.27	12.06	11.39	10.97	10.67	10.46	10.29	10.16	10.05
6	13.75	10.92	9.78	9.15	8.75	8.47	8.26	8.10	7.98	7.87
7	12.25	9.55	8.45	7.85	7.46	7.19	6.99	6.84	6.72	6.62
8	11.26	8.65	7.59	7.01	6.63	6.37	6.18	6.03	5.91	5.81
9	10.56	8.02	6.99	6.42	6.06	5.80	5.61	5.47	5.35	5.26
10	10.04	7.56	6.55	5.99	5.64	5.39	5.20	5.06	4.94	4.85
11	9.65	7.21	6.22	5.67	5.32	5.07	4.89	4.74	4.63	4.54
12	9.33	6.93	5.95	5.41	5.06	4.82	4.64	4.50	4.39	4.30
13	9.07	6.70	5.74	5.21	4.86	4.62	4.44	4.30	4.19	4.10
14	8.86	6.51	5.56	5.04	4.69	4.46	4.28	4.14	4.03	3.94
15	8.68	6.36	5.42	4.89	4.56	4.32	4.14	4.00	3.89	3.80
16	8.53	6.23	5.29	4.77	4.44	4.20	4.03	3.89	3.78	3.69
17	8.40	6.11	5.18	4.67	4.34	4.10	3.93	3.79	3.68	3.59
18	8.29	6.01	5.09	4.58	4.25	4.01	3.84	3.71	3.60	3.51
19	8.19	5.93	5.01	4.50	4.17	3.94	3.77	3.63	3.52	3.43
20	8.10	5.85	4.94	4.43	4.10	3.87	3.70	3.56	3.46	3.37
21	8.02	5.78	4.87	4.37	4.04	3.81	3.64	3.51	3.40	3.31
22	7.95	5.72	4.82	4.31	3.99	3.76	3.59	3.45	3.35	3.26
23	7.88	5.66	4.76	4.26	3.94	3.71	3.54	3.41	3.30	3.21
24	7.82	5.61	4.72	4.22	3.90	3.67	3.50	3.36	3.26	3.17
25	7.77	5.57	4.68	4.18	3.85	3.63	3.46	3.32	3.22	3.13
26	7.72	5.53	4.64	4.14	3.82	3.59	3.42	3.29	3.18	3.09
27	7.68	5.49	4.60	4.11	3.78	3.56	3.39	3.26	3.15	3.06
28	7.64	5.45	4.57	4.07	3.75	3.53	3.36	3.23	3.12	3.03
29	7.60	5.42	4.54	4.04	3.73	3.50	3.33	3.20	3.09	3.00
30	7.56	5.39	4.51	4.02	3.70	3.47	3.30	3.17	3.07	2.98
40	7.31	5.18	4.31	3.83	3.51	3.29	3.12	2.99	2.89	2.80
60	7.08	4.98	4.13	3.65	3.34	3.12	2.95	2.82	2.72	2.63
120	6.85	4.79	3.95	3.48	3.17	2.96	2.79	2.66	2.56	2.47
∞	6.63	4.61	3.78	3.32	3.02	2.80	2.64	2.51	2.41	2.32

F Distribution of .99 Quantiles (*continued*)

ν_2 (Denominator Degrees of Freedom)	ν_1 (Numerator Degrees of Freedom)								
	12	15	20	24	30	40	60	120	∞
1	6107	6157	6209	6235	6260	6287	6312	6339	6366
2	99.41	99.43	99.44	99.45	99.47	99.47	99.48	99.49	99.50
3	27.05	26.87	26.69	26.60	26.51	26.41	26.32	26.22	26.13
4	14.37	14.20	14.02	13.93	13.84	13.75	13.65	13.56	13.46
5	9.89	9.72	9.55	9.47	9.38	9.29	9.20	9.11	9.02
6	7.72	7.56	7.40	7.31	7.23	7.14	7.06	6.97	6.88
7	6.47	6.31	6.16	6.07	5.99	5.91	5.82	5.74	5.65
8	5.67	5.52	5.36	5.28	5.20	5.12	5.03	4.95	4.86
9	5.11	4.96	4.81	4.73	4.65	4.57	4.48	4.40	4.31
10	4.71	4.56	4.41	4.33	4.25	4.17	4.08	4.00	3.91
11	4.40	4.25	4.10	4.02	3.94	3.86	3.78	3.69	3.60
12	4.16	4.01	3.86	3.78	3.70	3.62	3.54	3.45	3.36
13	3.96	3.82	3.66	3.59	3.51	3.43	3.34	3.25	3.17
14	3.80	3.66	3.51	3.43	3.35	3.27	3.18	3.09	3.00
15	3.67	3.52	3.37	3.29	3.21	3.13	3.05	2.96	2.87
16	3.55	3.41	3.26	3.18	3.10	3.02	2.93	2.84	2.75
17	3.46	3.31	3.16	3.08	3.00	2.92	2.83	2.75	2.65
18	3.37	3.23	3.08	3.00	2.92	2.84	2.75	2.66	2.57
19	3.30	3.15	3.00	2.92	2.84	2.76	2.67	2.58	2.49
20	3.23	3.09	2.94	2.86	2.78	2.69	2.61	2.52	2.42
21	3.17	3.03	2.88	2.80	2.72	2.64	2.55	2.46	2.36
22	3.12	2.98	2.83	2.75	2.67	2.58	2.50	2.40	2.31
23	3.07	2.93	2.78	2.70	2.62	2.54	2.45	2.35	2.26
24	3.03	2.89	2.74	2.66	2.58	2.49	2.40	2.31	2.21
25	2.99	2.85	2.70	2.62	2.54	2.45	2.36	2.27	2.17
26	2.96	2.81	2.66	2.58	2.50	2.42	2.33	2.23	2.13
27	2.93	2.78	2.63	2.55	2.47	2.38	2.29	2.20	2.10
28	2.90	2.75	2.60	2.52	2.44	2.35	2.26	2.17	2.06
29	2.87	2.73	2.57	2.49	2.41	2.33	2.23	2.14	2.03
30	2.84	2.70	2.55	2.47	2.39	2.30	2.21	2.11	2.01
40	2.66	2.52	2.37	2.29	2.20	2.11	2.02	1.92	1.80
60	2.50	2.35	2.20	2.12	2.03	1.94	1.84	1.73	1.60
120	2.34	2.19	2.03	1.95	1.86	1.76	1.66	1.53	1.38
∞	2.18	2.04	1.88	1.79	1.70	1.59	1.47	1.32	1.00

This table was generated using MINITAB.

v_2 (Denominator Degrees of Freedom)	v_1 (Numerator Degrees of Freedom)									
	1	2	3	4	5	6	7	8	9	10
1	405261	499996	540349	562463	576409	585904	592890	598185	602359	605671
2	998.55	999.01	999.23	999.26	999.29	999.38	999.40	999.35	999.45	999.41
3	167.03	148.50	141.11	137.10	134.58	132.85	131.58	130.62	129.86	129.25
4	74.14	61.25	56.18	53.44	51.71	50.53	49.66	49.00	48.48	48.05
5	47.18	37.12	33.20	31.08	29.75	28.83	28.16	27.65	27.24	26.92
6	35.51	27.00	23.70	21.92	20.80	20.03	19.46	19.03	18.69	18.41
7	29.24	21.69	18.77	17.20	16.21	15.52	15.02	14.63	14.33	14.08
8	25.41	18.49	15.83	14.39	13.48	12.86	12.40	12.05	11.77	11.54
9	22.86	16.39	13.90	12.56	11.71	11.13	10.70	10.37	10.11	9.89
10	21.04	14.91	12.55	11.28	10.48	9.93	9.52	9.20	8.96	8.75
11	19.69	13.81	11.56	10.35	9.58	9.05	8.66	8.35	8.12	7.92
12	18.64	12.97	10.80	9.63	8.89	8.38	8.00	7.71	7.48	7.29
13	17.82	12.31	10.21	9.07	8.35	7.86	7.49	7.21	6.98	6.80
14	17.14	11.78	9.73	8.62	7.92	7.44	7.08	6.80	6.58	6.40
15	16.59	11.34	9.34	8.25	7.57	7.09	6.74	6.47	6.26	6.08
16	16.12	10.97	9.01	7.94	7.27	6.80	6.46	6.19	5.98	5.81
17	15.72	10.66	8.73	7.68	7.02	6.56	6.22	5.96	5.75	5.58
18	15.38	10.39	8.49	7.46	6.81	6.35	6.02	5.76	5.56	5.39
19	15.08	10.16	8.28	7.27	6.62	6.18	5.85	5.59	5.39	5.22
20	14.82	9.95	8.10	7.10	6.46	6.02	5.69	5.44	5.24	5.08
21	14.59	9.77	7.94	6.95	6.32	5.88	5.56	5.31	5.11	4.95
22	14.38	9.61	7.80	6.81	6.19	5.76	5.44	5.19	4.99	4.83
23	14.20	9.47	7.67	6.70	6.08	5.65	5.33	5.09	4.89	4.73
24	14.03	9.34	7.55	6.59	5.98	5.55	5.23	4.99	4.80	4.64
25	13.88	9.22	7.45	6.49	5.89	5.46	5.15	4.91	4.71	4.56
26	13.74	9.12	7.36	6.41	5.80	5.38	5.07	4.83	4.64	4.48
27	13.61	9.02	7.27	6.33	5.73	5.31	5.00	4.76	4.57	4.41
28	13.50	8.93	7.19	6.25	5.66	5.24	4.93	4.69	4.50	4.35
29	13.39	8.85	7.12	6.19	5.59	5.18	4.87	4.64	4.45	4.29
30	13.29	8.77	7.05	6.12	5.53	5.12	4.82	4.58	4.39	4.24
40	12.61	8.25	6.59	5.70	5.13	4.73	4.44	4.21	4.02	3.87
60	11.97	7.77	6.17	5.31	4.76	4.37	4.09	3.86	3.69	3.54
120	11.38	7.32	5.78	4.95	4.42	4.04	3.77	3.55	3.38	3.24
∞	10.83	6.91	5.42	4.62	4.10	3.74	3.47	3.27	3.10	2.96

v_2 (Denominator Degrees of Freedom)	v_1 (Numerator Degrees of Freedom)								
	12	15	20	24	30	40	60	120	∞
1	610644	615766	620884	623544	626117	628724	631381	634002	636619
2	999.46	999.40	999.44	999.45	999.47	999.49	999.50	999.52	999.50
3	128.32	127.37	126.42	125.94	125.45	124.96	124.47	123.97	123.47
4	47.41	46.76	46.10	45.77	45.43	45.09	44.75	44.40	44.05
5	26.42	25.91	25.40	25.13	24.87	24.60	24.33	24.06	23.79
6	17.99	17.56	17.12	16.90	16.67	16.44	16.21	15.98	15.75
7	13.71	13.32	12.93	12.73	12.53	12.33	12.12	11.91	11.70
8	11.19	10.84	10.48	10.30	10.11	9.92	9.73	9.53	9.33
9	9.57	9.24	8.90	8.72	8.55	8.37	8.19	8.00	7.81
10	8.45	8.13	7.80	7.64	7.47	7.30	7.12	6.94	6.76
11	7.63	7.32	7.01	6.85	6.68	6.52	6.35	6.18	6.00
12	7.00	6.71	6.40	6.25	6.09	5.93	5.76	5.59	5.42
13	6.52	6.23	5.93	5.78	5.63	5.47	5.30	5.14	4.97
14	6.13	5.85	5.56	5.41	5.25	5.10	4.94	4.77	4.60
15	5.81	5.54	5.25	5.10	4.95	4.80	4.64	4.47	4.31
16	5.55	5.27	4.99	4.85	4.70	4.54	4.39	4.23	4.06
17	5.32	5.05	4.78	4.63	4.48	4.33	4.18	4.02	3.85
18	5.13	4.87	4.59	4.45	4.30	4.15	4.00	3.84	3.67
19	4.97	4.70	4.43	4.29	4.14	3.99	3.84	3.68	3.51
20	4.82	4.56	4.29	4.15	4.01	3.86	3.70	3.54	3.38
21	4.70	4.44	4.17	4.03	3.88	3.74	3.58	3.42	3.26
22	4.58	4.33	4.06	3.92	3.78	3.63	3.48	3.32	3.15
23	4.48	4.23	3.96	3.82	3.68	3.53	3.38	3.22	3.05
24	4.39	4.14	3.87	3.74	3.59	3.45	3.29	3.14	2.97
25	4.31	4.06	3.79	3.66	3.52	3.37	3.22	3.06	2.89
26	4.24	3.99	3.72	3.59	3.44	3.30	3.15	2.99	2.82
27	4.17	3.92	3.66	3.52	3.38	3.23	3.08	2.92	2.75
28	4.11	3.86	3.60	3.46	3.32	3.18	3.02	2.86	2.69
29	4.05	3.80	3.54	3.41	3.27	3.12	2.97	2.81	2.64
30	4.00	3.75	3.49	3.36	3.22	3.07	2.92	2.76	2.59
40	3.64	3.40	3.14	3.01	2.87	2.73	2.57	2.41	2.23
60	3.32	3.08	2.83	2.69	2.55	2.41	2.25	2.08	1.89
120	3.02	2.78	2.53	2.40	2.26	2.11	1.95	1.77	1.54
∞	2.74	2.51	2.27	2.13	1.99	1.84	1.66	1.45	1.00

This table was generated using MINITAB.

Table B.7A
Factors for Two-Sided Tolerance Intervals for Normal Distributions

n	95% Confidence			99% Confidence		
	$p = .90$	$p = .95$	$p = .99$	$p = .90$	$p = .95$	$p = .99$
2		36.519	46.944		182.720	234.877
3	8.306	9.789	12.647	18.782	22.131	28.586
4	5.368	6.341	8.221	9.416	11.118	14.405
5	4.291	5.077	6.598	6.655	7.870	10.220
6	3.733	4.422	5.758	5.383	6.373	8.292
7	3.390	4.020	5.241	4.658	5.520	7.191
8	3.156	3.746	4.889	4.189	4.968	6.479
9	2.986	3.546	4.633	3.860	4.581	5.980
10	2.856	3.393	4.437	3.617	4.294	5.610
11	2.754	3.273	4.282	3.429	4.073	5.324
12	2.670	3.175	4.156	3.279	3.896	5.096
13	2.601	3.093	4.051	3.156	3.751	4.909
14	2.542	3.024	3.962	3.054	3.631	4.753
15	2.492	2.965	3.885	2.967	3.529	4.621
16	2.449	2.913	3.819	2.893	3.441	4.507
17	2.410	2.868	3.761	2.828	3.364	4.408
18	2.376	2.828	3.709	2.771	3.297	4.321
19	2.346	2.793	3.663	2.720	3.237	4.244
20	2.319	2.760	3.621	2.675	3.184	4.175
25	2.215	2.638	3.462	2.506	2.984	3.915
30	2.145	2.555	3.355	2.394	2.851	3.742
35	2.094	2.495	3.276	2.314	2.756	3.618
40	2.055	2.448	3.216	2.253	2.684	3.524
50	1.999	2.382	3.129	2.166	2.580	3.390
60	1.960	2.335	3.068	2.106	2.509	3.297
80	1.908	2.274	2.987	2.028	2.416	3.175
100	1.875	2.234	2.936	1.978	2.357	3.098
150	1.826	2.176	2.859	1.906	2.271	2.985
200	1.798	2.143	2.816	1.866	2.223	2.921
500	1.737	2.070	2.721	1.777	2.117	2.783
1000	1.709	2.036	2.676	1.736	2.068	2.718
∞	1.645	1.960	2.576	1.645	1.960	2.576

This table was computed using Mathcad.

Table B.7B
Factors for One-Sided Tolerance Intervals for Normal Distributions

n	95% Confidence			99% Confidence		
	$p = .90$	$p = .95$	$p = .99$	$p = .90$	$p = .95$	$p = .99$
2						
3	6.155	7.656	10.553	14.006	17.372	23.896
4	4.162	5.144	7.042	7.380	9.083	12.388
5	3.407	4.203	5.741	5.362	6.578	8.939
6	3.006	3.708	5.062	4.411	5.406	7.335
7	2.755	3.399	4.642	3.859	4.728	6.412
8	2.582	3.187	4.354	3.497	4.285	5.812
9	2.454	3.031	4.143	3.240	3.972	5.389
10	2.355	2.911	3.981	3.048	3.738	5.074
11	2.275	2.815	3.852	2.898	3.556	4.829
12	2.210	2.736	3.747	2.777	3.410	4.633
13	2.155	2.671	3.659	2.677	3.290	4.472
14	2.109	2.614	3.585	2.593	3.189	4.337
15	2.068	2.566	3.520	2.521	3.102	4.222
16	2.033	2.524	3.464	2.459	3.028	4.123
17	2.002	2.486	3.414	2.405	2.963	4.037
18	1.974	2.453	3.370	2.357	2.905	3.960
19	1.949	2.423	3.331	2.314	2.854	3.892
20	1.926	2.396	3.295	2.276	2.808	3.832
25	1.838	2.292	3.158	2.129	2.633	3.601
30	1.777	2.220	3.064	2.030	2.515	3.447
35	1.732	2.167	2.995	1.957	2.430	3.334
40	1.697	2.125	2.941	1.902	2.364	3.249
50	1.646	2.065	2.862	1.821	2.269	3.125
60	1.609	2.022	2.807	1.764	2.202	3.038
80	1.559	1.964	2.733	1.688	2.114	2.924
100	1.527	1.927	2.684	1.639	2.056	2.850
150	1.478	1.870	2.611	1.566	1.971	2.740
200	1.450	1.837	2.570	1.524	1.923	2.679
500	1.385	1.763	2.475	1.430	1.814	2.540
1000	1.354	1.727	2.430	1.385	1.762	2.475
∞	1.282	1.645	2.326	1.282	1.645	2.326

This table was computed using Mathcad.

Table B.8A
Factors for Simultaneous 95% Two-Sided Confidence Limits for Several Means

ν	Number of Means													
	1	2	3	4	5	6	7	8	9	10	12	14	16	32
2	4.303	5.571	6.340	6.886	7.306	7.645	7.929	8.172	8.385	8.573	8.894	9.162	9.390	10.529
3	3.182	3.960	4.430	4.764	5.023	5.233	5.410	5.562	5.694	5.812	6.015	6.184	6.328	7.055
4	2.776	3.382	3.745	4.003	4.203	4.366	4.503	4.621	4.725	4.817	4.975	5.107	5.221	5.794
5	2.571	3.091	3.399	3.619	3.789	3.928	4.044	4.145	4.233	4.312	4.447	4.560	4.657	5.150
6	2.447	2.916	3.193	3.389	3.541	3.664	3.769	3.858	3.937	4.008	4.129	4.230	4.317	4.760
7	2.365	2.800	3.055	3.236	3.376	3.489	3.585	3.668	3.740	3.805	3.916	4.009	4.090	4.498
8	2.306	2.718	2.958	3.127	3.258	3.365	3.454	3.532	3.600	3.660	3.764	3.852	3.927	4.310
9	2.262	2.657	2.885	3.046	3.171	3.272	3.357	3.430	3.494	3.552	3.650	3.733	3.805	4.169
10	2.228	2.609	2.829	2.983	3.103	3.199	3.281	3.351	3.412	3.467	3.562	3.641	3.710	4.058
11	2.201	2.571	2.784	2.933	3.048	3.142	3.220	3.288	3.347	3.400	3.491	3.568	3.634	3.969
12	2.179	2.540	2.747	2.892	3.004	3.095	3.171	3.236	3.294	3.345	3.433	3.507	3.571	3.897
13	2.160	2.514	2.717	2.858	2.967	3.055	3.129	3.193	3.249	3.299	3.385	3.457	3.519	3.836
14	2.145	2.493	2.691	2.830	2.936	3.022	3.095	3.157	3.212	3.260	3.344	3.415	3.475	3.784
15	2.131	2.474	2.669	2.805	2.909	2.994	3.065	3.126	3.180	3.227	3.309	3.378	3.438	3.740
16	2.120	2.458	2.650	2.784	2.886	2.969	3.039	3.099	3.152	3.199	3.279	3.347	3.405	3.701
17	2.110	2.444	2.633	2.765	2.866	2.948	3.017	3.076	3.127	3.173	3.253	3.319	3.376	3.668
18	2.101	2.432	2.619	2.749	2.849	2.929	2.997	3.055	3.106	3.151	3.229	3.295	3.351	3.638
19	2.093	2.421	2.606	2.734	2.833	2.912	2.979	3.037	3.087	3.132	3.209	3.273	3.329	3.611
20	2.086	2.411	2.594	2.721	2.819	2.897	2.963	3.020	3.070	3.114	3.190	3.254	3.308	3.587
24	2.064	2.380	2.558	2.681	2.775	2.851	2.914	2.969	3.016	3.059	3.132	3.193	3.246	3.513
30	2.042	2.350	2.522	2.641	2.732	2.805	2.866	2.918	2.964	3.005	3.075	3.133	3.184	3.439
36	2.028	2.331	2.499	2.615	2.704	2.775	2.834	2.885	2.930	2.970	3.038	3.094	3.143	3.391
40	2.021	2.321	2.488	2.602	2.690	2.760	2.819	2.869	2.913	2.952	3.019	3.075	3.123	3.367
60	2.000	2.292	2.454	2.564	2.649	2.716	2.772	2.821	2.863	2.900	2.964	3.018	3.064	3.295
120	1.980	2.264	2.420	2.527	2.608	2.673	2.727	2.773	2.814	2.849	2.910	2.961	3.005	3.225
144	1.977	2.259	2.415	2.521	2.602	2.666	2.720	2.766	2.806	2.841	2.902	2.952	2.996	3.214
∞	1.960	2.237	2.388	2.491	2.569	2.631	2.683	2.727	2.766	2.800	2.858	2.906	2.948	3.156

This table was prepared using a program written by Daniel L. Rose.

Table B.8B
Factors for Simultaneous 95% One-Sided Confidence Limits for Several Means

ν	1	2	3	4	5	6	7	8	9	10	12	14	16	32
						Number of Means								
2	2.920	4.075	4.834	5.397	5.842	6.208	6.516	6.781	7.014	7.220	7.573	7.867	8.118	9.364
3	2.353	3.090	3.551	3.888	4.154	4.372	4.557	4.717	4.858	4.983	5.199	5.380	5.535	6.315
4	2.132	2.722	3.080	3.340	3.544	3.711	3.852	3.974	4.082	4.179	4.345	4.484	4.604	5.212
5	2.015	2.532	2.840	3.062	3.234	3.376	3.495	3.599	3.690	3.772	3.912	4.031	4.132	4.650
6	1.943	2.417	2.696	2.894	3.049	3.175	3.282	3.374	3.455	3.528	3.653	3.758	3.849	4.312
7	1.895	2.340	2.599	2.783	2.925	3.041	3.139	3.224	3.299	3.365	3.480	3.577	3.660	4.085
8	1.860	2.285	2.530	2.703	2.837	2.946	3.038	3.117	3.187	3.250	3.357	3.447	3.525	3.923
9	1.833	2.243	2.479	2.644	2.772	2.875	2.962	3.038	3.104	3.163	3.265	3.351	3.424	3.801
10	1.812	2.211	2.439	2.598	2.720	2.820	2.904	2.976	3.039	3.096	3.193	3.275	3.346	3.707
11	1.796	2.186	2.407	2.561	2.680	2.776	2.857	2.927	2.988	3.042	3.136	3.215	3.283	3.631
12	1.782	2.164	2.380	2.531	2.647	2.740	2.819	2.886	2.946	2.999	3.090	3.166	3.232	3.569
13	1.771	2.147	2.359	2.506	2.619	2.710	2.787	2.853	2.911	2.962	3.051	3.126	3.190	3.517
14	1.761	2.132	2.340	2.485	2.596	2.685	2.760	2.825	2.881	2.932	3.018	3.091	3.154	3.473
15	1.753	2.119	2.324	2.467	2.576	2.663	2.737	2.800	2.856	2.905	2.990	3.062	3.123	3.436
16	1.746	2.108	2.311	2.451	2.558	2.645	2.717	2.779	2.834	2.883	2.966	3.036	3.096	3.403
17	1.740	2.099	2.299	2.437	2.543	2.628	2.700	2.761	2.815	2.863	2.945	3.014	3.073	3.375
18	1.734	2.090	2.288	2.425	2.530	2.614	2.684	2.745	2.798	2.845	2.926	2.994	3.052	3.349
19	1.729	2.083	2.279	2.415	2.518	2.601	2.671	2.731	2.783	2.830	2.910	2.977	3.034	3.327
20	1.725	2.076	2.271	2.405	2.507	2.590	2.659	2.718	2.770	2.816	2.895	2.961	3.018	3.307
24	1.711	2.055	2.245	2.375	2.474	2.554	2.621	2.678	2.728	2.772	2.848	2.912	2.967	3.244
30	1.697	2.034	2.219	2.346	2.442	2.519	2.584	2.639	2.687	2.730	2.803	2.864	2.917	3.183
36	1.688	2.020	2.202	2.327	2.421	2.496	2.559	2.613	2.660	2.702	2.773	2.833	2.884	3.142
40	1.684	2.014	2.194	2.317	2.410	2.485	2.547	2.600	2.647	2.688	2.758	2.817	2.868	3.122
60	1.671	1.993	2.169	2.289	2.379	2.451	2.511	2.563	2.607	2.647	2.715	2.771	2.820	3.063
120	1.658	1.974	2.145	2.261	2.349	2.418	2.476	2.526	2.569	2.607	2.672	2.726	2.773	3.005
144	1.656	1.971	2.141	2.257	2.344	2.413	2.471	2.520	2.563	2.601	2.665	2.719	2.765	2.995
∞	1.645	1.955	2.121	2.234	2.319	2.386	2.442	2.490	2.531	2.568	2.630	2.682	2.727	2.948

This table was prepared using a program written by Daniel L. Rose.

Table B.9A
.95 Quantiles of the Studentized Range Distribution

ν	Number of Means to Be Compared													
	2	3	4	5	6	7	8	9	10	11	12	13	15	20
5	3.64	4.60	5.22	5.67	6.03	6.33	6.58	6.80	6.99	7.17	7.32	7.47	7.72	8.21
6	3.46	4.34	4.90	5.30	5.63	5.90	6.12	6.32	6.49	6.65	6.79	6.92	7.14	7.59
7	3.34	4.16	4.68	5.06	5.36	5.61	5.82	6.00	6.16	6.30	6.43	6.55	6.76	7.17
8	3.26	4.04	4.53	4.89	5.17	5.40	5.60	5.77	5.92	6.05	6.18	6.29	6.48	6.87
9	3.20	3.95	4.41	4.76	5.02	5.24	5.43	5.59	5.74	5.87	5.98	6.09	6.28	6.64
10	3.15	3.88	4.33	4.65	4.91	5.12	5.30	5.46	5.60	5.72	5.83	5.93	6.11	6.47
11	3.11	3.82	4.26	4.57	4.82	5.03	5.20	5.35	5.49	5.61	5.71	5.81	5.98	6.33
12	3.08	3.77	4.20	4.51	4.75	4.95	5.12	5.27	5.39	5.51	5.61	5.71	5.88	6.21
13	3.06	3.73	4.15	4.45	4.69	4.88	5.05	5.19	5.32	5.43	5.53	5.63	5.79	6.11
14	3.03	3.70	4.11	4.41	4.64	4.83	4.99	5.13	5.25	5.36	5.46	5.55	5.71	6.03
15	3.01	3.67	4.08	4.37	4.59	4.78	4.94	5.08	5.20	5.31	5.40	5.49	5.65	5.96
16	3.00	3.65	4.05	4.33	4.56	4.74	4.90	5.03	5.15	5.26	5.35	5.44	5.59	5.90
17	2.98	3.63	4.02	4.30	4.52	4.70	4.86	4.99	5.11	5.21	5.31	5.39	5.54	5.84
18	2.97	3.61	4.00	4.28	4.49	4.67	4.82	4.96	5.07	5.17	5.27	5.35	5.50	5.79
19	2.96	3.59	3.98	4.25	4.47	4.65	4.79	4.92	5.04	5.14	5.23	5.31	5.46	5.75
20	2.95	3.58	3.96	4.23	4.45	4.62	4.77	4.90	5.01	5.11	5.20	5.28	5.43	5.71
24	2.92	3.53	3.90	4.17	4.37	4.54	4.68	4.81	4.92	5.01	5.10	5.18	5.32	5.59
30	2.89	3.49	3.85	4.10	4.30	4.46	4.60	4.72	4.82	4.92	5.00	5.08	5.21	5.47
40	2.86	3.44	3.79	4.04	4.23	4.39	4.52	4.63	4.73	4.82	4.90	4.98	5.11	5.36
60	2.83	3.40	3.74	3.98	4.16	4.31	4.44	4.55	4.65	4.73	4.81	4.88	5.00	5.24
120	2.80	3.36	3.68	3.92	4.10	4.24	4.36	4.47	4.56	4.64	4.71	4.78	4.90	5.13
∞	2.77	3.31	3.63	3.86	4.03	4.17	4.29	4.39	4.47	4.55	4.62	4.68	4.80	5.01

This table was computed using Mathcad.

Table B.9B
.99 Quantiles of the Studentized Range Distribution

ν	2	3	4	5	6	7	8	9	10	11	12	13	15	20
						Number of Means to Be Compared								
5	5.70	6.98	7.80	8.42	8.91	9.32	9.67	9.97	10.24	10.48	10.70	10.89	11.24	11.93
6	5.24	6.33	7.03	7.56	7.97	8.32	8.61	8.87	9.10	9.30	9.48	9.65	9.95	10.54
7	4.95	5.92	6.54	7.00	7.37	7.68	7.94	8.17	8.37	8.55	8.71	8.86	9.12	9.65
8	4.75	5.64	6.20	6.62	6.96	7.24	7.47	7.68	7.86	8.03	8.18	8.31	8.55	9.03
9	4.60	5.43	5.96	6.35	6.66	6.91	7.13	7.33	7.49	7.65	7.78	7.91	8.13	8.57
10	4.48	5.27	5.77	6.14	6.43	6.67	6.87	7.05	7.21	7.36	7.49	7.60	7.81	8.23
11	4.39	5.15	5.62	5.97	6.25	6.48	6.67	6.84	6.99	7.13	7.25	7.36	7.56	7.95
12	4.32	5.05	5.50	5.84	6.10	6.32	6.51	6.67	6.81	6.94	7.06	7.17	7.36	7.73
13	4.26	4.96	5.40	5.73	5.98	6.19	6.37	6.53	6.67	6.79	6.90	7.01	7.19	7.55
14	4.21	4.89	5.32	5.63	5.88	6.08	6.26	6.41	6.54	6.66	6.77	6.87	7.05	7.39
15	4.17	4.84	5.25	5.56	5.80	5.99	6.16	6.31	6.44	6.55	6.66	6.76	6.93	7.26
16	4.13	4.79	5.19	5.49	5.72	5.92	6.08	6.22	6.35	6.46	6.56	6.66	6.82	7.15
17	4.10	4.74	5.14	5.43	5.66	5.85	6.01	6.15	6.27	6.38	6.48	6.57	6.73	7.05
18	4.07	4.70	5.09	5.38	5.60	5.79	5.94	6.08	6.20	6.31	6.41	6.50	6.65	6.97
19	4.05	4.67	5.05	5.33	5.55	5.73	5.89	6.02	6.14	6.25	6.34	6.43	6.58	6.89
20	4.02	4.64	5.02	5.29	5.51	5.69	5.84	5.97	6.09	6.19	6.28	6.37	6.52	6.82
24	3.96	4.55	4.91	5.17	5.37	5.54	5.69	5.81	5.92	6.02	6.11	6.19	6.33	6.61
30	3.89	4.45	4.80	5.05	5.24	5.40	5.54	5.65	5.76	5.85	5.93	6.01	6.14	6.41
40	3.82	4.37	4.70	4.93	5.11	5.26	5.39	5.50	5.60	5.69	5.76	5.83	5.96	6.21
60	3.76	4.28	4.59	4.82	4.99	5.13	5.25	5.36	5.45	5.53	5.60	5.67	5.78	6.01
120	3.70	4.20	4.50	4.71	4.87	5.01	5.12	5.21	5.30	5.37	5.44	5.50	5.61	5.83
∞	3.64	4.12	4.40	4.60	4.76	4.88	4.99	5.08	5.16	5.23	5.29	5.35	5.45	5.65

This table was computed using Mathcad.

Answers to Section Exercises

Chapter 1

Section 1

1. Designing and improving complex products and systems often leads to situations where there is no known theory that can guide decisions. Engineers are then forced to experiment and collect data to find out how a system works, usually under time and monetary constraints. Engineers also collect data in order to monitor the quality of products and services. Statistical principles and methods can be used to find effective and efficient ways to collect and analyze such data.

2. The physical world is filled with variability. It comes from differences in raw materials, machinery, operators, environment, measuring devices, and other uncontrollable variables that change over time. This produces variability in engineering data, at least some of which is impossible to completely eliminate. Statistics must therefore address the reality of variability in data.

3. Descriptive statistics provides a way of summarizing patterns and major features of data. Inferential statistics uses a probability model to describe the process from which the data were obtained; data are then used to draw conclusions about the process by estimating parameters in the model and making predictions based on the model.

Section 2

1. Observational study—you might be interested in assessing the job satisfaction of a large number of manufacturing workers; you could administer a survey to measure various dimensions of job satisfaction. Experimental study—you might want to compare several different job routing schemes to see which one achieves the greatest throughput in a job shop.

2. Qualitative data—rating the quality of batches of ice cream as either poor, fair, good, or exceptional. Quantitative data—measuring the time (in hours) it takes for each of 1,000 integrated circuit chips to fail in a high-stress environment.

3. Any relationships between the variables x and y can only be derived from a bivariate sample.

4. You might want to compare two laboratories in their ability to determine percent impurities in rare metal specimens. Each specimen could be divided in two, with each half going to a different lab. Since each specimen is being measured twice for percent impurity, the data would be paired (according to specimen).

5. Full factorial data structure—tests are performed for all factor-level combinations:

Design	Paper	Loading Condition
delta	construction	with clip
t-wing	construction	with clip
delta	typing	with clip
t-wing	typing	with clip
delta	construction	without clip
t-wing	construction	without clip
delta	typing	without clip
t-wing	typing	without clip

Fractional factorial data structure—tests are performed for only some of the possible factor-level combinations. One possibility is to choose the following "half fraction":

Design	Paper	Loading Condition
delta	construction	without clip
t-wing	construction	with clip
delta	typing	with clip
t-wing	typing	without clip

6. Variables can be manipulated in an experiment. If changes in the response coincide with changes in factor levels, it is usually safe to infer that the changes in the factor caused the changes in the response (as long as other factors have been controlled and there is no source of bias). There is no control or manipulation in an observational study. Changes in the response may coincide with changes in another variable, but there is always the possibility that a *third* variable is causing the correlation. It is therefore risky to infer a cause-and-effect relationship between any variable and the response in an observational study.

Section 3

1. Even if a measurement system is accurate and precise, if it is not truly measuring the desired dimension or characteristic, then the measurements are useless. If a measurement system is valid and accurate, but imprecise, it may be useless because it produces too much variability (and this cannot be corrected by calibration). If a measurement system is valid and precise, but inaccurate, it might be easy to make it accurate (and thus useful) by calibrating it to a standard.

2. If the measurement system is not valid, then taking an average will still produce a measurement that is invalid. If the individual measurements are inaccurate, then the average will be inaccurate. Averaging many measurements only improves precision. Suppose that the long-run average yield of the process is stable over time. Imagine making 5 yield measurements every hour, for 24 hours. This produces 120 individual measurements, and 24 averages. Since the averages are "pulled" to the center, there will be less variability in the 24 averages than in the 120 individual measurements, so averaging improves precision.

3. Unstable measurement systems (e.g., instrument drift, multiple inconsistent devices) can lead to differences or changes in validity, precision, and accuracy. In a statistical engineering study, it is important to obtain valid, precise, and accurate measurements throughout the study. Changes or differences may create excessive variability, making it hard to draw conclusions. Changes or differences can also bias results by causing patterns in data that might incorrectly be attributed to factors in the experiment.

Section 4

1. Mathematical models can help engineers describe (in a relatively simple and concise way) how physical systems behave, or will behave. They are an integral part of designing and improving products and processes.

Chapter 2

Section 1

1. *Flight distance* might be defined as the horizontal distance that a plane travels after being launched from a mechanical slingshot. Specifically, the horizontal distance might be measured from the point on the floor directly below the slingshot to the point on the floor where any part of the plane first touches.

2. If all operators are trained to use measuring equipment in the same consistent way, this will result in better repeatability and reproducibility of measurements. The measurements will be more repeatable because individual operators will use the same technique from measurement to measurement, resulting in small variability among measurements of the same item by the same operator. The measurements will be more reproducible because all operators will be trained to use the same technique, resulting in small variability among measurements made by different operators.

3. This scheme will tend to "over-sample" larger lots and "under-sample" smaller lots, since the amount of information obtained about a large population from a particular sample size does not depend on the size of the population. To obtain the same amount of information from each lot, you should use an absolute (fixed) sample size instead of a relative one.

4. If the response variable is poorly defined, the data collected may not properly describe the characteristic of interest. Even if they do, operators may not be consistent in the way that they measure the response, resulting in more variation.

Section 2

1. Label the 38 runout values consecutively, 1–38, in the order given in Table 1.1 (smallest to largest). First sample labels: $\{12, 15, 5, 9, 11\}$; First sample runout values: $\{11, 11, 9, 10, 11\}$. Second sample labels: $\{34, 31, 36, 2, 14\}$; Second sample runout values: $\{17, 15, 18, 8, 11\}$. Third sample labels: $\{10, 35, 12, 27, 30\}$; Third sample runout values: $\{10, 17, 11, 14, 15\}$. Fourth sample labels: $\{15, 5, 19, 11, 8\}$; Fourth sample runout values: $\{11, 9, 12, 11, 10\}$. The samples are not identical. *Note:* the population mean is 12.63; the sample means are 10.4, 13.8, 13.4, and 10.6.

3. A simple random sample is not guaranteed to be representative of the population from which it is drawn. It gives every set of n items an equal chance of being selected, so there is always a chance that

the n items chosen will be "extreme" members of the population.

Section 3

1. Possible controlled variables: operator, launch angle, launch force, paper clip size, paper manufacturer, plane constructor, distance measurer, and wind. The response is Flight Distance and the experimental variables are Design, Paper Type, and Loading Condition. Concomitant variables might be wind speed and direction (if these cannot be controlled), ambient temperature, humidity, and atmospheric pressure.

2. Advantage: may reduce baseline variation (background noise) in the response, making it easier to see the effects of factors. Disadvantage: the variable may fluctuate in the real world, so controlling it makes the experiment more artificial—it will be harder to generalize conclusions from the experiment to the real world.

3. Treat "distance measurer" as an experimental (blocking) variable with 2 levels. For each level (team member), perform a full factorial experiment using the 3 primary factors. If there are differences in the way team members measure distance, then it will still be possible to unambiguously assess the effects of the primary factors within each "sub-experiment" (block).

4. List the tests for Mary in the same order given for Exercise 5 of Section 1.2. Then list the tests for Tom after Mary, again in the same order. Label the tests consecutively 1–16, in the order listed. Let the digits 01–05 refer to test 1, 06–10 to test 2, ..., and 76–80 to test 16. Move through Table B.1 choosing two digits at a time. Ignore previously chosen test labels or numbers between 81 and 00. Order the tests in the same order that their corresponding two-digit numbers are chosen from the table. Using this method (and starting from the upper-left of the table), the test labeled 3 (Mary, delta, typing, with clip) would be first, followed by the tests labeled 13, 9, 1, 2, 7, 10, 8, 14, 11, 6, 15, 4, 16, 12, and 5.

5. For the delta/construction/with clip condition (for example), flying the same plane twice would provide information about flight-to-flight variability for that particular plane. This would be useful if you are only interested in making conclusions about that particular plane. If you are interested in generalizing your conclusions to all delta design planes made with construction paper and loaded with a paper clip, then reflying the same airplane does not provide much more information. But making and flying two planes for this condition would give you some idea of variability among different planes of this type, and would therefore validate any general conclusions made from the study. This argument would be true for all 8 conditions, and would also apply to comparisons made among the 8 conditions.

6. Random sampling is used in enumerative studies. Its purpose is to choose a representative sample from some population of items. Randomization is used in analytical/experimental studies. Its purpose is to assign units to experimental conditions in an unbiased way, and to order procedures to prevent bias from unsupervised variables that may change over time.

7. Blocking is a way of controlling an extraneous variable: within each block, there may be less baseline variation (background noise) in the response than there would be if the variable were not controlled. This makes it easier to see the effects of the factors of interest within each block. Any effects of the extraneous variable can be isolated and distinguished from the effects of the factors of interest. Compared to holding the variable constant throughout the experiment, blocking also results in a more realistic experiment.

8. Replication is used to estimate the magnitude of baseline variation (background noise, experimental error) in the response, and thus helps sharpen and validate conclusions drawn from data. It provides verification that results are repeatable and establishes the limits of that repeatability.

9. It is not necessary to know exactly how the entire budget will be spent. Experimentation in engineering is usually sequential, and this requires some decisions to be made in the middle of the study. Although some may think that this is improper from a scientific/statistical point of view, it is only practical to base the design of later stages on the results of earlier stages.

Section 4

1. If you regard student as a blocking variable, then this would be a randomized complete block experiment. Otherwise, it would just be a completely randomized experiment (with a full factorial structure).

2. (a) Label the 24 runs as follows:

Labels	Level of A	Level of B	Level of C
1, 2, 3	1	1	1
4, 5, 6	2	1	1
7, 8, 9	1	2	1
10, 11, 12	2	2	1
13, 14, 15	1	1	2
16, 17, 18	2	1	2
19, 20, 21	1	2	2
22, 23, 24	2	2	2

Use the following coding for the test labels: table number 01–04 for test label 1, table number 05–08 for test label 2, ..., table number 93–96 for test number 24. Move through Table B.1 choosing two digits at a time, ignoring numbers between 97 and 00 and those corresponding to test labels that have already been picked. Order the tests in the same order that their corresponding two-digit numbers are picked from the table. Using this method, and starting from the upper-left corner of the table, the order would be 3, 4, 24, 16, 11, 2, 9, 12, 17, 8, 21, 1, 13, 7, 18, 5, 20, 14, 19, 15, 22, 23, 6, 10.   (b) Treat day as a blocking variable, and run each of the 8 factor-level combinations once on each day. Blocking allows comparisons among the factor-level combinations to be made within each day. If blocking were not used, differences among days might cause variation in the response which would cloud comparisons among the factor-level combinations. **(c)** List the 8 factor-level combinations separately for each day. For each day, label the runs as follows:

Label	Level of A	Level of B	Level of C
1	1	1	1
2	2	1	1
3	1	2	1
4	2	2	1
5	1	1	2
6	2	1	2
7	1	2	2
8	2	2	2

For each day, move through Table B.1 one digit at a time ignoring the digits 9 and 0 and any that have already been picked. Order the 8 runs in the same order that the numbers were picked from the table. Starting from where I left off in part (a), the order for day 1 is 5, 3, 8, 4, 1, 2, 6 (which implies that run 7 goes last). For day 2, the order is 5, 1, 8, 7, 2, 3, 6 (which implies that run 4 goes last). For day 3, the order is 1, 3, 2, 7, 4, 5, 8, (which implies that run 6 goes last).

The plan is summarized below:

Day	Level of A	Level of B	Level of C
1	1	1	2
1	1	2	1
1	2	2	2
1	2	2	1
1	1	1	1
1	2	1	1
1	2	1	2
1	1	2	2
2	1	1	2
2	1	1	1
2	2	2	2
2	1	2	2
2	2	1	1
2	1	2	1
2	2	1	2
2	2	2	1

Day	Level of A	Level of B	Level of C
3	1	1	1
3	1	2	1
3	2	1	1
3	1	2	2
3	2	2	1
3	1	1	2
3	2	2	2
3	2	1	2

Part (a) randomized all 24 runs together; here, each block of 8 runs is randomized separately.

3. The factor Person is the "block" variable.

Block	Design	Paper
Tom	delta	construction
Tom	t-wing	typing
Juanita	delta	typing
Juanita	t-wing	construction

4. Focusing on Design, you would want each person to test two delta-wing planes and two t-wing planes; this would allow you to clearly compare the two designs. You could separately compare the designs "within" each person. If possible, you would want a plan such that this is true for all three primary factors, simultaneously. This is possible by using the same pattern that is used in Table 2.6:

Person	Design	Paper	Loading Condition
Juanita	delta	construction	with clip
Tom	t-wing	construction	with clip
Tom	delta	typing	with clip
Juanita	t-wing	typing	with clip
Tom	delta	construction	without clip
Juanita	t-wing	construction	without clip
Juanita	delta	typing	without clip
Tom	t-wing	typing	without clip

This design also allows each person to test each Design/Paper combination once, each Design/

Loading combination once, and each Paper/Loading combination once.

5. This is an incomplete block experiment.

Section 5

1. A cause-and-effect diagram may be useful for representing a complex system in a relatively simple and visual way. It enables people to see how the components of the system interact, and may help identify areas which need the most attention/improvement.

Chapter 3

Section 1

1. One choice of intervals for the frequency table and histogram is 65.5–66.4, 66.5–67.4, . . . , 73.5–74.4. For this choice, the frequencies are 3, 2, 9, 5, 8, 6, 2, 3, 2; the relative frequencies are .075, .05, .225, .125, .2, .15, .05, .075, .05; the cumulative relative frequencies are .075, .125, .35, .475, .675, .825, .875, .95, 1. The plots reveal a fairly symmetric, bell-shaped distribution.

2. The plots show that the depths for 200 grain bullets are larger and have less variability than those for the 230 grain bullets.

3. (a) There are no obvious patterns. (b) The differences are -15, 0, -20, 0, -5, 0, -5, 0, -5, 20, -25, -5, -10, -20, and 0. The dot diagram shows that most of the differences are negative and "truncated" at zero. The exception is the tenth piece of equipment, with a difference of 20. This point does not fit in with the shape of the rest of the differences, so it is an outlier. Since most of the differences are negative, the bottom bolt generally required more torque than the top bolt.

Section 2

1. (a) For the lengthwise sample: $Median = .895$, $Q(.25) = .870$, $Q(.75) = .930$, $Q(.37) = .880$. For the crosswise sample: $Median = .775$, $Q(.25) = .690$, $Q(.75) = .800$, $Q(.37) = .738$. (b) On the whole, the impact strengths are larger and more consistent for lengthwise cuts. Each method also produced an unusual impact strength

value (outlier). (c) The nonlinearity of the Q–Q plot indicates that the overall shapes of these two data sets are not the same. The lengthwise cuts had an unusually large data point ("long right tail"), whereas the crosswise cuts had an unusually small data point ("long left tail"). Without these two outliers, the data sets would have similar shapes, since the rest of the Q–Q plot is fairly linear.

2. Use the $(i - .5)/n$ quantiles for the smaller data set. The plot coordinates are: $(.370, .907)$, $(.520, 1.22)$, $(.650, 1.47)$, $(.920, 1.70)$, $(2.89, 2.45)$, $(3.62, 5.89)$.

3. The first 3 plot coordinates are: $(65.6, -2.33)$, $(65.6, -1.75)$, $(66.2, -1.55)$. The normal plot is quite linear, indicating that the data are very bell-shaped.

4. Theoretical Q–Q plotting allows you to roughly check to see if a data set has a shape that is similar to some theoretical distribution. This can be useful in identifying a theoretical (probability) model to represent how the process is generating data. Such a model can then be used to make inferences (conclusions) about the process.

Section 3

1. For the lengthwise cuts: $\bar{x} = .919$, $Median = .895$, $R = .310$, $IQR = .060$, $s = .088$. For the crosswise cuts: $\bar{x} = .743$, $Median = .775$, $R = .430$, $IQR = .110$, $s = .120$. The sample means and medians show that the center of the distribution for lengthwise cuts is higher than the center for crosswise cuts. The sample ranges, interquartile ranges, and sample standard deviations show that there is less spread in the lengthwise data than in the crosswise data.

2. These values are statistics. They are summarizations of two samples of data, and do not represent exact summarizations of larger populations or theoretical (long-run) distributions.

4. In the first case, the sample mean and median increase by 1.3, but none of the measures of spread change; in the second case, all of the measures double.

Section 4

1. $\hat{p}$ = the proportion of part orders that are delivered on time to the factory floor. $\hat{u}$ = number of defects per shift produced on an assembly line. A measured value of 65% yield for a run of a chemical process is of neither form.

2. $\hat{p}_{\text{Laid}} = \frac{6}{38} = .158$. $\hat{p}_{\text{Hung}} = \frac{24}{39} = .615$. Most engineering situations call for minimizing variation. The $\hat{p}$ values do not give any indication of how much spread there is in each set of data, and would not be helpful in comparing the two methods with respect to variation.

3. Neither type. These rates represent continuous measurements on each specimen; there is no "counting" involved.

Chapter 4

Section 1

1. **(a)** $\hat{y} = 9.4 - 1.0x$ **(b)** $r = -.945$ **(c)** $r = .945$. This is the negative of the r in part (b), since the $\hat{y}$'s are perfectly negatively correlated with the x's. **(d)** $R^2 = .893 = r^2$ from both (b) and (c). **(e)** $-.4, .6, -.4, .6, -.4$. These are the vertical distances from each data point to the least squares line.

3. **(a)** $R^2 = .994$ **(b)** $\hat{y} = -3174.6 + 23.50x$. 23.5 **(c)** Residuals: 105.36, -21.13, -60.11, -97.58, 16.95, 14.48, 42.00, .02. **(d)** There is no replication (multiple experimental runs at a particular pot temperature). **(e)** For $x = 188$ °C, $\hat{y} = 1243.1$. For $x = 200$ °C, $\hat{y} = 1525.1$. It would not be wise to make a similar prediction at $x = 70$ °C because there is no evidence that the fitted relationship is correct for pot temperatures as low as $x = 70$ °C. Some data should be obtained around $x = 70$ °C.

4. **(a)** The scatterplot is not linear, so the given straight-line relationship does not seem appropriate. $R^2 = .723$. **(b)** This scatterplot is much more linear, and a straight-line relationship seems appropriate for the transformed variables. $R^2 = .965$. **(c)** $\widehat{\ln y} = 34.344 - 5.1857 \ln x$. For $x = 550$, $\widehat{\ln y} = 1.6229$ so $\hat{y} = e^{1.6229} = 5.07$ minutes.

The implied relationship between x and y is $y = e^{\beta_0} x^{\beta_1}$.

Section 2

1. $\hat{y} = -1315 + 5.6x + .04212x^2$. $R^2 = .996$. For the quadratic model, at $x = 200$ °C, $\hat{y} = 1487.2$, which is relatively close to 1525.1 from part (e) of Exercise 3 of Section 1.

2. **(a)** $\hat{y} = 6.0483 + .14167x_1 - .016944x_2$. $b_1 = .14167$ means that as x_1 increases by 1% (holding x_2 constant), y increases by roughly .142 cm^3/g. $b_2 = -.016944$ means that as x_2 increases by one minute (holding x_1 constant), y decreases by roughly .017 cm^3/g. $R^2 = .807$. **(b)** The residuals are $-.015, .143, .492, -.595, -.457, -.188$, .695, .143, $-.218$. **(c)** For $x_2 = 30$, the equation is $\hat{y} = 5.53998 + .14167x_1$. For $x_2 = 60$, the equation is $\hat{y} = 5.03166 + .14167x_1$. For $x_2 = 90$, the equation is $\hat{y} = 4.52334 + .14167x_1$. The fitted responses do not match up well, because the relationship between y and x_1 is not linear for any of the x_2 values. **(d)** At $x_1 = 10\%$ and $x_2 = 70$ minutes, $\hat{y} = 6.279$ cm^3/g. It would not be wise to make a similar prediction at $x_1 = 10\%$ and $x_2 = 120$ minutes because there is no evidence that the fitted relationship is correct under these conditions. Some data should be obtained around $x_1 = 10\%$ and $x_2 = 120$ minutes. **(e)** $\hat{y} = 4.98 + .260x_1 + .00081x_2 - .00197x_1x_2$, and $R^2 = .876$. The increase in R^2 from .807 to .876 is not very large; using the more complicated equation may not be desirable (this is subjective). **(f)** For $x_2 = 30$, the equation is $\hat{y} = 5.0076 + .20084x_1$. For $x_2 = 60$, the equation is $\hat{y} = 5.0319 + .14168x_1$. For $x_2 = 90$, the equation is $\hat{y} = 5.0562 + .08252x_1$. The new model allows there to be a different slope for different values of x_2, so these lines fit the data better than the lines in part (c). But they still do not account for the nonlinearity between x_1 and y. An x_1^2 term should be added to the model. **(g)** There is no replication (multiple experimental runs at a particular NaOH/Time combination). **(h)** These data have a complete (full) factorial structure. The straight-line least squares equation for x_1 is

$\hat{y} = 5.0317 + .14167x_1$ with a corresponding R^2 of .594. The straight-line least squares equation for x_2 is $\hat{y} = 7.3233 - .01694x_2$ with a corresponding R^2 of .212. The slopes in these one-variable linear equations are the same as the corresponding slopes in the two variable equation from (a). The R^2 value in (a) is the sum of the R^2 values from the two one-variable linear equations.

Section 3

1. **(a)** Labeling x_1 as A and x_2 as B, $a_1 = -.643, a_2 = -.413, a_3 = 1.057, b_1 = .537, b_2 = -.057, b_3 = -.480,$ $ab_{11} = -.250,$ $ab_{12} = -.007,$ $ab_{13} = .257, ab_{21} = -.210, ab_{22} = .013, ab_{23} = .197,$ $ab_{31} = .460, ab_{32} = -.007, ab_{33} = -.453.$ The fitted interactions ab_{31} and ab_{33} are large (relative to fitted main effects) indicating that the effect on y of changing NaOH from 9% to 15% depends on the Time (non-parallelism in the plot). It would not be wise to use the fitted main effects alone to summarize the data, since there may be an importantly large interaction.   **(b)** $\hat{y}_{11} = 6.20,$ $\hat{y}_{12} = 5.61, \hat{y}_{13} = 5.18, \hat{y}_{21} = 6.43, \hat{y}_{22} = 5.84,$ $\hat{y}_{23} = 5.41, \hat{y}_{31} = 7.90, \hat{y}_{32} = 7.31, \hat{y}_{33} = 6.88.$ Like the plot in part (c) and unlike the plot in (f) of Exercise 2 in Section 4.2, the fitted values for each level of B (x_2) must produce parallel plots; no interactions are allowed. However, unlike parts (c) and (f) of that exercise, the current model allows these fitted values to be nonlinear in x_1 (factorial models are generally more flexible than lines, curves, and surfaces).   **(c)** $R^2 = .914.$ The plots of residuals versus Time and residuals versus $\hat{y}_i$ both have patterns; these show that the "main effects only" model is not accounting for the apparent interaction between the two factors. Even though R^2 is higher than both of the models in Exercise 2 of Section 4.2, this model does not seem to be adequate.

2. **(a)** $\bar{y}_{...} = 20.792,$ $a_2 = .113,$ $b_2 = -13.807,$ $ab_{22} = -.086, c_2 = 7.081, ac_{22} = -.090, bc_{22} = -6.101, abc_{222} = .118.$ Other fitted effects can be obtained by appropriately changing the signs of the above. The simplest possible interpretation is that Diameter, Fluid, and their interaction

are the only effects on Time.   **(b)** $\bar{y}_{...} = 2.699,$ $a_2 = .006, b_2 = -.766, ab_{22} = -.003, c_2 = .271,$ $ac_{22} = -.003, bc_{22} = -.130, abc_{222} = .007.$ Yes, but the Diameter $\times$ Fluid interaction still seems to be important.   **(c)** In standard order, the fitted values are 3.19, 3.19, 1.66, 1.66, 3.74, 3.74, 2.20, 2.20. $R^2 = .974.$ For a model with all factorial effects $(\widehat{\ln y_{ijk}} = \overline{\ln y_{ijk}})$, $R^2 = .995.$ **(d)** $b_1 - b_2 = 1.532 \ln(\text{sec})$ decrease; divide the .188 raw drain time by $e^{1.532}$ to get the .314 drain time. This suggests that (.188 drain time/.314 drain time) = $e^{1.532} = 4.63$; the theory predicts this ratio to be 7.78.

3. Interpolation, and possibly some cautious extrapolation, is only possible using surface-fitting methods. In many engineering situations, an "optimal" setting of quantitative factors is sought. This can be facilitated by interpolation (or extrapolation) using a surface-fitting model.

Section 4

1. Transforming data can sometimes make relationships among variables simpler. Sometimes nonlinear relationships can be made linear, or factors and response can be transformed so that there are no interactions among the factors. Transformations can also potentially make the shape of a distribution simpler, allowing the use of statistical models that assume a particular distributional shape (such as the bell-shaped normal distribution).

2. In terms of the raw response, there will be interactions, since x_1 and x_2 are multiplied together in the power law. The suggested plot of raw y versus x_1 will have different slopes for different values of x_2. This means that the effect of changing x_1 depends on the setting of x_2, which is one way to define an interaction.

   In terms of the log of y, there will not be interactions, since x_1 and x_2 appear additively in the equation for $\ln y$. Therefore, the suggested plot of $\ln y$ versus x_1 will have the same slope for all values of x_2. This means that the effect of changing x_1 does not depend on the setting of x_2 (there are no interactions).

Section 5

1. A deterministic model is used to describe a situation where the outcome can be almost exactly predicted if certain variables are known. A stochastic/probabilistic model is used in situations where it is not possible to predict the exact outcome. This may happen when important variables are unknown, or when no known deterministic theory can describe the situation. An example of a deterministic model is the classical Economic Order Quantity (EOQ) model for inventory control. Given constant rate of demand R, order quantity X, ordering cost P, and per unit holding cost C, the total cost per time period is $Y = P\left(\frac{R}{X}\right) + C\left(\frac{X}{2}\right)$.

Chapter 5

Section 1

1. **(b)** 4.1; 1.136.

2. **(a)** X has a binomial distribution with $n = 10$ and $p = \frac{1}{3}$. Use equation (5.3) with $n = 10$ and $p = \frac{1}{3}$. $f(0)–f(10)$ are .0173, .0867, .1951, .2601, .2276, .1366, .0569, .0163, .0030, .0003, .0000. **(b)** Assuming that they are just guessing, the chance that 7 (or more) out of 10 subjects would be correct is $P(X \geq 7) = .0197$. Under the hypothesis that they are only guessing, this kind of extreme outcome would only happen about 1 in 50 times, so the outcome is strong evidence that they are not just guessing.

3. **(a)** Using equations (3.4) and (3.5), $\mu = 4$, $\sigma^2 = \frac{5}{3}$, and $\sigma = 1.291$.
   (b)

x	2	3	4	5	6
$P(X = x)$	$\frac{1}{6}$	$\frac{1}{6}$	$\frac{2}{6}$	$\frac{1}{6}$	$\frac{1}{6}$

Since all members of the population are equally likely to be chosen, the probability histogram for X is the same as the population relative frequency distribution. Using equations (5.1) and (5.2), $EX = 4$ and $\text{Var}X = \frac{5}{3}$. **(c)** Label the values 2, 3, $4_1, 4_2, 5, 6$.

First Item	Second Item	$\bar{x}$	s^2	Probability
2	3	2.5	.5	$\frac{1}{15}$
2	4_1	3.0	2.0	$\frac{1}{15}$
2	4_2	3.0	2.0	$\frac{1}{15}$
2	5	3.5	4.5	$\frac{1}{15}$
2	6	4.0	8.0	$\frac{1}{15}$
3	4_1	3.5	.5	$\frac{1}{15}$
3	4_2	3.5	.5	$\frac{1}{15}$
3	5	4.0	2.0	$\frac{1}{15}$
3	6	4.5	4.5	$\frac{1}{15}$
4_1	4_2	4.0	0	$\frac{1}{15}$
4_1	5	4.5	.5	$\frac{1}{15}$
4_1	6	5.0	2.0	$\frac{1}{15}$
4_2	5	4.5	.5	$\frac{1}{15}$
4_2	6	5.0	2.0	$\frac{1}{15}$
5	6	5.5	.5	$\frac{1}{15}$

Using the above table, the probability distribution for $\overline{X}$ is:

$\bar{x}$	2.5	3	3.5	4	4.5	5	5.5
$P(\overline{X} = \bar{x})$	$\frac{1}{15}$	$\frac{2}{15}$	$\frac{3}{15}$	$\frac{3}{15}$	$\frac{3}{15}$	$\frac{2}{15}$	$\frac{1}{15}$

Using equations (5.1) and (5.2), $E\overline{X} = 4$ and $\text{Var}\overline{X} = \frac{2}{3}$. As might be expected, the mean of $\overline{X}$ is the same as the mean of X, and the variance is smaller. The probability distribution for S^2 is

s^2	0	.5	2	4.5	8
$P(S^2 = s^2)$	$\frac{1}{15}$	$\frac{6}{15}$	$\frac{5}{15}$	$\frac{2}{15}$	$\frac{1}{15}$

4. For $p = .1$, $f(0)–f(5)$ are .59, .33, .07, .01, .00, .00; $\mu = np = .5$; $\sigma = \sqrt{np(1 - p)} = .67$. For $p = .3$, $f(0)–f(5)$ are .17, .36, .31, .13, .03, .00; $\mu = 1.5$; $\sigma = 1.02$. For $p = .5$, $f(0)–f(5)$ are .03, .16, .31, .31, .16, .03; $\mu = 2.5$; $\sigma = 1.12$. For

$p = .7$, $f(0)-f(5)$ are .00, .03, .13, .31, .36, .17; $\mu = 3.5$; $\sigma = 1.02$. For $p = .9$, $f(0)-f(5)$ are .00, .00, .01, .07, .33, .59; $\mu = 4.5$; $\sigma = .67$.

5. Binomial distribution: $n = 8$, $p = .20$. **(a)** .147 **(b)** .797 **(c)** $np = 1.6$ **(d)** $np(1 - p) = 1.28$ **(e)** 1.13

6. Geometric distribution: $p = .20$. **(a)** .08 **(b)** .59 **(c)** $1/p = 5$ **(d)** $(1 - p)/p^2 = 20$ **(e)** 4.47

7. For $\lambda = .5$, $f(0)$, $f(1)$, ... are .61, .30, .08, .01, .00, .00, ...; $\mu = \lambda = .5$; $\sigma = \sqrt{\lambda} = .71$. For $\lambda = 1.0$, $f(0)$, $f(1)$, ... are .37, .37, .18, .06, .02, .00, .00, ...; $\mu = 1.0$; $\sigma = 1.0$. For $\lambda = 2.0$, $f(0)$, $f(1)$, ... are .14, .27, .27, .18, .09, .04, .01, .00, .00, ...; $\mu = 2.0$; $\sigma = 1.41$. For $\lambda = 4.0$, $f(0)$, $f(1)$, ... are .02, .07, .15, .20, .20, .16, .10, .06, .03, .01, .00, .00, ...; $\mu = 4.0$; $\sigma = 2.0$.

8. **(a)** .323 **(b)** .368

9. **(a)** .0067 **(b)** $Y \sim$ Binomial($n = 4$, $p = .0067$); .00027

10. Probability is a mathematical system used to describe random phenomena. It is based on a set of axioms, and all the theory is deduced from the axioms. Once a model is specified, probability provides a deductive process that enables predictions to be made based on the theoretical model.

    Statistics uses probability theory to describe the source of variation seen in data. Statistics tries to create realistic probability models that have (unknown) parameters with meaningful interpretations. Then, based on observed data, statistical methods try to estimate the unknown parameters as accurately and precisely as possible. This means that statistics is inductive, using data to draw conclusions about the process or population from which the data came.

    Neither is a subfield of the other. Just as engineering uses calculus and differential equations to model physical systems, statistics uses probability to model variation in data. In each case the mathematics can stand alone as theory, so calculus is not a subfield of engineering and probability is not a subfield of statistics. Conversely, statistics is not a subfield of probability just as engineering is not a subfield of calculus; many simple statistical methods do not require the use of probability, and many engineering techniques do not require calculus.

11. A relative frequency distribution is based on *data*. A probability distribution is based on a theoretical model for probabilities. Since probability can be interpreted as long-run relative frequency, a relative frequency distribution approximates the underlying probability distribution, with the approximation getting better as the amount of data increases.

Section 2

1. **(a)** 2/9 **(c)** .5

$$
\textbf{(d)} \ F(x) = \begin{cases} 0 & \text{for } x \le 0 \\ \frac{10x - x^2}{9} & \text{for } 0 < x < 1 \\ 1 & \text{for } x \ge 1 \end{cases}
$$

(e) 13/27; .288

2. **(a)** .2676 **(b)** .1446 **(c)** .3393 **(d)** .3616 **(e)** .3524 **(f)** .9974 **(g)** 1.28 **(h)** 1.645 **(i)** 2.17

3. **(a)** .7291 **(b)** .3594 **(c)** .2794 **(d)** .4246 **(e)** .6384 **(f)** 48.922 **(g)** 44.872 **(h)** 7.056

4. **(a)** .4938 **(b)** Set μ to the midpoint of the specifications: $\mu = 2.0000$; .7888 **(c)** .0002551

5. **(a)** $P(X < 500) = .3934$; $P(X > 2000) = .1353$ **(b)** $Q(.05) = 51.29$; $Q(.90) = 2,302.58$

6. **(b)** $Median = 68.21 \times 10^6$ **(c)** $Q(.05) = 21.99 \times 10^6$; $Q(.95) = 128.9 \times 10^6$

Section 3

1. Data that are being generated from a particular distribution will have roughly the same shape as the density of the distribution, and this is more true for larger samples. Probability plotting provides a sensitive graphical way of deciding if the data have the same shape as a theoretical probability

distribution. If a distribution can be found that accurately describes the data generating process, one can then estimate probabilities and quantiles and make predictions about future process behavior based on the model.

2. Fit a line (by eye or some other method) through the points on the plot. The x-intercept is an approximate mean, and an approximate standard deviation is $\sigma \approx \dfrac{1}{\text{slope}} = \dfrac{\Delta x}{\Delta y} = \dfrac{\Delta \text{ data quantiles}}{\Delta \text{ std. normal quantiles}}$.

3. (b) $\mu \approx 69.5$; $\sigma \approx 1/\text{slope} = 2.1$

4. (a) First 3 coordinates of the normal plot of the raw data: $(17.88, -2.05)$, $(28.92, -1.48)$, $(33.00, -1.23)$. The normal plot is not linear, so a Gaussian (normal) distribution does not seem to fit these data. First 3 coordinates of the normal plot of the natural log of the data: $(2.884, -2.05)$, $(3.365, -1.48)$, $(3.497, -1.23)$. This normal plot is fairly linear, indicating that a lognormal distribution fits the data well. $\mu \approx 4.1504$, $\sigma \approx .5334$. 3.273; 26.391.   (b) The first 3 coordinates of the Weibull plot are $(2.88, -3.82)$, $(3.36, -2.70)$, $(3.50, -2.16)$. The Weibull plot is fairly linear, indicating that a Weibull distribution might be used to describe bearing load life. $\alpha \approx 81.12$, $\beta \approx 2.3$; 22.31.

5. (b) The exponential plot is fairly linear, indicating that an exponential distribution fits the data well. Since a line on the plot indicates that $Q(0) \approx 0$, no need for a threshold parameter greater than zero is indicated.

Section 4

1. If X and Y are independent, then observing the actual value of X does not in any way change probability assessments about the yet-to-be-observed Y, or vice-versa. Independence provides great mathematical simplicity in the description of the behavior of X and Y.

2. (a) For $x = 0, 1, 2$, $f_X(x) = .5, .4, .1$. For $y = 0, 1, 2, 3, 4$, $f_Y(y) = .21, .19, .26, .21, .13$.
(b) No, since $f(x, y) \neq f_X(x) f_Y(y)$.   (c) .6; .44
(d) 1.86; 1.74   (e) For $y = 0, 1, 2, 3, 4$, $f_{Y|X}(y \mid 0) = .3, .2, .2, .2, .1$; 1.6.

3. (a) For $y = 1, 2, 3, 4$, $f_{Y|X}(y \mid 0) = 0, 0, 0, 1$ and $f_{Y|X}(y \mid 1) = .25, .25, .25, .25$. $f(0, 1) = f(0, 2) = f(0, 3) = 0$, $f(0, 4) = p$, $f(1, 1) = f(1, 2) = f(1, 3) = f(1, 4) = .25(1 - p)$.
(b) $2.5 + 1.5p$ (c) $p > .143$

4. (a) Since X and Y are independent, $f(x, y) = f_X(x) f_Y(y)$ (Definition 27),

$$f(x, y) = \begin{cases} \frac{1}{.05}\frac{1}{.06} & \text{for } x \in (1.97, 2.02) \text{ and} \\ & y \in (2.00, 2.06) \\ 0 & \text{otherwise} \end{cases}$$

$$= \begin{cases} 333.33 & \text{for } x \in (1.97, 2.02) \text{ and} \\ & y \in (2.00, 2.06) \\ 0 & \text{otherwise} \end{cases}$$

(b) Find the volume below this density over the region in which $2.00 < y < x$ and $1.97 < x < 2.02$. This is .0667. (Using calculus, this is

$$\int_{2.00}^{2.02} \int_{2.00}^{x} 333.33 \, dy \, dx.)$$

5. (a)

$$f_X(x) = \begin{cases} 2x & \text{for } 0 \le x \le 1 \\ 0 & \text{otherwise} \end{cases} ;$$

$$f_Y(y) = \begin{cases} 2(1 - y) & \text{for } 0 \le y \le 1 \\ 0 & \text{otherwise} \end{cases} ;$$

$\mu = EX = 2/3$.
(b) Yes, since $f(x, y) = f_X(x) f_Y(y)$.   (c) .7083
(d) $E(X|Y = .5) = 2/3$

6. (a)

$$f(x, y) =$$

$$f_X(x) f_Y(y) = \begin{cases} e^{-x} e^{-y} & \text{if } x \ge 0 \text{ and } y \ge 0 \\ 0 & \text{otherwise} \end{cases}$$

(b) e^{-2t} (c) $f_T(t) = \begin{cases} 2e^{-2t} & \text{for } t \ge 0 \\ 0 & \text{otherwise} \end{cases}$.

This is an exponential distribution with mean .5.
(d) $(1 - e^{-t})^2$.

(e) $f_T(t) = \begin{cases} 2e^{-t}(1-e^{-t}) & \text{for } t \geq 0; \\ 0 & \text{otherwise} \end{cases}$.

$E(T) = 1.5$

Section 5

1. mean $= .75$ in.; standard deviation $= .0037$.

2. **(a)** Propagation of error formula gives 1.4159×10^{-6}.   **(b)** The lengths.

3. **(a)** $13/27$; $.0576$ **(b)** $\overline{X} \sim$ Normal with mean $13/27$ and standard deviation $.0576$.   **(c)** $.3745$ **(d)** $.2736$ **(e)** $13/27$, $.0288$; $\overline{X} \sim$ Normal with mean $13/27$ and standard deviation $.0288$; $.2611$; $.5098$.

4. $.7888$, $.9876$, 1.0000

5. Rearrange the relationship in terms of g to get $g = \frac{4\pi^2 L}{\tau^2}$. Take the given length and period to be approximately equal to the means of these input random variables. To use the propagation of error formula, the partial derivatives need to be evaluated at the means of the input random variables and $\frac{\partial g}{\partial L} = \frac{4\pi^2}{\tau^2} = 6.418837$ and $\frac{\partial g}{\partial \tau} = \frac{-8\pi^2 L}{\tau^3} = -25.8824089$. Then applying equation (5.59), $\text{Var}(g) \approx (6.418837)^2(.0208)^2 + (-25.8824089)^2 \times (.1)^2 = 6.7168$ ft^2/sec^4 so the approximate standard deviation of g is $\sqrt{6.7168} = 2.592$ ft/sec^2. The precision in the period measurement is the principal limitation on the precision of the derived g because its term (variance $\times$ squared partial derivative) contributes much more to the propagation of error formula than the length's term.

Chapter 6

Section 1

1. $[6.3, 7.9]$ ppm is a set of plausible values for the mean. The method used to construct this interval correctly contains the true mean in 95% of repeated applications. This particular interval either contains the mean or it doesn't (there is no probability involved). However, because the *method* is correct 95% of the time, we might say that we have 95% confidence that it was correct this time.

2. **(a)** $[111.0, 174.4]$ **(b)** $[105.0, 180.4]$ **(c)** 167.4 **(d)** 174.4 **(e)** $[111.0, 174.4]$ ppm is a set of plausible values for the mean aluminum content of samples of recycled PET plastic from the recycling pilot plant at Rutgers University. The method used to construct this interval correctly contains means in 90% of repeated applications. This particular interval either contains the mean or it doesn't (there is no probability involved). However, because the *method* is correct 90% of the time, we might say that we have 90% confidence that it was correct this time.

3. $n = 66$

4. **(a)** $\bar{x} = 4.6858$ and $s = .02900317$ **(b)** $= [4.676, 4.695]$ mm   **(c)** $[4.675, 4.696]$ mm. This interval is wider than the one in (b). To increase the confidence that μ is in the interval, you need to make the interval wider.   **(d)** The lower bound is 4.677 mm. This is larger than the lower endpoint of the interval in (b). Since the upper endpoint here is set to ∞, the lower endpoint must be increased to keep the confidence level the same.   **(e)** To make a 99% one-sided interval, construct a 98% two-sided interval and use the lower endpoint. This was done in part (a), and the resulting lower bound is 4.676. This is smaller than the value in (d); to increase the confidence, the interval must be made "wider."   **(f)** $[4.676, 4.695]$ ppm is a set of plausible values for the mean diameter of this type of screw as measured by this student with these calipers. The method used to construct this interval correctly contains means in 98% of repeated applications. This particular interval either contains the mean or it doesn't (there is no probability involved). However, because the *method* is correct 98% of the time, we might say that we have 98% confidence that it was correct this time.

Section 2

1. $H_0: \mu = 200$; $H_a: \mu > 200$; $z = -2.98$; p-value $\doteq .9986$. There is no evidence that the mean aluminum content for samples of recycled plastic is greater than 200 ppm.

2. (a) $H_0: \mu = .500$; $H_a: \mu \neq .500$; $z = 1.55$; p-value $\doteq .1212$. There is some (weak) evidence that the mean punch height is not .500 in. (The rounded $\bar{x}$ and s given produce a z that is quite a bit different from what the exact values produce. $\bar{x} = .005002395$ and $s = .002604151$, computed from the raw data, produce $z = 1.85$, and a p-value of $2(.0322) = .0644$.) **(b)** $[.49990, .50050]$ **(c)** If uniformity of stamps on the same piece of material is important, then the standard deviation (spread) of the distribution of punch heights will be important (in addition to the mean).

3. The mean of the punch heights is almost certainly not exactly equal to .50000000 inches. Given enough data, a hypothesis test would detect this as a "statistically significant" difference (and produce a small p-value). What is practically important is whether the mean is "close enough" to .500 inches. The confidence interval in part (b) answers this more practical question.

4. $H_0: \mu = 4.70$; $H_a: \mu \neq 4.70$; $z = -3.46$; p-value $\doteq .0006$. There is very strong evidence that the mean measured diameter differs from nominal.

5. Although there is evidence that the mean is not equal to nominal, the test does not say anything about how *far* the mean is from nominal. It may be "significantly" different from nominal, but the difference may be practically unimportant. A confidence interval is what is needed for determining how far the mean is from nominal.

Section 3

1. The normal distribution is bell-shaped and symmetric, with fairly "short" tails. The confidence interval methods depend on this regularity. If the distribution is skewed or prone to outliers/extreme observations, the normal-theory methods will not properly take this into account. The result is an interval whose real confidence level is different from the nominal value (and often lower than the nominal value).

2. (a) Independence among assemblies; normal distribution for top-bolt torques. **(b)** $H_0: \mu = 100$; $H_a: \mu \neq 100$; $t = 4.4$; p-value $\doteq .001$. There is

strong evidence that the mean torque is not 100 ft lb. **(c)** $[104.45, 117.55]$ **(d)** Independence among assemblies; normal distribution for differences. **(e)** $H_0: \mu_d = 0$; $H_a: \mu_d < 0$ (where differences are Top – Bottom); $t = -2.10$ on 14 df; $.025 < p$-value $< .05$. **(f)** $[-13.49, 1.49]$

3. (a) $[-0.0023, .0031]$ mm **(b)** $H_0: \mu_d = 0$; $H_a: \mu_d \neq 0$; $z = .24$; p-value $= .8104$. There is no evidence of a systematic difference between calipers. **(c)** The confidence interval in part (a) contains zero; in fact, zero is near the middle of the interval. This means that zero is a very plausible value for the mean difference—there is no evidence that the mean is not equal to zero. This is reflected by the large p-value in part (b).

4. (a) The data within each sample must be iid normal, and the two distributions must have the same variance σ^2. One way to check these assumptions is to normal plot both data sets on the same axes. For such small sample sizes, it is difficult to definitively verify the assumptions. But the plots are roughly linear with no outliers, indicating that the normal part of the assumption may be reasonable. The slopes are similar, indicating that the common variance assumption may be reasonable. **(b)** Label the Treaded data Sample 1 and the Smooth data Sample 2. $H_0: \mu_1 - \mu_2 = 0$; $H_a: \mu_1 - \mu_2 \neq 0$; $t = 2.49$; p-value is between .02 and .05. This is strong evidence of a difference in mean skid lengths. **(c)** $[2.65, 47.35]$ **(d)** $[2.3, 47.7]$

Section 4

1. (a) $[9.60, 37.73]$ **(b)** 57.58 **(c)** $H_0: \frac{\sigma_T^2}{\sigma_S^2} = 1$

$H_a: \frac{\sigma_T^2}{\sigma_S^2} \neq 1$; $f = .64$ on 5,5 df; p-value $> .50$

(d) $[.36, 1.80]$

2. (a) $[7.437, \infty)$ **(b)** $[44.662, \infty)$ **(c)** Top and bottom bolt torques for a given piece are probably not sensibly modeled as independent.

Section 5

1. (a) Conservative method: $[.562, .758]$; .578. Other method: $[.567, .753]$; .582. **(b)** $H_0: p = .55$; H_a:

$p > .55$; $z = 2.21$; p-value $= .0136$. **(c)** Conservative method: $[-.009, .269]$. Other method: $[-.005, .265]$. **(d)** $H_0: p_S - p_L = 0$; $H_a: p_S - p_L \neq 0$; $z = 1.87$; p-value $= .0614$.

2. 9604

3. Conservative method: $[.22, .35]$. Other method: $[.23, .34]$.

4. $H_0: p_1 - p_2 = 0$; $H_a: p_1 - p_2 \neq 0$; $z = -.97$; p-value $= .3320$.

Section 6

1. A consumer about to purchase a single auto would be most interested in a prediction bound, because the single auto that the consumer will purchase is likely to have mileage above the bound. This is not true for a confidence bound for the mean. That may be more useful for the EPA official, since this person wants to be sure that the manufacturer is producing cars that exceed some minimum average mileage. The design engineer would be most interested in a lower tolerance bound for most mileages, to be sure that a high percentage of the cars produced are able to cruise for at least 350 miles. A confidence for the mean or prediction bound does not answer this question.

2. **(a)** $[132.543, 297.656]$ **(b)** $[92.455, 337.745]$ **(c)** The tolerance interval is much wider than the prediction interval. The interval in (b) is meant to bracket 90% of all observations, while the the one from (a) is meant only to bracket a single additional observation.   **(d)** The confidence interval for mean lifetime is smaller than both the prediction interval and the tolerance interval. It is meant only to bracket the mean/center of the population, not additional observation(s).   **(e)** $[152.811, \infty)$ **(f)** $[113.969, \infty)$

3. **(a)** $[3.42, 6.38]$; $[30.6, 589.1]$ **(b)** $[3.87, 5.93]$; $[48.1, 375.0]$ **(c)** The intervals in (a) are wider than those in (b). This is usually true when applying tolerance intervals and prediction intervals in the same situation.

4. 92.6%; 74.9%

Chapter 7

Section 1

1. **(a)** The plot reveals two outliers. The assumptions of the one-way normal model appear to be less than perfectly met in this problem. (Both of the outliers come from the 8,000 psi condition. This is an indication that the common σ part of the one-way normal model may be less than perfect.)   **(b)** .02057. This measures the magnitude of baseline variation in any of the five treatments, assuming it is the same for all five treatments; $[.01521, .03277]$.

2. **(a)** The plot reveals one outlier/unusual residual (the 1.010 value from Van #1 produces the residual $-.0094$). One should proceed under the one-way model assumptions only with caution.   **(b)** The standardized residuals tell the same story told in part (a).   **(c)** $s_p = .0036$ measures the (supposedly common) variation in tilt angle for repeated measurement of a particular van; $[.0026, .0058]$.

Section 2

1. **(a)** .02646; 75%   **(b)** .03742   **(c)** $[-.0724, .0572]$ provides no convincing evidence of non-linearity over the range from 2,000 to 6,000, as it includes 0.

2. **(a)** The intervals in numerical order of the four vans are: $[1.0875, 1.0984]$, $[.9608, .9716]$, $[1.0145, 1.0242]$, $[.9968, 1.0076]$; at least 96% simultaneous confidence.   **(b)** $\Delta = .0077$; $\Delta = .0073$ **(c)** $[.013516, .02408]$

3. Before the data are collected, the probability is .05 that an individual 95% confidence interval will be in error—that it will not contain the quantity that it is supposed to contain. If several of these individual intervals are made, then the probability that *at least* one of the intervals is in error is greater than .05. (If each interval has a .05 chance of failing, then the overall chance of at least one failure is greater than .05.) When making several intervals, most people would like the overall or simultaneous error probability to be small. In order to make sure, for example, that the overall error probability is .05, the error probability associated with the individual intervals must be made smaller than .05. This is equivalent to increasing the individual confidences (above 95%), which makes the intervals wider.

Section 3

1. (a) .03682; it is larger.   (b) .05522; it is larger.

2. (a) $k_2^* = 2.88$ so the intervals in numerical order of the four vans are: [1.0878, 1.0982], [.9610, 9714], [1.0147, 1.0240], [.9970, 1.0074].   (b) $\Delta = .0097$; $\Delta = .0092$. These are larger than the earlier Δ's. The confidence level here is a simultaneous one while the earlier level was an individual one. The intervals here are doing a more ambitious job and must therefore be wider.

Section 4

1. (a) Small, since some means differ by more than the Δ there.   (b) $SSTr = .285135$, $MSTr = .071284$, df $= 4$; $SSE = .00423$, $MSE = .000423$, df $= 10$; $SSTot = .289365$, df $= 14$; $f = 168.52$ on 4,10 df; p-value $< .001$. $R^2 = .985$.

2. (a) Small, since some sample means differ by more than the Δ's there.   (b) $SSTr = .034134$, $MSTr = .011378$, df $= 3$; $SSE = .000175$, $MSE = .000013$, df $= 13$; $SSTot = .034308$, df $= 16$; $f = 847$ on 3,13 df; p-value $< .001$.

3. (a) To check that the μ_i's are normal, make a normal plot of the $\bar{y}_i$'s. To check that the ϵ_i's are normal, make a normal plot of the residuals. (Normal plotting each sample individually will not be very helpful because the sample sizes are so small.) Both plots are roughly linear, giving no evidence that the one-way random effects model assumptions are unreasonable.   (b) $SSTr = 9310.5$, $MSTr = 1862.1$, df $= 5$; $SSE = 194.0$, $MSE = 16.2$, df $= 12$; $SSTot = 9504.5$, df $= 17$; $f = 115.18$ on 5,12 df; p-value $< .001$. $\hat{\sigma} = 4.025$ measures variation in y from repeated measurements of the same rail; $\hat{\sigma}_\tau = 24.805$ measures the variation in y from differences among rails.   (c) [3.46, 13.38]

4. (a) Unstructured multisample data could also be thought of as data from one factor with r levels. In many situations, the specific levels of the factor included in the study are the levels of interest. For example, in comparing three drugs, the factor might be called "Treatment." It might have four levels: Drug 1, Drug 2, Drug 3, and Control. The experimenter is interested in comparing the specific drugs used in the study to each other and to the control. Sometimes the specific levels of the factor are not of interest in and of themselves, but only because they may represent (perhaps they are a random sample of) many different possible levels that could have been used in the study. A random effects analysis is appropriate in this situation. For an example, see part (b).   (b) If there are many technicians, and five of these were randomly chosen to be in the study, then interest is in the variation among all technicians, not just the five chosen for the study.   (c) $\hat{\sigma} = .00155$ in.; $\hat{\sigma}_\tau = .00071$ in.

Section 5

1. (a) Center line$_{\bar{x}} = 21.0$, $UCL_{\bar{x}} = 22.73$, $LCL_{\bar{x}} = 19.27$. Center line$_R = 1.693$, $UCL_R = 4.358$, no LCL_R.   (b) Center line$_s = .8862$, $UCL_s = 2.276$, no LCL_s.   (c) 1.3585; 1.3654; $s_p = 1.32$.   (d) Center line$_{\bar{x}} = 21.26$, $UCL_{\bar{x}} = 23.61$, $LCL_{\bar{x}} = 18.91$. Center line$_R = 2.3$, $UCL_R = 5.9202$, no LCL_R.   (e) Center line$_{\bar{x}} = 21.26$, $UCL_{\bar{x}} = 23.62$, $LCL_{\bar{x}} = 18.90$. Center line$_s = 1.21$, $UCL_s = 3.10728$, no LCL_s.

2. (a) $\dfrac{\bar{R}}{d_2} = \dfrac{4.052632}{2.326} = 1.742318 \times .001$ in.; $\dfrac{\bar{s}}{c_4} = \dfrac{1.732632}{.9400} = 1.843226 \times .001$ in. (b) For the R chart Center Line$_R = 2.326(1.843226) = 4.287344 \times .001$ in., $UCL_R = 4.918(1.843226) = 9.064985 \times .001$ in. and there is no lower control limit. For the s chart Center Line$_s = 1.732632 \times .001$ in. $UCL_s = 2.089(1.732632) = 3.619468 \times .001$ in. and there is again no lower control limit. Neither chart indicates that the short-term variability of the process (as measured by σ) was unstable.   (c) Use Center Line$_{\bar{x}} = 11.17895 \times .001$ in. above nominal, $LCL_{\bar{x}} = 11.17895 - 3\dfrac{1.843226}{\sqrt{5}} = 8.706 \times .001$

in. above nominal and $UCL_{\bar{x}} = 11.17895 + 3\frac{1.843226}{\sqrt{5}} = 13.65189 \times .001$ in. above nominal. $\bar{x}$ from sample 16 comes close to the upper control limit, but overall the process mean seems to have been stable over the time period. **(d)** The $\bar{x}$'s from samples 9 and 16 seem to have "jumped" from the previous $\bar{x}$. The coil change may be causing this jump, but it could also be explained by common cause variation. It may be something worth investigating. **(e)** Assuming that the mean could be adjusted (down), you need to look at one of the estimates of σ to answer this question about individual thread lengths. (You should not use control limits to answer this question!) If μ could be made equal to zero, then (assuming normally distributed thread lengths), almost all of the thread lengths would fall in the interval $\pm 3\sigma$. Using the estimate of σ based on $\bar{s}$ from part (a), this can be approximated by $3(1.843226) = 5.53 \times .001$ in. It does seem that the equipment is capable of producing thread lengths within .01 in. of nominal. If the equipment were not capable of meeting the given requirements, the company could invest in better equipment. This would "permanently" solve the problem, but it might not be feasible from a financial standpoint. A second option is to inspect the bolts and remove the ones that are not within .01 in. of nominal. This might be cheaper than investing in new equipment, but it will do nothing to improve the quality of the process in the long run. A third option is to study the process (through experimentation) to see if there might be some way of reducing the variability without making a large capital investment.

3. Control charting is used to monitor a process and detect changes (lack of stability) in a process. The focus is on detecting changes in a meaningful parameter such as μ, σ, p, or λ. Points that plot out of control are a signal that the process is not stable at the standard parameter value (for a standards given chart) or was not stable at any parameter value (for a retrospective chart). The overall goal is to reduce process variability by identifying assignable causes and taking action to eliminate them. Reducing variability increases the quality of the process output.

4. Shewhart control charts do not physically control a process in the sense of guiding or adjusting it. They only monitor the process, trying to detect process instability. There is an entirely different field dedicated to "engineering control"; this field uses feedback techniques that manipulate process variables to guide some response. Shewhart control charts simply monitor a response, and are not intended to be used to make "real time" adjustments.

5. Out-of-control points should be investigated. If the causes of such points can be determined and eliminated, this will reduce long-term variation from the process. There must be an active effort among those involved with the process to improve the quality; otherwise, control charts will do nothing to improve the process.

6. Control limits for an $\bar{x}$ chart are set so that, under the assumption that the process is stable, it would be very unusual for an $\bar{x}$ to plot outside the control limits. The chart recognizes that there will be *some* variation in the $\bar{x}$'s even if the process is stable, and prevents overadjustment by allowing the $\bar{x}$'s to vary "randomly" within the control limits. If the process mean or standard deviation changes, $\bar{x}$'s will be more likely to plot outside of the control limits, and sooner or later the alarm will sound. This provides an opportunity to investigate the cause of the change, and hopefully take steps to prevent it from happening again. In the long run, such troubleshooting may improve the process by making it less variable.

Section 6

1. **(a)** Center line$_{\hat{p}} = .02$, $UCL_{\hat{p}} = .0438$, no $LCL_{\hat{p}}$.
   (b) Center line$_{\hat{p}} = .0234$, $UCL_{\hat{p}} = .0491$, no $LCL_{\hat{p}}$.

2. Center line$_{\hat{u}_i} = .138$ for all i, $UCL_{\hat{u}_i} = .138 + 3\sqrt{\frac{.138}{k_i}}$, no $LCL_{\hat{u}_i}$ for all i.

3. (a) Center line$_{\hat{u}_i} = .714$ for all i, $UCL_{\hat{u}_i} = .714 +$ $3\sqrt{\frac{.714}{k_i}}$, no $LCL_{\hat{u}_i}$ for all i. The process seems to be stable.   **(b)** (i) if $k_i = 1$, .0078; if $k_i = 2$, .0033. (ii) if $k_i = 1$, .0959; if $k_i = 2$, .1133.

4. $\hat{p} = \frac{18}{250} = .072$, so Center Line$_{\hat{p}_i} = .072$. The control limits depend on the sample size n_i. For $n_i = 20$, $.072 - 3\sqrt{\frac{.072(1-.072)}{20}} = -.101399 < 0$, so there is no lower control limit, while $UCL_{\hat{p}_i} = .072 + 3\sqrt{\frac{.072(1-.072)}{20}} = .245399$. For $n_i = 30$, $.072 - 3\sqrt{\frac{.072(1-.072)}{30}} = -.06957966 < 0$, so there is no lower control limit, while $UCL_{\hat{p}_i} = .072 + 3\sqrt{\frac{.072(1-.072)}{30}} = .2135797$. For $n_i = 40$, $.072 - 3\sqrt{\frac{.072(1-.072)}{40}} = -.05061158 < 0$, so there is no lower control limit, while $UCL_{\hat{p}_i} = .072 + 3\sqrt{\frac{.072(1-.072)}{40}} = .1946116$. There is no evidence that the process fraction nonconforming was unstable (changing) over the time period studied.

5. If different data collectors have different ideas of exactly what a "nonconformance" is, then the data collected will not be consistent. A stable process may look unstable (according to the c chart) because of these inconsistencies.

6. It may indicate that the chart was not applied properly. For example, if hourly samples of size $m = 4$ are collected, it may or may not be reasonable to use a retrospective $\bar{x}$ chart with $m = 4$. If the 4 items sampled are from 4 different machines, 3 of which are stable at some mean and the 4th stable at a different mean, then the sample ranges and standard deviations will be inflated. This will make the control limits on the $\bar{x}$ chart too wide. Also, the $\bar{x}$'s will show very little variation about a center line somewhere between the two means. This is all a result of the fact that each sample is really coming from four different processes. Four different control charts should be used.

Chapter 8

Section 1

1. (a) Error bars: $\bar{y}_{ij} \pm 23.54$.   **(b)** $a_1 = 21.78$, $a_2 = -21.78$, $b_1 = -41.61$, $b_2 = 16.06$, $b_3 = 25.56$, $ab_{11} = -1.94$, $ab_{12} = 1.39$, $ab_{13} = .56$, $ab_{21} = 1.94$, $ab_{22} = -1.39$, $ab_{23} = -.56$. Interactions: $ab_{ij} \pm 9.52$. A main effects: $a_i \pm 6.73$. B main effects: $b_j \pm 9.52$. Interactions are not detectable, but main effects for both A and B are.   **(c)** $\bar{y}_{.j} - \bar{y}_{.j'} \pm 20.18$

2. (a) $s_p = 33.25$ measures baseline variation in y for each factor-level combination, assuming it is the same for all factor-level combinations.   **(b)** Error bars: $\bar{y}_{ij} \pm 27.36$.   **(d)** $a_1 = -2.77$, $a_2 = -17.4$, $a_3 = 20.17$, $b_1 = -13.33$, $b_2 = -1.20$, $b_3 = 14.53$, $ab_{11} = .033$, $ab_{12} = -5.40$, $ab_{13} = 5.37$, $ab_{21} = -2.13$, $ab_{22} = -.567$, $ab_{23} = 2.70$, $ab_{31} = 2.104$, $ab_{32} = 5.97$, $ab_{33} = -8.07$.   **(e)** 18.24. No.   **(f)** Use $(a_i - a_i') \pm 22.35$.   **(g)** Use $(a_i - a_i') \pm 26.88$.

Section 2

1. (a) $\hat{E} \pm .014$. B and C main effects, BC interaction.   **(b)** $s_{FE} = .0314$ with 20 df; close to $s_p = .0329$.   **(c)** Using few effects model: [3.037, 3.091]. Using general method: [3.005, 3.085].

2. (a) Only the main effect for A plots "off the line."   **(b)** Since the D main effect is almost as big (in absolute value) as the main effect for A, you might choose to include it. For this model, the fitted values are (in standard order): 16.375, 39.375, 16.375, 39.375, 16.375, 39.375, 16.375, 39.375, −4.125, 18.875, −4.125, 18.875, −4.125, 18.875, −4.125, 18.875.   **(c)** Set A low (unglazed) and D high (no clean). [0, 9.09].

3. (a) $\bar{y}_{....} = 3.594$, $a_2 = -.806$, $b_2 = .156$, $ab_{22} = -.219$, $c_2 = -.056$, $ac_{22} = -.031$, $bc_{22} = .081$, $abc_{222} = .031$, $d_2 = -.056$, $ad_{22} = -.156$, $bd_{22} = .006$, $abd_{222} = -.119$, $cd_{22} = -.031$, $acd_{222} = -.056$, $bcd_{222} = -.044$, $abcd_{2222} = .006$.   **(b)** It appears that only the main effect for A is detectably larger than the rest of the effects, since the point for a_2 is far away from the rest of the fitted effects.   **(c)** To minimize y, use A(+) (monks cloth) and B(+) (treatment Y).

Section 3

1. Since A ↔ BCDE, if both are large but opposite in sign, their estimated sum will be small.

2. (a) 8.23, .369, .256, −.056, .344, −.069, −.081, −.093, −.406, .181, .269, −.344, −.094, −.156, −.069, .019.   **(b)** .312. The sums $\alpha_2 + \beta\gamma\delta\epsilon_{2222}$, $\gamma_2 + \alpha\beta\delta\epsilon_{2222}$, $\delta_2 + \alpha\beta\gamma\epsilon_{2222}$, and $\alpha\beta\delta_{222} + \gamma\epsilon_{22}$ are detectable. Simplest explanation: A, C, D main effects and CE interaction are responsible for these large sums.   **(c)** A (+), C (+), D (−), and E (−). The abc combination, which did have the largest observed bond strength.

3. (b) (1), ad, bd, ab, cd, ac, bc, abcd. Estimated sums of effects: 3.600, −.850, .100, −.250, −.175, −.025, −, 075, −.025.   **(c)** The estimate of $\alpha_2 + \beta\gamma\delta_{222}$ plots off the line. Still, one might conclude that this is due to the main effect for A, but the conclusion here would be a little more tentative.

Section 4

1. The advantage of fractional factorial experiments is that the same number of factors can be studied using less experimental runs. This is important when there are a large number of factors, and/or experimental runs are expensive. The disadvantage is that there will be ambiguity in the results; only sums of effects can be estimated. The advantage of using a complete factorial experiment is that all means can be estimated, so all effects can be estimated.

2. It will be impossible to separate main effects from two-factor interactions. You would hope that any interactions are small compared to main effects; the results of the experiment can then be (tentatively) summarized in terms of main effects. (If all interactions are really zero, then it is possible to estimate all of the main effects.) Looking at Table 8.35, the best possible resolution is 3 (at most).

3. Those effects (or sums of effects) that are nearly zero will have corresponding estimates that are "randomly" scattered about zero. If all of the effects are nearly zero, then one might expect the

estimates from the Yates algorithm (excluding the one that includes the grand mean) to be bell-shaped around zero. A normal plot of these estimates would then be roughly linear. However, if there are effects (or sums of effects) that are relatively far from zero, the corresponding estimates will plot away from the rest (off the line), and may be considered more than just random noise. The principle of "sparsity of effects" says that in most situations, only a few of the many effects in a factorial experiment are dominant, and their estimates will then plot off the line on a normal plot.

4. (a) I ↔ ABCDF ↔ ABCEG ↔ DEFG **(b)** ABDF, ABEG, CDEFG   **(c)** +, +; −, −   **(d)** That only A, F, and their interaction are important in describing y.

5. 3.264

6. (a) I ↔ ABCE ↔ BCDE ↔ ADEF   **(b)** −, −; +, −   **(c)** .489

Chapter 9

Section 1

1. (a) $s_{LF} = 67.01$ measures the baseline variation in Average Molecular Weight for any particular Pot Temperature, assuming this variation is the same for all Pot Temperatures.   **(b)** Standardized residuals: 2.0131, −.3719, −.9998, −1.562, .2715, .2394, .7450, .0004   **(c)** [22.08, 24.91]   **(d)** [1761, 1853], [2630, 2770]   **(e)** [1745, 1869], [2605, 2795]   **(f)** 1705; 2590   **(g)** 1627; 2503   **(h)** $SSR = 4{,}676{,}798$, $MSR = 4{,}676{,}798$, df = 1; $SSE = 26{,}941$, $MSE = 4490$, df = 6; $SSTot = 4{,}703{,}739$, df = 7; $f = 1041.58$ on 1,6 df; p-value $< .001$

2. (a) $b_0 = 4345.9$, $b_1 = -3160.0$, $s_{LF} = 26.76$ (close to $s_p = 26.89$)   **(b)** Standardized residuals: 1.32, −.48, −.04, −.91, .52, −1.07, 1.94, −.04, −1.09.   **(c)** [−357.4, −274.64]   **(d)** $t = -14.47$ on 7 df, p-value $< .001$; or $f = 209.24$ on 1,7 df, p-value $< .001$.   **(e)** [2744.8, 2787.0]   **(f)** [2699.2, 2832.6]   **(g)** 2698.5

Section 2

1. (a) $s_{SF} = .04677$ measures variation in Elapsed Time for any particular Jetting Size, assuming this variation is the same for all Jetting Sizes. (b) Standardized residuals: $-.181, .649, -.794, -.747, 1.55, -1.26$. (c) [81.32, 126.66]; [$-3.17, -1.89$]; [.01344, .02245] (d) [14.462, 14.596]; [14.945, 15.145] (e) [14.415, 14.644]; [14.875, 15.215] (f) 14.440; 14.942 (g) 14.323; 14.816 (h) $SSR = .20639$, $MSR = .01319$, df $= 2$; $SSE = .00656$, $MSE = .00219$, df $= 3$; $SSTot = .21295$, df $= 5$; $f = 42.17$ on 2,3 df; p-value $= .005$. H_0 means that Elapsed Time is not related to Jetting Size. (i) $t = 9.38$; p-value $= .003$. $H_0: y \approx \beta_0 + \beta_1 x + 0$; i.e., Elapsed Time is related to Jetting Size only linearly (no curvature).

2. (a) $s_{SF} = .4851$ measures baseline variation in y for any (x_1, x_2) combination, assuming this variation is the same for all (x_1, x_2) combinations. (b) Standardized residuals: $-.041, .348, 1.36, -1.44, -1.00, -.457, 1.92, .348, -.604$. (c) [5.036, 7.060]; [.0775, .2058]; [$-.0298, -.0041$] (d) [5.992, 6.622]; [5.933, 6.625] (e) [5.798, 6.816]; [5.720, 6.838] (f) 5.571; 5.535 (g) 5.017; 4.970 (h) $SSR = 5.8854$, $MSR = 2.9427$, df $= 2$; $SSE = 1.4118$, $MSE = .2353$, df $= 6$; $SSTot = 7.2972$, df $= 8$; $f = 12.51$ on 2,6 df; p-value $= .007$.

Section 3

1. (a) $\hat{y} = 31.40 + 7.430 \ln x_1 - .08101 x_2 - .2760 (\ln x_1)^2 + .00004792 x_2^2 - .006596 x_2 \ln x_1$. $R^2 = .724$. $s_{SF} = 1.947$. $s_p = 2.136$, which is greater than s_{SF}, so there is no indication that the model is inappropriate. (b) Factor-level combinations have fitted values that differ by as much as .77. (d) (i) [.128, 2.781]. (ii) [$-2.693, 5.601$]. (iii) -2.332.

2. (a) Estimate of $\mu_{..} = .67407$; estimate of $\alpha_2 = .12407$; estimate of $\beta_2 = -.30926$. (b) There

is some hint of a pattern in the plot of Standardized Residuals versus levels of C, indicating that the amount of additive may be having a small effect that the model is not accounting for. Otherwise, the residuals do not provide any evidence that the model is inadequate. (c) $s_{FE} = .09623$. $s_p = .12247$. No; $s_{FE} < s_p$.

Appendix A (selected answers only)

Section 1

1. (a) .1865 (b) .6083

2. (a) .54 (b) .78

3. (a) .505 (b) .998

4. (a) .76 (b) .78 (c) .974

5. (a) .75 (b) .80 (c) .75 (d) Yes, since the answers to parts (a) and (c) are the same. (e) One such pair is "ring meets spec.s on first grind" and "ring is ground twice."

Section 2

1. $r = .99979$

2. $k = 2$

Section 3

1. (a) 1.7310×10^{13} (b) 2.2777×10^{12} (c) .1316

2. (a) .0000081 (b) .03402

3. (a) 1,757,600 (b) .00167 (c) .0167

4. (a) .5 (b) .167

Section 4

1. (a) 20; 15.81 (b) $1 - \frac{3}{2} \exp\left(-\frac{t}{15}\right) + \frac{1}{2} \exp\left(-\frac{t}{5}\right)$ (c) $f_T(t) = \frac{1}{10}\left(\exp\left(-\frac{t}{15}\right) - \exp\left(-\frac{t}{5}\right)\right)$ (d) $S_T(t) = \frac{3}{2}\exp\left(\frac{t}{15}\right) - \frac{1}{2}\exp\left(-\frac{t}{5}\right)$, $h_T(t) = \frac{1}{5}\left(\frac{\exp\left(-\frac{t}{15}\right) - \exp\left(-\frac{t}{5}\right)}{3\exp\left(-\frac{t}{15}\right) - \exp\left(-\frac{t}{5}\right)}\right)$. $h_T(t)$ is not constant. It starts at 0, and increases to an asymptote of 1/15.

Index

Note: **boldface** page numbers indicate definitions.